1998

$47.75

NATIONAL ELECTRICAL ESTIMATOR

Includes inside the back cover:
An estimating CD with all the costs in this book plus an interactive multimedia video that shows how to use this disk to compile construction cost estimates for your jobs.

by Edward J. Tyler, CPE

Craftsman Book Company
6058 Corte del Cedro / P.O. Box 6500 / Carlsbad, CA 92018

Acknowledgments

The author is a certified professional estimator, CPE. Certification from The American Society of Professional Estimators, ASPE, 11441 Georgia Avenue, Suite 412, Wheaton, MD 20902, (301) 929-8848, FAX (301) 929-0231.

The author wishes to thank the following individuals and companies for providing materials and information used in this book.

George H. Booth, Vice President Sales — Graybar Electric Company, Inc.

Steve Koundouriotis — P-W Western, Inc.

Don Geibel — Walker Division of Butler Manufacturing Company.

The tables on pages 411 and 412 are reprinted with permission from NFPA 70-1996, the *National Electrical Code*®, Copyright 1995, National Fire Protection Association, Quincy, MA 02269. This reprinted material is not the complete and official position of the National Fire Protection Association on the referenced subject, which is represented only by the standard in its entirety.

National Electrical Code® and *NEC*® are registered trademarks of the National Fire Protection Association, Inc. Quincy, MA 02269.

Cover photos by: *Tom Page Photography,* www.redrom.com/tom

Software resource: Robert W. Kramer III

Drawings by Ann Sweeney and the author

Contents

How to Use This Book

This manual is a guide to the cost of installing electrical work in buildings. It lists **costs to the electrical subcontractor** for a wide variety of electrical work.

Before using any estimate in this book, you should understand an important point about estimating electrical construction costs. Estimating is an art, not a science. There is no estimate that fits all work. The manhour estimates in this book will be accurate for many jobs. But no two crews will complete all tasks in exactly the same time. And no two jobs are identical. That's why electrical cost estimating requires an exercise of judgment. Every estimate has to be custom-made for the job, the crew and the contractor. No estimating reference, no computerized cost estimating system, no estimating service can consider all the variables that make every job unique.

This book is not intended as a replacement for well-informed judgment. But when supplemented with an estimator's professional evaluation, the figures in this manual will be a good aid in developing an informed opinion of the cost of electrical systems.

National **E**stimator **98** Inside the back cover of this book you'll find an envelope with a compact disk that you can install on your computer's hard drive. The disk has *National Estimator*, an easy-to-use estimating program, and all the cost data in this book. An interactive, multimedia video shows how to use the software to create electrical estimates for your jobs. Instructions for using National Estimator begin on page 492.

Labor Costs

The labor costs listed in this manual will apply on most jobs where the hourly wage in effect is the same or similar to the following rates:

Journeyman Electrician

Base Wage . . . .$19.86 per hr.

Fringe Benefits at 9.7% . . . .$1.92 per hr.

Taxes & Insurance at 25.69% . . . . .$5.10 per hr.

Total Labor Cost . . . .$26.88 per hr.

The total hourly cost includes the basic wage, taxable fringe benefits (vacation pay), workers' compensation insurance, liability insurance, taxes (state and federal unemployment, Social Security and Medicare), and typical non-taxable fringe benefits such as medical insurance.

If your hourly labor cost is much lower or much higher, costs of installation can be expected to be proportionately lower or higher than the installation costs listed in this book. If your total hourly labor cost is 25 percent less, for example, reduce the labor figures in the cost tables by 25 percent to find your local cost.

The *Craft@Hrs* column shows the recommended crew and manhours per unit for installation. For example, 2E in the *Craft@Hrs* column means that we recommend a crew of two electricians. 1E means that a crew of one electrician is recommended. Costs in the Labor Cost column are the result of multiplying the manhours per unit times the rate of $26.88 per hour.

For example, if the *Craft@Hrs* column shows 2E@.250, the *Labor Cost* column will show $6.72. That's .250 manhours multiplied by $26.88 per manhour and rounded to the nearest penny.

Divide the manhours per unit into 8 to find the number of units one electrician can install in one 8-hour day: 8 divided by .250 equals 32 units per day. Multiply that answer by the number of crew members to find the number of units the crew will install in an 8-hour day. For example, if the crew is two electricians, multiply 32 by 2 to find that the crew can be expected to install 64 units in an 8-hour day.

Some tasks can be expected to require less labor under certain conditions. For example, when conduit is run in groups, less labor will be required for each 100 linear feet of conduit. It's the estimator's responsibility to identify conditions that are likely to require more or less labor than the standard for the type of work being estimated.

This book lists both the labor cost per installed unit and the manhours required for installation. Manhours are listed in hundredths of an hour rather than minutes because it's easier to add, subtract, multiply and divide hundredths of a unit.

Material Costs

Material prices in this book are based on costs as of late 1997, and projected to mid 1998 by adding 3 to 8 percent. Prices are neither "retail" nor "wholesale" costs. Instead, they're intended to reflect what most electrical contractors who buy in moderate volume will be paying in spring 1998. Volume purchases of some materials will cost less because many dealers offer quantity discounts.

Material costs in this book include normal waste. If waste of materials or breakage is expected to exceed 3 to 5 percent of the materials used on the job, include a separate allowance for excessive waste.

The cost of delivering materials to the job site is not included in this book. Where the cost of delivery is significant and can be identified, add that cost to these figures.

Please note that the cost of some electrical materials is highly volatile. For example, copper wire prices have been known to move up or down 10 percent in one month. There's no reliable way to forecast price movements like this. Major price changes have not been assumed in this manual.

Add Sales Tax

No state or local sales tax is included in material prices listed here. Sales tax varies from area to area and may not be charged on purchases for some types of projects. Add the appropriate sales tax rate when sales tax will be charged on materials bought for the job.

Add Overhead and Profit

To complete the estimate, add your overhead and expected profit. Many contractors feel that adding 10 percent for **profit** yields an acceptable return on the money invested in the business. But there's no profit percentage that fits all jobs and all contractors.

For some electrical contractors, overhead may add as little as 10 percent to the labor and material cost. But routinely adding 10 percent for **overhead** is poor estimating practice. It's the estimator's responsibility to identify all overhead costs and include them in the estimate, either as a lump sum or as a percentage of the total labor and material cost.

Other Costs to Add

Some other costs are excluded from the figures in this manual: building electrical permits, special hoisting costs, freight costs not absorbed by the supplier, utility company charges for installation and service, special insurance and bonds, power equipment other than small tools, mobilization to remote sites, demobilization, nonproductive labor, and nonworking supervisors. Add these costs if they can be identified and if they will be a significant part of the job cost.

All Tables Assume "Good" Conditions

This means that there are few or no unusual conditions that would delay production. Conditions are good when work is performed during usual working hours in relatively clean surroundings and in readily accessible areas not over 12 feet above the finish floor. The temperature is between 50 and 85 degrees F. Electricians are working no more than eight hours a day and no more than five days a week.

Good conditions require that all tools and materials be available on the job site when needed. Tools, including power tools, are assumed to be in good working order. Where power tools are appropriate, it's assumed that temporary power is provided. Add the cost of temporary power when it will be furnished at your expense.

Proper supervision can make a big difference in labor productivity. The tables assume that there is adequate supervision but make no allowance for nonproductive labor — supervisors who direct but do no installation. If you plan to have nonproductive supervision on the job, add that cost to the figures in this manual.

Conditions are seldom "good" when the work area is confined or when a short construction schedule makes it necessary for many trades to work at the same time. The usual result will be stacks of material obstructing the work space and several tradesmen competing for access at the point of installation.

If the conditions on the job you're estimating aren't expected to be "good," adjust the labor figures in this book as appropriate. Occasionally larger jobs can be done faster because specialized equipment or specialized crews can be used to good advantage. This will usually reduce the installation cost. More often, conditions are less than "good." In that case, labor costs will be higher.

There's no accepted way to decide how much "bad" conditions will increase the labor hours needed. But it's accepted estimating practice to assign a cost multiplier of more than 1.0 to a job that can be expected to require more than the usual amount of labor per unit installed. For example, if conditions are less than "good" only in minor respects, you might multiply labor costs by 1.10. If conditions are very poor, a multiplier of 1.50 or more might be appropriate.

Other Factors That Affect Productivity

The tables in this book assume that the crew used for the job is the smallest crew appropriate for the work at hand. Usually this means that the crew is one journeyman electrician.

Most experts who have studied the productivity of construction trades will agree that the smallest crew that can do the job is usually the most efficient. For example, it would be foolish to have two men working together setting duplex receptacles — one handing tools and material to the other as needed. Only one of the two would be working at any given time. A more productive arrangement would be two one-man crews, each working independently.

Of course, there are exceptions. Sometimes a crew of one will take more than twice as long as a crew of two. When pulling feeder cable or setting floor-standing switchboards or motor control centers, more help will usually cut the labor cost per installed unit. Some jobs simply can't be done by a crew of one.

When work is done on a scaffold, there should be someone on the ground to chase parts and equipment and prepare lighting fixtures for hanging. It's a waste of manpower to have an electrician leave the scaffold and return every time parts or tools are needed. The scaffold installers should install one fixture while the "grunt" on the floor prepares the next. Conduit should be prefabricated on the ground from measurements taken by the electricians on the scaffold. The assistant on the ground should bend the conduit and hand it up to the installers.

Labor savings like this are obvious to anyone who has ever done this type of work, and are assumed in the labor tables in this book.

The Electrician

The labor hours in this book are intended to be typical of what a trained and motivated journeyman electrician with five years of experience will do on most jobs. Of course, it's assumed that the installer can read and follow plans and specifications and has the ability to lay out the work so that it complies with the code.

It shouldn't make any difference whether the work is in a hospital, a grocery store, a wood mill or a small convenience store. An experienced journeyman electrician should be able to handle the work at the rates shown here even though the materials and code requirements differ. But you'll have to make allowances if your installers are familiar with residential work only, and the job at hand is something else.

Improving Estimating Accuracy & Profit

It's been said that electrical estimators learn by making mistakes. The best estimators are the ones that have made the most mistakes. Once you've made every mistake possible, you're a real expert.

I can't subscribe 100 percent to that theory, but I know that there are plenty of pitfalls ready to trap unwary electrical estimators. This section is intended to suggest ways to spot potential problems before they become a major loss. It will also recommend some steps you can take to increase the profit on most jobs.

Labor Productivity

Improving output only slightly can result in major cost savings. Cutting only a minute or two off the installation time for each duplex receptacle or handy box can reduce the labor cost by several hundred or even a thousand dollars a job. Getting better (and predictably better) productivity from your electricians should be a primary concern for every electrical contractor.

Assuming that your electricians are experienced, well-trained, and have all the tools and materials they need to complete the work, the most significant increase in productivity will probably be through improved motivation.

The best form of motivation for most electricians is to encourage a sense of pride in the work they do. Encouraging self-respect is a powerful inducement for most skilled tradesmen. Every alert supervisor knows the value of recognition given for a job well done. Acknowledging good work builds confidence and encourages additional effort in the future.

Handling Inspections

The supervisor or foreman on site should have primary responsibility for dealing with all inspectors. Don't let others circumvent the supervisor's or foreman's position.

An inspector's only job is to see that the installation complies with the code. They aren't supervisors and don't direct the work. They can and do interpret the code and sometimes make mistakes. Encourage the foreman or supervisor to take issue promptly with a questionable interpretation. Ask the inspector to cite a code reference as his authority. If the inspector insists that his interpretation is correct, and if you feel that it's wrong, call the Building Official to initiate an appeal. Your trade association or the National Electrical Contractors' Association may be able to present persuasive arguments in your favor.

Some inspectors have a reputation for being impossible to deal with. Aggressive enforcement of questionable code interpretations can severely hurt project productivity. Following the code carefully will keep you out of most compliance problems. Every electrician and electrical supervisor has to know the code. Code classes are taught at night schools in many communities. Take a code class both to understand how the code is applied and to get current on code changes that have been enacted recently.

Mobilization and Demobilization

Many electrical subcontractors have job shacks and lockup boxes that can be moved onto the job for storing tools and materials. Some larger firms have trailer vans that can be moved from job to job. No matter what type of on-site storage you use, getting set up takes time. The bigger the job, the more time will probably be needed.

The first step is usually getting permission to set up your storage area somewhere on site. On some job sites storage space is at a premium. Some city projects literally have no storage space until parts of the building are completed. Occasionally tools and equipment will have to be stored off site. This will usually require daily mobilization and demobilization and increase the labor cost substantially. Be sure your estimate considers this important item.

Demobilization usually takes less time and costs less than mobilization. Removing the surplus material, tools and equipment can be done by helpers or material handlers rather than electricians.

One important mobilization item is temporary electrical service. Be sure you know who pays for installation of temporary power and who pays for power used on site during construction. It's common for the electrical contractor to cover the cost of electrical distribution and service. Installation will usually be done by your electricians and will have to pass inspection, of course.

Most communities require temporary electrical permits prior to starting work. Before applying for the permit, contact the electric utility and request a meeting with whoever coordinates extensions of service — usually the *planner*. Before having that meeting, know what size the service must be. The planner will let you know what voltage is available and where the point of connection will be. Don't end the meeting with the planner until you've covered every requirement and all procedures imposed by the electric utility.

Job Cleanup

It's hard to get good production when trash and debris obstruct access on the job site. That alone should be enough reason to dispose of accumulated waste regularly. Most specifications require that all subcontractors remove unused materials, cartons, wrappers and discarded equipment. On many jobs the general contractor will have the right to back-charge subs for removal of their discards if they don't clean the site themselves.

Encourage your crews to do their cleanup as installation is being done. For example, each time a fixture is removed from a carton, the tradesman should collapse the carton and throw it on the pile of discards. It takes little more time that way and the cleanup is less likely to be put off or ignored entirely.

Some contractors and subcontractors have a reputation for running a dirty job. You've probably seen jobs that are so cluttered that it's hard to understand how anyone can work efficiently. As the electrical contractor, you can't dictate to the general contractor or the other subcontractors, of course. But work habits of others affect your productivity and thus your profit.

My feeling is that it's entirely proper to discuss the matter with the general and the other subs when accumulated debris is slowing progress. Request a meeting right on the job in the middle of the clutter. This alone may do the trick.

If you don't insist on a clean job, the fire department probably will. A dirty job site costs everyone more time and money. A clean job is usually both more efficient and safer.

Production

No matter how simple and quick, most jobs will have some production problems. It's easy to explain why. Every job is unique. Every job brings together skilled tradesmen with varying preferences and habits. Most have never worked together before. Yet each must coordinate the work he does with those who precede him and those who follow. It's normal to expect that some adjustments will be needed before cooperation becomes routine.

Of course, the general contractor is the key to good cooperation among the trades. A general who schedules trades properly will have fewer problems and will help all subcontractors earn the profit they're hoping for. This is never automatic, of course. And some general contractors never seem to learn how to schedule properly. From the estimator's perspective, it's more expensive to work for a contractor who has scheduling problems than it is to work for a contractor who's more adept at coordination. Your estimates should reflect production problems when problems can be anticipated.

Good supervision will help avoid most production problems. Schedule material deliveries carefully. Have the right tools on hand when needed. Keep crews as small as possible. Don't work your crews more than 40 hours a week unless absolutely necessary. Too many bodies and too many hours will erode production.

If you're using a larger crew, don't have everyone work together. Instead, break the crew into two units and encourage friendly competition between the two, with a six-pack of beer going to the winning crew — to be consumed after hours and off the site, of course.

Corrections

The tables in this book assume that little or no time is spent making corrections after the work is done. Most electrical contractors have very few callbacks.

If you're often called back to replace faulty materials or correct defective workmanship, one of four things is happening. First, you could be working for some very particular contractors or owners or handling some very sensitive work. In that case, callbacks could be part of the job and should be included in each estimate. Second, you could be installing substandard materials. Third, your electricians could be doing haphazard work. Finally, your installation procedure could be omitting fixture and circuit tests that will locate problems before the owner finds them.

When qualified electricians install quality materials, the risk of a callback is small. Occasionally a ballast will fail after 10 or 20 hours in use. And sometimes an owner's negligence will damage a circuit or switch. When this happens, accept the service work as routine. Complete it promptly at no extra charge. Consider it cheap advertising, a chance to establish your credentials with the owner. You'll probably turn the service call into some extra work later.

Your Type of Work

Most electrical contractors prefer to handle certain types of work. Only a few have the capital, equipment and skills needed to handle the largest jobs. Most will do residential wiring because that's the most plentiful work available. Some prefer private work with as little government interference as possible. Others bid only government jobs.

The most profitable electrical contractors specialize in one type of work or one type of customer. The electrical construction field is too broad to try to do everything well. Select a part of the business that feels comfortable and concentrate on doing it as well or better than anyone else. Of course, some of the older and larger electrical shops will do almost any type of work. But nearly every electrical contractor prefers some class of job over all others — and would take only that work if there were enough available to stay busy.

Observe the electrical contractors in your area. Notice the companies that seem to be busiest and most profitable. See what class of customers they service or what type of work they do most. It's probably easier to follow the success of another contractor who's found a winning formula than it is to invent a new formula yourself.

Specialization lets you hire electricians who are specialists too. That tends to improve productivity, keeps costs down, and improves profits — as long as you're handling work that's in your specialty.

Coordination will be easier and the profits will usually be higher if you work for a limited number of general contractors. Some contractors seem to be masters at putting a project together. These same contractors probably pay promptly and treat their subs fairly. That makes your job easier and tends to fatten your bottom line. If you've found several contractors who make life more pleasant for you, keep them supplied with competitive bids that will bring more work into both your shop and theirs.

Most electrical contractors don't bid government work. It's a specialty that requires special knowledge: complying with detailed general conditions, observing regulations, anticipating inspection criteria and following administrative procedures. And every branch of federal, state and local government has its own requirements. Those who have mastered the procedures usually do quite well when work is plentiful. But there's a tide of government work that rises and falls just like the tide of construction activity in general.

Bid Shopping

Many contractors prefer projects that require subcontractor listings. The general contractor must list the subcontractors he plans to use, and must actually use the subs he lists. When listing of subs isn't required, the general can shop for lower subcontract bids right up to the time work begins, in some cases. Even if the general contractor has to list his subs in the contract with the owner, the general will usually have a month or two to shop bids after the contract is awarded.

When a general contractor uses your bid to land a job, it's just elementary fair play to expect that your company will get the contract. Giving all your competition a second look at the job is in no one's interest but the general contractor's. It's a waste of time to bid for general contractors who shop their sub bids. Neither is it good practice to undercut another electrical contractor whose estimate was used by the winning general contractor. Support the effort of reputable subcontractors who promote subcontractor listing at bid time.

Need More Help?

This book is concerned primarily with labor and material costs for electrical construction. You'll find only limited information here on how to compile an estimate. If you need a detailed explanation on how to make a material take-off and complete the bid, another book by this author may be helpful. An order form for *Estimating Electrical Construction* is bound into the last pages of this book.

Section 1: Conduit and Fittings

Every electrical estimator should be familiar with the *National Electrical Code®*. Nearly all inspection authorities follow *NEC®* recommendations on what is and what is not good electrical construction practice. Most inspection authorities accept electrical materials that comply with *NEC* standards. But some cities and counties have special requirements that supplement the current *NEC*. Others are still following an older edition of the *NEC*. The *NEC* is revised every three years to incorporate changes deemed necessary to keep the code up-to-date.

Be aware of the version of the *NEC* that applies at each job you're estimating, and stay current on special requirements that the inspection authority may impose.

Job specifications usually state that all work must comply with the *NEC*. But on many jobs the *NEC* sets only the minimum standard. Job specifications may prohibit what the *NEC* permits. For example, job specs might require specific installation methods or mandate specification grade fixtures.

The *National Electrical Code* classifies all enclosed channels intended to carry electrical conductors as "raceway." This includes conduit, busway and wireway. The most common raceway is electrical conduit. The code identifies the size and number of conductors that can be run through each size of conduit.

Conduit is intended to serve two purposes. First, it's a protective shield for the conductor it carries. It reduces the chance of accidental damage to the wire or insulation. Second, it protects people and property from accidental contact with the conductors. A ground or short is both a safety and a fire hazard.

Conduit is generally required in commercial and industrial buildings, hospitals, hotels, office buildings, stores and underground facilities. It's not generally used in wiring homes and apartments.

Several types of electrical conduit have been approved for electrical construction. Each is designed for a specific purpose or use. All conduit used in electrical construction as a raceway for conductors must bear a label issued by the Underwriter's Laboratories. The UL label indicates that the product has been approved for use under the *National Electrical Code.*

The *NEC* permits a maximum of four bends totaling 360 degrees between terminations in a run of conduit. Exposed conduit should be installed horizontal or vertical and should run parallel to building members. Concealed conduit should be run in the shortest direct line to reduce the length of run. Long runs waste materials, require excessive labor and, if long enough, can reduce the voltage available at the load end.

Electrical Metallic Tubing

EMT is also known as **thin wall** or **steel tube**. EMT conduit is nonferrous steel tubing sold in 10-foot lengths. Unlike water pipe, the ends aren't threaded. The conduit has a corrosion-resistant coating inside and outside. This coating may be hot-dipped galvanizing, electroplating, or some other material. The conduit sizes are ½", ¾", 1", 1¼", 1½", 2", 2½", 3", 3½" and 4".

Many types of EMT fittings are available. There are elbows, compression, set screw, indent and drive-on fittings which may be made of steel or die cast. Couplings and connectors are sold separately and not included in the price of the conduit. Various types of connectors may be purchased with or without insulated throats. The locknuts for the connectors are included in the cost of the connector.

Couplings are available for joining EMT to rigid metal conduit and to flexible conduit. These couplings are available in compression, set screw and drive-on type and are made of steel or die cast.

EMT conduit is sold without couplings. You have to figure the number of couplings needed and price them separately. To figure the number needed, allow one coupling for each 10 feet of conduit. Then add one coupling for each factory-made elbow.

EMT should be bent with a special conduit bender. The bender has a shoe that fits over and around about half of the conduit to keep the conduit from collapsing as it bends. With a bender it's easy to produce smooth, consistent bends up to 90 degrees. Hand benders are used on sizes from ½" to 1¼". EMT bending machines are available for all sizes of conduit. There are manual, hydraulic and electrically driven machines.

Offsets are made to take EMT conduit around obstructions, and when needed, to align the conduit at a box or cabinet. You can make offsets with a hand bender on sizes up to 1¼". Offsets in EMT conduit over 1¼" should be made with a machine.

In smaller sizes, conduit can be cut with a tubing cutter. Cut larger diameters with a hacksaw or by machine. Cut ends must be reamed to remove the burrs made while cutting. Burrs can damage insulation when wire is pulled through the conduit. Ream with a pocket knife or pliers on smaller sizes and with a metal file or pipe reamer on larger sizes.

EMT must be supported so it doesn't deflect on longer runs. Straps and nailers are the most common way of supporting EMT. Straps usually have one or two holes for securing to the building. Most inspection authorities won't let you support EMT on plumber's perforated metal tape. Straps come in thin steel, heavy duty steel or malleable types. There are special straps made of spring steel for supporting small sizes of EMT to hanger rods or drop ceiling wires.

EMT conduit should be supported at least every 10 feet with a strap or hanger and within 18 inches of every junction box or cabinet.

Other supports include beam clamps for attaching conduit to structural steel members and straps for mounting EMT on steel channel strut. These two-piece straps or clamps are inserted into the strut and bolted together to hold the conduit in place.

EMT can be installed inside or outside, in concrete or masonry, exposed or concealed in walls, floors or ceilings. But be sure to use the correct fittings in wet locations. EMT is not approved for most types of hazardous locations. Some specs limit the use of EMT to dry areas and don't allow placement in masonry or concrete. Conduit placed in concrete floor slab is generally placed below the reinforcing bar curtain or between curtains when two curtains are used. Tie the conduit to the rebar to prevent shifting as the concrete is placed.

Where conduit is turned up above the surface of the concrete, the radius of the turn must be concealed. Part of it can be concealed in a wall, but none should be visible after the building finish has been installed.

As with all types of conduit, EMT should be installed with a minimum of damage to the structure. Keep it clear of heating, ventilating and air conditioning ducts, fire sprinkler systems, plumbing lines, access doors, etc. When necessary, the installer will have to make offsets and bends so the conduit fits into devices, electrical boxes and cabinets.

Flexible Metal Conduit

There are several types of flex conduit: standard wall steel flex, reduced wall steel flex, and aluminum flex. It comes in diameters from 3/8" to 4" and is coiled in rolls of 100 feet in the small sizes and 25 feet in the larger sizes. Flex is usually used in concealed locations but never underground or in concrete. It's cut with a special flex cutter, a hand hacksaw, or with a power cutter such as a portable band saw. The inside cut edge must be reamed to remove cutting burrs which would damage insulation when wire is pulled through conduit.

Flex connectors are available with set screw, screw-in, clamp type, straight, or angled connectors. They're made of steel or die cast. Insulated connectors are also available. Die cast flex couplings are available for joining flex to flex, flex to EMT, or flex to threaded conduit. Support flex with conduit straps or nailers.

Most inspection authorities require that a bonding conductor be installed when electrical wiring is run in flex. Bonding ensures that there's electrical continuity in the flex from one end to the other.

Some specifications restrict the use of flex to short connections to equipment that is subject to vibration (such as motors and machinery) and for built-ins, recessed lighting, and lay-in lighting fixtures.

Flex conduit is popular in remodeling work where wiring in raceway has to be run through an existing cavity wall or in a ceiling cavity. With a little effort, your installer can fish the flex from point to point without opening the wall or ceiling.

Polyvinyl Chloride Conduit

PVC conduit is approved by the *NEC* for many types of applications. But there are some situations where it cannot replace metallic conduit. It's not approved for hazardous locations or in return air plenums. Check with the inspection authority for other restrictions. The standard length is 10 feet and sizes range from ½" to 6". Schedule 40 PVC is the standard weight. Schedule 80 has a heavier wall. PVC can be installed directly underground, concrete encased underground, exposed, in concrete walls, and in unit masonry.

One coupling is furnished with each length of conduit and is usually attached to the conduit. PVC must be bent with a special hot box which heats the conduit until it becomes pliable. Once heated to the right temperature, the tube is bent and then allowed to cool. PVC fittings fit both Schedule 40 and 80 conduit. Couplings, terminal adapters, female adapters, expansion fittings, end bells, caps, conduit bodies, pull boxes, outlet boxes and elbows require a special cement. The glue is air-drying and comes in half-pints, pints, quarts, and gallon containers. The smaller containers have a brush attached to the cap for applying the cement to the conduit or fittings. PVC conduit can join other types of conduit if you use the right fittings to tie the two types together.

PVC is nonconductive. That makes a bonding conductor necessary to ensure electrical continuity from the device to the service panel. You probably won't need a bonding conductor when PVC is used as communications conduit or in some application that doesn't include electrical wiring. When installed exposed, PVC requires extra support to keep it from sagging.

Some job specs restrict use of PVC to specific locations. One common restriction is to limit PVC to underground installations encased in a concrete envelope. Many specifications restrict its use to certain applications.

PVC conduit can be cut with a hand hacksaw, a wood crosscut saw, or with a power cutting machine. The inside cut edge should be reamed to remove the cutting burr. Use a pocket knife or a file.

Power and communications duct is usually called **P&C duct**. It's made of PVC in 25-foot lengths and in diameters from 1" to 6". There are two types of P&C duct. One is called **EB** for encased burial. The other is **DB** for direct burial. Fittings for P&C duct include couplings (one is furnished with each length), end bells, caps and plugs, terminal adapters, female adapters, elbows, and expansion fittings. The elbows are available in various shapes and with either long or short radii. Fittings can be used either on type EB or DB. Use a special cement to weld the fittings to the conduit.

Bend P&C duct with a hot box. It can be cut with the same tools as PVC conduit. The inside cut edge must be reamed to remove the cutting burr.

P&C duct is used for underground systems only, never above ground.

ABS underground duct is used and installed the same as PVC P&C duct. It requires a special ABS cement to weld the fittings to the conduit. The job specifications or the utility company may require either P&C, ABS or PVC duct, depending on the specific use.

Galvanized Rigid Conduit

GRS or **RSC** (for rigid steel conduit) is made with nonferrous steel and has a corrosion-resistant coating on the inside. The outer coating is either hot-dipped galvanizing or electroplate. It comes in diameters from ½" to 6" and in 10-foot lengths with a thread on each end. A coupling is furnished on one end of each length. GRS can be cut with a hand hacksaw, a pipe cutter, or with a cutting machine. The inner cut edge must be reamed to remove the burr. Use a pipe reamer or a file.

After the pipe has been cut and reamed, it can be threaded. Use a hand die for threading on a small job. Where there's more cutting and threading to be done, use a threading machine. Several types are available. Small portable electric threading tools cut sizes up to 2". Larger threading machines can cut, ream and thread conduit diameters up to 6". Another good choice for GRS up to 6" is a threading set that uses a tripod vise stand and a threading head that clamps to the pipe in the vise stand. The threading head is turned with a universal joint connected to a power vise. Another set uses a tripod vise stand to hold the conduit. The threading head clamped on the conduit is turned with a reduction gear assembly powered by an electric drill. This rig works well on diameters over 2".

Use enough cutting oil to keep the die cool and lubricated during thread cutting. Cutting oil comes in clear or dark and in small cans, gallons and barrels. Use an oil can to keep a film of oil ahead of the dies. Commercial oiling units hold about a gallon of cutting oil and recirculate oil back to the cutting teeth as oil drips into the catch basin. Most threading machines have automatic oilers that filter the oil as it's reused.

Elbows are available for all sizes of GRS. Long radius bends are available for the larger sizes. Some specifications require concentric bends for all exposed conduit installed parallel on a common hanging assembly or trapeze.

GRS fittings include couplings, locknuts, bushings, one-hole straps, two-hole straps, heavy duty two-hole straps, expansion fittings, threadless compression couplings, threadless set-screw couplings, threadless compression connectors, threadless set-screw connectors, three-piece union-type couplings, strut clamps, beam clamps, hanger clamps, condulets, split couplings, caps, and plugs.

Galvanized rigid conduit is bent about the same way as EMT except that the bender is made for bending rigid conduit. Hand benders are used on conduit up to 1". There are hand benders for 1¼" and 1½" rigid steel conduit, but it takes a lot of effort to make the bend. Power benders can be used on all sizes of conduit, even the ½".

There are three common types of rigid steel benders: one-shot benders create a single standard radius arc. Segment benders must be moved along the conduit as each few degrees of bend are made. The electric sidewinder bender has up to three bending shoes in place ready to bend any of three sizes of conduit. The sidewinder saves labor on larger rigid conduit jobs.

Supports for rigid conduit must be no more than 10 feet apart from support to support and within 18 inches of junction boxes or cabinets.

Trapeze hangers are often used to carry multiple runs of GRS conduit. Trapeze hangers can be made from strut, angle iron, or channel iron. The trapeze is supported from the structural frame of the building with threaded rod — usually either $^3/_8$" or ½" diameter. The upper part of the rod is attached to beam clamps or concrete anchors. The lower portion of the rod is run through the trapeze and is secured with double nuts and flat washers.

Like other hangers, trapezes have to be placed within 10 feet of each other and should be sized to support the total weight of the conduit and all cable. Trapeze hangers can be stacked one over the other with conduit clamped on each one.

IMC Conduit

Intermediate metal conduit (IMC) has a thinner wall than GRS. It comes in the same sizes and uses the same fittings as GRS. The same tools can be used for cutting, threading, and bending. It's made about the same way as GRS, comes in 10-foot lengths and is galvanized for corrosion resistance. The difference is that IMC is lighter and easier to install than GRS. Some specifications restrict its use to specific applications.

PVC Coated Conduit

Both GRS and IMC conduit come with a PVC coating for use in highly corrosive locations. Aluminum tubing also comes with a PVC coating, but applications are restricted to specific uses. The PVC coating is either 10, 20 or 40 mils thick, and is bonded directly to the conduit wall. Most fittings made for use with GRS are available with a PVC coating.

To thread PVC coated conduit, the PVC coating must be cut back away from the end to be threaded. When PVC coated conduit is put in a vise, be sure the coating is protected from the vise jaws. Also be careful when you're bending PVC coated conduit not to damage the coating. If the coating is damaged, patching material is available to restore the surface. The material comes in a spray can. Apply several thin layers to repair worn spots.

Conduit Take-Off

Here's how to calculate conduit quantities. First, scan the specs that cover conduit and conduit installation. Absorb all the information that relates to conduit. Then review the drawings for anything about conduit. The symbol list may include the engineer's design notations. Notes on the drawings or in the specs may set specific minimum conduit sizes. It's common for an engineer to require a minimum size conduit in the home run to the panel or cabinets or to specify a minimum size of ¾" throughout the job. It's also common practice to limit the maximum size of EMT to 2". Ignoring a note like that can be expensive.

For your quantity take-off, use any ruled 8½" by 11" tablet. Draw a pencil line down the left side of the sheet about an inch from the edge. Begin by looking for the smallest diameter of EMT. Write "EMT" at the top left of your take-off sheet. On the next line down, to the left of the vertical line, list the smallest EMT size found in the project — probably ½". To the right of the vertical line and on the same horizontal line as the size, you're going to list lengths of EMT of that diameter. Then you'll go to the next larger diameter, listing quantities until all EMT on the plans has been covered.

Check the plan scale before you start measuring conduit. If the plan has been reduced photographically to save paper, the scale will be inaccurate. Once you're sure of the correct scale, select the appropriate map measure or rule to compute conduit lengths.

Measure the length of each run of ½" EMT. Add enough conduit to include the run down to the wall switch, receptacle or panel. Write down the calculated length. As each run is listed on your take-off sheet, put a check mark on the plan over the line you just measured. Used an erasable color pencil and let each color stand for a particular conduit type. For example, red might be for GRS conduit. Follow the same color code on all estimates to avoid mistakes.

If there are more than two or three plan sheets, it's good practice to calculate the length of ½" EMT on each plan sheet and list that number separately on your take-off form. When you've finished taking off ½" EMT on the first plan sheet, list that quantity, and at the top of the column write in the plan sheet number. Then draw a vertical line to the right of that column and start accumulating lengths from the next plan sheet. As each plan page is taken off, enter the total and write the plan sheet number at the top of the column. Figure 1-1 shows what your take-off might look like if conduit and fittings are found on plan sheets E3 to E11.

When all of the smallest-diameter EMT has been listed, go on to the next larger size. Follow the same procedure.

After listing all EMT, begin with the fittings. Below the last horizontal line used for conduit, and to the left of the vertical line, write the word "Connectors." Below that, list all sizes of connectors needed for the job, again working from the smallest size to the largest. Don't bother to list the couplings. They'll be figured later from the total conduit length — one for each 10 feet and one for each elbow.

Count each connector needed for each conduit run on each plan sheet. Enter the total on your take-off form. When all connectors are counted, count EMT elbows from 1¼" to the largest size needed.

	E3	E4	E5	E6	E7	E8	E9	E10	E11	Total
Work Sheet					**Estimate No.:** M351					
Conduit / Fittings										
½" EMT	550	420	200	90	290	130	190	320		2190
¾"	20		30	20	80					150
1"			3		5		50			58
1¼"			30							30
1½"									90	90
2"					4				16	20
½" Conn	76	52	124	47	48	16	14	18		395
¾"	4		26	4	19	2				55
1"			4		5	2	2			13
1¼"			2							2
1½"									4	4
2"					2				4	6
1¼" Elb			2							2
1½"									3	3
2"									3	3
½" PVC			310	380	50					740
¾"			120	100	220	50				490
1"			40		320	40				400
1¼"						180				180
1½"				60					75	135
2"				10	25			70	75	180
4"								150		150
½" FA			45	30	4					79
¾"			4	4	12	2				22
1"			2		17	2				21
1¼"										0
1½"				4					2	6
2"				2	2				2	6
½" TA			5							5
¾"					4					4
1"					1					1
½" Elb			50	30	4					84
¾"			2	2	16	2				22

Figure 1-1

Follow this system for all estimates and for each item on every estimate. Keep it simple and uniform to avoid mistakes and omissions. When finished, your conduit and fitting take-off form might look like Figure 1-1. The right column is the sum of the columns to the left.

Many jobs limit the use of EMT to dry locations. So your EMT take-off will probably start with the lighting plans or the lighting portion of the plan.

Taking Off the Wire

Next, compute the quantity of wire needed. Head up another take-off form with the word "Wire" at the top. Put a vertical line down the left side of the page about an inch from the left edge. In this margin, list wire sizes from the smallest to the largest. To the right of the vertical line you'll list lengths for each wire gauge, on each plan sheet.

Start by measuring the length of ½" EMT with two #12 wires. Multiply by 2 to find the wire length. Then measure the length of ½" EMT with three #12 wires and multiply by 3. Keep following this procedure until the wire needed in all EMT has been computed. But watch for changes in the wire size on long runs. Sometimes the engineer will decide that a larger wire size is needed in the first portion of a run to reduce the voltage drop at the end of the line. This is common where the last device or fixture on a circuit is a long way from the panel.

Follow the same procedure for all conduit and wire. Record all of the measurements on the work sheets. Don't worry about waste of conduit or wire at this point. We'll include an allowance for waste after the totals are added and before figures are transferred to the pricing sheets.

Sometimes the specifications or a note on the plans will allow the use of aluminum feeder wire over a certain size, providing the ampacity of the wire is maintained and the conduit size is increased to accommodate the larger wire size. Be sure to observe these restrictions.

Taking Off Other Conduit

Some specifications permit the use of aluminum conduit in certain locations. The aluminum conduit is made in the same sizes as GRS. The fittings are identical except that they're made of aluminum instead of steel. Most specs prohibit the use of dissimilar metals in a conduit run and don't allow placing of aluminum conduit in concrete. Aluminum conduit saves time because it's lighter and easier to handle. But large wire sizes may be a little more difficult to pull in aluminum conduit. The insulation of the wire, the length of the conduit run, and the pulling lubricant used have an effect on pulling resistance.

When taking off the underground conduit, start a separate work sheet for trenching, surface cutting, breaking, and patching. List all excavation for underground pull boxes, handholes, manholes, poles, and light pole bases. Be sure the trenches are big enough for the number of duct they have to carry. If the specifications require concrete or sand encasement around underground duct, calculate the amount of concrete or sand as you compute measurements for each trench.

Be systematic. Follow the same procedure consistently on every take-off. If there are other estimators in your office, be sure they are using the same procedures. Being consistent reduces errors, minimizes omissions, and makes the work easier for others to check.

We've covered all common conduit. But some other types are used occasionally for special purposes:

Fiber duct is a paper and creosote duct. Type 1 is intended for concrete encasement and Type 2 is used for direct burial. Sizes range from 2" to 5". Lengths can be 5, 8 or 10 feet. End fittings are tapered. Ends that have been cut must be tapered with a duct lathe.

Transite duct is cement asbestos duct. Type 1 is for concrete encasement and Type 2 is for direct burial. Sizes range from 2" to 6". It's made in 5, 8 and 10-foot lengths. Transite is harder to cut and must have tapered ends for fittings.

Soapstone duct is made from a soapstone-like material in sizes from 2" to 4".

Wrought iron pipe comes in sizes from 2" to 4". It's used only for certain types of underground communications lines and has to be threaded on each end to accept fittings.

Clay conduit comes in sizes from 2" to 4". It's used for underground communication runs only.

These types of conduit are seldom specified today. You'll see them used only when an old duct line has to be extended. It may be hard to find a fitting that will join an existing duct system made with one type of duct to a new run of duct made from some other material. Sometimes an oversize plastic coupling can be used. In some cases an inside plastic coupling can be inserted into the old conduit. Then new conduit can be joined to start the new run.

Before extending an old underground duct system, check the old conduit with a mandrel to be sure the line is clean and clear. Old fiber duct that's been under water for a long time will swell, making the inside diameter too small to pull new cable.

Silicon-bronze conduit comes in sizes from ½" to 4". It's threaded like GRS and uses similar fittings, except that fittings are silicon-bronze also. It's used in extremely corrosive locations. This type of conduit will be available from your dealer on special request only. It's harder to bend, but can be bent with standard rigid bending tools. It threads very well with the standard threading tools and cutting oil.

Liquid-tight flexible metal conduit comes in sizes from ½" to 4". It's used to extend conduit to electrical equipment in damp or wet locations. Special fittings are available for connecting electrical systems and devices with this conduit. Your dealer probably stocks a limited supply of liquid-tight flex and will quote prices on request. The conduit can be cut with a hacksaw. Be sure to remove the cutting burr. Special connectors with grips are available to support the conduit and prevent any pulling strain.

Liquid-tight flexible non-metallic conduit comes in sizes from ½" to 1½". It's used in place of flexible metal conduit in concealed locations. Special fittings are available for making connections. Your dealer may have a limited supply in stock.

Flexible metallic tubing is available only in sizes from ⅜" to ¾". Special fittings are available for making connections. The tubing can be bent by hand and is cut with a hacksaw. The cutting burr must be removed before connectors are installed.

Other UL-approved raceways for electrical systems are covered in other sections of this book. See the sections on surface metal raceway, underfloor ducts, header ducts, cable tray, and wireway.

Using the Conduit Tables

The labor tables that follow are for conduit runs that average 50 feet. You'll note that there is no modification in the tables for shorter runs or longer runs of conduit. I agree that it takes more time per linear foot to install a 5-foot run of conduit than it does to install a 95-foot run of conduit. But I don't recommend that you tally shorter runs and longer runs separately and then compute labor separately for each. There's an easier way.

On most jobs the conduit runs average 50 feet. There will usually be about as many runs under 50 feet as there are runs over 50 feet. It's safe then, to use a 50-foot run as our benchmark. As long as the conduit runs on a job average close to 50 feet, there's no need to modify the figures in these tables. If conduit runs average well over 50 feet, consider reducing the cost per linear foot slightly. Increase the cost slightly if conduit runs average less than 50 feet.

The labor costs that follow include the labor needed to bore holes in wood stud walls. Where holes have to be cut through concrete or unit masonry, add these costs separately.

Typical conduit bending is included in the tables that follow. Usually you will have a bend or offset about every 20 feet. Labor needed to make bends and offsets is minor when installing the smaller sizes of conduit.

Concealed conduit is installed where it will be inaccessible once the structure or finish of the building is completed. *Exposed conduit* is attached to the surface where access is possible even after the building is completed. It's usually faster to run concealed conduit through wall and ceiling cavities that will be covered later by finish materials. Installing conduit on surfaces that won't be covered later usually takes more time.

If only a small percentage of the conduit is to be installed exposed, the cost difference will be minor and probably can be ignored. But if most of the job is exposed, add about 20 percent to the labor cost.

The conduit tables that follow assume that electricians are working from ladders and lifts up to 12 feet above the floor. Add to the labor cost for heights beyond 12 feet. If a large quantity of conduit has to be installed at 18 feet above the floor, for example, add 15 percent to the labor cost.

If there are conduit runs over 20 feet above the floor, check your labor contract for a **high time clause**. Some agreements require that electricians be paid time and one-half for heights from 20 to 50 feet and double time for heights beyond 50 feet. If high time must be paid, be sure the extra cost is covered in your bid.

Job Size Modifiers

It's seldom necessary to estimate lower productivity just because the job is small. If you're figuring a very small job with only four or five conduit runs, each with only a strap or two, you might want to use a higher hourly labor rate. On any other job that takes from two days to several years, you can use the labor units in the tables that follow. Of course, you'll still have to modify the figures for other than "good" conditions. And if you have long runs of feeder conduit with parallel runs on a common trapeze, you can reduce the labor units by as much as 40 percent.

Pitfalls

The most common error when estimating conduit is failing to read the plans and specs. Read carefully! Your profit depends on it. It's easy to miss a little note where the designer sets the minimum size for conduit at ¾" and 1" for all home runs to the panel. Look for a note on the plans that requires stub ups to ceiling cavities from power and lighting panels. The designer may require one ¾" conduit run for each three spare circuit breakers in a panel.

It's common for rigid conduit to be installed in a concrete floor slab. Where GRS is stubbed up out of the concrete for a wall switch, it's easier and cheaper to use EMT for the wall extension. The *NEC* permits making that extension in EMT. But some specs don't! Others require that a junction box be used to separate the two types of conduit. Failing to catch that note can be an expensive mistake.

You'll find all sorts of restrictions in specs and notes on the plans. That's why it's so important to read the plans and specs carefully. It's elementary, but it's so often overlooked.

Waste of Material

There will always be some waste on a job. Rounding off the conduit and wire needed to the next even 100 feet will usually allow enough extra material to cover all waste. But there are some cases where you can anticipate a waste problem. For example, suppose there will be 2 feet of waste for every 20 feet of conduit installed because of an unusual lighting pattern. Or suppose a row of junction boxes is spaced at 9 feet. Then a 10 percent waste allowance may be called for. That's almost certainly true if your job is installing the lighting only. There may be no chance to use waste materials on another part of the job.

Allowances

Be sure to make allowances for the vertical portion of every conduit run that stubs up or down in a wall. The floor plan doesn't show the 4 or 5 feet needed to run from the slab to the wall switch or panel. Even worse, if the job is a warehouse, the stub up to a switch or panel may be 15 to 20 feet. That's a wide miss! Watch for stub up.

Material	Craft@Hrs	Unit	Material Cost	Labor Cost	Installed Cost
EMT conduit in floor slab or multiple runs on a trapeze					
1/2"	1E@3.25	CLF	24.20	87.40	111.60
3/4"	1E@3.50	CLF	40.40	94.10	134.50
1"	1E@4.00	CLF	66.20	108.00	174.20
1-1/4"	1E@4.50	CLF	92.80	121.00	213.80
1-1/2"	1E@5.50	CLF	110.00	148.00	258.00
2"	1E@7.00	CLF	139.00	188.00	327.00
2-1/2"	1E@9.00	CLF	291.00	242.00	533.00
3"	1E@10.0	CLF	384.00	269.00	653.00
3-1/2"	1E@11.0	CLF	534.00	296.00	830.00
4"	1E@12.0	CLF	604.00	323.00	927.00
EMT conduit in concealed areas, walls and closed ceilings					
1/2"	1E@3.50	CLF	24.20	94.10	118.30
3/4"	1E@3.75	CLF	40.40	101.00	141.40
1"	1E@4.25	CLF	66.20	114.00	180.20
1-1/4"	1E@5.00	CLF	92.80	134.00	226.80
1-1/2"	1E@6.00	CLF	110.00	161.00	271.00
2"	1E@8.00	CLF	139.00	215.00	354.00
2-1/2"	1E@10.0	CLF	291.00	269.00	560.00
3"	1E@12.0	CLF	384.00	323.00	707.00
3-1/2"	1E@14.0	CLF	534.00	376.00	910.00
4"	1E@16.0	CLF	604.00	430.00	1,034.00
EMT conduit installed in exposed areas					
1/2"	1E@3.75	CLF	24.20	101.00	125.20
3/4"	1E@4.00	CLF	40.40	108.00	148.40
1"	1E@4.50	CLF	66.20	121.00	187.20
1-1/4"	1E@6.00	CLF	92.80	161.00	253.80
1-1/2"	1E@8.00	CLF	110.00	215.00	325.00
2"	1E@10.0	CLF	139.00	269.00	408.00
2-1/2"	1E@12.0	CLF	291.00	323.00	614.00
3"	1E@14.0	CLF	384.00	376.00	760.00
3-1/2"	1E@16.0	CLF	534.00	430.00	964.00
4"	1E@18.0	CLF	604.00	484.00	1,088.00

Use these figures to estimate the cost of EMT conduit installed in a building under the conditions described on pages 5 and 6. Costs listed are for each 100 linear feet installed. The crew is one electrician working at a labor cost of $26.88 per manhour. These costs include typical bending, boring out wood studs and joists (in concealed locations only), layout, material handling, and normal waste. Add for connectors, couplings, straps, boxes, wire, sales tax, delivery, supervision, mobilization, demobilization, cleanup, overhead and profit. Note: Conduit runs are assumed to be 50' long. Shorter runs will take more labor and longer runs will take less labor per linear foot.

EMT Fittings

Material	Craft@Hrs	Unit	Material Cost	Labor Cost	Installed Cost
EMT 45 degree elbows					
1"	1E@0.06	Ea	1.73	1.61	3.34
1-1/4"	1E@0.08	Ea	2.50	2.15	4.65
1-1/2"	1E@0.08	Ea	3.27	2.15	5.42
2"	1E@0.10	Ea	4.98	2.69	7.67
2-1/2"	1E@0.15	Ea	12.10	4.03	16.13
3"	1E@0.20	Ea	18.10	5.38	23.48
3-1/2"	1E@0.20	Ea	24.30	5.38	29.68
4"	1E@0.25	Ea	28.50	6.72	35.22
EMT 90 degree elbows					
1"	1E@0.08	Ea	1.73	2.15	3.88
1-1/4"	1E@0.10	Ea	2.50	2.69	5.19
1-1/2"	1E@0.10	Ea	3.27	2.69	5.96
2"	1E@0.15	Ea	4.98	4.03	9.01
2-/2"	1E@0.15	Ea	12.10	4.03	16.13
3"	1E@0.20	Ea	18.10	5.38	23.48
3-1/2"	1E@0.20	Ea	24.30	5.38	29.68
4"	1E@0.25	Ea	28.50	6.72	35.22

Use these figures to estimate the cost of EMT elbows installed on EMT conduit in a building under the conditions described on pages 5 and 6. Costs listed are for each elbow installed. The crew is one electrician working at a labor cost of $26.88 per manhour. These costs are for factory-made elbows and include layout, material handling, and normal waste. Add for field bending, couplings and connectors at the end of the run, sales tax, delivery, supervision, mobilization, demobilization, cleanup, overhead and profit. Note: Material costs assume purchase of full box quantities.

Conduit weight per 100 feet (in pounds)

Diameter	EMT steel	ENT plastic	PVC 40	Rigid steel	Intermediate rigid steel	Rigid aluminum
1/2"	30	11	18	79	57	30
3/4"	46	14	23	105	78	40
1"	66	20	35	153	112	59
1-1/4"	96	--	48	201	114	80
1-1/2"	112	--	57	249	176	96
2"	142	--	76	334	230	129
2-1/2"	230	--	125	527	393	205
3"	270	--	164	690	483	268
3-1/2"	350	--	198	831	561	321
4"	400	--	234	982	625	382
5"	--	--	317	1344	--	522
6"	--	--	412	1770	--	678

Material	Craft@Hrs	Unit	Material Cost	Labor Cost	Installed Cost
Indent EMT connectors					
1/2"	1E@0.05	Ea	.35	1.34	1.69
3/4"	1E@0.06	Ea	.64	1.61	2.25
1"	1E@0.08	Ea	1.01	2.15	3.16
Die cast set screw EMT connectors					
1/2"	1E@0.05	Ea	.21	1.34	1.55
3/4"	1E@0.06	Ea	.34	1.61	1.95
1"	1E@0.08	Ea	.56	2.15	2.71
1-1/4"	1E@0.10	Ea	.97	2.69	3.66
1-1/2"	1E@0.10	Ea	1.42	2.69	4.11
2"	1E@0.15	Ea	1.93	4.03	5.96
2-1/2"	1E@0.15	Ea	5.67	4.03	9.70
3"	1E@0.20	Ea	7.37	5.38	12.75
3-1/2"	1E@0.20	Ea	9.64	5.38	15.02
4"	1E@0.25	Ea	10.50	6.72	17.22
Die cast insulated throat set screw EMT connectors					
1/2"	1E@0.05	Ea	.33	1.34	1.67
3/4"	1E@0.06	Ea	.51	1.61	2.12
1"	1E@0.08	Ea	.79	2.15	2.94
1-1/4"	1E@0.10	Ea	1.65	2.69	4.34
1-1/2"	1E@0.10	Ea	2.13	2.69	4.82
2"	1E@0.15	Ea	3.23	4.03	7.26
2-1/2"	1E@0.15	Ea	11.30	4.03	15.33
3"	1E@0.20	Ea	14.20	5.38	19.58
3-1/2"	1E@0.20	Ea	18.40	5.38	23.78
4"	1E@0.25	Ea	21.00	6.72	27.72
Steel set screw EMT connectors					
1/2"	1E@0.05	Ea	.29	1.34	1.63
3/4"	1E@0.06	Ea	.47	1.61	2.08
1"	1E@0.08	Ea	.79	2.15	2.94
1-1/4"	1E@0.10	Ea	1.69	2.69	4.38
1-1/2"	1E@0.10	Ea	2.43	2.69	5.12
2"	1E@0.15	Ea	3.61	4.03	7.64
2-1/2"	1E@0.15	Ea	11.30	4.03	15.33
3"	1E@0.20	Ea	13.40	5.38	18.78
3-1/2"	1E@0.20	Ea	17.60	5.38	22.98
4"	1E@0.25	Ea	20.20	6.72	26.92

Use these figures to estimate the cost of EMT connectors installed on EMT conduit under the conditions described on pages 5 and 6. Costs listed are for each connector or expanded elbow installed. The crew is one electrician working at a labor cost of $26.88 per manhour. These costs include the connector locknut, removing the knockout, layout, material handling, and normal waste. Add for insulated bushings, sales tax, delivery, supervision, mobilization, demobilization, cleanup, overhead and profit. Note: Material costs assume purchase of full box quantities.

EMT Connectors

Material	Craft@Hrs	Unit	Material Cost	Labor Cost	Installed Cost
Steel insulated throat set screw EMT connectors					
1/2"	1E@0.05	Ea	.37	1.34	1.71
3/4"	1E@0.06	Ea	.60	1.61	2.21
1"	1E@0.08	Ea	.93	2.15	3.08
1-1/4"	1E@0.10	Ea	2.03	2.69	4.72
1-1/2"	1E@0.10	Ea	2.95	2.69	5.64
2"	1E@0.15	Ea	4.28	4.03	8.31
2-1/2"	1E@0.15	Ea	20.40	4.03	24.43
3"	1E@0.20	Ea	24.10	5.38	29.48
3-1/2"	1E@0.20	Ea	32.30	5.38	37.68
4"	1E@0.25	Ea	35.30	6.72	42.02

Material	Craft@Hrs	Unit	Material Cost	Labor Cost	Installed Cost
Die cast compression EMT connectors					
1/2"	1E@0.05	Ea	.33	1.34	1.67
3/4"	1E@0.06	Ea	.46	1.61	2.07
1"	1E@0.08	Ea	.76	2.15	2.91
1-1/4"	1E@0.10	Ea	1.41	2.69	4.10
1-1/2"	1E@0.10	Ea	2.05	2.69	4.74
2"	1E@0.15	Ea	3.01	4.03	7.04
2-1/2"	1E@0.15	Ea	8.64	4.03	12.67
3"	1E@0.20	Ea	11.70	5.38	17.08
3-1/2"	1E@0.20	Ea	16.70	5.38	22.08
4"	1E@0.25	Ea	19.00	6.72	25.72

Material	Craft@Hrs	Unit	Material Cost	Labor Cost	Installed Cost
Die cast insulated throat compression EMT connectors					
1/2"	1E@0.05	Ea	.41	1.34	1.75
3/4"	1E@0.06	Ea	.63	1.61	2.24
1"	1E@0.08	Ea	.97	2.15	3.12
1-1/4"	1E@0.10	Ea	1.80	2.69	4.49
1-1/2"	1E@0.10	Ea	2.76	2.69	5.45
2"	1E@0.15	Ea	4.03	4.03	8.06
2-1/2"	1E@0.15	Ea	18.20	4.03	22.23
3"	1E@0.20	Ea	22.80	5.38	28.18
3-1/2"	1E@0.20	Ea	30.80	5.38	36.18
4"	1E@0.25	Ea	34.80	6.72	41.52

Use these figures to estimate the cost of EMT connectors installed on EMT conduit under the conditions described on pages 5 and 6. Costs listed are for each connector installed. The crew is one electrician working at a labor cost of $26.88 per manhour. These costs include the connector locknut, removing the knockout, layout, material handling, and normal waste. Add for insulated bushings, sales tax, delivery, supervision, mobilization, demobilization, cleanup, overhead and profit. Note: Material costs assume purchase of full box quantities.

Material	Craft@Hrs	Unit	Material Cost	Labor Cost	Installed Cost
Steel compression EMT connectors					
1/2"	1E@0.05	Ea	.65	1.34	1.99
3/4"	1E@0.06	Ea	.90	1.61	2.51
1"	1E@0.08	Ea	1.35	2.15	3.50
1-1/4"	1E@0.10	Ea	2.70	2.69	5.39
1-1/2"	1E@0.10	Ea	3.94	2.69	6.63
2"	1E@0.15	Ea	5.68	4.03	9.71
2-1/2"	1E@0.15	Ea	17.80	4.03	21.83
3"	1E@0.20	Ea	24.60	5.38	29.98
3-1/2"	1E@0.20	Ea	37.10	5.38	42.48
4"	1E@0.25	Ea	37.90	6.72	44.62
Steel insulated throat compression EMT connectors					
1/2"	1E@0.05	Ea	.74	1.34	2.08
3/4"	1E@0.06	Ea	1.02	1.61	2.63
1"	1E@0.08	Ea	1.68	2.15	3.83
1-1/4"	1E@0.10	Ea	3.44	2.69	6.13
1-1/2"	1E@0.10	Ea	5.08	2.69	7.77
2"	1E@0.15	Ea	7.34	4.03	11.37
2-1/2"	1E@0.15	Ea	30.40	4.03	34.43
3"	1E@0.20	Ea	38.40	5.38	43.78
3-1/2"	1E@0.20	Ea	56.10	5.38	61.48
4"	1E@0.25	Ea	57.40	6.72	64.12
Die cast indent offset EMT connectors					
1/2"	1E@0.10	Ea	2.31	2.69	5.00
3/4"	1E@0.10	Ea	3.17	2.69	5.86
1"	1E@0.15	Ea	3.94	4.03	7.97
Die cast set screw offset EMT connectors					
1/2"	1E@0.10	Ea	3.48	2.69	6.17
3/4"	1E@0.10	Ea	4.62	2.69	7.31
1"	1E@0.15	Ea	5.96	4.03	9.99

Use these figures to estimate the cost of EMT connectors installed on EMT conduit under the conditions described on pages 5 and 6. Costs listed are for each connector installed. The crew is one electrician working at a labor cost of $26.88 per manhour. These costs include the connector locknut, removing the knockout, layout, material handling, and normal waste. Add for insulated bushings, sales tax, delivery, supervision, mobilization, demobilization, cleanup, overhead and profit. Note: Material costs assume purchase of full box quantities.

EMT Connectors and Couplings

Material	Craft@Hrs	Unit	Material Cost	Labor Cost	Installed Cost
Steel compression offset EMT connectors					
1/2"	1E@0.10	Ea	2.90	2.69	5.59
3/4"	1E@0.10	Ea	4.37	2.69	7.06
1"	1E@0.15	Ea	12.20	4.03	16.23
Indent EMT couplings					
1/2"	1E@0.05	Ea	.28	1.34	1.62
3/4"	1E@0.06	Ea	.44	1.61	2.05
1"	1E@0.08	Ea	.60	2.15	2.75
Die cast set screw EMT couplings					
1/2"	1E@0.05	Ea	.22	1.34	1.56
3/4"	1E@0.06	Ea	.37	1.61	1.98
1"	1E@0.08	Ea	.61	2.15	2.76
1-1/4"	1E@0.10	Ea	1.15	2.69	3.84
1-1/2"	1E@0.10	Ea	1.56	2.69	4.25
2"	1E@0.15	Ea	2.05	4.03	6.08
2-1/2"	1E@0.15	Ea	4.65	4.03	8.68
3"	1E@0.20	Ea	5.39	5.38	10.77
3-1/2"	1E@0.20	Ea	6.66	5.38	12.04
4"	1E@0.25	Ea	7.62	6.72	14.34

Use these figures to estimate the cost of EMT connectors and couplings installed on EMT conduit under the conditions described on pages 5 and 6. Costs listed are for each coupling or connector installed. The crew is one electrician working at a labor cost of $26.88 per manhour. These costs include the connector or coupling, layout, material handling, and normal waste. Add for conduit, sales tax, delivery, supervision, mobilization, demobilization, cleanup, overhead and profit. Note: Drive-on EMT fittings are rated as raintight and are also concrete tight. They are threaded with a standard electrical pipe thread and can be adapted easily to rigid conduit or other threaded fittings. Material costs assume purchase of full box quantities.

Material	Craft@Hrs	Unit	Material Cost	Labor Cost	Installed Cost
Set screw steel EMT couplings					
1/2"	1E@0.05	Ea	.37	1.34	1.71
3/4"	1E@0.06	Ea	.58	1.61	2.19
1"	1E@0.08	Ea	.90	2.15	3.05
1-1/4"	1E@0.10	Ea	1.97	2.69	4.66
1-1/2"	1E@0.10	Ea	2.87	2.69	5.56
2"	1E@0.15	Ea	3.85	4.03	7.88
2-1/2"	1E@0.15	Ea	9.32	4.03	13.35
3"	1E@0.20	Ea	10.50	5.38	15.88
3-1/2"	1E@0.20	Ea	12.30	5.38	17.68
4"	1E@0.25	Ea	14.10	6.72	20.82
Die cast compression EMT couplings					
1/2"	1E@0.05	Ea	.41	1.34	1.75
3/4"	1E@0.06	Ea	.57	1.61	2.18
1"	1E@0.08	Ea	.94	2.15	3.09
1-1/4"	1E@0.10	Ea	1.55	2.69	4.24
1-1/2"	1E@0.10	Ea	2.21	2.69	4.90
2"	1E@0.15	Ea	2.99	4.03	7.02
2-1/2"	1E@0.15	Ea	11.20	4.03	15.23
3"	1E@0.20	Ea	14.80	5.38	20.18
3-1/2"	1E@0.20	Ea	20.00	5.38	25.38
4"	1E@0.25	Ea	22.10	6.72	28.82
Steel compression EMT couplings					
1/2"	1E@0.05	Ea	.79	1.34	2.13
3/4"	1E@0.06	Ea	1.09	1.61	2.70
1"	1E@0.08	Ea	1.61	2.15	3.76
1-1/4"	1E@0.10	Ea	2.87	2.69	5.56
1-1/2"	1E@0.10	Ea	4.47	2.69	7.16
2"	1E@0.15	Ea	5.69	4.03	9.72
2-1/2"	1E@0.15	Ea	22.30	4.03	26.33
3"	1E@0.20	Ea	27.80	5.38	33.18
3-1/2"	1E@0.20	Ea	43.60	5.38	48.98
4"	1E@0.25	Ea	44.50	6.72	51.22

Use these figures to estimate the cost of EMT couplings installed on EMT conduit under the conditions described on pages 5 and 6. Costs listed are for each coupling installed. The crew is one electrician working at a labor cost of $26.88 per manhour. These costs include the coupling, layout, material handling, and normal waste. Add for conduit, sales tax, delivery, supervision, mobilization, demobilization, cleanup, overhead and profit. Note: Compression fittings are raintight and can be used in concrete. Material costs assume purchase of full box quantities.

EMT Couplings and Straps

Material	Craft@Hrs	Unit	Material Cost	Labor Cost	Installed Cost
Die cast EMT to flex couplings					
1/2"	1E@0.05	Ea	.93	1.34	2.27
3/4"	1E@0.05	Ea	1.24	1.34	2.58
1"	1E@0.06	Ea	1.74	1.61	3.35
Steel EMT to rigid compression couplings					
1/2"	1E@0.05	Ea	1.46	1.34	2.80
3/4"	1E@0.06	Ea	2.11	1.61	3.72
1"	1E@0.08	Ea	3.18	2.15	5.33
1-1/4"	1E@0.10	Ea	5.48	2.69	8.17
1-1/2"	1E@0.10	Ea	6.77	2.69	9.46
2"	1E@0.15	Ea	9.73	4.03	13.76
Steel EMT nail straps					
1/2"	1E@0.02	Ea	.04	.54	.58
3/4"	1E@0.03	Ea	.05	.81	.86
1"	1E@0.05	Ea	.07	1.34	1.41
Steel one hole EMT straps					
1/2"	1E@0.03	Ea	.12	.81	.93
3/4"	1E@0.04	Ea	.15	1.08	1.23
1"	1E@0.05	Ea	.23	1.34	1.57
1-1/4"	1E@0.06	Ea	.36	1.61	1.97
1-1/2"	1E@0.06	Ea	.56	1.61	2.17
2"	1E@0.10	Ea	.88	2.69	3.57
2-1/2"	1E@0.10	Ea	1.55	2.69	4.24
3"	1E@0.15	Ea	1.73	4.03	5.76
3-1/2"	1E@0.15	Ea	2.57	4.03	6.60
4"	1E@0.15	Ea	3.28	4.03	7.31

Use these figures to estimate the cost of EMT couplings and EMT straps installed on EMT conduit under the conditions described on pages 5 and 6. Costs listed are for each coupling and strap installed. The crew is one electrician working at a labor cost of $26.88 per manhour. These costs include cutting the EMT conduit, layout, material handling, and normal waste. Add the cost of conduit, sales tax, delivery, supervision, mobilization, demobilization, cleanup, overhead and profit. Note: Material costs assume purchase of full box quantities.

Material	Craft@Hrs	Unit	Material Cost	Labor Cost	Installed Cost

One hole heavy duty steel EMT straps

Material	Craft@Hrs	Unit	Material Cost	Labor Cost	Installed Cost
1/2"	1E@0.03	Ea	.34	.81	1.15
3/4"	1E@0.04	Ea	.45	1.08	1.53
1"	1E@0.05	Ea	.76	1.34	2.10
1-1/4"	1E@0.06	Ea	.98	1.61	2.59
1-1/2"	1E@0.06	Ea	1.54	1.61	3.15
2"	1E@0.10	Ea	2.44	2.69	5.13

One hole malleable EMT straps

Material	Craft@Hrs	Unit	Material Cost	Labor Cost	Installed Cost
1/2"	1E@0.03	Ea	.22	.81	1.03
3/4"	1E@0.04	Ea	.30	1.08	1.38
1"	1E@0.05	Ea	.44	1.34	1.78
1-1/4"	1E@0.06	Ea	.85	1.61	2.46
1-1/2"	1E@0.06	Ea	1.01	1.61	2.62
2"	1E@0.10	Ea	1.88	2.69	4.57
2-1/2"	1E@0.10	Ea	3.93	2.69	6.62
3"	1E@0.15	Ea	5.58	4.03	9.61
3-1/2"	1E@0.15	Ea	7.95	4.03	11.98
4"	1E@0.15	Ea	17.80	4.03	21.83

Two hole steel EMT straps

Material	Craft@Hrs	Unit	Material Cost	Labor Cost	Installed Cost
1/2"	1E@0.03	Ea	.06	.81	.87
3/4"	1E@0.04	Ea	.09	1.08	1.17
1"	1E@0.05	Ea	.14	1.34	1.48
1-1/4"	1E@0.06	Ea	.20	1.61	1.81
1-1/2"	1E@0.06	Ea	.24	1.61	1.85
2"	1E@0.10	Ea	.41	2.69	3.10
2-1/2"	1E@0.10	Ea	.72	2.69	3.41
3"	1E@0.15	Ea	.88	4.03	4.91
3-1/2"	1E@0.15	Ea	1.23	4.03	5.26
4"	1E@0.15	Ea	1.29	4.03	5.32

Use these figures to estimate the cost of EMT straps installed on EMT conduit under the conditions described on pages 5 and 6. Costs listed are for each strap installed. The crew is one electrician working at a labor cost of $26.88 per manhour. These costs include cutting the EMT conduit, layout, material handling, and normal waste. Add the cost of conduit, screws or nails to hold the straps, sales tax, delivery, supervision, mobilization, demobilization, cleanup, overhead and profit. Note: Material costs assume purchase of full box quantities.

EMT Straps, Hangers and Clips

Material	Craft@Hrs	Unit	Material Cost	Labor Cost	Installed Cost
Two hole heavy duty steel EMT straps					
1"	1E@0.05	Ea	.52	1.34	1.86
1-1/4"	1E@0.06	Ea	.78	1.61	2.39
1-1/2"	1E@0.06	Ea	.87	1.61	2.48
2"	1E@0.10	Ea	1.23	2.69	3.92
2-1/2"	1E@0.10	Ea	1.92	2.69	4.61
3"	1E@0.10	Ea	2.35	2.69	5.04
3-1/2"	1E@0.15	Ea	3.67	4.03	7.70
4"	1E@0.15	Ea	3.77	4.03	7.80
Steel EMT conduit hangers with bolt					
1/2"	1E@0.03	Ea	.18	.81	.99
3/4"	1E@0.04	Ea	.20	1.08	1.28
1"	1E@0.05	Ea	.21	1.34	1.55
1-1/4"	1E@0.06	Ea	.25	1.61	1.86
1-1/2"	1E@0.06	Ea	.41	1.61	2.02
2"	1E@0.10	Ea	.52	2.69	3.21
2-1/2"	1E@0.10	Ea	.61	2.69	3.30
3"	1E@0.15	Ea	.66	4.03	4.69
3-1/2"	1E@0.15	Ea	.75	4.03	4.78
4"	1E@0.15	Ea	.82	4.03	4.85
Beam clamp EMT conduit hanger assembly					
1/2"	1E@0.05	Ea	.74	1.34	2.08
3/4"	1E@0.06	Ea	.79	1.61	2.40
1"	1E@0.08	Ea	.85	2.15	3.00
1-1/4"	1E@0.10	Ea	.85	2.69	3.54
1-1/2"	1E@0.10	Ea	.88	2.69	3.57
2"	1E@0.15	Ea	1.04	4.03	5.07
Combination clips, EMT to strut					
1/2"	1E@0.06	Ea	.75	1.61	2.36
3/4"	1E@0.08	Ea	.84	2.15	2.99
1"	1E@0.10	Ea	.89	2.69	3.58
1-1/4"	1E@0.10	Ea	1.04	2.69	3.73
1-1/2"	1E@0.10	Ea	1.18	2.69	3.87
2"	1E@0.15	Ea	1.29	4.03	5.32

Use these figures to estimate the cost of EMT straps, hangers and clips installed on EMT conduit under the conditions described on pages 5 and 6. Costs listed are for each strap, hanger or clip installed. The crew is one electrician working at a labor cost of $26.88 per manhour. These costs include cutting the EMT conduit, layout, material handling, and normal waste. Add the cost of conduit, screws or nails to hold the straps, sales tax, delivery, supervision, mobilization, demobilization, cleanup, overhead and profit. Note: Material costs assume purchase of full box quantities.

EMT Clips, Adapters, Elbows, Caps and Benders

Material	Craft@Hrs	Unit	Material Cost	Labor Cost	Installed Cost
EMT clips for rod, wire, or steel flange					
1/2"	1E@0.04	Ea	.17	1.08	1.25
3/4"	1E@0.05	Ea	.17	1.34	1.51
1"	1E@0.06	Ea	.20	1.61	1.81
1-1/4"	1E@0.08	Ea	.28	2.15	2.43
EMT split adapters					
1/2"	1E@0.08	Ea	1.04	2.15	3.19
3/4"	1E@0.10	Ea	1.25	2.69	3.94
1"	1E@0.15	Ea	2.08	4.03	6.11
1-1/4"	1E@0.15	Ea	3.12	4.03	7.15
1-1/2"	1E@0.20	Ea	4.78	5.38	10.16
2"	1E@0.25	Ea	13.50	6.72	20.22
90 degree die cast set screw EMT elbows					
1/2"	1E@0.10	Ea	1.79	2.69	4.48
3/4"	1E@0.10	Ea	2.63	2.69	5.32
1"	1E@0.15	Ea	4.02	4.03	8.05
1-1/4"	1E@0.15	Ea	10.80	4.03	14.83
1-1/2"	1E@0.15	Ea	15.80	4.03	19.83
Die cast gasketed EMT pulling elbows					
1/2"	1E@0.10	Ea	2.27	2.69	4.96
3/4"	1E@0.10	Ea	3.11	2.69	5.80
1"	1E@0.15	Ea	7.19	4.03	11.22
1-1/4"	1E@0.15	Ea	10.80	4.03	14.83
Malleable EMT entrance caps					
1-1/4"	1E@0.15	Ea	4.83	4.03	8.86
1-1/2"	1E@0.20	Ea	8.61	5.38	13.99
2"	1E@0.30	Ea	16.00	8.06	24.06
2-1/2"	1E@0.70	Ea	41.50	18.80	60.30
EMT hand benders					
1/2"	--	Ea	15.80	--	--
3/4"	--	Ea	40.00	--	--
1"	--	Ea	65.20	--	--

Use these figures to estimate the cost of items shown above installed on EMT conduit under the conditions described on pages 5 and 6. Costs listed are for each item installed. The crew is one electrician working at a labor cost of $26.88 per manhour. These costs include the connector locknut, removing the knockout when required, layout, material handling, and normal waste. Add for conduit boxes, insulated bushings, sales tax, delivery, supervision, mobilization, demobilization, cleanup, overhead and profit. Note: Material costs assume purchase of full box quantities.

Flexible Conduit

Material	Craft@Hrs	Unit	Material Cost	Labor Cost	Installed Cost
Flex reduced wall steel conduit					
3/8"	1E@2.70	CLF	17.20	72.60	89.80
1/2"	1E@3.00	CLF	24.50	80.60	105.10
3/4"	1E@3.25	CLF	31.70	87.40	119.10
1"	1E@3.50	CLF	65.60	94.10	159.70
1-1/4"	1E@3.70	CLF	83.10	99.50	182.60
1-1/2"	1E@3.85	CLF	108.00	103.00	211.00
2"	1E@4.00	CLF	157.00	108.00	265.00
2-1/2"	1E@5.00	CLF	186.00	134.00	320.00
3"	1E@6.00	CLF	232.00	161.00	393.00
Flex reduced wall aluminum conduit					
3/8"	1E@2.50	CLF	21.00	67.20	88.20
1/2"	1E@2.70	CLF	30.00	72.60	102.60
3/4"	1E@3.00	CLF	42.00	80.60	122.60
1"	1E@3.30	CLF	86.10	88.70	174.80
1-1/4"	1E@3.50	CLF	110.00	94.10	204.10
1-1/2"	1E@3.70	CLF	141.00	99.50	240.50
2"	1E@3.80	CLF	205.00	102.00	307.00
2-1/2"	1E@4.25	CLF	261.00	114.00	375.00
3"	1E@6.00	CLF	302.00	161.00	463.00
3-1/2"	1E@8.00	CLF	400.00	215.00	615.00
4"	1E@10.0	CLF	476.00	269.00	745.00

Use these figures to estimate the cost of flexible conduit installed in a building, and for equipment hookup under the conditions described on pages 5 and 6. Costs listed are for each 100 linear feet installed. The crew is one electrician working at a labor cost of $26.88 per manhour. These costs include boring or notching wood studs and joists (in concealed locations), cutting flex conduit, layout, material handling, and normal waste. Add for connectors, couplings, straps, boxes, wire, bonding wire, sales tax, delivery, supervision, mobilization, demobilization, cleanup, overhead and profit. Note: Conduit runs are assumed to be 25' long. Labor costs per linear foot will be higher on shorter runs and lower on longer runs.

Material	Craft@Hrs	Unit	Material Cost	Labor Cost	Installed Cost
Die cast screw-in flex connectors					
3/8"	1E@0.03	Ea	.31	.81	1.12
1/2"	1E@0.03	Ea	.31	.81	1.12
3/4"	1E@0.05	Ea	.50	1.34	1.84
1"	1E@0.06	Ea	1.18	1.61	2.79
1-1/4"	1E@0.10	Ea	2.43	2.69	5.12
1-1/2"	1E@0.10	Ea	3.20	2.69	5.89
2"	1E@0.15	Ea	6.13	4.03	10.16
Die cast insulated throat screw-in flex connectors					
3/8"	1E@0.03	Ea	.42	.81	1.23
1/2"	1E@0.03	Ea	.42	.81	1.23
3/4"	1E@0.05	Ea	.63	1.34	1.97
1"	1E@0.06	Ea	1.86	1.61	3.47
1-1/4"	1E@0.10	Ea	4.42	2.69	7.11
1-1/2"	1E@0.10	Ea	5.50	2.69	8.19
2"	1E@0.15	Ea	10.40	4.03	14.43
Die cast duplex flex connectors					
3/8"	1E@0.05	Ea	.51	1.34	1.85
Die cast squeeze flex connectors					
3/8"	1E@0.05	Ea	.32	1.34	1.66
1/2"	1E@0.05	Ea	.42	1.34	1.76
3/4"	1E@0.06	Ea	.62	1.61	2.23
1"	1E@0.08	Ea	1.57	2.15	3.72
1-1/4"	1E@0.10	Ea	3.66	2.69	6.35
1-1/2"	1E@0.10	Ea	5.05	2.69	7.74
2"	1E@0.15	Ea	7.07	4.03	11.10

Use these figures to estimate the cost of flexible conduit connectors installed on flex conduit under the conditions described on pages 5 and 6. Costs listed are for each connector installed. The crew is one electrician working at a labor cost of $26.88 per manhour. These costs include the locknut, removing the knockout, layout, material handling, and normal waste. Add for connectors, couplings, straps, boxes, wire, bonding wire, sales tax, delivery, supervision, mobilization, demobilization, cleanup, overhead and profit. Note: Material costs assume purchase of full boxes.

Squeeze Flexible Conduit Connectors

Material	Craft@Hrs	Unit	Material Cost	Labor Cost	Installed Cost
Die cast insulated squeeze flex connectors					
1/2"	1E@0.05	Ea	.89	1.34	2.23
3/4"	1E@0.06	Ea	1.19	1.61	2.80
1"	1E@0.08	Ea	2.08	2.15	4.23
1-1/4"	1E@0.10	Ea	3.40	2.69	6.09
1-1/2"	1E@0.10	Ea	5.40	2.69	8.09
2"	1E@0.15	Ea	7.43	4.03	11.46
Malleable squeeze flex connectors					
3/8"	1E@0.05	Ea	.54	1.34	1.88
1/2"	1E@0.05	Ea	.87	1.34	2.21
3/4"	1E@0.06	Ea	.91	1.61	2.52
1"	1E@0.08	Ea	2.20	2.15	4.35
1-1/4"	1E@0.10	Ea	3.84	2.69	6.53
1-1/2"	1E@0.10	Ea	5.33	2.69	8.02
2"	1E@0.15	Ea	7.60	4.03	11.63
2-1/2"	1E@0.15	Ea	14.60	4.03	18.63
3"	1E@0.20	Ea	20.20	5.38	25.58
Malleable insulated squeeze flex connectors					
3/8"	1E@0.05	Ea	1.10	1.34	2.44
1/2"	1E@0.05	Ea	1.17	1.34	2.51
3/4"	1E@0.06	Ea	1.30	1.61	2.91
1"	1E@0.08	Ea	2.77	2.15	4.92
1-1/4"	1E@0.10	Ea	4.99	2.69	7.68
1-1/2"	1E@0.10	Ea	7.38	2.69	10.07
2"	1E@0.15	Ea	11.20	4.03	15.23
2-1/2"	1E@0.15	Ea	22.00	4.03	26.03
3"	1E@0.15	Ea	29.00	4.03	33.03

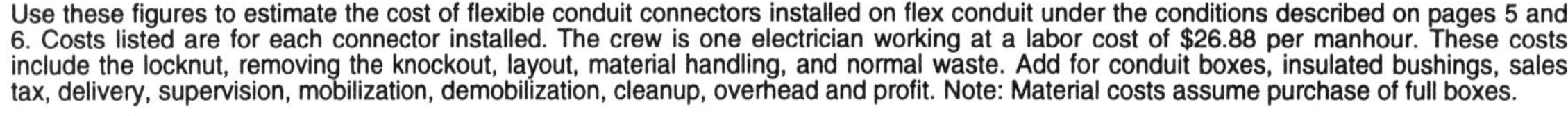

Use these figures to estimate the cost of flexible conduit connectors installed on flex conduit under the conditions described on pages 5 and 6. Costs listed are for each connector installed. The crew is one electrician working at a labor cost of $26.88 per manhour. These costs include the locknut, removing the knockout, layout, material handling, and normal waste. Add for conduit boxes, insulated bushings, sales tax, delivery, supervision, mobilization, demobilization, cleanup, overhead and profit. Note: Material costs assume purchase of full boxes.

Material	Craft@Hrs	Unit	Material Cost	Labor Cost	Installed Cost
45 degree die cast flex connectors					
3/8"	1E@0.05	Ea	1.28	1.34	2.62
1/2"	1E@0.05	Ea	2.01	1.34	3.35
3/4"	1E@0.06	Ea	2.76	1.61	4.37
45 degree malleable flex connectors					
3/8"	1E@0.05	Ea	1.28	1.34	2.62
1/2"	1E@0.06	Ea	2.01	1.61	3.62
90 degree die cast two screw flex connectors					
3/8"	1E@0.05	Ea	.86	1.34	2.20
1/2"	1E@0.05	Ea	1.46	1.34	2.80
3/4"	1E@0.06	Ea	2.29	1.61	3.90
1"	1E@0.08	Ea	3.96	2.15	6.11
1-1/4"	1E@0.10	Ea	8.05	2.69	10.74
1-1/2"	1E@0.15	Ea	13.00	4.03	17.03
2"	1E@0.20	Ea	16.20	5.38	21.58
2-1/2"	1E@0.25	Ea	46.80	6.72	53.52
3"	1E@0.25	Ea	62.30	6.72	69.02
90 degree die cast two screw insulated flex connectors					
3/8"	1E@0.05	Ea	.90	1.34	2.24
1/2"	1E@0.05	Ea	1.40	1.34	2.74
3/4"	1E@0.06	Ea	2.29	1.61	3.90
1"	1E@0.08	Ea	4.20	2.15	6.35
1-1/4"	1E@0.10	Ea	8.48	2.69	11.17
1-1/2"	1E@0.15	Ea	13.10	4.03	17.13
2"	1E@0.20	Ea	18.20	5.38	23.58
2-1/2"	1E@0.25	Ea	49.20	6.72	55.92
3"	1E@0.25	Ea	66.90	6.72	73.62

Use these figures to estimate the cost of flexible conduit connectors installed on flex conduit under the conditions described on pages 5 and 6. Costs listed are for each connector installed. The crew is one electrician working at a labor cost of $26.88 per manhour. These costs include the locknut, removing the knockout, layout, material handling, and normal waste. Add for conduit boxes, insulated bushings, sales tax, delivery, supervision, mobilization, demobilization, cleanup, overhead and profit. Note: Material costs assume purchase of full boxes.

Flexible Conduit Connectors and Couplings

Material	Craft@Hrs	Unit	Material Cost	Labor Cost	Installed Cost
90 degree malleable squeeze flex connectors					
3/8"	1E@0.05	Ea	1.00	1.34	2.34
1/2"	1E@0.05	Ea	1.55	1.34	2.89
3/4"	1E@0.06	Ea	2.55	1.61	4.16
1"	1E@0.08	Ea	4.67	2.15	6.82
1-1/4"	1E@0.10	Ea	9.42	2.69	12.11
1-1/2"	1E@0.15	Ea	14.50	4.03	18.53
2"	1E@0.20	Ea	20.20	5.38	25.58
2-1/2"	1E@0.25	Ea	54.70	6.72	61.42
3"	1E@0.25	Ea	74.30	6.72	81.02
3-1/2	1E@0.30	Ea	300.00	8.06	308.06
4"	1E@0.30	Ea	345.00	8.06	353.06
90 degree malleable insulated squeeze flex connectors					
3/8"	1E@0.05	Ea	1.17	1.34	2.51
1/2"	1E@0.05	Ea	1.80	1.34	3.14
3/4"	1E@0.06	Ea	2.88	1.61	4.49
1"	1E@0.08	Ea	4.63	2.15	6.78
1-1/4"	1E@0.10	Ea	8.88	2.69	11.57
1-1/2"	1E@0.15	Ea	15.90	4.03	19.93
2"	1E@0.20	Ea	20.80	5.38	26.18
2-1/2"	1E@0.25	Ea	55.80	6.72	62.52
3"	1E@0.25	Ea	76.30	6.72	83.02
3-1/2"	1E@0.30	Ea	300.00	8.06	308.06
4"	1E@0.30	Ea	345.00	8.06	353.06
Die cast screw-in flex couplings					
1/2"	1E@0.03	Ea	.45	.81	1.26
3/4"	1E@0.05	Ea	.65	1.34	1.99
1"	1E@0.06	Ea	.79	1.61	2.40
Die cast screw-in flex to EMT couplings					
3/8"	1E@0.05	Ea	.93	1.34	2.27
1/2"	1E@0.06	Ea	.93	1.61	2.54
3/4"	1E@0.08	Ea	1.24	2.15	3.39
1"	1E@0.10	Ea	1.74	2.69	4.43
Die cast set screw flex to rigid couplings					
1/2"	1E@0.05	Ea	2.01	1.34	3.35
3/4"	1E@0.06	Ea	3.07	1.61	4.68

Use these figures to estimate the cost of flexible conduit connectors and couplings installed on flex conduit under the conditions described on pages 5 and 6. Costs listed are for each connector or coupling installed. The crew is one electrician working at a labor cost of $26.88 per manhour. These costs include the locknut, removing the knockout, layout, material handling, and normal waste. Add for conduit boxes, insulated bushings, sales tax, delivery, supervision, mobilization, demobilization, cleanup, overhead and profit. Note: Material costs assume purchase of full boxes.

Material	Craft@Hrs	Unit	Material Cost	Labor Cost	Installed Cost
Type EF or Type LT flex steel conduit					
3/8"	1E@4.00	CLF	91.00	108.00	199.00
1/2"	1E@4.00	CLF	99.50	108.00	207.50
3/4"	1E@4.50	CLF	133.00	121.00	254.00
1"	1E@5.00	CLF	202.00	134.00	336.00
1-1/4"	1E@6.00	CLF	274.00	161.00	435.00
1-1/2"	1E@7.00	CLF	373.00	188.00	561.00
2"	1E@9.00	CLF	471.00	242.00	713.00
2-1/2"	1E@11.0	CLF	870.00	296.00	1,166.00
3"	1E@15.0	CLF	1,210.00	403.00	1,613.00
4"	1E@17.0	CLF	1,750.00	457.00	2,207.00
Type UA liquid-tight flex conduit					
3/8"	1E@4.00	CLF	128.00	108.00	236.00
1/2"	1E@4.00	CLF	150.00	108.00	258.00
3/4"	1E@4.50	CLF	209.00	121.00	330.00
1"	1E@5.00	CLF	324.00	134.00	458.00
1-1/4"	1E@6.00	CLF	459.00	161.00	620.00
1-1/2"	1E@7.00	CLF	480.00	188.00	668.00
2"	1E@9.00	CLF	598.00	242.00	840.00
2-1/2"	1E@11.0	CLF	1,070.00	296.00	1,366.00
3"	1E@15.0	CLF	1,510.00	403.00	1,913.00
4"	1E@17.0	CLF	2,130.00	457.00	2,587.00
Type OR liquid-tight flex conduit					
3/8"	1E@4.00	CLF	110.00	108.00	218.00
1/2"	1E@4.00	CLF	124.00	108.00	232.00
3/4"	1E@4.50	CLF	189.00	121.00	310.00
1"	1E@5.00	CLF	257.00	134.00	391.00
1-1/4"	1E@6.00	CLF	339.00	161.00	500.00
1-1/2"	1E@7.00	CLF	478.00	188.00	666.00
2"	1E@9.00	CLF	577.00	242.00	819.00
2-1/2"	1E@11.0	CLF	1,130.00	296.00	1,426.00
3"	1E@15.0	CLF	1,550.00	403.00	1,953.00
4"	1E@17.0	CLF	2,180.00	457.00	2,637.00
Construction grade liquid-tight flex conduit					
3/8"	1E@4.00	CLF	57.20	108.00	165.20
1/2"	1E@4.00	CLF	67.00	108.00	175.00
3/4"	1E@4.50	CLF	93.60	121.00	214.60
1"	1E@5.00	CLF	154.00	134.00	288.00
1-1/4"	1E@6.00	CLF	191.00	161.00	352.00
1-1/2"	1E@7.00	CLF	307.00	188.00	495.00
2"	1E@9.00	CLF	346.00	242.00	588.00

Use these figures to estimate the cost of liquid-tight flex conduit installed as part of equipment hookup under the conditions described on pages 5 and 6. Costs listed are for each linear foot installed. The crew is one electrician working at a labor cost of $26.88 per manhour. These costs include cutting conduit, layout, material handling, and normal waste. Add for connectors, boxes, straps, wire, bonding wire, sales tax, delivery, supervision, mobilization, demobilization, cleanup, overhead and profit. Note: Conduit runs are assumed to be 25' long. Labor costs per linear foot will be higher on shorter runs and lower on longer runs.

Liquid-tight Flexible Conduit and Connectors

Material	Craft@Hrs	Unit	Material Cost	Labor Cost	Installed Cost
Type HC liquid-tight extra flex conduit					
3/8"	1E@4.00	CLF	127.00	108.00	235.00
1/2"	1E@4.00	CLF	133.00	108.00	241.00
3/4"	1E@4.50	CLF	186.00	121.00	307.00
1"	1E@5.00	CLF	271.00	134.00	405.00
1-1/4"	1E@6.00	CLF	368.00	161.00	529.00
1-1/2"	1E@7.00	CLF	500.00	188.00	688.00
2"	1E@9.00	CLF	623.00	242.00	865.00
2-1/2"	1E@11.0	CLF	1,160.00	296.00	1,456.00
3"	1E@15.0	CLF	1,610.00	403.00	2,013.00
4"	1E@17.0	CLF	2,350.00	457.00	2,807.00
Type CN-P liquid-tight flex non-metallic conduit					
3/8"	1E@3.50	CLF	159.00	94.10	253.10
1/2"	1E@3.75	CLF	206.00	101.00	307.00
3/4"	1E@4.00	CLF	293.00	108.00	401.00
1"	1E@4.50	CLF	371.00	121.00	492.00
1-1/4"	1E@5.00	CLF	486.00	134.00	620.00
1-1/2"	1E@5.50	CLF	805.00	148.00	953.00
2"	1E@6.00	CLF	1,120.00	161.00	1,281.00
Malleable liquid-tight flex connectors					
3/8"	1E@0.10	Ea	1.61	2.69	4.30
1/2"	1E@0.10	Ea	1.61	2.69	4.30
3/4"	1E@0.10	Ea	2.29	2.69	4.98
1"	1E@0.15	Ea	3.36	4.03	7.39
1-1/4"	1E@0.20	Ea	5.77	5.38	11.15
1-1/2"	1E@0.20	Ea	8.22	5.38	13.60
2"	1E@0.25	Ea	15.10	6.72	21.82
2-1/2"	1E@0.25	Ea	69.70	6.72	76.42
3"	1E@0.30	Ea	78.30	8.06	86.36
4"	1E@0.30	Ea	98.80	8.06	106.86
Malleable insulated throat liquid-tight flex connectors					
3/8"	1E@0.10	Ea	1.93	2.69	4.62
1/2"	1E@0.10	Ea	1.93	2.69	4.62
3/4"	1E@0.10	Ea	2.83	2.69	5.52
1"	1E@0.15	Ea	4.37	4.03	8.40
1-1/4"	1E@0.20	Ea	7.08	5.38	12.46
1-1/2"	1E@0.20	Ea	10.10	5.38	15.48
2"	1E@0.25	Ea	19.00	6.72	25.72
2-1/2"	1E@0.25	Ea	104.00	6.72	110.72
3"	1E@0.30	Ea	116.00	8.06	124.06
4"	1E@0.30	Ea	139.00	8.06	147.06

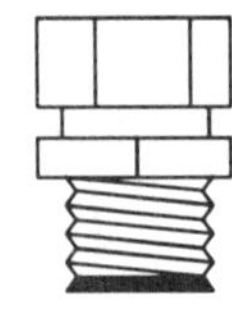

Use these figures to estimate the cost of liquid-tight flex conduit (two top tables) and connectors (two bottom tables) installed with equipment hookup under the conditions described on pages 5 and 6. Costs listed are for each 100 linear feet of conduit and each connector installed. The crew is one electrician working at a labor cost of $26.88 per manhour. These costs include cutting conduit, removal of the knockout for the connector, layout, material handling, and normal waste. Add for straps, boxes, wire, bonding wire, sales tax, delivery, supervision, mobilization, demobilization, cleanup, overhead and profit. Note: Connector costs assume the purchase of full box quantities.

Material	Craft@Hrs	Unit	Material Cost	Labor Cost	Installed Cost

Die cast liquid-tight flex connectors

Material	Craft@Hrs	Unit	Material Cost	Labor Cost	Installed Cost
3/8"	1E@0.10	Ea	1.42	2.69	4.11
1/2"	1E@0.10	Ea	1.42	2.69	4.11
3/4"	1E@0.10	Ea	2.05	2.69	4.74
1"	1E@0.15	Ea	2.98	4.03	7.01
1-1/4"	1E@0.20	Ea	5.10	5.38	10.48
1-1/2"	1E@0.20	Ea	7.26	5.38	12.64
2"	1E@0.25	Ea	13.20	6.72	19.92
2-1/2"	1E@0.25	Ea	64.10	6.72	70.82
3"	1E@0.30	Ea	72.40	8.06	80.46
4"	1E@0.30	Ea	82.40	8.06	90.46

Die cast insulated throat liquid-tight flex connectors

Material	Craft@Hrs	Unit	Material Cost	Labor Cost	Installed Cost
3/8"	1E@0.10	Ea	1.67	2.69	4.36
1/2"	1E@0.10	Ea	1.67	2.69	4.36
3/4"	1E@0.10	Ea	2.45	2.69	5.14
1"	1E@0.15	Ea	3.77	4.03	7.80
1-1/4"	1E@0.20	Ea	5.91	5.38	11.29
1-1/2"	1E@0.20	Ea	8.40	5.38	13.78
2"	1E@0.25	Ea	15.60	6.72	22.32
2-1/2"	1E@0.25	Ea	79.90	6.72	86.62
3"	1E@0.30	Ea	89.00	8.06	97.06
4"	1E@0.30	Ea	104.00	8.06	112.06

45 degree malleable liquid-tight flex connectors

Material	Craft@Hrs	Unit	Material Cost	Labor Cost	Installed Cost
3/8"	1E@0.12	Ea	2.60	3.23	5.83
1/2"	1E@0.12	Ea	2.60	3.23	5.83
3/4"	1E@0.15	Ea	3.94	4.03	7.97
1"	1E@0.15	Ea	8.13	4.03	12.16
1-1/4"	1E@0.20	Ea	12.50	5.38	17.88
1-1/2"	1E@0.20	Ea	15.50	5.38	20.88
2"	1E@0.25	Ea	22.70	6.72	29.42
2-1/2"	1E@0.25	Ea	102.00	6.72	108.72
3"	1E@0.30	Ea	123.00	8.06	131.06
4"	1E@0.30	Ea	136.00	8.06	144.06

45 degree malleable insulated throat liquid-tight flex connectors

Material	Craft@Hrs	Unit	Material Cost	Labor Cost	Installed Cost
3/8"	1E@0.12	Ea	3.22	3.23	6.45
1/2"	1E@0.12	Ea	3.22	3.23	6.45
3/4"	1E@0.15	Ea	4.85	4.03	8.88

Use these figures to estimate the cost of liquid-tight flex connectors installed on liquid-tight flex conduit under the conditions described on pages 5 and 6. Costs listed are for each connector installed. The crew is one electrician working at a labor cost of $26.88 per manhour. These costs include locknuts and removal of the knockout, layout, material handling, and normal waste. Add for conduit, insulating bushings, sales tax, delivery, supervision, mobilization, demobilization, cleanup, overhead and profit. Note: Material costs assume purchase of full box quantities.

Liquid-tight Flex Connectors and Couplings

Material	Craft@Hrs	Unit	Material Cost	Labor Cost	Installed Cost

45 degree malleable insulated throat liquid-tight flex connectors

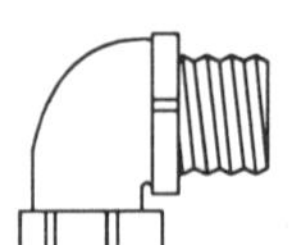

Material	Craft@Hrs	Unit	Material Cost	Labor Cost	Installed Cost
1"	1E@0.15	Ea	9.21	4.03	13.24
1-1/4"	1E@0.20	Ea	14.50	5.38	19.88
1-1/2"	1E@0.20	Ea	17.70	5.38	23.08
2"	1E@0.25	Ea	26.60	6.72	33.32
2-1/2"	1E@0.25	Ea	131.00	6.72	137.72
3"	1E@0.30	Ea	157.00	8.06	165.06
4"	1E@0.30	Ea	181.00	8.06	189.06

90 degree malleable liquid-tight flex connectors

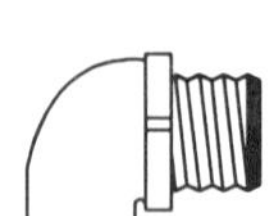

Material	Craft@Hrs	Unit	Material Cost	Labor Cost	Installed Cost
3/8"	1E@0.15	Ea	2.60	4.03	6.63
1/2"	1E@0.15	Ea	2.60	4.03	6.63
3/4"	1E@0.15	Ea	3.94	4.03	7.97
1"	1E@0.20	Ea	8.05	5.38	13.43
1-1/4"	1E@0.25	Ea	12.40	6.72	19.12
1-1/2"	1E@0.25	Ea	14.90	6.72	21.62
2"	1E@0.30	Ea	22.50	8.06	30.56
2-1/2"	1E@0.30	Ea	99.00	8.06	107.06
3"	1E@0.40	Ea	120.00	10.80	130.80
4"	1E@0.40	Ea	155.00	10.80	165.80

90 degree malleable insulated throat liquid-tight flex connectors

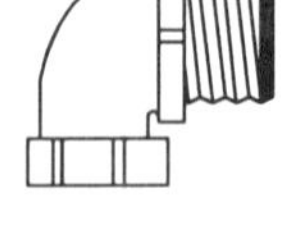

Material	Craft@Hrs	Unit	Material Cost	Labor Cost	Installed Cost
3/8"	1E@0.15	Ea	3.22	4.03	7.25
1/2"	1E@0.15	Ea	3.22	4.03	7.25
3/4"	1E@0.15	Ea	4.85	4.03	8.88
1"	1E@0.20	Ea	9.21	5.38	14.59
1-1/4"	1E@0.25	Ea	14.00	6.72	20.72
1-1/2"	1E@0.25	Ea	17.20	6.72	23.92
2"	1E@0.30	Ea	25.70	8.06	33.76
2-1/2"	1E@0.30	Ea	130.00	8.06	138.06
3"	1E@0.40	Ea	156.00	10.80	166.80
4"	1E@0.40	Ea	204.00	10.80	214.80

Malleable liquid-tight flex to rigid combination couplings

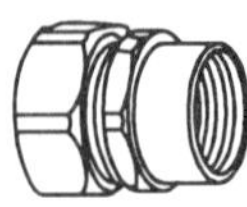

Material	Craft@Hrs	Unit	Material Cost	Labor Cost	Installed Cost
1/2"	1E@0.15	Ea	2.16	4.03	6.19
3/4"	1E@0.15	Ea	3.05	4.03	7.08
1"	1E@0.20	Ea	4.19	5.38	9.57
1-1/4"	1E@0.25	Ea	9.72	6.72	16.44
1-1/2"	1E@0.25	Ea	17.60	6.72	24.32
2"	1E@0.30	Ea	24.20	8.06	32.26
2-1/2"	1E@0.30	Ea	112.00	8.06	120.06
3"	1E@0.40	Ea	124.00	10.80	134.80
4"	1E@0.40	Ea	153.00	10.80	163.80

Use these figures to estimate the cost of liquid-tight flex connectors installed on liquid-tight flex conduit under the conditions described on pages 5 and 6. Costs listed are for each connector installed. The crew is one electrician working at a labor cost of $26.88 per manhour. These costs include locknuts and removal of the knockout, layout, material handling, and normal waste. Add for conduit, insulating bushings, sales tax, delivery, supervision, mobilization, demobilization, cleanup, overhead and profit. Note: Material costs assume purchase of full box quantities.

Material	Craft@Hrs	Unit	Material Cost	Labor Cost	Installed Cost

Schedule 40 PVC conduit, 10' lengths with coupling

Material	Craft@Hrs	Unit	Material Cost	Labor Cost	Installed Cost
1/2"	1E@3.10	CLF	9.00	83.30	92.30
3/4"	1E@3.20	CLF	12.00	86.00	98.00
1"	1E@3.30	CLF	18.00	88.70	106.70
1-1/4"	1E@3.40	CLF	24.00	91.40	115.40
1-1/2"	1E@3.45	CLF	29.00	92.70	121.70
2"	1E@3.50	CLF	36.00	94.10	130.10
2-1/2"	2E@3.60	CLF	65.00	96.80	161.80
3"	2E@3.75	CLF	80.00	101.00	181.00
3-1/2"	2E@3.85	CLF	94.00	103.00	197.00
4"	2E@4.00	CLF	108.00	108.00	216.00
5"	2E@4.25	CLF	151.00	114.00	265.00
6"	2E@4.50	CLF	196.00	121.00	317.00

Schedule 80 heavy wall PVC conduit, 10' lengths with coupling

Material	Craft@Hrs	Unit	Material Cost	Labor Cost	Installed Cost
1/2"	1E@3.20	CLF	16.00	86.00	102.00
3/4"	1E@3.30	CLF	22.00	88.70	110.70
1"	1E@3.40	CLF	32.00	91.40	123.40
1-1/4"	1E@3.50	CLF	44.00	94.10	138.10
1-1/2"	1E@3.60	CLF	53.00	96.80	149.80
2"	1E@3.70	CLF	73.00	99.50	172.50
2-1/2"	2E@3.90	CLF	109.00	105.00	214.00
3"	2E@4.00	CLF	142.00	108.00	250.00
4"	2E@4.50	CLF	208.00	121.00	329.00
5"	2E@5.00	CLF	300.00	134.00	434.00

30 degree Schedule 40 PVC elbows

Material	Craft@Hrs	Unit	Material Cost	Labor Cost	Installed Cost
1/2"	1E@0.05	Ea	1.26	1.34	2.60
3/4"	1E@0.06	Ea	1.31	1.61	2.92
1"	1E@0.08	Ea	1.59	2.15	3.74
1-1/4"	1E@0.10	Ea	2.23	2.69	4.92
1-1/2"	1E@0.10	Ea	3.06	2.69	5.75
2"	1E@0.15	Ea	4.82	4.03	8.85
2-1/2"	2E@0.15	Ea	9.13	4.03	13.16
3"	2E@0.20	Ea	15.50	5.38	20.88
3-1/2"	2E@0.20	Ea	17.60	5.38	22.98
4"	2E@0.25	Ea	25.90	6.72	32.62
5"	2E@0.30	Ea	41.60	8.06	49.66
6"	2E@0.50	Ea	48.50	13.40	61.90

Use these figures to estimate the cost of PVC conduit and elbows installed underground or in a building under the conditions described on pages 5 and 6. Costs listed are for 100 linear feet of conduit installed or for each elbow installed. The crew is one electrician for diameters to 2" and two electricians for 2-1/2" and larger conduit. The labor cost is $26.88 per manhour. These costs include making up joints with cement (glue), layout, material handling, and normal waste. Add for bends, connectors, end bell, spacers, wire, trenching, encasement, sales tax, delivery, supervision, mobilization, demobilization, cleanup, overhead and profit. Conduit runs are assumed to be 50' long. Shorter runs will take more labor and longer runs will take less labor per linear foot.

PVC Elbows and Couplings

Material	Craft@Hrs	Unit	Material Cost	Labor Cost	Installed Cost
45 degree Schedule 40 PVC elbows					
1/2"	1E@0.05	Ea	.86	1.34	2.20
3/4"	1E@0.06	Ea	.95	1.61	2.56
1"	1E@0.08	Ea	1.52	2.15	3.67
1-1/4"	1E@0.10	Ea	2.12	2.69	4.81
1-1/2"	1E@0.10	Ea	2.90	2.69	5.59
2"	1E@0.15	Ea	4.26	4.03	8.29
2-1/2"	1E@0.15	Ea	8.10	4.03	12.13
3"	1E@0.20	Ea	13.80	5.38	19.18
3-1/2"	1E@0.20	Ea	15.60	5.38	20.98
4"	1E@0.25	Ea	23.10	6.72	29.82
5"	1E@0.35	Ea	37.00	9.41	46.41
6"	1E@0.50	Ea	53.80	13.40	67.20
90 degree Schedule 40 PVC elbows					
1/2"	1E@0.05	Ea	.95	1.34	2.29
3/4"	1E@0.06	Ea	1.04	1.61	2.65
1"	1E@0.08	Ea	1.61	2.15	3.76
1-1/4"	1E@0.10	Ea	2.30	2.69	4.99
1-1/2"	1E@0.10	Ea	3.11	2.69	5.80
2"	1E@0.15	Ea	4.52	4.03	8.55
2-1/2"	1E@0.15	Ea	8.20	4.03	12.23
3"	1E@0.20	Ea	14.40	5.38	19.78
3-1/2"	1E@0.20	Ea	19.90	5.38	25.28
4"	1E@0.25	Ea	24.90	6.72	31.62
5"	1E@0.35	Ea	43.80	9.41	53.21
6"	1E@0.50	Ea	74.20	13.40	87.60
90 degree Schedule 80 PVC elbows					
1/2"	1E@0.06	Ea	1.49	1.61	3.10
3/4"	1E@0.08	Ea	2.12	2.15	4.27
1"	1E@0.10	Ea	3.22	2.69	5.91
1-1/4"	1E@0.15	Ea	4.35	4.03	8.38
1-1/2"	1E@0.15	Ea	6.41	4.03	10.44
2"	1E@0.20	Ea	8.67	5.38	14.05
2-1/2"	1E@0.20	Ea	15.30	5.38	20.68
3"	1E@0.25	Ea	29.40	6.72	36.12
4"	1E@0.30	Ea	47.90	8.06	55.96
Schedule 40 PVC couplings					
1/2"	1E@0.02	Ea	.24	.54	.78
3/4"	1E@0.03	Ea	.30	.81	1.11
1"	1E@0.05	Ea	.46	1.34	1.80
1-1/4"	1E@0.06	Ea	.60	1.61	2.21
1-1/2"	1E@0.08	Ea	.86	2.15	3.01

Use these figures to estimate the cost of PVC elbows and couplings installed on PVC conduit under the conditions described on pages 5 and 6. Costs listed are for each elbow or coupling installed. The crew is one electrician working at a labor cost of $26.88 per manhour. These costs include applying cement (glue), layout, material handling, and normal waste. Add for conduit, couplings, connectors, end bells, spacers, sales tax, delivery, supervision, mobilization, demobilization, cleanup, overhead and profit. Material costs assume purchase of full box quantities.

PVC Couplings, Adapters and Expansion Couplings

Material	Craft@Hrs	Unit	Material Cost	Labor Cost	Installed Cost

Schedule 40 PVC couplings

Material	Craft@Hrs	Unit	Material Cost	Labor Cost	Installed Cost
2"	1E@0.10	Ea	1.12	2.69	3.81
2-1/2"	1E@0.10	Ea	1.97	2.69	4.66
3"	1E@0.15	Ea	3.13	4.03	7.16
3-1/2"	1E@0.15	Ea	3.53	4.03	7.56
4"	1E@0.15	Ea	4.74	4.03	8.77
5"	1E@0.20	Ea	12.70	5.38	18.08
6"	1E@0.25	Ea	16.30	6.72	23.02

Type FA female PVC adapters

Material	Craft@Hrs	Unit	Material Cost	Labor Cost	Installed Cost
1/2"	1E@0.05	Ea	.34	1.34	1.68
3/4"	1E@0.06	Ea	.58	1.61	2.19
1"	1E@0.08	Ea	.80	2.15	2.95
1-1/4"	1E@0.10	Ea	1.03	2.69	3.72
1-1/2"	1E@0.10	Ea	1.11	2.69	3.80
2"	1E@0.15	Ea	1.53	4.03	5.56
2-1/2"	1E@0.15	Ea	2.89	4.03	6.92
3"	1E@0.20	Ea	4.40	5.38	9.78
3-1/2"	1E@0.20	Ea	5.74	5.38	11.12
4"	1E@0.25	Ea	5.78	6.72	12.50
5"	1E@0.30	Ea	14.70	8.06	22.76
6"	1E@0.40	Ea	19.30	10.80	30.10

Type TA terminal PVC adapters

Material	Craft@Hrs	Unit	Material Cost	Labor Cost	Installed Cost
1/2"	1E@0.05	Ea	.31	1.34	1.65
3/4"	1E@0.06	Ea	.57	1.61	2.18
1"	1E@0.08	Ea	.71	2.15	2.86
1-1/4"	1E@0.10	Ea	.90	2.69	3.59
1-1/2"	1E@0.10	Ea	1.11	2.69	3.80
2"	1E@0.15	Ea	1.59	4.03	5.62
2-1/2"	1E@0.15	Ea	2.69	4.03	6.72
3"	1E@0.20	Ea	3.91	5.38	9.29
3-1/2"	1E@0.20	Ea	5.13	5.38	10.51
4"	1E@0.25	Ea	6.72	6.72	13.44
5"	1E@0.30	Ea	13.20	8.06	21.26
6"	1E@0.40	Ea	15.90	10.80	26.70

2" range expansion PVC couplings

Material	Craft@Hrs	Unit	Material Cost	Labor Cost	Installed Cost
1/2"	1E@0.15	Ea	7.01	4.03	11.04
3/4"	1E@0.20	Ea	7.39	5.38	12.77
1"	1E@0.25	Ea	9.34	6.72	16.06
1-1/4"	1E@0.30	Ea	10.90	8.06	18.96
1-1/2"	1E@0.30	Ea	14.00	8.06	22.06
2"	1E@0.40	Ea	17.10	10.80	27.90

Use these figures to estimate the cost of PVC fittings installed on PVC conduit under the conditions described on pages 5 and 6. Costs listed are for each fitting installed. The crew is one electrician working at a labor cost of $26.88 per manhour. These costs include applying cement (glue), removal of knockouts, layout, material handling, and normal waste. Add for conduit, couplings, connectors, end bells, spacers, sales tax, delivery, supervision, mobilization, demobilization, cleanup, overhead and profit. Note: Material costs assume purchase of full box quantities.

PVC Expansion Couplings, End Bells, Caps and Plugs

Material	Craft@Hrs	Unit	Material Cost	Labor Cost	Installed Cost
6" range expansion PVC couplings					
1/2"	1E@0.15	Ea	12.40	4.03	16.43
3/4"	1E@0.20	Ea	12.70	5.38	18.08
1"	1E@0.25	Ea	12.80	6.72	19.52
1-1/4"	1E@0.30	Ea	12.80	8.06	20.86
1-1/2"	1E@0.30	Ea	12.90	8.06	20.96
2"	1E@0.40	Ea	14.50	10.80	25.30
2-1/2"	1E@0.40	Ea	19.60	10.80	30.40
3"	1E@0.50	Ea	25.50	13.40	38.90
3-1/2"	1E@0.50	Ea	29.70	13.40	43.10
4"	1E@0.60	Ea	36.80	16.10	52.90
5"	1E@0.70	Ea	54.30	18.80	73.10
6"	1E@0.75	Ea	73.10	20.20	93.30
PVC end bells					
1"	1E@0.10	Ea	1.54	2.69	4.23
1-1/4"	1E@0.15	Ea	1.86	4.03	5.89
1-1/2"	1E@0.15	Ea	1.86	4.03	5.89
2"	1E@0.20	Ea	2.89	5.38	8.27
2-1/2"	1E@0.20	Ea	3.03	5.38	8.41
3"	1E@0.25	Ea	3.38	6.72	10.10
3-1/2"	1E@0.25	Ea	3.64	6.72	10.36
4"	1E@0.30	Ea	4.04	8.06	12.10
5"	1E@0.35	Ea	6.21	9.41	15.62
6"	1E@0.40	Ea	6.81	10.80	17.61
PVC caps and plugs					
1/2" caps	1E@0.05	Ea	.71	1.34	2.05
3/4" caps	1E@0.06	Ea	.80	1.61	2.41
1" caps	1E@0.08	Ea	.90	2.15	3.05
1-1/4" caps	1E@0.10	Ea	1.28	2.69	3.97
1-1/2" plugs	1E@0.10	Ea	.51	2.69	3.20
2" plugs	1E@0.10	Ea	.86	2.69	3.55
2-1/2" plugs	1E@0.10	Ea	.67	2.69	3.36
3" plugs	1E@0.15	Ea	1.06	4.03	5.09
3-1/2" plugs	1E@0.15	Ea	1.26	4.03	5.29
4" plugs	1E@0.15	Ea	1.46	4.03	5.49
5" plugs	1E@0.20	Ea	1.93	5.38	7.31
6" plugs	1E@0.20	Ea	2.41	5.38	7.79

Use these figures to estimate the cost of PVC fittings installed on PVC conduit under the conditions described on pages 5 and 6. Costs listed are for each fitting installed. The crew is one electrician working at a labor cost of $26.88 per manhour. These costs include applying cement (glue), removal of knockouts, layout, material handling, and normal waste. Add for conduit, locknuts, insulated bushings, sales tax, delivery, supervision, mobilization, demobilization, cleanup, overhead and profit. Note: Material costs assume purchase of full box quantities.

PVC Reducing Bushings and Conduit Bodies

Material			Craft@Hrs	Unit	Material Cost	Labor Cost	Installed Cost

PVC reducing bushings

Material			Craft@Hrs	Unit	Material Cost	Labor Cost	Installed Cost
3/4"	to	1/2"	1E@0.03	Ea	.50	.81	1.31
1"	to	1/2"	1E@0.03	Ea	1.13	.81	1.94
1"	to	3/4"	1E@0.03	Ea	1.20	.81	2.01
1-1/4"	to	3/4"	1E@0.05	Ea	1.52	1.34	2.86
1-1/4"	to	1"	1E@0.05	Ea	1.52	1.34	2.86
1-1/2"	to	1"	1E@0.10	Ea	1.58	2.69	4.27
1-1/2"	to	1-1/4"	1E@0.10	Ea	1.58	2.69	4.27
2"	to	1-1/4"	1E@0.15	Ea	1.90	4.03	5.93
2-1/2"	to	2"	1E@0.15	Ea	6.13	4.03	10.16
3"	to	2"	1E@0.20	Ea	6.21	5.38	11.59
4"	to	3"	1E@0.25	Ea	8.45	6.72	15.17

Type C PVC conduit bodies

Material	Craft@Hrs	Unit	Material Cost	Labor Cost	Installed Cost
C 1/2"	1E@0.10	Ea	1.72	2.69	4.41
C 3/4"	1E@0.10	Ea	2.44	2.69	5.13
C 1"	1E@0.15	Ea	2.68	4.03	6.71
C 1-1/4"	1E@0.15	Ea	3.83	4.03	7.86
C 1-1/2"	1E@0.20	Ea	4.47	5.38	9.85
C 2"	1E@0.25	Ea	7.92	6.72	14.64

Type E PVC conduit bodies

Material	Craft@Hrs	Unit	Material Cost	Labor Cost	Installed Cost
E 1/2"	1E@0.10	Ea	1.72	2.69	4.41
E 3/4"	1E@0.10	Ea	2.44	2.69	5.13
E 1"	1E@0.15	Ea	2.68	4.03	6.71
E 1-1/4"	1E@0.15	Ea	3.83	4.03	7.86
E 1-1/2"	1E@0.20	Ea	4.47	5.38	9.85
E 2"	1E@0.25	Ea	7.92	6.72	14.64

Type LB PVC conduit bodies

Material	Craft@Hrs	Unit	Material Cost	Labor Cost	Installed Cost
LB 1/2"	1E@0.10	Ea	1.64	2.69	4.33
LB 3/4"	1E@0.10	Ea	2.32	2.69	5.01
LB 1"	1E@0.15	Ea	2.55	4.03	6.58
LB 1-1/4"	1E@0.15	Ea	3.65	4.03	7.68

Use these figures to estimate the cost of PVC fittings installed on PVC conduit under the conditions described on pages 5 and 6. Costs listed are for each fitting installed. The crew is one electrician working at a labor cost of $26.88 per manhour. These costs include applying cement (glue), removal of knockouts, layout, material handling, and normal waste. Add for conduit, locknuts, insulated bushings, sales tax, delivery, supervision, mobilization, demobilization, cleanup, overhead and profit. Note: Material costs assume purchase of full box quantities.

PVC Conduit Bodies and Service Entrance Caps

Material	Craft@Hrs	Unit	Material Cost	Labor Cost	Installed Cost
Type LB PVC conduit bodies					
LB 1-1/2"	1E@0.20	Ea	4.26	5.38	9.64
LB 2"	1E@0.25	Ea	7.54	6.72	14.26
LB 2-1/2"	1E@0.30	Ea	29.00	8.06	37.06
LB 3"	1E@0.30	Ea	30.00	8.06	38.06
LB 3-1/2"	1E@0.30	Ea	31.00	8.06	39.06
LB 4"	1E@0.40	Ea	32.90	10.80	43.70
Type LL PVC conduit bodies					
LL 1/2"	1E@0.10	Ea	1.72	2.69	4.41
LL 3/4"	1E@0.10	Ea	2.44	2.69	5.13
LL 1"	1E@0.15	Ea	2.68	4.03	6.71
LL 1-1/4"	1E@0.15	Ea	3.83	4.03	7.86
LL 1-1/2"	1E@0.20	Ea	4.47	5.38	9.85
LL 2"	1E@0.25	Ea	7.92	6.72	14.64
Type LR PVC conduit bodies					
LR 1/2"	1E@0.10	Ea	1.72	2.69	4.41
LR 3/4"	1E@0.10	Ea	2.44	2.69	5.13
LR 1"	1E@0.15	Ea	2.68	4.03	6.71
LR 1-1/4"	1E@0.15	Ea	3.83	4.03	7.86
LR 1-1/2"	1E@0.20	Ea	4.47	5.38	9.85
LR 2"	1E@0.25	Ea	7.92	6.72	14.64
Type T PVC conduit bodies					
T 1/2"	1E@0.10	Ea	2.18	2.69	4.87
T 3/4"	1E@0.15	Ea	2.72	4.03	6.75
T 1"	1E@0.15	Ea	2.75	4.03	6.78
T 1-1/4"	1E@0.20	Ea	4.47	5.38	9.85
T 1-1/2"	1E@0.25	Ea	5.89	6.72	12.61
T 2"	1E@0.30	Ea	8.53	8.06	16.59
PVC service entrance caps					
3/4"	1E@0.15	Ea	2.35	4.03	6.38
1"	1E@0.15	Ea	2.70	4.03	6.73
1-1/4"	1E@0.25	Ea	4.07	6.72	10.79
1-1/2"	1E@0.30	Ea	4.80	8.06	12.86
2"	1E@0.50	Ea	8.22	13.40	21.62
2-1/2"	1E@0.60	Ea	41.90	16.10	58.00
3"	1E@0.60	Ea	42.20	16.10	58.30
4"	1E@0.75	Ea	124.00	20.20	144.20

Use these figures to estimate the cost of PVC fittings installed on PVC conduit under the conditions described on pages 5 and 6. Costs listed are for each fitting installed. The crew is one electrician working at a labor cost of $26.88 per manhour. These costs include applying cement (glue), removal of knockouts, layout, material handling, and normal waste. Add for conduit, locknuts, insulated bushings, sales tax, delivery, supervision, mobilization, demobilization, cleanup, overhead and profit. Note: Material costs assume purchase of full box quantities.

PVC Boxes, Covers and Elbows

Material		Craft@Hrs	Unit	Material Cost	Labor Cost	Installed Cost

Type FS PVC boxes

Material		Craft@Hrs	Unit	Material Cost	Labor Cost	Installed Cost
FS1	1/2"	1E@0.20	Ea	3.85	5.38	9.23
FS2	3/4"	1E@0.20	Ea	4.01	5.38	9.39
FS3	1"	1E@0.25	Ea	4.36	6.72	11.08
FSC1	1/2"	1E@0.25	Ea	4.01	6.72	10.73
FSC2	3/4"	1E@0.25	Ea	4.36	6.72	11.08
FSC3	1"	1E@0.30	Ea	4.87	8.06	12.93
FSS1	1/2"	1E@0.25	Ea	4.01	6.72	10.73
FSS2	3/4"	1E@0.25	Ea	4.36	6.72	11.08
FSS3	1"	1E@0.30	Ea	4.87	8.06	12.93
FCSS1	1/2"	1E@0.30	Ea	4.36	8.06	12.42
FCSS2	3/4"	1E@0.30	Ea	4.87	8.06	12.93
FCSS3	1"	1E@0.35	Ea	5.36	9.41	14.77

Type FS, WP PVC box covers

Material	Craft@Hrs	Unit	Material Cost	Labor Cost	Installed Cost
1 gang blank	1E@0.10	Ea	1.58	2.69	4.27
1 gang single outlet	1E@0.10	Ea	4.30	2.69	6.99
1 gang duplex outlet	1E@0.10	Ea	2.81	2.69	5.50
1 gang single switch	1E@0.10	Ea	4.26	2.69	6.95
1 gang GFCI	1E@0.10	Ea	4.23	2.69	6.92

PVC junction boxes

Material	Craft@Hrs	Unit	Material Cost	Labor Cost	Installed Cost
4" x 4" x 2"	1E@0.25	Ea	6.15	6.72	12.87
4" x 4" x 4"	1E@0.25	Ea	10.20	6.72	16.92
4" x 4" x 6"	1E@0.30	Ea	10.40	8.06	18.46
5" x 5" x 2"	1E@0.30	Ea	10.80	8.06	18.86
6" x 6" x 4"	1E@0.35	Ea	12.40	9.41	21.81
6" x 6" x 6"	1E@0.40	Ea	14.50	10.80	25.30
8" x 8" x 4"	1E@0.40	Ea	23.40	10.80	34.20
8" x 8" x 7"	1E@0.50	Ea	31.20	13.40	44.60
12" x 12" x 4"	1E@0.70	Ea	32.00	18.80	50.80
12" x 12" x 6"	1E@0.75	Ea	33.00	20.20	53.20

30 degree sweeping PVC elbows

Material	Craft@Hrs	Unit	Material Cost	Labor Cost	Installed Cost
2" 30" radius	1E@0.15	Ea	7.82	4.03	11.85
2" 36" radius	1E@0.20	Ea	8.77	5.38	14.15
2" 48" radius	1E@0.25	Ea	10.00	6.72	16.72
3" 30" radius	1E@0.20	Ea	12.20	5.38	17.58
3" 36" radius	1E@0.25	Ea	13.70	6.72	20.42
3" 48" radius	1E@0.30	Ea	15.70	8.06	23.76
4" 30" radius	1E@0.25	Ea	19.10	6.72	25.82
4" 36" radius	1E@0.30	Ea	21.40	8.06	29.46
4" 48" radius	1E@0.40	Ea	24.50	10.80	35.30

Use these figures to estimate the cost of PVC fittings installed on PVC conduit under the conditions described on pages 5 and 6. Costs listed are for each fitting installed. The crew is one electrician working at a labor cost of $26.88 per manhour. These costs include applying cement (glue), removal of knockouts, layout, material handling, and normal waste. Add for conduit, locknuts, insulated bushings, sales tax, delivery, supervision, mobilization, demobilization, cleanup, overhead and profit. Note: Material costs assume purchase of full box quantities.

PVC Elbows

Material		Craft@Hrs	Unit	Material Cost	Labor Cost	Installed Cost
30 degree sweeping PVC elbows						
5"	30" radius	1E@0.30	Ea	29.60	8.06	37.66
5"	36" radius	1E@0.40	Ea	33.20	10.80	44.00
5"	48" radius	1E@0.50	Ea	37.20	13.40	50.60
6"	36" radius	1E@0.75	Ea	50.20	20.20	70.40
6"	48" radius	1E@1.00	Ea	56.20	26.90	83.10
45 degree sweeping PVC elbows						
2"	24" radius	1E@0.15	Ea	7.89	4.03	11.92
2"	30" radius	1E@0.15	Ea	8.23	4.03	12.26
2"	36" radius	1E@0.20	Ea	9.22	5.38	14.60
2"	48" radius	1E@0.30	Ea	10.60	8.06	18.66
2-1/2"	30" radius	1E@0.20	Ea	10.30	5.38	15.68
2-1/2"	36" radius	1E@0.25	Ea	11.50	6.72	18.22
2-1/2"	48" radius	1E@0.30	Ea	13.20	8.06	21.26
3"	24" radius	1E@0.25	Ea	12.30	6.72	19.02
3"	30" radius	1E@0.30	Ea	12.90	8.06	20.96
3"	36" radius	1E@0.30	Ea	14.80	8.06	22.86
3"	48" radius	1E@0.40	Ea	16.50	10.80	27.30
4"	24" radius	1E@0.30	Ea	19.30	8.06	27.36
4"	30" radius	1E@0.30	Ea	20.10	8.06	28.16
4"	36" radius	1E@0.35	Ea	22.50	9.41	31.91
4"	48" radius	1E@0.40	Ea	25.70	10.80	36.50
5"	30" radius	1E@0.35	Ea	31.20	9.41	40.61
5"	36" radius	1E@0.40	Ea	35.00	10.80	45.80
5"	48" radius	1E@0.50	Ea	39.10	13.40	52.50
6"	36" radius	1E@0.75	Ea	52.80	20.20	73.00
6"	48" radius	1E@1.00	Ea	59.20	26.90	86.10
90 degree sweeping PVC elbows						
2"	24" radius	1E@0.20	Ea	8.86	5.38	14.24
2"	30" radius	1E@0.25	Ea	9.25	6.72	15.97
2"	36" radius	1E@0.30	Ea	10.40	8.06	18.46
2"	48" radius	1E@0.35	Ea	11.90	9.41	21.31
2-1/2"	30" radius	1E@0.30	Ea	11.60	8.06	19.66
2-1/2"	36" radius	1E@0.35	Ea	12.80	9.41	22.21
2-1/2"	48" radius	1E@0.40	Ea	14.80	10.80	25.60

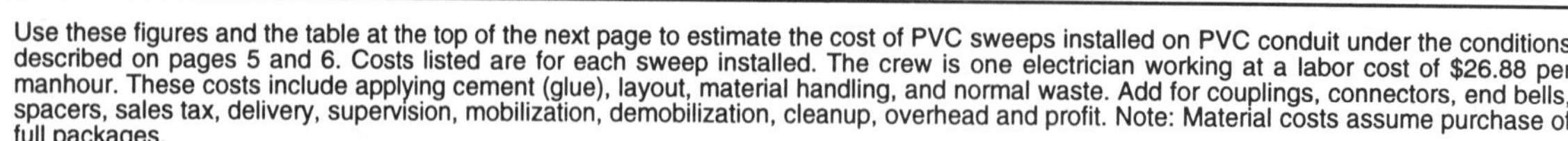

Use these figures and the table at the top of the next page to estimate the cost of PVC sweeps installed on PVC conduit under the conditions described on pages 5 and 6. Costs listed are for each sweep installed. The crew is one electrician working at a labor cost of $26.88 per manhour. These costs include applying cement (glue), layout, material handling, and normal waste. Add for couplings, connectors, end bells, spacers, sales tax, delivery, supervision, mobilization, demobilization, cleanup, overhead and profit. Note: Material costs assume purchase of full packages.

PVC Elbows, Power & Communication (P&C) Duct

Material		Craft@Hrs	Unit	Material Cost	Labor Cost	Installed Cost
90 degree sweeping PVC elbows						
3"	24" radius	2E@0.30	Ea	13.90	8.06	21.96
3"	30" radius	2E@0.35	Ea	14.50	9.41	23.91
3"	36" radius	2E@0.40	Ea	16.20	10.80	27.00
3"	48" radius	2E@0.50	Ea	18.50	13.40	31.90
4"	24" radius	2E@0.35	Ea	21.60	9.41	31.01
4"	30" radius	2E@0.40	Ea	22.60	10.80	33.40
4"	36" radius	2E@0.45	Ea	25.30	12.10	37.40
4"	48" radius	2E@0.55	Ea	28.90	14.80	43.70
5"	30" radius	2E@0.45	Ea	35.10	12.10	47.20
5"	36" radius	2E@0.50	Ea	39.30	13.40	52.70
5"	48" radius	2E@0.60	Ea	44.00	16.10	60.10
6"	36" radius	2E@0.60	Ea	59.40	16.10	75.50
6"	48" radius	2E@0.75	Ea	66.50	20.20	86.70
Type EB power and communication duct						
2"		2E@3.30	CLF	31.30	88.70	120.00
3"		2E@3.50	CLF	45.20	94.10	139.30
4"		2E@4.00	CLF	73.90	108.00	181.90
5"		2E@4.50	CLF	112.00	121.00	233.00
6"		2E@5.00	CLF	161.00	134.00	295.00
Type DB power and communication duct						
2"		2E@3.30	CLF	27.20	88.70	115.90
4"		2E@4.00	CLF	70.60	108.00	178.60
5"		2E@4.50	CLF	108.00	121.00	229.00
6"		2E@5.00	CLF	153.00	134.00	287.00
Type EB or DB power and communication duct couplings						
2"		1E@0.05	Ea	1.10	1.34	2.44
3"		1E@0.10	Ea	1.18	2.69	3.87
4"		1E@0.10	Ea	1.91	2.69	4.60
5"		1E@0.15	Ea	3.45	4.03	7.48
6"		1E@0.15	Ea	6.11	4.03	10.14
45 degree Type EB or DB power and communication duct elbows						
2"	24" radius	1E@0.15	Ea	4.08	4.03	8.11
3"	36" radius	1E@0.30	Ea	4.94	8.06	13.00
3"	48" radius	1E@0.40	Ea	6.37	10.80	17.17
4"	36" radius	1E@0.40	Ea	6.23	10.80	17.03
4"	48" radius	1E@0.75	Ea	8.01	20.20	28.21
5"	48" radius	1E@0.50	Ea	11.82	13.40	25.22

Use these figures to estimate the cost of PVC elbows (top table) and power and communication duct couplings and elbows (bottom tables). The footnote on the previous page applies to PVC sweep elbows. P&C duct is installed underground under the conditions described on pages 5 and 6. Costs listed are for each 100 linear feet installed. The crew is two electricians working at a labor cost of $26.88 per manhour. These costs include one coupling, applying cement (glue), multiple runs in the same trench, layout, material handling, and normal waste. Add for trenching, encasement, spacers and chairs, single duct runs, sales tax, delivery, supervision, mobilization, demobilization, cleanup, overhead and profit. Note: Encased burial requires spacers or chairs every 5 feet. Costs for spacers, chairs, encasement and trenching are listed elsewhere in this manual.

Power & Communication Duct Couplings, Elbows & Adapters

Material		Craft@Hrs	Unit	Material Cost	Labor Cost	Installed Cost

90 degree Type EB or DB power and communication duct elbows

Material		Craft@Hrs	Unit	Material Cost	Labor Cost	Installed Cost
2"	24" radius	1E@0.20	Ea	4.42	5.38	9.80
2"	36" radius	1E@0.30	Ea	4.62	8.06	12.68
2"	48" radius	1E@0.35	Ea	6.01	9.41	15.42
3"	24" radius	1E@0.30	Ea	5.04	8.06	13.10
3"	36" radius	1E@0.40	Ea	6.33	10.80	17.13
3"	48" radius	1E@0.50	Ea	9.99	13.40	23.39
4"	24" radius	1E@0.35	Ea	7.98	9.41	17.39
4"	36" radius	1E@0.45	Ea	8.98	12.10	21.08
4"	48" radius	1E@0.55	Ea	11.70	14.80	26.50
5"	36" radius	1E@0.50	Ea	14.00	13.40	27.40
5"	48" radius	1E@0.60	Ea	18.80	16.10	34.90
6"	48" radius	1E@0.75	Ea	26.20	20.20	46.40

5 degree power and communication bend couplings

Material	Craft@Hrs	Unit	Material Cost	Labor Cost	Installed Cost
2"	1E@0.05	Ea	4.00	1.34	5.34
3"	1E@0.10	Ea	5.26	2.69	7.95
4"	1E@0.10	Ea	6.22	2.69	8.91
5"	1E@0.15	Ea	6.81	4.03	10.84
6"	1E@0.15	Ea	6.97	4.03	11.00

Power and communication duct plugs

Material	Craft@Hrs	Unit	Material Cost	Labor Cost	Installed Cost
2"	1E@0.05	Ea	.82	1.34	2.16
3"	1E@0.10	Ea	1.01	2.69	3.70
4"	1E@0.10	Ea	1.39	2.69	4.08
5"	1E@0.15	Ea	1.85	4.03	5.88
6"	1E@0.15	Ea	2.31	4.03	6.34

Type FA female power and communication duct adapters

Material	Craft@Hrs	Unit	Material Cost	Labor Cost	Installed Cost
2"	1E@0.20	Ea	1.15	5.38	6.53
3"	1E@0.25	Ea	2.82	6.72	9.54
4"	1E@0.30	Ea	4.86	8.06	12.92
5"	1E@0.40	Ea	9.53	10.80	20.33
6"	1E@0.50	Ea	11.50	13.40	24.90

Type TA terminal power and communication adapters

Material	Craft@Hrs	Unit	Material Cost	Labor Cost	Installed Cost
2"	1E@0.20	Ea	1.15	5.38	6.53
3"	1E@0.25	Ea	2.82	6.72	9.54
4"	1E@0.30	Ea	4.86	8.06	12.92
5"	1E@0.40	Ea	9.53	10.80	20.33
6"	1E@0.50	Ea	11.50	13.40	24.90

Use these figures to estimate the cost of PVC fittings installed on PVC power and communication duct under the conditions described on pages 5 and 6. Costs listed are for each fitting installed. The crew is one electrician working at a labor cost of $26.88 per manhour. These costs include cutting and fitting, applying cement (glue), layout, material handling, and normal waste. Add for extra couplings, sales tax, delivery, supervision, mobilization, demobilization, cleanup, overhead and profit. Note: Material costs are based on purchase of full packages. All of these fittings can be used either on type EB or type DB duct.

Power & Communication Duct End Bells and Plastic Spacers

Material	Craft@Hrs	Unit	Material Cost	Labor Cost	Installed Cost
Power and communication duct end bells					
2"	1E@0.15	Ea	2.24	4.03	6.27
3"	1E@0.20	Ea	2.68	5.38	8.06
4"	1E@0.25	Ea	3.16	6.72	9.88
5"	1E@0.30	Ea	4.76	8.06	12.82
6"	1E@0.50	Ea	5.24	13.40	18.64
Base type plastic duct spacers					
2" 1-1/2" separation	1E@0.05	Ea	.73	1.34	2.07
3" 1-1/2" separation	1E@0.05	Ea	.80	1.34	2.14
4" 1-1/2" separation	1E@0.05	Ea	.89	1.34	2.23
5" 1-1/2" separation	1E@0.05	Ea	.95	1.34	2.29
6" 1-1/2" separation	1E@0.05	Ea	1.52	1.34	2.86
2" 2" separation	1E@0.05	Ea	.77	1.34	2.11
3" 2" separation	1E@0.05	Ea	.84	1.34	2.18
4" 2" separation	1E@0.05	Ea	.91	1.34	2.25
5" 2" separation	1E@0.05	Ea	.99	1.34	2.33
6" 2" separation	1E@0.05	Ea	1.61	1.34	2.95
2" 3" separation	1E@0.05	Ea	.87	1.34	2.21
3" 3" separation	1E@0.05	Ea	.96	1.34	2.30
4" 3" separation	1E@0.05	Ea	1.00	1.34	2.34
5" 3" separation	1E@0.05	Ea	1.64	1.34	2.98
6" 3" separation	1E@0.05	Ea	1.73	1.34	3.07
Intermediate type plastic duct spacers					
2" 1-1/2" separation	1E@0.05	Ea	.73	1.34	2.07
3" 1-1/2" separation	1E@0.05	Ea	.80	1.34	2.14
4" 1-1/2" separation	1E@0.05	Ea	.89	1.34	2.23
5" 1-1/2" separation	1E@0.05	Ea	.99	1.34	2.33
6" 1-1/2" separation	1E@0.05	Ea	1.61	1.34	2.95
2" 2" separation	1E@0.05	Ea	.77	1.34	2.11
3" 2" separation	1E@0.05	Ea	.84	1.34	2.18
4" 2" separation	1E@0.05	Ea	.91	1.34	2.25
5" 2" separation	1E@0.05	Ea	.99	1.34	2.33
6" 2" separation	1E@0.05	Ea	1.61	1.34	2.95
2" 3" separation	1E@0.05	Ea	.87	1.34	2.21
3" 3" separation	1E@0.05	Ea	.96	1.34	2.30
4" 3" separation	1E@0.05	Ea	1.00	1.34	2.34
5" 3" separation	1E@0.05	Ea	1.64	1.34	2.98
6" 3" separation	1E@0.05	Ea	1.73	1.34	3.07

Use these figures to estimate the cost of P&C end bell or plastic spacer installed with duct systems under the conditions described on pages 5 and 6. Costs listed are for each end bell or spacer installed. The crew is one electrician working at a labor cost of $26.88 per manhour. These costs include ganging spacers, tying duct to the spacer, layout, material handling, and normal waste. Add for duct, other fittings, sales tax, delivery, supervision, mobilization, demobilization, cleanup, overhead and profit. Note: Material costs are based on purchase of full packages. Tie wire should never be tied completely around the duct, it should be tied in a figure 8 pattern through open spaces in the side of the spacer and over the top part of the duct. Running wire completely around the duct will cause the wire to pick up an induction field from the current passing through the conductor, generating heat which will weaken the insulation.

ENT Conduit and Fittings

Material	Craft@Hrs	Unit	Material Cost	Labor Cost	Installed Cost
ENT conduit, non-metallic tubing					
1/2"	1E@2.15	CLF	17.50	57.80	75.30
3/4"	1E@2.25	CLF	24.10	60.50	84.60
1"	1E@2.50	CLF	39.00	67.20	106.20
ENT connectors					
1/2"	1E@0.03	Ea	.35	.81	1.16
3/4"	1E@0.04	Ea	.65	1.08	1.73
1"	1E@0.05	Ea	1.05	1.34	2.39
ENT couplings					
1/2"	1E@0.03	Ea	.40	.81	1.21
3/4"	1E@0.04	Ea	.51	1.08	1.59
1"	1E@0.05	Ea	.90	1.34	2.24
ENT male adapters					
1/2"	1E@0.03	Ea	.55	.81	1.36
3/4"	1E@0.04	Ea	1.10	1.08	2.18
1"	1E@0.05	Ea	1.55	1.34	2.89

Use these figures to estimate the cost of ENT conduit and fittings installed under the conditions described on pages 5 and 6. Costs listed are for each 100 linear feet installed and for each fitting installed. The crew is one electrician working at a labor cost of $26.88 per manhour. These costs include cutting and fitting, applying cement (glue), layout, material handling, and normal waste. Add for extra couplings, sales tax, delivery, supervision, mobilization, demobilization, cleanup, overhead and profit. Note: Material costs are based on purchase of full packages.

ENT conduit weight per 100 feet (in pounds)

Diameter	Weight
1/2"	11
3/4"	14
1"	20

Galvanized Rigid Steel (GRS) Conduit and Elbows

Material	Craft@Hrs	Unit	Material Cost	Labor Cost	Installed Cost
Standard wall galvanized rigid steel conduit					
1/2"	1E@4.00	CLF	93.00	108.00	201.00
3/4"	1E@4.50	CLF	117.00	121.00	238.00
1"	1E@5.00	CLF	173.00	134.00	307.00
1-1/4"	1E@7.00	CLF	221.00	188.00	409.00
1-1/2"	1E@8.00	CLF	270.00	215.00	485.00
2"	1E@10.0	CLF	361.00	269.00	630.00
2-1/2"	1E@12.0	CLF	593.00	323.00	916.00
3"	1E@14.0	CLF	760.00	376.00	1,136.00
3-1/2"	1E@16.0	CLF	925.00	430.00	1,355.00
4"	1E@18.0	CLF	1,100.00	484.00	1,584.00
5"	1E@25.0	CLF	2,200.00	672.00	2,872.00
6"	1E@30.0	CLF	3,100.00	806.00	3,906.00
45 degree galvanized rigid steel elbows					
1/2"	1E@0.10	Ea	2.16	2.69	4.85
3/4"	1E@0.10	Ea	2.60	2.69	5.29
1"	1E@0.12	Ea	3.85	3.23	7.08
1-1/4"	1E@0.15	Ea	5.52	4.03	9.55
1-1/2"	1E@0.15	Ea	7.17	4.03	11.20
2"	1E@0.20	Ea	10.40	5.38	15.78
2-1/2"	1E@0.25	Ea	18.90	6.72	25.62
3"	1E@0.25	Ea	27.50	6.72	34.22
3-1/2"	1E@0.30	Ea	43.10	8.06	51.16
4"	1E@0.30	Ea	49.30	8.06	57.36
5"	1E@0.50	Ea	117.00	13.40	130.40
6"	1E@1.00	Ea	176.00	26.90	202.90
90 degree galvanized rigid steel elbows					
1/2"	1E@0.10	Ea	2.16	2.69	4.85
3/4"	1E@0.10	Ea	2.60	2.69	5.29
1"	1E@0.12	Ea	3.85	3.23	7.08
1-1/4"	1E@0.15	Ea	5.52	4.03	9.55
1-1/2"	1E@0.15	Ea	7.17	4.03	11.20
2"	1E@0.20	Ea	10.40	5.38	15.78
2-1/2"	1E@0.25	Ea	18.90	6.72	25.62
3"	1E@0.25	Ea	27.50	6.72	34.22
3-1/2"	1E@0.30	Ea	43.10	8.06	51.16
4"	1E@0.30	Ea	49.30	8.06	57.36
5"	1E@0.50	Ea	117.00	13.40	130.40
6"	1E@1.00	Ea	176.00	26.90	202.90

Use these figures to estimate the cost of GRS conduit and elbows installed in buildings under the conditions described on pages 5 and 6. Costs listed are for each 100 linear feet of conduit or for each elbow installed. The crew is one electrician working at a labor cost of $26.88 per manhour. These costs include one coupling on each length of conduit, threading, cutting, straps, layout, material handling, and normal waste. Add for other fittings, boxes, wires, sales tax, delivery, supervision, mobilization, demobilization, cleanup, overhead and profit. Note: Couplings are not included with elbows. The elbows listed are factory made and have a standard radius. Conduit runs are assumed to be 50' long. Installation costs per linear foot will be less on longer runs and more on shorter runs.

Galvanized Rigid Steel Large Radius Elbows

Material		Craft@Hrs	Unit	Material Cost	Labor Cost	Installed Cost
90 degree galvanized rigid steel large radius elbows						
1"	12" radius	1E@0.10	Ea	12.30	2.69	14.99
1-1/4"	12" radius	1E@0.15	Ea	16.00	4.03	20.03
1-1/2"	12" radius	1E@0.15	Ea	20.00	4.03	24.03
2"	12" radius	1E@0.20	Ea	22.10	5.38	27.48
2-1/2"	12" radius	2E@0.20	Ea	43.00	5.38	48.38
1"	15" radius	1E@0.10	Ea	17.60	2.69	20.29
1-1/4"	15" radius	1E@0.15	Ea	20.70	4.03	24.73
1-1/2"	15" radius	1E@0.15	Ea	21.20	4.03	25.23
2"	15" radius	1E@0.20	Ea	25.50	5.38	30.88
2-1/2"	15" radius	2E@0.20	Ea	35.50	5.38	40.88
1"	18" radius	1E@0.10	Ea	17.80	2.69	20.49
1-1/4"	18" radius	1E@0.15	Ea	21.10	4.03	25.13
1-1/2"	18" radius	1E@0.15	Ea	24.60	4.03	28.63
2"	18" radius	1E@0.20	Ea	30.40	5.38	35.78
2-1/2"	18" radius	2E@0.20	Ea	42.00	5.38	47.38
3"	18" radius	2E@0.25	Ea	53.00	6.72	59.72
3-1/2"	18" radius	2E@0.25	Ea	76.30	6.72	83.02
4"	18" radius	2E@0.30	Ea	83.70	8.06	91.76
1"	24" radius	1E@0.10	Ea	20.90	2.69	23.59
1-1/4"	24" radius	1E@0.15	Ea	24.60	4.03	28.63
1-1/2"	24" radius	1E@0.15	Ea	29.00	4.03	33.03
2"	24" radius	1E@0.20	Ea	37.30	5.38	42.68
2-1/2"	24" radius	2E@0.20	Ea	50.80	5.38	56.18
3"	24" radius	2E@0.25	Ea	66.80	6.72	73.52
3-1/2"	24" radius	2E@0.25	Ea	90.00	6.72	96.72
4"	24" radius	2E@0.30	Ea	98.90	8.06	106.96
1"	30" radius	1E@0.15	Ea	23.20	4.03	27.23
1-1/4"	30" radius	1E@0.20	Ea	30.10	5.38	35.48
1-1/2"	30" radius	1E@0.20	Ea	35.10	5.38	40.48
2"	30" radius	1E@0.25	Ea	43.80	6.72	50.52
2-1/2"	30" radius	2E@0.25	Ea	59.20	6.72	65.92
3"	30" radius	2E@0.30	Ea	77.70	8.06	85.76
3-1/2"	30" radius	2E@0.30	Ea	104.00	8.06	112.06
4"	30" radius	2E@0.35	Ea	114.00	9.41	123.41
5"	30" radius	2E@0.50	Ea	153.00	13.40	166.40
1"	36" radius	1E@0.20	Ea	27.50	5.38	32.88
1-1/4"	36" radius	1E@0.25	Ea	33.70	6.72	40.42
1-1/2"	36" radius	1E@0.25	Ea	42.20	6.72	48.92
2"	36" radius	1E@0.30	Ea	50.40	8.06	58.46
2-1/2"	36" radius	2E@0.30	Ea	69.20	8.06	77.26
3"	36" radius	2E@0.35	Ea	91.10	9.41	100.51

Use these figures to estimate the cost of large radius GRS elbows installed on GRS conduit under the conditions described on pages 5 and 6. Costs listed are for each elbow installed. The crew is one electrician for size to 2" and two electricians for sizes over 2". The labor cost is $26.88 per manhour. These costs include layout, material handling, and normal waste. Add for other GRS fittings, conduit, field bending, sales tax, delivery, supervision, mobilization, demobilization, cleanup, overhead and profit. Note: All elbows are assumed to be factory made.

Galvanized Rigid Steel Elbows and Couplings

Material		Craft@Hrs	Unit	Material Cost	Labor Cost	Installed Cost
90 degree galvanized rigid steel large radius elbows						
3-1/2"	36" radius	2E@0.35	Ea	121.00	9.41	130.41
4"	36" radius	2E@0.40	Ea	133.00	10.80	143.80
5"	36" radius	2E@0.60	Ea	215.00	16.10	231.10
6"	36" radius	2E@1.00	Ea	235.00	26.90	261.90
1"	42" radius	1E@0.25	Ea	31.10	6.72	37.82
1-1/4"	42" radius	1E@0.30	Ea	37.70	8.06	45.76
1-1/2"	42" radius	1E@0.30	Ea	44.90	8.06	52.96
2"	42" radius	1E@0.35	Ea	56.00	9.41	65.41
2-1/2"	42" radius	2E@0.35	Ea	76.80	9.41	86.21
3"	42" radius	2E@0.40	Ea	101.00	10.80	111.80
3-1/2"	42" radius	2E@0.40	Ea	135.00	10.80	145.80
4"	42" radius	2E@0.50	Ea	163.00	13.40	176.40
5"	42" radius	2E@0.75	Ea	243.00	20.20	263.20
6"	42" radius	2E@1.25	Ea	256.00	33.60	289.60
1"	48" radius	1E@0.30	Ea	34.70	8.06	42.76
1-1/4"	48" radius	1E@0.35	Ea	42.00	9.41	51.41
1-1/2"	48" radius	1E@0.35	Ea	50.10	9.41	59.51
2"	48" radius	1E@0.40	Ea	63.50	10.80	74.30
2-1/2"	48" radius	2E@0.40	Ea	87.00	10.80	97.80
3"	48" radius	2E@0.50	Ea	114.00	13.40	127.40
3-1/2"	48" radius	2E@0.50	Ea	152.00	13.40	165.40
4"	48" radius	2E@0.70	Ea	184.00	18.80	202.80
5"	48" radius	2E@1.00	Ea	271.00	26.90	297.90
6"	48" radius	2E@1.50	Ea	289.00	40.30	329.30
Galvanized rigid steel couplings						
1/2"		1E@0.05	Ea	.78	1.34	2.12
3/4"		1E@0.06	Ea	.96	1.61	2.57
1"		1E@0.08	Ea	1.43	2.15	3.58
1-1/4"		1E@0.10	Ea	1.79	2.69	4.48
1-1/2"		1E@0.10	Ea	2.26	2.69	4.95
2"		1E@0.15	Ea	2.98	4.03	7.01
2-1/2"		2E@0.15	Ea	6.96	4.03	10.99
3"		2E@0.20	Ea	9.02	5.38	14.40
3-1/2"		2E@0.20	Ea	12.10	5.38	17.48
4"		2E@0.25	Ea	12.10	6.72	18.82
5"		2E@0.30	Ea	28.10	8.06	36.16
6"		2E@0.50	Ea	39.20	13.40	52.60

Use these figures to estimate the cost of large radius GRS elbows and couplings installed on GRS conduit under the conditions described on pages 5 and 6. Costs listed are for each elbow or coupling installed. The crew is one electrician for sizes to 2" and two electricians for sizes over 2". The labor cost is $26.88 per manhour. These costs include layout, material handling, and normal waste. Add for other GRS fittings, conduit, field bending, sales tax, delivery, supervision, mobilization, demobilization, cleanup, overhead and profit. Note: All elbows are assumed to be factory made.

GRS Terminations, Intermediate Metal Conduit (IMC) and Elbows

Material	Craft@Hrs	Unit	Material Cost	Labor Cost	Installed Cost
Galvanized rigid steel conduit terminations					
1/2"	1E@0.05	Ea	.37	1.34	1.71
3/4"	1E@0.06	Ea	.58	1.61	2.19
1"	1E@0.08	Ea	1.00	2.15	3.15
1-1/4"	1E@0.10	Ea	1.30	2.69	3.99
1-1/2"	1E@0.10	Ea	1.83	2.69	4.52
2"	1E@0.15	Ea	3.03	4.03	7.06
2-1/2"	2E@0.15	Ea	7.05	4.03	11.08
3"	2E@0.20	Ea	8.46	5.38	13.84
3-1/2"	2E@0.20	Ea	13.60	5.38	18.98
4"	2E@0.25	Ea	16.30	6.72	23.02
5"	2E@0.30	Ea	46.40	8.06	54.46
6"	2E@0.50	Ea	87.00	13.40	100.40
Intermediate metal conduit					
1/2"	1E@3.75	CLF	81.40	101.00	182.40
3/4"	1E@4.00	CLF	94.60	108.00	202.60
1"	1E@4.50	CLF	144.00	121.00	265.00
1-1/4"	1E@6.50	CLF	180.00	175.00	355.00
1-1/2"	1E@7.25	CLF	215.00	195.00	410.00
2"	1E@9.00	CLF	296.00	242.00	538.00
2-1/2"	2E@11.0	CLF	603.00	296.00	899.00
3"	2E@13.0	CLF	770.00	349.00	1,119.00
3-1/2"	2E@15.0	CLF	891.00	403.00	1,294.00
4"	2E@17.0	CLF	1,040.00	457.00	1,497.00
45 degree intermediate metal conduit elbows					
1/2"	1E@0.10	Ea	2.15	2.69	4.84
3/4"	1E@0.10	Ea	2.63	2.69	5.32
1"	1E@0.10	Ea	3.79	2.69	6.48
1-1/4"	1E@0.15	Ea	6.14	4.03	10.17
1-1/2"	1E@0.15	Ea	6.89	4.03	10.92
2"	1E@0.20	Ea	9.92	5.38	15.30
2-1/2"	2E@0.20	Ea	17.60	5.38	22.98
3"	2E@0.25	Ea	27.00	6.72	33.72
3-1/2"	2E@0.25	Ea	41.10	6.72	47.82
4"	2E@0.30	Ea	47.70	8.06	55.76
Galvanized rigid steel hand benders					
1/2"	--	Ea	20.00	--	--
3/4"	--	Ea	28.00	--	--
1"	--	Ea	46.00	--	--
1-1/4"	--	Ea	109.00	--	--

Use these figures to estimate the cost of GRS terminations, intermediate metal conduit and IMC elbows installed under the conditions described on pages 5 and 6. Costs listed are for each fitting or 100 linear feet installed. The crew is one electrician for GRS terminations and IMC to 2" and two electricians for GRS or IMC over 2". The labor cost is $26.88 per manhour. These costs include removing the knockout, field bending of the IMC and one coupling for each 10' length, layout, material handling, and normal waste. Add for straps and other fittings, sales tax, delivery, supervision, mobilization, demobilization, cleanup, overhead and profit. Note: Material cost is based on purchase of full packages. Conduit runs are assumed to be 50' long. Installation costs per linear foot will be less on longer runs and more on shorter runs.

IMC Elbows, Couplings and Running Thread

Material	Craft@Hrs	Unit	Material Cost	Labor Cost	Installed Cost
90 degree intermediate metal conduit elbows					
1/2"	1E@0.10	Ea	2.15	2.69	4.84
3/4"	1E@0.10	Ea	2.63	2.69	5.32
1"	1E@0.10	Ea	3.79	2.69	6.48
1-1/4"	1E@0.15	Ea	6.14	4.03	10.17
1-1/2"	1E@0.15	Ea	6.89	4.03	10.92
2"	1E@0.20	Ea	9.92	5.38	15.30
2-1/2"	1E@0.20	Ea	17.60	5.38	22.98
3"	1E@0.25	Ea	27.00	6.72	33.72
3-1/2"	1E@0.25	Ea	41.10	6.72	47.82
4"	1E@0.30	Ea	47.70	8.06	55.76
Rigid steel couplings (used on IMC)					
1/2"	1E@0.05	Ea	.71	1.34	2.05
3/4"	1E@0.06	Ea	.87	1.61	2.48
1"	1E@0.08	Ea	1.30	2.15	3.45
1-1/4"	1E@0.10	Ea	1.63	2.69	4.32
1-1/2"	1E@0.10	Ea	2.05	2.69	4.74
2"	1E@0.15	Ea	2.71	4.03	6.74
2-1/2"	1E@0.15	Ea	6.33	4.03	10.36
3"	1E@0.20	Ea	8.20	5.38	13.58
3-1/2"	1E@0.20	Ea	11.00	5.38	16.38
4"	1E@0.25	Ea	11.00	6.72	17.72
5"	1E@0.30	Ea	25.50	8.06	33.56
6"	1E@0.50	Ea	35.40	13.40	48.80
Steel running thread in 36" lengths					
1/2"	1E@0.15	Ea	12.40	4.03	16.43
3/4"	1E@0.15	Ea	13.60	4.03	17.63
1"	1E@0.20	Ea	22.90	5.38	28.28
1-1/4"	1E@0.20	Ea	26.00	5.38	31.38
1-1/2"	1E@0.25	Ea	28.20	6.72	34.92
2"	1E@0.25	Ea	39.40	6.72	46.12
2-1/2"	1E@0.30	Ea	59.80	8.06	67.86
3"	1E@0.30	Ea	77.10	8.06	85.16
3-1/2"	1E@0.35	Ea	92.50	9.41	101.91
4"	1E@0.40	Ea	108.00	10.80	118.80
5"	1E@0.50	Ea	233.00	13.40	246.40
6"	1E@0.75	Ea	235.00	20.20	255.20

Use these figures to estimate the cost of elbows, couplings and running thread installed on intermediate metal conduit under the conditions described on pages 5 and 6. Costs listed are for each fitting installed. The crew is one electrician working at a labor cost of $26.88 per manhour. These costs include cutting, removal of the knockout, layout, material handling, and normal waste. Add for elbow couplings, terminations, sales tax, delivery, supervision, mobilization, demobilization, cleanup, overhead and profit. Note: Elbows and running thread are factory made. Job specifications may prohibit the use of running thread.

Galvanized Steel Locknuts and Plastic or Insulated Bushings

Material	Craft@Hrs	Unit	Material Cost	Labor Cost	Installed Cost
Galvanized steel locknuts					
1/2"	1E@0.02	Ea	.13	.54	.67
3/4"	1E@0.02	Ea	.22	.54	.76
1"	1E@0.02	Ea	.35	.54	.89
1-1/4"	1E@0.03	Ea	.48	.81	1.29
1-1/2"	1E@0.03	Ea	.79	.81	1.60
2"	1E@0.05	Ea	1.15	1.34	2.49
2-1/2"	1E@0.05	Ea	3.20	1.34	4.54
3"	1E@0.07	Ea	4.11	1.88	5.99
3-1/2"	1E@0.07	Ea	6.94	1.88	8.82
4"	1E@0.09	Ea	8.65	2.42	11.07
5"	1E@0.10	Ea	18.50	2.69	21.19
6"	1E@0.20	Ea	31.60	5.38	36.98
Plastic bushings					
1/2"	1E@0.02	Ea	.13	.54	.67
3/4"	1E@0.02	Ea	.22	.54	.76
1"	1E@0.03	Ea	.36	.81	1.17
1-1/4"	1E@0.04	Ea	.54	1.08	1.62
1-1/2"	1E@0.04	Ea	.74	1.08	1.82
2"	1E@0.05	Ea	1.37	1.34	2.71
2-1/2"	1E@0.05	Ea	3.19	1.34	4.53
3"	1E@0.07	Ea	3.29	1.88	5.17
3-1/2"	1E@0.07	Ea	4.33	1.88	6.21
4"	1E@0.09	Ea	5.32	2.42	7.74
5"	1E@0.10	Ea	28.90	2.69	31.59
6"	1E@0.20	Ea	52.50	5.38	57.88
Insulated ground bushings					
1/2"	1E@0.10	Ea	1.30	2.69	3.99
3/4"	1E@0.10	Ea	1.80	2.69	4.49
1"	1E@0.10	Ea	2.19	2.69	4.88
1-1/4"	1E@0.15	Ea	2.47	4.03	6.50
1-1/2"	1E@0.15	Ea	2.70	4.03	6.73
2"	1E@0.20	Ea	3.95	5.38	9.33
2-1/2"	1E@0.20	Ea	6.22	5.38	11.60
3"	1E@0.25	Ea	8.12	6.72	14.84
3-1/2"	1E@0.25	Ea	9.01	6.72	15.73
4"	1E@0.30	Ea	11.60	8.06	19.66
5"	1E@0.40	Ea	53.00	10.80	63.80
6"	1E@0.50	Ea	78.80	13.40	92.20

Use these figures to estimate the cost of locknuts and bushings installed on GRS or IMC conduit under the conditions described on pages 5 and 6. Costs listed are for each locknut or bushing installed. The crew is one electrician working at a labor cost of $26.88 per manhour. These costs include removal of the knockout, layout, material handling, and normal waste. Add for conduit, sales tax, delivery, supervision, mobilization, demobilization, cleanup, overhead and profit. Note: Material costs assume purchase of full box quantities. The locknuts are steel for sizes up to 2" and malleable for sizes over 2". On conduit terminations at boxes or cabinets, one locknut is used inside the box and one locknut is used outside the box. A bushing is used at the end of each conduit run to protect the wire. An insulated ground bushing is used when connecting a ground wire to the conduit system.

Galvanized rigid steel nipples

Material		Craft@Hrs	Unit	Material Cost	Labor Cost	Installed Cost
1/2"	x close	1E@0.05	Ea	.43	1.34	1.77
1/2"	x 1-1/2"	1E@0.05	Ea	.53	1.34	1.87
1/2"	x 2"	1E@0.05	Ea	.59	1.34	1.93
1/2"	x 2-1/2"	1E@0.05	Ea	.64	1.34	1.98
1/2"	x 3"	1E@0.05	Ea	.69	1.34	2.03
1/2"	x 3-1/2"	1E@0.05	Ea	.76	1.34	2.10
1/2"	x 4"	1E@0.05	Ea	.82	1.34	2.16
1/2"	x 5"	1E@0.05	Ea	.92	1.34	2.26
1/2"	x 6"	1E@0.05	Ea	1.10	1.34	2.44
1/2"	x 8"	1E@0.05	Ea	1.91	1.34	3.25
1/2"	x 10"	1E@0.05	Ea	2.19	1.34	3.53
1/2"	x 12"	1E@0.05	Ea	2.52	1.34	3.86
3/4"	x close	1E@0.06	Ea	.57	1.61	2.18
3/4"	x 2"	1E@0.06	Ea	.66	1.61	2.27
3/4"	x 2-1/2"	1E@0.06	Ea	.75	1.61	2.36
3/4"	x 3"	1E@0.06	Ea	.81	1.61	2.42
3/4"	x 3-1/2"	1E@0.06	Ea	.84	1.61	2.45
3/4"	x 4"	1E@0.06	Ea	.96	1.61	2.57
3/4"	x 5"	1E@0.06	Ea	1.10	1.61	2.71
3/4"	x 6"	1E@0.06	Ea	1.28	1.61	2.89
3/4"	x 8"	1E@0.06	Ea	2.12	1.61	3.73
3/4"	x 10"	1E@0.06	Ea	2.54	1.61	4.15
3/4"	x 12"	1E@0.06	Ea	2.86	1.61	4.47
1"	x close	1E@0.08	Ea	.85	2.15	3.00
1"	x 2"	1E@0.08	Ea	.91	2.15	3.06
1"	x 2-1/2"	1E@0.08	Ea	1.08	2.15	3.23
1"	x 3"	1E@0.08	Ea	1.12	2.15	3.27
1"	x 3-1/2"	1E@0.08	Ea	1.28	2.15	3.43
1"	x 4"	1E@0.08	Ea	1.38	2.15	3.53
1"	x 5"	1E@0.08	Ea	1.55	2.15	3.70
1"	x 6"	1E@0.08	Ea	1.70	2.15	3.85
1"	x 8"	1E@0.08	Ea	2.68	2.15	4.83
1"	x 10"	1E@0.08	Ea	3.47	2.15	5.62
1"	x 12"	1E@0.08	Ea	3.93	2.15	6.08

Use these figures to estimate the cost of nipples installed on GRS conduit under the conditions described on pages 5 and 6. Costs listed are for each nipple installed. The crew is one electrician at a labor cost of $26.88 per manhour. These costs include removal of the knockout, layout, material handling, and normal waste. Add for terminations, couplings, sales tax, delivery, supervision, mobilization, demobilization, cleanup, overhead and profit. Note: Nipples are factory made, not field made. In many cases a coupling will be needed with a nipple.

Galvanized Rigid Steel Nipples

Material		Craft@Hrs	Unit	Material Cost	Labor Cost	Installed Cost
Galvanized rigid steel nipples						
1-1/4" x	close	1E@0.10	Ea	1.10	2.69	3.79
1-1/4" x	2"	1E@0.10	Ea	1.19	2.69	3.88
1-1/4" x	2-1/2"	1E@0.10	Ea	1.29	2.69	3.98
1-1/4" x	3"	1E@0.10	Ea	1.40	2.69	4.09
1-1/4" x	3-1/2"	1E@0.10	Ea	1.59	2.69	4.28
1-1/4" x	4"	1E@0.10	Ea	1.68	2.69	4.37
1-1/4" x	5"	1E@0.10	Ea	1.95	2.69	4.64
1-1/4" x	6"	1E@0.10	Ea	2.19	2.69	4.88
1-1/4" x	8"	1E@0.10	Ea	3.59	2.69	6.28
1-1/4" x	10"	1E@0.10	Ea	4.51	2.69	7.20
1-1/4" x	12"	1E@0.10	Ea	5.26	2.69	7.95
1-1/2" x	close	1E@0.10	Ea	1.33	2.69	4.02
1-1/2" x	2"	1E@0.10	Ea	1.41	2.69	4.10
1-1/2" x	2-1/2"	1E@0.10	Ea	1.58	2.69	4.27
1-1/2" x	3"	1E@0.10	Ea	1.70	2.69	4.39
1-1/2" x	3-1/2"	1E@0.10	Ea	1.96	2.69	4.65
1-1/2" x	4"	1E@0.10	Ea	2.13	2.69	4.82
1-1/2" x	5"	1E@0.10	Ea	2.39	2.69	5.08
1-1/2" x	6"	1E@0.10	Ea	2.94	2.69	5.63
1-1/2" x	8"	1E@0.10	Ea	4.49	2.69	7.18
1-1/2" x	10"	1E@0.10	Ea	5.41	2.69	8.10
1-1/2" x	12"	1E@0.10	Ea	5.87	2.69	8.56
2" x	close	1E@0.15	Ea	1.63	4.03	5.66
2" x	2-1/2"	1E@0.15	Ea	1.91	4.03	5.94
2" x	3"	1E@0.15	Ea	2.22	4.03	6.25
2" x	3-1/2"	1E@0.15	Ea	2.52	4.03	6.55
2" x	4"	1E@0.15	Ea	2.78	4.03	6.81
2" x	5"	1E@0.15	Ea	3.24	4.03	7.27
2" x	6"	1E@0.15	Ea	3.71	4.03	7.74
2" x	8"	1E@0.15	Ea	5.36	4.03	9.39
2" x	10"	1E@0.15	Ea	6.42	4.03	10.45
2" x	12"	1E@0.15	Ea	7.30	4.03	11.33
2-1/2" x	close	1E@0.15	Ea	4.51	4.03	8.54
2-1/2" x	3"	1E@0.15	Ea	4.57	4.03	8.60
2-1/2" x	3-1/2"	1E@0.15	Ea	5.40	4.03	9.43
2-1/2" x	4"	1E@0.15	Ea	5.66	4.03	9.69
2-1/2" x	5"	1E@0.15	Ea	6.69	4.03	10.72
2-1/2" x	6"	1E@0.15	Ea	7.59	4.03	11.62

Use these figures to estimate the cost of nipples installed on GRS conduit under the conditions described on pages 5 and 6. Costs listed are for each nipple installed. The crew is one electrician at a labor cost of $26.88 per manhour. These costs include removal of the knockout, layout, material handling, and normal waste. Add for terminations, couplings, sales tax, delivery, supervision, mobilization, demobilization, cleanup, overhead and profit. Note: Nipples are factory made, not field made. In many cases a coupling will be needed with a nipple.

Material		Craft@Hrs	Unit	Material Cost	Labor Cost	Installed Cost
Galvanized rigid steel nipples						
2-1/2" x	8"	1E@0.15	Ea	9.98	4.03	14.01
2-1/2" x	10"	1E@0.15	Ea	11.60	4.03	15.63
2-1/2" x	12"	1E@0.15	Ea	13.50	4.03	17.53
3"	x close	1E@0.20	Ea	5.30	5.38	10.68
3"	x 3"	1E@0.20	Ea	5.62	5.38	11.00
3"	x 3-1/2"	1E@0.20	Ea	6.34	5.38	11.72
3"	x 4"	1E@0.20	Ea	6.87	5.38	12.25
3"	x 5"	1E@0.20	Ea	7.98	5.38	13.36
3"	x 6"	1E@0.20	Ea	9.12	5.38	14.50
3"	x 8"	1E@0.20	Ea	11.90	5.38	17.28
3"	x 10"	1E@0.20	Ea	16.50	5.38	21.88
3"	x 12"	1E@0.20	Ea	17.00	5.38	22.38
3-1/2" x	close	1E@0.25	Ea	6.53	6.72	13.25
3-1/2" x	4"	1E@0.25	Ea	8.36	6.72	15.08
3-1/2" x	5"	1E@0.25	Ea	9.47	6.72	16.19
3-1/2" x	6"	1E@0.25	Ea	10.80	6.72	17.52
3-1/2" x	8"	1E@0.25	Ea	13.60	6.72	20.32
3-1/2" x	10"	1E@0.25	Ea	16.40	6.72	23.12
3-1/2" x	12"	1E@0.25	Ea	19.20	6.72	25.92
4"	x close	1E@0.25	Ea	7.72	6.72	14.44
4"	x 4"	1E@0.25	Ea	9.23	6.72	15.95
4"	x 5"	1E@0.25	Ea	11.00	6.72	17.72
4"	x 6"	1E@0.25	Ea	12.30	6.72	19.02
4"	x 8"	1E@0.25	Ea	15.30	6.72	22.02
4"	x 10"	1E@0.25	Ea	18.90	6.72	25.62
4"	x 12"	1E@0.25	Ea	22.50	6.72	29.22
5"	x close	1E@0.40	Ea	13.50	10.80	24.30
5"	x 5"	1E@0.40	Ea	19.70	10.80	30.50
5"	x 6"	1E@0.40	Ea	22.40	10.80	33.20
5"	x 8"	1E@0.40	Ea	25.50	10.80	36.30
5"	x 10"	1E@0.40	Ea	29.90	10.80	40.70
5"	x 12"	1E@0.40	Ea	36.20	10.80	47.00
6"	x close	1E@0.60	Ea	17.80	16.10	33.90
6"	x 5"	1E@0.60	Ea	24.60	16.10	40.70
6"	x 6"	1E@0.60	Ea	27.10	16.10	43.20
6"	x 8"	1E@0.60	Ea	31.60	16.10	47.70
6"	x 10"	1E@0.60	Ea	39.60	16.10	55.70
6"	x 12"	1E@0.60	Ea	43.90	16.10	60.00

Use these figures to estimate the cost of nipples installed on GRS conduit under the conditions described on pages 5 and 6. Costs listed are for each nipple installed. The crew is one electrician working at a cost of $26.88 per manhour. These costs include removal of the knockout, layout, material handling, and normal waste. Add for terminations, couplings, sales tax, delivery, supervision, mobilization, demobilization, cleanup, overhead and profit. Note: Nipples are factory made, not field made. In many cases a coupling will be needed with a nipple.

Aluminum Rigid Conduit (ARC), Elbows and Nipples

Material	Craft@Hrs	Unit	Material Cost	Labor Cost	Installed Cost
Aluminum rigid conduit					
1/2"	1E@3.75	CLF	83.70	101.00	184.70
3/4"	1E@4.00	CLF	111.00	108.00	219.00
1"	1E@4.50	CLF	159.00	121.00	280.00
1-1/4"	1E@6.00	CLF	208.00	161.00	369.00
1-1/2"	1E@7.00	CLF	258.00	188.00	446.00
2"	1E@8.50	CLF	345.00	228.00	573.00
2-1/2"	2E@10.0	CLF	546.00	269.00	815.00
3"	2E@12.0	CLF	717.00	323.00	1,040.00
3-1/2"	2E@14.0	CLF	861.00	376.00	1,237.00
4"	2E@16.0	CLF	1,020.00	430.00	1,450.00
5"	2E@20.0	CLF	1,460.00	538.00	1,998.00
6"	2E@25.0	CLF	1,920.00	672.00	2,592.00
90 degree aluminum rigid conduit elbows					
1/2"	1E@0.10	Ea	4.13	2.69	6.82
3/4"	1E@0.10	Ea	5.60	2.69	8.29
1"	1E@0.10	Ea	7.82	2.69	10.51
1-1/4"	1E@0.15	Ea	12.40	4.03	16.43
1-1/2"	1E@0.15	Ea	16.50	4.03	20.53
2"	1E@0.20	Ea	24.30	5.38	29.68
2-1/2"	2E@0.20	Ea	41.00	5.38	46.38
3"	2E@0.25	Ea	63.40	6.72	70.12
3-1/2"	2E@0.25	Ea	98.80	6.72	105.52
4"	2E@0.30	Ea	117.00	8.06	125.06
5"	2E@0.40	Ea	320.00	10.80	330.80
6"	2E@0.70	Ea	442.00	18.80	460.80
Aluminum rigid conduit nipples					
1/2" x close	1E@0.05	Ea	1.50	1.34	2.84
1/2" x 1-1/2"	1E@0.05	Ea	1.54	1.34	2.88
1/2" x 2"	1E@0.05	Ea	1.66	1.34	3.00
1/2" x 2-1/2"	1E@0.05	Ea	1.98	1.34	3.32
1/2" x 3"	1E@0.05	Ea	2.08	1.34	3.42
1/2" x 3-1/2"	1E@0.05	Ea	2.26	1.34	3.60
1/2" x 4"	1E@0.05	Ea	2.41	1.34	3.75
1/2" x 5"	1E@0.05	Ea	2.76	1.34	4.10
1/2" x 6"	1E@0.05	Ea	2.92	1.34	4.26
1/2" x 8"	1E@0.05	Ea	3.89	1.34	5.23
1/2" x 10"	1E@0.05	Ea	4.74	1.34	6.08
1/2" x 12"	1E@0.05	Ea	5.49	1.34	6.83
3/4" x close	1E@0.06	Ea	2.01	1.61	3.62
3/4" x 2"	1E@0.06	Ea	2.19	1.61	3.80
3/4" x 2-1/2"	1E@0.06	Ea	2.33	1.61	3.94

Use these figures to estimate the cost of aluminum rigid conduit, elbows and nipples installed in a building under the conditions described on pages 5 and 6. Costs listed are for each 100 linear feet of conduit or each fitting installed. The crew is one electrician for conduit sizes to 2" and two electricians for conduit over 2". The labor cost is $26.88 per manhour. These costs include conduit bending, one coupling for each length of conduit, layout, material handling, and normal waste. Add for extra couplings, straps, terminations, wire, sales tax, delivery, supervision, mobilization, demobilization, cleanup, overhead and profit. Note: Elbows and nipples are factory made. Do not install ARC in concrete or masonry construction. Conduit runs are assumed to be 50' long. Installation costs per linear foot will be less on longer runs and more on shorter runs.

Material	Craft@Hrs	Unit	Material Cost	Labor Cost	Installed Cost
Aluminum rigid conduit nipples					
3/4" x 3"	1E@0.06	Ea	2.81	1.61	4.42
3/4" x 3-1/2"	1E@0.06	Ea	2.92	1.61	4.53
3/4" x 4"	1E@0.06	Ea	3.07	1.61	4.68
3/4" x 5"	1E@0.06	Ea	3.69	1.61	5.30
3/4" x 6"	1E@0.06	Ea	4.18	1.61	5.79
3/4" x 8"	1E@0.06	Ea	5.50	1.61	7.11
3/4" x 10"	1E@0.06	Ea	6.44	1.61	8.05
3/4" x 12"	1E@0.06	Ea	7.90	1.61	9.51
1" x close	1E@0.08	Ea	2.76	2.15	4.91
1" x 2"	1E@0.08	Ea	3.06	2.15	5.21
1" x 2-1/2"	1E@0.08	Ea	3.36	2.15	5.51
1" x 3"	1E@0.08	Ea	3.61	2.15	5.76
1" x 3-1/2"	1E@0.08	Ea	4.09	2.15	6.24
1" x 4"	1E@0.08	Ea	4.44	2.15	6.59
1" x 5"	1E@0.08	Ea	5.26	2.15	7.41
1" x 6"	1E@0.08	Ea	6.22	2.15	8.37
1" x 8"	1E@0.08	Ea	7.71	2.15	9.86
1" x 10"	1E@0.08	Ea	9.68	2.15	11.83
1" x 12"	1E@0.08	Ea	11.40	2.15	13.55
1-1/4" x close	1E@0.10	Ea	3.70	2.69	6.39
1-1/4" x 2"	1E@0.10	Ea	3.78	2.69	6.47
1-1/4" x 2-1/2"	1E@0.10	Ea	4.16	2.69	6.85
1-1/4" x 3"	1E@0.10	Ea	4.71	2.69	7.40
1-1/4" x 3-1/2"	1E@0.10	Ea	5.36	2.69	8.05
1-1/4" x 4"	1E@0.10	Ea	5.90	2.69	8.59
1-1/4" x 5"	1E@0.10	Ea	6.93	2.69	9.62
1-1/4" x 6"	1E@0.10	Ea	8.04	2.69	10.73
1-1/4" x 8"	1E@0.10	Ea	10.20	2.69	12.89
1-1/4" x 10"	1E@0.10	Ea	12.50	2.69	15.19
1-1/4" x 12"	1E@0.10	Ea	14.60	2.69	17.29
1-1/2" x close	1E@0.10	Ea	4.61	2.69	7.30
1-1/2" x 2"	1E@0.10	Ea	4.75	2.69	7.44
1-1/2" x 2-1/2"	1E@0.10	Ea	5.05	2.69	7.74
1-1/2" x 3"	1E@0.10	Ea	5.74	2.69	8.43
1-1/2" x 3-1/2"	1E@0.10	Ea	7.19	2.69	9.88
1-1/2" x 4"	1E@0.10	Ea	7.23	2.69	9.92
1-1/2" x 5"	1E@0.10	Ea	8.28	2.69	10.97
1-1/2" x 6"	1E@0.10	Ea	9.57	2.69	12.26
1-1/2" x 8"	1E@0.10	Ea	12.30	2.69	14.99
1-1/2" x 10"	1E@0.10	Ea	15.10	2.69	17.79
1-1/2" x 12"	1E@0.10	Ea	17.60	2.69	20.29

Use these figures to estimate the cost of ARC nipples installed on ARC conduit under the conditions described on pages 5 and 6. Costs listed are for each nipple installed. The crew is one electrician at a labor cost of $26.88 per manhour. These costs include removing the knockout, layout, material handling, and normal waste. Add for extra couplings, straps, boxes, sales tax, delivery, supervision, mobilization, demobilization, cleanup, overhead and profit. Note: Material costs assume the purchase of full packages.

ARC Nipples

Material	Craft@Hrs	Unit	Material Cost	Labor Cost	Installed Cost
Aluminum rigid conduit nipples					
2" x close	1E@0.15	Ea	5.92	4.03	9.95
2" x 2-1/2"	1E@0.15	Ea	6.34	4.03	10.37
2" x 3"	1E@0.15	Ea	7.10	4.03	11.13
2" x 3-1/2"	1E@0.15	Ea	8.17	4.03	12.20
2" x 4"	1E@0.15	Ea	8.71	4.03	12.74
2" x 5"	1E@0.15	Ea	10.40	4.03	14.43
2" x 6"	1E@0.15	Ea	12.00	4.03	16.03
2" x 8"	1E@0.15	Ea	15.40	4.03	19.43
2" x 10"	1E@0.15	Ea	18.70	4.03	22.73
2" x 12"	1E@0.15	Ea	22.10	4.03	26.13
2-1/2" x close	1E@0.15	Ea	11.90	4.03	15.93
2-1/2" x 3"	1E@0.15	Ea	12.90	4.03	16.93
2-1/2" x 3-1/2"	1E@0.15	Ea	14.30	4.03	18.33
2-1/2" x 4"	1E@0.15	Ea	17.10	4.03	21.13
2-1/2" x 5"	1E@0.15	Ea	17.10	4.03	21.13
2-1/2" x 6"	1E@0.15	Ea	18.70	4.03	22.73
2-1/2" x 8"	1E@0.15	Ea	23.70	4.03	27.73
2-1/2" x 10"	1E@0.15	Ea	28.60	4.03	32.63
2-1/2" x 12"	1E@0.15	Ea	33.30	4.03	37.33
3" x close	1E@0.20	Ea	13.00	5.38	18.38
3" x 3"	1E@0.20	Ea	14.60	5.38	19.98
3" x 3-1/2"	1E@0.20	Ea	17.40	5.38	22.78
3" x 4"	1E@0.20	Ea	18.20	5.38	23.58
3" x 5"	1E@0.20	Ea	20.70	5.38	26.08
3" x 6"	1E@0.20	Ea	23.80	5.38	29.18
3" x 8"	1E@0.20	Ea	30.30	5.38	35.68
3" x 10"	1E@0.20	Ea	36.90	5.38	42.28
3" x 12"	1E@0.20	Ea	43.20	5.38	48.58
3-1/2" x close	1E@0.25	Ea	16.60	6.72	23.32
3-1/2" x 4"	1E@0.25	Ea	22.30	6.72	29.02
3-1/2" x 5"	1E@0.25	Ea	25.20	6.72	31.92
3-1/2" x 6"	1E@0.25	Ea	29.00	6.72	35.72
3-1/2" x 8"	1E@0.25	Ea	36.20	6.72	42.92
3-1/2" x 10"	1E@0.25	Ea	44.40	6.72	51.12
3-1/2" x 12"	1E@0.25	Ea	52.10	6.72	58.82

Use these figures to estimate the cost of ARC nipples installed on ARC conduit under the conditions described on pages 5 and 6. Costs listed are for each nipple installed. The crew is one electrician at a labor cost of $26.88 per manhour. These costs include removing the knockout, layout, material handling, and normal waste. Add for extra couplings, straps, boxes, sales tax, delivery, supervision, mobilization, demobilization, cleanup, overhead and profit. Note: Material costs assume the purchase of full packages. Nipples are factory made, not field made. In many cases a coupling will be needed with each nipple. Do not install aluminum fittings in concrete or masonry. The bending, cutting and threading tools for aluminum conduit are the same as used for GRS. Don't mix aluminum fittings with other types of fittings

ARC Nipples, Locknuts and Bushings

Material	Craft@Hrs	Unit	Material Cost	Labor Cost	Installed Cost
Aluminum rigid conduit nipples					
4" x close	1E@0.25	Ea	20.70	6.72	27.42
4" x 4"	1E@0.25	Ea	26.10	6.72	32.82
4" x 5"	1E@0.25	Ea	29.80	6.72	36.52
4" x 6"	1E@0.25	Ea	34.50	6.72	41.22
4" x 8"	1E@0.25	Ea	43.90	6.72	50.62
4" x 10"	1E@0.25	Ea	53.40	6.72	60.12
4" x 12"	1E@0.25	Ea	62.50	6.72	69.22
5" x close	1E@0.40	Ea	44.70	10.80	55.50
5" x 5"	1E@0.40	Ea	52.40	10.80	63.20
5" x 6"	1E@0.40	Ea	54.90	10.80	65.70
5" x 8"	1E@0.40	Ea	70.10	10.80	80.90
5" x 10"	1E@0.40	Ea	84.00	10.80	94.80
5" x 12"	1E@0.40	Ea	96.70	10.80	107.50
6" x close	1E@0.60	Ea	55.80	16.10	71.90
6" x 5"	1E@0.60	Ea	68.20	16.10	84.30
6" x 6"	1E@0.60	Ea	73.50	16.10	89.60
6" x 8"	1E@0.60	Ea	93.60	16.10	109.70
6" x 10"	1E@0.60	Ea	112.00	16.10	128.10
6" x 12"	1E@0.60	Ea	124.00	16.10	140.10
Aluminum locknuts					
1/2"	1E@0.02	Ea	.29	.54	.83
3/4"	1E@0.02	Ea	.43	.54	.97
1"	1E@0.02	Ea	.76	.54	1.30
1-1/4"	1E@0.03	Ea	1.09	.81	1.90
1-1/2"	1E@0.03	Ea	1.63	.81	2.44
2"	1E@0.05	Ea	2.32	1.34	3.66
2-1/2"	1E@0.05	Ea	4.60	1.34	5.94
3"	1E@0.07	Ea	5.95	1.88	7.83
3-1/2"	1E@0.07	Ea	12.80	1.88	14.68
4"	1E@0.09	Ea	16.00	2.42	18.42
5"	1E@0.10	Ea	35.40	2.69	38.09
6"	1E@0.20	Ea	62.80	5.38	68.18
Aluminum bushings					
1/2"	1E@0.02	Ea	.42	.54	.96
3/4"	1E@0.02	Ea	.62	.54	1.16
1"	1E@0.03	Ea	1.09	.81	1.90

Use these figures to estimate the cost of ARC nipples, locknuts and bushings installed on ARC conduit under the conditions described on pages 5 and 6. Costs listed are for each fitting installed. The crew is one electrician working at a labor cost of $26.88 per manhour. These costs include removing the knockout, layout, material handling, and normal waste. Add for extra couplings, sales tax, delivery, supervision, mobilization, demobilization, cleanup, overhead and profit. Note: Material costs are based on purchase of full packages. Nipples are factory made, not field made. In many cases a coupling will be needed with each nipple. Do not install aluminum fittings in concrete or masonry.

Aluminum Bushings and Terminations

Material	Craft@Hrs	Unit	Material Cost	Labor Cost	Installed Cost

Aluminum bushings

Material	Craft@Hrs	Unit	Material Cost	Labor Cost	Installed Cost
1-1/4"	1E@0.04	Ea	1.62	1.08	2.70
1-1/2"	1E@0.04	Ea	2.37	1.08	3.45
2"	1E@0.05	Ea	3.70	1.34	5.04
2-1/2"	1E@0.05	Ea	8.14	1.34	9.48
3"	1E@0.07	Ea	10.10	1.88	11.98
3-1/2"	1E@0.07	Ea	20.10	1.88	21.98
4"	1E@0.09	Ea	24.30	2.42	26.72
5"	1E@0.10	Ea	55.80	2.69	58.49
6"	1E@0.20	Ea	98.30	5.38	103.68

Aluminum insulated ground bushings

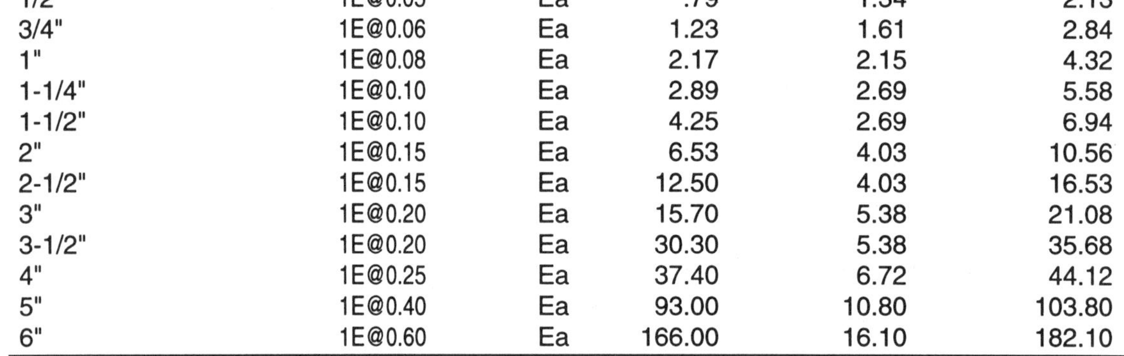

Material	Craft@Hrs	Unit	Material Cost	Labor Cost	Installed Cost
1/2"	1E@0.10	Ea	2.46	2.69	5.15
3/4"	1E@0.10	Ea	4.39	2.69	7.08
1"	1E@0.10	Ea	5.87	2.69	8.56
1-1/4"	1E@0.15	Ea	8.22	4.03	12.25
1-1/2"	1E@0.15	Ea	11.10	4.03	15.13
2"	1E@0.20	Ea	14.70	5.38	20.08
2-1/2"	1E@0.20	Ea	24.80	5.38	30.18
3"	1E@0.25	Ea	36.30	6.72	43.02
3-1/2"	1E@0.25	Ea	47.00	6.72	53.72
4"	1E@0.30	Ea	60.30	8.06	68.36
5"	1E@0.40	Ea	85.70	10.80	96.50
6"	1E@0.50	Ea	132.00	13.40	145.40

Conduit termination, two aluminum locknuts & one plastic bushing

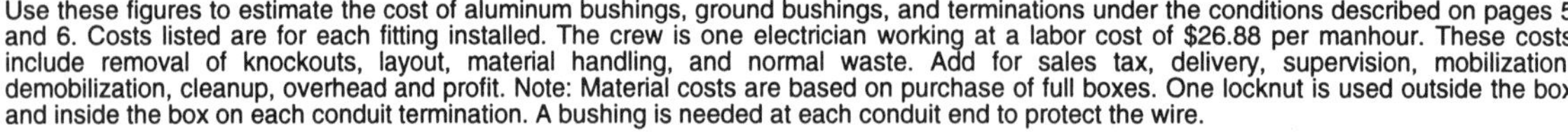

Material	Craft@Hrs	Unit	Material Cost	Labor Cost	Installed Cost
1/2"	1E@0.05	Ea	.79	1.34	2.13
3/4"	1E@0.06	Ea	1.23	1.61	2.84
1"	1E@0.08	Ea	2.17	2.15	4.32
1-1/4"	1E@0.10	Ea	2.89	2.69	5.58
1-1/2"	1E@0.10	Ea	4.25	2.69	6.94
2"	1E@0.15	Ea	6.53	4.03	10.56
2-1/2"	1E@0.15	Ea	12.50	4.03	16.53
3"	1E@0.20	Ea	15.70	5.38	21.08
3-1/2"	1E@0.20	Ea	30.30	5.38	35.68
4"	1E@0.25	Ea	37.40	6.72	44.12
5"	1E@0.40	Ea	93.00	10.80	103.80
6"	1E@0.60	Ea	166.00	16.10	182.10

Use these figures to estimate the cost of aluminum bushings, ground bushings, and terminations under the conditions described on pages 5 and 6. Costs listed are for each fitting installed. The crew is one electrician working at a labor cost of $26.88 per manhour. These costs include removal of knockouts, layout, material handling, and normal waste. Add for sales tax, delivery, supervision, mobilization, demobilization, cleanup, overhead and profit. Note: Material costs are based on purchase of full boxes. One locknut is used outside the box and inside the box on each conduit termination. A bushing is needed at each conduit end to protect the wire.

Cast Metal Entrance Elbows and Conduit Bodies

Material	Craft@Hrs	Unit	Material Cost	Labor Cost	Installed Cost

Cast metal Type SLB entrance elbows

Material	Craft@Hrs	Unit	Material Cost	Labor Cost	Installed Cost
1/2"	1E@0.10	Ea	3.36	2.69	6.05
3/4"	1E@0.15	Ea	4.38	4.03	8.41
1"	1E@0.15	Ea	7.11	4.03	11.14
1-1/4"	1E@0.20	Ea	9.67	5.38	15.05
1-1/2"	1E@0.20	Ea	22.60	5.38	27.98
2"	1E@0.25	Ea	32.60	6.72	39.32
2-1/2"	1E@0.30	Ea	83.70	8.06	91.76
3"	1E@0.40	Ea	104.00	10.80	114.80

Galvanized cast metal Types LB, LL or LR conduit bodies

Material	Craft@Hrs	Unit	Material Cost	Labor Cost	Installed Cost
1/2"	1E@0.10	Ea	6.75	2.69	9.44
3/4"	1E@0.15	Ea	8.13	4.03	12.16
1"	1E@0.20	Ea	12.00	5.38	17.38
1-1/4"	1E@0.25	Ea	18.10	6.72	24.82
1-1/2"	1E@0.25	Ea	22.40	6.72	29.12
2"	1E@0.30	Ea	37.30	8.06	45.36
2-1/2"	1E@0.40	Ea	75.10	10.80	85.90
3"	1E@0.50	Ea	97.10	13.40	110.50
3-1/2"	1E@0.70	Ea	155.00	18.80	173.80
4"	1E@1.00	Ea	181.00	26.90	207.90

Galvanized cast metal Type T conduit bodies

Material	Craft@Hrs	Unit	Material Cost	Labor Cost	Installed Cost
1/2"	1E@0.15	Ea	8.16	4.03	12.19
3/4"	1E@0.20	Ea	9.72	5.38	15.10
1"	1E@0.25	Ea	14.40	6.72	21.12
1-1/4"	1E@0.30	Ea	20.30	8.06	28.36
1-1/2"	1E@0.30	Ea	26.00	8.06	34.06
2"	1E@0.40	Ea	41.00	10.80	51.80
2-1/2"	1E@0.50	Ea	78.20	13.40	91.60
3"	1E@0.70	Ea	110.00	18.80	128.80
3-1/2"	1E@0.90	Ea	174.00	24.20	198.20
4"	1E@1.25	Ea	194.00	33.60	227.60

Galvanized cast metal Type X conduit bodies

Material	Craft@Hrs	Unit	Material Cost	Labor Cost	Installed Cost
1/2"	1E@0.20	Ea	10.20	5.38	15.58
3/4"	1E@0.25	Ea	12.10	6.72	18.82
1"	1E@0.30	Ea	16.70	8.06	24.76
1-1/4"	1E@0.40	Ea	22.90	10.80	33.70
1-1/2"	1E@0.40	Ea	28.70	10.80	39.50
2"	1E@0.50	Ea	51.70	13.40	65.10

Use these figures to estimate the cost of conduit bodies installed on EMT or GRS conduit under the conditions described on pages 5 and 6. Costs listed are for each body installed. The crew is one electrician working at a labor cost of $26.88 per manhour. These costs include layout, material handling, and normal waste. Add for conduit, nipples, boxes, covers, gaskets, sales tax, delivery, supervision, mobilization, demobilization, cleanup, overhead and profit. Note: Using a larger conduit body or a mogul size can reduce the installation time when wire sizes are larger.

Blank Conduit Body Covers

Material	Craft@Hrs	Unit	Material Cost	Labor Cost	Installed Cost
Steel blank conduit body covers					
1/2"	1E@0.05	Ea	1.49	1.34	2.83
3/4"	1E@0.05	Ea	1.81	1.34	3.15
1"	1E@0.05	Ea	2.50	1.34	3.84
1-1/4"	1E@0.10	Ea	2.99	2.69	5.68
1-1/2"	1E@0.10	Ea	2.99	2.69	5.68
2"	1E@0.10	Ea	5.27	2.69	7.96
2-1/2" - 3"	1E@0.15	Ea	8.27	4.03	12.30
2-1/2" - 4"	1E@0.20	Ea	9.72	5.38	15.10
Malleable blank conduit body covers					
1/2"	1E@0.05	Ea	2.31	1.34	3.65
3/4"	1E@0.05	Ea	2.80	1.34	4.14
1"	1E@0.10	Ea	4.45	2.69	7.14
1-1/4"	1E@0.10	Ea	5.93	2.69	8.62
1-1/2"	1E@0.10	Ea	5.93	2.69	8.62
2"	1E@0.15	Ea	8.75	4.03	12.78
2-1/2" - 3"	1E@0.20	Ea	10.20	5.38	15.58
2-1/2" - 4"	1E@0.25	Ea	16.00	6.72	22.72
Aluminum blank conduit body covers					
1/2"	1E@0.05	Ea	1.31	1.34	2.65
3/4"	1E@0.05	Ea	1.81	1.34	3.15
1"	1E@0.05	Ea	2.31	1.34	3.65
1-1/4"	1E@0.10	Ea	2.80	2.69	5.49
1-1/2"	1E@0.10	Ea	2.80	2.69	5.49
2"	1E@0.10	Ea	4.94	2.69	7.63
2-1/2" - 3"	1E@0.15	Ea	8.27	4.03	12.30
2-1/2" - 4"	1E@0.20	Ea	9.72	5.38	15.10

Use these figures to estimate the cost of blank conduit body covers installed on conduit bodies under the conditions described on pages 5 and 6. Costs listed are for each cover installed. The crew is one electrician working at a labor cost of $26.88 per manhour. These costs include layout, material handling, and normal waste. Add for conduit bodies, other fittings, sales tax, delivery, supervision, mobilization, demobilization, cleanup, overhead and profit. Note: These figures assume that the conduit body is readily accessible.

Conduit Body Gaskets, Conduit Bodies and Capped Elbows

Material	Craft@Hrs	Unit	Material Cost	Labor Cost	Installed Cost
Conduit body gaskets					
1/2"	1E@0.02	Ea	1.48	.54	2.02
3/4"	1E@0.02	Ea	1.48	.54	2.02
1"	1E@0.03	Ea	1.98	.81	2.79
1-1/4"	1E@0.05	Ea	2.50	1.34	3.84
1-1/2"	1E@0.05	Ea	2.50	1.34	3.84
2"	1E@0.07	Ea	2.97	1.88	4.85
2-1/2" - 3"	1E@0.10	Ea	4.94	2.69	7.63
2-1/2" - 4"	1E@0.15	Ea	6.27	4.03	10.30
Type LB, LL or LR aluminum conduit bodies					
1/2"	1E@0.10	Ea	6.05	2.69	8.74
3/4"	1E@0.15	Ea	7.62	4.03	11.65
1"	1E@0.15	Ea	11.10	4.03	15.13
1-1/4"	1E@0.20	Ea	16.00	5.38	21.38
1-1/2"	1E@0.20	Ea	20.00	5.38	25.38
2"	1E@0.25	Ea	33.30	6.72	40.02
2-1/2"	1E@0.30	Ea	70.40	8.06	78.46
3"	1E@0.40	Ea	90.30	10.80	101.10
Type LB, LL or LR mogul aluminum conduit bodies with covers & gaskets					
1"	1E@0.25	Ea	56.60	6.72	63.32
1-1/4"	1E@0.30	Ea	61.10	8.06	69.16
1-1/2"	1E@0.30	Ea	76.80	8.06	84.86
2"	1E@0.50	Ea	81.10	13.40	94.50
2-1/2"	1E@0.70	Ea	244.00	18.80	262.80
3"	1E@0.75	Ea	244.00	20.20	264.20
3-1/2"	1E@1.00	Ea	427.00	26.90	453.90
4"	1E@1.00	Ea	445.00	26.90	471.90
Galvanized capped elbows					
1/2"	1E@0.10	Ea	8.04	2.69	10.73
3/4"	1E@0.15	Ea	8.74	4.03	12.77
1"	1E@0.20	Ea	11.20	5.38	16.58
1-1/4"	1E@0.25	Ea	23.80	6.72	30.52
1-1/2"	1E@0.25	Ea	35.40	6.72	42.12

Use these figures to estimate the cost of conduit body gaskets, aluminum conduit bodies and capped elbows installed with covers and aluminum conduit under the conditions described on pages 5 and 6. Costs listed are for each fitting installed. The crew is one electrician working at a labor cost of $26.88 per manhour. These costs include layout, material handling, and normal waste. Add for covers, conduit, nipples, sales tax, delivery, supervision, mobilization, demobilization, cleanup, overhead and profit. Note: Standard conduit bodies do not include covers and gaskets. Cost of mogul bodies includes covers and gaskets.

Galvanized Cast Boxes

Material		Craft@Hrs	Unit	Material Cost	Labor Cost	Installed Cost
Galvanized cast boxes with threaded hubs						
FS-1	1/2" one gang	1E@0.20	Ea	9.95	5.38	15.33
FS-2	3/4" one gang	1E@0.25	Ea	10.50	6.72	17.22
FS-3	1" one gang	1E@0.30	Ea	11.40	8.06	19.46
FS-12	1/2" two gang	1E@0.25	Ea	17.80	6.72	24.52
FS-22	3/4" two gang	1E@0.30	Ea	18.30	8.06	26.36
FS-32	1" two gang	1E@0.35	Ea	20.00	9.41	29.41
FSC-1	1/2" one gang	1E@0.25	Ea	11.00	6.72	17.72
FSC-2	3/4" one gang	1E@0.30	Ea	12.10	8.06	20.16
FSC-3	1" one gang	1E@0.35	Ea	15.00	9.41	24.41
FSC-12	1/2" two gang	1E@0.30	Ea	19.70	8.06	27.76
FSC-22	3/4" two gang	1E@0.35	Ea	20.90	9.41	30.31
FSC-32	1" two gang	1E@0.40	Ea	23.50	10.80	34.30
FSCC-1	1/2" one gang	1E@0.35	Ea	13.50	9.41	22.91
FSCC-2	3/4" one gang	1E@0.40	Ea	17.80	10.80	28.60
FSCT-1	1/2" one gang	1E@0.35	Ea	13.50	9.41	22.91
FSCT-2	3/4" one gang	1E@0.40	Ea	16.80	10.80	27.60
FSL-1	1/2" one gang	1E@0.30	Ea	11.50	8.06	19.56
FSL-2	3/4" one gang	1E@0.35	Ea	12.60	9.41	22.01
FSLA-1	1/2" one gang	1E@0.35	Ea	13.50	9.41	22.91
FSLA-2	3/4" one gang	1E@0.40	Ea	17.80	10.80	28.60
FSR-1	1/2" one gang	1E@0.30	Ea	11.50	8.06	19.56
FSR-2	3/4" one gang	1E@0.35	Ea	12.60	9.41	22.01
FSS-1	1/2" one gang	1E@0.35	Ea	11.00	9.41	20.41
FSS-2	3/4" one gang	1E@0.40	Ea	12.10	10.80	22.90
FST-1	1/2" one gang	1E@0.35	Ea	17.50	9.41	26.91
FST-2	3/4" one gang	1E@0.40	Ea	19.60	10.80	30.40
FSX-1	1/2" one gang	1E@0.40	Ea	16.70	10.80	27.50
FSX-2	3/4" one gang	1E@0.45	Ea	20.60	12.10	32.70
FD-1	1/2" one gang	1E@0.25	Ea	11.50	6.72	18.22
FD-2	3/4" one gang	1E@0.30	Ea	12.60	8.06	20.66
FD-3	1" one gang	1E@0.35	Ea	13.50	9.41	22.91
FDC-1	1/2" one gang	1E@0.30	Ea	13.10	8.06	21.16
FDC-2	3/4" one gang	1E@0.35	Ea	14.10	9.41	23.51
FDC-3	1" one gang	1E@0.40	Ea	16.80	10.80	27.60

Use these figures to estimate the cost of galvanized cast boxes installed on conduit under the conditions described on pages 5 and 6. Costs listed are for each box installed. The crew is one electrician working at a labor cost of $26.88 per manhour. These costs include box mounting, layout, material handling, and normal waste. Add for covers, gaskets, sales tax, delivery, supervision, mobilization, demobilization, cleanup, overhead and profit. Note: Boxes are raintight or weatherproof when fitted with the proper cover. These figures assume that the boxes are surface mounted in accessible locations.

Covers for Galvanized Cast Boxes

Material	Craft@Hrs	Unit	Material Cost	Labor Cost	Installed Cost
Single gang stamped metal covers					
DS21 single receptacle	1E@0.05	Ea	2.49	1.34	3.83
DS23 duplex receptacle	1E@0.05	Ea	2.49	1.34	3.83
DS32 switch	1E@0.05	Ea	2.49	1.34	3.83
DS100 blank	1E@0.05	Ea	1.97	1.34	3.31
Two gang stamped metal covers					
S322 2 switches	1E@0.06	Ea	4.10	1.61	5.71
S1002 blank	1E@0.06	Ea	2.93	1.61	4.54
S32212 duplex	1E@0.06	Ea	7.09	1.61	8.70
S32232 Sw & duplex	1E@0.06	Ea	6.70	1.61	8.31
Single gang cast metal covers					
DS100G switch	1E@0.05	Ea	4.94	1.34	6.28
DS100G blank	1E@0.05	Ea	6.44	1.34	7.78
Two gang cast metal covers					
S322G 2 switches	1E@0.06	Ea	9.09	1.61	10.70
S1002G blank	1E@0.06	Ea	6.44	1.61	8.05
Single gang cast weatherproof covers					
DS128 Sw rod type	1E@0.10	Ea	22.22	2.69	24.91
DS181 Sw rocker type	1E@0.10	Ea	22.22	2.69	24.91
Two gang cast weatherproof covers					
DS1282 2 Sw rod type	1E@0.15	Ea	40.66	4.03	44.69
Single gang cast with hinged cover weatherproof					
WLRS-1 single recept	1E@0.10	Ea	19.62	2.69	22.31
WLRD-1 duplex recept	1E@0.10	Ea	23.22	2.69	25.91

Use these figures to estimate the cost of covers installed on galvanized boxes under the conditions described on pages 5 and 6. Costs listed are for each cover installed. The crew is one electrician working at a labor cost of $26.88 per manhour. These costs include the cover, mounting, layout, material handling, and normal waste. Add for sales tax, delivery, supervision, mobilization, demobilization, cleanup, overhead and profit. These figures assume that the boxes for the covers are surface mounted in accessible locations.

Galvanized Cast Expansion Fittings and Jumpers

Material	Craft@Hrs	Unit	Material Cost	Labor Cost	Installed Cost
Galvanized 4" cast expansion fitting					
1/2"	1E@0.25	Ea	29.50	6.72	36.22
3/4"	1E@0.30	Ea	34.40	8.06	42.46
1"	1E@0.40	Ea	40.60	10.80	51.40
1-1/4"	1E@0.50	Ea	54.80	13.40	68.20
1-1/2"	1E@0.50	Ea	75.20	13.40	88.60
2"	1E@0.60	Ea	112.00	16.10	128.10
2-1/2"	1E@0.70	Ea	176.00	18.80	194.80
3"	1E@0.70	Ea	217.00	18.80	235.80
3-1/2"	1E@0.80	Ea	320.00	21.50	341.50
4"	1E@1.00	Ea	380.00	26.90	406.90
5"	1E@1.25	Ea	609.00	33.60	642.60
Galvanized 8" cast expansion fittings					
1/2"	1E@0.30	Ea	30.40	8.06	38.46
3/4"	1E@0.40	Ea	36.70	10.80	47.50
1"	1E@0.50	Ea	39.90	13.40	53.30
1-1/4"	1E@0.60	Ea	51.40	16.10	67.50
1-1/2"	1E@0.60	Ea	93.60	16.10	109.70
2"	1E@0.70	Ea	112.00	18.80	130.80
2-1/2"	1E@0.80	Ea	172.00	21.50	193.50
3"	1E@1.00	Ea	291.00	26.90	317.90
3-1/2"	1E@1.25	Ea	500.00	33.60	533.60
4"	1E@1.30	Ea	583.00	34.90	617.90
5"	1E@1.50	Ea	954.00	40.30	994.30
4" bonding jumpers for galvanized cast expansion fitting					
1/2" - 3/4"	1E@0.15	Ea	19.30	4.03	23.33
1" - 1-1/4"	1E@0.20	Ea	20.70	5.38	26.08
1-1/2" - 2"	1E@0.30	Ea	20.70	8.06	28.76
2-1/2" - 3"	1E@0.40	Ea	31.40	10.80	42.20
3-1/2" - 4"	1E@0.50	Ea	35.60	13.40	49.00
5"	1E@0.70	Ea	70.00	18.80	88.80
8" bonding jumpers for galvanized cast expansion fitting					
1/2" - 3/4"	1E@0.15	Ea	18.60	4.03	22.63
1" - 1-1/4"	1E@0.25	Ea	20.60	6.72	27.32
1-1/2" - 2"	1E@0.35	Ea	20.60	9.41	30.01
2-1/2" - 3"	1E@0.45	Ea	33.30	12.10	45.40
3-1/2" - 4"	1E@0.60	Ea	36.40	16.10	52.50
5"	1E@0.80	Ea	70.10	21.50	91.60

Use these figures to estimate the cost of expansion fittings and bonding jumpers installed on conduit under the conditions described on pages 5 and 6. Costs listed are for each fitting installed. The crew is one electrician working at a labor cost of $26.88 per manhour. These costs include layout, material handling, and normal waste. Add for conduit, supports, sales tax, delivery, supervision, mobilization, demobilization, cleanup, overhead and profit. Note: These fittings are installed at construction expansion joints and are suitable for installation in concrete. The bonding jumper provides grounding continuity.

Material	Craft@Hrs	Unit	Material Cost	Labor Cost	Installed Cost
Steel or malleable reducing bushings					
3/4" - 1/2"	1E@0.05	Ea	.78	1.34	2.12
1" - 1/2"	1E@0.05	Ea	1.22	1.34	2.56
1" - 3/4"	1E@0.05	Ea	1.22	1.34	2.56
1-1/4" - 1/2"	1E@0.06	Ea	2.84	1.61	4.45
1-1/4" - 3/4"	1E@0.06	Ea	2.84	1.61	4.45
1-1/4" - 1"	1E@0.06	Ea	2.84	1.61	4.45
1-1/2" - 1/2"	1E@0.08	Ea	3.56	2.15	5.71
1-1/2" - 3/4"	1E@0.08	Ea	3.56	2.15	5.71
1-1/2" - 1"	1E@0.08	Ea	3.56	2.15	5.71
1-1/2" - 1-1/4"	1E@0.08	Ea	3.56	2.15	5.71
2" - 1/2"	1E@0.10	Ea	8.02	2.69	10.71
2" - 3/4"	1E@0.10	Ea	8.02	2.69	10.71
2" - 1"	1E@0.10	Ea	8.02	2.69	10.71
2" - 1-1/4"	1E@0.10	Ea	8.02	2.69	10.71
2" - 1-1/2"	1E@0.10	Ea	8.02	2.69	10.71
2-1/2" - 1-1/2"	1E@0.15	Ea	10.90	4.03	14.93
2-1/2" - 2"	1E@0.15	Ea	10.90	4.03	14.93
3" - 2-1/2"	1E@0.20	Ea	14.00	5.38	19.38
3-1/2" - 2"	1E@0.25	Ea	24.20	6.72	30.92
3-1/2" - 2-1/2"	1E@0.25	Ea	24.20	6.72	30.92
3-1/2" - 3"	1E@0.25	Ea	24.20	6.72	30.92
4" - 2-1/2"	1E@0.30	Ea	27.10	8.06	35.16
4" - 3"	1E@0.30	Ea	27.10	8.06	35.16
4" - 3-1/2"	1E@0.30	Ea	27.10	8.06	35.16
Aluminum reducing bushings					
3/4" - 1/2"	1E@0.05	Ea	1.78	1.34	3.12
1" - 1/2"	1E@0.05	Ea	2.00	1.34	3.34
1" - 3/4"	1E@0.05	Ea	2.00	1.34	3.34
1-1/4" - 1/2"	1E@0.06	Ea	3.07	1.61	4.68
1-1/4" - 3/4"	1E@0.06	Ea	3.07	1.61	4.68
1-1/4" - 1"	1E@0.06	Ea	3.07	1.61	4.68
1-1/2" - 1/2"	1E@0.08	Ea	5.82	2.15	7.97
1-1/2" - 3/4"	1E@0.08	Ea	5.82	2.15	7.97
1-1/2" - 1"	1E@0.08	Ea	5.82	2.15	7.97
1-1/2" - 1-1/4"	1E@0.08	Ea	5.82	2.15	7.97
2" - 1/2"	1E@0.10	Ea	8.82	2.69	11.51
2" - 3/4"	1E@0.10	Ea	8.82	2.69	11.51
2" - 1"	1E@0.10	Ea	8.55	2.69	11.24
2" - 1-1/4"	1E@0.10	Ea	8.55	2.69	11.24
2" - 1-1/2"	1E@0.10	Ea	8.55	2.69	11.24

Use these figures to estimate the cost of reducing bushings installed on conduit under the conditions described on pages 5 and 6. Costs listed are for each bushing installed. The crew is one electrician working at a labor cost of $26.88 per manhour. These costs include layout, material handling, and normal waste. Add for sales tax, delivery, supervision, mobilization, demobilization, cleanup, overhead and profit. Note: Material cost is based on purchase of full boxes. These bushings are used to reduce the threaded hub size in cast boxes when smaller conduit is used.

Reducing Bushings and Reducing Washers

Material	Craft@Hrs	Unit	Material Cost	Labor Cost	Installed Cost
Aluminum reducing bushings					
2-1/2" - 1"	1E@0.15	Ea	8.76	4.03	12.79
2-1/2" - 1-1/4"	1E@0.15	Ea	8.76	4.03	12.79
2-1/2" - 1-1/2"	1E@0.15	Ea	8.76	4.03	12.79
2-1/2" - 2"	1E@0.15	Ea	8.76	4.03	12.79
3" - 1-1/4"	1E@0.20	Ea	17.90	5.38	23.28
3" - 1-1/2"	1E@0.20	Ea	17.90	5.38	23.28
3" - 2"	1E@0.20	Ea	17.90	5.38	23.28
3" - 2-1/2"	1E@0.20	Ea	17.90	5.38	23.28
3-1/2" - 2"	1E@0.25	Ea	31.70	6.72	38.42
3-1/2" - 2-1/2"	1E@0.25	Ea	31.70	6.72	38.42
3-1/2" - 3"	1E@0.25	Ea	31.70	6.72	38.42
4" - 2"	1E@0.30	Ea	35.50	8.06	43.56
4" - 2-1/2"	1E@0.30	Ea	35.50	8.06	43.56
4" - 3"	1E@0.30	Ea	35.50	8.06	43.56
4" - 3-1/2"	1E@0.30	Ea	35.50	8.06	43.56
Steel reducing washers					
3/4" - 1/2"	1E@0.05	Pr	.31	1.34	1.65
1" - 1/2"	1E@0.06	Pr	.51	1.61	2.12
1" - 3/4"	1E@0.06	Pr	.51	1.61	2.12
1-1/4" - 1/2"	1E@0.08	Pr	.95	2.15	3.10
1-1/4" - 3/4"	1E@0.08	Pr	.86	2.15	3.01
1-1/4" - 1"	1E@0.08	Pr	.86	2.15	3.01
1-1/2" - 1/2"	1E@0.10	Pr	1.21	2.69	3.90
1-1/2" - 3/4"	1E@0.10	Pr	1.21	2.69	3.90
1-1/2" - 1"	1E@0.10	Pr	1.01	2.69	3.70
1-1/2" - 1-1/4"	1E@0.10	Pr	1.01	2.69	3.70
2" - 1/2"	1E@0.15	Pr	1.62	4.03	5.65
2" - 3/4"	1E@0.15	Pr	1.62	4.03	5.65
2" - 1"	1E@0.15	Pr	1.47	4.03	5.50
2" - 1-1/4"	1E@0.15	Pr	1.47	4.03	5.50
2" - 1-1/2"	1E@0.15	Pr	1.58	4.03	5.61
2-1/2" - 1"	1E@0.20	Pr	2.52	5.38	7.90
2-1/2" - 1-1/4"	1E@0.20	Pr	2.52	5.38	7.90
2-1/2" - 1-1/2"	1E@0.20	Pr	2.52	5.38	7.90
2-1/2" - 2"	1E@0.20	Pr	2.52	5.38	7.90
3" - 1-1/4"	1E@0.25	Pr	3.22	6.72	9.94
3" - 1-1/2"	1E@0.25	Pr	3.22	6.72	9.94
3" - 2"	1E@0.25	Pr	3.22	6.72	9.94
3" - 2-1/2"	1E@0.25	Pr	3.22	6.72	9.94
3-1/2" - 2"	1E@0.30	Pr	16.90	8.06	24.96
3-1/2" - 2-1/2"	1E@0.30	Pr	16.90	8.06	24.96
3-1/2" - 3"	1E@0.30	Pr	16.90	8.06	24.96
4" - 2"	1E@0.35	Pr	27.40	9.41	36.81
4" - 2-1/2"	1E@0.35	Pr	22.80	9.41	32.21
4" - 3"	1E@0.35	Pr	22.80	9.41	32.21
4" - 3-1/2"	1E@0.35	Pr	22.80	9.41	32.21

Use these figures to estimate the cost of Reducing bushings and reducing washers installed on conduit under the conditions described on pages 5 and 6. Costs for bushings are for each bushing installed. Costs for reducing washers are per pair of washers installed. The crew is one electrician working at a labor cost of $26.88 per manhour. These costs include layout, material handling, and normal waste. Add for sales tax, delivery, supervision, mobilization, demobilization, cleanup, overhead and profit. Note: Material cost is based on purchase of full boxes. These bushings are used to reduce the threaded hub size in cast boxes when smaller conduit is used.

Material	Craft@Hrs	Unit	Material Cost	Labor Cost	Installed Cost
Die cast bushed nipples					
1/2"	1E@0.05	Ea	.45	1.34	1.79
3/4"	1E@0.06	Ea	.63	1.61	2.24
1"	1E@0.08	Ea	1.27	2.15	3.42
1-1/4"	1E@0.10	Ea	2.18	2.69	4.87
1-1/2"	1E@0.10	Ea	2.83	2.69	5.52
2"	1E@0.15	Ea	4.33	4.03	8.36
2-1/2"	1E@0.20	Ea	10.70	5.38	16.08
3"	1E@0.20	Ea	15.70	5.38	21.08
3-1/2"	1E@0.25	Ea	32.60	6.72	39.32
4"	1E@0.25	Ea	53.90	6.72	60.62
Malleable bushed nipples					
1/2"	1E@0.05	Ea	.57	1.34	1.91
3/4"	1E@0.06	Ea	.79	1.61	2.40
1"	1E@0.08	Ea	1.61	2.15	3.76
1-1/4"	1E@0.10	Ea	2.76	2.69	5.45
1-1/2"	1E@0.10	Ea	3.59	2.69	6.28
2"	1E@0.15	Ea	5.50	4.03	9.53
2-1/2"	1E@0.20	Ea	13.60	5.38	18.98
3"	1E@0.20	Ea	19.90	5.38	25.28
3-1/2"	1E@0.25	Ea	41.40	6.72	48.12
4"	1E@0.25	Ea	68.40	6.72	75.12
Die cast insulated bushed nipples					
1/2"	1E@0.05	Ea	1.44	1.34	2.78
3/4"	1E@0.06	Ea	1.68	1.61	3.29
1"	1E@0.08	Ea	2.52	2.15	4.67
1-1/4"	1E@0.10	Ea	4.79	2.69	7.48
1-1/2"	1E@0.10	Ea	6.30	2.69	8.99
2"	1E@0.15	Ea	9.58	4.03	13.61
2-1/2"	1E@0.20	Ea	26.40	5.38	31.78
3"	1E@0.20	Ea	28.70	5.38	34.08
3-1/2"	1E@0.25	Ea	40.20	6.72	46.92
4"	1E@0.25	Ea	61.20	6.72	67.92
Malleable insulated bushed nipples					
1/2"	1E@0.05	Ea	1.83	1.34	3.17
3/4"	1E@0.06	Ea	2.13	1.61	3.74
1"	1E@0.08	Ea	3.20	2.15	5.35
1-1/4"	1E@0.10	Ea	6.07	2.69	8.76
1-1/2"	1E@0.10	Ea	8.01	2.69	10.70
2"	1E@0.15	Ea	12.20	4.03	16.23

Use these figures to estimate the cost of bushed nipples installed on conduit under the conditions described on pages 5 and 6. Costs listed are for each nipple installed. The crew is one electrician working at a labor cost of $26.88 per manhour. These costs include layout, material handling, and normal waste. Add for locknut, bushing, sales tax, delivery, supervision, mobilization, demobilization, cleanup, overhead and profit. Note: Material cost is based on purchase of full boxes. Bushed nipples are often used in threaded hubs.

Bushed Nipples, Couplings and Offset Nipples

Material	Craft@Hrs	Unit	Material Cost	Labor Cost	Installed Cost
Malleable insulated bushed nipples					
2-1/2"	1E@0.20	Ea	33.60	5.38	38.98
3"	1E@0.20	Ea	36.40	5.38	41.78
3-1/2"	1E@0.25	Ea	51.10	6.72	57.82
4"	1E@0.25	Ea	77.80	6.72	84.52
5"	1E@0.30	Ea	196.00	8.06	204.06
6"	1E@0.40	Ea	300.00	10.80	310.80
Malleable three piece couplings or unions					
1/2"	1E@0.10	Ea	2.55	2.69	5.24
3/4"	1E@0.10	Ea	3.28	2.69	5.97
1"	1E@0.15	Ea	6.30	4.03	10.33
1-1/4"	1E@0.20	Ea	12.70	5.38	18.08
1-1/2"	1E@0.20	Ea	16.00	5.38	21.38
2"	1E@0.25	Ea	32.10	6.72	38.82
2-1/2"	1E@0.30	Ea	73.90	8.06	81.96
3"	1E@0.30	Ea	112.00	8.06	120.06
3-1/2"	1E@0.50	Ea	181.00	13.40	194.40
4"	1E@0.50	Ea	211.00	13.40	224.40
5"	1E@1.00	Ea	438.00	26.90	464.90
6"	1E@1.25	Ea	588.00	33.60	621.60
Malleable offset nipples					
1/2"	1E@0.10	Ea	2.99	2.69	5.68
3/4"	1E@0.10	Ea	3.09	2.69	5.78
1"	1E@0.15	Ea	3.82	4.03	7.85
1-1/4"	1E@0.20	Ea	9.27	5.38	14.65
1-1/2"	1E@0.20	Ea	11.30	5.38	16.68
2"	1E@0.25	Ea	18.50	6.72	25.22
2-1/2"	1E@0.30	Ea	39.10	8.06	47.16
3"	1E@0.30	Ea	72.10	8.06	80.16
3-1/2"	1E@0.50	Ea	87.60	13.40	101.00
4"	1E@0.50	Ea	119.00	13.40	132.40
5"	1E@1.00	Ea	248.00	26.90	274.90
6"	1E@1.25	Ea	309.00	33.60	342.60
Cast metal male to male offset nipples					
1/2"	1E@0.10	Ea	1.52	2.69	4.21
3/4"	1E@0.15	Ea	1.84	4.03	5.87
1"	1E@0.20	Ea	2.78	5.38	8.16
1-1/4"	1E@0.25	Ea	4.07	6.72	10.79
1-1/2"	1E@0.25	Ea	6.62	6.72	13.34
2"	1E@0.30	Ea	6.72	8.06	14.78

Use these figures to estimate the cost of bushed nipples, unions, and offset nipples installed on conduit under the conditions described on pages 5 and 6. Costs listed are for each fitting installed. The crew is one electrician working at a labor cost of $26.88 per manhour. These costs include layout, material handling, and normal waste. Add for locknut, bushing, sales tax, delivery, supervision, mobilization, demobilization, cleanup, overhead and profit. Note: Material cost is based on purchase of full boxes. Three-piece couplings are made to fit the flat thread used on electrical fittings. Unions made for plumbing pipe should not be used in electrical systems.

Material	Craft@Hrs	Unit	Material Cost	Labor Cost	Installed Cost
Die cast offset nipples					
1/2"	1E@0.10	Ea	1.51	2.69	4.20
3/4"	1E@0.15	Ea	1.84	4.03	5.87
1"	1E@0.20	Ea	2.78	5.38	8.16
Malleable threadless connectors					
1/2"	1E@0.05	Ea	2.34	1.34	3.68
3/4"	1E@0.06	Ea	3.87	1.61	5.48
1"	1E@0.08	Ea	5.49	2.15	7.64
1-1/4"	1E@0.10	Ea	10.70	2.69	13.39
1-1/2"	1E@0.10	Ea	14.90	2.69	17.59
2"	1E@0.15	Ea	29.70	4.03	33.73
2-1/2"	1E@0.25	Ea	99.90	6.72	106.62
3"	1E@0.30	Ea	132.00	8.06	140.06
3-1/2"	1E@0.35	Ea	173.00	9.41	182.41
4"	1E@0.40	Ea	212.00	10.80	222.80
Malleable insulated threadless connectors					
1/2"	1E@0.05	Ea	2.94	1.34	4.28
3/4"	1E@0.06	Ea	4.73	1.61	6.34
1"	1E@0.08	Ea	6.85	2.15	9.00
1-1/4"	1E@0.10	Ea	13.60	2.69	16.29
1-1/2"	1E@0.10	Ea	18.90	2.69	21.59
2"	1E@0.15	Ea	46.90	4.03	50.93
2-1/2"	1E@0.25	Ea	163.00	6.72	169.72
3"	1E@0.30	Ea	214.00	8.06	222.06
3-1/2"	1E@0.35	Ea	251.00	9.41	260.41
4"	1E@0.40	Ea	304.00	10.80	314.80
Malleable threadless couplings					
1/2"	1E@0.10	Ea	3.81	2.69	6.50
3/4"	1E@0.10	Ea	5.62	2.69	8.31
1"	1E@0.15	Ea	9.52	4.03	13.55
1-1/4"	1E@0.20	Ea	15.70	5.38	21.08
1-1/2"	1E@0.20	Ea	19.70	5.38	25.08
2"	1E@0.25	Ea	44.90	6.72	51.62

Use these figures to estimate the cost of offset nipples, connectors, and couplings installed in conduit systems under the conditions described on pages 5 and 6. Costs listed are for each fitting installed. The crew is one electrician working at a labor cost of $26.88 per manhour. These costs include removing the knockout, layout, material handling, and normal waste. Add for locknuts, bushings, sales tax, delivery, supervision, mobilization, demobilization, cleanup, overhead and profit. Note: Material costs are based on purchase of full boxes. Threadless fittings are made for rigid conduit only and do not fit EMT conduit. They're rated for raintight or weatherproof applications.

Couplings and Connectors

Material	Craft@Hrs	Unit	Material Cost	Labor Cost	Installed Cost
Malleable threadless couplings					
2-1/2"	1E@0.30	Ea	141.00	8.06	149.06
3"	1E@0.40	Ea	186.00	10.80	196.80
3-1/2"	1E@0.50	Ea	248.00	13.40	261.40
4"	1E@0.60	Ea	317.00	16.10	333.10
Malleable set screw couplings					
1/2"	1E@0.10	Ea	2.59	2.69	5.28
3/4"	1E@0.10	Ea	3.36	2.69	6.05
1"	1E@0.15	Ea	5.38	4.03	9.41
1-1/4"	1E@0.20	Ea	9.21	5.38	14.59
1-1/2"	1E@0.20	Ea	11.80	5.38	17.18
2"	1E@0.25	Ea	26.60	6.72	33.32
2-1/2"	1E@0.30	Ea	65.90	8.06	73.96
3"	1E@0.40	Ea	79.00	10.80	89.80
3-1/2"	1E@0.50	Ea	112.00	13.40	125.40
4"	1E@0.60	Ea	147.00	16.10	163.10
Steel set screw connectors					
1/2"	1E@0.05	Ea	1.75	1.34	3.09
3/4"	1E@0.06	Ea	2.43	1.61	4.04
1"	1E@0.08	Ea	3.78	2.15	5.93
1-1/4"	1E@0.10	Ea	6.69	2.69	9.38
1-1/2"	1E@0.10	Ea	9.55	2.69	12.24
2"	1E@0.15	Ea	19.10	4.03	23.13
2-1/2"	1E@0.25	Ea	56.80	6.72	63.52
3"	1E@0.30	Ea	74.30	8.06	82.36
3-1/2"	1E@0.35	Ea	105.00	9.41	114.41
4"	1E@0.40	Ea	133.00	10.80	143.80
Insulated steel set screw connectors					
1/2"	1E@0.05	Ea	2.04	1.34	3.38
3/4"	1E@0.06	Ea	3.00	1.61	4.61
1"	1E@0.08	Ea	4.46	2.15	6.61
1-1/4"	1E@0.10	Ea	7.85	2.69	10.54
1-1/2"	1E@0.10	Ea	11.30	2.69	13.99
2"	1E@0.15	Ea	22.40	4.03	26.43
2-1/2"	1E@0.25	Ea	63.30	6.72	70.02
3"	1E@0.30	Ea	83.70	8.06	91.76
3-1/2"	1E@0.35	Ea	118.00	9.41	127.41
4"	1E@0.40	Ea	155.00	10.80	165.80

Use these figures to estimate the cost of couplings and connectors installed on conduit under the conditions described on pages 5 and 6. Costs listed are for each fitting installed. The crew is one electrician working at a labor cost of $26.88 per manhour. These costs include removing the knockout, the locknut, layout, material handling, and normal waste. Add for bushings, sales tax, delivery, supervision, mobilization, demobilization, cleanup, overhead and profit. Note: Material cost is based on purchase of full boxes.

Material	Craft@Hrs	Unit	Material Cost	Labor Cost	Installed Cost
Malleable set screw connectors					
1/2"	1E@0.05	Ea	1.75	1.34	3.09
3/4"	1E@0.06	Ea	2.43	1.61	4.04
1"	1E@0.08	Ea	3.78	2.15	5.93
1-1/4"	1E@0.10	Ea	6.69	2.69	9.38
1-1/2"	1E@0.10	Ea	9.55	2.69	12.24
2"	1E@0.15	Ea	19.10	4.03	23.13
2-1/2"	1E@0.25	Ea	56.80	6.72	63.52
3"	1E@0.30	Ea	74.30	8.06	82.36
3-1/2"	1E@0.35	Ea	105.00	9.41	114.41
4"	1E@0.40	Ea	133.00	10.80	143.80
Insulated malleable set screw connectors					
1/2"	1E@0.05	Ea	2.04	1.34	3.38
3/4"	1E@0.06	Ea	3.00	1.61	4.61
1"	1E@0.08	Ea	4.46	2.15	6.61
1-1/4"	1E@0.10	Ea	7.85	2.69	10.54
1-1/2"	1E@0.10	Ea	11.30	2.69	13.99
2"	1E@0.15	Ea	22.40	4.03	26.43
2-1/2"	1E@0.25	Ea	63.30	6.72	70.02
3"	1E@0.30	Ea	83.70	8.06	91.76
3-1/2"	1E@0.35	Ea	118.00	9.41	127.41
4"	1E@0.40	Ea	155.00	10.80	165.80
Steel one hole straps					
1/2"	1E@0.05	Ea	.22	1.34	1.56
3/4"	1E@0.06	Ea	.26	1.61	1.87
1"	1E@0.08	Ea	.54	2.15	2.69
1-1/4"	1E@0.10	Ea	.79	2.69	3.48
1-1/2"	1E@0.10	Ea	.94	2.69	3.63
2"	1E@0.10	Ea	1.67	2.69	4.36
2-1/2"	1E@0.15	Ea	2.54	4.03	6.57
3"	1E@0.20	Ea	2.83	5.38	8.21
3-1/2"	1E@0.20	Ea	4.20	5.38	9.58
4"	1E@0.20	Ea	5.35	5.38	10.73
Malleable one hole straps					
1/2"	1E@0.05	Ea	.32	1.34	1.66
3/4"	1E@0.06	Ea	.45	1.61	2.06
1"	1E@0.08	Ea	.68	2.15	2.83
1-1/4"	1E@0.10	Ea	1.34	2.69	4.03
1-1/2"	1E@0.10	Ea	1.58	2.69	4.27
2"	1E@0.10	Ea	3.07	2.69	5.76
2-1/2"	1E@0.15	Ea	6.81	4.03	10.84
3"	1E@0.20	Ea	9.67	5.38	15.05
3-1/2"	1E@0.20	Ea	13.80	5.38	19.18
4"	1E@0.25	Ea	30.70	6.72	37.42

Use these figures to estimate the cost of connectors and straps installed on conduit under the conditions described on pages 5 and 6. Costs listed are for each fitting installed. The crew is one electrician working at a labor cost of $26.88 per manhour. These costs include removing the knockout, the locknut, layout, material handling, and normal waste. Add for bushings, sales tax, delivery, supervision, mobilization, demobilization, cleanup, overhead and profit. Note: Material cost is based on purchase of full boxes.

Straps and Spacers

Material	Craft@Hrs	Unit	Material Cost	Labor Cost	Installed Cost
Steel two hole straps					
1/2"	1E@0.05	Ea	.10	1.34	1.44
3/4"	1E@0.06	Ea	.12	1.61	1.73
1"	1E@0.08	Ea	.18	2.15	2.33
1-1/4"	1E@0.10	Ea	.26	2.69	2.95
1-1/2"	1E@0.10	Ea	.32	2.69	3.01
2"	1E@0.10	Ea	.45	2.69	3.14
2-1/2"	1E@0.15	Ea	.78	4.03	4.81
3"	1E@0.20	Ea	.97	5.38	6.35
3-1/2"	1E@0.25	Ea	1.22	6.72	7.94
4"	1E@0.25	Ea	1.39	6.72	8.11
Aluminum one hole straps					
1/2"	1E@0.05	Ea	.46	1.34	1.80
3/4"	1E@0.06	Ea	.57	1.61	2.18
1"	1E@0.08	Ea	1.03	2.15	3.18
1-1/4"	1E@0.10	Ea	1.49	2.69	4.18
1-1/2"	1E@0.10	Ea	1.61	2.69	4.30
2"	1E@0.10	Ea	3.20	2.69	5.89
2-1/2"	1E@0.15	Ea	6.29	4.03	10.32
3"	1E@0.20	Ea	8.58	5.38	13.96
3-1/2"	1E@0.20	Ea	14.90	5.38	20.28
4"	1E@0.25	Ea	28.60	6.72	35.32
Malleable conduit spacers					
1/2" - 1"	1E@0.05	Ea	1.03	1.34	2.37
1-1/4" - 2"	1E@0.10	Ea	3.53	2.69	6.22
2-1/2" - 3"	1E@0.15	Ea	10.60	4.03	14.63
3-1/2" - 4"	1E@0.20	Ea	32.00	5.38	37.38
5" - 6"	1E@0.25	Ea	60.90	6.72	67.62
Aluminum conduit spacers					
1/2" - 1"	1E@0.05	Ea	1.13	1.34	2.47
1-1/4" - 2"	1E@0.10	Ea	3.09	2.69	5.78
2-1/2" - 3"	1E@0.15	Ea	13.80	4.03	17.83
3-1/2" - 4"	1E@0.20	Ea	41.50	5.38	46.88
5" - 6"	1E@0.25	Ea	138.00	6.72	144.72
Non-metallic conduit spacers					
1/2" - 1"	1E@0.05	Ea	.51	1.34	1.85
1-1/4" - 2"	1E@0.10	Ea	.75	2.69	3.44

Use these figures to estimate the cost of straps and spacers installed on conduit under the conditions described on pages 5 and 6. Costs listed are for each fitting installed. The crew is one electrician working at a labor cost of $26.88 per manhour. These costs include layout, material handling, and normal waste. Add for screws, bolts, anchors, sales tax, delivery, supervision, mobilization, demobilization, cleanup, overhead and profit. Note: Material cost is based on purchase of full boxes.

Material	Craft@Hrs	Unit	Material Cost	Labor Cost	Installed Cost
Rigid steel conduit clamps without bolts					
1/2"	1E@0.05	Ea	.23	1.34	1.57
3/4"	1E@0.06	Ea	.25	1.61	1.86
1"	1E@0.08	Ea	.34	2.15	2.49
1-1/4"	1E@0.10	Ea	.39	2.69	3.08
1-1/2"	1E@0.10	Ea	.48	2.69	3.17
2"	1E@0.10	Ea	.57	2.69	3.26
2-1/2"	1E@0.15	Ea	.67	4.03	4.70
3"	1E@0.15	Ea	.80	4.03	4.83
3-1/2"	1E@0.20	Ea	.98	5.38	6.36
4"	1E@0.20	Ea	2.32	5.38	7.70
Rigid steel conduit clamps with bolts					
1/2"	1E@0.05	Ea	.28	1.34	1.62
3/4"	1E@0.06	Ea	.32	1.61	1.93
1"	1E@0.08	Ea	.42	2.15	2.57
1-1/4"	1E@0.10	Ea	.51	2.69	3.20
1-1/2"	1E@0.10	Ea	.61	2.69	3.30
2"	1E@0.10	Ea	.69	2.69	3.38
2-1/2"	1E@0.15	Ea	.81	4.03	4.84
3"	1E@0.15	Ea	1.02	4.03	5.05
3-1/2"	1E@0.20	Ea	1.37	5.38	6.75
4"	1E@0.20	Ea	2.53	5.38	7.91
Clamp-type entrance caps					
1/2"	1E@0.15	Ea	5.09	4.03	9.12
3/4"	1E@0.20	Ea	6.00	5.38	11.38
1"	1E@0.25	Ea	7.31	6.72	14.03
1-1/4"	1E@0.30	Ea	8.70	8.06	16.76
1-1/2"	1E@0.30	Ea	15.50	8.06	23.56
2"	1E@0.50	Ea	28.80	13.40	42.20
2-1/2"	1E@0.60	Ea	74.60	16.10	90.70
3"	1E@0.75	Ea	88.20	20.20	108.40
3-1/2"	1E@1.00	Ea	94.60	26.90	121.50
4"	1E@1.25	Ea	127.00	33.60	160.60
Slip fitter entrance caps					
1/2"	1E@0.15	Ea	4.39	4.03	8.42
3/4"	1E@0.20	Ea	5.23	5.38	10.61
1"	1E@0.25	Ea	6.30	6.72	13.02
1-1/4"	1E@0.30	Ea	7.86	8.06	15.92
1-1/2"	1E@0.30	Ea	13.40	8.06	21.46
2"	1E@0.50	Ea	20.10	13.40	33.50
2-1/2"	1E@0.60	Ea	68.40	16.10	84.50
3"	1E@0.75	Ea	112.00	20.20	132.20
3-1/2"	1E@1.00	Ea	149.00	26.90	175.90
4"	1E@1.25	Ea	156.00	33.60	189.60

Use these figures to estimate the cost of clamps and entrance caps installed on conduit under the conditions described on pages 5 and 6. Costs listed are for each fitting installed. The crew is one electrician working at a labor cost of $26.88 per manhour. These costs include layout, material handling, and normal waste. Add for screws, bolts, anchors, sales tax, delivery, supervision, mobilization, demobilization, cleanup, overhead and profit. Note: Material cost is based on purchase of full boxes. Many other types of fittings are available. Those listed here are the most common.

PVC Coated Conduit, Elbows and Couplings

Material	Craft@Hrs	Unit	Material Cost	Labor Cost	Installed Cost
PVC coated steel conduit, 40 mil coating					
1/2"	1E@4.50	CLF	139.00	121.00	260.00
3/4"	1E@5.50	CLF	160.00	148.00	308.00
1"	1E@7.00	CLF	207.00	188.00	395.00
1-1/4"	1E@9.00	CLF	263.00	242.00	505.00
1-1/2"	1E@11.0	CLF	319.00	296.00	615.00
2"	1E@13.0	CLF	416.00	349.00	765.00
2-1/2"	2E@15.0	CLF	630.00	403.00	1,033.00
3"	2E@17.0	CLF	796.00	457.00	1,253.00
3-1/2"	2E@19.0	CLF	972.00	511.00	1,483.00
4"	2E@21.0	CLF	1,140.00	564.00	1,704.00
5"	2E@25.0	CLF	2,040.00	672.00	2,712.00
PVC coated steel 90 degree elbows					
1/2"	1E@0.10	Ea	5.11	2.69	7.80
3/4"	1E@0.10	Ea	5.30	2.69	7.99
1"	1E@0.15	Ea	6.08	4.03	10.11
1-1/4"	1E@0.20	Ea	7.47	5.38	12.85
1-1/2"	1E@0.20	Ea	9.18	5.38	14.56
2"	1E@0.25	Ea	12.80	6.72	19.52
2-1/2"	2E@0.30	Ea	24.20	8.06	32.26
3"	2E@0.35	Ea	38.60	9.41	48.01
3-1/2"	2E@0.40	Ea	50.60	10.80	61.40
4"	2E@0.50	Ea	53.60	13.40	67.00
5"	2E@0.75	Ea	129.00	20.20	149.20
PVC coated steel couplings					
1/2"	1E@0.05	Ea	1.46	1.34	2.80
3/4"	1E@0.06	Ea	1.52	1.61	3.13
1"	1E@0.08	Ea	1.98	2.15	4.13
1-1/4"	1E@0.10	Ea	2.31	2.69	5.00
1-1/2"	1E@0.10	Ea	2.75	2.69	5.44
2"	1E@0.15	Ea	4.02	4.03	8.05
2-1/2"	2E@0.20	Ea	9.96	5.38	15.34
3"	2E@0.20	Ea	12.10	5.38	17.48
3-1/2"	2E@0.25	Ea	15.50	6.72	22.22
4"	2E@0.25	Ea	18.20	6.72	24.92
5"	2E@0.30	Ea	59.10	8.06	67.16

Use these figures to estimate the cost of PVC coated conduit, elbows and couplings installed in corrosive areas under the conditions described on pages 5 and 6. Costs listed are for each 100 linear feet of conduit or for each fitting installed. The crew is one electrician for sizes up to 2" and two electricians for sizes over 2". The labor cost is $26.88 per manhour. These costs include cutting and threading, one coupling for each length of conduit, layout, material handling, and normal waste. Add for straps, locknuts, bushings, sales tax, delivery, supervision, mobilization, demobilization, cleanup, overhead and profit. Note: PVC patching material is available in spray cans for repairing any damaged PVC coating. Bending tools must be ground out when used on PVC conduit. Threading equipment must be modified for use on PVC conduit.

PVC Coated Straps and Clamps

Material	Craft@Hrs	Unit	Material Cost	Labor Cost	Installed Cost

PVC coated steel one hole straps

Material	Craft@Hrs	Unit	Material Cost	Labor Cost	Installed Cost
1/2"	1E@0.05	Ea	2.17	1.34	3.51
3/4"	1E@0.06	Ea	2.43	1.61	4.04
1"	1E@0.08	Ea	3.51	2.15	5.66
1-1/4"	1E@0.10	Ea	4.43	2.69	7.12
1-1/2"	1E@0.10	Ea	5.43	2.69	8.12
2"	1E@0.15	Ea	6.27	4.03	10.30
2-1/2"	1E@0.20	Ea	7.35	5.38	12.73
3"	1E@0.25	Ea	10.40	6.72	17.12
3-1/2"	1E@0.30	Ea	11.70	8.06	19.76
4"	1E@0.40	Ea	15.00	10.80	25.80

PVC coated malleable one hole straps

Material	Craft@Hrs	Unit	Material Cost	Labor Cost	Installed Cost
1/2"	1E@0.05	Ea	2.50	1.34	3.84
3/4"	1E@0.06	Ea	2.84	1.61	4.45
1"	1E@0.10	Ea	4.18	2.69	6.87
1-1/4"	1E@0.10	Ea	5.01	2.69	7.70
1-1/2"	1E@0.15	Ea	6.68	4.03	10.71
2"	1E@0.20	Ea	10.80	5.38	16.18
2-1/2"	1E@0.20	Ea	14.20	5.38	19.58
3"	1E@0.25	Ea	19.10	6.72	25.82
3-1/2"	1E@0.30	Ea	34.10	8.06	42.16
4"	1E@0.30	Ea	35.80	8.06	43.86

PVC coated right angle beam clamps

Material	Craft@Hrs	Unit	Material Cost	Labor Cost	Installed Cost
1/2"	1E@0.10	Ea	9.18	2.69	11.87
3/4"	1E@0.15	Ea	10.00	4.03	14.03
1"	1E@0.20	Ea	10.80	5.38	16.18
1-1/4"	1E@0.25	Ea	13.30	6.72	20.02
1-1/2"	1E@0.25	Ea	15.90	6.72	22.62
2"	1E@0.30	Ea	20.80	8.06	28.86
2-1/2"	1E@0.40	Ea	23.40	10.80	34.20
3"	1E@0.50	Ea	25.80	13.40	39.20
3-1/2"	1E@0.60	Ea	26.70	16.10	42.80
4"	1E@0.60	Ea	26.70	16.10	42.80

PVC coated parallel beam clamps

Material	Craft@Hrs	Unit	Material Cost	Labor Cost	Installed Cost
1/2"	1E@0.10	Ea	10.80	2.69	13.49
3/4"	1E@0.15	Ea	11.20	4.03	15.23
1"	1E@0.20	Ea	12.90	5.38	18.28
1-1/4"	1E@0.25	Ea	14.20	6.72	20.92
1-1/2"	1E@0.25	Ea	15.90	6.72	22.62
2"	1E@0.30	Ea	20.10	8.06	28.16
2-1/2"	1E@0.30	Ea	24.20	8.06	32.26
3"	1E@0.40	Ea	27.20	10.80	38.00
3-1/2"	1E@0.60	Ea	29.20	16.10	45.30
4"	1E@0.60	Ea	29.60	16.10	45.70

Use these figures to estimate the cost of PVC coated straps and clamps installed on PVC coated conduit under the conditions described on pages 5 and 6. Costs listed are for each fitting installed. The crew is one electrician working at a labor cost of $26.88 per manhour. These costs include screws, anchors, layout, material handling, and normal waste. Add for sales tax, delivery, supervision, mobilization, demobilization, cleanup, overhead and profit. Note: PVC patching material is available in spray cans for repairing any damaged PVC coating.

PVC Coated Clamps, U-bolts and Unions

Material	Craft@Hrs	Unit	Material Cost	Labor Cost	Installed Cost
PVC coated edge-type beam clamps					
1/2"	1E@0.10	Ea	10.00	2.69	12.69
3/4"	1E@0.15	Ea	10.80	4.03	14.83
1"	1E@0.20	Ea	12.50	5.38	17.88
1-1/4"	1E@0.25	Ea	15.40	6.72	22.12
1-1/2"	1E@0.25	Ea	22.50	6.72	29.22
2"	1E@0.30	Ea	24.20	8.06	32.26
PVC coated U-bolts					
1/2"	1E@0.10	Ea	2.50	2.69	5.19
3/4"	1E@0.15	Ea	2.50	4.03	6.53
1"	1E@0.20	Ea	2.50	5.38	7.88
1-1/4"	1E@0.25	Ea	3.34	6.72	10.06
1-1/2"	1E@0.25	Ea	3.76	6.72	10.48
2"	1E@0.30	Ea	4.18	8.06	12.24
2-1/2"	1E@0.35	Ea	7.94	9.41	17.35
3"	1E@0.35	Ea	8.11	9.41	17.52
3-1/2"	1E@0.40	Ea	8.35	10.80	19.15
4"	1E@0.50	Ea	11.70	13.40	25.10
5"	1E@0.60	Ea	16.70	16.10	32.80
PVC coated female conduit unions					
1/2"	1E@0.10	Ea	21.50	2.69	24.19
3/4"	1E@0.10	Ea	21.90	2.69	24.59
1"	1E@0.15	Ea	29.40	4.03	33.43
1-1/4"	1E@0.20	Ea	47.40	5.38	52.78
1-1/2"	1E@0.20	Ea	57.10	5.38	62.48
2"	1E@0.25	Ea	76.10	6.72	82.82
2-1/2"	1E@0.30	Ea	103.00	8.06	111.06
3"	1E@0.30	Ea	142.00	8.06	150.06
3-1/2"	1E@0.35	Ea	176.00	9.41	185.41
4"	1E@0.40	Ea	226.00	10.80	236.80
PVC coated male conduit unions					
1/2"	1E@0.10	Ea	22.30	2.69	24.99
3/4"	1E@0.10	Ea	24.70	2.69	27.39
1"	1E@0.15	Ea	30.90	4.03	34.93
1-1/4"	1E@0.20	Ea	50.00	5.38	55.38
1-1/2"	1E@0.20	Ea	60.80	5.38	66.18
2"	1E@0.25	Ea	76.10	6.72	82.82
2-1/2"	1E@0.30	Ea	121.00	8.06	129.06
3"	1E@0.30	Ea	160.00	8.06	168.06
3-1/2"	1E@0.35	Ea	208.00	9.41	217.41
4"	1E@0.40	Ea	258.00	10.80	268.80

Use these figures to estimate the cost of PVC coated clamps, U-bolts and unions installed on PVC coated conduit under the conditions described on pages 5 and 6. Costs listed are for each fitting installed. The crew is one electrician working at a labor cost of $26.88 per manhour. These costs include screws, bolts, nuts, layout, material handling, and normal waste. Add for sales tax, delivery, supervision, mobilization, demobilization, cleanup, overhead and profit. Note: PVC patching material is available in spray cans for repairing any damaged PVC coating. PVC conduit fittings are rigid conduit fittings that have a PVC bonded coating for corrosion protection.

PVC Coated Couplings and Conduit Bodies

Material	Craft@Hrs	Unit	Material Cost	Labor Cost	Installed Cost

PVC coated reducing couplings

Material	Craft@Hrs	Unit	Material Cost	Labor Cost	Installed Cost
3/4"- 1/2"	1E@0.05	Ea	14.20	1.34	15.54
1"- 1/2"	1E@0.06	Ea	14.60	1.61	16.21
1"- 3/4"	1E@0.06	Ea	17.40	1.61	19.01
1-1/4"- 3/4"	1E@0.08	Ea	21.70	2.15	23.85
1-1/4"- 1"	1E@0.08	Ea	22.90	2.15	25.05
1-1/2"- 3/4"	1E@0.10	Ea	18.40	2.69	21.09
1-1/2"- 1"	1E@0.10	Ea	21.70	2.69	24.39
1-1/2"- 1-1/4"	1E@0.10	Ea	35.80	2.69	38.49
2"- 3/4"	1E@0.15	Ea	39.20	4.03	43.23
2"- 1"	1E@0.15	Ea	37.90	4.03	41.93
2"- 1-1/4"	1E@0.15	Ea	40.10	4.03	44.13
2"- 1-1/2"	1E@0.20	Ea	45.90	5.38	51.28
3"- 2"	1E@0.20	Ea	55.90	5.38	61.28
3-1/2"- 2-1/2"	1E@0.25	Ea	84.20	6.72	90.92
4"- 3"	1E@0.30	Ea	111.00	8.06	119.06
5"- 4"	1E@0.40	Ea	138.00	10.80	148.80

PVC coated Type C conduit bodies

Material	Craft@Hrs	Unit	Material Cost	Labor Cost	Installed Cost
C-17 1/2"	1E@0.20	Ea	20.30	5.38	25.68
C-27 3/4"	1E@0.25	Ea	22.50	6.72	29.22
C-37 1"	1E@0.30	Ea	30.00	8.06	38.06
C-47 1-1/4"	1E@0.35	Ea	43.60	9.41	53.01
C-57 1-1/2"	1E@0.40	Ea	53.20	10.80	64.00
C-67 2"	1E@0.40	Ea	78.40	10.80	89.20
C-77 2-1/2"	1E@0.50	Ea	144.00	13.40	157.40
CLF-87 3"	1E@0.60	Ea	182.00	16.10	198.10
CLF-97 3-1/2"	1E@0.70	Ea	264.00	18.80	282.80
CLF-107 4"	1E@1.00	Ea	295.00	26.90	321.90

PVC coated Type LB conduit bodies

Material	Craft@Hrs	Unit	Material Cost	Labor Cost	Installed Cost
LB-17 1/2"	1E@0.20	Ea	20.30	5.38	25.68
LB-27 3/4"	1E@0.25	Ea	22.50	6.72	29.22
LB-37 1"	1E@0.30	Ea	29.00	8.06	37.06
LB-47 1-1/4"	1E@0.35	Ea	43.60	9.41	53.01
LB-57 1-1/2"	1E@0.35	Ea	53.20	9.41	62.61
LB-67 2"	1E@0.40	Ea	78.40	10.80	89.20
LB-77 2-1/2"	1E@0.50	Ea	144.00	13.40	157.40
LB-87 3"	1E@0.60	Ea	182.00	16.10	198.10
LB-97 3-1/2"	1E@0.70	Ea	264.00	18.80	282.80
LB-107 4"	1E@1.00	Ea	295.00	26.90	321.90

Use these figures to estimate the cost of PVC coated couplings and conduit bodies installed on PVC coated conduit under the conditions described on pages 5 and 6. Costs listed are for each fitting installed. The crew is one electrician working at a labor cost of $26.88 per manhour. These costs include covers, layout, material handling, and normal waste. Add for sales tax, delivery, supervision, mobilization, demobilization, cleanup, overhead and profit. Note: PVC patching material is available in spray cans for repairing any damaged PVC coating.

PVC Coated Couplings and Conduit Bodies

Material	Craft@Hrs	Unit	Material Cost	Labor Cost	Installed Cost
PVC coated Type T conduit bodies					
T-17 1/2"	1E@0.25	Ea	23.30	6.72	30.02
T-27 3/4"	1E@0.30	Ea	26.00	8.06	34.06
T-37 1"	1E@0.35	Ea	35.00	9.41	44.41
T-47 1-1/4"	1E@0.40	Ea	49.10	10.80	59.90
T-57 1-1/2"	1E@0.40	Ea	61.40	10.80	72.20
T-67 2"	1E@0.45	Ea	89.30	12.10	101.40
T-77 2-1/2"	1E@0.60	Ea	151.00	16.10	167.10
T-87 3"	1E@0.70	Ea	202.00	18.80	220.80
T-97 3-1/2"	1E@1.00	Ea	289.00	26.90	315.90
T-107 4"	1E@1.50	Ea	314.00	40.30	354.30
PVC coated Type TB conduit bodies					
TB-17 1/2"	1E@0.25	Ea	23.40	6.72	30.12
TB-27 3/4"	1E@0.30	Ea	30.00	8.06	38.06
TB-37 1"	1E@0.35	Ea	35.00	9.41	44.41
TB-47 1-1/4"	1E@0.40	Ea	49.10	10.80	59.90
TB-57 1-1/2"	1E@0.40	Ea	61.40	10.80	72.20
TB-67 2"	1E@0.45	Ea	88.40	12.10	100.50
PVC coated Type X conduit bodies					
X-17 1/2"	1E@0.30	Ea	24.80	8.06	32.86
X-27 3/4"	1E@0.35	Ea	29.00	9.41	38.41
X-37 1"	1E@0.40	Ea	37.90	10.80	48.70
X-47 1-1/4"	1E@0.45	Ea	47.90	12.10	60.00
X-57 1-1/2"	1E@0.45	Ea	56.20	12.10	68.30
X-67 2"	1E@0.50	Ea	85.90	13.40	99.30
PVC coated steel conduit body covers					
1/2"	1E@0.05	Ea	6.68	1.34	8.02
3/4"	1E@0.06	Ea	7.52	1.61	9.13
1"	1E@0.08	Ea	10.00	2.15	12.15
1-1/4"	1E@0.10	Ea	12.50	2.69	15.19
1-1/2"	1E@0.10	Ea	14.20	2.69	16.89
2"	1E@0.10	Ea	17.40	2.69	20.09
2-1/2" - 3"	1E@0.15	Ea	22.50	4.03	26.53
2-1/2" - 4"	1E@0.15	Ea	28.40	4.03	32.43

Use these figures to estimate the cost of PVC coated conduit bodies installed on PVC coated conduit and PVC coated body covers installed on conduit bodies under the conditions described on pages 5 and 6. Costs listed are for each fitting installed. The crew is one electrician working at a labor cost of $26.88 per manhour. These costs include layout, material handling, and normal waste. Add for sales tax, delivery, supervision, mobilization, demobilization, cleanup, overhead and profit. Note: PVC patching material is available in spray cans for repairing any damaged PVC coating.

Material	Craft@Hrs	Unit	Material Cost	Labor Cost	Installed Cost

PVC coated Type GUAB junction boxes with covers

Material	Craft@Hrs	Unit	Material Cost	Labor Cost	Installed Cost
1/2" - 2" dia.	1E@0.35	Ea	46.80	9.41	56.21
1/2" - 3" dia.	1E@0.40	Ea	56.90	10.80	67.70
3/4" - 2" dia.	1E@0.40	Ea	48.60	10.80	59.40
3/4" - 3" dia.	1E@0.45	Ea	56.00	12.10	68.10
1" - 3" dia.	1E@0.50	Ea	57.90	13.40	71.30
1-1/4" - 3-5/8" dia.	1E@0.60	Ea	90.20	16.10	106.30
1-1/2" - 5" dia.	1E@0.75	Ea	161.00	20.20	181.20
2" - 5" dia.	1E@1.00	Ea	169.00	26.90	195.90

PVC coated Type GUAC junction boxes with covers

Material	Craft@Hrs	Unit	Material Cost	Labor Cost	Installed Cost
1/2" - 2" dia.	1E@0.35	Ea	46.80	9.41	56.21
1/2" - 3" dia.	1E@0.40	Ea	56.90	10.80	67.70
3/4" - 2" dia.	1E@0.40	Ea	48.60	10.80	59.40
3/4" - 3" dia.	1E@0.45	Ea	56.00	12.10	68.10
1" - 3" dia.	1E@0.50	Ea	57.90	13.40	71.30
1-1/4" - 3-5/8" dia.	1E@0.60	Ea	90.20	16.10	106.30
1-1/2" - 5" dia.	1E@0.75	Ea	161.00	20.20	181.20
2" - 5" dia.	1E@1.00	Ea	169.00	26.90	195.90

PVC coated Type GUAL junction boxes with covers

Material	Craft@Hrs	Unit	Material Cost	Labor Cost	Installed Cost
1/2" - 2" dia.	1E@0.35	Ea	46.80	9.41	56.21
1/2" - 3" dia.	1E@0.40	Ea	56.90	10.80	67.70
3/4" - 2" dia.	1E@0.40	Ea	48.60	10.80	59.40
3/4" - 3" dia.	1E@0.45	Ea	56.00	12.10	68.10
1" - 3" dia.	1E@0.50	Ea	57.90	13.40	71.30
1-1/4" - 3-5/8" dia.	1E@0.60	Ea	90.20	16.10	106.30
1-1/2" - 5" dia.	1E@0.75	Ea	161.00	20.20	181.20
2" - 5" dia.	1E@1.00	Ea	169.00	26.90	195.90

PVC coated Type GUAN junction boxes with covers

Material	Craft@Hrs	Unit	Material Cost	Labor Cost	Installed Cost
1/2" - 2" dia.	1E@0.35	Ea	46.80	9.41	56.21
1/2" - 3" dia.	1E@0.40	Ea	56.90	10.80	67.70
3/4" - 2" dia.	1E@0.40	Ea	48.60	10.80	59.40
3/4" - 3" dia.	1E@0.45	Ea	56.00	12.10	68.10
1" - 3" dia.	1E@0.50	Ea	57.90	13.40	71.30
1-1/4" - 3-5/8" dia.	1E@0.60	Ea	90.20	16.10	106.30
1-1/2" - 5" dia.	1E@0.75	Ea	161.00	20.20	181.20
2" - 5" dia.	1E@1.00	Ea	169.00	26.90	195.90

PVC coated Type GUAW junction boxes with covers

Material	Craft@Hrs	Unit	Material Cost	Labor Cost	Installed Cost
1/2" - 2" dia.	1E@0.40	Ea	52.30	10.80	63.10
1/2" - 3" dia.	1E@0.45	Ea	58.80	12.10	70.90
3/4" - 2" dia.	1E@0.45	Ea	56.40	12.10	68.50
3/4" - 3" dia.	1E@0.50	Ea	62.50	13.40	75.90

Use these figures to estimate the cost of PVC coated junction boxes installed on PVC coated conduit under the conditions described on pages 5 and 6. Costs listed are for each fitting installed. The crew is one electrician working at a labor cost of $26.88 per manhour. These costs include the box cover, layout, material handling, and normal waste. Add for sales tax, delivery, supervision, mobilization, demobilization, cleanup, overhead and profit. Note: PVC patching material is available in spray cans for repairing damaged PVC coating.

PVC Coated Junction Boxes and Sealing Fittings

Material	Craft@Hrs	Unit	Material Cost	Labor Cost	Installed Cost
PVC coated Type GUAT junction boxes with covers					
1/2" - 2" dia.	1E@0.40	Ea	52.70	10.80	63.50
1/2" - 3" dia.	1E@0.45	Ea	59.10	12.10	71.20
3/4" - 2" dia.	1E@0.50	Ea	54.90	13.40	68.30
3/4" - 3" dia.	1E@0.55	Ea	61.70	14.80	76.50
1" - 3" dia.	1E@0.60	Ea	63.20	16.10	79.30
1-1/4" - 3-5/8" dia.	1E@0.65	Ea	95.70	17.50	113.20
1-1/2" - 5" dia.	1E@0.80	Ea	177.00	21.50	198.50
2" - 5" dia.	1E@1.10	Ea	186.00	29.60	215.60
PVC coated Type GUAX junction boxes with covers					
1/2" - 2" dia.	1E@0.45	Ea	52.30	12.10	64.40
1/2" - 3" dia.	1E@0.50	Ea	59.60	13.40	73.00
3/4" - 2" dia.	1E@0.55	Ea	55.10	14.80	69.90
3/4" - 3" dia.	1E@0.60	Ea	62.50	16.10	78.60
1" - 3" dia.	1E@0.65	Ea	68.90	17.50	86.40
1-1/4" - 3-5/8" dia.	1E@0.70	Ea	105.00	18.80	123.80
1-1/2" - 5" dia.	1E@0.90	Ea	184.00	24.20	208.20
2" - 5" dia.	1E@1.25	Ea	192.00	33.60	225.60
PVC coated Type EYD female sealing fittings					
1/2"	1E@0.35	Ea	57.90	9.41	67.31
3/4"	1E@0.40	Ea	59.60	10.80	70.40
1"	1E@0.45	Ea	67.00	12.10	79.10
1-1/4"	1E@0.50	Ea	73.50	13.40	86.90
1-1/2"	1E@0.60	Ea	92.60	16.10	108.70
2"	1E@0.75	Ea	101.00	20.20	121.20
PVC coated Type EYD male-female sealing fittings					
1/2"	1E@0.35	Ea	57.90	9.41	67.31
3/4"	1E@0.40	Ea	60.10	10.80	70.90
1"	1E@0.45	Ea	68.40	12.10	80.50
1-1/4"	1E@0.50	Ea	76.20	13.40	89.60
1-1/2"	1E@0.60	Ea	96.00	16.10	112.10
2"	1E@0.75	Ea	105.00	20.20	125.20
2-1/2"	1E@0.90	Ea	149.00	24.20	173.20
3"	1E@1.00	Ea	209.00	26.90	235.90
3-1/2"	1E@1.25	Ea	319.00	33.60	352.60
4"	1E@1.40	Ea	487.00	37.60	524.60
PVC coated Type EYS female sealing fittings					
1/2"	1E@0.30	Ea	26.60	8.06	34.66
3/4"	1E@0.35	Ea	27.50	9.41	36.91
1"	1E@0.40	Ea	32.10	10.80	42.90

Use these figures to estimate the cost of PVC coated junction boxes and sealing fittings installed on PVC coated conduit under the conditions described on pages 5 and 6. Costs listed are for each fitting installed. The crew is one electrician working at a labor cost of $26.88 per manhour. These costs include the cover, layout, material handling, and normal waste. Add for sales tax, delivery, supervision, mobilization, demobilization, cleanup, overhead and profit. Note: PVC patching material is available in spray cans for repairing damaged PVC coating.

Material	Craft@Hrs	Unit	Material Cost	Labor Cost	Installed Cost

PVC coated Type EYS female sealing fittings

Material	Craft@Hrs	Unit	Material Cost	Labor Cost	Installed Cost
1-1/4"	1E@0.50	Ea	38.50	13.40	51.90
1-1/2"	1E@0.50	Ea	52.30	13.40	65.70
2"	1E@0.60	Ea	63.30	16.10	79.40

PVC coated Type EYS male-female sealing fittings

Material	Craft@Hrs	Unit	Material Cost	Labor Cost	Installed Cost
1/2"	1E@0.30	Ea	26.60	8.06	34.66
3/4"	1E@0.35	Ea	28.50	9.41	37.91
1"	1E@0.40	Ea	33.90	10.80	44.70
1-1/4"	1E@0.50	Ea	40.40	13.40	53.80
1-1/2"	1E@0.50	Ea	55.10	13.40	68.50
2"	1E@0.60	Ea	67.00	16.10	83.10
2-1/2"	1E@0.75	Ea	98.50	20.20	118.70
3"	1E@0.90	Ea	123.00	24.20	147.20
3-1/2"	1E@1.00	Ea	303.00	26.90	329.90
4"	1E@1.25	Ea	468.00	33.60	501.60

PVC coated Type EZS female sealing fittings

Material	Craft@Hrs	Unit	Material Cost	Labor Cost	Installed Cost
1/2"	1E@0.30	Ea	42.20	8.06	50.26
3/4"	1E@0.35	Ea	43.20	9.41	52.61
1"	1E@0.40	Ea	48.60	10.80	59.40
1-1/4"	1E@0.50	Ea	55.10	13.40	68.50
1-1/2"	1E@0.50	Ea	80.90	13.40	94.30
2"	1E@0.60	Ea	104.00	16.10	120.10

PVC coated Type EZS male-female sealing fittings

Material	Craft@Hrs	Unit	Material Cost	Labor Cost	Installed Cost
1/2"	1E@0.30	Ea	43.20	8.06	51.26
3/4"	1E@0.35	Ea	44.10	9.41	53.51
1"	1E@0.40	Ea	51.50	10.80	62.30
1-1/4"	1E@0.50	Ea	55.10	13.40	68.50
1-1/2"	1E@0.50	Ea	84.20	13.40	97.60
2"	1E@0.60	Ea	108.00	16.10	124.10
2-1/2"	1E@0.70	Ea	172.00	18.80	190.80
3"	1E@0.90	Ea	258.00	24.20	282.20

Use these figures to estimate the cost of PVC coated sealing fittings installed on PVC coated conduit under the conditions described on pages 5 and 6. Costs listed are for each fitting installed. The crew is one electrician working at a labor cost of $26.88 per manhour. These costs include layout, material handling, and normal waste. Add for sales tax, delivery, supervision, mobilization, demobilization, cleanup, overhead and profit. Note: PVC patching material is available in spray cans for repairing damaged PVC coating.

Hanger Fittings

Material	Craft@Hrs	Unit	Material Cost	Labor Cost	Installed Cost
Plated threaded rod					
1/4-20 x 6'	1E@1.25	CLF	27.00	33.60	60.60
1/4-20 x 10'	1E@1.25	CLF	27.00	33.60	60.60
1/4-20 x 12'	1E@1.15	CLF	27.00	30.90	57.90
3/8-16 x 6'	1E@1.30	CLF	45.50	34.90	80.40
3/8-16 x 10'	1E@1.30	CLF	45.50	34.90	80.40
3/8-16 x 12'	1E@1.30	CLF	42.90	34.90	77.80
1/2-13 x 6'	1E@1.50	CLF	81.60	40.30	121.90
1/2-13 x 10'	1E@1.50	CLF	81.60	40.30	121.90
1/2-13 x 12'	1E@1.50	CLF	81.60	40.30	121.90
5/8-11 x 6'	1E@1.75	CLF	132.00	47.00	179.00
5/8-11 x 10'	1E@1.75	CLF	132.00	47.00	179.00
5/8-11 x 12'	1E@1.75	CLF	132.00	47.00	179.00
Rod couplings					
1/4-20	1E@0.05	Ea	.65	1.34	1.99
3/8-16	1E@0.05	Ea	1.01	1.34	2.35
1/2-13	1E@0.08	Ea	1.34	2.15	3.49
5/8-11	1E@0.10	Ea	2.15	2.69	4.84
Toggle bolts, wing nuts					
1/8 x 3"	1E@0.10	Ea	.19	2.69	2.88
3/16 x 3"	1E@0.10	Ea	.22	2.69	2.91
1/4 x 4"	1E@0.15	Ea	.29	4.03	4.32
3/8 x 4"	1E@0.20	Ea	.67	5.38	6.05
Expansion anchors, flush type					
1/4-20	1E@0.15	Ea	.58	4.03	4.61
3/8-16	1E@0.15	Ea	.16	4.03	4.19
1/2-13	1E@0.25	Ea	1.33	6.72	8.05
5/8-11	1E@0.30	Ea	2.14	8.06	10.20
Steel hex nuts					
1/4-20	1E@0.02	Ea	.04	.54	.58
3/8-16	1E@0.03	Ea	.07	.81	.88
1/2-13	1E@0.05	Ea	1.67	1.34	3.01
5/8-11	1E@0.10	Ea	2.83	2.69	5.52
Fender washers, 1-1/2" diameter					
1/4"	1E@0.02	Ea	.05	.54	.59
3/8"	1E@0.03	Ea	.06	.81	.87
1/2"	1E@0.04	Ea	.04	1.08	1.12

Use these figures to estimate the cost of installing steel hanger fittings for hanging or mounting conduit or electrical equipment under the conditions described on pages 5 and 6. Costs listed are for each 100 linear feet or steel channel strut, or each fitting installed. The crew is one electrical working at a labor cost of $26.88 per manhour. These costs include layout, material handling, and normal waste. Add for sales tax, delivery, supervision, mobilization, demobilization, cleanup, overhead and profit.

Material	Craft@Hrs	Unit	Material Cost	Labor Cost	Installed Cost
14 gauge steel channel					
13/16" x 1-5/8" painted	1E@4.00	CLF	164.00	108.00	272.00
13/16" x 1-5/8" galvanized	1E@4.00	CLF	175.00	108.00	283.00
1-5/8" x 1-5/8" painted	1E@6.00	CLF	223.00	161.00	384.00
1-5/8" x 1-5/8" galvanized	1E@6.00	CLF	230.00	161.00	391.00
14 gauge steel channel with 9/16" holes, 1-7/8" oc					
13/16" x 1-5/8" painted	1E@4.00	CLF	186.00	108.00	294.00
13/16" x 1-5/8" galvanized	1E@4.00	CLF	198.00	108.00	306.00
1-5/8" x 1-5/8" painted	1E@6.00	CLF	226.00	161.00	387.00
1-5/8" x 1-5/8" galvanized	1E@6.00	CLF	260.00	161.00	421.00
12 gauge steel channel					
13/16" x 1-5/8" painted	1E@4.00	CLF	204.00	108.00	312.00
13/16" x 1-5/8" galvanized	1E@4.00	CLF	225.00	108.00	333.00
1-5/8" x 1-5/8" painted	1E@6.00	CLF	248.00	161.00	409.00
1-5/8" x 1-5/8" galvanized	1E@6.00	CLF	251.00	161.00	412.00
12 gauge steel channel with 9/16" holes, 1-7/8" oc					
13/16" x 1-5/8" painted	1E@4.00	CLF	232.00	108.00	340.00
13/16" x 1-5/8" galvanized	1E@4.00	CLF	267.00	108.00	375.00
1-5/8" x 1-5/8" painted	1E@6.00	CLF	273.00	161.00	434.00
1-5/8" x 1-5/8" galvanized	1E@6.00	CLF	289.00	161.00	450.00
Channel nuts					
1/4-20 13/16" strut	1E@0.05	Ea	.50	1.34	1.84
3/8-16 13/16" strut	1E@0.05	Ea	.46	1.34	1.80
1/2-13 13/16" strut	1E@0.05	Ea	.51	1.34	1.85
1/4-20 1-5/8" strut	1E@0.05	Ea	.50	1.34	1.84
3/8-16 1-5/8" strut	1E@0.05	Ea	.46	1.34	1.80
1/2-13 1-5/8" strut	1E@0.05	Ea	.51	1.34	1.85
Channel spring nuts					
1/4-20 13/16" strut	1E@0.06	Ea	.53	1.61	2.14
3/8-16 13/16" strut	1E@0.06	Ea	.56	1.61	2.17
1/2-13 13/16" strut	1E@0.06	Ea	.62	1.61	2.23
1/4-20 1-5/8" strut	1E@0.06	Ea	.51	1.61	2.12
3/8-16 1-5/8" strut	1E@0.06	Ea	.54	1.61	2.15
1/2-13 1-5/8" strut	1E@0.06	Ea	.60	1.61	2.21

Use these figures to estimate the cost of installing steel channel strut and fittings for hanging or mounting conduit or electrical equipment under the conditions described on pages 5 and 6. Costs listed are for each 100 linear feet or steel channel strut, or each fitting installed. The crew is one electrical working at a labor cost of $26.88 per manhour. These costs include layout, material handling, and normal waste. Add for sales tax, delivery, supervision, mobilization, demobilization, cleanup, overhead and profit.

Section 2:
Wire and Cable

Wire and cable come in many types and sizes. Fortunately, only a few of these are of major concern to electrical estimators.

Article 310 of the *National Electrical Code* lists minimum conductor requirements for electrical systems. Tables in the *NEC* show installation standards that will apply in most communities where the *NEC* is followed. But be sure to use the version of the code that the inspectors are using. Even though the *NEC* is revised every three years, some cities and counties don't get around to adopting the current version until months or years after each revision.

In this book we're concerned with building construction exclusively. The scope of the *NEC* is much broader. We'll ignore *NEC* requirements that relate to ship building, aircraft manufacturing and the automobile industry.

In the construction industry, the most common wire types are identified as THW, THHN, THWN, XHHW, MTW, TF, TFF, TFFN, AF, and USE-RHH-RHW. These code letters identify the type of insulation on the wire.

Wire sizes used in building electrical systems range from #18 (called **18 gauge** under the American Wire Gauge system) to 1000 kcmil. The abbreviation kcmil stands for 1000 circular mils.

The smaller the gauge number, the larger the wire. #14 gauge is smaller than #12 gauge, for example. Gauges larger than 1 follow the "ought" scale, One ought (usually written as **1/0**) is larger than #1 gauge. Two ought (2/0) is larger still. Above four ought, wire size is identified in thousands of circular mils. A circular mil is the area of a circle that's one mil (1/1000th of an inch) in diameter. Wire that has a cross section area of 250,000 circular mils (250 kcmil), for example, weighs about 12 ounces per linear foot. 1000 kcmil wire weighs more than 3 pounds per linear foot.

Conductors in wire are either solid (one piece surrounded by insulation) or stranded. Stranded copper wire can be tinned as part of the manufacturing process and might be so specified.

Type THW

Type THW wire is Moisture- and Heat-Resistant Thermoplastic. It has a maximum operating temperature of 75 degrees C or 167 degrees F for dry or wet locations and 90 degrees C or 194 degrees F when used in electric discharge lighting equipment of 1000 open-circuit volts or less. The insulation is flame-retardant, moisture- and heat-resistant thermoplastic. It has no outer jacket or covering and is sized from #14 gauge with either solid or stranded conductors.

Type THHN

Type THHN wire is Heat-Resistant Thermoplastic and has a maximum operating temperature of 90 degrees C or 194 degrees F for dry locations. The insulation is flame-retardant, heat-resistant thermoplastic. It has a nylon jacket or the equivalent and is sized from #14 gauge with either solid or stranded conductors.

Type XHHW

Type XHHW wire is Moisture- and Heat-Resistant Thermoplastic and has a maximum operating temperature of 90 degrees C or 194 degrees F for dry locations, and 75 degrees C or 167 degrees F for wet locations. The insulation is flame-retardant cross-linked synthetic polymer. It has no outer jacket or covering and is sized from #14 gauge in either solid or stranded conductors.

Type USE

Type USE is Underground Service-Entrance Cable, Single Conductor, and has a maximum operating temperature of 75 degrees C or 167 degrees F for applications as listed in Article 338 - Service-Entrance Cable in the *NEC*. The insulation is heat- and moisture-resistant. It has a jacket or outer covering rated as moisture-resistant nonmetallic and is sized from #12 gauge in either solid or stranded conductors.

Type MTW

Type MTW is Moisture-, Heat-, and Oil-Resistant Thermoplastic and has a maximum operating temperature of 60 degrees C or 140 degrees F. It's used for machine tool wiring in wet locations and is covered by *NEC* Article 670, Metal-Working Machine Tools. The insulation is flame-retardant, moisture-, heat- and oil-resistant thermoplastic. It's sized from #14 gauge with stranded conductors. There are two thicknesses of insulation. One is a little thinner than the other. The thicker insulation, Type A, has no outer jacket or covering. The thinner insulation, Type B, has a nylon jacket or equivalent covering.

Type TF

Type TF is Thermoplastic-Covered Fixture Wire and has a maximum operating temperature of 60 degrees C or 140 degrees F. It's used for fixture wiring as permitted by the *NEC*. The insulation is thermoplastic. Sizes are #18 or #16 gauge with either solid or stranded conductors. It has no outer jacket or covering.

Type TFF

Type TFF is Thermoplastic-Covered Fixture Wire, Flexible Stranded. It has a maximum operating temperature of 60 degrees C or 140 degrees F and is used for fixture wiring, with some *NEC* restrictions. The insulation is thermoplastic. Sizes are either #18 or #16 gauge with stranded conductors. It has no outer jacket or covering.

Type TFFN

Type TFFN is Heat-Resistant Thermoplastic-Covered Fixture Wire, Flexible Stranded. It has a maximum operating temperature of 90 degrees C or 194 degrees F and is used for fixture wiring as permitted by the *NEC*. The insulation is thermoplastic and is sized at #18 or #16 gauge with solid or stranded conductors. It has a nylon or equivalent outer jacket.

Type AF

Type AF wire is Asbestos Covered Heat-Resistant Fixture Wire. It has a maximum operating temperature of 150 degrees C or 302 degrees F for fixture wiring. It's limited to 300 volts and indoor dry locations. The insulation is impregnated asbestos. Gauges run from #18 to #14 with stranded conductors. Moisture-resistant and impregnated asbestos insulation is sized from #12 to #10 gauge with stranded conductors.

Many other types of insulation are listed in the tables of the *NEC*. The types just listed are the ones most commonly used in building construction. Many other types of insulated wire were used in the past and have been discontinued. The *NEC* still lists some types that are seldom used and are no longer stocked by most dealers. For example, the smaller sizes of THW are no longer popular. Type THHN or THHN-THWN has become more popular as the industry standard changed to newer types of insulation. The insulation jacket on THHN-THWN is much thinner. Yet the wire has a higher ampacity than older THW. The *NEC* tables "Maximum Number of Conductors in Trade Sizes of Conduit or Tubing" permit more current-carrying capacity in conduit of a given size when using THHN wire than when using THW wire.

Flexible Cords

Many types of flexible power cords are listed in the *NEC*. Some of these apply to the construction industry and are important to the electrical estimator.

Type SJ

The trade name for Type SJ is Junior Hard Service Cord. It's made in sizes #18 or #10 with two, three or four conductors. The insulation is thermoset plastic with an outer jacket or covering of thermoset plastic. It's used for portable or pendant fixtures in damp locations where hard usage is expected. The insulation is rated at 300 volts for 60 degrees C.

Type S

The trade name for Type S is Hard Service Cord. It's made in gauges from #18 to #2 with two or more conductors. The insulation is thermoset plastic with an outer jacket or covering of thermoset plastic. It's used for portable or pendant fixtures in damp locations where extra hard use is expected. The insulation is rated at 600 volts for 60 degrees C.

Type SO

The trade name for Type SO is also Hard Service Cord. It's made in gauges from #18 to #2 with two or more conductors. The insulation is thermoset plastic with an outer jacket or covering of oil-resistant thermoset plastic. It's used for portable or pendant fixtures in damp locations where extra hard usage is expected. The insulation is rated at 600 volts for 60 degrees C.

Type STO

The trade name for Type STO is also Hard Service Cord. It's made in gauges from #18 to #2 with two or more conductors. The insulation is thermoplastic or thermoset with an outer jacket or covering of oil-resistant thermoplastic. It's used for portable or pendant fixtures in damp locations where extra hard usage is expected. The insulation is rated at 600 volts for 60 degrees C.

Type NM

The trade name for Type NM is Non-Metallic-Sheathed Cable. It's made in gauges from #14 to #2 with from two to four conductors and either with or without a ground wire. The ground wire may be either insulated or bare. The insulation is as listed in NEC Table 310-13. The cable has an outer jacket or covering of a flame-retardant and moisture-resistant plastic. Type NM is commonly used in house wiring. The insulation is rated at 600 volts for 60 degrees C.

Type UF

The trade name for Type UF Cable is Underground Feeder and Branch Circuit Cable. It's made in gauges from #14 to #4/0 with one, two, three or four conductors. Gauges from #14 to #10 have 60 mil insulation. Gauges from #8 to #2 have 80 mil insulation. Gauges from #1 to #4/0 have 95 mil insulation. It comes with or without a ground wire. Single conductor is rated for 600 volt, 60 degrees C or 140 degrees F. The insulation is moisture-resistant and is integral with the jacket or outer covering. Multiple conductor cable is rated at 600 volts and 75 degrees C or 167 degrees F. The insulation is moisture- and heat-resistant.

Type SEU

The trade name for Type SEU is Service-Entrance Cable. It's made in gauges from #12 to #4/0 with one or more conductors. The cable is rated 600 volts. The temperature rating of the wire is the same as the rating of the conductor itself. The cable has a flame-retardant, moisture-resistant jacket or covering.

Bare Copper Wire

Bare copper wire is made in gauges from #14 to #4/0 in either soft drawn, medium hard drawn or hard drawn temper. It's usually sold by the hundredweight, which is 100 pounds. It comes in solid strands in sizes up to #4 and stranded up to #4/0 gauge. The most common use for bare copper wire is in grounding electrical systems.

The *NEC* has a table that gives weights for 1000 linear feet of soft drawn bare copper wire. This simplifies the conversion of lengths to weights so you can convert the hundredweight price to a cost for the length you need.

Armored Cable

The trade name for Type AC or BX is Armored Cable. It consists of insulated conductors wrapped in a flexible metallic covering. The conductors have a moisture-resistant and flame-retardant fibrous cover. An internal bonding strip of copper or aluminum is in contact with the metallic covering for the entire length of the cable. It's rated for 600 volts and comes in gauges from #14 to #4 in single or multiple conductors. An approved insulating bushing or the equivalent must be installed at terminations in the cable between the conductor insulation and the outer metallic covering.

There are some restrictions on the use of armored cable. It's prohibited in theaters, places of assembly, motion picture studios, hazardous locations, where exposure to corrosive fumes or vapors is expected, on cranes or hoists, in storage battery rooms, in hoistways or elevators and in commercial garages. Check for exceptions in the *NEC* when working with AC or BX armored cable.

Other types of insulated conductors and cables are available and used occasionally in construction. But be sure you refer to the code and discuss the material with the inspection authority before including them in your bid. Some cities and counties prohibit certain conductors and cables.

High Voltage Wire and Cable

Medium and high voltage conductors and cables require special consideration. In most projects they're specified for a particular use or service. Many types of high voltage conductors are produced by some manufacturers. Some are stocked only by the factory. Others are not stocked at all and have to be made to order. These will be very expensive unless the factory run is quite large. In any case, be prepared to wait many weeks for delivery.

Terminations for high voltage conductors are also a specialty item. Installation should be done only by an experienced craftsman, usually called a cable splicer. Shielded conductors will require a stress cone for termination at the point of connection to switching equipment or some other device. Stress cones can be made as a kit that's adapted to the end of the cable to relieve stress from the shield.

Splicing cable is also a specialty that should be done by a qualified cable splicer. The splicer prepares the conductor and either uses a splice kit or tapes a splice. The process may take many hours to complete. Once a stress cone or a splice has been started, it should be completed as quickly as possible to keep moisture out of the splice.

Most professional electrical estimators have to figure the cost of extending an existing underground electrical system occasionally. For example, you may have to figure the cost of splicing new XLP cable to three conductor lead covered cable. An experienced cable splicer will be needed to make this splice properly. In an underground system that's subject to moisture or even continuous submersion in brackish water, this splice will have to be made correctly.

High voltage splicing requires experience and precision. If the cable is damaged or if the splice isn't made correctly, replacing the cable and the splice can be very expensive. If you can't locate an experienced cable splicer, consider subcontracting the work to a firm that specializes in this work. Having it done by a qualified expert can preserve many nights of sound sleep.

Most specs require a high-potential test on the cable after it's been installed. The test is done with a special portable test unit. An experienced operator is needed to run the test. Follow the test procedure recommended by the cable manufacturer.

There are good reasons for running this test. Of course, every reel of electrical cable gets a high potential test at the factory. But cable can be damaged between the factory and the job site. Your cable may have been cut from the original reel and re-coiled on a smaller reel for shipment to the job site. It probably sat in storage for at least several months and may have been exposed to adverse conditions for the entire time. When your electricians finally install the cable, it may be defective, even though it looks fine. But if moisture has entered from a cut end or if the cable was damaged in transit, it's much better to find out before the cable is energized. A fault in a high voltage cable can cause damage to other cable in the conduit or to other parts of the electrical system.

The high potential test is not a destructive test. If the cable is sound before the test begins, it will be unchanged by the test. The test proves that the insulation is in good condition and that the conductor will safely carry the rated voltage.

Here's the usual test procedure. First, identify the conductors by circuit and by phase (A, B, C, etc.). Voltage is applied for a few minutes and is increased to a certain value over a set amount of time. The operator records the voltage, time and amount of leakage up to a predetermined voltage. The maximum test voltage will be more than the rated voltage of the cable. The technician performing the test enters results on a special test form and submits it to the designing engineer for review. The form should then be filed with permanent project records.

If a fault shows up during the high potential test, the operator should have equipment available to locate the fault. Usually the problem will be at a stress cone or in a splice. One method of locating a fault in the cable is to install a ***thumper***, a large capacitor-type unit, in the line. The thumper gives off an electrical discharge every few seconds. The charge will create a pop or snap when it hits the fault, making location easy.

Repair the problem by resplicing. If the fault is in the cable, pull out the cable and examine it carefully. It was probably damaged during installation. If the cable jacket is good and doesn't have signs of damage, the cable itself may be at fault. The manufacturer may want to test the cable to determine the cause of failure. Most manufacturers will replace at no charge cable that fails a high potential test ***before*** being put into service. Once the cable is put into use, it's hard to get a free replacement, even if there is no apparent cable damage.

Aluminum Wire

The types of insulation listed for copper wire are the same as used on aluminum wire. There are two major differences between copper and aluminum wire: First, aluminum usually costs less for a similar gauge. Second, the ampacity of aluminum wire is less than that of the copper wire. That means that aluminum wire has to be heavier gauge to carry the same current. Tables in the *NEC* show the allowable current rating for aluminum wire.

Aluminum wire is much lighter than copper wire and can be installed in less time. Compare the difference in the labor units in this book for copper and aluminum wire. Many electricians prefer to install aluminum wire on some types of applications. But aluminum wire comes with a major problem: it oxidizes when not covered with insulation. When the wire is stripped and left bare, it develops a thin white oxide coating that resists the flow current. Eventually the wire begins to heat up at the point of connection. Heat expands the conductor. Cooling causes it to contract. Constant expansion and contraction makes the connection looser and accelerates oxidation. Eventually the conductor will fail. Sometimes overheating will cause a fire.

There are ways to reduce the problem of oxidation. An anti-oxidation material can be applied immediately after the cable is stripped and before it's terminated in the connector. The material can also be applied to the connector. Then the aluminum wire is inserted into the connector and tightened. A good practice is to retighten the connector after a few days and again after the circuit has been in operation for about a year. Some wire and cable connectors are made with an oxidation inhibitor material inside the area where the wire will be inserted. Only connectors rated for aluminum wire should be used on aluminum circuits.

Today, aluminum wire is seldom used on branch circuits to convenience receptacles or lighting switches. But it's common on feeder circuits such as runs from distribution switchboards to lighting or power panels.

When you decide to change a feeder from copper wire to aluminum wire, be sure to compare the carrying capacity of the two materials. You'll find that it takes larger aluminum wire to carry the same amperage as a copper wire. This may mean that you'll have to use larger conduit size to accommodate the larger wire diameter. But using aluminum on feeder circuits will probably still reduce the labor and material cost.

Wire Connectors

Many types of wire connectors and lugs are available. Some are insulated, made for multiple conductors, or intended to be watertight. Be sure the connections you price will comply with the code and job specs. Not all types of connections are listed in this manual. But you'll find a representative selection here that gives you a good idea of how to price most of the common wire connectors. It would take a 200-page book just to list labor and material costs for all the connectors and lugs that are available.

Pulling Wire in Conduit

The best way to thread wire through conduit depends on the type of wire and the type of conduit. Some conduit runs are several hundred feet long. On a job like that, power equipment is recommended. If the circuit length is closer to 50 feet and the wire gauge is #6 or less, a fish tape will be the only tool you need.

Pulling wire is the essence of simplicity: Thread a line through the full length of conduit from box to box, bend wire around the end of that line and pull it back the other way.

Several kinds of fish tape are made for pulling building wire. Tape used on EMT, PVC and rigid steel conduit is usually ⅛" spring steel and will be 100 feet long. Lengths of 200 feet are also available. For larger conduit, use ¼" spring steel fish tape in 200 feet lengths.

Form a pulling eye in spring steel tape by bending back the last several inches of tape. Heat the tape before bending to keep the steel from breaking as you make the eye.

For flexible conduit, it's easier to use a special fish tape made specifically for flex. The end of the tape comes with an eye to receive the building wire.

Before attaching building wire to the pulling eye, always strip insulation off the last 3 or 4 inches of wire. Feed the bare conductors through the eye, bend the conductors back tightly and wrap the ends with plastic tape. Then start the pulling.

Any time you pull wire, be sure wire is feeding smoothly into the box or conduit at the far end. If there's too much pulling resistance, stop pulling. Something's wrong. You don't want to strip insulation off the wire. Coat the wire with special pulling compound if necessary.

Other Pulling Methods

Some electricians prefer to use a small fishing vacuum unit that draws string through the conduit. A small sponge (called a *mouse*) sized for the conduit is attached to a fine plastic string. The mouse has a steel eye fitted at one or both ends. String is tied to the mouse eye and then the mouse is sucked through the conduit with the vacuum.

The mouse can also be blown through conduit with air pressure rather than a vacuum. This is the common way of pulling wire in long feeder conduit and underground ducts between manholes or handholes. Some electricians prefer using a parachute in place of the mouse. The parachute is blown through the duct with a plastic line attached.

When pulling heavy wire or pulling wire on longer runs, the thin line first pulled through the conduit may not be strong enough to pull the wire. In that case, attach the first line pulled to some stronger string. Pull that string through the entire length. Then pull the wire with the stronger string.

On underground runs it's good practice to pull a mandrel and wire brush through the duct before pulling any wire. This guarantees that the conduit is free of debris. The mandrel should be at least 4" long and ¼" smaller than the duct. Tie a heavy line to both ends of the mandrel and brush when it's pulled. If the mandrel gets hung up, you want to be able to pull it back.

Don't pull wire in conduit that you know is damaged. Obstructions or breaks in the duct will probably damage any wire you pull. Make repairs first. Then pull the wire. That's the easiest way.

Getting Set Up

Start each pull by running fish tape through the conduit. Set the rolls of wire where they can be handled into place with minimum effort. When pulling starts, it should continue without interruption until complete.

Pulling is easier if you buy smaller gauge building wire (sizes from AWG 18 to 6) on small disposable reels. Each has from 500 to 2500 feet of continuous wire. Mount these reels on a reel frame or dolly that will hold as many reels as there are conductors in the pull. Set the reel dolly at the feed end of the conduit.

Larger reels have to be set on reel jacks. Run conduit or pipe through the center axle of the reel and rest the pipe on the jacks. Then lift the reel so it's off the floor. The reel should be mounted so wire feeds from the top of the coil. That makes it easier to handle the wire when necessary. More bending or stooping is needed if wire feeds from the bottom of the reel. As wire is fed into the conduit or duct, someone should apply pulling compound to ease the pulling effort.

Power Pulling Equipment

A pulling machine is usually required when pulling heavy wire for feeder circuits. Power pulling rigs are available in several sizes. A truck-mounted winch can also be used for outdoor work. Whatever power equipment is used, be sure to have good communication between both ends of the conduit. If the pull has to be stopped quickly, the operator should know right away.

The best pull rate when using power equipment is between 10 and 15 feet per minute. It's easy to pull faster, until something goes wrong. Then, the faster the pull, the more likely damage will result before pulling can be stopped.

On feeder pulls, a nylon pulling rope may be the best choice if the pulling rig has a capstan. Wrap two to four turns of pulling rope around the capstan. Have the electrician at the feed end hold the rope taut as the slack is being taken out. Begin the pull. If the load increases during the pull, have the operator pull a little slack and let the rope slip on the capstan. That slows the pull to avoid damaging the wire.

There are other ways to pull wire. But those listed here are the most common methods for conduit and duct. Use whatever method helps you get the pulling done most efficiently on the job at hand. No single pulling method is ideal for all situations. Most electrical contractors have several types of pulling equipment and use the method that's most appropriate for job conditions.

Adjusting the Cost Tables

The tables in this section show labor and material costs and labor hours per 1000 linear feet of wire when three conductors are pulled at the same time. That's the most common case on most jobs. The labor cost will be higher per 1000 linear feet of wire when only two conductors are pulled at once. It will usually be lower when more than three conductors are pulled at once.

When pulling three conductors simultaneously, use the tables in this section without modification. If there are two, three and four wire pulls on a job, and if the average pull is three wires, use the tables without modification. Only if two or four wire pulls predominate will it be necessary to adjust the cost tables.

If you find that most pulls on a job will be only two wires, add about 10 percent to the labor cost per 1000 linear feet of wire. If most pulls will be more than three wires, reduce the labor cost by about 10 percent for each conductor over three.

Increasing Productivity

A good supervisor can improve pulling efficiency by planning the job before work begins. Think about the best place to set up the reels. Setting up at the panels is usually a good idea because you can serve many pulls from there. But it may be easier to fish from other locations back to the panel. Avoid pulling uphill. Do whatever is necessary to make the pull easier. Be sure there's enough workspace at the pull end to make pulling possible.

String used with mouse fishing can be reused many times if it's handled properly. Some electrical contractors feel that string is cheaper than the labor needed to coil it for reuse. For most contractors it's better to save the string.

When the wire is pulled, leave plenty of extra wire at both ends to make the connection. Leave 10 feet at full size panels, 10 feet to 15 feet at motor control centers, and 18 inches to 24 inches at outlets and fixtures. Your take-off should reflect these allowances, of course.

Flexible Cords

The tables that follow include prices for flexible cords. These cords aren't installed in conduit. They're used to connect portable equipment, to extend power temporarily, or with certain types of lighting fixtures. Power cords are available on reels with lengths from 25 feet to 100 feet. Larger reels can be ordered also. Flexible cords must be protected from physical damage. Check the *NEC* for restrictions on each type of cord.

Taking Off Wire

Wire take-off should be based on the conduit take-off. Add the length of all conduit runs that have the same number of conductors and use the same wire gauge. Then multiply the computed length by the number of conductors and add for the extra wire needed at each outlet box and panel. Don't worry about the colors at this point. Get the total quantity so you can price the wire and labor.

If you get the job, the first step is to check the panel schedules. Find out which circuits are single phase and which are three phase. If the panels are three phase, wire colors will be black, red, blue and white for 120/208V and brown, orange, yellow and white for 277/480V. If the panels are 120/240V, wire colors will be black, red and white. You may want to set aside another color for switch legs.

When figuring the quantity of each color, figure about 80 percent of the wire will be the primary colors noted above. Divide the 80 percent equally among the colors needed. The other 20 percent will be white. Wire colors for switch legs can be figured separately.

Copper Building Wire

Material	Craft@Hrs	Unit	Material Cost	Labor Cost	Installed Cost
Type THW 600 volt solid copper building wire					
# 14	2E@6.00	KLF	51.20	161.00	212.20
# 12	2E@7.00	KLF	77.60	188.00	265.60
# 10	2E@8.00	KLF	122.00	215.00	337.00
Type THW 600 volt stranded copper building wire					
# 14	2E@6.00	KLF	59.40	161.00	220.40
# 12	2E@7.00	KLF	89.10	188.00	277.10
# 10	2E@8.00	KLF	134.00	215.00	349.00
# 8	2E@9.00	KLF	230.00	242.00	472.00
# 6	2E@10.0	KLF	326.00	269.00	595.00
# 4	2E@12.0	KLF	505.00	323.00	828.00
# 2	3E@13.0	KLF	778.00	349.00	1,127.00
# 1	3E@14.0	KLF	1,010.00	376.00	1,386.00
# 1/0	3E@15.0	KLF	1,230.00	403.00	1,633.00
# 2/0	3E@16.0	KLF	1,510.00	430.00	1,940.00
# 3/0	3E@17.0	KLF	1,890.00	457.00	2,347.00
# 4/0	3E@18.0	KLF	2,340.00	484.00	2,824.00
# 250 KCMIL	4E@20.0	KLF	2,790.00	538.00	3,328.00
# 300 KCMIL	4E@23.0	KLF	3,320.00	618.00	3,938.00
# 350 KCMIL	4E@24.0	KLF	3,880.00	645.00	4,525.00
# 400 KCMIL	4E@25.0	KLF	4,350.00	672.00	5,022.00
# 500 KCMIL	4E@26.0	KLF	5,490.00	699.00	6,189.00
# 600 KCMIL	4E@30.0	KLF	7,010.00	806.00	7,816.00
# 750 KCMIL	4E@32.0	KLF	8,680.00	860.00	9,540.00
#1000 KCMIL	4E@36.0	KLF	12,400.00	968.00	13,368.00
Type THHN 600 volt solid copper building wire					
# 14	2E@6.00	KLF	51.20	161.00	212.20
# 12	2E@7.00	KLF	77.60	188.00	265.60
# 10	2E@8.00	KLF	122.00	215.00	337.00

Use these figures to estimate the cost of copper THW solid and stranded, and THHN solid wire installed in conduit under the conditions described on pages 5 and 6. Costs listed are for each 1,000 linear feet installed. The crew is two electricians for sizes up to #4, three electricians for sizes over #4 to #4/0 and four electricians for sizes over #4/0. The labor cost per manhour is $26.88. These costs include fishing string, circuit make-up, splices on wire up to #6, reel set-up, pulling compound, phase identification, circuit testing, layout, material handling, and normal waste. Add for a bonding wire run in non-metallic or flexible conduit, sales tax, delivery, supervision, mobilization, demobilization, cleanup, overhead and profit. Note: These costs are for conduit runs of less than 100 feet and assume that three wires of the same size are pulled at the same time. A deposit is often required on larger-wire reels. These costs assume that larger reels are returned for credit. Order large-size wire in longest lengths available to reduce waste.

See Simple Conversion Table On Page 491
Copper to Aluminum Wire Ampacities

Copper Building Wire

Type THHN 600 volt stranded copper building wire

Material	Craft@Hrs	Unit	Material Cost	Labor Cost	Installed Cost
# 14	2E@6.00	KLF	59.40	161.00	220.40
# 12	2E@7.00	KLF	89.10	188.00	277.10
# 10	2E@8.00	KLF	134.00	215.00	349.00
# 8	2E@9.00	KLF	230.00	242.00	472.00
# 6	2E@10.0	KLF	326.00	269.00	595.00
# 4	2E@12.0	KLF	505.00	323.00	828.00
# 2	3E@13.0	KLF	778.00	349.00	1,127.00
# 1	3E@14.0	KLF	1,010.00	376.00	1,386.00
# 1/0	3E@15.0	KLF	1,230.00	403.00	1,633.00
# 2/0	3E@16.0	KLF	1,510.00	430.00	1,940.00
# 3/0	3E@17.0	KLF	1,890.00	457.00	2,347.00
# 4/0	3E@18.0	KLF	2,340.00	484.00	2,824.00
#250 KCMIL	4E@20.0	KLF	2,790.00	538.00	3,328.00
#300 KCMIL	4E@23.0	KLF	3,320.00	618.00	3,938.00
#350 KCMIL	4E@24.0	KLF	3,880.00	645.00	4,525.00
#400 KCMIL	4E@25.0	KLF	4,350.00	672.00	5,022.00
#500 KCMIL	4E@26.0	KLF	5,490.00	699.00	6,189.00

Type XHHW 600 volt solid copper building wire

Material	Craft@Hrs	Unit	Material Cost	Labor Cost	Installed Cost
# 14	2E@6.00	KLF	118.00	161.00	279.00
# 12	2E@7.00	KLF	155.00	188.00	343.00
# 10	2E@8.00	KLF	215.00	215.00	430.00

Type XHHW 600 volt stranded copper building wire

Material	Craft@Hrs	Unit	Material Cost	Labor Cost	Installed Cost
# 14	2E@6.00	KLF	125.00	161.00	286.00
# 12	2E@7.00	KLF	158.00	188.00	346.00
# 10	2E@8.00	KLF	221.00	215.00	436.00
# 8	2E@9.00	KLF	286.00	242.00	528.00
# 6	2E@10.0	KLF	345.00	269.00	614.00
# 4	2E@12.0	KLF	528.00	323.00	851.00
# 2	3E@13.0	KLF	821.00	349.00	1,170.00
# 1	3E@14.0	KLF	1,020.00	376.00	1,396.00
# 1/0	3E@15.0	KLF	1,270.00	403.00	1,673.00
# 2/0	3E@16.0	KLF	1,560.00	430.00	1,990.00
# 3/0	3E@17.0	KLF	1,900.00	457.00	2,357.00
# 4/0	3E@18.0	KLF	2,400.00	484.00	2,884.00
# 250 KCMIL	4E@20.0	KLF	2,920.00	538.00	3,458.00
# 300 KCMIL	4E@23.0	KLF	3,470.00	618.00	4,088.00
# 350 KCMIL	4E@24.0	KLF	4,050.00	645.00	4,695.00
# 400 KCMIL	4E@25.0	KLF	4,490.00	672.00	5,162.00
# 500 KCMIL	4E@26.0	KLF	5,560.00	699.00	6,259.00
# 600 KCMIL	4E@30.0	KLF	7,280.00	806.00	8,086.00
# 750 KCMIL	4E@32.0	KLF	9,030.00	860.00	9,890.00
#1000 KCMIL	4E@36.0	KLF	13,500.00	968.00	14,468.00

Use these figures to estimate the cost of copper THHN stranded, XHHW solid and stranded wire installed in conduit under the conditions described on pages 5 and 6. Costs listed are for each 1,000 linear feet installed. The crew is two electricians for sizes up to #4, three electricians for sizes over #4 to #4/0 and four electricians for sizes over #4/0. The labor cost per manhour is $26.88. These costs include fishing string, circuit make-up, splices on wire up to #6, reel set-up, pulling compound, phase identification, circuit testing, layout, material handling, and normal waste. Add for a bonding wire run in non-metallic or flexible conduit, sales tax, delivery, supervision, mobilization, demobilization, cleanup, overhead and profit. Note: These costs are for conduit runs of less than 100 feet and assume that three wires of the same size are pulled at the same time. A deposit is often required on larger-wire reels. These costs assume that larger reels are returned for credit. Order large-size wire in longest lengths available to reduce waste.

Material	Craft@Hrs	Unit	Material Cost	Labor Cost	Installed Cost

Type USE, RHH-RHW 600 volt solid copper building wire

Material	Craft@Hrs	Unit	Material Cost	Labor Cost	Installed Cost
# 12	2E@7.00	KLF	121.00	188.00	309.00
# 10	2E@8.00	KLF	157.00	215.00	372.00
# 8	2E@9.00	KLF	217.00	242.00	459.00

Type USE, RHH-RHW 600 volt stranded copper building wire

Material	Craft@Hrs	Unit	Material Cost	Labor Cost	Installed Cost
# 14	2E@6.00	KLF	141.00	161.00	302.00
# 12	2E@7.00	KLF	164.00	188.00	352.00
# 10	2E@8.00	KLF	228.00	215.00	443.00
# 8	2E@9.00	KLF	326.00	242.00	568.00
# 6	2E@10.0	KLF	360.00	269.00	629.00
# 4	2E@12.0	KLF	553.00	323.00	876.00
# 2	3E@13.0	KLF	851.00	349.00	1,200.00
# 1	3E@14.0	KLF	1,030.00	376.00	1,406.00
# 1/0	3E@15.0	KLF	1,310.00	403.00	1,713.00
# 2/0	3E@16.0	KLF	1,980.00	430.00	2,410.00
# 3/0	3E@17.0	KLF	2,010.00	457.00	2,467.00
# 4/0	3E@18.0	KLF	2,430.00	484.00	2,914.00
# 250 KCMIL	4E@20.0	KLF	3,040.00	538.00	3,578.00
# 300 KCMIL	4E@23.0	KLF	3,560.00	618.00	4,178.00
# 350 KCMIL	4E@24.0	KLF	4,140.00	645.00	4,785.00
# 400 KCMIL	4E@25.0	KLF	4,590.00	672.00	5,262.00
# 500 KCMIL	4E@26.0	KLF	5,600.00	699.00	6,299.00
# 600 KCMIL	4E@30.0	KLF	7,610.00	806.00	8,416.00
# 750 KCMIL	4E@32.0	KLF	9,250.00	860.00	10,110.00
#1000 KCMIL	4E@36.0	KLF	13,900.00	968.00	14,868.00

Type MTW 600 volt 90 degree stranded copper building wire

Material	Craft@Hrs	Unit	Material Cost	Labor Cost	Installed Cost
# 18	2E@5.00	KLF	68.30	134.00	202.30
# 16	2E@5.50	KLF	78.50	148.00	226.50
# 14	2E@6.00	KLF	105.00	161.00	266.00
# 12	2E@7.00	KLF	151.00	188.00	339.00
# 10	2E@8.00	KLF	237.00	215.00	452.00
# 8	2E@9.00	KLF	509.00	242.00	751.00

Type MTW heavy insulation 600 volt 90 degree stranded copper building wire

Material	Craft@Hrs	Unit	Material Cost	Labor Cost	Installed Cost
# 14	2E@6.00	KLF	138.00	161.00	299.00
# 12	2E@7.00	KLF	191.00	188.00	379.00
# 10	2E@8.00	KLF	270.00	215.00	485.00
# 8	2E@9.00	KLF	410.00	242.00	652.00
# 6	2E@10.0	KLF	818.00	269.00	1,087.00
# 4	2E@12.0	KLF	1,210.00	323.00	1,533.00

Use these figures to estimate the cost of copper USE RHH-RHW solid and stranded, MTW stranded 90-degree and MTW heavy insulation stranded 90-degree wire installed in conduit under the conditions described on pages 5 and 6. Costs listed are for each 1,000 linear feet installed. The crew is two electricians for sizes up to #4, three electricians for sizes over #4 to #4/0 and four electricians for sizes over #4/0. The labor cost per manhour is $26.88. These costs include fishing string, circuit make-up, splices on wire up to #6, reel set-up, pulling compound, phase identification, circuit testing, layout, material handling, and normal waste. Add for a bonding wire run in non-metallic or flexible conduit, sales tax, delivery, supervision, mobilization, demobilization, cleanup, overhead and profit. Note: These costs are for conduit runs of less than 100 feet and assume that three wires of the same size are pulled at the same time. Type MTW is machine tool wire and can be used for control wiring. A deposit is often required on larger-wire reels. These costs assume that larger reels are returned for credit. Order large-size wire in longest lengths available to reduce waste.

Copper Flexible Wire and Flexible Cords

Material	Craft@Hrs	Unit	Material Cost	Labor Cost	Installed Cost
Type TFFN 90 degree stranded copper wire, 600 volt					
#18, 16 strand	2E@5.00	KLF	41.10	134.00	175.10
#16, 26 strand	2E@5.25	KLF	55.80	141.00	196.80
Type AWN, 90 degree copper appliance wire, 1,000 volt					
#14, 19 strand	2E@5.00	KLF	156.00	134.00	290.00
Type AWN, 105 degree copper appliance wire, 600 volt					
#18 stranded	2E@5.00	KLF	70.00	134.00	204.00
#16 stranded	2E@5.25	KLF	80.00	141.00	221.00
#14 stranded	2E@8.00	KLF	110.00	215.00	325.00
#12 stranded	2E@10.0	KLF	152.00	269.00	421.00
#10 stranded	2E@12.0	KLF	240.00	323.00	563.00
# 8 stranded	2E@14.0	KLF	510.00	376.00	886.00
Type SJ 300 volt 60 degree flexible copper cord					
# 18-2	2E@8.00	KLF	245.00	215.00	460.00
# 16-2	2E@9.00	KLF	270.00	242.00	512.00
# 14-2	2E@10.0	KLF	375.00	269.00	644.00
# 18-3	2E@8.25	KLF	305.00	222.00	527.00
# 16-3	2E@9.25	KLF	380.00	249.00	629.00
# 14-3	2E@10.3	KLF	520.00	277.00	797.00

Use these figures to estimate the cost of copper building wire installed in conduit systems and for flexible cords under the conditions described on pages 5 and 6. Costs listed are for each 1,000 linear feet installed. The crew is two electricians working at a labor cost of $26.88 per manhour. These costs include fishing string, circuit make-up, stripping, phase identification, circuit testing, layout, material handling, and normal waste. Add for a cord connectors, supports, sales tax, delivery, supervision, mobilization, demobilization, cleanup, overhead and profit. Note: Costs for wire pulled in conduit assume conduit runs of less than 100 feet and assume that three wires of the same size are pulled at the same time. Small-gauge wire is usually packed on spools in 500-foot lengths. Some manufacturers offer lengths of 1,500 and 2,500 feet. Flexible cord is usually sold in 250-foot lengths. The *NEC* places limits on how flexible cords can be used. Never pull flexible cord in conduit.

Material	Craft@Hrs	Unit	Material Cost	Labor Cost	Installed Cost

Type SJ 300 volt 60 degree flexible cord

Material	Craft@Hrs	Unit	Material Cost	Labor Cost	Installed Cost
# 18-4	1E@8.50	KLF	483.00	228.00	711.00
# 16-4	1E@9.50	KLF	600.00	255.00	855.00
# 14-4	1E@10.5	KLF	947.00	282.00	1,229.00

Type S 600 volt 60 degree flexible cord

Material	Craft@Hrs	Unit	Material Cost	Labor Cost	Installed Cost
# 18-2	1E@8.00	KLF	343.00	215.00	558.00
# 16-2	1E@9.00	KLF	397.00	242.00	639.00
# 14-2	1E@10.0	KLF	626.00	269.00	895.00
# 12-2	1E@12.0	KLF	815.00	323.00	1,138.00
# 10-2	1E@14.0	KLF	979.00	376.00	1,355.00
# 18-3	1E@8.25	KLF	407.00	222.00	629.00
# 16-3	1E@9.25	KLF	483.00	249.00	732.00
# 14-3	1E@10.3	KLF	788.00	277.00	1,065.00
# 12-3	1E@12.3	KLF	1,010.00	331.00	1,341.00
# 10-3	1E@14.3	KLF	1,300.00	384.00	1,684.00
# 18-4	1E@8.50	KLF	544.00	228.00	772.00
# 16-4	1E@9.50	KLF	680.00	255.00	935.00
# 14-4	1E@10.5	KLF	951.00	282.00	1,233.00
# 12-4	1E@12.5	KLF	1,250.00	336.00	1,586.00
# 10-4	1E@14.5	KLF	1,580.00	390.00	1,970.00

Type SJO 300 volt 60 degree flexible cord

Material	Craft@Hrs	Unit	Material Cost	Labor Cost	Installed Cost
# 18-2	1E@8.00	KLF	334.00	215.00	549.00
# 16-2	1E@9.00	KLF	528.00	242.00	770.00
# 14-2	1E@10.0	KLF	708.00	269.00	977.00
# 18-3	1E@8.25	KLF	361.00	222.00	583.00
# 16-3	1E@9.25	KLF	452.00	249.00	701.00
# 14-3	1E@10.3	KLF	596.00	277.00	873.00

Use these figures to estimate the cost of copper flexible cords installed exposed in buildings under the conditions described on pages 5 and 6. Costs listed are for each 1,000 linear feet installed. The crew is one electrician working at a labor cost of $26.88 per manhour. These costs include circuit make-up, reel set-up, phase identification, circuit testing, stripping, layout, material handling, and normal waste. Add for cord connectors, supports, sales tax, delivery, supervision, mobilization, demobilization, cleanup, overhead and profit. Note: The *NEC* places limits on how flexible cords can be used. Never install flexible cord in conduit.

Flexible Cords

Material	Craft@Hrs	Unit	Material Cost	Labor Cost	Installed Cost
Type SJO 300 volt 60 degree flexible cord					
# 18-4	1E@8.50	KLF	555.00	228.00	783.00
# 16-4	1E@9.50	KLF	646.00	255.00	901.00
# 14-4	1E@10.5	KLF	920.00	282.00	1,202.00
Type SO 600 volt 60 degree flexible cord					
# 18-2	1E@8.00	KLF	475.00	215.00	690.00
# 16-2	1E@9.00	KLF	543.00	242.00	785.00
# 14-2	1E@10.0	KLF	865.00	269.00	1,134.00
# 12-2	1E@12.0	KLF	1,150.00	323.00	1,473.00
# 10-2	1E@14.0	KLF	1,410.00	376.00	1,786.00
# 8-2	1E@15.0	KLF	2,370.00	403.00	2,773.00
# 6-2	1E@16.0	KLF	2,790.00	430.00	3,220.00
# 18-3	1E@8.25	KLF	559.00	222.00	781.00
# 16-3	1E@9.25	KLF	621.00	249.00	870.00
# 14-3	1E@10.3	KLF	1,020.00	277.00	1,297.00
# 12-3	1E@12.3	KLF	1,380.00	331.00	1,711.00
# 10-3	1E@14.3	KLF	1,730.00	384.00	2,114.00
# 8-3	1E@15.3	KLF	2,950.00	411.00	3,361.00
# 6-3	1E@16.3	KLF	3,910.00	438.00	4,348.00
# 18-4	1E@8.50	KLF	772.00	228.00	1,000.00
# 16-4	1E@9.50	KLF	865.00	255.00	1,120.00
# 14-4	1E@10.5	KLF	1,280.00	282.00	1,562.00
# 12-4	1E@12.5	KLF	1,570.00	336.00	1,906.00
# 10-4	1E@14.5	KLF	1,990.00	390.00	2,380.00
# 8-4	1E@15.5	KLF	3,720.00	417.00	4,137.00
# 6-4	1E@16.5	KLF	4,780.00	444.00	5,224.00

Use these figures to estimate the cost of copper flexible cords installed exposed in buildings under the conditions described on pages 5 and 6. Costs listed are for each 1,000 linear feet installed. The crew is one electrician working at a labor cost of $26.88 per manhour. These costs include circuit make-up, reel set-up, phase identification, circuit testing, stripping, layout, material handling, and normal waste. Add for cord connectors, supports, sales tax, delivery, supervision, mobilization, demobilization, cleanup, overhead and profit. Note: The *NEC* places limits on how flexible cords can be used. Never install flexible cord in conduit.

Material	Craft@Hrs	Unit	Material Cost	Labor Cost	Installed Cost

Type STO 600 volt 60 degree oil-resistant flexible cord

Material	Craft@Hrs	Unit	Material Cost	Labor Cost	Installed Cost
# 18-2	1E@8.00	KLF	348.00	215.00	563.00
# 16-2	1E@9.00	KLF	665.00	242.00	907.00
# 14-2	1E@10.0	KLF	1,020.00	269.00	1,289.00
# 12-2	1E@12.0	KLF	1,400.00	323.00	1,723.00
# 10-2	1E@14.0	KLF	1,740.00	376.00	2,116.00
# 18-3	1E@8.25	KLF	680.00	222.00	902.00
# 16-3	1E@9.25	KLF	566.00	249.00	815.00
# 14-3	1E@10.3	KLF	945.00	277.00	1,222.00
# 12-3	1E@12.3	KLF	999.00	331.00	1,330.00
# 10-3	1E@14.3	KLF	1,570.00	384.00	1,954.00
# 18-4	1E@8.50	KLF	938.00	228.00	1,166.00
# 16-4	1E@9.50	KLF	783.00	255.00	1,038.00
# 14-4	1E@10.5	KLF	1,180.00	282.00	1,462.00
# 12-4	1E@12.5	KLF	1,440.00	336.00	1,776.00
# 10-4	1E@14.5	KLF	2,440.00	390.00	2,830.00

Type NM 600 volt non-metallic sheathed cable (Romex) without ground

Material	Craft@Hrs	Unit	Material Cost	Labor Cost	Installed Cost
# 14-2	1E@6.00	KLF	134.00	161.00	295.00
# 12-2	1E@6.50	KLF	184.00	175.00	359.00
# 10-2	1E@7.00	KLF	286.00	188.00	474.00
# 8-2	1E@9.00	KLF	592.00	242.00	834.00
# 6-2	1E@12.0	KLF	1,020.00	323.00	1,343.00
# 14-3	1E@6.50	KLF	226.00	175.00	401.00
# 12-3	1E@7.00	KLF	325.00	188.00	513.00
# 10-3	1E@7.50	KLF	483.00	202.00	685.00
# 8-3	1E@10.0	KLF	960.00	269.00	1,229.00
# 6-3	1E@14.0	KLF	1,420.00	376.00	1,796.00
# 4-3	1E@16.0	KLF	2,320.00	430.00	2,750.00

Use these figures to estimate the cost of copper flexible cords and type NM cable installed in buildings under the conditions described on pages 5 and 6. Costs listed are for each 1,000 linear feet installed. The crew is one electrician working at a labor cost of $26.88 per manhour. These costs include circuit make-up, reel set-up, phase identification, circuit testing, stripping, boring wood studs, layout, material handling, and normal waste. Add for supports, staples, connectors, sales tax, delivery, supervision, mobilization, demobilization, cleanup, overhead and profit. Note: The *NEC* places limits on how flexible cords and NM cable can be used. Non-metallic (NM) cable is limited to residences. Never install flexible cord or non-metallic cable in conduit.

Copper 600 Volt Non-metallic Sheathed Cable

Material	Craft@Hrs	Unit	Material Cost	Labor Cost	Installed Cost
Type NM copper 600 volt non-metallic sheathed cable (Romex) with ground					
#14-2	1E@6.25	KLF	147.00	168.00	315.00
#12-2	1E@6.75	KLF	218.00	181.00	399.00
#10-2	1E@7.25	KLF	354.00	195.00	549.00
# 8-2	1E@9.25	KLF	751.00	249.00	1,000.00
# 6-2	1E@12.5	KLF	1,080.00	336.00	1,416.00
#14-3	1E@6.75	KLF	245.00	181.00	426.00
#12-3	1E@7.00	KLF	352.00	188.00	540.00
#10-3	1E@7.75	KLF	559.00	208.00	767.00
# 8-3	1E@10.3	KLF	1,120.00	277.00	1,397.00
# 6-3	1E@14.0	KLF	1,550.00	376.00	1,926.00
# 4-3	1E@16.0	KLF	2,450.00	430.00	2,880.00
# 2-3	1E@18.0	KLF	3,530.00	484.00	4,014.00
#14-4	1E@7.00	KLF	469.00	188.00	657.00
#12-4	1E@7.25	KLF	704.00	195.00	899.00
#10-4	1E@8.00	KLF	932.00	215.00	1,147.00
Type UF 600 volt copper solid direct burial cable					
#14	1E@4.50	KLF	108.00	121.00	229.00
#12	1E@4.75	KLF	157.00	128.00	285.00
#10	1E@5.00	KLF	239.00	134.00	373.00
# 8	1E@6.00	KLF	334.00	161.00	495.00
Type UF 600 volt copper stranded direct burial cable					
# 8	1E@6.00	KLF	316.00	161.00	477.00
# 6	1E@7.00	KLF	389.00	188.00	577.00
# 4	1E@8.00	KLF	570.00	215.00	785.00
# 2	1E@10.0	KLF	867.00	269.00	1,136.00
# 1	1E@12.0	KLF	1,280.00	323.00	1,603.00
# 1/0	2E@14.0	KLF	1,440.00	376.00	1,816.00
# 2/0	2E@15.0	KLF	1,850.00	403.00	2,253.00
# 3/0	2E@16.0	KLF	2,220.00	430.00	2,650.00
# 4/0	2E@18.0	KLF	2,690.00	484.00	3,174.00

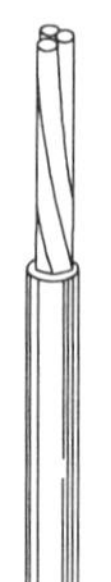

Use these figures to estimate the cost of copper Type NM cable installed in residences and UF cable installed underground under the conditions described on pages 5 and 6. Costs listed are for each 1,000 linear feet installed. The crew is one electrician except for UF cable over #1 when two electricians are needed. The labor cost per manhour is $26.88. These costs include circuit make-up, reel set-up, phase identification, circuit testing, stripping, layout, material handling, and normal waste. Add for connectors, staples, trenching, sales tax, delivery, supervision, mobilization, demobilization, cleanup, overhead and profit. Note: Always bury a warning tape about 12 inches above UF cable.

Copper 600 Volt Non-metallic Sheathed Cable

Material	Craft@Hrs	Unit	Material Cost	Labor Cost	Installed Cost

Type UF 600 volt copper stranded direct burial cable without ground

Material	Craft@Hrs	Unit	Material Cost	Labor Cost	Installed Cost
#14-2	1E@4.75	KLF	179.00	128.00	307.00
#12-2	1E@5.00	KLF	227.00	134.00	361.00
#10-2	1E@5.50	KLF	345.00	148.00	493.00
#14-3	1E@5.00	KLF	250.00	134.00	384.00
#12-3	1E@5.25	KLF	355.00	141.00	496.00
#10-3	1E@5.75	KLF	520.00	155.00	675.00
# 8-3	1E@6.50	KLF	1,410.00	175.00	1,585.00
# 6-3	1E@7.50	KLF	2,070.00	202.00	2,272.00

Type UF 600 volt copper stranded direct burial cable with ground

Material	Craft@Hrs	Unit	Material Cost	Labor Cost	Installed Cost
#14-2	1E@5.00	KLF	194.00	134.00	328.00
#12-2	1E@5.25	KLF	268.00	141.00	409.00
#10-2	1E@6.00	KLF	405.00	161.00	566.00
# 8-2	1E@7.00	KLF	1,180.00	188.00	1,368.00
# 6-2	1E@8.00	KLF	1,780.00	215.00	1,995.00
#14-3	1E@5.25	KLF	284.00	141.00	425.00
#12-3	1E@5.50	KLF	398.00	148.00	546.00
#10-3	1E@6.25	KLF	581.00	168.00	749.00
# 8-3	1E@7.25	KLF	1,550.00	195.00	1,745.00
# 6-3	1E@8.25	KLF	2,240.00	222.00	2,462.00

Type SEU 600 volt copper stranded service entrance cable

Material	Craft@Hrs	Unit	Material Cost	Labor Cost	Installed Cost
# 8-2	1E@10.0	KLF	594.00	269.00	863.00
# 6-2	1E@12.0	KLF	834.00	323.00	1,157.00
# 4-2	1E@14.0	KLF	1,210.00	376.00	1,586.00
#12-3	1E@6.00	KLF	616.00	161.00	777.00
#10-3	1E@8.50	KLF	736.00	228.00	964.00
# 8-3	1E@11.0	KLF	932.00	296.00	1,228.00
# 6-2 & 8	1E@12.5	KLF	1,150.00	336.00	1,486.00
# 6-3	1E@13.0	KLF	1,270.00	349.00	1,619.00
# 4-2 & 6	1E@14.0	KLF	1,660.00	376.00	2,036.00
# 4-3	1E@14.5	KLF	1,870.00	390.00	2,260.00
# 3-2 & 5	1E@15.5	KLF	2,050.00	417.00	2,467.00

Use these figures to estimate the cost of copper UF and SEU cable installed underground, and SEU in buildings, under the conditions described on pages 5 and 6. Costs listed are for each 1,000 linear feet installed. The crew is one electrician working at a labor cost of $26.88 per manhour. These costs include circuit make-up, reel set-up, phase identification, circuit testing, stripping, layout, material handling, and normal waste. Add for trenching supports, connectors, the SEU service entrance cap, sales tax, delivery, supervision, mobilization, demobilization, cleanup, overhead and profit. Note: Always bury a warning tape about 12 inches above UF cable.

Copper 600 Volt Non-metallic Sheathed Cable

Material	Craft@Hrs	Unit	Material Cost	Labor Cost	Installed Cost
Type SEU 600 volt copper service entrance cable					
# 3-3	1E@16.0	KLF	2,360.00	430.00	2,790.00
# 2-2 & 4	1E@17.5	KLF	2,710.00	470.00	3,180.00
# 2-3	1E@18.0	KLF	2,940.00	484.00	3,424.00
# 1-3	1E@20.0	KLF	4,520.00	538.00	5,058.00
# 1/0-3	1E@22.0	KLF	5,030.00	591.00	5,621.00
# 2/0-3	1E@24.0	KLF	5,950.00	645.00	6,595.00
# 3/0-3	1E@26.0	KLF	7,340.00	699.00	8,039.00
# 4/0-3	1E@28.0	KLF	8,220.00	753.00	8,973.00
Soft drawn solid bare copper wire					
#14	1E@4.00	KLF	45.00	108.00	153.00
#12	1E@4.25	KLF	66.00	114.00	180.00
#10	1E@4.50	KLF	98.00	121.00	219.00
# 8	1E@4.75	KLF	185.00	128.00	313.00
# 6	1E@5.00	KLF	275.00	134.00	409.00
# 4	1E@6.00	KLF	415.00	161.00	576.00
Soft drawn stranded bare copper wire					
# 8	1E@4.75	KLF	190.00	128.00	318.00
# 6	1E@5.00	KLF	285.00	134.00	419.00
# 4	1E@6.00	KLF	445.00	161.00	606.00
# 2	1E@8.00	KLF	695.00	215.00	910.00
# 1	1E@10.0	KLF	910.00	269.00	1,179.00
# 1/0	1E@11.0	KLF	1,160.00	296.00	1,456.00
# 2/0	1E@12.0	KLF	1,450.00	323.00	1,773.00
# 3/0	1E@13.0	KLF	1,850.00	349.00	2,199.00
# 4/0	1E@14.0	KLF	2,300.00	376.00	2,676.00

Use these figures to estimate the cost of copper SEU service entrance cable and bare copper wire installed in buildings, under the conditions described on pages 5 and 6. Costs listed are for each 1,000 linear feet installed. The crew is one electrician working at a labor cost of $26.88 per manhour. These costs include circuit make-up, reel set-up, phase identification, circuit testing, stripping, layout, material handling, and normal waste. Add for connectors, supports, sales tax, delivery, supervision, mobilization, demobilization, cleanup, overhead and profit. Note: The *NEC* permits installation of bare copper wire in conduit with conductors. But bare copper wire installed outside conduit must be protected against damage. Some dealers sell bare copper wire by weight rather than length. The table at the bottom of the page shows weights per 1,000 linear feet.

Approximate weight per 1000' for soft drawn bare copper wire (pounds)

Size		Weight	Size		Weight
#14	solid	12.40	#2	stranded	204.90
#12	solid	19.80	#1	stranded	258.40
#10	solid	31.43	#1/0	stranded	325.80
#8	stranded	50.97	#2/0	stranded	410.90
#6	stranded	81.05	#3/0	stranded	518.10
#4	stranded	128.90	#4/0	stranded	653.30

Copper solid armored (BX) cable

Material	Craft@Hrs	Unit	Material Cost	Labor Cost	Installed Cost
#14-2	1E@8.50	KLF	350.00	228.00	578.00
#12-2	1E@9.50	KLF	361.00	255.00	616.00
#10-2	1E@11.0	KLF	637.00	296.00	933.00
#14-3	1E@9.50	KLF	454.00	255.00	709.00
#12-3	1E@10.5	KLF	528.00	282.00	810.00
#10-3	1E@14.0	KLF	831.00	376.00	1,207.00
#14-4	1E@11.0	KLF	610.00	296.00	906.00
#12-4	1E@13.0	KLF	741.00	349.00	1,090.00
#10-4	1E@15.0	KLF	1,250.00	403.00	1,653.00

Copper stranded armored (BX) cable

Material	Craft@Hrs	Unit	Material Cost	Labor Cost	Installed Cost
# 8-2	1E@12.0	KLF	1,210.00	323.00	1,533.00
# 6-2	1E@15.0	KLF	1,650.00	403.00	2,053.00
# 8-3	1E@15.0	KLF	1,530.00	403.00	1,933.00
# 6-3	1E@17.0	KLF	2,210.00	457.00	2,667.00
# 4-3	1E@20.0	KLF	3,190.00	538.00	3,728.00
# 2-3	1E@24.0	KLF	4,540.00	645.00	5,185.00
# 8-4	1E@17.0	KLF	2,120.00	457.00	2,577.00
# 6-4	1E@20.0	KLF	2,850.00	538.00	3,388.00
# 4-4	1E@24.0	KLF	4,240.00	645.00	4,885.00

Copper solid armored (BX) cable with bare copper ground

Material	Craft@Hrs	Unit	Material Cost	Labor Cost	Installed Cost
# 8	1E@10.0	KLF	474.00	269.00	743.00
# 6	1E@12.0	KLF	599.00	323.00	922.00

Copper stranded armored (BX) cable with bare copper ground

Material	Craft@Hrs	Unit	Material Cost	Labor Cost	Installed Cost
# 4	1E@14.0	KLF	879.00	376.00	1,255.00

Use these figures to estimate the cost of copper BX cable installed in buildings under the conditions described on pages 5 and 6. Costs listed are for each 1,000 linear feet installed. The crew is one electrician working at a labor cost of $26.88 per manhour. These costs include boring, notching, stripping, circuit make-up, anti-short bushings, layout, material handling, and normal waste. Add for connectors, supports, sales tax, delivery, supervision, mobilization, demobilization, cleanup, overhead and profit. Note: The *NEC* and your local code restrict the ways BX can be used. Be sure to include anti-short bushings for the cable to be used. These bushings are inserted between the armor and the wire.

Single Conductor Copper Power Cable

Material	Craft@Hrs	Unit	Material Cost	Labor Cost	Installed Cost

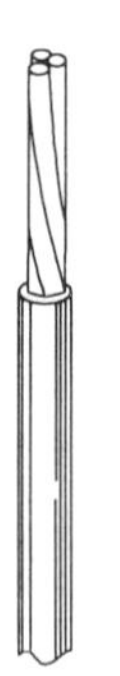

5000 volt cross-linked polyethylene non-shielded copper (XLP) power cable

Material	Craft@Hrs	Unit	Material Cost	Labor Cost	Installed Cost
# 6	2E@14.0	KLF	500.00	376.00	876.00
# 4	2E@16.0	KLF	600.00	430.00	1,030.00
# 2	2E@18.0	KLF	720.00	484.00	1,204.00
# 1/0	3E@22.0	KLF	1,100.00	591.00	1,691.00
# 2/0	3E@24.0	KLF	1,300.00	645.00	1,945.00
# 4/0	3E@28.0	KLF	1,900.00	753.00	2,653.00
#350 KCMIL	3E@34.0	KLF	2,400.00	914.00	3,314.00
#500 KCMIL	3E@36.0	KLF	2,800.00	968.00	3,768.00

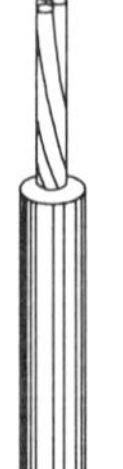

5000 volt cross-linked polyethylene tape shielded copper (XLP) power cable

Material	Craft@Hrs	Unit	Material Cost	Labor Cost	Installed Cost
# 6	2E@14.0	KLF	800.00	376.00	1,176.00
# 4	2E@16.0	KLF	960.00	430.00	1,390.00
# 2	2E@18.0	KLF	1,200.00	484.00	1,684.00
# 1/0	3E@22.0	KLF	1,800.00	591.00	2,391.00
# 2/0	3E@24.0	KLF	2,100.00	645.00	2,745.00
# 4/0	3E@28.0	KLF	3,100.00	753.00	3,853.00
# 350 KCMIL	3E@34.0	KLF	3,900.00	914.00	4,814.00
# 500 KCMIL	3E@36.0	KLF	4,600.00	968.00	5,568.00

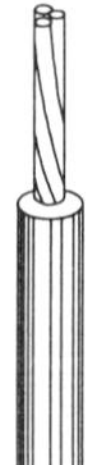

15000 volt ethylene-propylene-rubber tape shielded copper (EPR) power cable

Material	Craft@Hrs	Unit	Material Cost	Labor Cost	Installed Cost
# 2	2E@20.0	KLF	1,800.00	538.00	2,338.00
# 1/0	3E@26.0	KLF	2,000.00	699.00	2,699.00
# 2/0	3E@28.0	KLF	2,100.00	753.00	2,853.00
# 4/0	3E@32.0	KLF	3,300.00	860.00	4,160.00
# 350 KCMIL	3E@38.0	KLF	4,400.00	1,020.00	5,420.00
# 500 KCMIL	3E@42.0	KLF	5,300.00	1,130.00	6,430.00

Use these figures to estimate the cost of medium-voltage copper power cable used for the primary service under the conditions described on pages 5 and 6. Costs listed are for each 1,000 linear feet installed. The crew is two electricians for sizes to #1 and three or four electricians for sizes over #1. Cost per manhour is $26.88. These costs include fishing string, reel set-up, pulling gear set-up, phase identification, pulling compound, layout, material handling, and normal waste. Add for terminations, splices, fire-proofing, high-potential testing, sales tax, delivery, supervision, mobilization, demobilization, cleanup, overhead and profit. Note: These figures assume that cable is pulled in conduit runs of 100 feet or less and that three conductors are pulled at the same time. Keep medium-voltage cable sealed against moisture at all times.

Material	Craft@Hrs	Unit	Material Cost	Labor Cost	Installed Cost

Type THW 600 volt 75 degree stranded aluminum wire

Material	Craft@Hrs	Unit	Material Cost	Labor Cost	Installed Cost
# 6	2E@9.00	KLF	248.00	242.00	490.00
# 4	2E@10.0	KLF	309.00	269.00	578.00
# 2	2E@11.0	KLF	419.00	296.00	715.00
# 1	2E@12.0	KLF	612.00	323.00	935.00
# 1/0	3E@13.0	KLF	713.00	349.00	1,062.00
# 2/0	3E@14.0	KLF	850.00	376.00	1,226.00
# 3/0	3E@15.0	KLF	1,050.00	403.00	1,453.00
# 4/0	3E@16.0	KLF	1,180.00	430.00	1,610.00
# 250 KCMIL	3E@18.0	KLF	1,400.00	484.00	1,884.00
# 300 KCMIL	3E@19.0	KLF	1,940.00	511.00	2,451.00
# 350 KCMIL	3E@20.0	KLF	1,970.00	538.00	2,508.00
# 400 KCMIL	3E@21.0	KLF	2,310.00	564.00	2,874.00
# 500 KCMIL	3E@22.0	KLF	2,540.00	591.00	3,131.00
# 600 KCMIL	3E@23.0	KLF	3,220.00	618.00	3,838.00
# 700 KCMIL	3E@24.0	KLF	3,720.00	645.00	4,365.00
# 750 KCMIL	3E@25.0	KLF	3,750.00	672.00	4,422.00
#1000 KCMIL	3E@30.0	KLF	5,590.00	806.00	6,396.00

Type THHN-THWN 600 volt 90 degree stranded aluminum wire

Material	Craft@Hrs	Unit	Material Cost	Labor Cost	Installed Cost
# 6	2E@9.00	KLF	248.00	242.00	490.00
# 4	2E@10.0	KLF	309.00	269.00	578.00
# 2	2E@11.0	KLF	419.00	296.00	715.00
# 1	2E@12.0	KLF	612.00	323.00	935.00
# 1/0	3E@13.0	KLF	713.00	349.00	1,062.00
# 2/0	3E@14.0	KLF	850.00	376.00	1,226.00
# 3/0	3E@15.0	KLF	1,050.00	403.00	1,453.00
# 4/0	3E@16.0	KLF	1,180.00	430.00	1,610.00
# 250 KCMIL	3E@18.0	KLF	1,400.00	484.00	1,884.00
# 300 KCMIL	3E@19.0	KLF	1,940.00	511.00	2,451.00
# 350 KCMIL	3E@20.0	KLF	1,970.00	538.00	2,508.00
# 400 KCMIL	3E@21.0	KLF	2,310.00	564.00	2,874.00
# 500 KCMIL	3E@22.0	KLF	2,540.00	591.00	3,131.00
# 600 KCMIL	3E@23.0	KLF	3,220.00	618.00	3,838.00
# 700 KCMIL	3E@24.0	KLF	3,720.00	645.00	4,365.00
# 750 KCMIL	3E@25.0	KLF	3,750.00	672.00	4,422.00
#1000 KCMIL	3E@30.0	KLF	5,590.00	806.00	6,396.00

Use these figures to estimate the cost of aluminum THW and THHN-THWN wire installed in conduit under the conditions described on pages 5 and 6. Costs listed are for each 1,000 linear feet installed. The crew is two electricians for sizes to #1 and three electricians for sizes over #1. The labor cost per manhour is $26.88. These costs include fishing string, reel set-up, pulling gear set-up, phase identification, pulling compound, layout, material handling, and normal waste. Add for splicing, anti-oxidation compound, sales tax, delivery, supervision, mobilization, demobilization, cleanup, overhead and profit. Note: These figures assume that wire is pulled in conduit runs of 100 feet or less and that three conductors are pulled at the same time. Use anti-oxidation compound on all aluminum connections. The ampacity of copper wire is greater than the ampacity of aluminum wire of the same size. Check the *NEC* before substituting aluminum wire for copper wire. Terminations for aluminum wire must be made with approved fittings.

See Simple Conversion Table on Page 491

Copper To Aluminum Wire Ampacities

600 Volt Aluminum Wire

Material	Craft@Hrs	Unit	Material Cost	Labor Cost	Installed Cost

Type XHHW cross-linked polyethylene 600 volt 90 degree (XLP) aluminum wire

Material	Craft@Hrs	Unit	Material Cost	Labor Cost	Installed Cost
# 6	2E@9.00	KLF	248.00	242.00	490.00
# 4	2E@10.0	KLF	309.00	269.00	578.00
# 2	2E@11.0	KLF	419.00	296.00	715.00
# 1	2E@12.0	KLF	612.00	323.00	935.00
# 1/0	3E@13.0	KLF	713.00	349.00	1,062.00
# 2/0	3E@14.0	KLF	850.00	376.00	1,226.00
# 3/0	3E@15.0	KLF	1,050.00	403.00	1,453.00
# 4/0	3E@16.0	KLF	1,180.00	430.00	1,610.00
# 250 KCMIL	3E@18.0	KLF	1,400.00	484.00	1,884.00
# 300 KCMIL	3E@19.0	KLF	1,940.00	511.00	2,451.00
# 350 KCMIL	3E@20.0	KLF	1,970.00	538.00	2,508.00
# 400 KCMIL	3E@21.0	KLF	2,310.00	564.00	2,874.00
# 500 KCMIL	3E@22.0	KLF	2,540.00	591.00	3,131.00
# 600 KCMIL	3E@23.0	KLF	3,220.00	618.00	3,838.00
# 700 KCMIL	3E@24.0	KLF	3,720.00	645.00	4,365.00
# 750 KCMIL	3E@25.0	KLF	3,750.00	672.00	4,422.00
#1000 KCMIL	3E@30.0	KLF	5,590.00	806.00	6,396.00

Type USE, RHH-RHW 600 volt cross-linked polyethylene (XLP) aluminum wire

Material	Craft@Hrs	Unit	Material Cost	Labor Cost	Installed Cost
# 6	2E@9.00	KLF	284.00	242.00	526.00
# 4	2E@10.0	KLF	340.00	269.00	609.00
# 2	2E@11.0	KLF	455.00	296.00	751.00
# 1	2E@12.0	KLF	656.00	323.00	979.00
# 1/0	3E@13.0	KLF	755.00	349.00	1,104.00
# 2/0	3E@14.0	KLF	884.00	376.00	1,260.00
# 3/0	3E@15.0	KLF	1,090.00	403.00	1,493.00
# 4/0	3E@16.0	KLF	1,210.00	430.00	1,640.00
# 250 KCMIL	3E@18.0	KLF	1,540.00	484.00	2,024.00
# 300 KCMIL	3E@19.0	KLF	2,100.00	511.00	2,611.00
# 350 KCMIL	3E@20.0	KLF	2,150.00	538.00	2,688.00
# 400 KCMIL	3E@21.0	KLF	2,500.00	564.00	3,064.00
# 500 KCMIL	3E@22.0	KLF	2,750.00	591.00	3,341.00
# 600 KCMIL	3E@23.0	KLF	3,510.00	618.00	4,128.00
# 700 KCMIL	3E@24.0	KLF	4,040.00	645.00	4,685.00
# 750 KCMIL	3E@25.0	KLF	4,100.00	672.00	4,772.00
#1000 KCMIL	3E@30.0	KLF	6,010.00	806.00	6,816.00

Use these figures to estimate the cost of aluminum XHHW wire installed in conduit or USE RHH-RHW wire installed under the conditions described on pages 5 and 6. Costs listed are for each 1,000 linear feet installed. The crew is two electricians for sizes to #1 and three electricians for sizes over #1. The labor cost per manhour is $26.88. These costs include fishing string, reel set-up, pulling gear set-up, phase identification, pulling compound, layout, material handling, and normal waste. Add for splicing, anti-oxidation compound, sales tax, delivery, supervision, mobilization, demobilization, cleanup, overhead and profit. Note: The figures for XHHW wire assume that wire is pulled in conduit runs of 100 feet or less and that three conductors are pulled at the same time. Use anti-oxidation compound on all aluminum connections. Type USE, RHH-RHW wire is approved for use overhead, direct burial or in underground duct systems. The ampacity of copper wire is greater than the ampacity of aluminum wire of the same size. Check the *NEC* before substituting aluminum wire for copper wire. Terminations for aluminum wire must be made with approved fittings.

Material	Craft@Hrs	Unit	Material Cost	Labor Cost	Installed Cost

Type SEU 600 volt plastic jacket service entrance aluminum cable

Material	Craft@Hrs	Unit	Material Cost	Labor Cost	Installed Cost
# 8-3	2E@10.0	KLF	1,190.00	269.00	1,459.00
# 6-3	2E@11.0	KLF	1,250.00	296.00	1,546.00
# 4-2 & 6	2E@12.0	KLF	1,580.00	323.00	1,903.00
# 4-3	2E@13.0	KLF	1,660.00	349.00	2,009.00
# 2-2 & 4	2E@14.0	KLF	1,990.00	376.00	2,366.00
# 1-2 & 3	2E@16.0	KLF	2,780.00	430.00	3,210.00
# 1-3	2E@17.0	KLF	2,930.00	457.00	3,387.00
# 1/0-2 & 2	2E@18.0	KLF	3,050.00	484.00	3,534.00
# 1/0-3	2E@19.0	KLF	3,190.00	511.00	3,701.00
# 2/0-2 & 1	2E@20.0	KLF	3,450.00	538.00	3,988.00
# 2/0-3	2E@21.0	KLF	3,710.00	564.00	4,274.00
# 3/0-2 & 1/0	2E@22.0	KLF	4,380.00	591.00	4,971.00
# 3/0-3	2E@23.0	KLF	4,770.00	618.00	5,388.00
# 4/0-2 & 2/0	2E@24.0	KLF	4,920.00	645.00	5,565.00
# 4/0-3	2E@25.0	KLF	5,170.00	672.00	5,842.00

Type SE-SER 600 volt plastic jacket aluminum cable

Material		Craft@Hrs	Unit	Material Cost	Labor Cost	Installed Cost
# 8-3	#2 Gr	2E@11.0	KLF	1,450.00	296.00	1,746.00
# 6-3	#6 Gr	2E@12.0	KLF	1,730.00	323.00	2,053.00
# 4-3	#6 Gr	2E@14.0	KLF	1,990.00	376.00	2,366.00
# 2-3	#4 Gr	2E@16.0	KLF	2,780.00	430.00	3,210.00
# 1-3	#3 Gr	2E@18.0	KLF	4,110.00	484.00	4,594.00
# 1/0-3	#2 Gr	2E@20.0	KLF	4,250.00	538.00	4,788.00
# 2/0-3	#2 Gr	2E@22.0	KLF	4,920.00	591.00	5,511.00
# 3/0-3	#1/0 Gr	2E@24.0	KLF	6,120.00	645.00	6,765.00
# 4/0-3	#2/0 Gr	2E@26.0	KLF	7,170.00	699.00	7,869.00

Weatherproof polyethylene 600 volt solid aluminum wire

	Code Name	Craft@Hrs	Unit	Material Cost	Labor Cost	Installed Cost
#6	Apple	2E@8.00	KLF	185.00	215.00	400.00
#4	Pear	2E@9.00	KLF	185.00	242.00	427.00
#2	Cherry	2E@10.0	KLF	311.00	269.00	580.00

Use these figures to estimate the cost of aluminum service entrance cable installed under the conditions described on pages 5 and 6. Costs listed are for each 1,000 linear feet installed. The crew is two electricians working at the cost of $26.88 per manhour. These costs include stripping, phase identification, layout, material handling, and normal waste. Add for service entrance cap, anti-oxidation compound, supports, sales tax, delivery, supervision, mobilization, demobilization, cleanup, overhead and profit. Note: Use anti-oxidation compound on all aluminum connections. The ampacity of copper wire is greater than the ampacity of aluminum wire of the same size. Check the *NEC* before substituting aluminum wire for copper wire. Terminations for aluminum wire must be made with approved fittings.

600 Volt Aluminum Wire

Material			Craft@Hrs	Unit	Material Cost	Labor Cost	Installed Cost

Weatherproof polyethylene 600 volt stranded aluminum wire

Material	Code Name	Strands	Craft@Hrs	Unit	Material Cost	Labor Cost	Installed Cost
#6	Plum	7	2E@8.00	KLF	245.00	215.00	460.00
#4	Apricot	7	2E@9.00	KLF	215.00	242.00	457.00
#2	Peach	7	2E@9.25	KLF	325.00	249.00	574.00
#1/0	Quince	7	4E@10.0	KLF	504.00	269.00	773.00
#2/0	Orange	7	4E@10.3	KLF	609.00	277.00	886.00
#3/0	Fig	7	4E@12.0	KLF	800.00	323.00	1,123.00
#4/0	Olive	7	4E@13.0	KLF	886.00	349.00	1,235.00

600 volt aluminum conductor steel reinforced (ACSR) aluminum wire

Material	Code Name	Strands	Craft@Hrs	Unit	Material Cost	Labor Cost	Installed Cost
#6	Walnut	6	4E@8.00	KLF	222.00	215.00	437.00
#4	Butternut	6	4E@9.00	KLF	273.00	242.00	515.00
#4	Hickory	7	4E@9.00	KLF	293.00	242.00	535.00
#2	Pignut	6	4E@10.0	KLF	394.00	269.00	663.00
#2	Beech	7	4E@10.0	KLF	435.00	269.00	704.00
#1/0	Almond	6	5E@12.0	KLF	616.00	323.00	939.00
#2/0	Pecan	6	5E@13.0	KLF	754.00	349.00	1,103.00
#3/0	Filbert	6	5E@14.0	KLF	915.00	376.00	1,291.00
#4/0	Buckeye	6	5E@15.0	KLF	1,170.00	403.00	1,573.00

600 volt cross-linked polyethylene (XLP) solid aluminum wire

Material	Code Name	Craft@Hrs	Unit	Material Cost	Labor Cost	Installed Cost
#6	Apple-XLP	2E@8.00	KLF	196.00	215.00	411.00
#4	Pear-XLP	2E@9.00	KLF	219.00	242.00	461.00
#2	Cherry-XLP	2E@10.0	KLF	409.00	269.00	678.00

Use these figures to estimate the cost of aluminum wire installed on overhead supports under the conditions described on pages 5 and 6. Costs listed are for each 1,000 linear feet installed. For wire other than ACSR, the crew is two electricians for wire up to #2 and four electricians for wire over #2. For ACSR wire on overhead poles, use four electricians on wire to #2 and five electricians on wire over #2. The labor cost is $26.88 per manhour. These costs include reel set-up, pulling gear set-up, tensioning, layout, material handling, and normal waste. Add for insulators, terminations, line hardware, anti-oxidation compound, pre-formed ties, sales tax, delivery, supervision, mobilization, demobilization, cleanup, overhead and profit. Note: Use anti-oxidation compound on all aluminum connections, fittings and lugs. The ampacity of copper wire is greater than the ampacity of aluminum wire of the same size. Check the *NEC* before substituting aluminum wire for copper wire. Terminations for aluminum wire must be made with approved fittings.

Material		Craft@Hrs	Unit	Material Cost	Labor Cost	Installed Cost

600 volt cross-linked polyethylene (XLP) stranded aluminum wire

Material	Code Name	Strands	Craft@Hrs	Unit	Material Cost	Labor Cost	Installed Cost
#6	Plum-XLP	7	2E@8.00	KLF	178.00	215.00	393.00
#4	Apricot-XLP	7	2E@9.00	KLF	246.00	242.00	488.00
#2	Peach-XLP	7	2E@10.0	KLF	353.00	269.00	622.00
#1/0	Quince-XLP	7	4E@12.0	KLF	532.00	323.00	855.00
#2/0	Orange-XLP	7	4E@13.0	KLF	630.00	349.00	979.00
#3/0	Fig-XLP	7	4E@14.0	KLF	825.00	376.00	1,201.00
#4/0	Olive-XLP	7	4E@15.0	KLF	925.00	403.00	1,328.00

600 volt cross-linked polyethylene ACSR aluminum wire

Material	Code Name	Strands	Craft@Hrs	Unit	Material Cost	Labor Cost	Installed Cost
#6	Walnut-XLP	6	4E@8.00	KLF	299.00	215.00	514.00
#4	Butternut-XLP	6	4E@9.00	KLF	331.00	242.00	573.00
#4	Hickory-XLP	7	4E@9.00	KLF	347.00	242.00	589.00
#2	Pignut-XLP	6	4E@10.0	KLF	474.00	269.00	743.00
#2	Beech-XLP	7	4E@10.0	KLF	578.00	269.00	847.00
#1/0	Almond-XLP	6	5E@12.0	KLF	670.00	323.00	993.00
#2/0	Pecan-XLP	6	5E@13.0	KLF	813.00	349.00	1,162.00
#3/0	Filbert-XLP	6	5E@14.0	KLF	1,090.00	376.00	1,466.00
#4/0	Buckeye-XLP	6	5E@15.0	KLF	1,320.00	403.00	1,723.00

600 volt polyethylene duplex aluminum service drop wire

Material	Code Name	Strands	Craft@Hrs	Unit	Material Cost	Labor Cost	Installed Cost
#6	Pekingese	1	2E@12.0	KLF	311.00	323.00	634.00
#6	Collie	7	2E@11.0	KLF	351.00	296.00	647.00
#4	Spaniel	7	2E@14.0	KLF	454.00	376.00	830.00
#2	Doberman	7	2E@16.0	KLF	736.00	430.00	1,166.00
#1/0	Malamute	19	4E@18.0	KLF	1,180.00	484.00	1,664.00

600 volt triplex aluminum service drop wire, ACSR

Material	Code Name	Strands	Craft@Hrs	Unit	Material Cost	Labor Cost	Installed Cost
#4	Oyster	7	4E@14.0	KLF	696.00	376.00	1,072.00
#2	Clam	7	4E@16.0	KLF	948.00	430.00	1,378.00
#1/0	Murex	7	5E@18.0	KLF	1,420.00	484.00	1,904.00
#2/0	Nassa	7	5E@20.0	KLF	1,710.00	538.00	2,248.00
#4/0	Portunas	19	5E@24.0	KLF	2,610.00	645.00	3,255.00

Use these figures to estimate the cost of aluminum wire installed on overhead supports under the conditions described on pages 5 and 6. Costs listed are for each 1,000 linear feet installed. For wire other than ACSR, the crew is two electricians for wire up to #2 and four electricians for wire over #2. For ACSR wire on overhead poles, use four electricians on wire to #2 and five electricians on wire over #2. The labor cost is $26.88 per manhour. These costs include reel set-up, pulling gear set-up, tensioning, layout, material handling, and normal waste. Add for insulators, terminations, line hardware, anti-oxidation compound, pre-formed ties, sales tax, delivery, supervision, mobilization, demobilization, cleanup, overhead and profit. Note: Use anti-oxidation compound on all aluminum connections, fittings and lugs. The ampacity of copper wire is greater than the ampacity of aluminum wire of the same size. Check the *NEC* before substituting aluminum wire for copper wire. Terminations for aluminum wire must be made with approved fittings.

600 Volt Aluminum Service Drop Wire

Material			Craft@Hrs	Unit	Material Cost	Labor Cost	Installed Cost

600 volt triplex aluminum service drop wire, ACSR

Material	Code Name	Strands	Craft@Hrs	Unit	Material Cost	Labor Cost	Installed Cost
#6	Paludina	1	3E@14.0	KLF	443.00	376.00	819.00
#6	Voluta	7	3E@14.0	KLF	558.00	376.00	934.00
#4	Periwinkle	7	3E@16.0	KLF	733.00	430.00	1,163.00
#2	Conch	7	3E@18.0	KLF	981.00	484.00	1,465.00
#1/0	Neritina	7	3E@20.0	KLF	1,510.00	538.00	2,048.00
#1/0	Cenia	19	3E@20.0	KLF	1,570.00	538.00	2,108.00
#2/0	Runcina	7	3E@24.0	KLF	1,800.00	645.00	2,445.00
#4/0	Zurara	19	3E@28.0	KLF	2,770.00	753.00	3,523.00

600 volt triplex aluminum service drop wire with reduced neutral

Material	Code Name	Strands	Craft@Hrs	Unit	Material Cost	Labor Cost	Installed Cost
#4	Scallop	1	3E@16.0	KLF	650.00	430.00	1,080.00
#4	Strombus	7	3E@16.0	KLF	677.00	430.00	1,107.00
#2	Cockle	7	3E@18.0	KLF	893.00	484.00	1,377.00
#1/0	Janthina	7	3E@20.0	KLF	1,350.00	538.00	1,888.00
#2/0	Clio	19	3E@24.0	KLF	1,660.00	645.00	2,305.00
#4/0	Cerapus	19	3E@28.0	KLF	2,430.00	753.00	3,183.00

600 volt quadruplex aluminum service drop wire

Material	Code Name	Strands	Craft@Hrs	Unit	Material Cost	Labor Cost	Installed Cost
#4	Pinto	7	3E@18.0	KLF	928.00	484.00	1,412.00
#2	Mustang	7	3E@20.0	KLF	1,280.00	538.00	1,818.00
#1/0	Criollo	7	3E@24.0	KLF	2,120.00	645.00	2,765.00
#2/0	Percheron	19	3E@26.0	KLF	2,510.00	699.00	3,209.00
#4/0	Oldenberg	19	3E@30.0	KLF	3,660.00	806.00	4,466.00

Use these figures to estimate the cost of aluminum service drop wire installed on overhead supports under the conditions described on pages 5 and 6. Costs listed are for each 1,000 linear feet installed. The crew is three electricians working at a labor rate of $26.88 per manhour. These costs include reel set-up, pulling gear set-up, tensioning, layout, material handling, and normal waste. Add for insulators, terminations, line hardware, anti-oxidation compound, pre-formed ties, sales tax, delivery, supervision, mobilization, demobilization, cleanup, overhead and profit. Note: Use anti-oxidation compound on all aluminum connections, fittings and lugs. The ampacity of copper wire is greater than the ampacity of aluminum wire of the same size. Check the *NEC* before substituting aluminum wire for copper wire. Terminations for aluminum wire must be made with approved fittings. Multi-conductor service drop is factory-twisted and sold on spools of 500 and 1,000 feet. Many suppliers will sell service drop wire in special lengths at a modest additional charge.

Material		Craft@Hrs	Unit	Material Cost	Labor Cost	Installed Cost

600 volt quadruplex aluminum service drop wire, ACSR neutral

	Code Name	Strands	Craft@Hrs	Unit	Material Cost	Labor Cost	Installed Cost
#6	Chola	7	3E@16.0	KLF	762.00	430.00	1,192.00
#4	Hackney	7	3E@18.0	KLF	958.00	484.00	1,442.00
#2	Palomino	7	3E@20.0	KLF	1,310.00	538.00	1,848.00
#1/0	Costena	19	3E@24.0	KLF	2,200.00	645.00	2,845.00
#2/0	Grullo	19	3E@26.0	KLF	2,610.00	699.00	3,309.00
#3/0	Suffolk	19	3E@28.0	KLF	3,200.00	753.00	3,953.00
#4/0	Appaloosa	19	3E@30.0	KLF	3,860.00	806.00	4,666.00

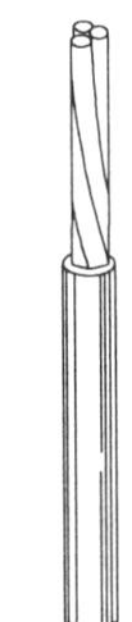

600 volt cross-linked polyethylene duplex aluminum service drop wire with aluminum neutral

	Code Name	Strands	Craft@Hrs	Unit	Material Cost	Labor Cost	Installed Cost
#6	Pekingese	1	3E@13.0	KLF	385.00	349.00	734.00
#6	Collie	7	3E@14.0	KLF	417.00	376.00	793.00
#4	Spaniel	7	3E@15.0	KLF	486.00	403.00	889.00
#2	Doberman	7	3E@17.0	KLF	767.00	457.00	1,224.00
#1/0	Malamute	19	3E@19.0	KLF	1,240.00	511.00	1,751.00

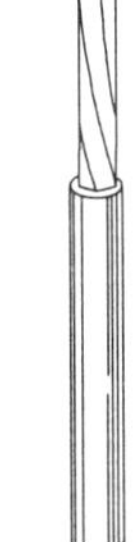

600 volt cross-linked polyethylene duplex aluminum service drop wire with ACSR neutral

	Code Name	Strands	Craft@Hrs	Unit	Material Cost	Labor Cost	Installed Cost
#6	Setter	1	3E@13.0	KLF	328.00	349.00	677.00
#6	Shepherd	7	3E@14.0	KLF	442.00	376.00	818.00
#4	Terrier	7	3E@15.0	KLF	507.00	403.00	910.00
#2	Chow	7	3E@17.0	KLF	873.00	457.00	1,330.00

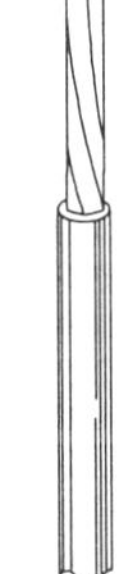

600 volt cross-linked polyethylene triplex aluminum service drop wire with aluminum neutral

	Code Name	Strands	Craft@Hrs	Unit	Material Cost	Labor Cost	Installed Cost
#4	Oyster	7	3E@16.0	KLF	767.00	430.00	1,197.00
#2	Clam	7	3E@18.0	KLF	1,010.00	484.00	1,494.00
#1/0	Murex	7	3E@20.0	KLF	1,470.00	538.00	2,008.00
#2/0	Nassa	7	3E@24.0	KLF	1,770.00	645.00	2,415.00
#4/0	Portunas	19	3E@28.0	KLF	2,700.00	753.00	3,453.00

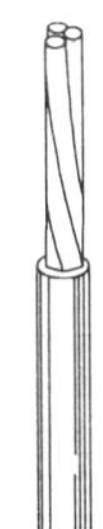

Use these figures to estimate the cost of aluminum service drop wire installed on overhead supports under the conditions described on pages 5 and 6. Costs listed are for each 1,000 linear feet installed. The crew is three electricians working at a labor rate of $26.88 per manhour. These costs include reel set-up, pulling gear set-up, tensioning, layout, material handling, and normal waste. Add for insulators, terminations, line hardware, anti-oxidation compound, pre-formed ties, sales tax, delivery, supervision, mobilization, demobilization, cleanup, overhead and profit. Note: Use anti-oxidation compound on all aluminum connections, fittings and lugs. The ampacity of copper wire is greater than the ampacity of aluminum wire of the same size. Check the *NEC* before substituting aluminum wire for copper wire. Terminations for aluminum wire must be made with approved fittings. Multi-conductor service drop is factory-twisted and sold on spools of 500 and 1,000 feet. Many suppliers will sell service drop wire in special lengths at a modest additional charge.

600 Volt Aluminum Service Drop Wire

Material		Craft@Hrs	Unit	Material Cost	Labor Cost	Installed Cost

Cross-linked polyethylene (XLP) triplex aluminum service drop wire with ACSR neutral

Material	Code Name	Strands	Craft@Hrs	Unit	Material Cost	Labor Cost	Installed Cost
#6	Paludina	1	3E@14.0	KLF	561.00	376.00	937.00
#6	Voluta	7	3E@14.0	KLF	623.00	376.00	999.00
#4	Periwinkle	7	3E@16.0	KLF	796.00	430.00	1,226.00
#2	Conch	7	3E@18.0	KLF	1,050.00	484.00	1,534.00
#1/0	Neritina	7	3E@20.0	KLF	1,540.00	538.00	2,078.00
#1/0	Cenia	19	3E@20.0	KLF	1,630.00	538.00	2,168.00
#2/0	Runcina	7	3E@24.0	KLF	1,850.00	645.00	2,495.00
#4/0	Zuzara	19	3E@28.0	KLF	2,810.00	753.00	3,563.00

Cross-linked polyethylene (XLP) triplex aluminum service drop wire with reduced neutral

Material	Code Name	Strands	Craft@Hrs	Unit	Material Cost	Labor Cost	Installed Cost
#4	Scallop	6	3E@16.0	KLF	690.00	430.00	1,120.00
#4	Strombus	6	3E@16.0	KLF	763.00	430.00	1,193.00
#2	Cockle	4	3E@18.0	KLF	975.00	484.00	1,459.00
#1/0	Janthina	2	3E@20.0	KLF	1,400.00	538.00	1,938.00
#2/0	Clio	1	3E@24.0	KLF	1,740.00	645.00	2,385.00
#4/0	Cerapus	2/0	3E@28.0	KLF	2,530.00	753.00	3,283.00

Cross-linked polyethylene (XLP) quadruplex aluminum service drop wire with aluminum neutral

Material	Code Name	Strands	Craft@Hrs	Unit	Material Cost	Labor Cost	Installed Cost
#4	Pinto	4	3E@18.0	KLF	990.00	484.00	1,474.00
#2	Mustang	4	3E@20.0	KLF	1,350.00	538.00	1,888.00
#1/0	Criollo	1/0	3E@24.0	KLF	2,220.00	645.00	2,865.00
#2/0	Percheron	2/0	3E@26.0	KLF	2,550.00	699.00	3,249.00
#4/0	Oldenberg	4/0	3E@30.0	KLF	3,760.00	806.00	4,566.00

Cross-linked polyethylene (XLP) quadruplex aluminum service drop wire with ACSR neutral

Material	Code Name	Strands	Craft@Hrs	Unit	Material Cost	Labor Cost	Installed Cost
#6	Chola	6	3E@16.0	KLF	850.00	430.00	1,280.00
#4	Hackney	4	3E@18.0	KLF	1,020.00	484.00	1,504.00
#2	Palomino	2	3E@20.0	KLF	1,390.00	538.00	1,928.00
#1/0	Costena	1/0	3E@24.0	KLF	2,300.00	645.00	2,945.00
#2/0	Grullo	2/0	3E@26.0	KLF	2,630.00	699.00	3,329.00
#3/0	Suffolk	3/0	3E@28.0	KLF	3,250.00	753.00	4,003.00
#4/0	Appaloosa	4/0	3E@30.0	KLF	3,940.00	806.00	4,746.00

Use these figures to estimate the cost of aluminum service drop wire installed on overhead supports under the conditions described on pages 5 and 6. Costs listed are for each 1,000 linear feet installed. The crew is three electricians working at a labor rate of $26.88 per manhour. These costs include reel set-up, pulling gear set-up, tensioning, layout, material handling, and normal waste. Add for insulators, terminations, line hardware, anti-oxidation compound, pre-formed ties, sales tax, delivery, supervision, mobilization, demobilization, cleanup, overhead and profit. Note: Use anti-oxidation compound on all aluminum connections, fittings and lugs. The ampacity of copper wire is greater than the ampacity of aluminum wire of the same size. Check the *NEC* before substituting aluminum wire for copper wire. Terminations for aluminum wire must be made with approved fittings. Multi-conductor service drop is factory-twisted and sold on spools of 500 and 1,000 feet. Many suppliers will sell service drop wire in special lengths at a modest additional charge.

Aluminum Type URD 600 Volt Underground Distribution Cable

Material			Craft@Hrs	Unit	Material Cost	Labor Cost	Installed Cost

Type URD 600 volt aluminum underground cable, two phase conductors

Material	Code Name	Strands	Craft@Hrs	Unit	Material Cost	Labor Cost	Installed Cost
#4	Taft	4	4E@10.0	KLF	1,030.00	269.00	1,299.00
#2	Wells	4	4E@11.0	KLF	1,220.00	296.00	1,516.00
#2	Juilliard	2	4E@11.0	KLF	1,320.00	296.00	1,616.00
#1/0	Marion	2	4E@12.0	KLF	1,780.00	323.00	2,103.00
#1/0	Montclair	1/0	4E@13.0	KLF	2,050.00	349.00	2,399.00
#2/0	Bliss	1/0	4E@14.0	KLF	2,230.00	376.00	2,606.00
#2/0	Bloomfield	2/0	4E@15.0	KLF	2,380.00	403.00	2,783.00
#3/0	Whittier	1/0	4E@16.0	KLF	2,620.00	430.00	3,050.00
#3/0	Pace	3/0	4E@17.0	KLF	2,930.00	457.00	3,387.00
#4/0	Regis	2/0	4E@18.0	KLF	3,050.00	484.00	3,534.00
#4/0	Manhattan	4/0	4E@19.0	KLF	3,390.00	511.00	3,901.00
#250	Adelphi	3/0	4E@20.0	KLF	3,840.00	538.00	4,378.00
#350	Concordia	4/0	4E@24.0	KLF	4,850.00	645.00	5,495.00

Type URD aluminum underground cable, triplex, black ground with yellow stripe

Material	Code Name	Strands	Craft@Hrs	Unit	Material Cost	Labor Cost	Installed Cost
#4	Vassar	4	4E@12.0	KLF	966.00	323.00	1,289.00
#2	Stephens	4	4E@13.0	KLF	1,180.00	349.00	1,529.00
#2	Ramapo	2	4E@14.0	KLF	1,250.00	376.00	1,626.00
#1/0	Brenau	2	4E@15.0	KLF	1,730.00	403.00	2,133.00
#1/0	Bergen	1/0	4E@16.0	KLF	1,960.00	430.00	2,390.00
#2/0	Converse	1	4E@17.0	KLF	2,090.00	457.00	2,547.00
#2/0	Hunter	2/0	4E@18.0	KLF	2,200.00	484.00	2,684.00
#3/0	Hollins	1/0	4E@19.0	KLF	2,350.00	511.00	2,861.00
#3/0	Rockland	3/0	4E@20.0	KLF	2,510.00	538.00	3,048.00
#4/0	Sweetbriar	4/0	4E@21.0	KLF	2,760.00	564.00	3,324.00
#4/0	Monmouth	4/0	4E@22.0	KLF	3,090.00	591.00	3,681.00
#250	Pratt	3/0	4E@24.0	KLF	3,500.00	645.00	4,145.00
#250	Wesleyan	4/0	4E@26.0	KLF	4,460.00	699.00	5,159.00

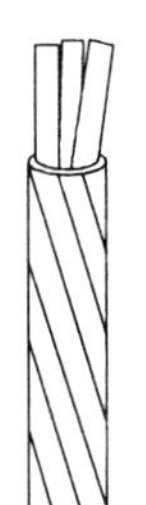

Use these figures to estimate the cost of aluminum underground distribution cable installed in an open trench under the conditions described on pages 5 and 6. Costs listed are for each 1,000 linear feet installed. The crew is four electricians working at a labor rate of $26.88 per manhour. These costs include reel set-up in a pickup truck, layout, material handling, and normal waste. Add for terminations, splicing, warning tape, hardware, anti-oxidation compound, trenching, sales tax, delivery, supervision, mobilization, demobilization, cleanup, overhead and profit. Note: Use anti-oxidation compound on all aluminum connections. The ampacity of copper wire is greater than the ampacity of aluminum wire of the same size. Check the *NEC* before substituting aluminum wire for copper wire. Terminations for aluminum wire must be made in approved fittings.

Aluminum Wire and Steel Messenger Strand

Material			Craft@Hrs	Unit	Material Cost	Labor Cost	Installed Cost

Bare aluminum conductor

	Code Name	Strands	Craft@Hrs	Unit	Material Cost	Labor Cost	Installed Cost
#4	Swan	6	2E@5.50	KLF	343.00	148.00	491.00
#4	Swanate	7	2E@5.50	KLF	328.00	148.00	476.00
#2	Sparrow	6	2E@6.50	KLF	308.00	175.00	483.00
#2	Sparate	7	2E@6.50	KLF	305.00	175.00	480.00
#1/0	Raven	6	4E@9.00	KLF	299.00	242.00	541.00
#2/0	Quail	6	4E@10.0	KLF	296.00	269.00	565.00
#3/0	Pigeon	6	4E@11.0	KLF	294.00	296.00	590.00
#4/0	Penguin	6	4E@12.7	KLF	293.00	341.00	634.00
266, #800	Partridge	6	4E@13.0	KLF	361.00	349.00	710.00
336, #400	Merlin	18	4E@15.0	KLF	323.00	403.00	726.00
336, #400	Linnet	26	4E@16.0	KLF	332.00	430.00	762.00

Galvanized steel messenger strand

Common Grade	Tensile Strength	Craft@Hrs	Unit	Material Cost	Labor Cost	Installed Cost
1/4"	3150	2E@6.00	KLF	141.00	161.00	302.00
5/16"	5350	2E@7.00	KLF	195.00	188.00	383.00
3/8"	6950	2E@8.00	KLF	221.00	215.00	436.00
7/16"	9350	2E@9.00	KLF	340.00	242.00	582.00
1/2"	12100	2E@10.0	KLF	429.00	269.00	698.00
High Strength	**Tensile Strength**					
1/4"	4750	2E@6.25	KLF	143.00	168.00	311.00
5/16"	8000	2E@7.25	KLF	196.00	195.00	391.00
3/8"	10800	2E@8.25	KLF	222.00	222.00	444.00
7/16"	14500	2E@9.25	KLF	344.00	249.00	593.00
1/2"	18800	2E@10.3	KLF	436.00	277.00	713.00
Extra High Strength	**Tensile Strength**					
1/4"	6650	2E@6.50	KLF	143.00	175.00	318.00
5/16"	11200	2E@7.50	KLF	196.00	202.00	398.00
3/8"	15400	2E@8.50	KLF	222.00	228.00	450.00
7/16"	20800	2E@9.50	KLF	344.00	255.00	599.00
1/2"	26900	2E@10.5	KLF	436.00	282.00	718.00
Utility Grade	**Tensile Strength**					
5/16"	6000	2E@8.00	KLF	211.00	215.00	426.00
3/8"	11500	2E@9.00	KLF	227.00	242.00	469.00
7/16"	18000	2E@10.0	KLF	351.00	269.00	620.00
1/2"	25000	2E@12.0	KLF	443.00	323.00	766.00

Use these figures to estimate the cost of bare aluminum wire and galvanized steel messenger strand installed in overhead line construction under the conditions described on pages 5 and 6. Costs listed are for each 1,000 linear feet installed. The crew is two electricians for wire up to #2 and four electricians for wire over #2. The labor cost per manhour is $26.88. These costs include reel set-up, pulling gear set-up, tensioning, layout, material handling, and normal waste. Add for insulators, line hardware, anti-oxidation compound, pre-formed ties, sales tax, delivery, supervision, mobilization, demobilization, cleanup, overhead and profit. Note: Use anti-oxidation compound on all aluminum connections. The ampacity of copper wire is greater than the ampacity of aluminum wire of the same size. Check the *NEC* before substituting aluminum wire for copper wire. Terminations for aluminum wire must be made in approved fittings.

Weights of galvanized steel messenger strand per 1000 feet for common grade, high strength, extra high strength and utility grades are: 1/4" - 121 pounds; 5/16" - 205 pounds; 3/8" - 273 pounds; 7/16" - 399 pounds; and 1/2" - 517 pounds.

Material	Craft@Hrs	Unit	Material Cost	Labor Cost	Installed Cost
Insulated screw-on connectors					
Wire Size					
#22 - #12	1E@3.00	100	7.43	80.60	88.03
#22 - #12	1E@15.0	500	29.60	403.00	432.60
#18 - #10	1E@3.00	100	9.72	80.60	90.32
#18 - #10	1E@15.0	500	38.70	403.00	441.70
#14 - # 8	1E@3.00	100	20.30	80.60	100.90
#14 - # 8	1E@15.0	500	83.30	403.00	486.30
#12 - # 6	1E@3.00	100	33.90	80.60	114.50
Insulated screw-on self stripping connectors					
Wire Size					
#22 - #18	1E@3.00	100	12.30	80.60	92.90
#22 - #18	1E@15.0	500	52.80	403.00	455.80
#18 - #14	1E@3.00	100	12.90	80.60	93.50
#18 - #14	1E@15.0	500	55.10	403.00	458.10
#18 - #14	1E@30.0	1000	88.40	806.00	894.40
#12 - #10	1E@3.00	100	16.00	80.60	96.60
Insulated crimp sleeves					
Wire Size					
#22 - #14	1E@3.00	100	10.50	80.60	91.10
#22 - #14	1E@30.0	1000	59.90	806.00	865.90
#18 - #10	1E@1.50	50	16.10	40.30	56.40
#18 - #10	1E@15.0	500	40.80	403.00	443.80
Uninsulated crimp sleeves					
Wire Size					
#18 - #10	1E@3.00	100	6.32	80.60	86.92
#18 - #10	1E@30.0	1000	60.60	806.00	866.60

Use these figures to estimate the cost of wire connectors installed under the conditions described on pages 5 and 6. Costs listed are for 50, 100, 500 or 1,000 units. The crew is one electrician working at a labor cost of $26.88 per manhour. These costs include wire stripping, layout, material handling, and normal waste. Add for sales tax, delivery, supervision, mobilization, demobilization, cleanup, overhead and profit.

Wire Connectors

Material	Craft@Hrs	Unit	Material Cost	Labor Cost	Installed Cost

Two way wire connectors

Wire Size					
# 8	1E@0.03	Ea	1.10	.81	1.91
# 6	1E@0.04	Ea	1.91	1.08	2.99
# 4	1E@0.05	Ea	2.16	1.34	3.50
# 2	1E@0.08	Ea	2.73	2.15	4.88
# 1	1E@0.10	Ea	3.14	2.69	5.83
# 1/0	1E@0.10	Ea	3.39	2.69	6.08
# 2/0	1E@0.15	Ea	3.62	4.03	7.65
# 3/0	1E@0.15	Ea	4.21	4.03	8.24
# 4/0	1E@0.15	Ea	4.59	4.03	8.62
# 250 KCMIL	1E@0.20	Ea	5.47	5.38	10.85
# 300 KCMIL	1E@0.20	Ea	6.09	5.38	11.47
# 350 KCMIL	1E@0.20	Ea	6.20	5.38	11.58
# 400 KCMIL	1E@0.25	Ea	7.36	6.72	14.08
# 500 KCMIL	1E@0.25	Ea	9.08	6.72	15.80
# 600 KCMIL	1E@0.30	Ea	13.90	8.06	21.96
# 700 KCMIL	1E@0.30	Ea	16.30	8.06	24.36
# 750 KCMIL	1E@0.35	Ea	16.30	9.41	25.71
#1000 KCMIL	1E@0.40	Ea	22.50	10.80	33.30

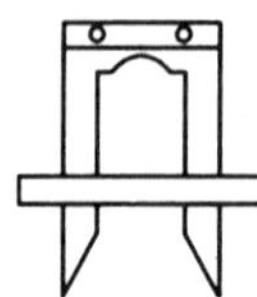

Split bolt connectors for copper wire only

Wire Size					
# 10	1E@0.20	Ea	1.95	5.38	7.33
# 8	1E@0.20	Ea	2.08	5.38	7.46
# 6	1E@0.25	Ea	2.27	6.72	8.99
# 4	1E@0.30	Ea	2.73	8.06	10.79
# 2	1E@0.30	Ea	3.84	8.06	11.90
# 1	1E@0.40	Ea	4.14	10.80	14.94
# 1/0	1E@0.50	Ea	5.13	13.40	18.53
# 2/0	1E@0.50	Ea	8.30	13.40	21.70
# 3/0	1E@0.60	Ea	11.60	16.10	27.70
# 4/0	1E@0.60	Ea	14.10	16.10	30.20
# 250 KCMIL	1E@0.70	Ea	14.10	18.80	32.90
# 350 KCMIL	1E@0.70	Ea	25.80	18.80	44.60
# 500 KCMIL	1E@0.75	Ea	33.70	20.20	53.90
# 750 KCMIL	1E@1.00	Ea	56.70	26.90	83.60
#1000 KCMIL	1E@1.25	Ea	77.50	33.60	111.10

Use these figures to estimate the cost of wire connectors installed under the conditions described on pages 5 and 6. Costs listed are for each connector installed. The crew is one electrician working at a labor cost of $26.88 per manhour. These costs include cutting the wire, insulation stripping, taping for 600-volt capacity, layout, material handling, and normal waste. Add for insulating the connector, special encapsulating kits, sales tax, delivery, supervision, mobilization, demobilization, cleanup, overhead and profit. Note: Crimping and high pressure tools are needed when installing connectors.

Material	Craft@Hrs	Unit	Material Cost	Labor Cost	Installed Cost

Split bolt connectors with spacer for copper or aluminum wire

Wire Size					
# 8	1E@0.20	Ea	2.32	5.38	7.70
# 6	1E@0.20	Ea	2.81	5.38	8.19
# 4	1E@0.25	Ea	3.94	6.72	10.66
# 2	1E@0.30	Ea	4.22	8.06	12.28
# 1	1E@0.40	Ea	5.38	10.80	16.18
# 1/0	1E@0.50	Ea	8.72	13.40	22.12
# 4/0	1E@0.50	Ea	14.80	13.40	28.20
# 350 KCMIL	1E@0.70	Ea	25.00	18.80	43.80
# 500 KCMIL	1E@0.75	Ea	33.50	20.20	53.70
# 750 KCMIL	1E@1.00	Ea	56.70	26.90	83.60
#1000 KCMIL	1E@1.25	Ea	78.60	33.60	112.20

Two bolt connectors for copper wire only

Material	Craft@Hrs	Unit	Material Cost	Labor Cost	Installed Cost
# 4/0	1E@0.70	Ea	10.40	18.80	29.20
# 350 KCMIL	1E@1.00	Ea	20.40	26.90	47.30
# 500 KCMIL	1E@1.25	Ea	23.80	33.60	57.40
# 800 KCMIL	1E@1.40	Ea	32.90	37.60	70.50
#1000 KCMIL	1E@1.50	Ea	46.50	40.30	86.80

Two bolt connectors for copper or aluminum wire

Material	Craft@Hrs	Unit	Material Cost	Labor Cost	Installed Cost
# 2/0	1E@0.50	Ea	8.33	13.40	21.73
# 250 KCMIL	1E@0.70	Ea	13.80	18.80	32.60
# 350 KCMIL	1E@0.70	Ea	27.10	18.80	45.90
# 500 KCMIL	1E@0.75	Ea	31.60	20.20	51.80
# 800 KCMIL	1E@1.50	Ea	61.80	40.30	102.10

Two bolt connectors with spacer for copper or aluminum wire

Material	Craft@Hrs	Unit	Material Cost	Labor Cost	Installed Cost
# 2/0	1E@0.55	Ea	9.93	14.80	24.73
# 250 KCMIL	1E@0.75	Ea	10.60	20.20	30.80
# 350 KCMIL	1E@0.75	Ea	16.50	20.20	36.70
# 500 KCMIL	1E@0.80	Ea	37.60	21.50	59.10
# 800 KCMIL	1E@1.10	Ea	52.10	29.60	81.70
#1000 KCMIL	1E@1.60	Ea	73.60	43.00	116.60

Use these figures to estimate the cost of split bolt and two bolt connectors installed under the conditions described on pages 5 and 6. Costs listed are for each connector installed. The crew is one electrician working at a labor cost of $26.88 per manhour. These costs include cutting the wire, insulation, stripping, taping for 600-volt capacity, forming the wire in a pull box or panel, layout, material handling, and normal waste. Add for insulating the connector, sales tax, delivery, supervision, mobilization, demobilization, cleanup, overhead and profit. Note: Split bolt connectors are installed with pump pliers. Tap the connector with a hammer after tightening and then retighten.

Copper Wire Connector Lugs

| | | | | Material | Labor | Installed |
| Material | Craft@Hrs | Unit | | Cost | Cost | Cost |

One hole solder type connector lugs for copper wire

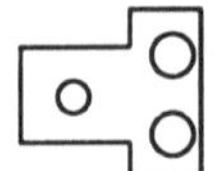

Material		Craft@Hrs	Unit	Material Cost	Labor Cost	Installed Cost
# 10		1E@0.10	Ea	.38	2.69	3.07
# 8		1E@0.15	Ea	.48	4.03	4.51
# 6		1E@0.15	Ea	.56	4.03	4.59
# 4		1E@0.20	Ea	.59	5.38	5.97
# 2		1E@0.20	Ea	.78	5.38	6.16
# 1/0		1E@0.25	Ea	1.04	6.72	7.76
# 2/0		1E@0.25	Ea	1.26	6.72	7.98
# 3/0		1E@0.30	Ea	1.72	8.06	9.78
# 4/0		1E@0.30	Ea	2.21	8.06	10.27
# 250	KCMIL	1E@0.40	Ea	4.21	10.80	15.01
# 400	KCMIL	1E@0.50	Ea	6.26	13.40	19.66
# 500	KCMIL	1E@0.50	Ea	9.86	13.40	23.26
# 600	KCMIL	1E@0.50	Ea	10.50	13.40	23.90
# 800	KCMIL	1E@0.60	Ea	17.50	16.10	33.60
#1000	KCMIL	1E@0.70	Ea	19.20	18.80	38.00
#1500	KCMIL	1E@0.75	Ea	34.30	20.20	54.50

One hole solderless type connector lugs for copper wire

Material		Craft@Hrs	Unit	Material Cost	Labor Cost	Installed Cost
# 10		1E@0.05	Ea	.60	1.34	1.94
# 6		1E@0.08	Ea	.63	2.15	2.78
# 4		1E@0.10	Ea	.87	2.69	3.56
# 2		1E@0.10	Ea	.84	2.69	3.53
# 1/0		1E@0.15	Ea	1.68	4.03	5.71
# 3/0		1E@0.20	Ea	3.31	5.38	8.69
# 4/0		1E@0.20	Ea	3.76	5.38	9.14
# 350	KCMIL	1E@0.30	Ea	6.99	8.06	15.05
# 500	KCMIL	1E@0.40	Ea	10.20	10.80	21.00
#1000	KCMIL	1E@0.60	Ea	25.40	16.10	41.50

One hole double conductor solderless type connector lugs for copper wire

Material	Craft@Hrs	Unit	Material Cost	Labor Cost	Installed Cost
# 2	1E@0.20	Ea	16.40	5.38	21.78
# 1	1E@0.25	Ea	20.10	6.72	26.82
# 2/0	1E@0.30	Ea	29.50	8.06	37.56
# 4/0	1E@0.40	Ea	38.40	10.80	49.20

Two hole three conductor solderless type connector lugs for copper wire

Material		Craft@Hrs	Unit	Material Cost	Labor Cost	Installed Cost
# 4		1E@0.30	Ea	19.50	8.06	27.56
# 1		1E@0.50	Ea	23.80	13.40	37.20
# 2/0		1E@0.75	Ea	34.40	20.20	54.60
# 4/0		1E@1.00	Ea	45.50	26.90	72.40
# 500	KCMIL	1E@1.50	Ea	88.40	40.30	128.70

Use these figures to estimate the cost of conductor lugs installed under the conditions described on pages 5 and 6. Costs listed are for each lug installed. The crew is one electrician working at a labor cost of $26.88 per manhour. These costs include the lug only (and no other material) wire cutting, insulation, stripping for the termination, layout, material handling, and normal waste. Add for insulating the lug (if needed), sales tax, delivery, supervision, mobilization, demobilization, cleanup, overhead and profit. Note: Always recheck solderless lugs for tightness.

Section 3: Outlet Boxes

In construction wiring, switches and electrical receptacles (known as wiring devices) are mounted in boxes which are secured to the wall, ceiling or floor. The box protects the device from damage and gives access to the wiring even after construction is complete. Once the box is installed and wired, the device is mounted and connected to the conductors. Finally, a box cover is installed to prevent accidental contact with conductors in the box.

Manufacturers of electrical equipment make boxes for every conceivable application. Some are used only in residences. Others are intended for commercial and industrial jobs. The *National Electrical Code* and your local code dictate the type of boxes that may be used in your community.

The size of the outlet box or junction box is determined by the size and number of conductors to be installed, and the size of the wiring device needed. When a plaster ring or switch ring is to be set against the box, the box size, the type, and the number of wiring devices to be used determine the size of the ring. The depth of the plaster ring or the switch ring will vary with the thickness of the final wall finish.

Boxes with plaster rings can be installed in ceilings and walls in wood construction, unit masonry, tile or concrete. Special outlet boxes and special plaster and switch rings are made for unit masonry and tile work. The ring depth is greater so the ring lip is flush with the surface of the finished wall. Check the architectural details of the plan to be sure you're pricing the right box and plaster ring. Some types cost considerably more than the standard grade used on most jobs.

This section does not deal with outlet boxes used in hazardous installations. Hazardous conditions require special outlet boxes and fittings that are approved for special environments.

Handy Boxes

Handy boxes are generally intended for surface mounting, though some are made with brackets attached that permit flush mounting. Handy boxes are 2" wide by 4" high. They come in 1½", 1⅞" 2⅛" and 2¹³/₁₆" depths. The box is one-piece sheet metal with knockouts on all sides and the back. The tapped ears for adapting devices are turned inward into the box.

The boxes and covers for handy boxes are plated steel. The covers are stamped. Cover types include blank, single switch, single receptacle, duplex receptacle, four-wire twist-lock receptacle, and despard device with three knockout squares. Price labor and material cost for the wiring device separately.

Handy boxes have punched holes for mounting with round-head wood screws, round-head machine screws, dome-head stove bolts, tie wire or masonry anchors.

Sectional Switch Boxes

Sectional switch boxes are generally installed flush with the wall in wood frame walls. The finish surface can be either plaster, drywall, insulating board or paneling. The boxes can be ganged together by removing side panels. They're made of plated steel with punched holes for mounting. Boxes can have knockouts in all sides and back or may have Romex or BX clamps for holding the cable in place.

Sectional boxes are made 2", 2½" and 2¾" deep. The tapped mounting ears are turned outward away from the box opening and the ears are spaced for standard wiring devices. Sectional switch boxes can be mounted in the wall or ceiling with round head wood screws, round head machine screws, dome head stove bolts, tie wire or masonry anchors. Be sure your electricians set the box flush with the level of the finished surface.

Welded Switch Boxes

Welded switch boxes look like sectional boxes. The difference is that they can't be ganged by removing the side panels. They're made with Romex or BX clamps for holding the cable in place.

Welded switch boxes are available with mounting brackets attached to the sides for rough-in mounting to the construction framing. The boxes are stamped, welded, punched, plated steel. The mounting ears are turned outward away from the box opening and are spaced for mounting standard wiring devices. The box can be mounted with nails, round-head wood screws, round-head machine screws, dome-head stove bolts, tie wire or masonry anchors.

Welded switch boxes come in single gang or two gang. They're 2½", 2¹³/₁₆" or 3¹³/₁₆" deep.

Octagon Boxes

Octagon boxes are one-piece pressed plated steel boxes. They're punched for knockouts and mounting and are available with mounting brackets that are attached to the box for flush mounting in a wall or ceiling.

The box's diameter is either three (called *3-0*) or four (called *4-0*) inches. The mounting ears are turned inward into the box for mounting covers or fixtures. The boxes can be mounted on fixed or adjustable bar hangers which are nailed to the framing. The box is attached to the bar hanger with dome-head stove bolts or round-head machine screws.

Some bar hangers have a ⅜" fixture stud for securing the lighting fixture directly to the bar hanger for support. The center knockout is removed from the box, the fixture stud is inserted into the box through the removed knockout, and a ⅜" locknut is used to secure the box to the stud.

Octagon boxes are 1½" or 2⅛" deep. A very shallow box called a *pancake box* is also made in 3-0 and 4-0 sizes. The pancake box is used when space is restricted and only one circuit is needed.

Another octagon box is used in concrete construction. Octagon concrete rings are stamped plated steel and are generally used in overhead concrete construction. The rings have ½" and ¾" knockouts and mounting ears that turn into the box for adapting a top cover and finish cover or lighting fixture. The box has tabs for nailing to concrete forms.

The octagon rings are 2½", 3", 3½", 4", 5" or 6" deep. The top cover can be blank with ½" or ¾" knockouts or the cover can be purchased with a fixture stud attached for mounting a lighting fixture directly to the stud. Octagon concrete rings are 4-0.

Square Boxes

The smallest square box is 4" x 4" x 1½" deep. It can be either a one-piece unit or welded, stamped plated steel. The box has ½" or ¾" knockouts or comes with a combination of ½" and ¾" knockouts. The 4-S box can purchased with mounting brackets and with Romex or BX clamps.

4-S boxes are also made 2⅛" deep with the same knockouts as a 4-S by 1½" box.

The tapped ears are in diagonal corners and have 8-32 tapped holes for attaching a cover or an extension ring. There are punched holes for mounting the box to the framing. Machine screws come with the box.

The extension ring is 1½" deep with ½" or ¾" or a combination of ½" and ¾" knockouts. The bottom diagonal corners have cutouts for mounting to the 4-S box. The top has tapped ears in diagonal corners with 8-32 tapped holes for attaching a cover. Again, machine screws come with the box.

4-S boxes are also made with two sets of mounting ears and 6-32 tapped holes for mounting two wiring devices. These two-ganged 4-S boxes can be purchased with mounting brackets for flush construction. 4-S boxes make good junction boxes in conduit systems. They're used for interior, dry, accessible locations and can be used in concrete walls or ceilings if the mounting holes are covered to keep liquid concrete from filling the box. When used in concrete construction, the point where conduit enters the box should be sealed to keep concrete out of the conduit.

Another square box is the 4¹¹/₁₆" x 4¹¹/₁₆" box. It's made the same as the 4-S box except that knockouts in the 2⅛" deep box can be ½", ¾", 1" or 1¼". Combinations of knockouts are ½" and ¾", 1" and 1¼". It has four tapped tabs for mounting a cover or an extension ring. The tapped holes are for 10-32 machine screws, which come with the box.

4¹¹/₁₆" boxes can be purchased with flush mounting brackets. They're used for interior, dry, accessible locations.

Bar Hangers

Bar hangers support an outlet box that has to be placed between two framing members. They're offset to allow for the depth of the box. Two lengths are available: one for nailing to studs 16" on center; the other for nailing to studs 24" on center.

There are two types of bar hangers. One comes with stove bolts for attaching the box to the bar hanger. The bar hanger has one long slot for adjustment or centering the box. The stove bolts are passed through the box and through the slot in the bar hanger. A square nut is used for tightening each bolt, thus securing the box to the bar hanger in the correct position.

The other type bar hanger has a ⅜" fixture stud that can be moved on the bar hanger. This permits adjustment of the box position. The center knockout is removed from the back of the outlet box, a ⅜" fixture stud is inserted into the box and a ⅜" locknut is used to secure the box to the bar hanger in the position desired.

Outlet Box Covers

Covers for outlet boxes mentioned so far in this section are made of plated steel. They're stamped with mounting holes lined up to fit the outlet box. Some covers are raised plaster rings or raised switch rings that extend the box rough-in to the wall or ceiling surface. The switch rings are made in single- or two-gang for one or two wiring devices, either lighting switches or convenience receptacles.

The covers for 3" octagon boxes are:

  3-0 flat blank covers

  3-0 duplex receptacle covers

  3-0 flat with ½" knockout covers

  3-0 single receptacle covers

The covers for 4" octagon boxes are:

  4-0 flat blank covers

  4-0 duplex receptacle covers

  4-0 flat with ½" knockout covers

  4-0 flat with ¾" knockout covers

  4-0 single receptacle covers

  4-0 twist-lock receptacle covers

  4-0 raised sign receptacle covers

  4-0 ¼" raised plaster ring

  4-0 ½" raised plaster ring

  4-0 ⅝" raised plaster ring

  4-0 ¾" raised plaster ring

  4-0 1" raised plaster ring

  4-0 1¼" raised plaster ring

  4-0 ½" raised blank cover

  4-0 ½" raised one device

Covers for 4-S outlet boxes are similar to 4-0 covers except that they're square and many more types are available.

Square covers for 4¹¹/₁₆" boxes also come in types that will meet almost any application.

Special Outlet Boxes

Special outlet boxes are available to cover applications not mentioned so far in this section. Ganged boxes are used when more than one wiring device is to be installed.

Multi-ganged boxes can be one-piece or welded plated steel. They have ½" and ¾" knockouts and the mounting holes are punched. Most have two 8-32 machine screws at one end and two holding tabs at the other end to hold the switch ring. The switch ring is usually raised ¾".

The multi-ganged box ranges from two-gang to ten-gang. They're generally used for multiple switches. But occasionally you'll see them used for multiple receptacles.

Another special outlet box is the masonry box. It's made of plated steel, is stamped with knockouts and has mounting holes punched. The boxes are usually single- or two-gang and are deep enough so that they can be flush mounted and still intercept conduit that's run through the cavity of a masonry unit. The top and bottom edge of the box are turned inward and tapped for 6-32 machine screws that hold the wiring device in place.

There are through boxes for masonry work that permit mounting devices back-to-back on both sides of a wall. They measure 3½" high and 7½" deep. Both the front and back of the box are made to receive wiring devices.

Non-Metallic Outlet Boxes

Non-metallic outlet boxes are made of either fiberglass or PVC in shapes and sizes like sheet metal boxes.

Fiberglass outlet boxes are generally used in residential concealed wiring. Many come with nails in place in a bracket for attaching to studs or joists. They have threaded metal inserts placed for standard wiring devices. Both 3" and 4" diameter boxes are available. These boxes use fiberglass covers that are similar to steel covers made for steel boxes.

Fiberglass switch boxes are available in single-gang, two-gang, three-gang and four-gang. They all can be purchased with or without mounting brackets and Romex clamps.

Fiberglass 4-S boxes are either 1⅝", 2¼" or 2½" deep. Many covers and plaster rings are available. These are also made of fiberglass.

Another series of non-metallic outlet boxes is made of PVC plastic. These are generally used for surface wiring but can be flush mounted. The switch boxes resemble FS condulets except that they don't have threaded hubs. PVC hubs are threadless because PVC conduit is glued or cemented in place. Other PVC boxes are made as junction boxes. They can be installed in concrete and are approved for corrosive conditions. But they're not acceptable for hazardous environments.

Special covers are available for PVC boxes. Most are weatherproof. There are blank covers, single receptacle covers, duplex receptacle covers, single switch covers and ground fault interrupter (GFI) covers.

I've described many of the common types of outlet boxes. But other types are available. What's listed here is representative of what's stocked by most dealers. Unless you handle a lot of exotic applications, this section has covered nearly all the boxes you use regularly.

Taking Off Outlet Boxes

Many estimators don't count the outlet boxes when making their take-off. Instead, they figure the number of wiring devices that will need a single-gang box, a two-gang box, etc. They use this device count to indicate the number of boxes needed. If you follow this method, be sure to add one outlet box for each single fixture and one for each row of fixtures.

I can't recommend this shortcut. Each box type and size has its own labor and material cost. Counting boxes on the plans is the only way to find the right number of each type of box and the right number of plaster rings.

When making your count, be sure to figure the right size box at each location. NEC Article 370, **Outlet, Switch and Junction Boxes and Fittings** limits the number of wires allowed in each box. To be accurate, your take-off must consider the number of conductors in each box. Even if the specs allow a 4" square by 1½" box, you may have to use a larger box to comply with the code. Either an extension ring or a 4$^{11}/_{16}$" square box may be required.

It's also worth your time to make an accurate count of plaster and switch rings. The architectural plans should show the wall treatment at each box. On masonry walls, use 1½" to 2" deep rings. If the wall has ½" thick drywall over ¼" plywood, use a ¾" deep ring.

Take the time to make your box estimate as accurate as possible. Little omissions compounded many times can make a big difference. The material cost of a box isn't very much. But add in the labor cost and multiply by a hundred boxes or more. The result will usually be more than your profit on most jobs. You can see why counting boxes is worth your time.

Labor for Outlet Box Installation

The labor costs in this section are based on standardized labor units. Using these units should both speed and simplify your box estimates. There are too many different types of boxes and installation conditions to develop an estimate for each case.

Instead, use the labor standard that applies to the situation that's closest to the box being estimated.

The labor standards in this section include all the time needed for layout, handling and installation of a box intended for a duplex receptacle and a wall switch next to a door or fixture outlet. Labor for the ring and cover aren't included, of course. See the tables for rings and covers.

I realize that the labor units in this section are higher than many electrical contractors use. But I find that they're accurate when the time required to receive, store, move and handle each box is considered. And most jobs have at least one box that's a real problem. You may set the first fifty boxes faster than the times listed in this section. But the fifty-first unit is going to take 20 minutes if your electrician has no room to work and has to chip out space for the box.

Watch for These Opportunities

Some specs permit a 2" x 4" outlet box as the end device on a circuit. At the end of the circuit only one conduit will enter the box. Usually this will be noted on the plan. Check the symbol list also. Sometimes the symbol list will specify outlet box minimum sizes.

Any time an outlet box has nearly as many conductors as permitted by the code, it may save labor to install the next larger size box. That makes it easier to tuck wires back into the box and install the device.

Handy Boxes and Switch Boxes

Material	Craft@Hrs	Unit	Material Cost	Labor Cost	Installed Cost
Handy boxes					
1-1/2" deep 1/2" KO	1E@0.15	Ea	1.11	4.03	5.14
1-7/8" deep 1/2" KO	1E@0.17	Ea	1.13	4.57	5.70
1-7/8" deep 3/4" KO	1E@0.17	Ea	1.52	4.57	6.09
1-7/8" flat bracket	1E@0.17	Ea	1.87	4.57	6.44
1-7/8" extension	1E@0.17	Ea	1.51	4.57	6.08
2-1/2" deep 1/2" KO	1E@0.20	Ea	1.46	5.38	6.84
2-1/2" deep 3/4" KO	1E@0.20	Ea	2.07	5.38	7.45
2-1/8" angle bracket	1E@0.20	Ea	2.45	5.38	7.83
2-1/8" flat bracket	1E@0.20	Ea	2.03	5.38	7.41
Handy box covers					
Blank	1E@0.03	Ea	.39	.81	1.20
Switch	1E@0.03	Ea	.62	.81	1.43
Single receptacle	1E@0.03	Ea	.62	.81	1.43
Duplex receptacle	1E@0.03	Ea	.38	.81	1.19
4 wire twistlock	1E@0.03	Ea	.82	.81	1.63
Handy boxes, large size					
1-5/8" deep 1/2" KO	1E@0.20	Ea	2.02	5.38	7.40
2-3/16" deep 1/2" KO	1E@0.25	Ea	2.06	6.72	8.78
Handy box covers, large size					
Blank	1E@0.03	Ea	.92	.81	1.73
Blank with 1/2" KO	1E@0.03	Ea	1.21	.81	2.02
Switch	1E@0.03	Ea	.92	.81	1.73
Single receptacle	1E@0.03	Ea	1.11	.81	1.92
Duplex receptacle	1E@0.03	Ea	.92	.81	1.73
4 wire twistlock	1E@0.03	Ea	1.11	.81	1.92
Sectional switch boxes, gangable with ears					
1-1/2" deep 1/2" KO	1E@0.15	Ea	1.33	4.03	5.36
2" deep 1/2" KO	1E@0.15	Ea	1.75	4.03	5.78
2-1/2" deep 1/2" KO	1E@0.17	Ea	1.39	4.57	5.96
2-3/4" deep 1/2" KO	1E@0.20	Ea	1.54	5.38	6.92
2-3/4" deep 3/4" KO	1E@0.20	Ea	2.19	5.38	7.57
3-1/2" deep 1/2" KO	1E@0.25	Ea	1.85	6.72	8.57
3-1/2" deep 3/4" KO	1E@0.25	Ea	2.40	6.72	9.12

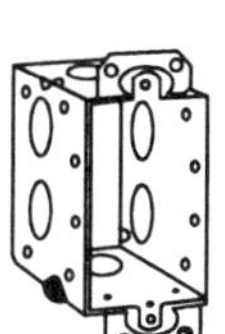

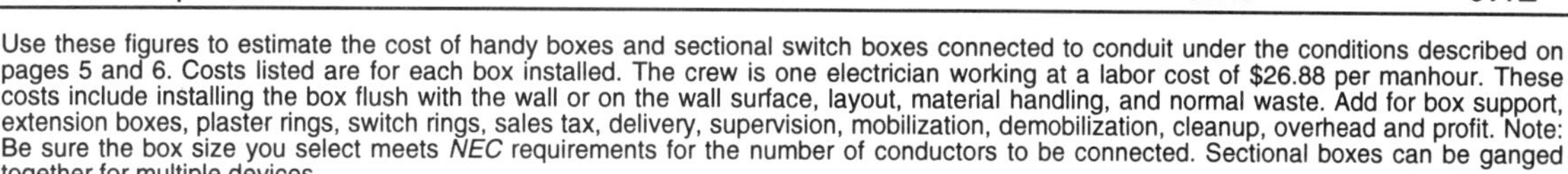

Use these figures to estimate the cost of handy boxes and sectional switch boxes connected to conduit under the conditions described on pages 5 and 6. Costs listed are for each box installed. The crew is one electrician working at a labor cost of $26.88 per manhour. These costs include installing the box flush with the wall or on the wall surface, layout, material handling, and normal waste. Add for box support, extension boxes, plaster rings, switch rings, sales tax, delivery, supervision, mobilization, demobilization, cleanup, overhead and profit. Note: Be sure the box size you select meets *NEC* requirements for the number of conductors to be connected. Sectional boxes can be ganged together for multiple devices.

Switch Boxes

Material	Craft@Hrs	Unit	Material Cost	Labor Cost	Installed Cost
Sectional switch boxes, gangable, no ears					
2" deep 1/2" KO	1E@0.15	Ea	1.55	4.03	5.58
2-1/2" deep 1/2" KO	1E@0.17	Ea	1.43	4.57	6.00
2-3/4" deep 1/2" KO	1E@0.20	Ea	2.14	5.38	7.52
3-1/2" deep 1/2" KO	1E@0.25	Ea	2.12	6.72	8.84
3-1/2" deep 3/4" KO	1E@0.25	Ea	2.68	6.72	9.40
Old work box with ears and side cleats					
1-1/2" KO	1E@0.30	Ea	3.73	8.06	11.79
Switch boxes for conduit					
2-1/2" with long bracket	1E@0.17	Ea	2.08	4.57	6.65
2-1/2" with flat bracket	1E@0.17	Ea	1.79	4.57	6.36
2-3/4" with flat bracket	1E@0.20	Ea	2.18	5.38	7.56
Thru boxes, plaster walls, with brackets					
2-1/8" x 5" x 4"	1E@0.25	Ea	5.74	6.72	12.46
Thru boxes, dry wall, with brackets					
2-1/8" x 4-5/16" x 4"	1E@0.25	Ea	4.78	6.72	11.50
Sectional boxes, beveled corners, for Romex					
2-1/4" deep no ears	1E@0.15	Ea	1.16	4.03	5.19
2-1/4" deep with ears	1E@0.15	Ea	1.33	4.03	5.36
2-1/4" with long bracket	1E@0.15	Ea	2.13	4.03	6.16
2-1/4" with flat bracket	1E@0.15	Ea	2.02	4.03	6.05
2-1/4" 2-gang, flat bracket	1E@0.20	Ea	3.94	5.38	9.32
Switch boxes, beveled corners, non-gangable, for Romex, with ears					
2-1/4" deep	1E@0.15	Ea	1.48	4.03	5.51
2-1/4" deep with nails	1E@0.15	Ea	1.68	4.03	5.71
Switch boxes, square corners, non-gangable, for Romex					
1-1/2" deep with ears	1E@0.15	Ea	1.27	4.03	5.30
3-1/2" deep with ears	1E@0.20	Ea	3.10	5.38	8.48
2-1/2" deep no ears	1E@0.15	Ea	1.61	4.03	5.64
2-7/8" deep no ears	1E@0.17	Ea	1.77	4.57	6.34
2-1/2" deep with nails	1E@0.17	Ea	1.93	4.57	6.50
2-3/4" deep with nails	1E@0.20	Ea	1.83	5.38	7.21
3-1/8" deep with nails	1E@0.20	Ea	3.53	5.38	8.91
3-13/32" deep with nails	1E@0.25	Ea	3.90	6.72	10.62

Use these figures to estimate the cost of sectional switch boxes and welded switch boxes connected to conduit under the conditions described on pages 5 and 6. Costs listed are for each box installed. The crew is one electrician working at a labor cost of $26.88 per manhour. These costs include installing the box flush with the wall or on the wall surface, layout, material handling, and normal waste. Add for box support, extension boxes, plaster rings, switch rings, sales tax, delivery, supervision, mobilization, demobilization, cleanup, overhead and profit. Note: Be sure the box size you select meets *NEC* requirements for the number of conductors to be connected. Sectional boxes can be ganged together for multiple devices. Romex and BX boxes have special clamps to hold the cable in place.

Switch Boxes and Octagon Boxes

Material	Craft@Hrs	Unit	Material Cost	Labor Cost	Installed Cost
Switch boxes, square corners, non-gangable, for Romex					
2-13/16" d with flat bracket	1E@0.17	Ea	2.68	4.57	7.25
3-1/8" d with flat bracket	1E@0.20	Ea	3.37	5.38	8.75
Switch boxes, square corners, non-gangable, for BX					
2" deep with ears	1E@0.15	Ea	1.95	4.03	5.98
2-1/2" deep with ears	1E@0.17	Ea	1.66	4.57	6.23
2-3/4" deep with ears	1E@0.17	Ea	2.54	4.57	7.11
3-1/2" deep with ears	1E@0.20	Ea	3.37	5.38	8.75
2-1/2" deep no ears	1E@0.17	Ea	2.47	4.57	7.04
2-1/2" with long bracket	1E@0.17	Ea	2.37	4.57	6.94
2-1/2" with flat bracket	1E@0.17	Ea	1.95	4.57	6.52
2-3/4" with flat bracket	1E@0.17	Ea	2.23	4.57	6.80
3-1/2" with flat bracket	1E@0.20	Ea	3.11	5.38	8.49
3" octagon boxes, 1-1/2" deep except pancake box					
3-0 pancake 1/2" deep	1E@0.15	Ea	1.18	4.03	5.21
3-0 for Romex	1E@0.15	Ea	2.42	4.03	6.45
3-0 1/2" KO	1E@0.15	Ea	1.60	4.03	5.63
3-0 extension 1/2" KO	1E@0.15	Ea	2.50	4.03	6.53
4" octagon boxes, 1-1/2" deep except pancake box					
4-0 pancake 1/2" deep	1E@0.17	Ea	1.68	4.57	6.25
4-0 1/2" KO	1E@0.20	Ea	1.05	5.38	6.43
4-0 3/4" KO	1E@0.20	Ea	1.43	5.38	6.81
4-0 1/2" & 3/4" KO	1E@0.20	Ea	1.30	5.38	6.68
4-0 BX flat bracket 1/2" KO	1E@0.20	Ea	2.03	5.38	7.41
4-0 for Romex	1E@0.20	Ea	1.46	5.38	6.84
4-0 Romex with ears	1E@0.20	Ea	2.45	5.38	7.83
4-0 Romex J-bracket	1E@0.20	Ea	2.09	5.38	7.47
4-0 Romex flat bracket	1E@0.20	Ea	2.15	5.38	7.53
4-0 extension ring 1/2" KO	1E@0.20	Ea	1.78	5.38	7.16
4-0 ext. 1/2" & 3/4" KO	1E@0.20	Ea	1.77	5.38	7.15

Use these figures to estimate the cost of welded switch boxes and octagon boxes connected to conduit under the conditions described on pages 5 and 6. Costs listed are for each box installed. The crew is one electrician working at a labor cost of $26.88 per manhour. These costs include installing the box flush with the wall or on the wall surface, layout, material handling, and normal waste. Add for box support, extension boxes, plaster rings, switch rings, sales tax, delivery, supervision, mobilization, demobilization, cleanup, overhead and profit. Note: Be sure the box size you select meets *NEC* requirements for the number of conductors to be connected. Sectional boxes can be ganged together for multiple devices.

Octagon Boxes, Concrete Rings and Box Hangers

Material	Craft@Hrs	Unit	Material Cost	Labor Cost	Installed Cost
4" octagon boxes 2-1/8" deep					
4-0 1/2" KO	1E@0.20	Ea	1.87	5.38	7.25
4-0 3/4" KO	1E@0.20	Ea	2.04	5.38	7.42
4-0 1/2" & 3/4" KO	1E@0.20	Ea	2.12	5.38	7.50
4-0 1" KO	1E@0.20	Ea	2.38	5.38	7.76
4-0 with clamps	1E@0.20	Ea	2.69	5.38	8.07
4-0 with clamps & J-bkt	1E@0.20	Ea	3.24	5.38	8.62
4-0 clamps & flat bkt	1E@0.20	Ea	3.24	5.38	8.62
4" octagon boxes 1-1/2" deep with adjustable hanger and stud					
4-0 1/2" KO	1E@0.25	Ea	4.51	6.72	11.23
4" octagon boxes 1-1/2" deep with adjustable hanger and clip					
4-0 1/2" KO	1E@0.25	Ea	4.79	6.72	11.51
4-3/8" octagon rings 1/2" and 3/4" KO					
4-3/8" 2" deep	1E@0.25	Ea	3.14	6.72	9.86
4-3/8" 2-1/2" deep	1E@0.25	Ea	3.22	6.72	9.94
4-3/8" 3" deep	1E@0.25	Ea	3.26	6.72	9.98
4-3/8" 3-1/2" deep	1E@0.30	Ea	3.42	8.06	11.48
4-3/8" 4" deep	1E@0.30	Ea	4.47	8.06	12.53
4-3/8" 5" deep	1E@0.35	Ea	5.90	9.41	15.31
4-3/8" 6" deep	1E@0.35	Ea	7.00	9.41	16.41
Concrete ring plates					
4-3/8" with 1/2" & 3/4" KO	1E@0.05	Ea	.94	1.34	2.28
4-3/8" with 3/8" stud	1E@0.05	Ea	1.60	1.34	2.94
Drop ceiling boxes with 1/2" and 3/4" KO					
4-0 with 18" hanger bars	1E@0.25	Ea	6.00	6.72	12.72
4-0 with 24" hanger bars	1E@0.25	Ea	6.30	6.72	13.02
4-0 with 30" hanger bars	1E@0.25	Ea	6.33	6.72	13.05
Adjustable bar hangers with 3.8" stud					
11-1/2" to 18-1/2"	1E@0.05	Ea	1.47	1.34	2.81
19-1/2" to 26-1/2"	1E@0.05	Ea	1.91	1.34	3.25
Octagon concrete box covers					
Flat blank	1E@0.05	Ea	.95	1.34	2.29
Flat blank with 3/8 stud	1E@0.05	Ea	1.38	1.34	2.72

Use these figures to estimate the cost of octagon boxes and rings installed on conduit under the conditions described on pages 5 and 6. Costs listed are for each box installed. The crew is one electrician working at a labor cost of $26.88 per manhour. These costs include installing the box flush with the wall or on the wall surface, layout, material handling, and normal waste. Add for box support, extension boxes, plaster rings, switch rings, sales tax, delivery, supervision, mobilization, demobilization, cleanup, overhead and profit. Note: Be sure the box size you select meets *NEC* requirements for the number of conductors to be connected.

Material	Craft@Hrs	Unit	Material Cost	Labor Cost	Installed Cost
4" x 4" x 1-1/4" deep square boxes					
4-S 1/2 KO	1E@0.25	Ea	2.12	6.72	8.84
4-S bracket	1E@0.25	Ea	3.08	6.72	9.80
4" x 4" x 1-1/2" deep square boxes					
4-S 1/2 KO	1E@0.25	Ea	1.53	6.72	8.25
4-S 3/4 KO	1E@0.25	Ea	1.35	6.72	8.07
4-S 1/2 & 3/4 KO	1E@0.25	Ea	1.23	6.72	7.95
4-S bracket	1E@0.25	Ea	2.29	6.72	9.01
4-S Romex	1E@0.25	Ea	2.04	6.72	8.76
4-S Romex, bracket	1E@0.25	Ea	2.99	6.72	9.71
4-S BX	1E@0.25	Ea	2.33	6.72	9.05
4-S BX, bracket	1E@0.25	Ea	2.98	6.72	9.70
4-S ext. 1/2 KO	1E@0.15	Ea	1.66	4.03	5.69
4-S ext. 3/4 KO	1E@0.15	Ea	2.31	4.03	6.34
4-S ext. 1/2 & 3/4 KO	1E@0.15	Ea	1.69	4.03	5.72
4-S thru 1/2 KO	1E@0.25	Ea	3.16	6.72	9.88
4-S thru 3/4 KO	1E@0.25	Ea	3.38	6.72	10.10
4-S thru 1/2 & 3/4 KO	1E@0.25	Ea	3.38	6.72	10.10
4" x 4" x 2-1/8" deep square boxes					
4-S 1/2 KO	1E@0.27	Ea	2.15	7.26	9.41
4-S 3/4 KO	1E@0.27	Ea	2.22	7.26	9.48
4-S 1 KO	1E@0.27	Ea	2.20	7.26	9.46
4-S 1/2 & 3/4 KO	1E@0.27	Ea	2.22	7.26	9.48
4-S 1/2 bracket	1E@0.27	Ea	3.47	7.26	10.73
4-S 3/4 bracket	1E@0.27	Ea	3.47	7.26	10.73
4-S 1/2 & 3/4 bracket	1E@0.27	Ea	3.47	7.26	10.73
4-S Romex	1E@0.27	Ea	3.99	7.26	11.25
4-S Romex, bracket	1E@0.27	Ea	3.83	7.26	11.09
4-11/16" x 4-11/16" x 1-1/2" deep square boxes					
4-11/16 1/2 KO	1E@0.30	Ea	3.61	8.06	11.67
4-11/16 3/4 KO	1E@0.30	Ea	3.82	8.06	11.88
4-11/16 1/2 & 3/4 KO	1E@0.30	Ea	3.40	8.06	11.46
4-11/16 ext 1/2 KO	1E@0.25	Ea	4.17	6.72	10.89
4-11/16 ext 1/2 & 3/4 KO	1E@0.25	Ea	4.17	6.72	10.89

Use these figures to estimate the cost of square boxes connected to conduit under the conditions described on pages 5 and 6. Costs listed are for each box installed. The crew is one electrician working at a labor cost of $26.88 per manhour. These costs include installing the box flush with the wall or on the wall surface, layout, material handling, and normal waste. Add for box support, extension boxes, plaster rings, switch rings, sales tax, delivery, supervision, mobilization, demobilization, cleanup, overhead and profit. Note: Be sure the box size you select meets *NEC* requirements for the number of conductors to be connected.

Square Boxes and Covers

Material	Craft@Hrs	Unit	Material Cost	Labor Cost	Installed Cost
4-11/16" x 4-11/16" x 2-1/8" deep square boxes					
4-11/16 1/2 KO	1E@0.33	Ea	3.17	8.87	12.04
4-11/16 3/4 KO	1E@0.33	Ea	3.17	8.87	12.04
4-11/16 1/2&3/4 KO	1E@0.33	Ea	2.78	8.87	11.65
4-11/16 3/4 & 1 KO	1E@0.33	Ea	3.23	8.87	12.10
4-11/16 1 KO	1E@0.33	Ea	3.32	8.87	12.19
4-11/16 1 1/4 KO	1E@0.33	Ea	4.13	8.87	13.00
4-11/16 1/2 & 3/4 bkt	1E@0.33	Ea	4.60	8.87	13.47
4-11/16 1 bracket	1E@0.33	Ea	4.60	8.87	13.47
4-11/16 ext 1/2 KO	1E@0.25	Ea	4.14	6.72	10.86
4-11/16 ext 3/4 KO	1E@0.25	Ea	4.14	6.72	10.86
4-11/16 ext 1/2 & 3/4 KO	1E@0.25	Ea	4.14	6.72	10.86
4-11/16 ext 1 KO	1E@0.25	Ea	5.46	6.72	12.18
Outlet box covers					
3-O flat blank	1E@0.03	Ea	.58	.81	1.39
3-O 1/2 KO	1E@0.03	Ea	.58	.81	1.39
3-O duplex	1E@0.03	Ea	1.13	.81	1.94
3-O single outlet	1E@0.03	Ea	.58	.81	1.39
4-O flat blank	1E@0.05	Ea	.51	1.34	1.85
4-O 1/2 KO	1E@0.05	Ea	.41	1.34	1.75
4-O 3/4 KO	1E@0.05	Ea	.41	1.34	1.75
4-O duplex	1E@0.05	Ea	1.28	1.34	2.62
4-O single outlet	1E@0.05	Ea	1.27	1.34	2.61
4-O T/L recept.	1E@0.05	Ea	1.27	1.34	2.61
4-O plaster ring 1/4	1E@0.05	Ea	1.40	1.34	2.74
4-O plaster ring 1/2	1E@0.05	Ea	.92	1.34	2.26
4-O plaster ring 5/8	1E@0.05	Ea	1.15	1.34	2.49
4-O plaster ring 3/4	1E@0.05	Ea	1.69	1.34	3.03
4-O plaster ring 1	1E@0.05	Ea	2.81	1.34	4.15
4-O plas. ring 1-1/4	1E@0.05	Ea	3.22	1.34	4.56
4-O raised blank 1/2	1E@0.05	Ea	1.13	1.34	2.47
4-S flat blank	1E@0.05	Ea	.47	1.34	1.81
4-S flat blank 1/2 KO	1E@0.05	Ea	.51	1.34	1.85
4-S plaster ring 1/4	1E@0.05	Ea	1.31	1.34	2.65
4-S plaster ring 1/2	1E@0.05	Ea	.99	1.34	2.33
4-S plaster ring 5/8	1E@0.05	Ea	.92	1.34	2.26
4-S plaster ring 3/4	1E@0.05	Ea	1.26	1.34	2.60
4-S plaster ring 1	1E@0.05	Ea	3.36	1.34	4.70
4-S plaster ring 1-1/4	1E@0.05	Ea	3.36	1.34	4.70

Use these figures to estimate the cost of square boxes and covers installed on conduit under the conditions described on pages 5 and 6. Costs listed are for each unit installed. The crew is one electrician working at a labor cost of $26.88 per manhour. These costs include installing the unit, layout, material handling, and normal waste. Add for box support, extension boxes, plaster rings, switch rings, sales tax, delivery, supervision, mobilization, demobilization, cleanup, overhead and profit. Note: Be sure the box size you select meets *NEC* requirements for the number of conductors to be connected.

Material	Craft@Hrs	Unit	Material Cost	Labor Cost	Installed Cost
Single gang 4" square switch rings					
4-S sw ring flat	1E@0.05	Ea	.92	1.34	2.26
4-S sw ring 1/4	1E@0.05	Ea	.84	1.34	2.18
4-S sw ring 1/2	1E@0.05	Ea	.65	1.34	1.99
4-S sw ring 5/8	1E@0.05	Ea	.75	1.34	2.09
4-S sw ring 3/4	1E@0.05	Ea	.92	1.34	2.26
4-S sw ring 1	1E@0.05	Ea	1.24	1.34	2.58
4-S sw ring 1-1/4	1E@0.05	Ea	1.49	1.34	2.83
4-S tile 1/2	1E@0.05	Ea	2.11	1.34	3.45
4-S tile 3/4	1E@0.05	Ea	2.20	1.34	3.54
4-S tile 1	1E@0.05	Ea	2.62	1.34	3.96
4-S tile 1-1/4	1E@0.06	Ea	2.76	1.61	4.37
4-S tile 1-1/2	1E@0.06	Ea	2.20	1.61	3.81
4-S tile 2	1E@0.06	Ea	2.32	1.61	3.93
Two gang 4" square switch rings					
4-S sw ring flat	1E@0.06	Ea	1.36	1.61	2.97
4-S sw ring 1/4	1E@0.06	Ea	1.72	1.61	3.33
4-S sw ring 1/2	1E@0.06	Ea	1.11	1.61	2.72
4-S sw ring 5/8	1E@0.06	Ea	1.24	1.61	2.85
4-S sw ring 3/4	1E@0.06	Ea	1.44	1.61	3.05
4-S sw ring 1	1E@0.06	Ea	2.01	1.61	3.62
4-S sw ring 1-1/4	1E@0.06	Ea	2.12	1.61	3.73
4-S tile 1/2	1E@0.06	Ea	2.87	1.61	4.48
4-S tile 3/4	1E@0.06	Ea	3.01	1.61	4.62
4-S tile 1	1E@0.06	Ea	3.08	1.61	4.69
4-S tile 1-1/4	1E@0.08	Ea	3.03	2.15	5.18
4-S tile 1-1/2	1E@0.08	Ea	2.73	2.15	4.88
4-S tile 2	1E@0.08	Ea	2.90	2.15	5.05
Adjustable bar hangers with 3.8" stud					
11-1/2" to 18-1/2"	1E@0.05	Ea	1.47	1.34	2.81
19-1/2" to 26-1/2"	1E@0.05	Ea	1.91	1.34	3.25
Adjustable bar hanger with clip					
11-1/2" to 18-1/2"	1E@0.05	Ea	1.53	1.34	2.87
19-1/2" to 26-1/2"	1E@0.05	Ea	1.72	1.34	3.06
Bar hangers with stove bolts					
20" straight	1E@0.05	Ea	2.36	1.34	3.70
24" straight	1E@0.05	Ea	2.58	1.34	3.92
21" 1" offset	1E@0.05	Ea	2.95	1.34	4.29
22" 1-1/2" offset	1E@0.05	Ea	2.45	1.34	3.79

Use these figures to estimate the cost of square switch rings installed on outlet boxes under the conditions described on pages 5 and 6. Costs listed are for each switch ring installed. The crew is one electrician working at a labor cost of $26.88 per manhour. These costs include adapting the cover to the box, layout, material handling, and normal waste. Add for the outlet box, wiring devices, sales tax, delivery, supervision, mobilization, demobilization, cleanup, overhead and profit. Note: Be sure to select plaster and switch rings as specified and that are appropriate for the wall depth. Many types are available.

Square Switch Rings and Surface Covers

Material	Craft@Hrs	Unit	Material Cost	Labor Cost	Installed Cost
Single gang 4-11/16" square switch rings					
4-11/16 sw ring 1/4	1E@0.06	Ea	3.22	1.61	4.83
4-11/16 sw ring 1/2	1E@0.06	Ea	2.22	1.61	3.83
4-11/16 sw ring 5/8	1E@0.06	Ea	2.25	1.61	3.86
4-11/16 sw ring 3/4	1E@0.06	Ea	2.22	1.61	3.83
4-11/16 sw ring 1	1E@0.06	Ea	2.31	1.61	3.92
4-11/16 sw ring 1-1/4	1E@0.06	Ea	2.51	1.61	4.12
4-11/16 tile 1/2	1E@0.06	Ea	5.08	1.61	6.69
4-11/16 tile 3/4	1E@0.06	Ea	5.08	1.61	6.69
4-11/16 tile 1	1E@0.06	Ea	5.08	1.61	6.69
4-11/16 tile 1-1/4	1E@0.08	Ea	5.18	2.15	7.33
4-11/16 tile 1-1/2	1E@0.08	Ea	4.99	2.15	7.14
4-11/16 tile 2	1E@0.08	Ea	5.13	2.15	7.28
Two gang 4-11/16" square switch rings					
4-11/16 sw ring 1/2	1E@0.08	Ea	2.44	2.15	4.59
4-11/16 sw ring 5/8	1E@0.08	Ea	2.25	2.15	4.40
4-11/16 sw ring 3/4	1E@0.08	Ea	2.53	2.15	4.68
4-11/16 sw ring 1	1E@0.08	Ea	3.57	2.15	5.72
4-11/16 sw ring 1-1/4	1E@0.08	Ea	3.65	2.15	5.80
4-11/16 tile 1/2	1E@0.10	Ea	5.21	2.69	7.90
4-11/16 tile 3/4	1E@0.10	Ea	5.21	2.69	7.90
4-11/16 tile 1	1E@0.10	Ea	5.44	2.69	8.13
4-11/16 tile 1-1/4	1E@0.10	Ea	5.44	2.69	8.13
4-11/16 tile 1-1/2	1E@0.10	Ea	5.42	2.69	8.11
4-11/16 tile 2	1E@0.10	Ea	5.58	2.69	8.27
4" square surface covers, raised 1/4"					
4 S blank	1E@0.05	Ea	1.91	1.34	3.25
4 S one switch	1E@0.05	Ea	1.56	1.34	2.90
4 S two switch	1E@0.05	Ea	1.62	1.34	2.96
4 S one single recept	1E@0.05	Ea	1.47	1.34	2.81
4 S two single recept	1E@0.05	Ea	1.81	1.34	3.15
4 S one duplex recept	1E@0.05	Ea	1.20	1.34	2.54
4 S two duplex recept	1E@0.05	Ea	1.16	1.34	2.50
4 S sw & single recept	1E@0.05	Ea	1.57	1.34	2.91
4 S sw & duplex recept	1E@0.05	Ea	1.33	1.34	2.67
4 S single & duplex recep	1E@0.05	Ea	1.66	1.34	3.00
4 S four wire recept	1E@0.05	Ea	1.64	1.34	2.98

Use these figures to estimate the cost of square switch rings and surface covers installed on outlet boxes under the conditions described on pages 5 and 6. Costs listed are for each ring or cover installed. The crew is one electrician working at a labor cost of $26.88 per manhour. These costs include adapting the cover to the box, layout, material handling, and normal waste. Add for outlet box, wiring devices, sales tax, delivery, supervision, mobilization, demobilization, cleanup, overhead and profit. Note: Be sure to select plaster and switch rings as specified and that are appropriate to the wall depth. Many types are available.

Material	Craft@Hrs	Unit	Material Cost	Labor Cost	Installed Cost

4" square surface box covers, raised 1/2"

Material	Craft@Hrs	Unit	Material Cost	Labor Cost	Installed Cost
4-S blank	1E@0.05	Ea	1.91	1.34	3.25
4-S one switch	1E@0.05	Ea	1.56	1.34	2.90
4-S two switch	1E@0.05	Ea	1.62	1.34	2.96
4-S one single recept	1E@0.05	Ea	1.47	1.34	2.81
4-S two single recept	1E@0.05	Ea	1.81	1.34	3.15
4-S one duplex recept	1E@0.05	Ea	1.20	1.34	2.54
4-S two duplex recept	1E@0.05	Ea	1.16	1.34	2.50
4-S sw & single recept	1E@0.05	Ea	1.57	1.34	2.91
4-S sw & duplex recept	1E@0.05	Ea	1.33	1.34	2.67
4-S single & duplex recept	1E@0.05	Ea	1.33	1.34	2.67
4-S 3 w T/L recept	1E@0.05	Ea	1.64	1.34	2.98
4-S 4 w T/L recept	1E@0.05	Ea	1.64	1.34	2.98
4-S 3w recept	1E@0.05	Ea	1.64	1.34	2.98
4-S 4w recept	1E@0.05	Ea	1.64	1.34	2.98
4-S 3 despard	1E@0.05	Ea	2.92	1.34	4.26
4-S 6 despard	1E@0.05	Ea	3.50	1.34	4.84

4-11/16" square box covers

Material	Craft@Hrs	Unit	Material Cost	Labor Cost	Installed Cost
4-11/16 flat blank	1E@0.06	Ea	.88	1.61	2.49
4-11/16 blank 1/2 KO	1E@0.06	Ea	1.40	1.61	3.01
4-11/16 pl ring 1/2	1E@0.06	Ea	2.69	1.61	4.30
4-11/16 pl ring 5/8	1E@0.06	Ea	2.85	1.61	4.46
4-11/16 pl ring 1	1E@0.06	Ea	6.35	1.61	7.96
4-11/16 pl ring 1-1/4	1E@0.06	Ea	6.51	1.61	8.12

4-11/16" square surface box covers, raised 1/2"

Material	Craft@Hrs	Unit	Material Cost	Labor Cost	Installed Cost
Blank	1E@0.06	Ea	4.75	1.61	6.36
One switch	1E@0.06	Ea	4.19	1.61	5.80
Two switches	1E@0.06	Ea	4.39	1.61	6.00
One duplex	1E@0.06	Ea	4.30	1.61	5.91
Two duplex	1E@0.06	Ea	4.30	1.61	5.91
One switch & duplex	1E@0.06	Ea	4.43	1.61	6.04
One single receptacle	1E@0.06	Ea	4.19	1.61	5.80
Two single receptacle	1E@0.06	Ea	4.43	1.61	6.04
One 30A T/L receptacle	1E@0.06	Ea	4.19	1.61	5.80
One 60A T/L receptacle	1E@0.06	Ea	4.31	1.61	5.92

Use these figures to estimate the cost of square and octagon box covers installed on outlet boxes under the conditions described on pages 5 and 6. Costs listed are for each cover installed. The crew is one electrician working at a labor cost of $26.88 per manhour. These costs include adapting the cover to the box, layout, material handling, and normal waste. Add for outlet box, wiring devices, sales tax, delivery, supervision, mobilization, demobilization, cleanup, overhead and profit. Note: Be sure to select plaster and switch rings as specified and that are appropriate to the wall depth. Many types are available.

Gang Switch Boxes and Masonry Boxes

Material	Craft@Hrs	Unit	Material Cost	Labor Cost	Installed Cost
Gang boxes 1-5/8" deep with 1/2" and 3/4" KO					
2 gang	1E@0.25	Ea	8.28	6.72	15.00
3 gang	1E@0.30	Ea	9.96	8.06	18.02
4 gang	1E@0.40	Ea	13.70	10.80	24.50
5 gang	1E@0.50	Ea	18.90	13.40	32.30
6 gang	1E@0.60	Ea	35.10	16.10	51.20
Gang switch rings, 1/2" raised					
2 gang	1E@0.05	Ea	3.97	1.34	5.31
3 gang	1E@0.10	Ea	4.62	2.69	7.31
4 gang	1E@0.10	Ea	6.30	2.69	8.99
5 gang	1E@0.10	Ea	8.23	2.69	10.92
6 gang	1E@0.10	Ea	13.90	2.69	16.59
One gang masonry boxes, 1/2" and 3/4" knockouts					
2-1/2" deep, 1 gang	1E@0.20	Ea	3.48	5.38	8.86
3-1/2" deep, 1 gang	1E@0.25	Ea	3.03	6.72	9.75
3-1/2" deep, thru box	1E@0.30	Ea	5.39	8.06	13.45
5-1/2" deep, thru box	1E@0.40	Ea	5.89	10.80	16.69
7-1/2" deep, thru box	1E@0.50	Ea	6.86	13.40	20.26
Multi-gang masonry boxes, 1/2" and 3/4" knockouts					
2-1/2" deep, 2 gang	1E@0.30	Ea	4.72	8.06	12.78
3-1/2" deep, 2 gang	1E@0.35	Ea	4.75	9.41	14.16
2-1/2" deep, 3 gang	1E@0.40	Ea	6.31	10.80	17.11
3-1/2" deep, 3 gang	1E@0.40	Ea	6.58	10.80	17.38
2-1/2" deep, 4 gang	1E@0.50	Ea	7.10	13.40	20.50
3-1/2" deep, 4 gang	1E@0.50	Ea	7.31	13.40	20.71
Tomic bolt hangers					
5/16" x 4"	1E@0.06	Ea	1.09	1.61	2.70
5/16" x 5"	1E@0.06	Ea	1.23	1.61	2.84

Use these figures to estimate the cost of knockout boxes and masonry boxes installed in buildings under the conditions described on pages 5 and 6. Costs listed are for each box installed. The crew is one electrician working at a labor cost of $26.88 per manhour. These costs include switch rings, layout, material handling, and normal waste. Add for box support, wiring devices, sales tax, delivery, supervision, mobilization, demobilization, cleanup, overhead and profit. Note: Ganged switch boxes are made in one piece.

Boxes made for masonry construction come as a single piece and have provisions for attaching devices to the box. But there's no separate switch ring. The knockout for conduit is near the back of the box so conduit can be run from the masonry cells to the box with little or no bending. Usually a short piece of rebar is used to secure the box in place. All other holes in the box should be sealed with masking tape to prevent mortar from entering the box. Be sure that conduit entering the bottom of the box is also sealed.

Material	Craft@Hrs	Unit	Material Cost	Labor Cost	Installed Cost

3-1/2" round by 1-3/4" deep fiberglass outlet boxes

Material	Craft@Hrs	Unit	Material Cost	Labor Cost	Installed Cost
With flash ring	1E@0.15	Ea	.59	4.03	4.62
With ground strap	1E@0.15	Ea	.83	4.03	4.86
With bracket	1E@0.15	Ea	.59	4.03	4.62
With bracket & ground	1E@0.15	Ea	.83	4.03	4.86

3-1/2" round by 2" deep fiberglass outlet boxes

Material	Craft@Hrs	Unit	Material Cost	Labor Cost	Installed Cost
With flash ring	1E@0.15	Ea	.80	4.03	4.83
With ground strap	1E@0.15	Ea	1.06	4.03	5.09
With bracket	1E@0.15	Ea	.80	4.03	4.83
With bracket & ground	1E@0.15	Ea	1.06	4.03	5.09
With bracket & clamps	1E@0.15	Ea	.80	4.03	4.83
With bkt. clamps, ground	1E@0.15	Ea	1.06	4.03	5.09
With ears	1E@0.15	Ea	1.06	4.03	5.09
With bracket & clamp	1E@0.15	Ea	1.27	4.03	5.30
With snap-in bracket	1E@0.17	Ea	1.27	4.57	5.84

3-1/2" round by 2-7/8" deep fiberglass outlet boxes

Material	Craft@Hrs	Unit	Material Cost	Labor Cost	Installed Cost
With nails	1E@0.15	Ea	.99	4.03	5.02
With nails & clamps	1E@0.15	Ea	1.25	4.03	5.28
With nails & ground	1E@0.15	Ea	1.25	4.03	5.28
With nails, strap & ground	1E@0.15	Ea	1.25	4.03	5.28
With adj. bar hanger	1E@0.17	Ea	2.13	4.57	6.70
With hanger & ground	1E@0.17	Ea	2.40	4.57	6.97
With hanger, grd. & clamps	1E@0.17	Ea	2.40	4.57	6.97
With flash ring	1E@0.15	Ea	.99	4.03	5.02

4" round by 1-5/8" deep fiberglass outlet boxes

Material	Craft@Hrs	Unit	Material Cost	Labor Cost	Installed Cost
Outlet box	1E@0.20	Ea	.81	5.38	6.19
With wire clamps	1E@0.20	Ea	1.04	5.38	6.42
With 16" adj. hanger	1E@0.25	Ea	2.25	6.72	8.97
With 24" adj. hanger	1E@0.25	Ea	2.38	6.72	9.10
With hanger & clamps	1E@0.20	Ea	2.38	5.38	7.76
With hanger & ground	1E@0.20	Ea	2.50	5.38	7.88
With nails	1E@0.20	Ea	.99	5.38	6.37
With nails & clamps	1E@0.20	Ea	1.25	5.38	6.63
With nails & ground	1E@0.20	Ea	1.25	5.38	6.63

Use these figures to estimate the cost of round fiberglass outlet boxes installed in residential buildings under the conditions described on pages 5 and 6. Costs listed are for each box installed. The crew is one electrician working at a labor cost of $26.88 per manhour. These costs include layout, material handling, and normal waste. Add for wiring devices, covers, sales tax, delivery, supervision, mobilization, demobilization, cleanup, overhead and profit. Note: Be sure the box size you select meets *NEC* requirements for the number of conductors to be connected.

Fiberglass Outlet Boxes

Material	Craft@Hrs	Unit	Material Cost	Labor Cost	Installed Cost
4" round by 2-5/16" deep fiberglass outlet boxes					
With wire clamps	1E@0.20	Ea	.85	5.38	6.23
With bracket	1E@0.20	Ea	1.26	5.38	6.64
With bracket & clamps	1E@0.20	Ea	1.26	5.38	6.64
With 16" adj. hanger	1E@0.25	Ea	1.50	6.72	8.22
With 16" adj. clamps	1E@0.25	Ea	2.23	6.72	8.95
With 16" adj. ground	1E@0.25	Ea	2.19	6.72	8.91
With 24" adj. hanger	1E@0.25	Ea	1.96	6.72	8.68
With 24" adj. clamps	1E@0.25	Ea	2.23	6.72	8.95
With 24" adj. ground	1E@0.25	Ea	2.19	6.72	8.91
4" round by 2-3/8" deep fiberglass outlet boxes					
With nails	1E@0.20	Ea	1.05	5.38	6.43
With nails & clamps	1E@0.20	Ea	1.29	5.38	6.67
With nails, ground & strap	1E@0.20	Ea	1.55	5.38	6.93
With nails, ground & clamps	1E@0.20	Ea	1.55	5.38	6.93
4" square by 1-5/8" deep fiberglass outlet boxes					
4/S box	1E@0.20	Ea	1.07	5.38	6.45
4/S bracket	1E@0.20	Ea	1.49	5.38	6.87
4" square by 2-1/4" deep fiberglass outlet boxes					
4/S box	1E@0.20	Ea	1.12	5.38	6.50
4/S clamps	1E@0.20	Ea	1.38	5.38	6.76
4/S bracket	1E@0.20	Ea	1.57	5.38	6.95
4" square by 2-1/2" deep fiberglass outlet boxes					
With nails	1E@0.20	Ea	1.67	5.38	7.05
With wire clamps	1E@0.20	Ea	1.93	5.38	7.31

Use these figures to estimate the cost of fiberglass outlet boxes installed in residential buildings under the conditions described on pages 5 and 6. Costs listed are for each box installed. The crew is one electrician working at a labor cost of $26.88 per manhour. These costs include layout, material handling, and normal waste. Add for wiring devices, covers, sales tax, delivery, supervision, mobilization, demobilization, cleanup, overhead and profit. Note: These boxes are generally used with Romex (type NM) cable in residences. Ground the box by attaching the Romex ground conductor to the ground terminal on the wiring device. Round boxes are generally used for lighting fixtures and junction boxes. They're usually installed flush with the wall or ceiling surface. A lighting fixture can be attached directly to the box with 8-32 machine screws. Conductors must be connected in the box with approved connectors. Be sure the outlet box is secured to the building frame well enough to support the weight of the lighting fixture. The outlet box can either be nailed to a stud or fastened with wood screws.

Fiberglass Covers and Outlet Boxes

Material	Craft@Hrs	Unit	Material Cost	Labor Cost	Installed Cost
3-1/2" round fiberglass box covers					
Blank	1E@0.05	Ea	.74	1.34	2.08
4" round fiberglass box covers					
Blank	1E@0.05	Ea	.58	1.34	1.92
Plaster ring, 7/16"	1E@0.05	Ea	.76	1.34	2.10
4" square fiberglass box covers					
Blank	1E@0.05	Ea	.72	1.34	2.06
Plaster ring, 1-G	1E@0.05	Ea	.76	1.34	2.10
Plaster ring, 2-G	1E@0.05	Ea	.85	1.34	2.19
Handy box covers					
Blank	1E@0.05	Ea	.46	1.34	1.80
Switch cover	1E@0.05	Ea	.47	1.34	1.81
Duplex cover	1E@0.05	Ea	.45	1.34	1.79
Handy boxes					
1-5/8 deep	1E@0.20	Ea	2.51	5.38	7.89
Single gang switch boxes, 1-5/8" deep					
With mounting ears	1E@0.15	Ea	.69	4.03	4.72
With bracket	1E@0.15	Ea	1.26	4.03	5.29
Single gang switch boxes, 1-1/2" deep					
With mounting ears	1E@0.15	Ea	.68	4.03	4.71
Single gang switch boxes, 1-3/4" deep					
With mounting ears	1E@0.15	Ea	.71	4.03	4.74
With ears, clamps	1E@0.15	Ea	1.26	4.03	5.29
With nails	1E@0.15	Ea	.84	4.03	4.87

Use these figures to estimate the cost of fiberglass outlet boxes and covers installed in residential buildings under the conditions described on pages 5 and 6. Costs listed are for each box or cover installed. The crew is one electrician working at a labor cost of $26.88 per manhour. These costs include layout, material handling, and normal waste. Add for wiring devices, covers, sales tax, delivery, supervision, mobilization, demobilization, cleanup, overhead and profit. Note: These boxes are generally used with Romex (type NM) cable in residences. Ground the box by attaching the Romex ground conductor to the ground terminal on the wiring device. Round boxes are generally used for lighting fixtures and junction boxes. They're usually installed flush with the wall or ceiling surface. A lighting fixture can be attached directly to the box with 8-32 machine screws. Conductors must be connected in the box with approved connectors. Be sure the outlet box is secured to the building frame well enough to support the weight of the lighting fixture. The outlet box can either be nailed to a stud or fastened with wood screws.

Fiberglass Switch Boxes

Material	Craft@Hrs	Unit	Material Cost	Labor Cost	Installed Cost
Single gang fiberglass switch boxes, 2-1/4" deep					
With nails	1E@0.15	Ea	.84	4.03	4.87
With bracket	1E@0.15	Ea	1.14	4.03	5.17
Single gang fiberglass switch boxes, 2-1/2" deep					
With mounting ears	1E@0.15	Ea	.84	4.03	4.87
With ears, clamps	1E@0.15	Ea	1.02	4.03	5.05
With ears, bracket	1E@0.15	Ea	1.02	4.03	5.05
With bracket, 4 clamps	1E@0.15	Ea	1.02	4.03	5.05
With bracket, 8 clamps	1E@0.15	Ea	1.37	4.03	5.40
Single gang fiberglass switch boxes, 2-9/16" deep					
With nails	1E@0.15	Ea	.83	4.03	4.86
Single gang fiberglass switch boxes, 2-3/4" deep					
With nails	1E@0.15	Ea	.66	4.03	4.69
With speed clips	1E@0.15	Ea	.85	4.03	4.88
With bracket	1E@0.15	Ea	.96	4.03	4.99
Single gang fiberglass switch boxes, 2-7/8" deep					
With nails	1E@0.15	Ea	.78	4.03	4.81
With mounting ears	1E@0.15	Ea	.78	4.03	4.81
With ears, clamps	1E@0.15	Ea	1.28	4.03	5.31
With ears, bracket	1E@0.15	Ea	1.11	4.03	5.14
With ears, bkt, 4 clamps	1E@0.15	Ea	.99	4.03	5.02
With ears, bkt, 8 clamps	1E@0.15	Ea	1.34	4.03	5.37
Single gang fiberglass switch boxes, 3" deep					
With nails	1E@0.15	Ea	.46	4.03	4.49
With nails speed clips	1E@0.15	Ea	.68	4.03	4.71
With bracket	1E@0.15	Ea	.85	4.03	4.88
Single gang fiberglass switch boxes, 3-1/4" deep					
With nails	1E@0.15	Ea	.46	4.03	4.49
With bracket	1E@0.15	Ea	.85	4.03	4.88

Use these figures to estimate the cost of fiberglass switch boxes installed in residential buildings under the conditions described on pages 5 & 6. Costs listed are for each box installed. The crew is one electrician working at a labor cost of $26.88 per manhour. These costs include layout, material handling, and normal waste. Add for wiring devices, sales tax, delivery, supervision, mobilization, demobilization, cleanup, overhead and profit. Note: These boxes are generally used with Romex (type NM) cable in residences. Ground the box by attaching the Romex ground conductor to the ground terminal on the wiring device. Fiberglass single-gang boxes can be used for lighting switches, duplex receptacles, single receptacles, clock outlets and other wiring devices that fit into the box and match the mounts in the box.

Material	Craft@Hrs	Unit	Material Cost	Labor Cost	Installed Cost
Two gang fiberglass switch boxes, 1-3/4" deep					
Box with mounting ears	1E@0.25	Ea	1.50	6.72	8.22
Two gang fiberglass switch boxes, 2-1/2" deep					
With mounting ears	1E@0.25	Ea	1.45	6.72	8.17
With ears, clamps	1E@0.25	Ea	1.71	6.72	8.43
With ears, bracket	1E@0.25	Ea	1.65	6.72	8.37
With ears, bracket, clamps	1E@0.25	Ea	1.94	6.72	8.66
Two gang fiberglass switch boxes, 3" deep					
With nails	1E@0.25	Ea	1.34	6.72	8.06
With nails & 2 clamps	1E@0.25	Ea	1.60	6.72	8.32
With nails & 4 clamps	1E@0.25	Ea	1.85	6.72	8.57
With nails & speed clips	1E@0.25	Ea	1.34	6.72	8.06
With hanger	1E@0.25	Ea	1.70	6.72	8.42
With hanger & 2 clamps	1E@0.25	Ea	2.00	6.72	8.72
With hanger & 4 clamps	1E@0.25	Ea	2.35	6.72	9.07
Three gang switch boxes, 3" deep					
With nails	1E@0.30	Ea	1.90	8.06	9.96
With nails & 3 clamps	1E@0.30	Ea	2.26	8.06	10.32
With nails & 6 clamps	1E@0.30	Ea	2.70	8.06	10.76
With nails & hanger	1E@0.30	Ea	2.39	8.06	10.45
With bracket	1E@0.30	Ea	2.25	8.06	10.31
With nails & speed clamp	1E@0.30	Ea	2.80	8.06	10.86
With offset bracket	1E@0.30	Ea	2.25	8.06	10.31
With bracket, hanger	1E@0.30	Ea	2.74	8.06	10.80
With bkt, hgr. & 3 clamps	1E@0.30	Ea	3.15	8.06	11.21
With offset brackets	1E@0.30	Ea	2.39	8.06	10.45
With brackets & 3 clamps	1E@0.30	Ea	2.94	8.06	11.00
Four gang switch boxes, 3" deep					
With nails	1E@0.50	Ea	2.70	13.40	16.10
With nails & 4 clamps	1E@0.50	Ea	3.26	13.40	16.66
With nails & 8 clamps	1E@0.50	Ea	3.86	13.40	17.26
With nails & bracket	1E@0.50	Ea	3.24	13.40	16.64
With nails, bkt & 4 clamps	1E@0.50	Ea	3.81	13.40	17.21
With nails, bkt & 6 clamps	1E@0.50	Ea	4.32	13.40	17.72
With offset bracket	1E@0.50	Ea	3.06	13.40	16.46
With adj. bar hanger	1E@0.50	Ea	3.62	13.40	17.02
With hanger & 4 clamps	1E@0.50	Ea	3.99	13.40	17.39
With 2 offset brackets	1E@0.50	Ea	3.06	13.40	16.46

Use these figures to estimate the cost of two gang fiberglass switch boxes installed in residential buildings under the conditions described on pages 5 & 6. Costs listed are for each box installed. The crew is one electrician working at a labor cost of $26.88 per manhour. These costs include layout, material handling, and normal waste Add for wiring devices, sales tax, delivery, supervision, mobilization, demobilization, cleanup, overhead and profit. Note: These boxes are generally used with Romex (type NM) cable in residences. Ground the box by attaching the Romex ground conductor to the ground terminal on the wiring device. Fiberglass two gang boxes can be used for lighting switches, duplex receptacles, single receptacles, clock outlets and other wiring devices that fit into the box and match the mounts in the box. These boxes are installed flush with the surface of the wall.

NEMA Class 1 Sheet Metal Pull Boxes

Material	Craft@Hrs	Unit	Material Cost	Labor Cost	Installed Cost
NEMA class 1 sheet metal surface or flush mounted screw cover pull boxes					
4 x 4 x 4	1E@0.25	Ea	4.86	6.72	11.58
4 x 6 x 4	1E@0.30	Ea	5.81	8.06	13.87
6 x 6 x 3	1E@0.35	Ea	7.08	9.41	16.49
6 x 6 x 4	1E@0.35	Ea	6.89	9.41	16.30
6 x 8 x 4	1E@0.40	Ea	7.97	10.80	18.77
6 x 12 x 4	1E@0.45	Ea	11.30	12.10	23.40
8 x 8 x 4	1E@0.45	Ea	9.59	12.10	21.69
8 x 10 x 4	1E@0.50	Ea	10.60	13.40	24.00
8 x 12 x 4	1E@0.50	Ea	12.10	13.40	25.50
10 x 10 x 4	1E@0.50	Ea	15.90	13.40	29.30
10 x 12 x 4	1E@0.55	Ea	18.40	14.80	33.20
12 x 12 x 4	1E@0.55	Ea	20.70	14.80	35.50
12 x 16 x 4	1E@0.60	Ea	28.90	16.10	45.00
12 x 18 x 4	1E@0.65	Ea	30.10	17.50	47.60
12 x 24 x 4	1E@0.70	Ea	39.60	18.80	58.40
18 x 18 x 4	1E@0.70	Ea	40.40	18.80	59.20
18 x 24 x 4	1E@0.75	Ea	72.70	20.20	92.90
18 x 30 x 4	1E@0.80	Ea	76.90	21.50	98.40
24 x 24 x 4	1E@0.80	Ea	76.60	21.50	98.10
24 x 30 x 4	1E@1.00	Ea	90.90	26.90	117.80
24 x 36 x 4	1E@1.25	Ea	100.00	33.60	133.60
30 x 30 x 4	1E@1.25	Ea	108.00	33.60	141.60
6 x 6 x 6	1E@0.40	Ea	8.15	10.80	18.95
6 x 8 x 6	1E@0.45	Ea	9.59	12.10	21.69
6 x 12 x 6	1E@0.50	Ea	12.20	13.40	25.60
8 x 8 x 6	1E@0.50	Ea	11.40	13.40	24.80
8 x 10 x 6	1E@0.50	Ea	12.70	13.40	26.10
8 x 12 x 6	1E@0.55	Ea	14.50	14.80	29.30
10 x 10 x 6	1E@0.55	Ea	15.00	14.80	29.80
10 x 12 x 6	1E@0.60	Ea	16.60	16.10	32.70
12 x 12 x 6	1E@0.60	Ea	18.70	16.10	34.80
12 x 16 x 6	1E@0.65	Ea	25.50	17.50	43.00
12 x 18 x 6	1E@0.70	Ea	27.50	18.80	46.30
12 x 24 x 6	1E@0.75	Ea	34.10	20.20	54.30
15 x 18 x 6	1E@0.75	Ea	27.10	20.20	47.30
16 x 16 x 6	1E@0.75	Ea	31.30	20.20	51.50
18 x 18 x 6	1E@0.75	Ea	33.90	20.20	54.10
18 x 24 x 6	1E@0.80	Ea	61.90	21.50	83.40
18 x 30 x 6	1E@0.90	Ea	83.10	24.20	107.30
24 x 24 x 6	1E@1.00	Ea	77.50	26.90	104.40
24 x 30 x 6	1E@1.25	Ea	88.90	33.60	122.50
24 x 36 x 6	1E@1.25	Ea	111.00	33.60	144.60
30 x 30 x 6	1E@1.50	Ea	133.00	40.30	173.30

Use these figures to estimate the cost of sheet metal screw cover pull boxes installed on conduit under the conditions described on pages 5 and 6. Costs listed are for each pull box installed. The crew is one electrician working at a labor cost of $26.88 per manhour. These costs include the cover, layout, material handling, and normal waste. Add for supports, sales tax, delivery, supervision, mobilization, demobilization, cleanup, overhead and profit. Note: Pull boxes are used as junction boxes when a smaller box would be too crowded with wire and when a long conduit run must be divided into two pulls. Special sizes of pull boxes are available when a box must fit in a limited space. Allow several weeks for delivery of special size boxes. Many inspection authorities want to see the UL (Underwriters' Laboratory) label on items like pull boxes.

Material	Craft@Hrs	Unit	Material Cost	Labor Cost	Installed Cost

NEMA class 1 sheet metal surface or flush mounted hinged cover pull boxes

Material	Craft@Hrs	Unit	Material Cost	Labor Cost	Installed Cost
6 x 6 x 4	1E@0.35	Ea	7.83	9.41	17.24
8 x 8 x 4	1E@0.45	Ea	10.20	12.10	22.30
10 x 10 x 4	1E@0.50	Ea	13.40	13.40	26.80
10 x 12 x 4	1E@0.55	Ea	14.90	14.80	29.70
12 x 12 x 4	1E@0.55	Ea	16.60	14.80	31.40
12 x 18 x 4	1E@0.65	Ea	26.90	17.50	44.40
6 x 6 x 6	1E@0.40	Ea	8.73	10.80	19.53
8 x 8 x 6	1E@0.50	Ea	12.20	13.40	25.60
12 x 12 x 6	1E@0.60	Ea	20.00	16.10	36.10
12 x 18 x 6	1E@0.70	Ea	29.20	18.80	48.00
8 x 8 x 8	1E@0.55	Ea	22.50	14.80	37.30
12 x 12 x 8	1E@0.65	Ea	31.20	17.50	48.70
12 x 16 x 8	1E@0.70	Ea	40.80	18.80	59.60
12 x 18 x 8	1E@0.75	Ea	43.90	20.20	64.10
12 x 24 x 8	1E@0.80	Ea	48.90	21.50	70.40
18 x 18 x 8	1E@0.80	Ea	59.70	21.50	81.20
18 x 24 x 8	1E@1.00	Ea	82.60	26.90	109.50
24 x 24 x 8	1E@1.25	Ea	104.00	33.60	137.60
24 x 30 x 8	1E@1.50	Ea	106.00	40.30	146.30
24 x 36 x 8	1E@1.60	Ea	122.00	43.00	165.00
30 x 30 x 8	1E@1.75	Ea	163.00	47.00	210.00
12 x 18 x 10	1E@0.80	Ea	58.70	21.50	80.20
12 x 24 x 10	1E@0.90	Ea	65.80	24.20	90.00
18 x 18 x 10	1E@1.00	Ea	66.20	26.90	93.10
18 x 24 x 10	1E@1.25	Ea	108.00	33.60	141.60
24 x 24 x 10	1E@1.50	Ea	131.00	40.30	171.30
24 x 30 x 10	1E@1.75	Ea	158.00	47.00	205.00
24 x 36 x 10	1E@2.00	Ea	179.00	53.80	232.80
30 x 30 x 10	1E@2.25	Ea	225.00	60.50	285.50
12 x 24 x 12	1E@1.00	Ea	73.10	26.90	100.00
18 x 18 x 12	1E@1.00	Ea	87.60	26.90	114.50
18 x 24 x 12	1E@1.50	Ea	102.00	40.30	142.30
24 x 24 x 12	1E@1.75	Ea	143.00	47.00	190.00
24 x 30 x 12	1E@2.00	Ea	187.00	53.80	240.80
24 x 36 x 12	1E@2.25	Ea	199.00	60.50	259.50
30 x 36 x 12	1E@2.50	Ea	298.00	67.20	365.20
36 x 36 x 12	1E@2.75	Ea	431.00	73.90	504.90

Use these figures to estimate the cost of sheet metal hinged cover pull boxes installed on conduit under the conditions described on pages 5 and 6. Costs listed are for each pull box installed. The crew is one electrician working at a labor cost of $26.88 per manhour. These costs include the cover, layout, material handling, and normal waste. Add for supports, sales tax, delivery, supervision, mobilization, demobilization, cleanup, overhead and profit. Note: Pull boxes are used as junction boxes when a smaller box would be too crowded with wire and when a long conduit run must be divided into two pulls. Special sizes of pull boxes are available when a box must fit in a limited space. Allow several weeks for delivery of special size boxes. Many inspection authorities want to see the UL (Underwriters' Laboratory) label on items like pull boxes.

NEMA Class 3R (Raintight) Pull Boxes

NEMA class 3R (raintight) pull boxes

Material	Craft@Hrs	Unit	Material Cost	Labor Cost	Installed Cost
4 x 4 x 4	1E@0.30	Ea	8.46	8.06	16.52
4 x 6 x 4	1E@0.35	Ea	9.45	9.41	18.86
6 x 6 x 4	1E@0.35	Ea	11.10	9.41	20.51
6 x 8 x 4	1E@0.40	Ea	12.60	10.80	23.40
8 x 8 x 4	1E@0.45	Ea	14.50	12.10	26.60
8 x 10 x 4	1E@0.50	Ea	16.40	13.40	29.80
8 x 12 x 4	1E@0.50	Ea	18.50	13.40	31.90
10 x 10 x 4	1E@0.50	Ea	19.20	13.40	32.60
10 x 12 x 4	1E@0.55	Ea	21.60	14.80	36.40
12 x 12 x 4	1E@0.55	Ea	23.40	14.80	38.20
12 x 16 x 4	1E@0.50	Ea	33.40	13.40	46.80
18 x 18 x 4	1E@0.70	Ea	44.90	18.80	63.70
6 x 6 x 6	1E@0.40	Ea	14.70	10.80	25.50
6 x 8 x 6	1E@0.45	Ea	14.90	12.10	27.00
8 x 8 x 6	1E@0.50	Ea	14.90	13.40	28.30
8 x 10 x 6	1E@0.50	Ea	19.50	13.40	32.90
8 x 12 x 6	1E@0.55	Ea	22.10	14.80	36.90
10 x 10 x 6	1E@0.55	Ea	22.80	14.80	37.60
10 x 12 x 6	1E@0.60	Ea	25.20	16.10	41.30
12 x 12 x 6	1E@0.60	Ea	28.10	16.10	44.20
12 x 16 x 6	1E@0.65	Ea	35.50	17.50	53.00
12 x 18 x 6	1E@0.70	Ea	36.90	18.80	55.70
12 x 24 x 6	1E@0.75	Ea	66.60	20.20	86.80
18 x 18 x 6	1E@0.75	Ea	54.90	20.20	75.10
18 x 24 x 6	1E@0.80	Ea	69.80	21.50	91.30
24 x 24 x 6	1E@1.00	Ea	94.90	26.90	121.80
12 x 12 x 8	1E@0.65	Ea	32.20	17.50	49.70
12 x 18 x 8	1E@0.75	Ea	41.90	20.20	62.10
18 x 18 x 8	1E@0.80	Ea	61.30	21.50	82.80
18 x 24 x 8	1E@1.00	Ea	84.90	26.90	111.80
24 x 24 x 8	1E@1.25	Ea	106.00	33.60	139.60
18 x 18 x 10	1E@1.00	Ea	68.10	26.90	95.00
18 x 24 x 10	1E@1.25	Ea	110.00	33.60	143.60
24 x 30 x 10	1E@1.75	Ea	163.00	47.00	210.00
24 x 24 x 12	1E@2.00	Ea	147.00	53.80	200.80

Use these figures to estimate the cost of raintight screw cover pull boxes installed on conduit under the conditions described on pages 5 and 6. Costs listed are for each pull box installed. The crew is one electrician working at a labor cost of $26.88 per manhour. These costs include the cover, layout, material handling, and normal waste. Add for supports, sales tax, delivery, supervision, mobilization, demobilization, cleanup, overhead and profit. Note: Pull boxes are used as junction boxes when a smaller box would be too crowded with wire and when a long conduit run must be divided into two pulls. Raintight pull boxes usually have knockouts on the box bottom. Special sizes of pull boxes are available when a box must fit in a limited space. Allow several weeks for delivery of special size boxes. Many inspection authorities want to see the UL (Underwriters' Laboratory) label on items like pull boxes.

Material	Craft@Hrs	Unit	Material Cost	Labor Cost	Installed Cost

NEMA class 3R (raintight) galvanized sheet metal hinged pull boxes

Material	Craft@Hrs	Unit	Material Cost	Labor Cost	Installed Cost
6 x 6 x 4	1E@0.35	Ea	19.10	9.41	28.51
8 x 8 x 4	1E@0.45	Ea	24.50	12.10	36.60
10 x 10 x 4	1E@0.50	Ea	32.00	13.40	45.40
12 x 12 x 4	1E@0.55	Ea	35.70	14.80	50.50
12 x 12 x 6	1E@0.65	Ea	41.30	17.50	58.80
12 x 18 x 6	1E@0.70	Ea	46.30	18.80	65.10
24 x 24 x 6	1E@1.00	Ea	134.00	26.90	160.90
18 x 18 x 8	1E@0.80	Ea	86.40	21.50	107.90
24 x 24 x 10	1E@1.50	Ea	136.00	40.30	176.30

J 3R

JIC wiring boxes, NEMA 4, hinged clamping cover, no panels

Material	Craft@Hrs	Unit	Material Cost	Labor Cost	Installed Cost
4 x 4 x 3	1E@0.30	Ea	29.40	8.06	37.46
6 x 4 x 3	1E@0.40	Ea	33.10	10.80	43.90
8 x 6 x 3-1/2	1E@0.50	Ea	41.00	13.40	54.40
6 x 4 x 4	1E@0.30	Ea	40.10	8.06	48.16
6 x 8 x 4	1E@0.40	Ea	38.80	10.80	49.60
10 x 8 x 4	1E@0.60	Ea	51.00	16.10	67.10
12 x 10 x 5	1E@0.70	Ea	65.10	18.80	83.90
14 x 12 x 6	1E@0.80	Ea	81.90	21.50	103.40
16 x 14 x 6	1E@0.90	Ea	96.50	24.20	120.70

J 3R

JIC wiring boxes, NEMA 12, hinged clamping cover, no panels

Material	Craft@Hrs	Unit	Material Cost	Labor Cost	Installed Cost
4 x 4 x 3	1E@0.40	Ea	20.60	10.80	31.40
6 x 4 x 3	1E@0.50	Ea	23.60	13.40	37.00
8 x 6 x 3-1/2	1E@0.60	Ea	41.00	16.10	57.10
4 x 4 x 4	1E@0.40	Ea	21.60	10.80	32.40
6 x 4 x 4	1E@0.45	Ea	24.80	12.10	36.90
6 x 6 x 4	1E@0.50	Ea	27.50	13.40	40.90
10 x 8 x 4	1E@0.70	Ea	36.00	18.80	54.80
12 x 10 x 5	1E@0.80	Ea	46.30	21.50	67.80
14 x 12 x 6	1E@0.90	Ea	58.40	24.20	82.60
16 x 14 x 6	1E@1.00	Ea	68.50	26.90	95.40

J 3R

JIC panels

Material	Craft@Hrs	Unit	Material Cost	Labor Cost	Installed Cost
6 x 4 x 3	1E@0.05	Ea	1.96	1.34	3.30
8 x 6 x 3-1/2	1E@0.05	Ea	2.70	1.34	4.04
6 x 4 x 4	1E@0.05	Ea	1.96	1.34	3.30
6 x 6 x 4	1E@0.05	Ea	2.16	1.34	3.50
10 x 8 x 4	1E@0.05	Ea	3.58	1.34	4.92
12 x 10 x 5	1E@0.06	Ea	4.73	1.61	6.34
14 x 12 x 6	1E@0.06	Ea	6.28	1.61	7.89
16 x 14 x 6	1E@0.08	Ea	8.31	2.15	10.46

J 3R

Use these figures to estimate the cost of raintight hinged cover pull boxes and weatherproof gasketed pull boxes installed on conduit under the conditions described on pages 5 and 6. Costs listed are for each pull box installed. The crew is one electrician working at a labor cost of $26.88 per manhour. These costs include the cover, layout, material handling, and normal waste. Add for supports, sales tax, delivery, supervision, mobilization, demobilization, cleanup, overhead and profit. Note: Pull boxes are used as junction boxes when a smaller box would be too crowded with wire and when a long conduit run must be divided into two pulls. Raintight pull boxes usually have knockouts on the box bottom. Special sizes of pull boxes are available when a box must fit in a limited space. Allow several weeks for delivery of special size boxes. Many inspection authorities want to see the UL (Underwriters' Laboratory) label on items like pull boxes.

Round Floor Outlet Boxes

Material	Craft@Hrs	Unit	Material Cost	Labor Cost	Installed Cost

Adjustable sheet metal round floor outlet boxes

Opening	Depth	Craft@Hrs	Unit	Material Cost	Labor Cost	Installed Cost
4-3/16"	3"	1E@1.00	Ea	34.80	26.90	61.70
4-3/16"	4"	1E@1.00	Ea	44.60	26.90	71.50

Semi-adjustable sheet metal round floor outlet boxes

Opening	Depth	Craft@Hrs	Unit	Material Cost	Labor Cost	Installed Cost
3-5/16"	2-1/2"	1E@1.00	Ea	37.10	26.90	64.00
3-5/16"	3"	1E@1.00	Ea	46.40	26.90	73.30

Adjustable cast iron round floor outlet boxes

Opening	Depth	Craft@Hrs	Unit	Material Cost	Labor Cost	Installed Cost
4-3/16"	2-5/8"	1E@1.00	Ea	38.80	26.90	65.70
4-3/16"	2-13/16"	1E@1.00	Ea	27.70	26.90	54.60
4-3/16"	3"	1E@1.00	Ea	49.40	26.90	76.30
4-3/16"	3-1/16"	1E@1.00	Ea	32.40	26.90	59.30
4-3/16	3-3/4"	1E@1.00	Ea	55.20	26.90	82.10
4-3/16"	4-3/4"	1E@1.25	Ea	86.80	33.60	120.40

Semi-adjustable cast iron round floor outlet boxes

Opening	Depth	Craft@Hrs	Unit	Material Cost	Labor Cost	Installed Cost
3-3/4"	2"	1E@1.00	Ea	46.50	26.90	73.40

Non-adjustable cast iron round floor outlet boxes

Opening	Depth	Craft@Hrs	Unit	Material Cost	Labor Cost	Installed Cost
4-3/16"	3-3/4"	1E@1.00	Ea	32.10	26.90	59.00

Use these figures to estimate the cost of floor boxes installed in building floors under the conditions described on pages 5 and 6. Costs listed are for each floor box installed. The crew is one electrician working at a labor cost of $26.88 per manhour. These costs include layout, material handling, and normal waste. Add for supports, wiring devices, special adapters, covers, sales tax, delivery, supervision, mobilization, demobilization, cleanup, overhead and profit. Note: The labor cost assumes that floor boxes are being installed in a concrete floor slab on grade. Deduct 10% for formed concrete floor slabs installed above grade. Many other sizes of floor boxes are available. Cast iron boxes have threaded hubs of various sizes. Some boxes can be installed in wood floor framing.

Adjustable Floor Boxes and Covers

Material	Craft@Hrs	Unit	Material Cost	Labor Cost	Installed Cost

Adjustable round floor box covers

Diameter	Use					
3-5/16"	3/4" hole	1E@0.10	Ea	24.70	2.69	27.39
3-5/16"	1" hole	1E@0.10	Ea	26.50	2.69	29.19
3-5/16"	SR	1E@0.10	Ea	25.10	2.69	27.79
3-3/4"	3/4" hole	1E@0.10	Ea	46.90	2.69	49.59
3-3/4"	1" hole	1E@0.10	Ea	30.30	2.69	32.99
3-3/4"	SR	1E@0.10	Ea	25.20	2.69	27.89
4-3/16"	3/4" hole	1E@0.10	Ea	24.50	2.69	27.19
4-3/16"	1" hole	1E@0.10	Ea	31.00	2.69	33.69
4-3/16"	2-1/8 S R	1E@0.10	Ea	21.60	2.69	24.29
4-3/16"	2-3/8 S R	1E@0.10	Ea	43.90	2.69	46.59
4-3/16"	DR	1E@0.10	Ea	32.40	2.69	35.09
4-3/16"	duplex flap	1E@0.10	Ea	38.40	2.69	41.09

Adjustable rectangle sheet metal floor boxes

Opening	Depth					
3-3/16 x 4-3/8	3"	1E@1.25	Ea	53.00	33.60	86.60
3-3/16 x 4-3/8	4"	1E@1.25	Ea	42.30	33.60	75.90

Adjustable rectangle one gang cast iron floor boxes

Opening	Depth					
3-3/16 x 4-3/8	2-7/8"	1E@1.25	Ea	45.60	33.60	79.20
3-3/16 x 4-3/8	3-1/8"	1E@1.25	Ea	49.30	33.60	82.90

Adjustable rectangle two gang cast iron floor boxes

Opening	Depth					
6-3/8 x 4-3/8	3-3/8"	1E@1.50	Ea	101.00	40.30	141.30

Adjustable rectangle three gang cast iron floor boxes

Opening	Depth					
9-9/16 x 4-3/8	3-3/8"	1E@1.75	Ea	143.00	47.00	190.00

Use these figures to estimate the cost of floor boxes installed in building floors under the conditions described on pages 5 and 6. Costs listed are for each floor box installed. The crew is one electrician working at a labor cost of $26.88 per manhour. These costs include layout, material handling, and normal waste. Add for supports, wiring devices, special adapters, covers, sales tax, delivery, supervision, mobilization, demobilization, cleanup, overhead and profit. Note: The labor cost assumes that floor boxes are being installed in a concrete floor slab on grade. Deduct 10% for formed concrete floor slabs installed above grade. Many other sizes of floor boxes are available. Cast iron boxes have threaded hubs of various sizes. Some boxes can be installed in wood floor framing. SR is single receptacle, DR is duplex receptacle. Semi-adjustable rectangle one gang cast iron floor boxes

Semi-adjustable Floor Boxes and Covers

Material	Craft@Hrs	Unit	Material Cost	Labor Cost	Installed Cost
Semi-adjustable rectangle one gang cast iron floor boxes					
Opening ___ Depth					
3-3/16 x 4-3/8 2"	1E@1.25	Ea	45.60	33.60	79.20
Semi-adjustable rectangle two gang cast iron floor boxes					
Opening ___ Depth					
6-3/8 x 4-3/8 2"	1E@1.50	Ea	73.40	40.30	113.70
Semi-adjustable rectangle three gang cast iron floor boxes					
Opening ___ Depth					
9-9/16 x 4-3/8 2"	1E@1.75	Ea	119.00	47.00	166.00
Semi-adjustable rectangle floor box covers					
Opening					
2-1/8 - 3/4" combination	1E@0.15	Ea	26.40	4.03	30.43
2-1/8 - 1" combination	1E@0.15	Ea	29.60	4.03	33.63
2-1/8 - single	1E@0.15	Ea	26.20	4.03	30.23
2-3/8 - single	1E@0.15	Ea	60.00	4.03	64.03
Duplex	1E@0.15	Ea	30.80	4.03	34.83
Duplex, flip	1E@0.15	Ea	56.30	4.03	60.33
Pedestal 3/4 x 3"	1E@0.10	Ea	22.00	2.69	24.69
Pedestal 3/4 x 6"	1E@0.10	Ea	36.60	2.69	39.29
Pedestal duplex 3/4"	1E@0.25	Ea	124.00	6.72	130.72
Pedestal duplex 1"	1E@0.25	Ea	124.00	6.72	130.72
Pedestal duplex 2G 3/4"	1E@0.35	Ea	122.00	9.41	131.41
Pedestal duplex 2G 1"	1E@0.25	Ea	122.00	6.72	128.72
Pedestal telephone	1E@0.25	Ea	73.00	6.72	79.72
Pedestal single reception	1E@0.25	Ea	80.10	6.72	86.82
Pedestal flange, 3/4"	1E@0.15	Ea	36.30	4.03	40.33
Pedestal flange, 1"	1E@0.15	Ea	46.50	4.03	50.53
Mound, telephone, 1 side	1E@0.25	Ea	57.90	6.72	64.62
Mound, telephone, 2 sides	1E@0.25	Ea	58.10	6.72	64.82
Mound, duplex, 1 side	1E@0.25	Ea	59.50	6.72	66.22
Mound, duplex 2 sides	1E@0.25	Ea	65.90	6.72	72.62

Use these figures to estimate the cost of floor boxes and accessories installed in building floors under the conditions described on pages 5 and 6. Costs listed are for each unit installed. The crew is one electrician working at a labor cost of $26.88 per manhour. These costs include layout, material handling, and normal waste. Add for supports, wiring devices, special adapters, sales tax, delivery, supervision, mobilization, demobilization, cleanup, overhead and profit.

Material	Craft@Hrs	Unit	Material Cost	Labor Cost	Installed Cost

Round brass carpet flanges for floor boxes

Carpet thickness

Material	Craft@Hrs	Unit	Material Cost	Labor Cost	Installed Cost
3/8"	1E@0.10	Ea	39.00	2.69	41.69
1/2"	1E@0.10	Ea	41.90	2.69	44.59
5/8"	1E@0.10	Ea	44.70	2.69	47.39
3/4"	1E@0.10	Ea	47.70	2.69	50.39

Clear polycarbonate carpet flanges for floor boxes

Material	Craft@Hrs	Unit	Material Cost	Labor Cost	Installed Cost
Round	1E@0.10	Ea	10.60	2.69	13.29
Rectangle, one gang	1E@0.15	Ea	11.30	4.03	15.33
Rectangle, two gang	1E@0.17	Ea	13.80	4.57	18.37
Rectangle, three gang	1E@0.20	Ea	17.50	5.38	22.88

Brass crown plugs for floor boxes

Material	Craft@Hrs	Unit	Material Cost	Labor Cost	Installed Cost
3/4"	1E@0.05	Ea	5.28	1.34	6.62
1"	1E@0.06	Ea	8.18	1.61	9.79

Brass flush plugs for floor boxes

Material	Craft@Hrs	Unit	Material Cost	Labor Cost	Installed Cost
3/4"	1E@0.05	Ea	9.60	1.34	10.94
1"	1E@0.06	Ea	16.00	1.61	17.61
1-1/2"	1E@0.10	Ea	12.80	2.69	15.49
2-1/8"	1E@0.10	Ea	10.90	2.69	13.59
2-3/8"	1E@0.10	Ea	18.40	2.69	21.09

Use these figures to estimate the cost of floor box accessories installed in building floors under the conditions described on pages 5 and 6. Costs listed are for each unit installed. The crew is one electrician working at a labor cost of $26.88 per manhour. These costs include layout, material handling, and normal waste. Add for supports, wiring devices, special adapters, sales tax, delivery, supervision, mobilization, demobilization, cleanup, overhead and profit. Note: Some manufacturers offer aluminum finish floor box parts that fit their floor boxes. The labor for these accessories will be the same as for the brass accessories listed above. Adjustable floor boxes are usually installed in concrete floors. Non-adjustable boxes are usually installed in wood floors. The boxes and accessories listed in this section are only a few of the many items available.

Section 4:
Lighting Fixtures

If you handle many commercial jobs, you'll spend a lot of time estimating lighting fixtures. Light fixtures are a major cost item on many commercial jobs. Designers scour manufacturer's catalogs to find exactly the right fixture that shows off their client's merchandise to best advantage. When you see a particular fixture specified for an upscale store, you can bet that thought and consideration went into its selection.

There's a vogue in lighting fixtures, just as there is in hair styles and Paris gowns. And, from time to time, a new edition of the *National Electrical Code* changes the way some lighting fixtures are made and installed. That's why lighting fixtures become obsolete and are replaced by more modern, efficient or attractive units.

In the last 10 years *NEC* restrictions and state law have set limits on lighting in residential and commercial buildings. The purpose of these regulations is to promote the conservation of electricity. Standards have been set for energy use in most work environments. Lighting levels have been decreased and better controlled in many areas. In response, manufacturers have designed energy-efficient lighting fixtures for both new construction and retrofit jobs. Don't underestimate the care needed when estimating electrical lighting fixtures. Start with a complete take-off of the lighting system as it's designed. Then review the installation details for each fixture. Count all the special components that will be required: stem and canopy hangers, ceiling spacers, end-of-row caps, special support material, the type and quantity of lamps, mounting height, safety clips, safety wires or cables, guards, finish color, etc. There are so many choices and components that it's easy to overlook something.

Stem and Canopies

Many lighting fixtures are suspended on a stem device. The stem is usually attached to the outlet box and both supports the fixture and serves as the wireway from the outlet box to the fixture. Some engineers require that the stem include a swivel so the fixture can swing free in an earthquake or if hit by something. Otherwise, the stem might break, dropping the fixture on whatever is directly below. The swivel is usually a ball aligner fitting that's attached to an outlet box fitted with a threaded hub to receive the stem. Ball aligner fittings come in several sizes that fit most of the popular stem sizes.

Ball aligners are made for a certain degree of movement. The maximum movement permitted varies with the manufacturer and with the style of swivel. Be sure the aligner you install meets job specs.

The stem canopy is an escutcheon that serves as trim between the connecting part of the stem and the outlet box. Some canopy assemblies have a swivel built in. Some are referred to as *earthquake stem and canopies*. Some stem and canopy sets also have a swivel fitting at the point of connection to the lighting fixture. This dual swivel system is common on fluorescent fixtures where more than one stem is used on the same fixture, or when fluorescent fixtures are suspended in rows of two or more.

Four-foot-long fluorescent fixtures mounted in rows may need only one stem and canopy per fixture, plus an additional stem and canopy for the end fixture. For example, if you find a row of three fixtures on the plans, figure on using four stem and canopy sets.

Your supplier will quote a separate price for stem and canopy sets. The price is probably for a standard stem length of two feet, with an additional cost per foot for stems over two feet. If you use long stems, it might be less expensive to buy the stem material yourself and cut individual stems to order in your shop.

Most inspection authorities don't permit couplings in a stem. A coupling becomes the weak spot that would be the point of rupture under stress.

Stem and canopy sets are usually painted with white enamel at the factory. Any other color will probably cost extra. Enamel white is the standard finish because it matches the standard color for fluorescent fixtures.

Ceiling Spacers

Incorrect mounting of some types of lighting fixtures can be a fire hazard. For example, some fixtures have a ballast that's in direct contact with the top of the fixture housing. The housing is going to get pretty hot. You can see that mounting that fixture flush against a combustible ceiling material is a bad idea, even if the inspector doesn't catch the problem.

The code requires spacers that provide a gap between the fixture and the ceiling. The spacer is usually 1½" deep. It's painted white to match the fixture. It attaches to the outlet box and joins it to the fixture. Some spacers can be attached directly to the ceiling material. Your supplier will quote prices on several common types of fixture spacers.

Ideally, support for the fixture is an integral part of the ceiling, such as in a suspended grid ceiling. The grid is designed to include ceiling fixtures and may provide all the support needed. But in many cases you'll have to provide additional support for the fixtures you install.

Usually wood screws are used to attach the fixture or a suspension hanger to the ceiling frame or blocking. Be sure the screws extend through the ceiling surface and into the framing. In an emergency, your support should hold up both the fixture and the weight of a man.

Ganging Fluorescent Fixtures

You'll need some additional parts when fluorescent fixtures are installed in rows of two or more. An end-of-row cap is needed at the end of each row and joiners are needed between fixtures. Usually these parts have to be ordered separately and will be an additional cost item.

When you've finished counting all the fluorescent lighting fixtures during the take-off, go back and count the joiners and end caps. And don't forget the additional labor required to install these fittings.

In some cases your supplier will offer fixtures that are factory-built into rows of fixtures. The fixture cost will be higher, but the installed cost will probably be lower than assemblies of individual fixtures built-up on the job. Long rows of fixtures can be made with fixture channels that reduce the number of ballasts needed and eliminate separate joiners and caps.

Special Support Material

Recessed lighting fixtures in plaster ceilings require plaster frames. These frames are part of the rough-in because they have to be installed before the lath is hung. The plaster frame provides a finished opening for the lighting fixture. When you order recessed fixtures, order the plaster frames at the same time. Plaster frames cost extra, so don't forget to list them separately on your takeoff sheet.

Some recessed fixtures intended for recessed mounting in drywall ceilings also require a factory-made frame that has to be set before the drywall is hung. The drywall installer cuts an opening in the drywall to fit around the frame.

Lighting fixtures installed in some types of drop ceilings may need more support than the drop ceiling offers. Usually furring channels are used to support the ceiling adjacent to the fixture. The support spreads the weight of the fixture over a wider area and helps hold the fixture in position. When a fixture is installed in a concealed spline suspended ceiling

system, the fixture will be attached directly to the additional support.

Some fluorescent fixtures have **bat wings** attached to the housing sides. When the fixture is installed, the bat wings are extended to provide support. Adjusting screws in the bat wings are used to level the fixture flush with the ceiling tile. When this type of fixture is needed, the designer probably specified a fixture with bat wings. But check the exact mounting method to be sure that you're pricing what's needed.

Type and Quantity of Lamps

The lighting fixtures aren't finished until lamps are installed. Check the lighting fixture schedule on the plans or in the specs to see what lamps are needed. The lamps are usually ordered separately — though some fixtures come with lamps. When lamps are included in the fixture price, the supplier's catalog should note that fact.

Sometimes it will save money to have the supplier ship the fixtures with lamps already installed. Consider asking for a quote on this service if you're installing a large quantity of fluorescent fixtures. Some suppliers can provide fixtures with both lamps and a flex conduit tail already installed. This speeds field hookup and may reduce your installed cost on a larger job.

Your take-off of fixtures should leave space for listing the lamps required. Note the quantity of each type of fixture and find the desired lamp. Later, you'll add material prices and installation labor for both fixtures and lamps.

Don't omit the labor for lamp installation. Every lamp has to be ordered, received, stored, handled, distributed on the job and installed. And there will be occasional breakage and waste. If you don't include an estimate of the time required for installing lamps, and include this cost in your bid, you're running a charity, not a business. True, you could include lamping labor in the labor time for fixture installation. But this manual doesn't. Some fixtures arrive on the job with lamps already installed. And occasionally you may have a job where the lamps are not included in the electrical work. That's why it's best to estimate the lamping labor separately.

Mounting Height

Somewhere in the plans, the fixture schedule, or the specs, you'll find the mounting height for suspended and wall-mounted lighting fixtures. Usually the installation height is noted on the plans in the area where the fixture is to be mounted. Note that the mounting height has an effect on installation cost. The longer the suspension system, the more material required and the higher your cost. Higher ceilings have the same effect. Working from the floor is fastest. Using a 6 foot step ladder takes more time. The greater the height, the more equipment needed and the more up and down time required, even if you use a modern power lift.

Safety Clips

Safety clips may be required on fluorescent fixtures installed in suspended ceilings. The clips probably won't show up in the plans or specs. But they're a cost item that should be in your estimate. Safety clips are usually required in exposed "T" bar grid ceilings. The ceiling tile rests in exposed runners. Where a lighting fixture is required, the fixture replaces one of the ceiling tiles. The fixture should have a safety clip attached to either the fixture or the exposed runners. The clip keeps the fixture from falling if the grid is displaced or damaged in some way. In seismic zones these clips are called earthquake clips.

Safety Wire and Cable

On many jobs, safety wires or cable supports will be required in addition to the regular fixture support. Safety wires are often used on fluorescent fixtures mounted in an exposed "T" bar grid ceiling. The wires run from the fixture to an anchor point above the suspended ceiling. Small cable is sometimes used for this purpose. Usually only two wires are needed per fixture. The wires run from diagonal corners of the fixture and should have a little slack so the fixture doesn't lift the "T" bar runners. Safety wires are usually required even though safety clips are also installed. That may seem like wearing both a belt and suspenders to keep your pants up. But I guess you can't be too careful.

Safety cables may also be required on high bay lighting fixtures. If the fixture is mounted on a hook arrangement, the safety cable is attached to the fixture housing and runs either to the framing above or to the conduit system.

Guards

Guards protect either a fixture or the lamp from damage. Some guards are like a simple wire basket. Others are more substantial. Some outdoor wall bracket fixtures have guards to protect the glass globe from damage. Storage room lights may have guards to prevent accidental damage to the lamps. Auditorium fixtures usually have guards to protect the public if glass in the fixture should shatter. Watch for notes in the plans or specs on what guards are required. Most fixtures are available both with and without guards. Omitting the guard from your estimate can be an expensive oversight.

Finish Color

Many designers specify a certain finish or color for the exposed parts of fixtures. Finishes include anodizing, colored enamel, simulated wood finishes or real wood frames. Note that the finish specified may not be the standard finish or may come at additional cost. In some cases, you may have to arrange for custom finishing or do the finishing yourself. Special finishes are expensive. Be sure you understand what finish is required before pricing each fixture.

Estimating Lighting Fixtures

Many electrical estimators start each labor and material take-off by counting the lighting fixtures. Scanning the plans for fixtures is a quick and easy introduction to the project. Count and list each type of lighting fixture on your work sheet. On larger jobs, there may be several sheets with fixtures. As you count the fixtures, study the type of ceiling they will be mounted on or in. Look for the mounting height of suspended and wall fixtures. Look for the fluorescents that are in rows and can be tandem mounted. Watch the voltage and switching controls and look for dimmers. If fluorescent fixtures are controlled from a dimmer, you'll need to price fixtures with special dimming ballasts.

Some specs require lighting fixtures with radio suppressors. The manufacturer will install suppressors, but you have to order the right material and include the extra cost in the estimate.

Some fixtures have remote ballasts that are mounted away from the fixture. Be sure to list these units separately because the labor cost is higher and more materials are needed.

Watch out for lighting fixtures that have to be installed in coving built by the contractor. Also note carefully any special diffusers required on lighting fixtures. They're sure to be an additional cost item. Sometimes the diffuser is supplied by the contractor who builds the coving. That may or may not be the electrical contractor. Be sure you know what your work includes.

When you've taken off the lighting fixtures throughout the project, you know a lot about the building. That makes fixture takeoff a good introduction to the project.

Using a Pricing Service

Some electrical suppliers have the fixture takeoff done by a fixture estimating service. The supplier may quote a competitive package price for all fixtures on the job. The service lists all fixtures — and may even remember to include most of the additional parts and pieces required for installation.

Some of these fixtures services are reasonably accurate. And the package price may be attractive because they substitute the lowest-cost fixtures whenever permitted by the plans and specs. But it's hard for you to check their work and verify fixture prices. And, since the supplier won't be doing the installation, they may omit important items like supports and clips. Your own fixture estimate is more likely to include everything that's needed to comply with the code and do the job.

Doing Your Take-off

Use a separate take-off form to list all lighting fixtures and the accessories. Across the top of the page list the names of each supplier you plan to get fixture quotes from. When you've assembled all quotes and extended figures to a total, select the most competitive supplier. But watch out for those alternates some suppliers like to slip in. It's their duty to advise you when they offer an alternate to what's specified. It's always a gamble to accept an alternate bid without first checking with the designing engineer. If the engineer rejects the alternate, you may have to supply the item specified at your own cost. The supplier won't make good on the difference if the alternate was rejected.

Work carefully when taking off lighting fixtures. Counting the fixtures is easy — deceptively so. Remembering to include everything needed for the lighting system is much harder.

Figuring the Manhours

The labor units listed in this section are based on good installation conditions as defined in the section "How to Use This Book" at the front of this manual. If you expect unusual job conditions to delay production, increase the labor time accordingly.

Here are some conditions that will increase your labor cost:

- A work area littered with construction debris
- Fixtures stored in unmarked boxes
- Fixtures stored at a considerable distance from the points of installation
- Extra high ceilings
- Restricted work area
- Smaller room size
- Working around furniture and equipment in an occupied room. Remodeling work almost always takes more labor per unit than new construction.

It takes judgment to spot conditions that will slow the work. Every job isn't "standard." Don't expect that the manhours in this manual will cover every situation without modification. Be ready to increase (or reduce) the figures offered here when appropriate. It's better to be cautious when in doubt. There are old electrical estimators and there are bold electrical estimators. But there are no *old and bold* electrical estimators.

Surface Mounted Incandescent Light Fixtures

Material		Craft@Hrs	Unit	Material Cost	Labor Cost	Installed Cost
Surface mounted porcelain keyless incandescent light fixtures						
250V 3-3/4" dia.		1E@0.20	Ea	2.01	5.38	7.39
250V 4-1/2" dia.		1E@0.20	Ea	1.88	5.38	7.26
250V 5-1/2" dia.		1E@0.20	Ea	3.08	5.38	8.46
600V 4-1/2" dia.		1E@0.20	Ea	2.42	5.38	7.80
Surface mounted porcelain chain pull incandescent light fixtures						
250V 4-1/2" dia.		1E@0.20	Ea	4.06	5.38	9.44
250V 5-1/4" dia.		1E@0.20	Ea	4.23	5.38	9.61
Cover mounted keyless incandescent light fixtures						
600V 3-1/2" dia.		1E@0.25	Ea	2.78	6.72	9.50
600V 4-1/8" dia.		1E@0.25	Ea	3.48	6.72	10.20
Ceiling square glass incandescent light fixtures, two 60-watt lamps						
White	12"	1E@0.30	Ea	5.79	8.06	13.85
White	14"	1E@0.30	Ea	9.50	8.06	17.56
Satin white	12"	1E@0.30	Ea	8.22	8.06	16.28
Satin white	14"	1E@0.30	Ea	12.50	8.06	20.56
Gold applique	12"	1E@0.30	Ea	8.69	8.06	16.75
Gold applique	14"	1E@0.30	Ea	10.50	8.06	18.56
White textured	12"	1E@0.30	Ea	8.69	8.06	16.75
White patterned	12"	1E@0.30	Ea	5.79	8.06	13.85
White patterned	14"	1E@0.30	Ea	8.86	8.06	16.92
Ceiling square glass incandescent light fixtures, three 60-watt lamps						
White, scroll	13"	1E@0.35	Ea	15.60	9.41	25.01
White patterned	16"	1E@0.35	Ea	15.70	9.41	25.11

Use these figures to estimate the cost of lighting fixtures installed in buildings under the conditions described on pages 5 and 6. Costs listed are for each fixture installed. The crew is one electrician working at a labor cost of $26.88 per manhour. These costs include layout, material handling, and normal waste. Add for lamp, accessories, sales tax, delivery, supervision, mobilization, demobilization, cleanup, overhead and profit. Note: These costs are based on internally pre-wired fixtures installed in new construction not over 10 feet above the floor. Many types of glass diffusers are available. Many have textured glass with gold designs that can be matched with larger or smaller diffusers.

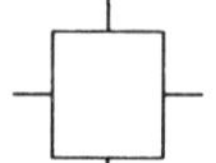

Ceiling Mounted Incandescent Light Fixtures

Material		Craft@Hrs	Unit	Material Cost	Labor Cost	Installed Cost

Ceiling round glass incandescent light fixtures, two 60-watt lamps

Material		Craft@Hrs	Unit	Material Cost	Labor Cost	Installed Cost
Floral textured	12"	1E@0.30	Ea	8.92	8.06	16.98
Satin white	12"	1E@0.30	Ea	9.85	8.06	17.91
Silver applique	13"	1E@0.30	Ea	11.90	8.06	19.96
White floral	13"	1E@0.30	Ea	11.90	8.06	19.96

Ceiling round glass incandescent light fixtures, three 60-watt lamps

Material		Craft@Hrs	Unit	Material Cost	Labor Cost	Installed Cost
Satin white	15"	1E@0.35	Ea	16.50	9.41	25.91
White scroll	15"	1E@0.35	Ea	17.40	9.41	26.81
Satin floral	15"	1E@0.35	Ea	17.70	9.41	27.11
White floral	15"	1E@0.35	Ea	18.20	9.41	27.61
Silver applique	15"	1E@0.35	Ea	18.90	9.41	28.31
Colorful clown	18"	1E@0.35	Ea	26.90	9.41	36.31
Needlepoint	15"	1E@0.35	Ea	27.50	9.41	36.91
Scalloped	16"	1E@0.35	Ea	29.20	9.41	38.61
Leaf & floral	14"	1E@0.35	Ea	51.20	9.41	60.61
Etched floral	14"	1E@0.35	Ea	52.40	9.41	61.81

Ceiling round glass incandescent light fixtures, four 60-watt lamps

Material		Craft@Hrs	Unit	Material Cost	Labor Cost	Installed Cost
White prismatic	18"	1E@0.40	Ea	27.50	10.80	38.30
White textured	18"	1E@0.40	Ea	33.60	10.80	44.40

Ceiling round glass incandescent light fixtures, two 60-watt lamps, with canopy

Material		Craft@Hrs	Unit	Material Cost	Labor Cost	Installed Cost
White canopy	12"	1E@0.40	Ea	13.90	10.80	24.70
White canopy	14"	1E@0.40	Ea	15.90	10.80	26.70
White canopy	11"	1E@0.40	Ea	19.90	10.80	30.70
Brass canopy	11"	1E@0.40	Ea	21.40	10.80	32.20
Brass canopy	12"	1E@0.40	Ea	22.00	10.80	32.80

Use these figures to estimate the cost for lighting fixtures installed in buildings under the conditions described on pages 5 and 6. Costs listed are for each fixture installed. The crew is one electrician working at labor cost of $26.88 per manhour. These costs include layout, material handling, and normal waste. Add for the lamp, accessories, sales tax, delivery, supervision, mobilization, demobilization, cleanup, overhead, and profit. Note: These costs are based on internally pre wired fixtures installed in new construction not over 10 feet above the floor. Many types of glass diffusers are available. Many have textured glass with gold designs that can be matched with larger or smaller diffusers.

Incandescent Light Fixtures

Material		Craft@Hrs	Unit	Material Cost	Labor Cost	Installed Cost

Ceiling round incandescent light fixtures, three 60-watt lamps, with canopy

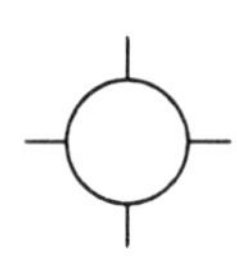

Material		Craft@Hrs	Unit	Material Cost	Labor Cost	Installed Cost
Brass canopy	13"	1E@0.40	Ea	20.50	10.80	31.30
Brass canopy	14"	1E@0.40	Ea	28.10	10.80	38.90
Brass canopy	15"	1E@0.40	Ea	28.40	10.80	39.20
White canopy	13"	1E@0.40	Ea	26.60	10.80	37.40
Brass/jeweled	15"	1E@0.40	Ea	32.40	10.80	43.20
Wagon wheel	16"	1E@0.40	Ea	36.50	10.80	47.30
Patterned brass	18"	1E@0.40	Ea	36.50	10.80	47.30
Almond canopy	14"	1E@0.40	Ea	34.70	10.80	45.50
Peach floral	17"	1E@0.40	Ea	42.50	10.80	53.30
Cathedral	15"	1E@0.40	Ea	50.00	10.80	60.80

Ceiling white glass incandescent light fixtures, 60- or 100-watt lamp, exposed canopy

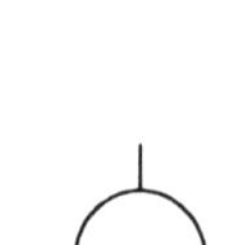

Material		Craft@Hrs	Unit	Material Cost	Labor Cost	Installed Cost
Polished brass	5"	1E@0.35	Ea	6.08	9.41	15.49
Antique brass	5"	1E@0.35	Ea	7.30	9.41	16.71
Polished brass	6"	1E@0.35	Ea	8.92	9.41	18.33
Antique brass	6"	1E@0.35	Ea	8.92	9.41	18.33
Polished brass	7"	1E@0.35	Ea	9.27	9.41	18.68
Antique brass	7"	1E@0.35	Ea	9.27	9.41	18.68
White globe	7"	1E@0.40	Ea	10.40	10.80	21.20
Chrome globe	7"	1E@0.40	Ea	11.00	10.80	21.80
Polished brass	9"	1E@0.40	Ea	13.00	10.80	23.80
Antique brass	9"	1E@0.40	Ea	13.00	10.80	23.80
White globe	10"	1E@0.45	Ea	16.50	12.10	28.60
Polished brass	10"	1E@0.45	Ea	17.10	12.10	29.20
Antique brass	10"	1E@0.45	Ea	17.10	12.10	29.20

Ceiling opal glass utility incandescent light fixtures, 60-watt lamp

Material		Craft@Hrs	Unit	Material Cost	Labor Cost	Installed Cost
Globe	5"	1E@0.40	Ea	13.90	10.80	24.70
Globe	6"	1E@0.40	Ea	16.50	10.80	27.30

Use these figures to estimate the cost of lighting fixtures installed in buildings under the conditions described on pages 5 and 6. Costs listed are for each fixture installed. The crew is one electrician working at a labor cost of $26.88 per manhour. These costs include layout, material handling, and normal waste. Add for lamp, accessories, sales tax, delivery, supervision, mobilization, demobilization, cleanup, overhead and profit. Note: These costs are based on internally pre-wired fixtures installed in new construction not over 10 feet above the floor. The fixture body for utility lights is satin finish cast aluminum. The opal globe screws into the fixture body.

Incandescent Light Fixtures

Material		Craft@Hrs	Unit	Material Cost	Labor Cost	Installed Cost

Opal glass utility incandescent light fixtures, 100-watt lamp

Material		Craft@Hrs	Unit	Material Cost	Labor Cost	Installed Cost
Wall mount sphere	5"	1E@0.40	Ea	16.20	10.80	27.00
Wall mount sphere	6"	1E@0.40	Ea	18.50	10.80	29.30
Double unit	5"	1E@0.50	Ea	35.30	13.40	48.70

Polycarbonate diffuser utility incandescent light fixtures, 60-watt lamp

Material		Craft@Hrs	Unit	Material Cost	Labor Cost	Installed Cost
Ceiling	5"	1E@0.40	Ea	20.30	10.80	31.10
Wall	5"	1E@0.40	Ea	21.40	10.80	32.20
Double unit	5"	1E@0.50	Ea	40.20	13.40	53.60

Wall mounted adjustable flood incandescent light fixtures, one lamp

Material		Craft@Hrs	Unit	Material Cost	Labor Cost	Installed Cost
Satin aluminum	100W	1E@0.40	Ea	16.50	10.80	27.30
Satin brass	100W	1E@0.40	Ea	16.50	10.80	27.30
White	50W	1E@0.40	Ea	17.40	10.80	28.20
White sphere	50W	1E@0.40	Ea	30.80	10.80	41.60
White sphere	75W	1E@0.40	Ea	33.50	10.80	44.30
White cylinder	75W	1E@0.40	Ea	32.30	10.80	43.10
White cylinder	150W	1E@0.40	Ea	35.00	10.80	45.80

Wall mounted adjustable flood incandescent light fixtures, two lamps

Material		Craft@Hrs	Unit	Material Cost	Labor Cost	Installed Cost
Satin aluminum	100W	1E@0.50	Ea	29.20	13.40	42.60
Satin brass	100W	1E@0.50	Ea	29.20	13.40	42.60
White	100W	1E@0.50	Ea	30.40	13.40	43.80
Antique brass	100W	1E@0.50	Ea	30.40	13.40	43.80

Wraparound glass bathroom incandescent fixtures, two or four 75-watt lamps

Material		Craft@Hrs	Unit	Material Cost	Labor Cost	Installed Cost
White glass	13"	1E@0.40	Ea	8.69	10.80	19.49
White with outlet	13"	1E@0.40	Ea	9.85	10.80	20.65
White glass	24"	1E@0.45	Ea	15.30	12.10	27.40
White with outlet	24"	1E@0.45	Ea	17.00	12.10	29.10
Ribbed glass	13"	1E@0.40	Ea	11.20	10.80	22.00
Ribbed glass	24"	1E@0.45	Ea	18.80	12.10	30.90

Use these figures to estimate the cost of lighting fixtures installed in buildings under the conditions described on pages 5 and 6. Costs listed are for each fixture installed. The crew is one electrician working at a labor cost of $26.88 per manhour. These costs include layout, material handling, and normal waste. Add for lamp, accessories, sales tax, delivery, supervision, mobilization, demobilization, cleanup, overhead and profit. Note: These costs are based on internally pre-wired fixtures installed in new construction not over 10 feet above the floor.

Incandescent Drum Light Fixtures and Entrance Lights

Material			Craft@Hrs	Unit	Material Cost	Labor Cost	Installed Cost

White glass utility incandescent drum light fixtures with canopy

Material			Craft@Hrs	Unit	Material Cost	Labor Cost	Installed Cost
White	7"	60W	1E@0.40	Ea	8.57	10.80	19.37
Chrome	7"	60W	1E@0.40	Ea	9.27	10.80	20.07
Antique brass	7"	60W	1E@0.40	Ea	9.27	10.80	20.07
White	9"	100W	1E@0.40	Ea	12.70	10.80	23.50
Chrome	9"	100W	1E@0.40	Ea	13.30	10.80	24.10
Antique brass	9"	100W	1E@0.40	Ea	13.30	10.80	24.10
White	11"	2-60W	1E@0.45	Ea	16.80	12.10	28.90
Chrome	11"	2-60W	1E@0.45	Ea	17.60	12.10	29.70
Antique brass	11"	2-60W	1E@0.45	Ea	17.60	12.10	29.70

Opal glass utility incandescent drum light fixtures with canopy

Material			Craft@Hrs	Unit	Material Cost	Labor Cost	Installed Cost
Chrome	9"	100W	1E@0.40	Ea	20.80	10.80	31.60
Chrome	11"	2-75W	1E@0.45	Ea	28.10	12.10	40.20
Chrome	13"	3-60W	1E@0.45	Ea	37.30	12.10	49.40
Chrome	14"	3-60W	1E@0.45	Ea	48.60	12.10	60.70

Round opal glass utility incandescent drum light fixtures with canopy

Material			Craft@Hrs	Unit	Material Cost	Labor Cost	Installed Cost
Concealed can	8"	60W	1E@0.40	Ea	20.00	10.80	30.80
White	11"	100W	1E@0.40	Ea	26.90	10.80	37.70
White	12"	2-60W	1E@0.45	Ea	37.00	12.10	49.10

Square opal glass utility incandescent drum light fixtures with canopy

Material			Craft@Hrs	Unit	Material Cost	Labor Cost	Installed Cost
Concealed can	8"	60W	1E@0.40	Ea	22.60	10.80	33.40
White	11"	100W	1E@0.40	Ea	29.50	10.80	40.30
White	12"	2-60W	1E@0.45	Ea	38.80	12.10	50.90

Satin black exterior entrance fixtures, 100-watt lamp

Material		Craft@Hrs	Unit	Material Cost	Labor Cost	Installed Cost
Clear glass	7" H	1E@0.40	Ea	4.92	10.80	15.72
Clear glass	8" H	1E@0.40	Ea	5.79	10.80	16.59
Textured glass	9" H	1E@0.40	Ea	8.11	10.80	18.91
Satin white	7" H	1E@0.40	Ea	9.85	10.80	20.65
Stippled glass	8" H	1E@0.40	Ea	7.82	10.80	18.62
White glass	8" H	1E@0.40	Ea	11.60	10.80	22.40

Use these figures to estimate the cost of lighting fixtures installed in buildings under the conditions described on pages 5 and 6. Costs listed are for each fixture installed. The crew is one electrician working at a labor cost of $26.88 per manhour. These costs include layout, material handling, and normal waste. Add for lamp, accessories, sales tax, delivery, supervision, mobilization, demobilization, cleanup, overhead and profit. Note: These costs are based on internally pre-wired fixtures installed in new construction not over 10 feet above the floor.

Recessed Lighting Fixtures with Pre-wired Housings

Material	Craft@Hrs	Unit	Material Cost	Labor Cost	Installed Cost
Square white glass recessed light fixtures with metal trim					
Chrome	8" 60W	1E@0.60	Ea	19.70	16.10	35.80
White	9" 100W	1E@0.60	Ea	20.00	16.10	36.10
Chrome	9" 100W	1E@0.60	Ea	20.40	16.10	36.50
White	11" 150W	1E@0.60	Ea	22.70	16.10	38.80
Chrome	11" 150W	1E@0.60	Ea	23.10	16.10	39.20
Square recessed light fixtures with white trim					
Flat glass	9" 100W	1E@0.60	Ea	19.70	16.10	35.80
Drop opal	9" 100W	1E@0.60	Ea	25.50	16.10	41.60
Flat glass	11" 150W	1E@0.60	Ea	25.50	16.10	41.60
Drop opal	11" 150W	1E@0.60	Ea	31.30	16.10	47.40
Flat glass	13" 200W	1E@0.60	Ea	45.10	16.10	61.20
Drop glass	13" 200W	1E@0.60	Ea	49.80	16.10	65.90
Round recessed light fixtures with white trim					
Downlight	7" 150R40	1E@0.60	Ea	17.60	16.10	33.70
Shower	7" 75W	1E@0.60	Ea	22.50	16.10	38.60
Louver	6" 75R30	1E@0.60	Ea	25.00	16.10	41.10
Cone	6" 75R30	1E@0.60	Ea	26.90	16.10	43.00
Baffle	6" 75R30	1E@0.60	Ea	28.90	16.10	45.00
Scoop	8" 100W	1E@0.60	Ea	28.90	16.10	45.00
Baffle	6" 50R20	1E@0.60	Ea	30.40	16.10	46.50
Cone	8" 150R40	1E@0.60	Ea	30.40	16.10	46.50
Louver	8" 150R40	1E@0.60	Ea	31.10	16.10	47.20
Baffle	8" 150R40	1E@0.60	Ea	32.30	16.10	48.40
Downlight	6" 50R20	1E@0.60	Ea	32.70	16.10	48.80
Fresnel	7" 100W	1E@0.60	Ea	35.00	16.10	51.10
Eyeball	6" 50R20	1E@0.60	Ea	37.30	16.10	53.40
Opal	8" 100W	1E@0.60	Ea	42.00	16.10	58.10
Scoop	8" 75R30	1E@0.60	Ea	44.30	16.10	60.40
Eyeball	8" 50R20	1E@0.60	Ea	46.50	16.10	62.60
Adj accent	8" 75R30	1E@0.60	Ea	49.70	16.10	65.80
Downlight	8" 75R30	1E@0.60	Ea	49.70	16.10	65.80
Cone	8" 150W	1E@0.60	Ea	52.70	16.10	68.80
Eyeball	9" 150R40	1E@0.60	Ea	53.10	16.10	69.20
Baffle	7" 150R40	1E@0.60	Ea	53.50	16.10	69.60
Downlight	9" 150R40	1E@0.60	Ea	62.80	16.10	78.90

Use these figures to estimate the cost of lighting fixtures installed in buildings under the conditions described on pages 5 and 6. Costs listed are for each fixture installed. The crew is one electrician working at a labor cost of $26.88 per manhour. These costs include layout, material handling, and normal waste. Add for lamp, accessories, sales tax, delivery, supervision, mobilization, demobilization, cleanup, overhead and profit. Note: These costs are based on internally pre-wired fixtures installed in new construction not over 10 feet above the floor.

Recessed Lighting Fixtures with Pre-wired Housings

Material			Craft@Hrs	Unit	Material Cost	Labor Cost	Installed Cost
Round recessed light fixtures with white trim							
Scoop	7"	150R40	1E@0.60	Ea	65.80	16.10	81.90
Reflector	9"	300W	1E@0.60	Ea	67.80	16.10	83.90
Wall washer	7"	150W	1E@0.60	Ea	78.90	16.10	95.00
Wall washer	9"	300W	1E@0.60	Ea	83.20	16.10	99.30
Round recessed light fixtures with white trim and heat guard							
Downlight	7"	150R40	1E@0.60	Ea	19.60	16.10	35.70
Shower	7"	75W	1E@0.60	Ea	19.90	16.10	36.00
Louver	6"	75R30	1E@0.60	Ea	27.00	16.10	43.10
Cone	6"	75R30	1E@0.60	Ea	28.90	16.10	45.00
Baffle	6"	75R30	1E@0.60	Ea	30.90	16.10	47.00
Wall washer	8"	150R40	1E@0.60	Ea	30.90	16.10	47.00
Baffle	8"	50R20	1E@0.60	Ea	32.40	16.10	48.50
Cone	8"	150R40	1E@0.60	Ea	32.40	16.10	48.50
Louver	8"	150R40	1E@0.60	Ea	33.20	16.10	49.30
Baffle	8"	150R40	1E@0.60	Ea	34.30	16.10	50.40
Downlight	6"	50R20	1E@0.60	Ea	34.70	16.10	50.80
Fresnel	7"	100W	1E@0.60	Ea	37.00	16.10	53.10
Eyeball	6"	50R20	1E@0.60	Ea	39.40	16.10	55.50
Baffle	9"	300R40	1E@0.60	Ea	41.30	16.10	57.40
Eyeball	8"	75R30	1E@0.60	Ea	44.00	16.10	60.10
Opal	8"	100W	1E@0.60	Ea	44.00	16.10	60.10
Wall washer	8"	75R30	1E@0.60	Ea	45.50	16.10	61.60
Eyeball	8"	50R20	1E@0.60	Ea	48.60	16.10	64.70
Adj accent	8"	75R30	1E@0.60	Ea	51.70	16.10	67.80
Downlight	8"	75R30	1E@0.60	Ea	51.70	16.10	67.80
Cone	8"	150W	1E@0.60	Ea	54.80	16.10	70.90
Eyeball	9"	150R40	1E@0.60	Ea	55.10	16.10	71.20
Baffle	7"	150R40	1E@0.60	Ea	55.50	16.10	71.60
Downlight	9"	150R40	1E@0.60	Ea	65.10	16.10	81.20
Baffle	9"	300R40	1E@0.60	Ea	65.90	16.10	82.00
Wall washer	7"	150R40	1E@0.60	Ea	67.80	16.10	83.90
Reflector	9"	300W	1E@0.60	Ea	69.70	16.10	85.80
Wall washer	7"	150W	1E@0.60	Ea	81.00	16.10	97.10
Wall washer	9"	300W	1E@0.60	Ea	85.10	16.10	101.20

Use these figures to estimate the cost of lighting fixtures installed in buildings under the conditions described on pages 5 and 6. Costs listed are for each fixture installed. The crew is one electrician working at a labor cost of $26.88 per manhour. These costs include layout, material handling, and normal waste. Add for lamp, accessories, sales tax, delivery, supervision, mobilization, demobilization, cleanup, overhead and profit. Note: These costs are based on internally pre-wired fixtures installed in new construction not over 10 feet above the floor.

Material	Craft@Hrs	Unit	Material Cost	Labor Cost	Installed Cost

Surface mounted single circuit track light, 120 volt

Material	Craft@Hrs	Unit	Material Cost	Labor Cost	Installed Cost
2' starter, bronze	1E@0.30	Ea	18.50	8.06	26.56
2' starter, white	1E@0.30	Ea	18.50	8.06	26.56
2' starter, black	1E@0.30	Ea	18.50	8.06	26.56
4' starter, bronze	1E@0.40	Ea	27.30	10.80	38.10
4' starter, white	1E@0.40	Ea	27.30	10.80	38.10
4' starter, black	1E@0.40	Ea	27.30	10.80	38.10
8' starter, bronze	1E@0.55	Ea	43.10	14.80	57.90
8' starter, white	1E@0.55	Ea	43.10	14.80	57.90
8' starter, black	1E@0.55	Ea	43.10	14.80	57.90
8' joiner, bronze	1E@0.45	Ea	37.00	12.10	49.10
8' joiner, white	1E@0.45	Ea	37.00	12.10	49.10
8' joiner, black	1E@0.45	Ea	37.00	12.10	49.10

Track light universal lampholder for medium base lamp

Material		Craft@Hrs	Unit	Material Cost	Labor Cost	Installed Cost
Bronze	5"	1E@0.20	Ea	20.00	5.38	25.38
White	5"	1E@0.20	Ea	20.00	5.38	25.38
Black	5"	1E@0.20	Ea	20.00	5.38	25.38
Chrome	5"	1E@0.20	Ea	20.00	5.38	25.38

Track light shielded universal lampholder for R30 lamp

Material		Craft@Hrs	Unit	Material Cost	Labor Cost	Installed Cost
Bronze	10"	1E@0.20	Ea	9.27	5.38	14.65
White	10"	1E@0.20	Ea	9.27	5.38	14.65
Black	10"	1E@0.20	Ea	9.27	5.38	14.65
Chrome	10"	1E@0.20	Ea	9.27	5.38	14.65

Track light shielded universal lampholder for R40 or Par38 lamp

Material		Craft@Hrs	Unit	Material Cost	Labor Cost	Installed Cost
Bronze	11"	1E@0.20	Ea	9.27	5.38	14.65
White	11"	1E@0.20	Ea	9.27	5.38	14.65
Black	11"	1E@0.20	Ea	9.27	5.38	14.65
Chrome	11"	1E@0.20	Ea	9.27	5.38	14.65

Track light continental lampholder for R20 lamp

Material		Craft@Hrs	Unit	Material Cost	Labor Cost	Installed Cost
Bronze	6"	1E@0.20	Ea	21.20	5.38	26.58
White	6"	1E@0.20	Ea	21.20	5.38	26.58
Black	6"	1E@0.20	Ea	21.20	5.38	26.58
Chrome	6"	1E@0.20	Ea	21.20	5.38	26.58

Use these figures to estimate the cost of lighting fixtures installed in buildings under the conditions described on pages 5 and 6. Costs listed are for each fixture installed. The crew is one electrician working at a labor cost of $26.88 per manhour. These costs include layout, material handling, and normal waste. Add for lamp, accessories, sales tax, delivery, supervision, mobilization, demobilization, cleanup, overhead and profit. Note: These costs are based on internally pre-wired fixtures installed in new construction not over 10 feet above the floor. Many types of fittings are available for track lighting.

Track Light Fixtures

Material		Craft@Hrs	Unit	Material Cost	Labor Cost	Installed Cost

Track light continental lampholder for R20 lamp

Material		Craft@Hrs	Unit	Material Cost	Labor Cost	Installed Cost
Polished brass	7"	1E@0.20	Ea	35.80	5.38	41.18
Bronze	7"	1E@0.20	Ea	30.40	5.38	35.78
White	7"	1E@0.20	Ea	30.40	5.38	35.78
Black	7"	1E@0.20	Ea	30.40	5.38	35.78

Track light continental lampholder for R30 lamp

Bronze	10"	1E@0.20	Ea	31.20	5.38	36.58
White	10"	1E@0.20	Ea	31.20	5.38	36.58
Black	10"	1E@0.20	Ea	31.20	5.38	36.58

Track light continental lampholder for R40 lamp

Bronze	12"	1E@0.20	Ea	35.80	5.38	41.18
White	12"	1E@0.20	Ea	35.80	5.38	41.18
Black	12"	1E@0.20	Ea	35.80	5.38	41.18
Bronze	9"	1E@0.20	Ea	29.90	5.38	35.28
White	9"	1E@0.20	Ea	29.90	5.38	35.28
Black	9"	1E@0.20	Ea	29.90	5.38	35.28

Track light petite cylinder lampholder for R14 lamp

Bronze	5"	1E@0.20	Ea	22.30	5.38	27.68
White	5"	1E@0.20	Ea	22.30	5.38	27.68
Black	5"	1E@0.20	Ea	22.30	5.38	27.68

Track light petite round cylinder lampholder for R20 lamp

Bronze	5"	1E@0.20	Ea	23.80	5.38	29.18
White	5"	1E@0.20	Ea	23.80	5.38	29.18
Black	5"	1E@0.20	Ea	23.80	5.38	29.18

Use these figures to estimate the cost of lighting fixtures installed in buildings under the conditions described on pages 5 and 6. Costs listed are for each fixture installed. The crew is one electrician working at a labor cost of $26.88 per manhour. These costs include layout, material handling, and normal waste. Add for lamp, accessories, sales tax, delivery, supervision, mobilization, demobilization, cleanup, overhead and profit. Note: These costs are based on internally pre-wired fixtures installed in new construction not over 10 feet above the floor.

Material	Craft@Hrs	Unit	Material Cost	Labor Cost	Installed Cost
Decorator track light fixture for R20 or 100-watt lamp					
Square 6"	1E@0.20	Ea	30.80	5.38	36.18
Cone 16"	1E@0.20	Ea	55.50	5.38	60.88
Scoop 5"	1E@0.20	Ea	60.10	5.38	65.48
Track light cylinder lampholder for various lamp sizes					
Bronze 5"	1E@0.20	Ea	24.30	5.38	29.68
White 5"	1E@0.20	Ea	24.30	5.38	29.68
Black 5"	1E@0.20	Ea	24.30	5.38	29.68
Bronze 7"	1E@0.20	Ea	27.70	5.38	33.08
White 7"	1E@0.20	Ea	27.70	5.38	33.08
Black 7"	1E@0.20	Ea	27.70	5.38	33.08
Antique brass 7"	1E@0.20	Ea	28.10	5.38	33.48
Bronze 8"	1E@0.20	Ea	32.00	5.38	37.38
White 8"	1E@0.20	Ea	32.00	5.38	37.38
Black 8"	1E@0.20	Ea	32.00	5.38	37.38
Antique brass 8"	1E@0.20	Ea	32.00	5.38	37.38
Bronze 9"	1E@0.20	Ea	43.10	5.38	48.48
White 9"	1E@0.20	Ea	43.10	5.38	48.48
Black 9"	1E@0.20	Ea	43.10	5.38	48.48
Bronze 10"	1E@0.20	Ea	51.60	5.38	56.98
White 10"	1E@0.20	Ea	51.60	5.38	56.98
Black 10"	1E@0.20	Ea	51.60	5.38	56.98
Track light spherical lampholder for various lamp sizes					
Polished alum. 5"	1E@0.20	Ea	39.20	5.38	44.58
Bronze 5"	1E@0.20	Ea	34.70	5.38	40.08
White 5"	1E@0.20	Ea	34.70	5.38	40.08
Black 5"	1E@0.20	Ea	34.70	5.38	40.08
Polished alum. 6"	1E@0.20	Ea	42.40	5.38	47.78
Bronze 6"	1E@0.20	Ea	37.70	5.38	43.08
White 6"	1E@0.20	Ea	37.70	5.38	43.08
Black 6"	1E@0.20	Ea	37.70	5.38	43.08
Polished alum. 7"	1E@0.20	Ea	54.60	5.38	59.98
Bronze 7"	1E@0.20	Ea	44.70	5.38	50.08
White 7"	1E@0.20	Ea	44.70	5.38	50.08
Black 7"	1E@0.20	Ea	44.70	5.38	50.08

Use these figures to estimate the cost of lighting fixtures installed in buildings under the conditions described on pages 5 and 6. Costs listed are for each fixture installed. The crew is one electrician working at a labor cost of $26.88 per manhour. These costs include layout, material handling, and normal waste. Add for lamp, accessories, sales tax, delivery, supervision, mobilization, demobilization, cleanup, overhead and profit. Note: These costs are based on internally pre-wired fixtures installed in new construction not over 10 feet above the floor.

Track Light Fixtures and Exit Lighting Fixtures with Universal Arrows

Material		Craft@Hrs	Unit	Material Cost	Labor Cost	Installed Cost
Track light cylinder lampholders with rounded base, track lighting						
Bronze	5"	1E@0.20	Ea	32.40	5.38	37.78
White	5"	1E@0.20	Ea	32.40	5.38	37.78
Black	5"	1E@0.20	Ea	32.40	5.38	37.78
Antique brass	5"	1E@0.20	Ea	32.40	5.38	37.78
Bronze	6"	1E@0.20	Ea	37.70	5.38	43.08
White	6"	1E@0.20	Ea	37.70	5.38	43.08
Black	6"	1E@0.20	Ea	37.70	5.38	43.08
Antique brass	6"	1E@0.20	Ea	37.70	5.38	43.08
Bronze	7"	1E@0.20	Ea	53.90	5.38	59.28
White	7"	1E@0.20	Ea	53.90	5.38	59.28
Black	7"	1E@0.20	Ea	53.90	5.38	59.28
Antique brass	7"	1E@0.20	Ea	53.90	5.38	59.28
Bronze	8"	1E@0.20	Ea	67.80	5.38	73.18
White	8"	1E@0.20	Ea	67.80	5.38	73.18
Black	8"	1E@0.20	Ea	67.80	5.38	73.18
Antique brass	8"	1E@0.20	Ea	67.80	5.38	73.18
Track light stepped base lampholders for R30 or R40 lamps						
Bronze	8"	1E@0.20	Ea	61.60	5.38	66.98
White	8"	1E@0.20	Ea	61.60	5.38	66.98
Black	8"	1E@0.20	Ea	61.60	5.38	66.98
Polished brass	8"	1E@0.20	Ea	69.30	5.38	74.68
Single face exit fixtures with universal arrows, 20-watt T6 1/2 lamps						
Ceiling mounted, red		1E@0.50	Ea	18.41	13.40	31.81
End mounted, red		1E@0.50	Ea	18.41	13.40	31.81
Wall mounted, red		1E@0.50	Ea	18.41	13.40	31.81
Ceiling mounted, green		1E@0.50	Ea	18.41	13.40	31.81
End mounted, green		1E@0.50	Ea	18.41	13.40	31.81
Wall mounted, green		1E@0.50	Ea	18.41	13.40	31.81

Use these figures to estimate the cost of lighting fixtures installed in buildings under the conditions described on pages 5 and 6. Costs listed are for each fixture installed. The crew is one electrician working at a labor cost of $26.88 per manhour. These costs include layout, material handling, and normal waste. Add for lamp, accessories, sales tax, delivery, supervision, mobilization, demobilization, cleanup, overhead and profit. Note: These costs are based on internally pre-wired fixtures installed in new construction not over 10 feet above the floor.

Exit Lighting Fixtures with Universal Arrows

Material	Craft@Hrs	Unit	Material Cost	Labor Cost	Installed Cost
Double face exit fixtures, 20-watt T6 1/2 lamps					
Ceiling mounted, red	1E@0.55	Ea	19.53	14.80	34.33
End mounted, red	1E@0.55	Ea	19.53	14.80	34.33
Ceiling mounted, green	1E@0.55	Ea	19.53	14.80	34.33
End mounted, green	1E@0.55	Ea	19.53	14.80	34.33
Aluminum housing single face exit fixtures, 20-watt T6 1/2 lamps					
Ceiling mounted, red	1E@0.50	Ea	33.74	13.40	47.14
End mounted, red	1E@0.50	Ea	33.74	13.40	47.14
Wall mounted, red	1E@0.50	Ea	33.74	13.40	47.14
Ceiling mounted, green	1E@0.50	Ea	33.74	13.40	47.14
End mounted, green	1E@0.50	Ea	33.74	13.40	47.14
Wall mounted, green	1E@0.50	Ea	33.74	13.40	47.14
Aluminum housing double face exit fixtures, 20-watt T6 1/2 lamps					
Ceiling mounted, red	1E@0.55	Ea	138.39	14.80	153.19
End mounted, red	1E@0.55	Ea	138.39	14.80	153.19
Ceiling mounted, green	1E@0.55	Ea	138.39	14.80	153.19
End mounted, green	1E@0.55	Ea	138.39	14.80	153.19
1 or 2 face exit fixtures with rechargeable batteries					
Ceiling mounted, red	1E@0.75	Ea	115.00	20.20	135.20
End mounted, red	1E@0.75	Ea	115.00	20.20	135.20
Wall mounted, red	1E@0.75	Ea	115.00	20.20	135.20
Ceiling mounted, green	1E@0.75	Ea	115.00	20.20	135.20
End mounted, green	1E@0.75	Ea	115.00	20.20	135.20
Wall mounted, green	1E@0.75	Ea	115.00	20.20	135.20

Use these figures to estimate the cost of exit lighting fixtures installed in buildings under the conditions described on pages 5 and 6. Costs listed are for each fixture installed. The crew is one electrician working at a labor cost of $26.88 per manhour. These costs include layout, material handling, and normal waste. Add for lamp, accessories, sales tax, delivery, supervision, mobilization, demobilization, cleanup, overhead and profit. Note: These costs are based on internally pre-wired fixtures installed in new construction not over 10 feet above the floor. Check the job specifications carefully when selecting emergency fixtures. Those listed here are standard grade and may not meet rigid specifications. Get advice from your supplier on the fixtures needed.

Surface Mounted 120 Volt Fluorescent Lighting Fixtures

Material	Craft@Hrs	Unit	Material Cost	Labor Cost	Installed Cost
Surface mounted 120 volt fluorescent single-lamp strip lighting fixtures					
18"	1E@0.25	Ea	10.00	6.72	16.72
24"	1E@0.30	Ea	14.70	8.06	22.76
36"	1E@0.35	Ea	15.30	9.41	24.71
48"	1E@0.40	Ea	15.30	10.80	26.10
48" energy saving	1E@0.40	Ea	15.30	10.80	26.10
72" slimline	1E@0.50	Ea	30.40	13.40	43.80
96" slimline	1E@0.60	Ea	32.10	16.10	48.20
96" 2-48" tandem	1E@0.70	Ea	50.60	18.80	69.40
Surface mounted 120 volt fluorescent two-lamp strip lighting fixtures					
24"	1E@0.35	Ea	14.70	9.41	24.11
36"	1E@0.40	Ea	43.20	10.80	54.00
48"	1E@0.45	Ea	36.30	12.10	48.40
48" energy saving	1E@0.45	Ea	36.30	12.10	48.40
72" slimline	1E@0.55	Ea	30.40	14.80	45.20
96" slimline	1E@0.65	Ea	34.30	17.50	51.80
96" 2-48" tandem	1E@0.75	Ea	73.20	20.20	93.40
Side-mounted 120 volt fluorescent single-lamp lighting fixtures					
18"	1E@0.25	Ea	20.50	6.72	27.22
24"	1E@0.30	Ea	20.50	8.06	28.56
36"	1E@0.35	Ea	23.50	9.41	32.91
48"	1E@0.40	Ea	23.50	10.80	34.30
Side-mounted 120 volt fluorescent single-lamp on both sides fixtures					
18"	1E@0.30	Ea	30.00	8.06	38.06
24"	1E@0.35	Ea	30.00	9.41	39.41
36"	1E@0.40	Ea	34.00	10.80	44.80
48"	1E@0.45	Ea	34.00	12.10	46.10

Use these figures to estimate the cost of lighting fixtures installed in buildings under the conditions described on pages 5 and 6. Costs listed are for each fixture installed. The crew is one electrician working at a labor cost of $26.88 per manhour. These costs include layout, material handling, and normal waste. Add for lamp, accessories, sales tax, delivery, supervision, mobilization, demobilization, cleanup, overhead and profit. Note: These costs are based on internally pre-wired fixtures installed in new construction not over 10 feet above the floor.

Material	Craft@Hrs	Unit	Material Cost	Labor Cost	Installed Cost

Two-lamp industrial fluorescent lighting fixtures with enameled reflector

Material	Craft@Hrs	Unit	Material Cost	Labor Cost	Installed Cost
24"	1E@0.35	Ea	18.60	9.41	28.01
48"	1E@0.45	Ea	20.80	12.10	32.90
48" energy saving	1E@0.45	Ea	20.80	12.10	32.90
72" slimline	1E@0.55	Ea	53.00	14.80	67.80
96" slimline	1E@0.65	Ea	53.00	17.50	70.50
96" slimline, energy saving	1E@0.65	Ea	53.00	17.50	70.50

Two-lamp industrial fluorescent lighting fixtures with 15 percent uplight

Material	Craft@Hrs	Unit	Material Cost	Labor Cost	Installed Cost
48"	1E@0.45	Ea	40.10	12.10	52.20
48" high output	1E@0.45	Ea	64.20	12.10	76.30
48" very high output	1E@0.45	Ea	97.50	12.10	109.60
48" powergroove	1E@0.45	Ea	83.70	12.10	95.80
72" slimline	1E@0.55	Ea	66.50	14.80	81.30
96" slimline	1E@0.65	Ea	66.50	17.50	84.00
96" high output	1E@0.65	Ea	81.40	17.50	98.90
96" 2-48" tandem	1E@0.65	Ea	81.40	17.50	98.90
96" powergroove	1E@0.65	Ea	107.00	17.50	124.50

Three-lamp industrial fluorescent lighting fixtures with 15 percent uplight

Material	Craft@Hrs	Unit	Material Cost	Labor Cost	Installed Cost
48"	1E@0.50	Ea	60.80	13.40	74.20
96" slimline	1E@0.70	Ea	95.20	18.80	114.00
96" high output	1E@0.70	Ea	127.00	18.80	145.80
96" powergroove	1E@0.70	Ea	138.00	18.80	156.80

Four-lamp industrial fluorescent lighting fixtures with 15 percent uplight

Material	Craft@Hrs	Unit	Material Cost	Labor Cost	Installed Cost
48"	1E@0.55	Ea	60.80	14.80	75.60
96" slimline	1E@0.75	Ea	98.60	20.20	118.80

Use these figures to estimate the cost of lighting fixtures installed in buildings under the conditions described on pages 5 and 6. Costs listed are for each fixture installed. The crew is one electrician working at a labor cost of $26.88 per manhour. These costs include layout, material handling, and normal waste. Add for lamp, accessories, sales tax, delivery, supervision, mobilization, demobilization, cleanup, overhead and profit. Note: These costs are based on internally pre-wired fixtures installed in new construction not over 10 feet above the floor. Industrial fluorescent fixtures are made with many uplight hues and paint finishes. On some jobs additional accessories will be needed: joiners, hanging fittings, stems and canopies.

Fluorescent Lighting Fixtures

Material	Craft@Hrs	Unit	Material Cost	Labor Cost	Installed Cost
Two-lamp industrial fluorescent lighting, enameled reflector, 25% uplight					
48"	1E@0.40	Ea	59.60	10.80	70.40
96" slimline	1E@0.65	Ea	103.00	17.50	120.50
96" high output	1E@0.65	Ea	119.00	17.50	136.50
96" very high output	1E@0.65	Ea	154.00	17.50	171.50
96" 2-48" tandem	1E@0.65	Ea	118.00	17.50	135.50
Two-lamp enclosed-gasketed fluorescent fixtures for damp locations					
48"	1E@0.50	Ea	67.70	13.40	81.10
48" high output	1E@0.50	Ea	111.00	13.40	124.40
96" slimline	1E@0.70	Ea	120.00	18.80	138.80
96" high output	1E@0.70	Ea	161.00	18.80	179.80
96" 2-48" tandem	1E@0.70	Ea	135.00	18.80	153.80
Two-lamp enclosed-gasketed fluorescent lighting fixtures for wet locations					
48"	1E@0.50	Ea	71.10	13.40	84.50
48" high output	1E@0.50	Ea	117.00	13.40	130.40
96" slimline	1E@0.70	Ea	124.00	18.80	142.80
96" high output	1E@0.70	Ea	143.00	18.80	161.80
96" 2-48" tandem	1E@0.70	Ea	136.00	18.80	154.80
Wall mounted single-lamp fluorescent lighting fixtures with acrylic lens					
24"	1E@0.40	Ea	36.70	10.80	47.50
36"	1E@0.50	Ea	47.00	13.40	60.40
48"	1E@0.55	Ea	47.00	14.80	61.80

Use these figures to estimate the cost of lighting fixtures installed in buildings under the conditions described on pages 5 and 6. Costs listed are for each fixture installed. The crew is one electrician working at a labor cost of $26.88 per manhour. These costs include layout, material handling, and normal waste. Add for lamp, accessories, sales tax, delivery, supervision, mobilization, demobilization, cleanup, overhead and profit. Note: These costs are based on internally pre-wired fixtures installed in new construction not over 10 feet above the floor. Industrial fluorescent fixtures are made with many uplight hues and paint finishes. On some jobs additional accessories will be needed: joiners, hanging fittings, stems and canopies.

Material	Craft@Hrs	Unit	Material Cost	Labor Cost	Installed Cost

Wall mounted two-lamp fluorescent lighting fixtures with acrylic lens

Material	Craft@Hrs	Unit	Material Cost	Labor Cost	Installed Cost
24"	1E@0.45	Ea	44.70	12.10	56.80
36"	1E@0.55	Ea	55.00	14.80	69.80
48"	1E@0.60	Ea	52.70	16.10	68.80

Two-lamp wraparound injection-molded corridor fluorescent fixtures

Material	Craft@Hrs	Unit	Material Cost	Labor Cost	Installed Cost
48"	1E@0.60	Ea	45.90	16.10	62.00
96" 2-48" tandem	1E@0.75	Ea	91.70	20.20	111.90

Four-lamp wraparound injection-molded corridor fluorescent fixtures

Material	Craft@Hrs	Unit	Material Cost	Labor Cost	Installed Cost
48"	1E@0.70	Ea	80.30	18.80	99.10
96" 2-48" tandem	1E@0.90	Ea	161.00	24.20	185.20

Two-lamp wraparound ceiling fluorescent lighting fixtures

Material	Craft@Hrs	Unit	Material Cost	Labor Cost	Installed Cost
48"	1E@0.60	Ea	43.60	16.10	59.70
48" energy saving	1E@0.60	Ea	56.20	16.10	72.30
96" 2-48" tandem	1E@0.75	Ea	91.70	20.20	111.90
96" 2-48" energy saving	1E@0.75	Ea	115.00	20.20	135.20

Four-lamp wraparound ceiling fluorescent lighting fixtures

Material	Craft@Hrs	Unit	Material Cost	Labor Cost	Installed Cost
48"	1E@0.70	Ea	76.80	18.80	95.60
48" energy savings	1E@0.70	Ea	102.00	18.80	120.80
96" 2-48" tandem	1E@0.90	Ea	114.00	24.20	138.20
96" 2-48" energy saving	1E@0.90	Ea	138.00	24.20	162.20

Use these figures to estimate the cost of lighting fixtures installed in buildings under the conditions described on pages 5 and 6. Costs listed are for each fixture installed. The crew is one electrician working at a labor cost of $26.88 per manhour. These costs include layout, material handling, and normal waste. Add for lamp, accessories, sales tax, delivery, supervision, mobilization, demobilization, cleanup, overhead and profit. Note: These costs are based on internally pre-wired fixtures installed in new construction not over 10 feet above the floor. Many styles of surface mounted fluorescent fixtures are available. Prices range from economy to architectural quality.

Framed Troffer Lay-in T-Bar Fluorescent Fixtures

Material	Craft@Hrs	Unit	Material Cost	Labor Cost	Installed Cost
24"-wide lay-in T-bar fluorescent fixtures, steel frame					
48" 2 lamp	1E@0.70	Ea	71.10	18.80	89.90
48" 3 lamp	1E@0.75	Ea	80.30	20.20	100.50
48" 4 lamp	1E@0.80	Ea	79.10	21.50	100.60
48" 2 lamp, energy saver	1E@0.70	Ea	83.70	18.80	102.50
48" 3 lamp, energy saver	1E@0.75	Ea	103.00	20.20	123.20
48" 4 lamp, energy saver	1E@0.80	Ea	101.00	21.50	122.50
Regressed 24"-wide lay-in T-bar fluorescent fixtures, steel frame					
48" 2 lamp	1E@0.70	Ea	75.70	18.80	94.50
48" 3 lamp	1E@0.75	Ea	86.00	20.20	106.20
48" 4 lamp	1E@0.80	Ea	84.80	21.50	106.30
48" 2 lamp, energy saver	1E@0.70	Ea	86.00	18.80	104.80
48" 3 lamp, energy saver	1E@0.75	Ea	108.00	20.20	128.20
48" 4 lamp, energy saver	1E@0.80	Ea	105.00	21.50	126.50
24"-wide lay-in T-bar fluorescent fixtures, aluminum frame					
48" 2 lamp	1E@0.70	Ea	72.20	18.80	91.00
48" 3 lamp	1E@0.75	Ea	81.40	20.20	101.60
48" 4 lamp	1E@0.80	Ea	87.10	21.50	108.60
48" 2 lamp, energy saver	1E@0.70	Ea	87.10	18.80	105.90
48" 3 lamp, energy saver	1E@0.75	Ea	105.00	20.20	125.20
48" 4 lamp, energy saver	1E@0.80	Ea	103.00	21.50	124.50
Regressed 24"-wide lay-in T-bar fluorescent fixtures, aluminum frame					
48" 2 lamp	1E@0.70	Ea	76.80	18.80	95.60
48" 3 lamp	1E@0.75	Ea	87.10	20.20	107.30
48" 4 lamp	1E@0.80	Ea	87.10	21.50	108.60
48" 2 lamp, energy saver	1E@0.70	Ea	89.40	18.80	108.20
48" 3 lamp, energy saver	1E@0.75	Ea	110.00	20.20	130.20
48" 4 lamp, energy saver	1E@0.80	Ea	108.00	21.50	129.50

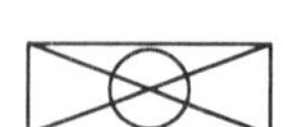

Use these figures to estimate the cost of fluorescent fixtures installed in buildings under the conditions described on pages 5 and 6. Costs listed are for each fixture installed. The crew is one electrician working at a labor cost of $26.88 per manhour. These costs include layout, material handling, and normal waste. Add for lamp, accessories, sales tax, delivery, supervision, mobilization, demobilization, cleanup, overhead and profit. Note: These costs are based on internally pre-wired fixtures installed in new construction not over 10 feet above the floor. These fixtures can be ordered with the lamps installed. Be sure to specify the lamp color needed. Lay-in fixtures are available both with flex tails ready for make-up in J-boxes and for soft wiring systems. Soft wiring is a special plug-in system used when wiring fixtures in a suspended ceiling system. The inter-connecting cables are factory assembled in various lengths with circuiting to meet most job requirements. Soft wiring systems carry the UL label and are accepted in most code jurisdictions. Energy saving (ES) ballasts are available and should be installed at the factory to save field labor hours. When ES fluorescent lighting fixtures are installed, be sure to use ES lamps. Special surface mounted and semi-flush fluorescent fixtures with wood siding are available for designer applications. You can also order parabolic louvers with colored anodized finishes.

Framed Troffer Lay-in T-Bar Fluorescent Fixtures

Material	Craft@Hrs	Unit	Material Cost	Labor Cost	Installed Cost
24"-wide air-handling lay-in T-bar fluorescent fixtures					
48" 2 lamp	1E@0.70	Ea	94.80	18.80	113.60
48" 3 lamp	1E@0.75	Ea	113.00	20.20	133.20
48" 4 lamp	1E@0.80	Ea	113.00	21.50	134.50
48" 2 lamp, energy saver	1E@0.70	Ea	107.00	18.80	125.80
48" 3 lamp, energy saver	1E@0.75	Ea	120.00	20.20	140.20
48" 4 lamp, energy saver	1E@0.80	Ea	120.00	21.50	141.50
48"-wide air-handling lay-in T-bar fluorescent fixtures					
48" 4 lamp	1E@1.25	Ea	334.00	33.60	367.60
48" 6 lamp	1E@1.30	Ea	353.00	34.90	387.90
48" 8 lamp	1E@1.35	Ea	372.00	36.30	408.30
48" 4 lamp, energy saver	1E@1.25	Ea	367.00	33.60	400.60
48" 6 lamp, energy saver	1E@1.30	Ea	391.00	34.90	425.90
48" 8 lamp, energy saver	1E@1.35	Ea	422.00	36.30	458.30
12"-wide heat-recovery lay-in T-bar fluorescent fixtures					
48" 1 lamp	1E@0.55	Ea	77.20	14.80	92.00
48" 2 lamp	1E@0.60	Ea	77.20	16.10	93.30
48" 3 lamp	1E@0.65	Ea	95.70	17.50	113.20
48" 1 lamp, energy saver	1E@0.55	Ea	89.80	14.80	104.60
48" 2 lamp, energy saver	1E@0.60	Ea	89.80	16.10	105.90
48" 3 lamp, energy saver	1E@0.65	Ea	102.00	17.50	119.50

Use these figures to estimate the cost of fluorescent fixtures installed in buildings under the conditions described on pages 5 and 6. Costs listed are for each fixture installed. The crew is one electrician working at a labor cost of $26.88 per manhour. These costs include layout, material handling, and normal waste. Add for lamp, accessories, sales tax, delivery, supervision, mobilization, demobilization, cleanup, overhead and profit. Note: These costs are based on internally pre-wired fixtures installed in new construction not over 10 feet above the floor. These fixtures can be ordered with the lamps installed. Be sure to specify the lamp color needed. Lay-in fixtures are available both with flex tails ready for make-up in J-boxes and for soft wiring systems. Soft wiring is a special plug-in system used when wiring fixtures in a suspended ceiling system. The inter-connecting cables are factory assembled in various lengths with circuiting to meet most job requirements. Soft wiring systems carry the UL label and are accepted in most code jurisdictions. Energy saving (ES) ballasts are available and should be installed at the factory to save field labor hours. When ES fluorescent lighting fixtures are installed, be sure to use ES lamps. Special surface mounted and semi-flush fluorescent fixtures with wood siding are available for designer applications. You can also order parabolic louvers with colored anodized finishes.

Framed Troffer Lay-in T-Bar Fluorescent Fixtures

Material	Craft@Hrs	Unit	Material Cost	Labor Cost	Installed Cost
20"-wide heat-recovery lay-in T-bar fluorescent fixtures					
48" 2 lamp	1E@0.65	Ea	84.50	17.50	102.00
48" 3 lamp	1E@0.70	Ea	103.00	18.80	121.80
48" 4 lamp	1E@0.75	Ea	115.00	20.20	135.20
48" 2 lamp, energy saver	1E@0.65	Ea	97.10	17.50	114.60
48" 3 lamp, energy saver	1E@0.70	Ea	128.00	18.80	146.80
48" 4 lamp, energy saver	1E@0.75	Ea	128.00	20.20	148.20
24"-square heat-recovery lay-in T-bar fluorescent fixtures					
F40 U bent, 1 lamp	1E@0.50	Ea	77.20	13.40	90.60
F40 U bent, 2 lamp	1E@0.55	Ea	77.20	14.80	92.00
F40 U bent, 3 lamp	1E@0.60	Ea	95.70	16.10	111.80
F20 4 lamp	1E@0.60	Ea	106.00	16.10	122.10
24"-wide heat-recovery lay-in T-bar fluorescent fixtures					
48" 2 lamp	1E@0.70	Ea	84.50	18.80	103.30
48" 3 lamp	1E@0.75	Ea	103.00	20.20	123.20
48" 4 lamp	1E@0.80	Ea	103.00	21.50	124.50
48" 2 lamp, energy saver	1E@0.70	Ea	97.10	18.80	115.90
48" 3 lamp, energy saver	1E@0.75	Ea	128.00	20.20	148.20
48" 4 lamp, energy saver	1E@0.80	Ea	128.00	21.50	149.50
24"-wide return-air-handling lay-in T-bar fluorescent fixtures					
48" 2 lamp	1E@0.70	Ea	78.80	18.80	97.60
48" 3 lamp	1E@0.75	Ea	94.20	20.20	114.40
48" 4 lamp	1E@0.80	Ea	94.20	21.50	115.70
48" 2 lamp, energy saver	1E@0.70	Ea	91.40	18.80	110.20
48" 3 lamp, energy saver	1E@0.75	Ea	119.00	20.20	139.20
48" 4 lamp, energy saver	1E@0.80	Ea	119.00	21.50	140.50

Use these figures to estimate the cost of fluorescent fixtures installed in buildings under the conditions described on pages 5 and 6. Costs listed are for each fixture installed. The crew is one electrician working at a labor cost of $26.88 per manhour. These costs include layout, material handling, and normal waste. Add for lamp, accessories, sales tax, delivery, supervision, mobilization, demobilization, cleanup, overhead and profit. Note: These costs are based on internally pre-wired fixtures installed in new construction not over 10 feet above the floor. These fixtures can be ordered with the lamps installed. Be sure to specify the lamp color needed. Lay-in fixtures are available both with flex tails ready for make-up in J-boxes and for soft wiring systems. Soft wiring is a special plug-in system used when wiring fixtures in a suspended ceiling system. The inter-connecting cables are factory assembled in various lengths with circuiting to meet most job requirements. Soft wiring systems carry the UL label and are accepted in most code jurisdictions. Energy saving (ES) ballasts are available and should be installed at the factory to save field labor hours. When ES fluorescent lighting fixtures are installed, be sure to use ES lamps. Special surface-mounted and semi-flush fluorescent fixtures with wood siding are available for designer applications. You can also order parabolic louvers with colored anodized finishes.

Framed Troffer Lay-in T-Bar Fluorescent Fixtures

Material	Craft@Hrs	Unit	Material Cost	Labor Cost	Installed Cost

12"-wide lay-in T-bar fluorescent fixture with vandal resistant lens

Material	Craft@Hrs	Unit	Material Cost	Labor Cost	Installed Cost
48" 1 lamp	1E@0.55	Ea	103.00	14.80	117.80
48" 2 lamp	1E@0.60	Ea	103.00	16.10	119.10
48" 3 lamp	1E@0.65	Ea	121.00	17.50	138.50
48" 1 lamp, energy saver	1E@0.55	Ea	115.00	14.80	129.80
48" 2 lamp, energy saver	1E@0.60	Ea	115.00	16.10	131.10
48" 3 lamp, energy saver	1E@0.65	Ea	128.00	17.50	145.50

24"-square lay-in T-bar fluorescent fixture with vandal resistant lens

Material	Craft@Hrs	Unit	Material Cost	Labor Cost	Installed Cost
F40 U bent, 1 lamp	1E@0.50	Ea	103.00	13.40	116.40
F40 U bent, 2 lamp	1E@0.55	Ea	103.00	14.80	117.80
F40 U bent, 3 lamp	1E@0.60	Ea	121.00	16.10	137.10
F20 4 lamp	1E@0.60	Ea	132.00	16.10	148.10

24"-wide lay-in T-bar fluorescent fixture with vandal resistant lens

Material	Craft@Hrs	Unit	Material Cost	Labor Cost	Installed Cost
48" 2 lamp	1E@0.70	Ea	114.00	18.80	132.80
48" 3 lamp	1E@0.75	Ea	132.00	20.20	152.20
48" 4 lamp	1E@0.80	Ea	132.00	21.50	153.50
48" 2 lamp, energy saver	1E@0.70	Ea	126.00	18.80	144.80
48" 3 lamp, energy saver	1E@0.75	Ea	157.00	20.20	177.20
48" 4 lamp, energy saver	1E@0.80	Ea	157.00	21.50	178.50

Use these figures to estimate the cost of fluorescent fixtures installed in buildings under the conditions described on pages 5 and 6. Costs listed are for each fixture installed. The crew is one electrician working at a labor cost of $26.88 per manhour. These costs include layout, material handling, and normal waste. Add for lamp, accessories, sales tax, delivery, supervision, mobilization, demobilization, cleanup, overhead and profit. Note: These costs are based on internally pre-wired fixtures installed in new construction not over 10 feet above the floor. These fixtures can be ordered with the lamps installed. Be sure to specify the lamp color needed. Lay-in fixtures are available both with flex tails ready for make-up in J-boxes and for soft wiring systems. Soft wiring is a special plug-in system used when wiring fixtures in a suspended ceiling system. The inter-connecting cables are factory assembled in various lengths with circuiting to meet most job requirements. Soft wiring systems carry the UL label and are accepted in most code jurisdictions. Energy saving (ES) ballasts are available and should be installed at the factory to save field labor hours. When ES fluorescent lighting fixtures are installed, be sure to use ES lamps. Special surface-mounted and semi-flush fluorescent fixtures with wood siding are available for designer applications. You can also order parabolic louvers with colored anodized finishes.

Recessed High Intensity Discharge Fixtures

Material		Craft@Hrs	Unit	Material Cost	Labor Cost	Installed Cost

24"-square recessed HID open fixtures

Material		Craft@Hrs	Unit	Material Cost	Labor Cost	Installed Cost
175W	Mercury vapor	1E@0.55	Ea	132.00	14.80	146.80
400W	Mercury vapor	1E@0.55	Ea	115.00	14.80	129.80
400W	Metal halide	1E@0.55	Ea	170.00	14.80	184.80
100W	Hi pres sodium	1E@0.55	Ea	191.00	14.80	205.80
250W	Hi pres sodium	1E@0.55	Ea	208.00	14.80	222.80

24"-square recessed HID prismatic lens fixtures

Material		Craft@Hrs	Unit	Material Cost	Labor Cost	Installed Cost
250W	Mercury vapor	1E@0.60	Ea	125.00	16.10	141.10
400W	Mercury vapor	1E@0.60	Ea	130.00	16.10	146.10
400W	Metal halide	1E@0.60	Ea	153.00	16.10	169.10
150W	Hi pres sodium	1E@0.60	Ea	220.00	16.10	236.10
250W	Hi pres sodium	1E@0.60	Ea	302.00	16.10	318.10

24"-square regressed HID door fixtures

Material		Craft@Hrs	Unit	Material Cost	Labor Cost	Installed Cost
400W	Mercury vapor	1E@0.60	Ea	252.00	16.10	268.10
175W	Metal halide	1E@0.60	Ea	193.00	16.10	209.10
400W	Metal halide	1E@0.60	Ea	279.00	16.10	295.10
100W	Hi pres sodium	1E@0.60	Ea	227.00	16.10	243.10
150W	Hi pres sodium	1E@0.60	Ea	232.00	16.10	248.10

24"-square regressed HID ventilated door fixtures

Material		Craft@Hrs	Unit	Material Cost	Labor Cost	Installed Cost
400W	Mercury vapor	1E@0.60	Ea	262.00	16.10	278.10
175W	Metal halide	1E@0.60	Ea	203.00	16.10	219.10
400W	Metal halide	1E@0.60	Ea	289.00	16.10	305.10
100W	Hi pres sodium	1E@0.60	Ea	237.00	16.10	253.10
150W	Hi pres sodium	1E@0.60	Ea	242.00	16.10	258.10

Use these figures to estimate the cost of lighting fixtures installed in buildings under the conditions described on pages 5 and 6. Costs listed are for each fixture installed. The crew is one electrician working at a labor cost of $26.88 per manhour. These costs include layout, material handling, and normal waste. Add for lamp, accessories, sales tax, delivery, supervision, mobilization, demobilization, cleanup, overhead and profit. Note: These costs are based on internally pre-wired fixtures installed in new construction not over 16 feet above the floor. These fixtures are designed for interior locations and for use in suspended ceiling systems.

Recessed High Intensity Discharge Fixtures

Material	Craft@Hrs	Unit	Material Cost	Labor Cost	Installed Cost
14"-high bay recessed HID open reflector fixtures					
400W MV, 480V	1E@0.65	Ea	140.00	17.50	157.50
400W MV, Taps	1E@0.65	Ea	147.00	17.50	164.50
400W MV, Taps	1E@0.65	Ea	161.00	17.50	178.50
400W MH, Taps	1E@0.65	Ea	166.00	17.50	183.50
250W HPS, 480V	1E@0.65	Ea	168.00	17.50	185.50
250W HPS, Taps	1E@0.65	Ea	176.00	17.50	193.50
400W HPS, 480V	1E@0.65	Ea	189.00	17.50	206.50
400W HPS, Taps	1E@0.65	Ea	194.00	17.50	211.50
18"-high bay recessed HID open reflector fixtures					
250W MV, Taps	1E@0.70	Ea	190.00	18.80	208.80
400W MV, Taps	1E@0.70	Ea	192.00	18.80	210.80
400W MH, Taps	1E@0.70	Ea	292.00	18.80	310.80
200W HPS, Taps	1E@0.70	Ea	250.00	18.80	268.80
250W HPS, Taps	1E@0.70	Ea	296.00	18.80	314.80
310W HPS, Taps	1E@0.70	Ea	272.00	18.80	290.80
400W HPS, Taps	1E@0.70	Ea	279.00	18.80	297.80
23"-high bay recessed HID open reflector fixtures					
1000W MV, Taps	1E@0.75	Ea	376.00	20.20	396.20
1000W MH, Taps	1E@0.75	Ea	442.00	20.20	462.20
1000W HPS, Taps	1E@0.75	Ea	517.00	20.20	537.20
11"-low bay recessed HID open reflector retrofit fixtures					
35W HPS, MSB	1E@0.15	Ea	98.00	4.03	102.03
50W HPS, MSB	1E@0.15	Ea	100.00	4.03	104.03
70W HPS, MSB	1E@0.15	Ea	103.00	4.03	107.03
35W HPS, CM	1E@0.35	Ea	98.00	9.41	107.41
50W HPS, CM	1E@0.35	Ea	100.00	9.41	109.41
70W HPS, CM	1E@0.35	Ea	103.00	9.41	112.41
35W HPS, OBM	1E@0.25	Ea	98.00	6.72	104.72
50W HPS, OBM	1E@0.25	Ea	100.00	6.72	106.72
70W HPS, OBM	1E@0.25	Ea	103.00	6.72	109.72

Use these figures to estimate the cost of lighting fixtures installed in buildings under the conditions described on pages 5 and 6. Costs listed are for each fixture installed. The crew is one electrician working at a labor cost of $26.88 per manhour. These costs include layout, material handling, and normal waste. Add for lamp, accessories, sales tax, delivery, supervision, mobilization, demobilization, cleanup, overhead and profit. Note: These costs are based on internally pre-wired fixtures installed in new construction not over 16 feet above the floor. Abbreviations: MV indicates mercury vapor, MH is metal halide, HPS is high pressure sodium.

Recessed High Intensity Discharge Fixtures

Material	Craft@Hrs	Unit	Material Cost	Labor Cost	Installed Cost
14"-low bay recessed HID open reflector fixtures					
100W MV	1E@0.65	Ea	136.00	17.50	153.50
175W MV	1E@0.65	Ea	137.00	17.50	154.50
250W MV	1E@0.65	Ea	138.00	17.50	155.50
175W MH	1E@0.65	Ea	163.00	17.50	180.50
50W HPS	1E@0.65	Ea	176.00	17.50	193.50
70W HPS	1E@0.65	Ea	179.00	17.50	196.50
100W HPS	1E@0.65	Ea	181.00	17.50	198.50
150W HPS	1E@0.65	Ea	184.00	17.50	201.50
10"-diameter recessed high intensity work area enclosed fixtures					
50W HPS	1E@0.50	Ea	140.00	13.40	153.40
70W HPS	1E@0.50	Ea	142.00	13.40	155.40
10"-diameter recessed high intensity work fixture for retrofit					
50W HPS	1E@0.15	Ea	156.00	4.03	160.03
70W HPS	1E@0.15	Ea	158.00	4.03	162.03
18"-diameter recessed HID industrial enclosed fixtures					
100W MV	1E@0.70	Ea	154.00	18.80	172.80
150W MV	1E@0.70	Ea	156.00	18.80	174.80
250W MV	1E@0.70	Ea	157.00	18.80	175.80
175W MH	1E@0.70	Ea	159.00	18.80	177.80
100W HPS	1E@0.70	Ea	199.00	18.80	217.80
150W HPS	1E@0.70	Ea	204.00	18.80	222.80

Use these figures to estimate the cost of lighting fixtures installed in buildings under the conditions described on pages 5 and 6. Costs listed are for each fixture installed. The crew is one electrician working at a labor cost of $26.88 per manhour. These costs include layout, material handling, and normal waste. Add for lamp, accessories, sales tax, delivery, supervision, mobilization, demobilization, cleanup, overhead and profit. Note: These costs are based on internally pre-wired fixtures installed in new construction not over 16 feet above the floor. Abbreviations: Mercury vapor is MV, metal halide is MH, high pressure sodium is HPS. Some high bay high intensity fixtures are available with a variety of mounting accessories: power hook assemblies for quick servicing, soft wiring connectors, and flexible tails for connections to J-boxes. Some manufacturers offer a computerized lighting design plan to help select the right fixture and the best location. In some cases it may be necessary to have high intensity light fixtures made to order to meet specific requirements.

Interior Industrial High Intensity Discharge Luminaires

Material	Craft@Hrs	Unit	Material Cost	Labor Cost	Installed Cost

16"-diameter enclosed 120 volt interior industrial HID fixtures

Material	Craft@Hrs	Unit	Material Cost	Labor Cost	Installed Cost
250W MH	1E@0.70	Ea	241.00	18.80	259.80
200W HPS	1E@0.70	Ea	282.00	18.80	300.80
250W HPS	1E@0.70	Ea	293.00	18.80	311.80

29"-diameter enclosed 120, 208, 240 or 277 volt interior industrial HID fixtures

Material	Craft@Hrs	Unit	Material Cost	Labor Cost	Installed Cost
400W MV	1E@0.75	Ea	222.00	20.20	242.20
250W MH	1E@0.75	Ea	224.00	20.20	244.20
400W MH	1E@0.75	Ea	288.00	20.20	308.20
200W HPS	1E@0.75	Ea	315.00	20.20	335.20
250W HPS	1E@0.75	Ea	319.00	20.20	339.20
310W HPS	1E@0.75	Ea	325.00	20.20	345.20
400W HPS	1E@0.75	Ea	328.00	20.20	348.20

16"-diameter enclosed 120, 208, 240 or 277 volt interior industrial HID fixtures for fast installation

Material	Craft@Hrs	Unit	Material Cost	Labor Cost	Installed Cost
400W MV	1E@0.60	Ea	276.00	16.10	292.10
250W MH	1E@0.60	Ea	303.00	16.10	319.10
400W MH	1E@0.60	Ea	306.00	16.10	322.10
200W HPS	1E@0.60	Ea	343.00	16.10	359.10
250W HPS	1E@0.60	Ea	349.00	16.10	365.10
310W HPS	1E@0.60	Ea	355.00	16.10	371.10
400W HPS	1E@0.60	Ea	361.00	16.10	377.10

Use these figures to estimate the cost of lighting fixtures installed in buildings under the conditions described on pages 5 and 6. Costs listed are for each fixture installed. The crew is one electrician working at a labor cost of $26.88 per manhour. These costs include layout, material handling, and normal waste. Add for lamp, accessories, sales tax, delivery, supervision, mobilization, demobilization, cleanup, overhead and profit. Note: These costs are based on internally pre-wired fixtures installed in new construction not over 16 feet above the floor. Abbreviations: Mercury vapor is MV, metal halide is MH, high pressure sodium is HPS. Some high bay high intensity fixtures are available with a variety of mounting accessories: power hook assemblies for quick servicing, soft wiring connectors, and flexible tails for connections to J-boxes. Some manufacturers offer a computerized lighting design plan to help select the right fixture and the best location. In some cases it may be necessary to have high intensity light fixtures made to order to meet specific requirements.

HID Bracket-Mounted Exterior Floodlights

Material	Craft@Hrs	Unit	Material Cost	Labor Cost	Installed Cost
HID bracket-mounted exterior floodlights with lamp base down					
250W Mercury vapor	1E@1.25	Ea	276.00	33.60	309.60
400W Mercury vapor	1E@1.25	Ea	289.00	33.60	322.60
1000W Mercury vapor	1E@1.25	Ea	394.00	33.60	427.60
400W Metal halide	1E@1.25	Ea	315.00	33.60	348.60
1000W Metal halide	1E@1.25	Ea	426.00	33.60	459.60
200W Hi pres sodium	1E@1.25	Ea	328.00	33.60	361.60
250W Hi pres sodium	1E@1.25	Ea	342.00	33.60	375.60
400W Hi pres sodium	1E@1.25	Ea	361.00	33.60	394.60
1000W Hi pres sodium	1E@1.25	Ea	525.00	33.60	558.60
HID bracket-mounted exterior floodlights with lamp base horizontal					
250W Mercury vapor	1E@1.25	Ea	276.00	33.60	309.60
400W Mercury vapor	1E@1.25	Ea	302.00	33.60	335.60
1000W Mercury vapor	1E@1.25	Ea	394.00	33.60	427.60
400W Metal halide	1E@1.25	Ea	315.00	33.60	348.60
1000W Metal halide	1E@1.25	Ea	426.00	33.60	459.60
200W Hi pres sodium	1E@1.25	Ea	328.00	33.60	361.60
250W Hi pres sodium	1E@1.25	Ea	342.00	33.60	375.60
400W Hi pres sodium	1E@1.25	Ea	361.00	33.60	394.60
1000W Hi pres sodium	1E@1.25	Ea	525.00	33.60	558.60
Small HID bracket-mounted exterior floodlights					
35W Hi pres sodium	1E@1.00	Ea	129.00	26.90	155.90
50W Hi pres sodium	1E@1.00	Ea	131.00	26.90	157.90
70W Hi pres sodium	1E@1.00	Ea	133.00	26.90	159.90
100W Hi pres sodium	1E@1.00	Ea	135.00	26.90	161.90
Heavy duty HID bracket-mounted floodlights with lamp base horizontal					
400W Mercury vapor	1E@1.30	Ea	597.00	34.90	631.90
1000W Mercury vapor	1E@1.30	Ea	779.00	34.90	813.90
400W Metal halide	1E@1.30	Ea	597.00	34.90	631.90
1000W Metal halide	1E@1.30	Ea	779.00	34.90	813.90

Use these figures to estimate the cost of bracket-mounted area lighting installed under the conditions described on pages 5 and 6. Costs listed are for each fixture installed. The crew is one electrician working at a labor cost of $26.88 per manhour. These costs include layout, material handling, and normal waste. Add for lamp, accessories, sales tax, delivery, supervision, mobilization, demobilization, cleanup, overhead and profit. Note: Many of these lighting fixtures can be mounted in clusters to reduce the number of poles needed. Some manufacturers offer a computerized layout service that will find the most efficient combination of lamps and locations that meets job requirements.

HID Lamp Base Horizontal Pole-Mounted Floodlights

Material	Craft@Hrs	Unit	Material Cost	Labor Cost	Installed Cost

HID pole-mounted floodlights with lamp base horizontal without poles

Material	Craft@Hrs	Unit	Material Cost	Labor Cost	Installed Cost
100W Mercury vapor	1E@0.70	Ea	224.00	18.80	242.80
175W Mercury vapor	1E@0.70	Ea	290.00	18.80	308.80
250W Mercury vapor	1E@0.70	Ea	306.00	18.80	324.80
400W Mercury vapor	1E@0.70	Ea	324.00	18.80	342.80
1000W Mercury vapor	1E@0.70	Ea	472.00	18.80	490.80
250W Metal halide	1E@0.70	Ea	347.00	18.80	365.80
400W Metal halide	1E@0.70	Ea	352.00	18.80	370.80
1000W Metal halide	1E@0.70	Ea	498.00	18.80	516.80
175W Super metalarc	1E@0.70	Ea	335.00	18.80	353.80
250W Super metalarc	1E@0.70	Ea	357.00	18.80	375.80
400W Super metalarc	1E@0.70	Ea	362.00	18.80	380.80
100W Hi pres sodium	1E@0.70	Ea	329.00	18.80	347.80
150W Hi pres sodium	1E@0.70	Ea	341.00	18.80	359.80
200W Hi pres sodium	1E@0.70	Ea	365.00	18.80	383.80
250W Hi pres sodium	1E@0.70	Ea	374.00	18.80	392.80
400W Hi pres sodium	1E@0.70	Ea	415.00	18.80	433.80
1000W Hi pres sodium	1E@0.70	Ea	590.00	18.80	608.80

HID floodlights with lamp base horizontal, including typical poles

Material	Craft@Hrs	Unit	Material Cost	Labor Cost	Installed Cost
175W Mercury vapor	1E@0.70	Ea	375.00	18.80	393.80
250W Mercury vapor	1E@0.70	Ea	441.00	18.80	459.80
400W Mercury vapor	1E@0.70	Ea	608.00	18.80	626.80
1000W Mercury vapor	1E@0.80	Ea	1,030.00	21.50	1,051.50
250W Metal halide	1E@0.70	Ea	596.00	18.80	614.80
400W Metal halide	1E@0.70	Ea	529.00	18.80	547.80
1000W Metal halide	1E@0.80	Ea	753.00	21.50	774.50
175W Super metalarc	1E@0.70	Ea	446.00	18.80	464.80
250W Super metalarc	1E@0.70	Ea	613.00	18.80	631.80
400W Super metalarc	1E@0.70	Ea	658.00	18.80	676.80
18W Lo pres sodium	1E@0.70	Ea	357.00	18.80	375.80
35W Lo pres sodium	1E@0.70	Ea	418.00	18.80	436.80
100W Hi pres sodium	1E@0.70	Ea	391.00	18.80	409.80
150W Hi pres sodium	1E@0.70	Ea	474.00	18.80	492.80
250W Hi pres sodium	1E@0.70	Ea	646.00	18.80	664.80
400W Hi pres sodium	1E@0.70	Ea	619.00	18.80	637.80
1000W Hi pres sodium	1E@0.80	Ea	845.00	21.50	866.50

Use these figures to estimate the cost of pole-mounted area lighting installed under the conditions described on pages 5 and 6. Costs listed are for each fixture installed. The crew is one electrician working at a labor cost of $26.88 per manhour. These costs include layout, material handling, and normal waste. Add for lamp, pole, accessories, sales tax, delivery, supervision, mobilization, demobilization, cleanup, overhead and profit. Note: Many of these lighting fixtures can be mounted in clusters to reduce the number of poles needed. Some manufacturers offer a computerized layout service that will find the most efficient combination of lamps and locations that meets job requirements.

HID Exterior Floodlights

Material	Craft@Hrs	Unit	Material Cost	Labor Cost	Installed Cost
Pole-mounted sphere HID exterior floodlights, lamp base down					
75W Mercury vapor	1E@0.70	Ea	188.00	18.80	206.80
100W Mercury vapor	1E@0.70	Ea	190.00	18.80	208.80
175W Mercury vapor	1E@0.70	Ea	192.00	18.80	210.80
175W Metal halide	1E@0.70	Ea	194.00	18.80	212.80
35W Hi pres sodium	1E@0.70	Ea	200.00	18.80	218.80
50W Hi pres sodium	1E@0.70	Ea	204.00	18.80	222.80
70W Hi pres sodium	1E@0.70	Ea	209.00	18.80	227.80
100W Hi pres sodium	1E@0.70	Ea	214.00	18.80	232.80
Wall-mounted cylinder HID exterior floodlights, lamp base down					
75W Mercury vapor	1E@0.80	Ea	216.00	21.50	237.50
100W Mercury vapor	1E@0.80	Ea	219.00	21.50	240.50
175W Mercury vapor	1E@0.80	Ea	222.00	21.50	243.50
175W Metal halide	1E@0.80	Ea	225.00	21.50	246.50
35W Hi pres sodium	1E@0.80	Ea	237.00	21.50	258.50
50W Hi pres sodium	1E@0.80	Ea	241.00	21.50	262.50
70W Hi pres sodium	1E@0.80	Ea	248.00	21.50	269.50
100W Hi pres sodium	1E@0.80	Ea	254.00	21.50	275.50
150W Hi pres sodium	1E@0.80	Ea	258.00	21.50	279.50
Wall-mounted round HID exterior floodlights, lamp base horizontal					
175W Mercury vapor	1E@0.75	Ea	444.00	20.20	464.20
250W Mercury vapor	1E@0.75	Ea	462.00	20.20	482.20
400W Mercury vapor	1E@0.75	Ea	518.00	20.20	538.20
250W Metal halide	1E@0.75	Ea	523.00	20.20	543.20
400W Metal halide	1E@0.75	Ea	523.00	20.20	543.20
175W Super metalarc	1E@0.75	Ea	480.00	20.20	500.20
250W Super metalarc	1E@0.75	Ea	525.00	20.20	545.20
400W Super metalarc	1E@0.75	Ea	525.00	20.20	545.20
70W Hi pres sodium	1E@0.75	Ea	468.00	20.20	488.20
100W Hi pres sodium	1E@0.75	Ea	474.00	20.20	494.20
150W Hi pres sodium	1E@0.75	Ea	499.00	20.20	519.20
250W Hi pres sodium	1E@0.75	Ea	618.00	20.20	638.20
400W Hi pres sodium	1E@0.75	Ea	618.00	20.20	638.20

Use these figures to estimate the cost of area lighting installed under the conditions described on pages 5 and 6. Costs listed are for each fixture installed. The crew is one electrician working at a labor cost of $26.88 per manhour. These costs include layout, material handling, and normal waste. Add for lamp, pole, accessories, sales tax, delivery, supervision, mobilization, demobilization, cleanup, overhead and profit. Note: Many of these lighting fixtures can be mounted in clusters to reduce the number of poles needed. Some manufacturers offer a computerized layout service that will find the most efficient combination of lamps and locations that meets job requirements.

HID Off-Street Area Lighting, Lamp Base Up

Material	Craft@Hrs	Unit	Material Cost	Labor Cost	Installed Cost

HID off-street 120 volt area lighting with photocell control, lamp base up

Material	Craft@Hrs	Unit	Material Cost	Labor Cost	Installed Cost
175W Mercury vapor	1E@0.60	Ea	77.40	16.10	93.50
50W Hi pres sodium	1E@0.60	Ea	167.00	16.10	183.10
70W Hi pres sodium	1E@0.60	Ea	169.00	16.10	185.10
100W Hi pres sodium	1E@0.60	Ea	171.00	16.10	187.10
150W Hi pres sodium	1E@0.60	Ea	173.00	16.10	189.10

HID off-street area lighting 120, 208, 240, 277 or 480 volt open luminaires

Material	Craft@Hrs	Unit	Material Cost	Labor Cost	Installed Cost
1000W Mercury vapor	1E@1.25	Ea	529.00	33.60	562.60
1000W Metal halide	1E@1.25	Ea	529.00	33.60	562.60
400W Hi pres sodium	1E@1.25	Ea	435.00	33.60	468.60
1000W Hi pres sodium	1E@1.25	Ea	614.00	33.60	647.60

HID off-street area lighting 120, 208, 240, 277 or 480 volt flat lens

Material	Craft@Hrs	Unit	Material Cost	Labor Cost	Installed Cost
1000W Mercury vapor	1E@1.30	Ea	547.00	34.90	581.90
1000W Metal halide	1E@1.30	Ea	547.00	34.90	581.90
400W Hi pres sodium	1E@1.30	Ea	490.00	34.90	524.90
1000W Hi pres sodium	1E@1.30	Ea	725.00	34.90	759.90

HID off-street area lighting 120, 208, 240, 277 or 480 volt closed asymmetric lens

Material	Craft@Hrs	Unit	Material Cost	Labor Cost	Installed Cost
1000W Mercury vapor	1E@1.30	Ea	632.00	34.90	666.90
1000W Metal halide	1E@1.30	Ea	632.00	34.90	666.90
400W Hi pres sodium	1E@1.30	Ea	721.00	34.90	755.90
1000W Hi pres sodium	1E@1.30	Ea	766.00	34.90	800.90

Use these figures to estimate the cost of bracket-mounted area lighting installed under the conditions described on pages 5 and 6. Costs listed are for each fixture installed. The crew is one electrician working at a labor cost of $26.88 per manhour. These costs include layout, material handling, and normal waste. Add for lamp, accessories, sales tax, delivery, supervision, mobilization, demobilization, cleanup, overhead and profit. Note: Many of these lighting fixtures can be mounted in clusters to reduce the number of poles needed. Some manufacturers offer a computerized layout service that will find the most efficient combination of lamps and locations that meets job requirements. The luminaires on this page are intended for mounting at 60' above ground level. All high intensity discharge luminaires must go through a warm-up period when first energized and a cool-down period when de-energized. Always burn-in the lamp for at least an hour on the first turn-on. Some HID lamps (especially Metal halide) will not restart quickly after being disconnected from their power source. They must go through cool-down and warm-up all over again.

HID Exterior Pole-Mounted Floodlights

Material	Craft@Hrs	Unit	Material Cost	Labor Cost	Installed Cost
HID exterior pole-mounted rectangular architectural horizontal floodlights					
100W Mercury vapor	1E@0.85	Ea	422.00	22.80	444.80
175W Mercury vapor	1E@0.85	Ea	426.00	22.80	448.80
250W Mercury vapor	1E@0.85	Ea	439.00	22.80	461.80
400W Mercury vapor	1E@0.85	Ea	453.00	22.80	475.80
400W Metal halide	1E@0.85	Ea	473.00	22.80	495.80
175W Super metalarc	1E@0.85	Ea	460.00	22.80	482.80
250W Super metalarc	1E@0.85	Ea	479.00	22.80	501.80
400W Super metalarc	1E@0.85	Ea	486.00	22.80	508.80
70W Hi pressure sodium	1E@0.85	Ea	479.00	22.80	501.80
100W Hi pressure sodium	1E@0.85	Ea	488.00	22.80	510.80
150W Hi pressure sodium	1E@0.85	Ea	493.00	22.80	515.80
200W Hi pressure sodium	1E@0.85	Ea	522.00	22.80	544.80
250W Hi pressure sodium	1E@0.85	Ea	528.00	22.80	550.80
310W Hi pressure sodium	1E@0.85	Ea	532.00	22.80	554.80
400W Hi pressure sodium	1E@0.85	Ea	535.00	22.80	557.80
HID exterior pole-mounted architectural horizontal floodlights, 45° cutoff					
400W Metal halide	1E@1.00	Ea	626.00	26.90	652.90
1000W Metal halide	1E@1.00	Ea	835.00	26.90	861.90
400W Hi pressure sodium	1E@1.00	Ea	722.00	26.90	748.90
1000W Hi pressure sodium	1E@1.00	Ea	926.00	26.90	952.90
HID exterior pole-mounted architectural square floodlights, lamp base down					
1000W Mercury vapor	1E@1.00	Ea	653.00	26.90	679.90
400W Metal halide	1E@1.00	Ea	619.00	26.90	645.90
1000W Metal halide	1E@1.00	Ea	653.00	26.90	679.90
250W Hi pressure sodium	1E@1.00	Ea	641.00	26.90	667.90
400W Hi pressure sodium	1E@1.00	Ea	658.00	26.90	684.90

Use these figures to estimate the cost of pole-mounted area lighting installed under the conditions described on pages 5 and 6. Costs listed are for each fixture installed. The crew is one electrician working at a labor cost of $26.88 per manhour. These costs include layout, material handling, and normal waste. Add for the pole, lamp, excavation, accessories, sales tax, delivery, supervision, mobilization, demobilization, cleanup, overhead and profit. Note: Many of these lighting fixtures can be mounted in clusters to reduce the number of poles needed. Some manufacturers offer a computerized layout service that will find the most efficient combination of lamps and locations that meets job requirements.

Material	Craft@Hrs	Unit	Material Cost	Labor Cost	Installed Cost

HID exterior square walkway 120 volt bollards, lamp base down

Material	Craft@Hrs	Unit	Material Cost	Labor Cost	Installed Cost
75W Mercury vapor	1E@1.75	Ea	500.00	47.00	547.00
100W Mercury vapor	1E@1.75	Ea	507.00	47.00	554.00
35W Hi pres sodium	1E@1.75	Ea	433.00	47.00	480.00
50W Hi pres sodium	1E@1.75	Ea	441.00	47.00	488.00
70W Hi pres sodium	1E@1.75	Ea	540.00	47.00	587.00
100W Hi pres sodium	1E@1.75	Ea	553.00	47.00	600.00

HID exterior round walkway 120 volt bollards, lamp base down

Material	Craft@Hrs	Unit	Material Cost	Labor Cost	Installed Cost
70W Mercury vapor	1E@1.75	Ea	500.00	47.00	547.00
100W Mercury vapor	1E@1.75	Ea	507.00	47.00	554.00
35W Hi pres sodium	1E@1.75	Ea	433.00	47.00	480.00
50W Hi pres sodium	1E@1.75	Ea	441.00	47.00	488.00
70W Hi pres sodium	1E@1.75	Ea	540.00	47.00	587.00
100W Hi pres sodium	1E@1.75	Ea	553.00	47.00	600.00

HID exterior prismatic lens 120 to 480 volt wall luminaires, lamp base horizontal

Material	Craft@Hrs	Unit	Material Cost	Labor Cost	Installed Cost
100W Mercury vapor	1E@1.00	Ea	181.00	26.90	207.90
175W Mercury vapor	1E@1.00	Ea	182.00	26.90	208.90
175W Metal halide	1E@1.00	Ea	186.00	26.90	212.90
18W Lo pres sodium	1E@1.00	Ea	208.00	26.90	234.90
35W Hi pres sodium	1E@1.00	Ea	191.00	26.90	217.90
50W Hi pres sodium	1E@1.00	Ea	196.00	26.90	222.90
70W Hi pres sodium	1E@1.00	Ea	217.00	26.90	243.90
100W Hi pres sodium	1E@1.00	Ea	222.00	26.90	248.90

HID exterior clear lens 120 to 480 volt wall luminaires, lamp base horizontal

Material	Craft@Hrs	Unit	Material Cost	Labor Cost	Installed Cost
100W Mercury vapor	1E@1.00	Ea	210.00	26.90	236.90
175W Mercury vapor	1E@1.00	Ea	212.00	26.90	238.90
175W Metal halide	1E@1.00	Ea	220.00	26.90	246.90
18W Lo pres sodium	1E@1.00	Ea	243.00	26.90	269.90
35W Hi pres sodium	1E@1.00	Ea	217.00	26.90	243.90
50W Hi pres sodium	1E@1.00	Ea	221.00	26.90	247.90
70W Hi pres sodium	1E@1.00	Ea	251.00	26.90	277.90
100W Hi pres sodium	1E@1.00	Ea	256.00	26.90	282.90

Use these figures to estimate the cost of bracket mounted area lighting installed under the conditions described on pages 5 and 6. Costs listed are for each fixture installed. The crew is one electrician working at a labor cost of $26.88 per manhour. These costs include layout, material handling, and normal waste. Add for lamp, accessories, sales tax, delivery, supervision, mobilization, demobilization, cleanup, overhead and profit. Note: Many of these lighting fixtures can be mounted in clusters to reduce the number of poles needed. Some manufacturers offer a computerized layout service that will find the most efficient combination of lamps and locations that meets job requirements.

Pole-Mounted HID Exterior Floodlights

Material	Craft@Hrs	Unit	Material Cost	Labor Cost	Installed Cost
HID exterior pole-mounted flat lens square arm 120 volt luminaires					
1000W Mercury vapor	1E@1.50	Ea	703.00	40.30	743.30
1000W Metal halide	1E@1.50	Ea	737.00	40.30	777.30
1000W Hi pres sodium	1E@1.50	Ea	933.00	40.30	973.30
HID pole-mounted square exterior walkway 120 volt luminaires, base down					
100W Mercury vapor	1E@0.70	Ea	775.00	18.80	793.80
175W Mercury vapor	1E@0.70	Ea	797.00	18.80	815.80
250W Mercury vapor	1E@0.70	Ea	804.00	18.80	822.80
400W Mercury vapor	1E@0.70	Ea	878.00	18.80	896.80
175W Metal halide	1E@0.70	Ea	855.00	18.80	873.80
250W Metal halide	1E@0.70	Ea	884.00	18.80	902.80
400W Metal halide	1E@0.70	Ea	891.00	18.80	909.80
70W Hi pres sodium	1E@0.70	Ea	793.00	18.80	811.80
100W Hi pres sodium	1E@0.70	Ea	804.00	18.80	822.80
150W Hi pres sodium	1E@0.70	Ea	878.00	18.80	896.80
250W Hi pres sodium	1E@0.70	Ea	931.00	18.80	949.80
400W Hi pres sodium	1E@0.70	Ea	954.00	18.80	972.80
HID pole-mounted round exterior walkway 120 volt luminaires, base down					
100W Mercury vapor	1E@0.70	Ea	793.00	18.80	811.80
175W Mercury vapor	1E@0.70	Ea	816.00	18.80	834.80
400W Mercury vapor	1E@0.70	Ea	901.00	18.80	919.80
175W Metal halide	1E@0.70	Ea	873.00	18.80	891.80
250W Metal halide	1E@0.70	Ea	902.00	18.80	920.80
400W Metal halide	1E@0.70	Ea	914.00	18.80	932.80
70W Hi pres sodium	1E@0.70	Ea	811.00	18.80	829.80
100W Hi pres sodium	1E@0.70	Ea	821.00	18.80	839.80
150W Hi pres sodium	1E@0.70	Ea	896.00	18.80	914.80
250W Hi pres sodium	1E@0.70	Ea	953.00	18.80	971.80
400W Hi pres sodium	1E@0.70	Ea	976.00	18.80	994.80

Use these figures to estimate the cost of pole-mounted area lighting installed under the conditions described on pages 5 and 6. Costs listed are for each fixture installed. The crew is one electrician working at a labor cost of $26.88 per manhour. These costs include layout, material handling, and normal waste. Add for lamppole, accessories, sales tax, delivery, supervision, mobilization, demobilization, cleanup, overhead and profit. Note: Many of these lighting fixtures can be mounted in clusters to reduce the number of poles needed. Some manufacturers offer a computerized layout service that will find the most efficient combination of lamps and locations that meets job requirements.

Pole-Mounted HID Square Luminaires, 480 Volts

Material	Craft@Hrs	Unit	Material Cost	Labor Cost	Installed Cost
HID pole-mounted 480 volt square luminaires with flat glass lens					
1000W Mercury vapor	1E@1.50	Ea	600.00	40.30	640.30
400W Metal halide	1E@1.50	Ea	595.00	40.30	635.30
1000W Metal halide	1E@1.50	Ea	600.00	40.30	640.30
250W Hi pres sodium	1E@1.50	Ea	617.00	40.30	657.30
400W Hi pres sodium	1E@1.50	Ea	633.00	40.30	673.30
1000W Hi pres sodium	1E@1.50	Ea	703.00	40.30	743.30
HID pole-mounted 480 volt square luminaires with convex glass lens					
1000W Mercury vapor	1E@1.50	Ea	654.00	40.30	694.30
400W Metal halide	1E@1.50	Ea	658.00	40.30	698.30
1000W Metal halide	1E@1.50	Ea	633.00	40.30	673.30
250W Hi pres sodium	1E@1.50	Ea	649.00	40.30	689.30
400W Hi pres sodium	1E@1.50	Ea	658.00	40.30	698.30
1000W Hi pres sodium	1E@1.50	Ea	751.00	40.30	791.30
HID pole-mounted 480 volt round luminaires with flat glass lens					
1000W Mercury vapor	1E@1.50	Ea	600.00	40.30	640.30
400W Metal halide	1E@1.50	Ea	574.00	40.30	614.30
1000W Metal halide	1E@1.50	Ea	600.00	40.30	640.30
250W Hi pres sodium	1E@1.50	Ea	595.00	40.30	635.30
400W Hi pres sodium	1E@1.50	Ea	611.00	40.30	651.30
1000W Hi pres sodium	1E@1.50	Ea	703.00	40.30	743.30
HID pole-mounted 480 volt round luminaires with convex glass lens					
1000W Mercury vapor	1E@1.50	Ea	633.00	40.30	673.30
400W Metal halide	1E@1.50	Ea	590.00	40.30	630.30
1000W Metal halide	1E@1.50	Ea	633.00	40.30	673.30
250W Hi pres sodium	1E@1.50	Ea	627.00	40.30	667.30
400W Hi pres sodium	1E@1.50	Ea	633.00	40.30	673.30
1000W Hi pres sodium	1E@1.50	Ea	729.00	40.30	769.30

Use these figures to estimate the cost of pole-mounted area lighting installed under the conditions described on pages 5 and 6. Costs listed are for each fixture installed. The crew is one electrician working at a labor cost of $26.88 per manhour. These costs include layout, material handling, and normal waste. Add for lamp pole, accessories, sales tax, delivery, supervision, mobilization, demobilization, cleanup, overhead and profit. Note: These luminaires are intended to be mounted on 40' to 60' poles.

HID Street and Roadway Luminaires

Material	Craft@Hrs	Unit	Material Cost	Labor Cost	Installed Cost
HID street and roadway luminaires, 28"-long housing					
100W Mercury vapor	1E@0.80	Ea	125.00	21.50	146.50
175W Mercury vapor	1E@0.80	Ea	149.00	21.50	170.50
250W Mercury vapor	1E@0.80	Ea	172.00	21.50	193.50
50W Hi pres sodium	1E@0.80	Ea	146.00	21.50	167.50
70W Hi pres sodium	1E@0.80	Ea	174.00	21.50	195.50
100W Hi pres sodium	1E@0.80	Ea	149.00	21.50	170.50
150W Hi pres sodium	1E@0.80	Ea	161.00	21.50	182.50
200W Hi pres sodium	1E@0.80	Ea	219.00	21.50	240.50
250W Hi pres sodium	1E@0.80	Ea	233.00	21.50	254.50
HID street and roadway luminaires, 29"-long housing					
100W Mercury vapor	1E@0.85	Ea	153.00	22.80	175.80
175W Mercury vapor	1E@0.85	Ea	155.00	22.80	177.80
250W Mercury vapor	1E@0.85	Ea	173.00	22.80	195.80
70W Hi pres sodium	1E@0.85	Ea	155.00	22.80	177.80
100W Hi pres sodium	1E@0.85	Ea	158.00	22.80	180.80
150W Hi pres sodium	1E@0.85	Ea	164.00	22.80	186.80
200W Hi pres sodium	1E@0.85	Ea	211.00	22.80	233.80
250W Hi pres sodium	1E@0.85	Ea	223.00	22.80	245.80
HID street and roadway luminaires, 33"-long housing					
400W Mercury vapor	1E@1.00	Ea	163.00	26.90	189.90
400W Metal halide	1E@1.00	Ea	217.00	26.90	243.90
200W Hi pres sodium	1E@1.00	Ea	230.00	26.90	256.90
250W Hi pres sodium	1E@1.00	Ea	237.00	26.90	263.90
310W Hi pres sodium	1E@1.00	Ea	253.00	26.90	279.90
400W Hi pres sodium	1E@1.00	Ea	256.00	26.90	282.90

Use these figures to estimate the cost of street and roadway luminaires installed under the conditions described on pages 5 and 6. Costs listed are for each unit installed. The crew is one electrician working at a labor cost of $26.88 per manhour. These costs include layout, material handling, and normal waste. Add for pole, arm, lamp, accessories, sales tax, delivery, supervision, mobilization, demobilization, cleanup, overhead and profit. Note: These luminaires mount on an arm that supports the fixture housing in a horizontal position. Usually the arm is attached to an 18' to 30' pole or to a wall mounting bracket. The housing has provision for installing a photo cell control switch.

Material		Craft@Hrs	Unit	Material Cost	Labor Cost	Installed Cost

HID street luminaires, 33"-long housing, lamp base horizontal

Material		Craft@Hrs	Unit	Material Cost	Labor Cost	Installed Cost
400W	Mercury vapor	1E@1.25	Ea	248.00	33.60	281.60
200W	Hi pres sodium	1E@1.25	Ea	302.00	33.60	335.60
250W	Hi pres sodium	1E@1.25	Ea	316.00	33.60	349.60
310W	Hi pres sodium	1E@1.25	Ea	335.00	33.60	368.60
400W	Hi pres sodium	1E@1.25	Ea	344.00	33.60	377.60

HID street luminaires, 40"-long housing, lamp base horizontal

Material		Craft@Hrs	Unit	Material Cost	Labor Cost	Installed Cost
1000W	Mercury vapor	1E@1.50	Ea	265.00	40.30	305.30
1000W	Metal halide	1E@1.50	Ea	542.00	40.30	582.30
1000W	Hi pres sodium	1E@1.50	Ea	372.00	40.30	412.30

Enclosed indoor mercury vapor ballasts

Material		Craft@Hrs	Unit	Material Cost	Labor Cost	Installed Cost
400W	MV 1 lamp	1E@0.40	Ea	154.00	10.80	164.80
400W	MV 2 lamp	1E@0.40	Ea	206.00	10.80	216.80
1000W	MV 1 lamp	1E@0.45	Ea	263.00	12.10	275.10
1000W	MV 2 lamp	1E@0.45	Ea	467.00	12.10	479.10

Exterior weatherproof potted high intensity ballasts

Material		Craft@Hrs	Unit	Material Cost	Labor Cost	Installed Cost
175W	MV 1 lamp	1E@0.50	Ea	166.00	13.40	179.40
250W	MV 1 lamp	1E@0.50	Ea	188.00	13.40	201.40
400W	MV 1 lamp	1E@0.50	Ea	188.00	13.40	201.40
400W	MV 2 lamp	1E@0.60	Ea	424.00	16.10	440.10
1000W	MV 1 lamp	1E@0.55	Ea	354.00	14.80	368.80
1000W	MV 2 lamp	1E@0.60	Ea	653.00	16.10	669.10
400W	MH 1 lamp	1E@0.50	Ea	250.00	13.40	263.40
400W	MH 2 lamp	1E@0.60	Ea	467.00	16.10	483.10
1000W	MH 1 lamp	1E@0.55	Ea	510.00	14.80	524.80
250W	HPS 1 lamp	1E@0.50	Ea	335.00	13.40	348.40
400W	HPS 1 lamp	1E@0.50	Ea	401.00	13.40	414.40
1000W	HPS 1 lamp	1E@0.60	Ea	744.00	16.10	760.10

Use these figures to estimate the cost of street and roadway luminaires and ballasts installed under the conditions described on pages 5 and 6. Costs listed are for each unit installed. The crew is one electrician working at a labor cost of $26.88 per manhour. These costs include layout, material handling, and normal waste. Add for pole, arm, lamp, special accessories, sales tax, delivery, supervision, mobilization, demobilization, cleanup, overhead and profit. Note: These luminaires mount on an arm that supports the fixture housing in a horizontal position. Usually the arm is attached to an 18' to 30' pole or to a wall mounting bracket. The housing has provision for installing a photo cell control switch.

Steel Light Poles with Square Mounting Base

Material	Craft@Hrs	Unit	Material Cost	Labor Cost	Installed Cost
Round steel street light poles with square mounting base					
8' high, 4" diameter	2E@0.70	Ea	223.00	18.80	241.80
8' high, 6" diameter	2E@0.80	Ea	353.00	21.50	374.50
10' high, 4" diameter	2E@0.75	Ea	234.00	20.20	254.20
10' high, 6" diameter	2E@0.85	Ea	376.00	22.80	398.80
12' high, 4" diameter	2E@0.90	Ea	250.00	24.20	274.20
12' high, 6" diameter	2E@0.95	Ea	400.00	25.50	425.50
14' high, 4" diameter	2E@1.00	Ea	287.00	26.90	313.90
14' high, 6" diameter	2E@1.10	Ea	435.00	29.60	464.60
16' high, 4" diameter	2E@1.20	Ea	310.00	32.30	342.30
16' high, 6" diameter	2E@1.30	Ea	494.00	34.90	528.90
18' high, 4" diameter	2E@1.50	Ea	323.00	40.30	363.30
18' high, 6" diameter	2E@1.60	Ea	561.00	43.00	604.00
Round tapered steel street light poles with square mounting base					
20' high	2E@1.25	Ea	615.00	33.60	648.60
25' high	2E@1.50	Ea	700.00	40.30	740.30
30' high	2E@1.75	Ea	801.00	47.00	848.00
35' high	2E@2.00	Ea	990.00	53.80	1,043.80
35' high, heavy duty	2E@2.50	Ea	1,230.00	67.20	1,297.20
39' high	2E@2.50	Ea	1,120.00	67.20	1,187.20
39' high, heavy duty	2E@3.00	Ea	1,390.00	80.60	1,470.60
50' high	2E@5.00	Ea	2,170.00	134.00	2,304.00
Square steel street light poles with square mounting base					
10' high	2E@0.75	Ea	280.00	20.20	300.20
12' high	2E@0.90	Ea	314.00	24.20	338.20
14' high	2E@1.00	Ea	346.00	26.90	372.90
16' high	2E@1.20	Ea	369.00	32.30	401.30
20' high	2E@1.60	Ea	542.00	43.00	585.00
24' high	2E@1.75	Ea	739.00	47.00	786.00
30' high	2E@2.00	Ea	800.00	53.80	853.80
35' high	2E@2.50	Ea	939.00	67.20	1,006.20

Use these figures to estimate the cost of street and yard light poles installed under the conditions described on pages 5 and 6. Costs listed are for each pole installed. The crew is two electricians working at a labor cost of $26.88 per manhour. These costs include four anchor bolts, layout, material handling, and normal waste. Add for the fixture, lamps, pole foundation, brackets, arms, excavation, sales tax, delivery, supervision, mobilization, demobilization, cleanup, overhead and profit. Note: Hoisting equipment may be necessary on the larger poles. Read the specs carefully when pricing poles. Some jobs will require items that are not included in these costs: decorative bases, anchor bolt covers, ballast bases, weatherproof receptacles, pole vibration dampers, or lowering devices.

Light Poles with Square Mounting Base

Material	Craft@Hrs	Unit	Material Cost	Labor Cost	Installed Cost
Square tapered steel light poles with square mounting base					
30' high	2E@2.00	Ea	727.00	53.80	780.80
30' high, heavy duty	2E@2.50	Ea	1,030.00	67.20	1,097.20
35' high	2E@2.50	Ea	985.00	67.20	1,052.20
35' high, heavy duty	2E@3.00	Ea	1,240.00	80.60	1,320.60
39' high	2E@3.00	Ea	1,010.00	80.60	1,090.60
39' high, heavy duty	2E@4.00	Ea	1,290.00	108.00	1,398.00
50' high	2E@6.00	Ea	1,770.00	161.00	1,931.00
50' high, heavy duty	2E@7.00	Ea	2,070.00	188.00	2,258.00
60' high	2E@7.00	Ea	2,250.00	188.00	2,438.00
60' high, heavy duty	2E@8.00	Ea	2,510.00	215.00	2,725.00
Hinged square steel light poles with square mounting base					
20' high	2E@1.60	Ea	720.00	43.00	763.00
24' high	2E@1.80	Ea	873.00	48.40	921.40
30' high	2E@2.20	Ea	1,040.00	59.10	1,099.10
35' high	2E@3.00	Ea	1,260.00	80.60	1,340.60
39' high	2E@3.50	Ea	1,600.00	94.10	1,694.10
Hinged square tapered steel light poles with square mounting base					
30' high	2E@2.25	Ea	1,200.00	60.50	1,260.50
35' high	2E@2.55	Ea	1,450.00	68.50	1,518.50
39' high	2E@3.10	Ea	1,850.00	83.30	1,933.30
Round tapered aluminum light poles with square mounting base					
6' high	2E@0.50	Ea	143.00	13.40	156.40
6' high, heavy duty	2E@0.60	Ea	157.00	16.10	173.10
8' high	2E@0.65	Ea	163.00	17.50	180.50
8' high, heavy duty	2E@0.70	Ea	182.00	18.80	200.80
10' high	2E@0.70	Ea	184.00	18.80	202.80
10' high, heavy duty	2E@0.75	Ea	211.00	20.20	231.20
12' high	2E@0.80	Ea	267.00	21.50	288.50
12' high, heavy duty	2E@0.85	Ea	308.00	22.80	330.80
15' high	2E@0.85	Ea	230.00	22.80	252.80
15' high, heavy duty	2E@0.90	Ea	403.00	24.20	427.20
18' high	2E@1.00	Ea	310.00	26.90	336.90
18' high, heavy duty	2E@1.15	Ea	452.00	30.90	482.90
20' high	2E@1.15	Ea	373.00	30.90	403.90
20' high, heavy duty	2E@1.25	Ea	350.00	33.60	383.60
25' high	2E@1.30	Ea	591.00	34.90	625.90
25' high, heavy duty	2E@1.35	Ea	1,140.00	36.30	1,176.30

Use these figures to estimate the cost of street and yard light poles installed under the conditions described on pages 5 and 6. Costs listed are for each pole installed. The crew is two electricians working at a labor cost of $26.88 per manhour. These costs include four anchor bolts, layout, material handling, and normal waste. Add for the fixture, lamps, pole foundation, brackets, arms, excavation, sales tax, delivery, supervision, mobilization, demobilization, cleanup, overhead and profit. Note: Hoisting equipment may be necessary on the larger poles. Read the specs carefully when pricing poles. Some jobs will require items that are not included in these costs: decorative bases, anchor bolt covers, ballast bases, weatherproof receptacles, pole vibration dampers, or lowering devices.

Round Tapered Aluminum Light Poles

Material	Craft@Hrs	Unit	Material Cost	Labor Cost	Installed Cost

Round tapered aluminum light poles without arms, square base

Material	Craft@Hrs	Unit	Material Cost	Labor Cost	Installed Cost
30' high	2E@1.40	Ea	815.00	37.60	852.60
30' high, heavy duty	2E@1.50	Ea	1,710.00	40.30	1,750.30
35' high	2E@1.75	Ea	1,010.00	47.00	1,057.00
35' high, heavy duty	2E@1.90	Ea	2,480.00	51.10	2,531.10
40' high	2E@2.25	Ea	1,350.00	60.50	1,410.50
40' high, heavy duty	2E@2.50	Ea	2,640.00	67.20	2,707.20
45' high	2E@2.75	Ea	1,570.00	73.90	1,643.90
45' high, heavy duty	2E@3.00	Ea	2,940.00	80.60	3,020.60
50' high	2E@3.50	Ea	1,850.00	94.10	1,944.10
50' high, heavy duty	2E@3.75	Ea	3,100.00	101.00	3,201.00

Round tapered aluminum light poles with one arm, square base

Material	Craft@Hrs	Unit	Material Cost	Labor Cost	Installed Cost
20' pole, 4' arm	2E@1.30	Ea	425.00	34.90	459.90
20' pole, 6' arm	2E@1.35	Ea	450.00	36.30	486.30
20' pole, 8' arm	2E@1.40	Ea	461.00	37.60	498.60
25' pole, 4' arm	2E@1.40	Ea	584.00	37.60	621.60
25' pole, 6' arm	2E@1.45	Ea	608.00	39.00	647.00
25' pole, 8' arm	2E@1.50	Ea	620.00	40.30	660.30
30' pole, 4' arm	2E@1.50	Ea	774.00	40.30	814.30
30' pole, 6' arm	2E@1.55	Ea	800.00	41.70	841.70
30' pole, 8' arm	2E@1.60	Ea	812.00	43.00	855.00
35' pole, 4' arm	2E@1.80	Ea	925.00	48.40	973.40
35' pole, 6' arm	2E@1.85	Ea	961.00	49.70	1,010.70
35' pole, 8' arm	2E@1.90	Ea	1,100.00	51.10	1,151.10

Round tapered aluminum light poles with two arms, square base

Material	Craft@Hrs	Unit	Material Cost	Labor Cost	Installed Cost
20' pole, 4' arms	2E@1.40	Ea	500.00	37.60	537.60
20' pole, 6' arms	2E@1.50	Ea	551.00	40.30	591.30
20' pole, 8' arms	2E@1.60	Ea	574.00	43.00	617.00
25' pole, 4' arms	2E@1.60	Ea	660.00	43.00	703.00
25' pole, 6' arms	2E@1.70	Ea	709.00	45.70	754.70
25' pole, 8' arms	2E@1.80	Ea	731.00	48.40	779.40
30' pole, 4' arms	2E@1.80	Ea	836.00	48.40	884.40
30' pole, 6' arms	2E@1.90	Ea	885.00	51.10	936.10
30' pole, 8' arms	2E@2.00	Ea	908.00	53.80	961.80
35' pole, 4' arms	2E@2.00	Ea	987.00	53.80	1,040.80
35' pole, 6' arms	2E@2.10	Ea	1,040.00	56.40	1,096.40
35' pole, 8' arms	2E@2.20	Ea	1,060.00	59.10	1,119.10

Use these figures to estimate the cost of street and yard light poles installed under the conditions described on pages 5 and 6. Costs listed are for each pole installed. The crew is two electricians working at a labor cost of $26.88 per manhour. These costs include four anchor bolts, layout, material handling, and normal waste. Add for the fixture, lamps, pole foundation, brackets, arms (except as noted above), excavation, sales tax, delivery, supervision, mobilization, demobilization, cleanup, overhead and profit. Note: Hoisting equipment may be necessary on the larger poles. Read the specs carefully when pricing poles. Some jobs will require items that are not included in these costs: decorative bases, anchor bolt covers, ballast bases, weatherproof receptacles, pole vibration dampers, or lowering devices.

Square Base Aluminum Light Poles

Material	Craft@Hrs	Unit	Material Cost	Labor Cost	Installed Cost
Round tapered aluminum light poles with three arms, square base					
30' pole, 4' arms	2E@1.90	Ea	977.00	51.10	1,028.10
30' pole, 6' arms	2E@2.15	Ea	1,030.00	57.80	1,087.80
30' pole, 8' arms	2E@2.30	Ea	1,060.00	61.80	1,121.80
35' pole, 4' arms	2E@2.30	Ea	1,100.00	61.80	1,161.80
35' pole, 6' arms	2E@2.45	Ea	1,180.00	65.90	1,245.90
35' pole, 8' arms	2E@2.60	Ea	1,210.00	69.90	1,279.90
Round tapered aluminum light poles with four arms, square base					
30' pole, 4' arms	2E@2.00	Ea	1,050.00	53.80	1,103.80
30' pole, 6' arms	2E@2.30	Ea	1,150.00	61.80	1,211.80
30' pole, 8' arms	2E@2.60	Ea	1,190.00	69.90	1,259.90
35' pole, 4' arms	2E@2.60	Ea	1,200.00	69.90	1,269.90
35' pole, 6' arms	2E@2.75	Ea	1,300.00	73.90	1,373.90
35' pole, 8' arms	2E@2.90	Ea	1,350.00	78.00	1,428.00
Square aluminum light poles without arms					
8' high	2E@0.65	Ea	336.00	17.50	353.50
10' high	2E@0.70	Ea	382.00	18.80	400.80
12' high	2E@0.80	Ea	394.00	21.50	415.50
15' high	2E@0.85	Ea	435.00	22.80	457.80
15' high, heavy duty	2E@0.90	Ea	537.00	24.20	561.20
18' high	2E@1.00	Ea	707.00	26.90	733.90
18' high, heavy duty	2E@1.15	Ea	938.00	30.90	968.90
20' high	2E@1.15	Ea	641.00	30.90	671.90
20' high, heavy duty	2E@1.25	Ea	1,180.00	33.60	1,213.60
25' high	2E@1.30	Ea	969.00	34.90	1,003.90
25' high, heavy duty	2E@1.35	Ea	1,540.00	36.30	1,576.30
30' high	2E@1.40	Ea	1,610.00	37.60	1,647.60
30' high, heavy duty	2E@1.50	Ea	1,780.00	40.30	1,820.30

Use these figures to estimate the cost of street and yard light poles installed under the conditions described on pages 5 and 6. Costs listed are for each pole installed. The crew is two electricians working at a labor cost of $26.00 per manhour. These costs include four anchor bolts, layout, material handling, and normal waste. Add for the fixture, lamps, pole foundation, brackets, arms (except as noted above), excavation, sales tax, delivery, supervision, mobilization, demobilization, cleanup, overhead and profit. Note: Hoisting equipment may be necessary on the larger poles. Read the specs carefully when pricing poles. Some jobs will require items that are not included in these costs: decorative bases, anchor bolt covers, ballast bases, weatherproof receptacles, pole vibration dampers, or lowering devices.

Fixed Tenon Fiberglass Light Poles

Material	Craft@Hrs	Unit	Material Cost	Labor Cost	Installed Cost
Round tapered fixed tenon fiberglass light poles, square base without arms					
10' high	2E@0.60	Ea	441.00	16.10	457.10
12' high	2E@0.65	Ea	459.00	17.50	476.50
14' high	2E@0.70	Ea	484.00	18.80	502.80
16' high	2E@0.80	Ea	540.00	21.50	561.50
18' high	2E@0.90	Ea	591.00	24.20	615.20
20' high	2E@1.00	Ea	647.00	26.90	673.90
25' high	2E@1.15	Ea	765.00	30.90	795.90
30' high	2E@1.30	Ea	1,080.00	34.90	1,114.90
35' high	2E@1.45	Ea	1,280.00	39.00	1,319.00
39' high	2E@1.60	Ea	1,480.00	43.00	1,523.00
Round tapered direct burial fixed tenon fiberglass light poles without arms					
10' high	2E@0.60	Ea	251.00	16.10	267.10
12' high	2E@0.65	Ea	270.00	17.50	287.50
14' high	2E@0.70	Ea	283.00	18.80	301.80
16' high	2E@0.80	Ea	402.00	21.50	423.50
18' high	2E@0.90	Ea	502.00	24.20	526.20
20' high	2E@1.00	Ea	516.00	26.90	542.90
25' high	2E@1.15	Ea	704.00	30.90	734.90
30' high	2E@1.30	Ea	854.00	34.90	888.90
35' high	2E@1.45	Ea	1,070.00	39.00	1,109.00
Round tapered direct burial fixed tenon fiberglass light poles with two arms					
20' high	2E@1.30	Ea	704.00	34.90	738.90
25' high	2E@1.45	Ea	905.00	39.00	944.00
30' high	2E@1.70	Ea	1,180.00	45.70	1,225.70
35' high	2E@1.75	Ea	1,420.00	47.00	1,467.00
39' high	2E@2.00	Ea	1,630.00	53.80	1,683.80
Square tapered fixed tenon fiberglass light poles, square base, two arms					
16' high	2E@0.80	Ea	665.00	21.50	686.50
18' high	2E@0.90	Ea	744.00	24.20	768.20
20' high	2E@1.00	Ea	889.00	26.90	915.90
25' high	2E@1.15	Ea	1,140.00	30.90	1,170.90
30' high	2E@1.30	Ea	1,310.00	34.90	1,344.90

Use these figures to estimate the cost of street and yard light poles installed under the conditions described on pages 5 and 6. Costs listed are for each pole installed. The crew is two electricians working at a labor cost of $26.88 per manhour. These costs include four anchor bolts, layout, material handling, and normal waste. Add for the fixture, lamps, pole foundation, brackets, arms, excavation, sales tax, delivery, supervision, mobilization, demobilization, cleanup, overhead and profit. Note: Read the specs carefully when pricing poles. Some jobs will require items that are not included in these costs: decorative bases, anchor bolt covers, ballast bases, weatherproof receptacles, pole vibration dampers, or lowering devices.

Material		Craft@Hrs	Unit	Material Cost	Labor Cost	Installed Cost

Standard voltage incandescent lamps, ES = Energy-Saving

Material		Craft@Hrs	Unit	Material Cost	Labor Cost	Installed Cost
25A19/W	120V	1E@0.05	Ea	1.23	1.34	2.57
25A19	130V	1E@0.05	Ea	1.64	1.34	2.98
25A/CL	120V	1E@0.05	Ea	1.28	1.34	2.62
25A/CL	130V	1E@0.05	Ea	1.70	1.34	3.04
25A19/RS	120V	1E@0.05	Ea	2.78	1.34	4.12
25A19/RS	130V	1E@0.05	Ea	3.76	1.34	5.10
30R20	120V	1E@0.06	Ea	5.37	1.61	6.98
30R20	130V	1E@0.06	Ea	7.51	1.61	9.12
34A19/ES	120V	1E@0.05	Ea	.95	1.34	2.29
34A19/ES	130V	1E@0.05	Ea	1.25	1.34	2.59
40A19	120V	1E@0.05	Ea	.95	1.34	2.29
40A19	130V	1E@0.05	Ea	1.25	1.34	2.59
40A19/W	120V	1E@0.05	Ea	1.26	1.34	2.40
40A19/CL	120V	1E@0.05	Ea	1.28	1.34	2.62
40A19/CL	130V	1E@0.05	Ea	1.70	1.34	3.04
50A/RS	120V	1E@0.05	Ea	2.56	1.34	3.90
50A/RS	130V	1E@0.05	Ea	3.46	1.34	4.80
50PAR/FL	120V	1E@0.08	Ea	7.61	2.15	9.76
50PAR/FL	130V	1E@0.08	Ea	9.70	2.15	11.85
50PAR/SP	120V	1E@0.08	Ea	8.43	2.15	10.58
50PAR/SP	130V	1E@0.08	Ea	10.80	2.15	12.95
50R20	120V	1E@0.06	Ea	5.37	1.61	6.98
50R20	130V	1E@0.06	Ea	7.51	1.61	9.12
52A/ES	120V	1E@0.05	Ea	.76	1.34	2.10
52A/ES	130V	1E@0.05	Ea	1.12	1.34	2.46
60A19	120V	1E@0.05	Ea	.76	1.34	2.10
60A19	130V	1E@0.05	Ea	1.01	1.34	2.35
60A19/CL	120V	1E@0.05	Ea	1.28	1.34	2.62
67A/ES	120V	1E@0.05	Ea	1.94	1.34	3.28
67A/ES	125V	1E@0.05	Ea	2.28	1.34	3.62
67A/ES	130V	1E@0.05	Ea	2.61	1.34	3.95
75A	120V	1E@0.05	Ea	.76	1.34	2.10
75A	130V	1E@0.05	Ea	1.00	1.34	2.34
75A/CL	120V	1E@0.05	Ea	1.28	1.34	2.62
75PAR/FL/ES	120V	1E@0.08	Ea	7.67	2.15	9.82
75PAR/FL/ES	130V	1E@0.08	Ea	8.82	2.15	10.97
75PAR/SP/ES	120V	1E@0.08	Ea	7.37	2.15	9.52
75PAR/SP/ES	130V	1E@0.08	Ea	9.78	2.15	11.93
75R20	120V	1E@0.08	Ea	5.95	2.15	8.10
75R30	120V	1E@0.08	Ea	10.40	2.15	12.55

Use these figures to estimate the cost of lamps installed in lighting fixtures under the conditions described on pages 5 and 6. Costs listed are for each lamp installed. The crew is one electrician working at a labor cost of $26.88 per manhour. These costs include layout, material handling, and normal waste. Add for the lighting fixture, sales tax, delivery, supervision, mobilization, demobilization, cleanup, overhead and profit. Note: Some special voltage fixtures come with the right lamp packaged in the fixture box. If the lamp isn't supplied with the fixture, be sure to select the lamp with the right voltage and wattage. If the specifications require a lamp in a fixture, only a certain lamp will meet job requirements.

Incandescent Lamps

Material		Craft@Hrs	Unit	Material Cost	Labor Cost	Installed Cost

Standard voltage lamps

Material		Craft@Hrs	Unit	Material Cost	Labor Cost	Installed Cost
100A	120V	1E@0.05	Ea	.76	1.34	2.10
100A	130V	1E@0.05	Ea	1.00	1.34	2.34
100A21	120V	1E@0.05	Ea	1.39	1.34	2.73
100A21	130V	1E@0.05	Ea	1.86	1.34	3.20
100A23	120V	1E@0.05	Ea	2.17	1.34	3.51
100A/RS	120V	1E@0.05	Ea	1.79	1.34	3.13
100A/RS	130V	1E@0.05	Ea	2.44	1.34	3.78
100PAR/FL	120V	1E@0.08	Ea	6.78	2.15	8.93
100PAR/FL	130V	1E@0.08	Ea	9.70	2.15	11.85
100PAR/SP	120V	1E@0.08	Ea	8.43	2.15	10.58
100PAR/SP	130V	1E@0.08	Ea	10.80	2.15	12.95
120ER40/ES	120V	1E@0.08	Ea	7.08	2.15	9.23
120ER40/ES	130V	1E@0.08	Ea	9.42	2.15	11.57
135A/ES	120V	1E@0.05	Ea	1.39	1.34	2.73
135A/ES	130V	1E@0.05	Ea	2.04	1.34	3.38
150A	120V	1E@0.05	Ea	1.41	1.34	2.75
150A	130V	1E@0.05	Ea	2.04	1.34	3.38
150A/CL	120V	1E@0.05	Ea	1.86	1.34	3.20
150A21/RS	120V	1E@0.05	Ea	3.97	1.34	5.31
150A21/RS	130V	1E@0.05	Ea	5.91	1.34	7.25
150A23/RS	120V	1E@0.05	Ea	2.07	1.34	3.41
150A23/RS	130V	1E@0.05	Ea	2.80	1.34	4.14
200A	120V	1E@0.05	Ea	1.97	1.34	3.31
200A	130V	1E@0.05	Ea	2.63	1.34	3.97
200/CL	120V	1E@0.05	Ea	3.23	1.34	4.57
200/CL	130V	1E@0.05	Ea	4.37	1.34	5.71
200A/CL	120V	1E@0.05	Ea	2.26	1.34	3.60
200A/CL	130V	1E@0.05	Ea	3.06	1.34	4.40
200IF	120V	1E@0.05	Ea	3.36	1.34	4.70
200IF	130V	1E@0.05	Ea	4.52	1.34	5.86
200PAR/MFL	120V	1E@0.08	Ea	24.40	2.15	26.55
200PST/MFL	130V	1E@0.08	Ea	32.80	2.15	34.95
200PAR/NSP	120V	1E@0.08	Ea	24.40	2.15	26.55
200PAR/NSP	130V	1E@0.08	Ea	32.80	2.15	34.95
200PS30	120V	1E@0.05	Ea	5.67	1.34	7.01
200PS30	130V	1E@0.05	Ea	7.54	1.34	8.88
250R40	120V	1E@0.08	Ea	5.88	2.15	8.03
300M/PS25	120V	1E@0.06	Ea	3.27	1.61	4.88
300M/PS25	130V	1E@0.06	Ea	4.38	1.61	5.99
300M/PS25/IF	120V	1E@0.06	Ea	3.54	1.61	5.15
300M/PS25/IF	130V	1E@0.06	Ea	4.83	1.61	6.44

Use these figures to estimate the cost of lamps installed in lighting fixtures under the conditions described on pages 5 and 6. Costs listed are for each lamp installed. The crew is one electrician working at a labor cost of $26.88 per manhour. These costs include layout, material handling, and normal waste. Add for the lighting fixture, sales tax, delivery, supervision, mobilization, demobilization, cleanup, overhead and profit. Note: Some special voltage fixtures come with the right lamp packaged in the fixture box. If the lamp isn't supplied with the fixture, be sure to select the lamp with the right voltage and wattage. If the specifications require a lamp in a fixture, only a certain lamp will meet job requirements.

Material		Craft@Hrs	Unit	Material Cost	Labor Cost	Installed Cost
Standard voltage lamps						
300/PS35/CL	120V	1E@0.06	Ea	4.74	1.61	6.35
300/PS35/CL	130V	1E@0.06	Ea	6.31	1.61	7.92
300/PS35/IF	120V	1E@0.06	Ea	5.03	1.61	6.64
300/PS35/IF	130V	1E@0.06	Ea	6.66	1.61	8.27
300/PS35/RS	130V	1E@0.06	Ea	12.30	1.61	13.91
300R/FL	120V	1E@0.10	Ea	13.20	2.69	15.89
300R/FL	130V	1E@0.10	Ea	17.60	2.69	20.29
300R/SP	120V	1E@0.10	Ea	13.20	2.69	15.89
300R/3FL	120V	1E@0.10	Ea	17.60	2.69	20.29
300R/3FL	130V	1E@0.10	Ea	23.70	2.69	26.39
400R/FL	120V	1E@0.10	Ea	23.80	2.69	26.49
500/PS35/CL	120V	1E@0.06	Ea	7.86	1.61	9.47
500/PS35/CL	130V	1E@0.06	Ea	10.40	1.61	12.01
500/PS35/IF	120V	1E@0.06	Ea	8.22	1.61	9.83
500/PS35/IF	130V	1E@0.06	Ea	11.10	1.61	12.71
500/PS40	120V	1E@0.08	Ea	8.50	2.15	10.65
500PAR64/MFL	120V	1E@0.10	Ea	52.70	2.69	55.39
500PAR64/WFL	120V	1E@0.10	Ea	52.70	2.69	55.39
500PAR64/NSP	120V	1E@0.10	Ea	52.70	2.69	55.39
500R/FL	120V	1E@0.10	Ea	20.10	2.69	22.79
500R/FL	130V	1E@0.10	Ea	27.20	2.69	29.89
500R52	120V	1E@0.10	Ea	25.00	2.69	27.69
500R52	130V	1E@0.10	Ea	33.70	2.69	36.39
750/PS52/CL	120V	1E@0.10	Ea	18.20	2.69	20.89
750/PS52/CL	130V	1E@0.10	Ea	24.60	2.69	27.29
750R52	120V	1E@0.10	Ea	29.90	2.69	32.59
750R52	130V	1E@0.10	Ea	40.50	2.69	43.19
1000/PS52/CL	120V	1E@0.15	Ea	18.70	4.03	22.73
1000/PS52/CL	130V	1E@0.15	Ea	25.20	4.03	29.23
1500/PS52/CL	120V	1E@0.15	Ea	30.30	4.03	34.33
1500/PS52/CL	130V	1E@0.15	Ea	40.70	4.03	44.73

Material		Craft@Hrs	Unit	Material Cost	Labor Cost	Installed Cost
Special voltage incandescent lamps						
25A	12V	1E@0.05	Ea	2.71	1.34	4.05
25A	34V	1E@0.05	Ea	2.78	1.34	4.12
25A17/RS	75V	1E@0.05	Ea	3.40	1.34	4.74
25A	230V	1E@0.05	Ea	3.20	1.34	4.54
25PAR36	5.5V	1E@0.08	Ea	13.00	2.15	15.15
25PAR36/NSP	12V	1E@0.08	Ea	11.20	2.15	13.35
25PAR36/WFL	12V	1E@0.08	Ea	11.20	2.15	13.35
25PAR36/VWFL	12V	1E@0.08	Ea	11.20	2.15	13.35
25PAR46	5.5V	1E@0.08	Ea	13.00	2.15	15.15

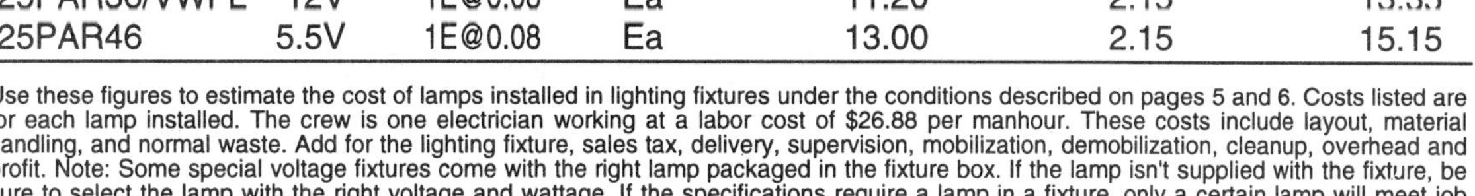

Use these figures to estimate the cost of lamps installed in lighting fixtures under the conditions described on pages 5 and 6. Costs listed are for each lamp installed. The crew is one electrician working at a labor cost of $26.88 per manhour. These costs include layout, material handling, and normal waste. Add for the lighting fixture, sales tax, delivery, supervision, mobilization, demobilization, cleanup, overhead and profit. Note: Some special voltage fixtures come with the right lamp packaged in the fixture box. If the lamp isn't supplied with the fixture, be sure to select the lamp with the right voltage and wattage. If the specifications require a lamp in a fixture, only a certain lamp will meet job requirements.

Incandescent Lamps

Material		Craft@Hrs	Unit	Material Cost	Labor Cost	Installed Cost

Special voltage incandescent lamps

Material		Craft@Hrs	Unit	Material Cost	Labor Cost	Installed Cost
50A19/RS	75V	1E@0.05	Ea	3.43	1.34	4.77
50A19	250V	1E@0.05	Ea	2.97	1.34	4.31
50A19/RS	250V	1E@0.05	Ea	3.97	1.34	5.31
50A21	12V	1E@0.05	Ea	3.12	1.34	4.46
50A21	30V	1E@0.05	Ea	3.27	1.34	4.61
50A21	34V	1E@0.05	Ea	2.45	1.34	3.79
50A21	300V	1E@0.05	Ea	3.85	1.34	5.19
60A21	230V	1E@0.05	Ea	3.44	1.34	4.78
60PAR46	38V	1E@0.08	Ea	21.30	2.15	23.45
75A21	12V	1E@0.05	Ea	3.54	1.34	4.88
100A21	230V	1E@0.05	Ea	3.05	1.34	4.39
100A21	250V	1E@0.05	Ea	3.05	1.34	4.39
100A21/RS	250V	1E@0.05	Ea	3.36	1.34	4.70
100A21	277V	1E@0.05	Ea	4.57	1.34	5.91
100PAR38/FL	12V	1E@0.08	Ea	25.80	2.15	27.95
100PAR46	60V	1E@0.10	Ea	26.50	2.69	29.19
100R30/CL	12V	1E@0.08	Ea	14.60	2.15	16.75
200PS30	250V	1E@0.05	Ea	5.37	1.34	6.71
240PAR56/MFL	12V	1E@0.08	Ea	25.80	2.15	27.95
240PAR56/WFL	12V	1E@0.08	Ea	25.80	2.15	27.95
240PAR56/NSP	12V	1E@0.08	Ea	25.80	2.15	27.95
300R40/FL	250V	1E@0.10	Ea	23.70	2.69	26.39
500PAR64/MFL	230V	1E@0.10	Ea	66.50	2.69	69.19
500PAR64/WFL	230V	1E@0.10	Ea	66.50	2.69	69.19
500R40/FL	250V	1E@0.10	Ea	29.30	2.69	31.99
500R52	250V	1E@0.10	Ea	34.60	2.69	37.29
1000PS52	250V	1E@0.15	Ea	23.20	4.03	27.23
1000PS52	277V	1E@0.15	Ea	35.60	4.03	39.63

Standard voltage incandescent halogen lamps, Energy-Saving

Material		Craft@Hrs	Unit	Material Cost	Labor Cost	Installed Cost
45PAR38/FL	120V	1E@0.08	Ea	10.10	2.15	12.25
45PAR38/FL	130V	1E@0.08	Ea	10.90	2.15	13.05
45PAR38/SP	120V	1E@0.08	Ea	10.10	2.15	12.25
45PAR38/SP	130V	1E@0.08	Ea	12.10	2.15	14.25
50PAR38/FL	120V	1E@0.08	Ea	7.61	2.15	9.76
50PAR38/FL	130V	1E@0.08	Ea	9.70	2.15	11.85
50PAR38/SP	120V	1E@0.08	Ea	8.43	2.15	10.58
50PAR38/SP	130V	1E@0.08	Ea	10.80	2.15	12.95

Use these figures to estimate the cost of special purpose incandescent lamps installed in lighting fixtures under the conditions described on pages 5 and 6. Costs listed are for each lamp installed. The crew is one electrician working at a labor cost of $26.88 per manhour. These costs include layout, material handling, and normal waste. Add for the lighting fixture, sales tax, delivery, supervision, mobilization, demobilization, cleanup, overhead and profit. Note: Some special voltage fixtures come with the right lamp packaged in the fixture box. If the lamp isn't supplied with the fixture, be sure to select the lamp with the right voltage and wattage. If the specifications require a lamp in a fixture, only a certain lamp will meet job requirements.

Material	Craft@Hrs	Unit	Material Cost	Labor Cost	Installed Cost
Standard voltage incandescent halogan lamps, Energy-Saving, this table only					
90PAR38/FL	120V	1E@0.08	Ea	9.14	2.15	11.29
90PAR38/FL	130V	1E@0.08	Ea	10.90	2.15	13.05
90PAR38/SP	120V	1E@0.08	Ea	9.14	2.15	11.29
90PAR38/SP	130V	1E@0.08	Ea	10.90	2.15	13.05
100PAR38/FL	120V	1E@0.08	Ea	6.78	2.15	8.93
100PAR38/FL	130V	1E@0.08	Ea	9.70	2.15	11.85
100PAR38/SP	120V	1E@0.08	Ea	8.43	2.15	10.58
100PAR38/SP	130V	1E@0.08	Ea	10.80	2.15	12.95

Tubular quartz lamps

Material	Craft@Hrs	Unit	Material Cost	Labor Cost	Installed Cost
100T3/CL	120V	1E@0.06	Ea	12.30	1.61	13.91
150T3/CL	120V	1E@0.06	Ea	12.80	1.61	14.41
200T3/CL	120V	1E@0.06	Ea	37.60	1.61	39.21
250T3/CL	120V	1E@0.06	Ea	11.60	1.61	13.21
300T3/CL	120V	1E@0.06	Ea	12.30	1.61	13.91
425T3/CL	120V	1E@0.06	Ea	27.60	1.61	29.21
500T3/CL	120V	1E@0.06	Ea	10.30	1.61	11.91
500T3/CL	130V	1E@0.06	Ea	11.30	1.61	12.91
1000T3/CL	220V	1E@0.08	Ea	49.50	2.15	51.65
1000T3/CL	240V	1E@0.08	Ea	49.50	2.15	51.65
1500T3/CL	208V	1E@0.10	Ea	39.20	2.69	41.89
1500T3/CL	220V	1E@0.10	Ea	39.20	2.69	41.89
1500T3/CL	240V	1E@0.10	Ea	15.20	2.69	17.89
1500T3/CL	277V	1E@0.10	Ea	16.00	2.69	18.69
6000T3/CL	480V	1E@0.15	Ea	57.60	4.03	61.63

Tubular quartz heat lamps

Material	Craft@Hrs	Unit	Material Cost	Labor Cost	Installed Cost
300T3/CL	120V	1E@0.06	Ea	33.70	1.61	35.31
375T3/CL	120V	1E@0.06	Ea	32.70	1.61	34.31
500T3/CL	120V	1E@0.06	Ea	51.90	1.61	53.51
1000T3/CL	230V	1E@0.08	Ea	41.70	2.15	43.85
1600T3/CL	230V	1E@0.10	Ea	87.60	2.69	90.29
1600T3/CL	277V	1E@0.10	Ea	64.70	2.69	67.39

Use these figures to estimate the cost of special purpose incandescent lamps installed in lighting fixtures under the conditions described on pages 5 and 6. Costs listed are for each lamp installed. The crew is one electrician working at a labor cost of $26.88 per manhour. These costs include layout, material handling, and normal waste. Add for the lighting fixture, sales tax, delivery, supervision, mobilization, domobilization, cleanup, overhead and profit. Note: Some special voltage fixtures come with the right lamp packaged in the fixture box. If the lamp isn't supplied with the fixture, be sure to select the lamp with the right voltage and wattage. If the specifications require a lamp in a fixture, only a certain lamp will meet job requirements. Counsel your installers that they have to keep their fingers off the socket metal of quartz lamps. Oil from their skin increases electrical resistance which makes the point of contact overheat.

High Intensity Discharge Lamps

Material	Craft@Hrs	Unit	Material Cost	Labor Cost	Installed Cost
Mercury vapor lamps					
HR40/50/B17/DX	1E@0.10	Ea	52.60	2.69	55.29
HR75/B17/DX	1E@0.10	Ea	52.60	2.69	55.29
HR100/A23/CL	1E@0.10	Ea	48.20	2.69	50.89
HR100/B17/DX	1E@0.10	Ea	50.70	2.69	53.39
HR100/ED23/CL	1E@0.10	Ea	31.90	2.69	34.59
HR100/ED23/DX	1E@0.10	Ea	28.40	2.69	31.09
HR100/ED23/WDX	1E@0.10	Ea	50.30	2.69	52.99
HR100/PAR38/FL	1E@0.06	Ea	88.40	1.61	90.01
HR100/PAR38/SP	1E@0.06	Ea	88.40	1.61	90.01
HR100/R40/FL	1E@0.06	Ea	75.40	1.61	77.01
HR100/R40/DX/WFL	1E@0.06	Ea	77.60	1.61	79.21
HR175/ED28/CL	1E@0.10	Ea	21.60	2.69	24.29
HR175/ED28/DX	1E@0.10	Ea	21.10	2.69	23.79
HR175/ED28/WDX	1E@0.10	Ea	47.50	2.69	50.19
HR175/R40/FL	1E@0.06	Ea	78.30	1.61	79.91
HR175/R40/RFL	1E@0.06	Ea	64.40	1.61	66.01
HR250/ED28/CL	1E@0.10	Ea	41.80	2.69	44.49
HR250/ED28/DX	1E@0.10	Ea	37.40	2.69	40.09
HR250/ED28/WDX	1E@0.10	Ea	72.30	2.69	74.99
HR400/BT37/DX	1E@0.10	Ea	43.60	2.69	46.29
HR400/ED37/CL	1E@0.10	Ea	27.80	2.69	30.49
HR400/ED37/DX	1E@0.10	Ea	30.00	2.69	32.69
HR400/ED37/WDX	1E@0.10	Ea	58.10	2.69	60.79
HR400/R52/DX	1E@0.10	Ea	90.00	2.69	92.69
HR400/R60/DX/WFL	1E@0.10	Ea	124.00	2.69	126.69
HR400/R60/SP	1E@0.10	Ea	113.00	2.69	115.69
HR400/T16/CL	1E@0.10	Ea	108.00	2.69	110.69
HR1000/BT65/CL	1E@0.10	Ea	80.40	2.69	83.09
HR1000/BT56/DX	1E@0.10	Ea	68.70	2.69	71.39
HR175/ED28/DX	1E@0.10	Ea	64.00	2.69	66.69
HR250/ED28/DX	1E@0.10	Ea	78.90	2.69	81.59
HR400/ED28/DX	1E@0.10	Ea	71.50	2.69	74.19
Self-ballasted lamps (retrofit)					
HSB160/ED24/DX, 120V	1E@0.10	Ea	77.70	2.69	80.39
HSB250/ED28/DX, 120V	1E@0.10	Ea	124.00	2.69	126.69
HSB250/ED28/DX, 130V	1E@0.10	Ea	124.00	2.69	126.69
HSB450/BT37/DX, 120V	1E@0.10	Ea	221.00	2.69	223.69
HSB750/R57/DX, 120V	1E@0.10	Ea	325.00	2.69	327.69

Use these figures to estimate the cost of quartz lamps installed in lighting fixtures under the conditions described on pages 5 and 6. Costs listed are for each lamp installed. The crew is one electrician working at a labor cost of $26.88 per manhour. These costs include layout, material handling, and normal waste. Add for the lighting fixture, sales tax, delivery, supervision, mobilization, demobilization, cleanup, overhead and profit. Note: Some special voltage fixtures .

Metal Halide High Intensity Discharge Lamps

Material	Craft@Hrs	Unit	Material Cost	Labor Cost	Installed Cost

Metal halide high intensity discharge lamps, ES = Energy-Saving

Material	Craft@Hrs	Unit	Material Cost	Labor Cost	Installed Cost
MXR50/BD17/CL	1E@0.10	Ea	59.90	2.69	62.59
MXR50/BD17/CTD	1E@0.10	Ea	63.60	2.69	66.29
MXR70/BD17/CL	1E@0.10	Ea	56.70	2.69	59.39
MXR70/BD17/CTD	1E@0.10	Ea	60.40	2.69	63.09
MXR70/ED17/CL	1E@0.10	Ea	65.50	2.69	68.19
MXR70/ED17/CTD	1E@0.10	Ea	69.30	2.69	71.99
MXR70/PAR38/FL	1E@0.08	Ea	65.80	2.15	67.95
MXR70/PAR38/WFL	1E@0.08	Ea	65.80	2.15	67.95
MXR70/PAR38/SP	1E@0.08	Ea	65.80	2.15	67.95
MXR100/BD17/CL	1E@0.10	Ea	56.70	2.69	59.39
MXR100/BD17/CTD	1E@0.10	Ea	60.40	2.69	63.09
MXR100/ED17/CL	1E@0.10	Ea	65.50	2.69	68.19
MXR100/ED17/CTD	1E@0.10	Ea	69.30	2.69	71.99
MXR100/PAR38/FL	1E@0.10	Ea	65.80	2.69	68.49
MXR100/PAR38/WFL	1E@0.10	Ea	65.80	2.69	68.49
MXR100/PAR38/SP	1E@0.10	Ea	65.80	2.69	68.49
MXR150/BD17/CL	1E@0.10	Ea	62.70	2.69	65.39
MXR150/BD17/CTD	1E@0.10	Ea	66.20	2.69	68.89
MVR175/BD17/CL	1E@0.10	Ea	55.90	2.69	58.59
MVR175/BD17/CTD	1E@0.10	Ea	58.70	2.69	61.39
MVR175/ED28/CL	1E@0.10	Ea	41.00	2.69	43.69
MVR175/ED28/CTD	1E@0.10	Ea	43.70	2.69	46.39
MVR250/ED28/CL	1E@0.10	Ea	46.30	2.69	48.99
MVR250/ED28/CTD	1E@0.10	Ea	49.00	2.69	51.69
MVR360/ED37/CL/ES	1E@0.10	Ea	55.30	2.69	57.99
MVR360/ED37/CTD/ES	1E@0.10	Ea	58.20	2.69	60.89
MVR400/ED28/CL	1E@0.10	Ea	68.30	2.69	70.99
MVR400/ED28/CTD	1E@0.10	Ea	74.60	2.69	77.29
MVR400/ED37/CL	1E@0.10	Ea	39.40	2.69	42.09
MVR400/ED37/CTD	1E@0.10	Ea	49.90	2.69	52.59
MVR1000/BT56/CL	1E@0.10	Ea	105.00	2.69	107.69
MVR1000/BT56/CTD	1E@0.10	Ea	116.00	2.69	118.69

Metal halide high intensity discharge ballasts, encased

Material	Craft@Hrs	Unit	Material Cost	Labor Cost	Installed Cost
175W, 120V to 480V	1E@1.00	Ea	156.00	26.90	182.90
250W, 120V to 480V	1E@1.00	Ea	192.00	26.90	218.90
400W, 120V to 480V	1E@1.00	Ea	188.00	26.90	214.90
1000W, 120V to 480V	1E@1.00	Ea	431.00	26.90	457.90

Use these figures to estimate the cost of high intensity discharge lamps installed in lighting fixtures under the conditions described on pages 5 and 6. Costs listed are for each lamp installed. The crew is one electrician working at a labor cost of $26.88 per manhour. These costs include layout, material handling, and normal waste. Add for the lighting fixture, sales tax, delivery, supervision, mobilization, demobilization, cleanup, overhead and profit. Note: The lamps above are identified by both wattage and by the shape of the bulb.

High Intensity Discharge Ballasts

Material	Craft@Hrs	Unit	Material Cost	Labor Cost	Installed Cost
Mercury vapor single lamp indoor ballasts					
100 W 120/208/240277V	1E@1.00	Ea	91.70	26.90	118.60
100W 480V	1E@1.00	Ea	91.70	26.90	118.60
175W 120/208/240/277V	1E@1.00	Ea	119.00	26.90	145.90
175W 480V	1E@1.00	Ea	119.00	26.90	145.90
250W 120/208/240/277V	1E@1.00	Ea	132.00	26.90	158.90
250W 480V	1E@1.00	Ea	132.00	26.90	158.90
400W 120/208/240/277V	1E@1.00	Ea	130.00	26.90	156.90
400W 480V	1E@1.00	Ea	130.00	26.90	156.90
1000W 120/208/240/277V	1E@1.00	Ea	232.00	26.90	258.90
1000W 480V	1E@1.00	Ea	261.00	26.90	287.90

Material	Craft@Hrs	Unit	Material Cost	Labor Cost	Installed Cost
Weatherproof high intensity lighting ballasts					
100W 120/208/240277V	1E@1.00	Ea	165.00	26.90	191.90
100W 480V	1E@1.00	Ea	165.00	26.90	191.90
175W 120/208/240/277V	1E@1.00	Ea	173.00	26.90	199.90
175W 480V	1E@1.00	Ea	173.00	26.90	199.90
250W 120/208/240/277V	1E@1.00	Ea	185.00	26.90	211.90
250W 480V	1E@1.00	Ea	185.00	26.90	211.90
400W 120/208/240/277V	1E@1.00	Ea	171.00	26.90	197.90
400W 480V	1E@1.00	Ea	191.00	26.90	217.90
1000W 120/208/240/277V	1E@1.00	Ea	378.00	26.90	404.90
1000W 480V	1E@1.00	Ea	378.00	26.90	404.90

Material	Craft@Hrs	Unit	Material Cost	Labor Cost	Installed Cost
High intensity ballast core and coil with capacitor					
100 W 120/208/240277V	1E@1.25	Ea	45.40	33.60	79.00
100W 480V	1E@1.25	Ea	59.50	33.60	93.10
175W 120/208/240/277V	1E@1.25	Ea	55.00	33.60	88.60
175W 480V	1E@1.25	Ea	63.80	33.60	97.40
250W 120/208/240/277V	1E@1.25	Ea	66.10	33.60	99.70
250W 480V	1E@1.25	Ea	85.70	33.60	119.30
400W 120/208/240/277V	1E@1.25	Ea	75.10	33.60	108.70
400W 480V	1E@1.25	Ea	86.00	33.60	119.60
1000W 120/208/240/277V	1E@1.25	Ea	117.00	33.60	150.60
1000W 480V	1E@1.25	Ea	120.00	33.60	153.60

Use these figures to estimate the cost of installing high intensity discharge ballasts under the conditions described on pages 5 and 6. Costs listed are for each ballast installed. The crew is one electrician working at a labor cost of $26.88 per manhour. These costs include layout, material handling, and normal waste. Add for the wiring connections, enclosures if needed, sales tax, delivery, supervision, mobilization, demobilization, cleanup, overhead and profit.

Metal Halide High Intensity Discharge Lamps and Ballasts

Material		Craft@Hrs	Unit	Material Cost	Labor Cost	Installed Cost

Metal halide high intensity lamps

Material		Craft@Hrs	Unit	Material Cost	Labor Cost	Installed Cost
MVR175/C/U	E28	1E@0.10	Ea	42.10	2.69	44.79
MVR175/C	E28	1E@0.10	Ea	39.60	2.69	42.29
MVR250/C/U	E28	1E@0.10	Ea	47.30	2.69	49.99
MVR250U	E28	1E@0.10	Ea	44.80	2.69	47.49
MVR325/C/I/U/ES	E28	1E@0.10	Ea	48.40	2.69	51.09
MVR325/I/U/ES	E37	1E@0.10	Ea	61.30	2.69	63.99
MVT325/I/U/ES	E37	1E@0.10	Ea	96.10	2.69	98.79
MVT325/C/I/U/ES	E37	1E@0.10	Ea	101.00	2.69	103.69
MVR400/VBU	E37	1E@0.10	Ea	52.00	2.69	54.69
MVR400/C/I/U	E37	1E@0.10	Ea	68.10	2.69	70.79
MVR400/C/U	E37	1E@0.10	Ea	48.20	2.69	50.89
MVR400/C/VBD	E37	1E@0.10	Ea	59.20	2.69	61.89
MVR400/C/VBU	E37	1E@0.10	Ea	51.30	2.69	53.99
MVR400/I/U	E37	1E@0.10	Ea	62.90	2.69	65.59
MVR400/U	E37	1E@0.10	Ea	38.60	2.69	41.29
MVR400V BD	E37	1E@0.10	Ea	60.00	2.69	62.69
MVR950/I/VBD	BT56	1E@0.10	Ea	154.00	2.69	156.69
MVR960/I/VBU	BT56	1E@0.10	Ea	154.00	2.69	156.69
MVR1000/C/U	BT56	1E@0.10	Ea	110.00	2.69	112.69
MVR1000/U	BT56	1E@0.10	Ea	99.60	2.69	102.29
MVR1500/HBD/E	BT56	1E@0.10	Ea	125.00	2.69	127.69
MVR1500/HBU/E	BT56	1E@0.10	Ea	113.00	2.69	115.69

Metal halide high intensity indoor ballasts, encased

Material		Craft@Hrs	Unit	Material Cost	Labor Cost	Installed Cost
175W	120/208/240/277V	1E@1.00	Ea	151.00	26.90	177.90
175W	480V	1E@1.00	Ea	151.00	26.90	177.90
250W	120/208/240/277V	1E@1.00	Ea	186.00	26.90	212.90
250W	480V	1E@1.00	Ea	186.00	26.90	212.90
400W	120/208/240/277V	1E@1.00	Ea	183.00	26.90	209.90
400W	480V	1E@1.00	Ea	183.00	26.90	209.90
1000W	120/208/240/277V	1E@1.00	Ea	418.00	26.90	444.90
1000W	480V	1E@1.00	Ea	418.00	26.90	444.90

Use these figures to estimate the cost of metal halide discharge lamps and ballasts installed in or with lighting fixtures under the conditions described on pages 5 and 6. Costs listed are for each lamp or ballast installed. The crew is one electrician working at a labor cost of $26.88 per manhour. These costs include layout, material handling, and normal waste. Add for the lighting fixture, sales tax, delivery, supervision, mobilization, demobilization, cleanup, overhead and profit. Note: Metal halide lamps have a high rated lamp life when installed in some fixtures.

High Intensity Discharge Lamps

Material		Craft@Hrs	Unit	Material Cost	Labor Cost	Installed Cost
Low pressure sodium lamps						
18W	T16	1E@0.15	Ea	45.70	4.03	49.73
35W	T16	1E@0.15	Ea	50.70	4.03	54.73
55W	T16	1E@0.15	Ea	55.60	4.03	59.63
90W	T21	1E@0.15	Ea	64.50	4.03	68.53
135W	T21	1E@0.15	Ea	82.50	4.03	86.53
High pressure sodium lamps, ES = Energy-Saving						
35W	B17	1E@0.10	Ea	43.30	2.69	45.99
35W/D	B17	1E@0.10	Ea	44.30	2.69	46.99
50W	B17	1E@0.10	Ea	43.30	2.69	45.99
50W/D	B17	1E@0.10	Ea	44.70	2.69	47.39
50W	ED23.5	1E@0.10	Ea	52.70	2.69	55.39
50W/D	ED23.5	1E@0.10	Ea	63.50	2.69	66.19
70W	B17	1E@0.10	Ea	43.30	2.69	45.99
70W/D	B17	1E@0.10	Ea	44.70	2.69	47.39
70W	ED23.5	1E@0.10	Ea	46.90	2.69	49.59
70W/D	ED23.5	1E@0.10	Ea	63.50	2.69	66.19
100W	B17	1E@0.10	Ea	49.10	2.69	51.79
100W/D	B17	1E@0.10	Ea	50.40	2.69	53.09
100W	ED23.5	1E@0.10	Ea	49.00	2.69	51.69
100W/D	ED23.5	1E@0.10	Ea	66.20	2.69	68.89
110W/ES	ED23.5	1E@0.10	Ea	79.30	2.69	81.99
150W	B17	1E@0.10	Ea	96.90	2.69	99.59
150W/D	B17	1E@0.10	Ea	51.90	2.69	54.59
150W	ED23.5	1E@0.10	Ea	53.40	2.69	56.09
150W/D	ED23.5	1E@0.10	Ea	50.60	2.69	53.29
150W/ES	ED28	1E@0.10	Ea	73.00	2.69	75.69
150W	ED28	1E@0.10	Ea	69.10	2.69	71.79
200W	ED18	1E@0.10	Ea	70.50	2.69	73.19
215W/ES	ED28	1E@0.10	Ea	78.40	2.69	81.09
250W	ED18	1E@0.10	Ea	53.80	2.69	56.49
250W/D	ED18	1E@0.10	Ea	71.80	2.69	74.49
250W/S	ED18	1E@0.10	Ea	81.80	2.69	84.49
250W/SBY	ED18	1E@0.10	Ea	112.00	2.69	114.69
310W	ED18	1E@0.10	Ea	89.80	2.69	92.49
400W	ED18	1E@0.10	Ea	49.20	2.69	51.89
400W/SBY	ED18	1E@0.10	Ea	118.00	2.69	120.69
400W/D	ED37	1E@0.10	Ea	75.80	2.69	78.49
400W	T7-R7s	1E@0.10	Ea	113.00	2.69	115.69
1000W	E25	1E@0.10	Ea	150.00	2.69	152.69
1000W	T7-R7s	1E@0.10	Ea	158.00	2.69	160.69

Use these figures to estimate the cost of high Intensity discharge lamps installed in lighting fixtures under the conditions described on pages 5 and 6. Costs listed are for each lamp installed. The crew is one electrician working at a labor cost of $26.88 per manhour. These costs include layout, material handling, and normal waste. Add for the lighting fixture, sales tax, delivery, supervision, mobilization, demobilization, cleanup, overhead and profit. Note: Be sure the lamp selected is compatible with the installed ballast.

Material	Craft@Hrs	Unit	Material Cost	Labor Cost	Installed Cost

High intensity discharge core and coil with capacitor ballasts

Material	Craft@Hrs	Unit	Material Cost	Labor Cost	Installed Cost
175W 120/208/240/277V	1E@1.25	Ea	146.00	33.60	179.60
175W 480V	1E@1.25	Ea	189.00	33.60	222.60
250W 120/208/240/277V	1E@1.25	Ea	189.00	33.60	222.60
250W 480V	1E@1.25	Ea	189.00	33.60	222.60
400W 120/208/240/277V	1E@1.25	Ea	213.00	33.60	246.60
400W 480V	1E@1.25	Ea	213.00	33.60	246.60
1000W 120/208/240/277V	1E@1.25	Ea	294.00	33.60	327.60
1000W 480V	1E@1.25	Ea	294.00	33.60	327.60

High pressure sodium indoor encased ballasts

Material	Craft@Hrs	Unit	Material Cost	Labor Cost	Installed Cost
250W 120/208/240/277V	1E@1.00	Ea	333.00	26.90	359.90
250W 480V	1E@1.00	Ea	263.00	26.90	289.90
400W 120/208/240/277V	1E@1.00	Ea	415.00	26.90	441.90
400W 480V	1E@1.00	Ea	415.00	26.90	441.90

High pressure sodium weatherproof ballasts

Material	Craft@Hrs	Unit	Material Cost	Labor Cost	Installed Cost
250W 120/208/240/277V	1E@1.00	Ea	407.00	26.90	433.90
250W 480V	1E@1.00	Ea	301.00	26.90	327.90
400W 120/208/240/277V	1E@1.00	Ea	515.00	26.90	541.90
400W 480V	1E@1.00	Ea	515.00	26.90	541.90

WP

Use these figures to estimate the cost of installing high intensity discharge lamps and ballasts under the conditions described on pages 5 and 6. Costs listed are for each lamp or ballast installed. The crew is one electrician working at a labor cost of $26.88 per manhour. These costs include layout, material handling, and normal waste. Add for the wiring connections, enclosures if needed, sales tax, delivery, supervision, mobilization, demobilization, cleanup, overhead and profit. Note: Be sure the lamp selected is compatible with the installed ballast.

Fluorescent Lamps

Material	Craft@Hrs	Unit	Material Cost	Labor Cost	Installed Cost
Preheat fluorescent lamps for use with starters					
F4T5/CW	1E@0.02	Ea	5.09	.54	5.63
F6T5/CW	1E@0.02	Ea	4.59	.54	5.13
F6T5/D	1E@0.02	Ea	6.65	.54	7.19
F8T5/CW	1E@0.03	Ea	4.59	.81	5.40
F8T5/WW	1E@0.03	Ea	6.80	.81	7.61
F8T5/D	1E@0.03	Ea	6.65	.81	7.46
F13T5/CW	1E@0.03	Ea	5.70	.81	6.51
F13T5/WW	1E@0.03	Ea	7.66	.81	8.47
F13T8/CW	1E@0.04	Ea	7.39	1.08	8.47
F14T8/CW	1E@0.04	Ea	7.93	1.08	9.01
F14T8/D	1E@0.04	Ea	8.64	1.08	9.72
F15T8/CW	1E@0.04	Ea	6.16	1.08	7.24
F15T8/WW	1E@0.04	Ea	5.45	1.08	6.53
F15T8/D	1E@0.04	Ea	5.60	1.08	6.68
F30T8/CW	1E@0.04	Ea	6.38	1.08	7.46
F30T8/WW	1E@0.04	Ea	7.15	1.08	8.23
F30T8/D	1E@0.04	Ea	7.58	1.08	8.66
F14T12/CW	1E@0.04	Ea	5.10	1.08	6.18
F14T12/D	1E@0.04	Ea	6.16	1.08	7.24
F15T12/CW	1E@0.04	Ea	5.03	1.08	6.11
F15T12/WW	1E@0.04	Ea	5.88	1.08	6.96
F15T12/D	1E@0.04	Ea	6.16	1.08	7.24
F20T12/CW	1E@0.04	Ea	4.30	1.08	5.38
F20T12/WW	1E@0.04	Ea	4.63	1.08	5.71
F20T12/D	1E@0.04	Ea	5.20	1.08	6.28
Rapid start fluorescent lamps, ES = Energy-Saving					
F30T12/CW/ES	1E@0.04	Ea	6.52	1.08	7.60
F30T12/WW/ES	1E@0.04	Ea	7.82	1.08	8.90
F30T12/CW	1E@0.04	Ea	5.46	1.08	6.54
F30T12/WW	1E@0.04	Ea	6.98	1.08	8.06
F30T12/D	1E@0.04	Ea	8.02	1.08	9.10
F40T12/CW/ES	1E@0.05	Ea	3.01	1.34	4.35
F40T12/WW/ES	1E@0.05	Ea	3.47	1.34	4.81
F40T12/W/ES	1E@0.05	Ea	3.25	1.34	4.59
F40T12/DX/ES	1E@0.05	Ea	5.06	1.34	6.40
F40T12/CW/U/3/ES	1E@0.05	Ea	13.40	1.34	14.74
F40T12/WW/U/3/ES	1E@0.06	Ea	13.40	1.61	15.01
F40T12/W/U/3/ES	1E@0.06	Ea	13.90	1.61	15.51
F40T12/CW/U/6/ES	1E@0.06	Ea	12.10	1.61	13.71
F40T12/WW/U/6/ES	1E@0.06	Ea	12.10	1.61	13.71
F40T12/W/U/6/ES	1E@0.06	Ea	12.60	1.61	14.21

Use these figures to estimate the cost of fluorescent lamps installed in lighting fixtures under the conditions described on pages 5 and 6. Costs listed are for each lamp installed. The crew is one electrician working at a labor cost of $26.88 per manhour. These costs include layout, material handling, and normal waste. Add for the lighting fixture, sales tax, delivery, supervision, mobilization, demobilization, cleanup, overhead and profit. Note: These figures are typical minimums for most fluorescent fixtures. If the fixture must be dismantled for lamping or if access to the fixture is difficult, the labor cost will be higher.

Material	Craft@Hrs	Unit	Material Cost	Labor Cost	Installed Cost

Slimline fluorescent lamps, ES = Energy-Saving

Material	Craft@Hrs	Unit	Material Cost	Labor Cost	Installed Cost
F42T6/CW	1E@0.05	Ea	14.50	1.34	15.84
F42T6/WW	1E@0.05	Ea	15.90	1.34	17.24
F64T6/CW	1E@0.08	Ea	14.70	2.15	16.85
F64T6/WW	1E@0.08	Ea	16.20	2.15	18.35
F72T8/CW	1E@0.08	Ea	15.00	2.15	17.15
F72T8/WW	1E@0.08	Ea	16.20	2.15	18.35
F96T8/CW	1E@0.10	Ea	13.40	2.69	16.09
F24T12/CW	1E@0.05	Ea	11.40	1.34	12.74
F36T12/CW	1E@0.05	Ea	11.40	1.34	12.74
F42T12/CW	1E@0.05	Ea	11.40	1.34	12.74
F48T12/CW/ES	1E@0.05	Ea	9.07	1.34	10.41
F48T12/WW/ES	1E@0.05	Ea	9.85	1.34	11.19
F48T12/W/ES	1E@0.05	Ea	9.85	1.34	11.19
F48T12/CW	1E@0.05	Ea	7.42	1.34	8.76
F48T12/D	1E@0.05	Ea	9.28	1.34	10.62
F60T12/CW	1E@0.08	Ea	10.50	2.15	12.65
F60T12/D	1E@0.08	Ea	12.00	2.15	14.15
F64T12/CW	1E@0.08	Ea	10.60	2.15	12.75
F64T12/D	1E@0.08	Ea	11.80	2.15	13.95
F72T12/CW	1E@0.10	Ea	8.50	2.69	11.19
F72T12/CWX	1E@0.10	Ea	11.70	2.69	14.39
F72T12/WW	1E@0.10	Ea	10.60	2.69	13.29
F72T12/D	1E@0.10	Ea	10.60	2.69	13.29
F84T12/CW	1E@0.10	Ea	11.10	2.69	13.79
F96T12/CW/ES	1E@0.10	Ea	6.39	2.69	9.08
F96T12/WW/ES	1E@0.10	Ea	8.54	2.69	11.23
F96T12/W/ES	1E@0.10	Ea	7.88	2.69	10.57

High-Output (HO) fluorescent lamps, ES = Energy-Saving

Material	Craft@Hrs	Unit	Material Cost	Labor Cost	Installed Cost
F18T12/CW/HO	1E@0.04	Ea	11.50	1.08	12.58
F24T12/CW/HO	1E@0.04	Ea	11.50	1.08	12.58
F24T12/D/HO	1E@0.04	Ea	13.50	1.08	14.58
F30T12/CW/HO	1E@0.04	Ea	13.70	1.08	14.78
F36T12/CW/HO	1E@0.04	Ea	12.20	1.08	13.28
F36T12/D/HO	1E@0.04	Ea	13.60	1.08	14.68
F42T12/CW/HO	1E@0.05	Ea	12.80	1.34	14.14
F42T12/D/HO	1E@0.05	Ea	14.10	1.34	15.44
F48T12/CW/HO	1E@0.05	Ea	11.60	1.34	12.94
F48T12/WW/HO	1E@0.05	Ea	11.60	1.34	12.94
F48T12/D/HO	1E@0.05	Ea	8.50	1.34	9.84
F48T12/W/HO/ES	1E@0.05	Ea	10.80	1.34	12.14
F64T12/CW/HO	1E@0.08	Ea	10.80	2.15	12.95
F64T12/D/HO	1E@0.08	Ea	12.50	2.15	14.65

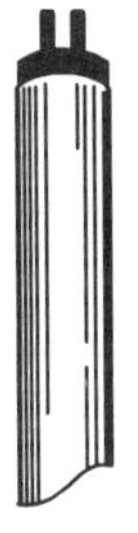

Use these figures to estimate the cost of fluorescent lamps installed in lighting fixtures under the conditions described on pages 5 and 6. Costs listed are for each lamp installed. The crew is one electrician working at a labor cost of $26.88 per manhour. These costs include layout, material handling, and normal waste. Add for the lighting fixture, sales tax, delivery, supervision, mobilization, demobilization, cleanup, overhead and profit. Note: These figures are typical minimums for most fluorescent fixtures. If the fixture must be dismantled for lamping or if access to the fixture is difficult, the labor cost will be higher.

Fluorescent Lamps

Material	Craft@Hrs	Unit	Material Cost	Labor Cost	Installed Cost
High-Output (HO) fluorescent lamps, ES = Energy-Saving, continued					
F72T12/CW/HO	1E@0.08	Ea	8.75	2.15	10.90
F72T12/WW/HO	1E@0.08	Ea	11.80	2.15	13.95
F72T12/D/HO	1E@0.08	Ea	11.60	2.15	13.75
F84T12/CW/HO	1E@0.08	Ea	9.33	2.15	11.48
F84T12/D/HO	1E@0.08	Ea	11.80	2.15	13.95
F96T12/CW/HO/ES	1E@0.10	Ea	8.75	2.69	11.44
F96T12/WW/HO/ES	1E@0.10	Ea	11.10	2.69	13.79
F96T12/W/HO/ES	1E@0.10	Ea	9.46	2.69	12.15
Very High-Output (VHO) fluorescent lamps, ES = Energy-Saving					
F48T12/CW/VHO	1E@0.05	Ea	16.90	1.34	18.24
F48T12/WW/VHO	1E@0.05	Ea	19.10	1.34	20.44
F72T12/CW/VHO	1E@0.08	Ea	17.60	2.15	19.75
F96T12/CW/VHO/ES	1E@0.10	Ea	16.70	2.69	19.39
F96T12/W/VHO/ES	1E@0.10	Ea	20.00	2.69	22.69
F96T12/CW/VHO	1E@0.10	Ea	16.20	2.69	18.89
F96T12/WW/VHO	1E@0.10	Ea	20.80	2.69	23.49
F96T12/D/VHO	1E@0.10	Ea	20.80	2.69	23.49
Power Groove (PG) fluorescent lamps, ES = Energy-Saving					
F48PG17/CW/ES	1E@0.05	Ea	20.80	1.34	22.14
F48PG17/CW	1E@0.05	Ea	19.20	1.34	20.54
F48PG17/D	1E@0.05	Ea	22.40	1.34	23.74
F72PG17/CW	1E@0.08	Ea	20.30	2.15	22.45
F96PG17/CW/ES	1E@0.10	Ea	19.60	2.69	22.29
F96PG17/W/ES	1E@0.10	Ea	25.60	2.69	28.29
F96PG17/CW	1E@0.10	Ea	18.10	2.69	20.79
F96PG17/D	1E@0.10	Ea	23.00	2.69	25.69
Circular fluorescent lamps					
FC6T9/CW	1E@0.08	Ea	7.34	2.15	9.49
FC6T9/WW	1E@0.08	Ea	8.03	2.15	10.18
FC8T9/CW	1E@0.08	Ea	6.25	2.15	8.40
FC8T9/WW	1E@0.08	Ea	7.79	2.15	9.94
FC12T9/CW	1E@0.10	Ea	6.71	2.69	9.40
FC12T9/WW	1E@0.10	Ea	8.86	2.69	11.55
FC12T9/D	1E@0.10	Ea	9.96	2.69	12.65
FC16T9/CW	1E@0.10	Ea	9.75	2.69	12.44
FC16T9/WW	1E@0.10	Ea	11.80	2.69	14.49
FC16T9/D	1E@0.10	Ea	12.50	2.69	15.19

Use these figures to estimate the cost of fluorescent lamps installed in lighting fixtures under the conditions described on pages 5 and 6. Costs listed are for each lamp installed. The crew is one electrician working at a labor cost of $26.88 per manhour. These costs include layout, material handling, and normal waste. Add for the lighting fixture, sales tax, delivery, supervision, mobilization, demobilization, cleanup, overhead and profit. Note: These figures are typical minimums for most fluorescent fixtures. If the fixture must be dismantled for lamping or if access to the fixture is difficult, the labor cost will be higher.

Material	Craft@Hrs	Unit	Material Cost	Labor Cost	Installed Cost

Rapid start energy-saving fluorescent ballasts, 430 M.A.-HPF

Material	Craft@Hrs	Unit	Material Cost	Labor Cost	Installed Cost
1 F40T12-120V	1E@0.40	Ea	18.00	10.80	28.80
1 F40T12-277V	1E@0.40	Ea	49.00	10.80	59.80
2 F40T12-120V	1E@0.40	Ea	17.00	10.80	27.80
2 F40T12-277V	1E@0.40	Ea	20.00	10.80	30.80
3 F40T12-120V	1E@0.40	Ea	35.00	10.80	45.80
3 F40T12-277V	1E@0.40	Ea	68.00	10.80	78.80

Rapid start energy-saving fluorescent ballasts, 460 M.A.-HPF

Material	Craft@Hrs	Unit	Material Cost	Labor Cost	Installed Cost
3 F40T12 (34W)-120V	1E@0.40	Ea	52.00	10.80	62.80
3 F40T12 (34W)-277V	1E@0.40	Ea	55.00	10.80	65.80

Rapid start energy-saving fluorescent ballasts, 800 M.A.-HPF

Material	Craft@Hrs	Unit	Material Cost	Labor Cost	Installed Cost
2 F72T12-120V	1E@0.50	Ea	68.00	13.40	81.40
2 F96T12-120V	1E@0.50	Ea	68.00	13.40	81.40
2 F72T12-277V	1E@0.50	Ea	70.00	13.40	83.40
2 F96T12-277V	1E@0.50	Ea	70.00	13.40	83.40

Electronic energy-saving fluorescent ballasts

Material	Craft@Hrs	Unit	Material Cost	Labor Cost	Installed Cost
1 F30T12-120V	1E@0.40	Ea	60.00	10.80	70.80
1 F40T12-120V	1E@0.40	Ea	60.00	10.80	70.80
1 F40T10-120V	1E@0.40	Ea	60.00	10.80	70.80
1 F30T12-277V	1E@0.40	Ea	64.00	10.80	74.80
1 F40T12-277V	1E@0.40	Ea	64.00	10.80	74.80
1 F40T10-277V	1E@0.40	Ea	64.00	10.80	74.80
2 F40T12-120V	1E@0.40	Ea	98.00	10.80	108.80
2 F40T10-120V	1E@0.40	Ea	98.00	10.80	108.80
2 F40T12-277V	1E@0.40	Ea	99.00	10.80	109.80
2 F40T10-277V	1E@0.40	Ea	99.00	10.80	109.80
2 F48T12-120V	1E@0.40	Ea	79.20	10.80	90.00
2 F72T12-120V	1E@0.50	Ea	79.20	13.40	92.60
2 F96T12-120V	1E@0.50	Ea	79.20	13.40	92.60
2 F48T10-277V	1E@0.50	Ea	82.30	13.40	95.70
2 F72T12-277V	1E@0.50	Ea	82.30	13.40	95.70
2 F96T12-277V	1E@0.50	Ea	82.30	13.40	95.70

Slimline energy-saving fluorescent ballasts, 425 M.A.-HPF

Material	Craft@Hrs	Unit	Material Cost	Labor Cost	Installed Cost
2 F72T12-120V	1E@0.50	Ea	26.00	13.40	39.40
2 F96T10-120V	1E@0.50	Ea	26.00	13.40	39.40
2 F72T12-277V	1E@0.50	Ea	61.00	13.40	74.40
2 F96T12-277V	1E@0.50	Ea	72.00	13.40	85.40

Use these figures to estimate the cost of installing fluorescent ballasts in lighting fixtures under the conditions described on pages 5 and 6. Costs listed are for each ballast installed. The crew is one electrician working at a labor cost of $26.88 per manhour. These costs include layout, material handling, and normal waste. Add for the lighting fixture, wire connections, sales tax, delivery, supervision, mobilization, demobilization, cleanup, overhead and profit. These costs are typical minimums for most fluorescent fixtures. If the ballast compartment requires substantial dismantling and reassembly or if the fixture is difficult to reach, the labor cost will be higher.

Fluorescent Ballasts, Standard

Material	Craft@Hrs	Unit	Material Cost	Labor Cost	Installed Cost
Rapid start standard fluorescent ballasts, 430 M.A.-LPF					
1 F30T12-120V	1E@0.40	Ea	11.00	10.80	21.80
1 F40T12-120V	1E@0.40	Ea	11.00	10.80	21.80

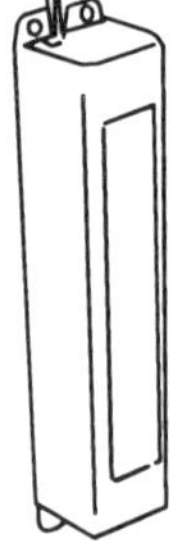

Material	Craft@Hrs	Unit	Material Cost	Labor Cost	Installed Cost
Rapid start standard fluorescent ballasts, 460 M.A.-HPF					
1 F30T12-120V	1E@0.40	Ea	34.60	10.80	45.40
1 F30T12-277V	1E@0.40	Ea	36.70	10.80	47.50
1 F40T12-120V	1E@0.40	Ea	16.80	10.80	27.60
1 F40T12-277V	1E@0.40	Ea	19.20	10.80	30.00
2 F30T12-120V	1E@0.40	Ea	30.60	10.80	41.40
2 F30T12-277V	1E@0.40	Ea	33.00	10.80	43.80
2 F40T12-120V	1E@0.40	Ea	16.30	10.80	27.10
2 F40T12-277V	1E@0.40	Ea	19.20	10.80	30.00

Material	Craft@Hrs	Unit	Material Cost	Labor Cost	Installed Cost
Rapid start standard fluorescent ballasts, 800 M.A.-HPF					
1 F24T12-120V	1E@0.40	Ea	70.80	10.80	81.60
1 F48T12-120V	1E@0.40	Ea	70.80	10.80	81.60
1 F72T12-120V	1E@0.50	Ea	59.80	13.40	73.20
1 F96T12-120V	1E@0.50	Ea	59.80	13.40	73.20
1 F24T12-277V	1E@0.40	Ea	69.60	10.80	80.40
1 F48T12-277V	1E@0.40	Ea	69.60	10.80	80.40
1 F72T12-277V	1E@0.50	Ea	72.00	13.40	85.40
1 F96T12-277V	1E@0.50	Ea	72.00	13.40	85.40
2 F24T12-120V	1E@0.40	Ea	67.40	10.80	78.20
2 F48T12-120V	1E@0.40	Ea	67.40	10.80	78.20
2 F72T12-120V	1E@0.50	Ea	50.80	13.40	64.20
2 F96T12-120V	1E@0.50	Ea	50.80	13.40	64.20
2 F24T12-277V	1E@0.40	Ea	71.80	10.80	82.60
2 F48T12-277V	1E@0.40	Ea	71.80	10.80	82.60
2 F72T12-277V	1E@0.50	Ea	53.90	13.40	67.30
2 F96T12-277V	1E@0.50	Ea	53.90	13.40	67.30

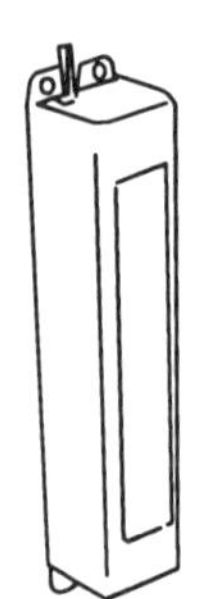

Material	Craft@Hrs	Unit	Material Cost	Labor Cost	Installed Cost
Rapid start standard fluorescent ballasts, 1,500 M.A.-HPF					
1 F48T12-120V	1E@0.40	Ea	80.20	10.80	91.00
1 F72T12-120V	1E@0.50	Ea	80.20	13.40	93.60
1 F96T12-120V	1E@0.50	Ea	80.20	13.40	93.60
1 F48T12-277V	1E@0.40	Ea	89.40	10.80	100.20
1 F72T12-277V	1E@0.50	Ea	89.40	13.40	102.80
1 F96T12-277V	1E@0.50	Ea	89.40	13.40	102.80
2 F48T12-120V	1E@0.40	Ea	80.20	10.80	91.00
2 F72T12-120V	1E@0.50	Ea	91.80	13.40	105.20
2 F96T12-120V	1E@0.50	Ea	91.80	13.40	105.20
2 F48T12-277V	1E@0.40	Ea	89.40	10.80	100.20
2 F72T12-277V	1E@0.50	Ea	102.00	13.40	115.40
2 F96T12-277V	1E@0.50	Ea	102.00	13.40	115.40

Use these figures to estimate the cost of installing fluorescent ballasts in lighting fixtures under the conditions described on pages 5 and 6. Costs listed are for each ballast installed. The crew is one electrician working at a labor cost of $26.88 per manhour. These costs include layout, material handling, and normal waste. Add for lighting fixture, wire connections, sales tax, delivery, supervision, mobilization, demobilization, cleanup, overhead and profit. These costs are typical minimums for most fluorescent fixtures. If the ballast compartment requires substantial dismantling and reassembly or if the fixture is difficult to reach, the labor cost will be higher.

Material	Craft@Hrs	Unit	Material Cost	Labor Cost	Installed Cost
Instant start standard fluorescent ballasts, slimline, 200 M.A.-LPF					
1 F42T6-120V	1E@0.40	Ea	33.20	10.80	44.00
Instant start standard fluorescent ballasts, slimline, 200 M.A.-HPF					
1 F42T6-120V	1E@0.40	Ea	32.50	10.80	43.30
1 F64T6-120V	1E@0.40	Ea	46.20	10.80	57.00
1 F72T8-120V	1E@0.50	Ea	46.50	13.40	59.90
1 F96T8-120V	1E@0.50	Ea	46.50	13.40	59.90
2 F42T6-120V	1E@0.40	Ea	76.70	10.80	87.50
2 F64T6-120V	1E@0.40	Ea	50.70	10.80	61.50
2 F72T8-120V	1E@0.50	Ea	53.40	13.40	66.80
2 F96T8-120V	1E@0.50	Ea	53.40	13.40	66.80
Instant start standard fluorescent ballasts, slimline, 425 M.A.-HPF					
1 F24T12-120V	1E@0.40	Ea	36.30	10.80	47.10
1 F40T12-120V	1E@0.40	Ea	36.30	10.80	47.10
1 F48T12-120V	1E@0.40	Ea	36.30	10.80	47.10
1 F60T12-120V	1E@0.40	Ea	31.40	10.80	42.20
1 F64T12-120V	1E@0.40	Ea	31.40	10.80	42.20
1 F72T12-120V	1E@0.50	Ea	31.40	13.40	44.80
1 F96T12-120V	1E@0.50	Ea	31.40	13.40	44.80
1 F24T12-277V	1E@0.40	Ea	43.50	10.80	54.30
1 F40T12-277V	1E@0.40	Ea	43.50	10.80	54.30
1 F48T12-277V	1E@0.40	Ea	43.50	10.80	54.30
1 F60T12-277V	1E@0.40	Ea	31.40	10.80	42.20
1 F64T12-277V	1E@0.40	Ea	31.40	10.80	42.20
1 F72T12-277V	1E@0.50	Ea	33.40	13.40	46.80
1 F96T12-277V	1E@0.50	Ea	33.40	13.40	46.80
2 F24T12-120V	1E@0.40	Ea	30.50	10.80	41.30
2 F40T12-120V	1E@0.40	Ea	30.50	10.80	41.30
2 F48T12-120V	1E@0.40	Ea	30.50	10.80	41.30
2 F60T12-120V	1E@0.40	Ea	34.50	10.80	45.30
2 F64T12-120V	1E@0.40	Ea	34.50	10.80	45.30
2 F72T12-120V	1E@0.50	Ea	26.80	13.40	40.20
2 F96T12-120V	1E@0.50	Ea	26.80	13.40	40.20
2 F40T12-277V	1E@0.40	Ea	34.50	10.80	45.30
2 F48T12-277V	1E@0.40	Ea	34.50	10.80	45.30
2 F60T12-277V	1E@0.40	Ea	37.80	10.80	48.60
2 F64T12-277V	1E@0.40	Ea	37.80	10.80	48.60
2 F72T12-277V	1E@0.50	Ea	73.80	13.40	87.20
2 F96T12-277V	1E@0.50	Ea	73.80	13.40	87.20

Use these figures to estimate the cost of installing fluorescent ballasts in lighting fixtures under the conditions described on pages 5 and 6. Costs listed are for each ballast installed. The crew is one electrician working at a labor cost of $26.88 per manhour. These costs include layout, material handling, and normal waste. Add for lighting fixture, wire connections, sales tax, delivery, supervision, mobilization, demobilization, cleanup, overhead and profit. These cost are typical minimums for most fluorescent fixtures. If the ballast compartment requires substantial dismantling and reassembly or if the fixture is difficult to reach, the labor cost will be higher.

Fluorescent Ballasts

Material	Craft@Hrs	Unit	Material Cost	Labor Cost	Installed Cost
Preheat fluorescent ballasts, LPF					
1 F4T5-120V	1E@0.35	Ea	3.96	9.41	13.37
1 F6T5-120V	1E@0.35	Ea	3.96	9.41	13.37
1 F8T5-120V	1E@0.35	Ea	3.96	9.41	13.37
1 F14T12-120V	1E@0.35	Ea	3.30	9.41	12.71
1 F20T12-120V	1E@0.40	Ea	3.30	10.80	14.10
2 F15T8-120V	1E@0.35	Ea	3.30	9.41	12.71
2 F25T12-120V	1E@0.40	Ea	7.11	10.80	17.91
Preheat fluorescent ballasts, LPF, Class P					
1 F30T8-120V	1E@0.40	Ea	10.50	10.80	21.30
2 F40T12-120V	1E@0.40	Ea	10.50	10.80	21.30
Preheat fluorescent ballasts, LPF, Class P					
2 F40T12-120V	1E@0.40	Ea	29.30	10.80	40.10
2 F90T17-120V	1E@0.50	Ea	88.30	13.40	101.70
2 F100T17-120V	1E@0.50	Ea	88.30	13.40	101.70
Trigger start fluorescent ballasts, LPF, Class P					
1 F14T12-120V	1E@0.40	Ea	10.50	10.80	21.30
1 F20T12-120V	1E@0.40	Ea	10.50	10.80	21.30
1 F15T8-120V	1E@0.40	Ea	10.50	10.80	21.30
2 F14T12-120V	1E@0.40	Ea	15.20	10.80	26.00
2 F20T12-120V	1E@0.40	Ea	15.20	10.80	26.00
2 F15T8-120V	1E@0.40	Ea	15.20	10.80	26.00
2 F15T12-120V	1E@0.40	Ea	15.20	10.80	26.00
2 F20T12-120V	1E@0.40	Ea	15.20	10.80	26.00
2 F15T8-120V	1E@0.40	Ea	15.20	10.80	26.00
Trigger start fluorescent ballasts, HPF, Class P					
1 F14T12-120V	1E@0.40	Ea	37.00	10.80	47.80
1 F20T12-120V	1E@0.40	Ea	37.00	10.80	47.80
1 F15T8-120V	1E@0.40	Ea	37.00	10.80	47.80
1 F14T12-277V	1E@0.40	Ea	39.20	10.80	50.00
1 F20T12-277V	1E@0.40	Ea	39.20	10.80	50.00
1 F15T8-277V	1E@0.40	Ea	39.20	10.80	50.00
2 F14T12-120V	1E@0.40	Ea	32.60	10.80	43.40
2 F20T12-120V	1E@0.40	Ea	32.60	10.80	43.40
2 F15T8-120V	1E@0.40	Ea	32.60	10.80	43.40
2 F14T12-277V	1E@0.40	Ea	34.50	10.80	45.30
2 F20T12-277V	1E@0.40	Ea	34.50	10.80	45.30
2 F15T8-277V	1E@0.40	Ea	34.50	10.80	45.30

Use these figures to estimate the cost of installing fluorescent ballasts in lighting fixtures under the conditions described on pages 5 and 6. Costs listed are for each ballast installed. The crew is one electrician working at a labor cost of $26.88 per manhour. These costs include layout, material handling, and normal waste. Add for lighting fixture, wire connections, sales tax, delivery, supervision, mobilization, demobilization, cleanup, overhead and profit. These costs are typical minimums for most fluorescent fixtures. If the ballast compartment requires substantial dismantling and reassembly or if the fixture is difficult to reach, the labor cost will be higher.

Section 5: Wiring Devices

The term *wiring devices* refers to the circuit control switches and convenience outlets in an electrical system. The switches we use to turn lights on and off and the convenience receptacles we use to get power for electrical appliances are wiring devices.

Why the Cost of Devices Varies

Estimating the cost of wiring devices may seem to be a simple matter of accurate counting. On some jobs it is. But pricing the number of devices counted can be more difficult. Many types of wiring devices are available. Some look about the same, serve the same function, but vary widely in price. It's not always easy to be sure you're pricing the right device.

This section describes the more common devices and suggests what to watch out for when pricing the device section of your estimate.

Standard-grade devices used in residences will be the least expensive because they're made in great quantity and carried by every electrical supplier. Less common devices are made to meet special requirements and constructed of special materials. For example, switches and trim plates made of nylon are intended for heavy use. They cost more than standard residential-grade devices and may not be stocked by your local supplier.

Every modern wiring device has a voltage and amperage rating. This rating has an effect on cost. Usually, the greater the ampacity and voltage rating, the higher the cost.

Devices intended to meet exacting specifications or to serve in special applications are said to be *specification grade*. Usually these devices are made with more expensive contact material, are designed with special features, or are built to be particularly durable. That increases the cost, often to several times the cost of the standard-grade device. But the added cost may be a good investment from the owner's standpoint. For example, one specification-grade convenience outlet has contact material with better shape-retention characteristics. That ensures better electrical contact after years of hard use. You've probably seen duplex outlets that will hardly hold an electrical plug in place after a year or two of service.

Most manufacturers offer wiring devices with smooth surfaces that stay cleaner longer — and are easier to clean when they do get soiled. Older switch covers had decorative grooves that accumulated smudges and were hard to clean.

Standard colors for wiring devices are ivory, white and brown. Other colors are available but will cost more.

Some devices have built-in grounding. There are two common types of grounded devices. One has a screw or terminal for attaching a ground conductor. The other has a special spring on one of the mounting screws. This spring provides a positive ground through the device frame to the mounting screw and on to a metal-backed outlet box. Automatic grounding devices cost a little more but install a little faster than the type with a grounding terminal or grounding screw, because the electrician saves one connection.

Wiring Device Descriptions

Duplex receptacles are the most common wiring device. Many types are available. The traditional duplex outlet has screw terminals around which wire is wrapped to make contact. Then the screw is tightened. Some have holes designed to receive and hold a stripped wire conductor. That eliminates wrapping the wire around a screw contact and tightening the screw. Other receptacles have both screw terminals and insert holes. Some are side wired. Others are wired at the back.

Many duplex receptacles have a removable tab between the screw terminals on each side. Remove the tab to isolate one part of the receptacle from the other. These are commonly used for switched outlets. Some residential rooms are wired so that a switch by the entrance controls one or more outlets where table lamps will be installed. Outlets with tabs can also be wired to two different circuits.

Clock hanger receptacles are usually mounted high on a wall. They have a recessed single outlet for a clock. They also have a tab to hold the clock. This is a decorative item, so many colors and styles are available. Clock hanger outlets are usually connected to an unswitched circuit.

Locking receptacles come in a variety of sizes and configurations. Each service load capacity has a different size and configuration as set by NEMA (National Electrical Manufacturers' Association) standards. They're made that way so that appliances that use the receptacles can use only receptacles rated at the correct capacity. Locking receptacles can be wall-mounted in a box or may be attached to an extension cord.

Be careful when selecting trim plates for locking receptacles. The size of the receptacle face varies with the service rating. A 20-amp, 4-wire receptacle is larger than a 20-amp, 3-wire receptacle. Locking receptacles are commonly called *twist-lock* devices because they must be rotated or twisted when inserted and removed.

Switches are classified by the number of poles. Figure 5-1 shows the most simple type, the one-pole switch. Notice that the switch interrupts only one conductor when open. A one-pole switch can control only one circuit.

Two-pole switches can control two separate circuits at the same time. For example, a large room may need two lighting circuits. If only one switch is needed, use a single two-pole switch. Several two-pole switches can be ganged to control a number of circuits. This reduces the labor, switch, and box cost.

Many offices require two-level lighting. Regulations designed to reduce energy consumption sometimes require two-level lighting in all office buildings. Each four-lamp fluorescent fixture will have two lamps controlled from one switch and two lamps controlled by another switch located immediately beside the first. When there are more than two circuits involved, two-pole switches can be used in the same two-gang switch box.

Three-way switches allow switching a lighting circuit from two separate locations. For example, in a hall you want lighting controlled from switches at either end. On a stairway you want lights controlled from both upstairs and downstairs. Figure 5-2 shows that both switches 1 and 2 can control the circuit no matter what position the other switch is in.

Four-way switches are used where three switches need control of the circuit. The four-way switch is connected into the switching circuits between two three-way switches. In Figure 5-3, switch 3 can open or close the circuit no matter what position switches 1 and 2 are in.

There are two types of switches with illuminated handles. One lights when the switch is in the *on* position. The other lights when the switch is in the *off* position. Either can be used as a pilot light. But be sure you order the right switch. They're not interchangeable and can't be modified on site.

The switch that lights when on is used where the switch and the fixture or appliance are in different rooms — such as the light for a storage room or closet.

The switch that lights when the switch is off is used in rooms that are normally dark when unlighted. When the switch is on, the handle isn't illuminated.

Quiet (silent) switches are a good choice for better-quality homes and offices. You'll hear these called *mercury switches*, but they may or may not be made with mercury contacts. Quiet switches come in voltages to 277V.

Heavy duty "T" rated switches are made for heavy loads. The circuit is closed by a contact armature that moves very fast to minimize arcing when the switch changes position.

Decorative wiring devices are offered by some manufacturers. Unlike standard devices, decorative devices are generally rectangular. The trim plate is designed to fit only a decorative style device. And note that trim plates made by some manufacturers for decorative switches won't fit decorative duplex receptacles by the same manufacturer. That creates a problem when placing a decorative switch and a decorative duplex receptacle in the same box.

Horsepower-rated switches are used to control small motor loads, such as exhaust fans, where manual motor starters are not required. The motor has a built-in overload protector.

Key-operated switches are common in public areas where lighting shouldn't be controlled by the public — for example, in a public restroom where the light has to be on while the building is open to the public. Key switches come with either a simple stamped 2-prong key or a conventional lockset.

Duplex switches have two switches mounted in a housing shaped like a duplex outlet. They can be wired to control two separate circuits or to control one circuit with two loads. Duplex switches can be used to control two-level lighting in an office.

Momentary-contact switches provide current only for an instant. When not being activated, the switch is always open. Momentary switches are commonly used to throw another switch like a lighting contactor that controls a large bank of fluorescent lights. If the lighting contactor has a close and open set of coils, a momentary-center-off switch is used. The contactor is closed when the momentary switch handle is lifted to the *on* position. When the switch handle is released, it returns to the center position. To de-energize the lighting contactor, the momentary switch is pushed down to the *off* position. Any time the momentary switch handle is released, it returns to the center position.

Maintain-contact switches are used when current is to flow in both the *on* and *off* positions. There is a center *off* position. But the switch handle must be moved to that center position. It doesn't return to center by just releasing the handle. One application for this type of switch is a conveyor that can be operated in two directions but also has to be stopped when not in use. There are many other applications.

Simple Switch Control Schemes

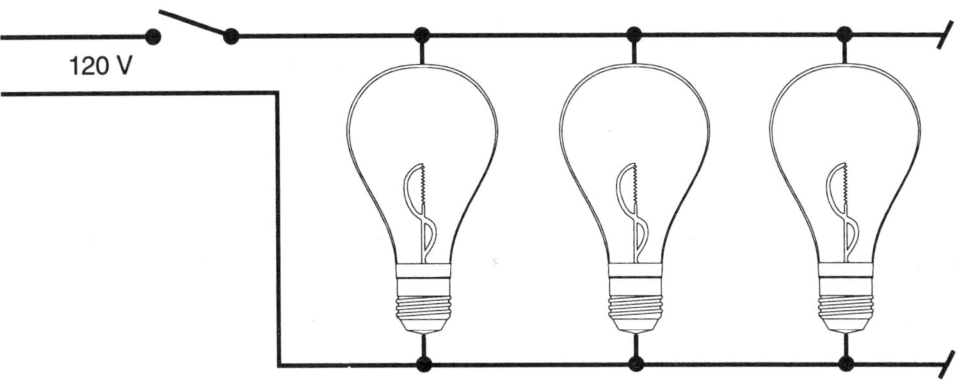

Figure 5-1
Single Pole Switch

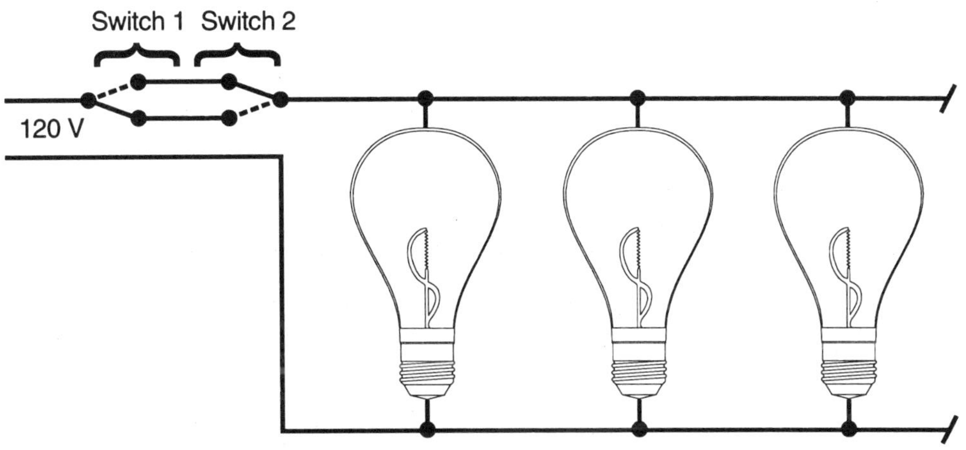

Figure 5-2
Three Way Switch

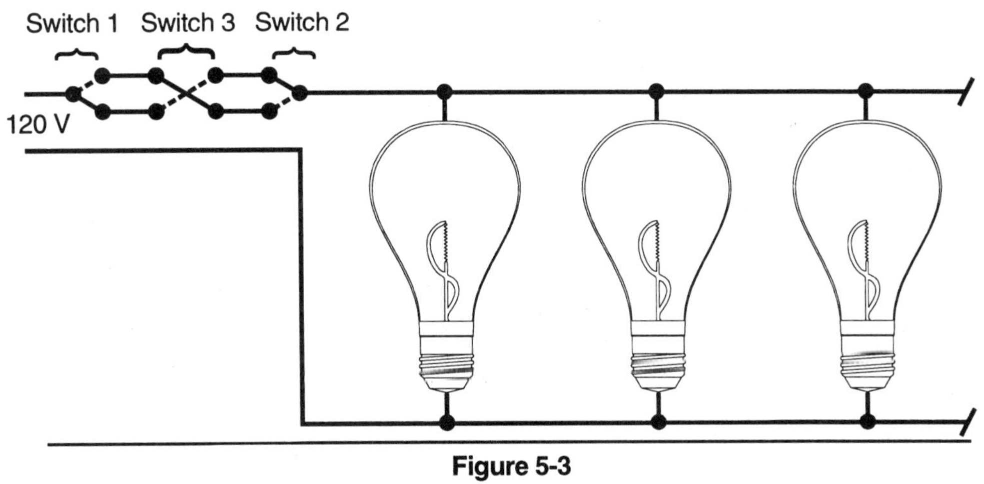

Figure 5-3
Four Way Switch

Dimming switches have to be sized for the anticipated lighting load. Incandescent lighting loads and incandescent dimmers are rated in watts. Incandescent dimmers can't be used with fluorescent lighting. Fluorescent dimmers are rated by the number of lamps in the dimming circuit. Fluorescent lighting fixtures need special dimming ballasts. Standard fluorescent ballasts won't work on a dimming circuit.

There are many types of dimming systems, from a simple manual wheel to large motorized units. Large dimming systems, such as stage lighting, are custom engineered for the lighting load expected.

Photoelectric switches are made to energize a circuit from dusk until dawn. Exterior lighting is usually controlled with photoelectric switches built into the luminaire. Photoelectric switches can also be used to control latches on entryways or gates. When the light beam is interrupted, the switch operates.

Device plates come in as many sizes, shapes and colors and are made of as many materials as the devices they cover. But if you need a special plate, order it early. Some plates are stocked or made only on special order.

Electrical material catalogs list many plates that are not included in this section. Most are made for specific uses. Some are not interchangeable with any other plate and may require a specific outlet box.

Every month new wiring devices and trim plates are introduced and old devices and plates are discontinued. Keep your catalog current for quick reference.

Many remodeling contractors use semi-jumbo and jumbo device trim plates. Oversize plates hide mistakes and make it unnecessary to repaint just because a device was changed. The standard device trim plate doesn't provide much cover beyond the edge of a recessed outlet box.

Deep-cut trim plates are made with a more concave shape so they fit flush against a wall even if the box protrudes slightly beyond the wall surface. That's a big advantage when a box is being cut into an existing wall. Deep-cut trim plates allow more room for the mounting screws that hold the box to the wall surface.

When a standard box is cut into an existing wall, the mounting yoke or ears at the top and bottom of the box set flush with the wall surface. Adding screws to attach the box to the wall will hold a standard trim plate away from the wall surface. That's unsightly. The easy solution is to use a deep-cut plate. That's much easier than recessing the mounting ears into the finish wall surface.

Galvanized or plated trim plates are used to cover an outlet box that's surface mounted. You'll hear these plates called "industrial" covers or "handy box" covers. They're made for switches, receptacles and combinations of both. Material and labor costs for this item are listed in Section 3 of this manual.

Weatherproof plates also come in a wide variety of sizes and descriptions. There are spring-loaded types with a flip cover that returns to the closed position when released. The locking type can be opened only with a key. Some are made for use in corrosive environments. some are intended for use with vertically-mounted duplex receptacles. Others are intended for horizontal mounting.

Take-off and Pricing Wiring Devices

The remainder of this section points out some important considerations when preparing a take-off of wiring devices.

The cost tables in this section cover all the common wiring devices. But many more devices are available. Your supplier can provide a more complete list of the devices that will meet any conceivable need.

Your device take-off should start with the switches. I recommend using an 8½" x 11" lined tablet to list all devices. Draw a line down the left side of the page. The line should be about 1½ inches from the left edge. At the top of the sheet and just to the right of the line you've drawn, write in the number of the first plan sheet you're using. For example, the first plan sheet might be E2. To the left of the vertical line, enter a device symbol, such as **S** for a single pole switch. Under E2, enter the number of single-pole switches counted on sheet E2. Then go on to the next type of switch.

On the next line, to the left of the vertical line, and on the next line below **S**, write in **SS** for two single-pole switches. Continue listing types of switches down the left column until you've covered all of the lighting switches found on sheet E2. Then go on to sheet E3. Head the next column E3 and list each type of switch found on that sheet.

You'll usually find three-way switches (S3 or S3S3 is the common symbol) in pairs. When you find one, look for the mate across the room, down the hall or up a flight of stairs. When you see S3S3, there will most likely be another S3S3 in the circuit across the room.

Switching combinations may be SS3 or SS3S3 or SSS3 or S3SS3 or S3S3S. Custom-designed lighting may include some exotic switching. But the owner and designer probably intend exactly what the plans require. Be sure you install what's on the plans.

Any uncommon symbols should be defined in a schedule on one of the plan sheets. But don't be surprised if you find switch symbols on the plans that aren't on the schedule or symbols on the schedule that aren't on the plans. Electrical designers usually use standard switch schedules without considering what applies or doesn't apply to the job at hand. The schedule you see is probably no more than a standard guide.

When all the lighting switches on all sheets are counted and recorded, start counting duplex receptacles. Count and record all the receptacles. Watch for notes on the plans about special devices such as locking receptacles, special voltages and amperages. Watch for specifications that require ivory duplex receptacles in some areas and brown duplex receptacles in other areas.

Continuing the Take-off

The specifications should list the type of device plate for the switches and receptacles. For example, you may need stainless steel plates in some rooms and smooth plastic plates in other rooms. The plans and specifications shouldn't leave any doubt about the grade or type of wiring device needed.

The specifications may list a wiring device by manufacturer and catalog number only. Look in that manufacturer's catalog to find the grade and color of the device specified. If the specs indicate "or equal" after the catalog number, be very sure any substitute you use is a true equal. Some owners and architects watch this very carefully.

When designer-grade wiring devices are specified, the specs will list the devices to install at each location. Designer devices include both switches and receptacles.

Count devices accurately. Mistakes in counting make problems for the installer. Have enough devices to complete the job without leaving a surplus that has to go back to the shop. Leftovers tend to become misfits that take up valuable storage space and may never be used. When doing your take-off, don't assume that changes will add more devices or that some devices will be lost on the jobsite. Do a good job of counting in the beginning to get the job started off right.

Improving Labor Productivity

Your electricians can't work efficiently if they don't have the right materials. But don't burden the job with excess materials. For example, don't ship finish materials until it's nearly time to install finish materials.

Tell your installers how many hours are estimated for conduit, pulling wire, installing devices, etc. If they can't meet your expectations, you should know why. If your estimate is too high, you should be informed of that fact.

On medium to large jobs, have a good installer set the devices. A quick electrician can really pick up speed if there are several hundred devices to set. But be sure each device is installed flush with the wall. The plate should fit properly. And don't leave any dirty fingerprints or scrapes behind. A dirty or messy job reflects on your company as well as the installer.

On a large job like a hospital, some contractors have the trim crew use a lockable push cart that can be wheeled from room to room. Load the wiring devices, plates, grounding jumpers, extra mounting screws, a selection of trim screws and a trash box on the cart. Be sure the cart is narrow enough to pass easily through a doorway. The cart should be locked when left unattended.

On a large job with a good trim electrician installing devices, reduce the labor costs in this section by 5 to 10 percent. If there are few devices and if they're spaced farther apart, increase the labor cost by 5 percent.

Residential Grade, 15 Amp AC Quiet Switches

Material	Craft@Hrs	Unit	Material Cost	Labor Cost	Installed Cost
15 amp switches 120 volt push-in and side-wire, residential grade					
Single pole, brown	1E@0.25	Ea	.66	6.72	7.38
Single pole, ivory	1E@0.25	Ea	.66	6.72	7.38
Single pole, white	1E@0.25	Ea	.66	6.72	7.38
Three-way, brown	1E@0.30	Ea	1.11	8.06	9.17
Three-way, ivory	1E@0.30	Ea	1.11	8.06	9.17
Three-way, white	1E@0.30	Ea	1.11	8.06	9.17
15 amp switches 120v, push-in, side-wire, ground screw, residential					
Single pole, brown	1E@0.25	Ea	.68	6.72	7.40
Single pole, ivory	1E@0.25	Ea	.68	6.72	7.40
Single pole, white	1E@0.25	Ea	.68	6.72	7.40
Three-way, brown	1E@0.30	Ea	1.13	8.06	9.19
Three-way, ivory	1E@0.30	Ea	1.13	8.06	9.19
Three-way, white	1E@0.30	Ea	1.13	8.06	9.19
15 amp switches 120v, push-in, side-wire, ground screw, no ears, residential					
Single pole, ivory	1E@0.25	Ea	1.14	6.72	7.86
Single pole, white	1E@0.25	Ea	1.14	6.72	7.86
Three-way, ivory	1E@0.30	Ea	1.93	8.06	9.99
Three-way, white	1E@0.30	Ea	1.93	8.06	9.99
15 amp switches 120/277v, push-in, side-wire, no-ground, residential					
Single pole, brown	1E@0.25	Ea	5.08	6.72	11.80
Single pole, ivory	1E@0.25	Ea	5.08	6.72	11.80
Three-way, brown	1E@0.30	Ea	8.33	8.06	16.39
Three-way, ivory	1E@0.30	Ea	8.33	8.06	16.39

Use these figures to estimate the cost of lighting switches installed in outlet boxes under the conditions described on pages 5 and 6. Costs listed are for each switch installed. The crew is one electrician working at a labor cost of $26.88 per manhour. These costs include mounting screws, layout, material handling, and normal waste. Add for the outlet box, trim plate, plaster ring, sales tax, delivery, supervision, mobilization, demobilization, cleanup, overhead and profit. Labor costs will be slightly lower when switches are ganged or when many switches are mounted on the same wall. Note: Be careful to select the switch size and type appropriate for the intended load and use. Switch designations are as follows: single pole is SP, double pole is DP, three-way is 3W, four-way is 4W, brown is BR, ivory is IV, white is W, red is R, gray is GY, locking is L.

Industrial Grade, 15 Amp AC Quiet Switches

Material	Craft@Hrs	Unit	Material Cost	Labor Cost	Installed Cost
15 amp switches 120/277v, shallow side-wire, ground screw, industrial					
Single pole, brown	1E@0.25	Ea	4.35	6.72	11.07
Single pole, ivory	1E@0.25	Ea	4.35	6.72	11.07
Single pole, white	1E@0.25	Ea	4.35	6.72	11.07
Double pole, brown	1E@0.30	Ea	7.40	8.06	15.46
Double pole, ivory	1E@0.30	Ea	7.40	8.06	15.46
Double pole, white	1E@0.30	Ea	7.40	8.06	15.46
Three-way, brown	1E@0.30	Ea	6.39	8.06	14.45
Three-way, ivory	1E@0.30	Ea	6.39	8.06	14.45
Three-way, white	1E@0.30	Ea	6.39	8.06	14.45
Four-way, brown	1E@0.40	Ea	20.60	10.80	31.40
Four-way, ivory	1E@0.40	Ea	20.60	10.80	31.40
Four-way, white	1E@0.40	Ea	20.60	10.80	31.40

S

S_2

S_3

S_4

Material	Craft@Hrs	Unit	Material Cost	Labor Cost	Installed Cost
15 amp switches 120/277 volt back- and side-wire, industrial					
Single pole, brown	1E@0.25	Ea	6.10	6.72	12.82
Single pole ivory	1E@0.25	Ea	6.10	6.72	12.82
Single pole, white	1E@0.25	Ea	6.10	6.72	12.82
Double pole, brown	1E@0.30	Ea	10.30	8.06	18.36
Double pole ivory	1E@0.30	Ea	10.30	8.06	18.36
Double pole white	1E@0.30	Ea	10.30	8.06	18.36
Three-way, brown	1E@0.30	Ea	9.24	8.06	17.30
Three-way ivory	1E@0.30	Ea	9.24	8.06	17.30
Three-way, white	1E@0.30	Ea	9.24	8.06	17.30
Four-way, brown	1E@0.40	Ea	24.50	10.80	35.30
Four-way ivory	1E@0.40	Ea	24.50	10.80	35.30
Four-way, white	1E@0.40	Ea	24.50	10.80	35.30

S

S_2

S_3

S_4

Use these figures to estimate the cost of lighting switches installed in outlet boxes under the conditions described on pages 5 and 6. Costs listed are for each switch installed. The crew is one electrician working at a labor cost of $26.88 per manhour. These costs include mounting screws, layout, material handling, and normal waste. Add for the outlet box, trim plate, plaster ring, sales tax, delivery, supervision, mobilization, demobilization, cleanup, overhead and profit. Labor costs will be slightly lower when switches are ganged or when many switches are mounted on the same wall. Note: Be careful to select the switch size and type appropriate for the intended load and use. Switch designations are as follows: single pole is SP, double pole is DP, three-way is 3W, four-way is 4W, brown is BR, ivory is IV, white is W, red is R, gray is GY, locking is L.

Industrial Grade, 20 Amp AC Quiet Switches

Material	Craft@Hrs	Unit	Material Cost	Labor Cost	Installed Cost

20A switches 120/277v, back & side-wire, 1 HP at 120v, 2 HP at 240v, industrial

Material	Craft@Hrs	Unit	Material Cost	Labor Cost	Installed Cost
Single pole, brown	1E@0.25	Ea	6.70	6.72	13.42
Single pole, ivory	1E@0.25	Ea	6.70	6.72	13.42
Single pole, white	1E@0.25	Ea	6.70	6.72	13.42
Double pole, brown	1E@0.30	Ea	8.05	8.06	16.11
Double pole, ivory	1E@0.30	Ea	8.05	8.06	16.11
Double pole, white	1E@0.30	Ea	8.05	8.06	16.11
Three-way, brown	1E@0.30	Ea	8.05	8.06	16.11
Three-way, ivory	1E@0.30	Ea	8.05	8.06	16.11
Three-way, white	1E@0.30	Ea	8.05	8.06	16.11
Four-way, brown	1E@0.40	Ea	24.70	10.80	35.50
Four-way, ivory	1E@0.40	Ea	24.70	10.80	35.50
Four-way, white	1E@0.40	Ea	24.70	10.80	35.50

20A switches 120/277v, back & side-wire, 1 HP at 120v, 2 HP at 240v, framed toggle, industrial

Material	Craft@Hrs	Unit	Material Cost	Labor Cost	Installed Cost
Single pole, brown	1E@0.25	Ea	9.92	6.72	16.64
Single pole, ivory	1E@0.25	Ea	9.92	6.72	16.64
Single pole, white	1E@0.25	Ea	9.92	6.72	16.64
Double pole, brown	1E@0.30	Ea	13.40	8.06	21.46
Double pole, ivory	1E@0.30	Ea	13.40	8.06	21.46
Double pole, white	1E@0.30	Ea	13.40	8.06	21.46
Three-way, brown	1E@0.30	Ea	10.60	8.06	18.66
Three-way ivory	1E@0.30	Ea	10.60	8.06	18.66
Three-way, white	1E@0.30	Ea	10.60	8.06	18.66
Four-way, brown	1E@0.40	Ea	28.80	10.80	39.60
Four-way ivory	1E@0.40	Ea	28.80	10.80	39.60
Four-way, white	1E@0.40	Ea	28.80	10.80	39.60

Use these figures to estimate the cost of lighting switches installed in outlet boxes under the conditions described on pages 5 and 6. Costs listed are for each switch installed. The crew is one electrician working at a labor cost of $26.88 per manhour. These costs include mounting screws, layout, material handling, and normal waste. Add for the outlet box, trim plate, plaster ring, sales tax, delivery, supervision, mobilization, demobilization, cleanup, overhead and profit. Labor costs will be slightly lower when switches are ganged or when many switches are mounted on the same wall. Note: The switches listed above are side wired. Be careful to select the switch size and type appropriate for the intended load and use. Switch designations are as follows: single pole is SP, double pole is DP, three-way is 3W, four-way is 4W, brown is BR, ivory is IV, white is W, red is R, gray is GY, locking is L.

Commercial Grade, 15 Amp AC Quiet Switches

Material	Craft@Hrs	Unit	Material Cost	Labor Cost	Installed Cost	

15A switches 120/277v, side-wire, ground screw, 1/2 HP at 120v, 2 HP at 240v, commercial

Material	Craft@Hrs	Unit	Material Cost	Labor Cost	Installed Cost	
Single pole, brown	1E@0.25	Ea	1.79	6.72	8.51	
Single pole, ivory	1E@0.25	Ea	1.79	6.72	8.51	S
Single pole, white	1E@0.25	Ea	1.79	6.72	8.51	
Double pole, brown	1E@0.30	Ea	5.43	8.06	13.49	
Double pole, ivory	1E@0.30	Ea	5.43	8.06	13.49	S_2
Double pole, white	1E@0.30	Ea	5.43	8.06	13.49	
Three-way, brown	1E@0.30	Ea	3.20	8.06	11.26	
Three-way, ivory	1E@0.30	Ea	3.20	8.06	11.26	
Three-way, white	1E@0.30	Ea	3.20	8.06	11.26	S_3
Four-way, brown	1E@0.40	Ea	8.01	10.80	18.81	
Four-way, ivory	1E@0.40	Ea	8.01	10.80	18.81	
Four-way, white	1E@0.40	Ea	8.01	10.80	18.81	S_4

15A switches 120/277v, back- & side-wire, ground screw, framed toggle, commercial

Material	Craft@Hrs	Unit	Material Cost	Labor Cost	Installed Cost	
Single pole, brown	1E@0.25	Ea	2.53	6.72	9.25	
Single pole, ivory	1E@0.25	Ea	2.53	6.72	9.25	S
Single pole, white	1E@0.25	Ea	2.53	6.72	9.25	
Double pole, brown	1E@0.30	Ea	10.90	8.06	18.96	
Double pole, ivory	1E@0.30	Ea	10.90	8.06	18.96	S_2
Double pole, white	1E@0.30	Ea	10.90	8.06	18.96	
Three-way, brown	1E@0.30	Ea	3.45	8.06	11.51	
Three-way, ivory	1E@0.30	Ea	3.45	8.06	11.51	
Three-way, white	1E@0.30	Ea	3.45	8.06	11.51	S_3
Four-way, brown	1E@0.40	Ea	20.90	10.80	31.70	
Four-way, ivory	1E@0.40	Ea	20.90	10.80	31.70	
Four-way, white	1E@0.40	Ea	20.90	10.80	31.70	S_4

15A switches 120/277v, lighted handle toggle, commercial grade

Material	Craft@Hrs	Unit	Material Cost	Labor Cost	Installed Cost	
Single pole, ivory	1E@0.25	Ea	6.92	6.72	13.64	S
Single pole, clear	1E@0.25	Ea	6.92	6.72	13.64	
Three-way, ivory	1E@0.30	Ea	11.10	8.06	19.16	
Three-way, clear	1E@0.30	Ea	11.10	8.06	19.16	S_3

Use these figures to estimate the cost of lighting switches installed in outlet boxes under the conditions described on pages 5 and 6. Costs listed are for each switch installed. The crew is one electrician working at a labor cost of $26.88 per manhour. These costs include mounting screws, layout, material handling, and normal waste. Add for the outlet box, trim plate, plaster ring, sales tax, delivery, supervision, mobilization, demobilization, cleanup, overhead and profit. Labor costs will be slightly lower when switches are ganged or when many switches are mounted on the same wall. Note: The copper and aluminum rated switches listed above are side wired. Be careful to select the switch size and type appropriate for the intended load and use. Switch designations are as follows: single pole is SP, double pole is DP, three-way is 3W, four-way is 4W, brown is BR, ivory is IV, white is W, red is R, gray is GY, locking is L.

Commercial Grade, 20 Amp AC Quiet Switches

Material	Craft@Hrs	Unit	Material Cost	Labor Cost	Installed Cost

20A switches 120/277v, side-wire, ground screw, 1 HP at 120v, 2 HP at 240v, commercial

Material	Craft@Hrs	Unit	Material Cost	Labor Cost	Installed Cost
Single pole, brown	1E@0.25	Ea	2.77	6.72	9.49
Single pole, ivory	1E@0.25	Ea	2.77	6.72	9.49
Single pole, white	1E@0.25	Ea	2.77	6.72	9.49
Double pole, brown	1E@0.30	Ea	8.32	8.06	16.38
Double pole, ivory	1E@0.30	Ea	8.32	8.06	16.38
Double pole, white	1E@0.30	Ea	8.32	8.06	16.38
Three-way, brown	1E@0.30	Ea	4.01	8.06	12.07
Three-way, ivory	1E@0.30	Ea	4.01	8.06	12.07
Three-way, white	1E@0.30	Ea	4.01	8.06	12.07
Four-way, brown	1E@0.40	Ea	15.70	10.80	26.50
Four-way, ivory	1E@0.40	Ea	15.70	10.80	26.50
Four-way, white	1E@0.40	Ea	15.70	10.80	26.50

20A switches 120/277v, side-wire, gr. screw, 1 HP at 120v, 2 HP at 240v, framed toggle, commercial

Material	Craft@Hrs	Unit	Material Cost	Labor Cost	Installed Cost
Single pole, brown	1E@0.25	Ea	2.90	6.72	9.62
Single pole, ivory	1E@0.25	Ea	2.90	6.72	9.62
Single pole, white	1E@0.25	Ea	2.90	6.72	9.62
Double pole, brown	1E@0.30	Ea	8.47	8.06	16.53
Double pole, ivory	1E@0.30	Ea	8.47	8.06	16.53
Double pole, white	1E@0.30	Ea	8.47	8.06	16.53
Three way, brown	1E@0.30	Ea	4.13	8.06	12.19
Three way, ivory	1E@0.30	Ea	4.13	8.06	12.19
Three way, white	1E@0.30	Ea	4.13	8.06	12.19
Four-way, brown	1E@0.40	Ea	15.90	10.80	26.70
Four-way, ivory	1E@0.40	Ea	15.90	10.80	26.70
Four-way, white	1E@0.40	Ea	15.90	10.80	26.70

20A switches 120/277v, back- & side-wire, 1 HP at 120v and 2 HP at 240v, commercial

Material	Craft@Hrs	Unit	Material Cost	Labor Cost	Installed Cost
Single pole, brown	1E@0.25	Ea	3.33	6.72	10.05
Single pole, ivory	1E@0.25	Ea	3.33	6.72	10.05
Single pole, white	1E@0.25	Ea	3.33	6.72	10.05
Double pole, brown	1E@0.30	Ea	12.20	8.06	20.26
Double pole, ivory	1E@0.30	Ea	12.20	8.06	20.26
Double-pole white	1E@0.30	Ea	12.20	8.06	20.26
Three-way, brown	1E@0.30	Ea	4.13	8.06	12.19
Three-way, ivory	1E@0.30	Ea	4.13	8.06	12.19
Three-way, white	1E@0.30	Ea	4.13	8.06	12.19
Four-way, brown	1E@0.40	Ea	22.50	10.80	33.30
Four-way, ivory	1E@0.40	Ea	22.50	10.80	33.30
Four-way, white	1E@0.40	Ea	22.50	10.80	33.30

Use these figures to estimate the cost of lighting switches installed in outlet boxes under the conditions described on pages 5 and 6. Costs listed are for each switch installed. The crew is one electrician working at a labor cost of $26.88 per manhour. These costs include mounting screws, layout, material handling, and normal waste. Add for the outlet box, trim plate, plaster ring, sales tax, delivery, supervision, mobilization, demobilization, cleanup, overhead and profit. Labor costs will be slightly lower when switches are ganged or when many switches are mounted on the same wall. Note: Prewired switches have pigtails that are ready for attachment to the lighting circuit. Be careful to select the switch size and type appropriate for the intended load and use. Switch designations are as follows: single pole is SP, double pole is DP, three-way is 3W, four-way is 4W, brown is BR, ivory is IV, white is W, red is R, gray is GY, locking is L.

Material	Craft@Hrs	Unit	Material Cost	Labor Cost	Installed Cost
15 amp AC key operated switches					
Single pole	1E@0.25	Ea	8.21	6.72	14.93
Three-way	1E@0.30	Ea	12.00	8.06	20.06
20 amp AC key operated switches					
Single pole	1E@0.25	Ea	12.60	6.72	19.32
Three pole	1E@0.30	Ea	14.30	8.06	22.36
30 amp AC key operated switches					
Single pole	1E@0.25	Ea	26.10	6.72	32.82
Three-way	1E@0.30	Ea	31.00	8.06	39.06
Four-way	1E@0.35	Ea	66.70	9.41	76.11
15 amp AC key operated switches, grounded					
Single pole	1E@0.25	Ea	13.60	6.72	20.32
Double pole	1E@0.30	Ea	18.40	8.06	26.46
Three-way	1E@0.30	Ea	16.00	8.06	24.06
Four-way	1E@0.35	Ea	37.90	9.41	47.31
20 amp AC key operated switches, grounded					
Single pole	1E@0.25	Ea	18.00	6.72	24.72
Double pole	1E@0.30	Ea	20.40	8.06	28.46
Three-way	1E@0.30	Ea	19.90	8.06	27.96
Four-way	1E@0.35	Ea	40.60	9.41	50.01
30 amp AC key operated switches, grounded					
Single pole	1E@0.25	Ea	26.10	6.72	32.82
Double pole	1E@0.30	Ea	31.60	8.06	39.66
Three-way	1E@0.30	Ea	31.00	8.06	39.06
Four-way	1E@0.35	Ea	66.70	9.41	76.11

Use these figures to estimate the cost of lighting switches installed in outlet boxes under the conditions described on pages 5 and 6. Costs listed are for each switch installed. The crew is one electrician working at a labor cost of $26.88 per manhour. These costs include mounting screws, layout, material handling, and normal waste. Add for the outlet box, trim plate, plaster ring, sales tax, delivery, supervision, mobilization, demobilization, cleanup, overhead and profit. Labor costs will be slightly lower when switches are ganged or when many switches are mounted on the same wall. Note: Be careful to select the switch size and type appropriate for the intended load and use. Switch designations are as follows: single pole is SP, double pole is DP, three-way is 3W, four-way is 4W, brown is BR, ivory is IV, white is W, red is R, gray is GY, locking is L.

AC Specialty Switches

Material	Craft@Hrs	Unit	Material Cost	Labor Cost	Installed Cost
15 amp duplex AC switches, single pole					
Brown	1E@0.30	Ea	7.98	8.06	16.04
Ivory	1E@0.30	Ea	8.02	8.06	16.08
White	1E@0.30	Ea	9.88	8.06	17.94
20 amp duplex AC switches, single pole					
Brown	1E@0.30	Ea	11.80	8.06	19.86
Ivory	1E@0.30	Ea	11.90	8.06	19.96
White	1E@0.30	Ea	13.20	8.06	21.26
15 amp duplex AC switches, three-way					
Brown	1E@0.35	Ea	14.10	9.41	23.51
Ivory	1E@0.35	Ea	14.20	9.41	23.61
White	1E@0.35	Ea	15.70	9.41	25.11
15 amp duplex AC switches, single pole and three-way					
Brown	1E@0.35	Ea	11.30	9.41	20.71
Ivory	1E@0.35	Ea	11.40	9.41	20.81
White	1E@0.35	Ea	12.60	9.41	22.01
15 amp duplex AC decorator switches					
Brown, single pole	1E@0.30	Ea	10.90	8.06	18.96
Ivory, single pole	1E@0.30	Ea	10.90	8.06	18.96
White, single pole	1E@0.30	Ea	10.90	8.06	18.96
Gray, single pole	1E@0.30	Ea	11.90	8.06	19.96
Brown, three-way	1E@0.35	Ea	16.50	9.41	25.91
Ivory, three-way	1E@0.35	Ea	16.50	9.41	25.91
White, three-way	1E@0.35	Ea	16.50	9.41	25.91
Gray, three-way	1E@0.35	Ea	16.90	9.41	26.31
Brown, SP & three-way	1E@0.35	Ea	16.40	9.41	25.81
Ivory, SP & three-way	1E@0.35	Ea	16.40	9.41	25.81
White, SP & three-way	1E@0.35	Ea	16.40	9.41	25.81
Gray, SP & three-way	1E@0.35	Ea	16.90	9.41	26.31

Use these figures to estimate the cost of lighting switches installed in outlet boxes under the conditions described on pages 5 and 6. Costs listed are for each switch installed. The crew is one electrician working at a labor cost of $26.88 per manhour. These costs include mounting screws, layout, material handling, and normal waste. Add for the outlet box, trim plate, plaster ring, sales tax, delivery, supervision, mobilization, demobilization, cleanup, overhead and profit. Labor costs will be slightly lower when switches are ganged or when many switches are mounted on the same wall. Note: Duplex switches require a duplex receptacle cover and can be wired for two separate circuits. Decorator switches require a special trim plate that fits both decorator trim plates and decorator receptacles. Be careful to select the switch size and type appropriate for the intended load and use. Switch designations are as follows: single pole is SP, double pole is DP, three-way is 3W, four-way is 4W, brown is BR, ivory is IV, white is W, red is R, gray is GY, locking is L.

Material	Craft@Hrs	Unit	Material Cost	Labor Cost	Installed Cost

30 amp, 120/277 volt momentary contact 3-position AC switches

Material	Craft@Hrs	Unit	Material Cost	Labor Cost	Installed Cost
Brown	1E@0.35	Ea	47.90	9.41	57.31
Ivory	1E@0.35	Ea	43.60	9.41	53.01

30 amp, 120/277 volt maintain contact 3-position AC switches

Material	Craft@Hrs	Unit	Material Cost	Labor Cost	Installed Cost
Brown	1E@0.35	Ea	53.50	9.41	62.91
Ivory	1E@0.35	Ea	53.70	9.41	63.11

20 amp, 120/277 volt interchangeable AC switches

Material	Craft@Hrs	Unit	Material Cost	Labor Cost	Installed Cost
Single pole, brown	1E@0.20	Ea	13.10	5.38	18.48
Single pole, ivory	1E@0.20	Ea	13.50	5.38	18.88
Double pole, brown	1E@0.25	Ea	20.90	6.72	27.62
Double pole, ivory	1E@0.25	Ea	21.50	6.72	28.22
Three-way, brown	1E@0.30	Ea	15.50	8.06	23.56
Three-way, ivory	1E@0.30	Ea	15.90	8.06	23.96
Four-way, brown	1E@0.30	Ea	37.60	8.06	45.66
Four-way, ivory	1E@0.30	Ea	37.10	8.06	45.16

15 amp, 120/277 volt AC decorator switches

Material	Craft@Hrs	Unit	Material Cost	Labor Cost	Installed Cost
Single pole, brown	1E@0.25	Ea	9.86	6.72	16.58
Single pole, ivory	1E@0.25	Ea	9.86	6.72	16.58
Single pole, white	1E@0.25	Ea	9.86	6.72	16.58
Single pole, gray	1E@0.25	Ea	9.86	6.72	16.58
Single pole, black	1E@0.25	Ea	10.80	6.72	17.52
Double pole, brown	1E@0.30	Ea	18.20	8.06	26.26
Double pole, ivory	1E@0.30	Ea	18.20	8.06	26.26
Double pole, white	1E@0.30	Ea	18.20	8.06	26.26
Double pole, gray	1E@0.30	Ea	18.20	8.06	26.26
Double pole, black	1E@0.30	Ea	18.20	8.06	26.26
Three-way, brown	1E@0.35	Ea	12.40	9.41	21.81
Three-way, ivory	1E@0.35	Ea	13.50	9.41	22.91
Three-way, white	1E@0.35	Ea	14.90	9.41	24.31
Three-way, gray	1E@0.35	Ea	14.90	9.41	24.31
Three-way, black	1E@0.35	Ea	14.90	9.41	24.31
Four-way, brown	1E@0.35	Ea	28.60	9.41	38.01
Four-way, ivory	1E@0.35	Ea	28.60	9.41	38.01
Four-way, white	1E@0.35	Ea	28.60	9.41	38.01
Four-way, gray	1E@0.35	Ea	28.60	9.41	38.01
Four-way, black	1E@0.35	Ea	28.60	9.41	38.01

Use these figures to estimate the cost of lighting switches installed in outlet boxes under the conditions described on pages 5 and 6. Costs listed are for each switch installed. The crew is one electrician working at a labor cost of $26.88 per manhour. These costs include mounting screws, layout, material handling, and normal waste. Add for the outlet box, trim plate, plaster ring, sales tax, delivery, supervision, mobilization, demobilization, cleanup, overhead and profit. Labor costs will be slightly lower when switches are ganged or when many switches are mounted on the same wall. Note: Be careful to select the switch size and type appropriate for the intended load and use. Switch designations are as follows: single pole is SP, double pole is DP, Three-way is 3W, Four-way is 4W, brown is BR, ivory is IV, white is W, red is R, gray is GY, locking is L, normally closed is NC, normally open is NO.

AC Specialty Switches

Material	Craft@Hrs	Unit	Material Cost	Labor Cost	Installed Cost
15 amp, 120 volt lighted handle AC decorator switches					
Single pole, ivory	1E@0.25	Ea	6.11	6.72	12.83
Single pole, white	1E@0.25	Ea	6.80	6.72	13.52
Three-way, ivory	1E@0.30	Ea	8.05	8.06	16.11
Three-way, white	1E@0.30	Ea	8.93	8.06	16.99
20 amp, 120 volt lighted handle AC decorator switches					
Single pole, ivory	1E@0.25	Ea	18.00	6.72	24.72
Single pole, white	1E@0.25	Ea	18.00	6.72	24.72
Three-way, ivory	1E@0.30	Ea	25.60	8.06	33.66
Three-way, white	1E@0.30	Ea	25.60	8.06	33.66
20 amp, 120/277 volt AC decorator switches					
Single pole, brown	1E@0.25	Ea	9.83	6.72	16.55
Single pole, ivory	1E@0.25	Ea	9.83	6.72	16.55
Single pole, white	1E@0.25	Ea	9.83	6.72	16.55
Double pole, brown	1E@0.30	Ea	18.20	8.06	26.26
Double pole, ivory	1E@0.30	Ea	18.20	8.06	26.26
Double pole, white	1E@0.30	Ea	18.20	8.06	26.26
Double pole, gray	1E@0.30	Ea	18.20	8.06	26.26
Double pole, black	1E@0.30	Ea	18.20	8.06	26.26
Three-way, brown	1E@0.30	Ea	4.38	8.06	12.44
Three-way, ivory	1E@0.30	Ea	4.38	8.06	12.44
Three-way, white	1E@0.30	Ea	4.38	8.06	12.44
Four-way brown	1E@0.35	Ea	17.50	9.41	26.91
Four-way, ivory	1E@0.35	Ea	17.50	9.41	26.91
Four-way, white	1E@0.35	Ea	17.50	9.41	26.91
Four-way, gray	1E@0.35	Ea	17.50	9.41	26.91
Four-way, black	1E@0.35	Ea	17.50	9.41	26.91
15 amp, 120 volt lighted handle AC touch switches					
Single pole, ivory	1E@0.25	Ea	6.11	6.72	12.83
Single pole, white	1E@0.25	Ea	6.80	6.72	13.52
Three-way, ivory	1E@0.30	Ea	8.05	8.06	16.11
Three-way, white	1E@0.30	Ea	8.93	8.06	16.99

Use these figures to estimate the cost of lighting switches installed in outlet boxes under the conditions described on pages 5 and 6. Costs listed are for each switch installed. The crew is one electrician working at a labor cost of $26.88 per manhour. These costs include mounting screws, layout, material handling, and normal waste. Add for the outlet box, trim plate, plaster ring, sales tax, delivery, supervision, mobilization, demobilization, cleanup, overhead and profit. Labor costs will be slightly lower when switches are ganged or when many switches are mounted on the same wall. Note: A standard trim plate will fit over a push switch. These switches can be ganged the same as any standard lighting switch. Be careful to select the switch size and type appropriate for the intended load and use. Switch designations are as follows: single pole is SP, double pole is DP, three-way is 3W, four-way is 4W, brown is BR, ivory is IV, white is W, red is R, gray is GY, locking is L.

Material	Craft@Hrs	Unit	Material Cost	Labor Cost	Installed Cost
15 amp, 120 volt duplex AC switches with red neon pilot					
Single pole, brown	1E@0.25	Ea	6.88	6.72	13.60
Single pole, ivory	1E@0.25	Ea	7.70	6.72	14.42
Single pole, white	1E@0.25	Ea	9.45	6.72	16.17
Three-way, brown	1E@0.30	Ea	10.80	8.06	18.86
Three-way, ivory	1E@0.30	Ea	10.90	8.06	18.96
Three-way, white	1E@0.30	Ea	11.00	8.06	19.06
15 amp, 120 volt duplex AC switches and receptacle					
Single pole, brown	1E@0.30	Ea	7.63	8.06	15.69
Single pole, ivory	1E@0.30	Ea	7.70	8.06	15.76
Single pole, white	1E@0.30	Ea	9.45	8.06	17.51
Three-way, brown	1E@0.35	Ea	12.30	9.41	21.71
Three-way, ivory	1E@0.35	Ea	12.50	9.41	21.91
Three-way, white	1E@0.35	Ea	12.50	9.41	21.91
20 amp duplex AC switches with grounding receptacle					
Single pole, brown	1E@0.30	Ea	11.30	8.06	19.36
Single pole, white	1E@0.30	Ea	12.60	8.06	20.66
15 amp, 125 volt ground fault interrupter duplex AC receptacles					
Ivory	1E@0.40	Ea	20.20	10.80	31.00
White	1E@0.40	Ea	20.20	10.80	31.00
Red, hospital	1E@0.40	Ea	54.80	10.80	65.60
15 amp, 125 volt hospital grade GFI AC receptacles					
Ivory	1E@0.40	Ea	40.30	10.80	51.10
White	1E@0.40	Ea	40.30	10.80	51.10
Gray	1E@0.40	Ea	40.30	10.80	51.10
Red	1E@0.40	Ea	40.30	10.80	51.10
20 amp, 125 volt hospital grade GFI AC receptacles					
Ivory	1E@0.40	Ea	42.90	10.80	53.70
White	1E@0.40	Ea	42.90	10.80	53.70
Gray	1E@0.40	Ea	42.90	10.80	53.70
Red	1E@0.40	Ea	43.80	10.80	54.60

Use these figures to estimate the cost of lighting switches installed in outlet boxes under the conditions described on pages 5 and 6. Costs listed are for each switch installed. The crew is one electrician working at a labor cost of $26.88 per manhour. These costs include mounting screws, layout, material handling, and normal waste. Add for the outlet box, trim plate, plaster ring, sales tax, delivery, supervision, mobilization, demobilization, cleanup, overhead and profit. Labor costs will be slightly lower when switches are ganged or when many switches are mounted on the same wall. Note: Duplex switch and receptacle combination devices can be wired so the switch controls the receptacle. Be careful to select the switch size and type appropriate for the intended load and use. Switch designations are as follows: single pole is SP, double pole is DP, three-way is 3W, four-way is 4W, brown is BR, ivory is IV, white is W, red is R, gray is GY, locking is L.

AC Specialty Switches

Material	Craft@Hrs	Unit	Material Cost	Labor Cost	Installed Cost
15 amp decorator AC switches with pilot light					
Single pole, brown	1E@0.25	Ea	11.90	6.72	18.62
Single pole, ivory	1E@0.25	Ea	11.90	6.72	18.62
Single pole, white	1E@0.25	Ea	11.90	6.72	18.62
Single pole, gray	1E@0.25	Ea	12.20	6.72	18.92
Three-way, brown	1E@0.30	Ea	13.90	8.06	21.96
Three-way, ivory	1E@0.30	Ea	13.90	8.06	21.96
Three-way, white	1E@0.30	Ea	13.90	8.06	21.96
Three-way, gray	1E@0.30	Ea	14.30	8.06	22.36
15 amp, 120 volt decorator duplex AC switches and receptacle					
Single pole, brown	1E@0.30	Ea	10.40	8.06	18.46
Single pole, ivory	1E@0.30	Ea	10.40	8.06	18.46
Single pole, white	1E@0.30	Ea	10.40	8.06	18.46
Single pole, gray	1E@0.30	Ea	10.40	8.06	18.46
Rotary 60 Hz incandescent AC dimmers					
Single pole, 600W	1E@0.25	Ea	27.40	6.72	34.12
Single pole, 1000W	1E@0.25	Ea	53.70	6.72	60.42
Single pole, 1500W	1E@0.25	Ea	115.00	6.72	121.72
Single pole, 2000W	1E@0.25	Ea	172.00	6.72	178.72
Rotary 60 Hz incandescent AC dimmers, push on-off					
Single pole, 600W	1E@0.25	Ea	30.20	6.72	36.92
Single pole, 1000W	1E@0.25	Ea	73.30	6.72	80.02
Three-way, 600W	1E@0.30	Ea	42.90	8.06	50.96
Three-way, 1000W	1E@0.30	Ea	84.10	8.06	92.16
Rotary 60 Hz low voltage incandescent AC dimmers					
Single pole, 600W	1E@0.25	Ea	112.00	6.72	118.72
Single pole, 1000W	1E@0.25	Ea	146.00	6.72	152.72
Single pole, 1500W	1E@0.25	Ea	244.00	6.72	250.72
Rotary 60 Hz lighted knob incandescent AC dimmers					
Single pole, 600W	1E@0.25	Ea	62.50	6.72	69.22
Single pole, 1000W	1E@0.25	Ea	41.00	6.72	47.72
Three-way, 600W	1E@0.30	Ea	68.40	8.06	76.46

Use these figures to estimate the cost of decorator duplex switches and receptacles installed in outlet boxes under the conditions described on pages 5 and 6. Costs listed are for each switch installed. The crew is one electrician working at a labor cost of $26.88 per manhour. These costs include mounting screws, layout, material handling, and normal waste. Add for the outlet box, trim plate, plaster ring, sales tax, delivery, supervision, mobilization, demobilization, cleanup, overhead and profit. Labor costs will be slightly lower when switches are ganged or when many switches are mounted on the same wall. Note: Decorator dimming switches can be ganged together. Be careful to select the switch size and type appropriate for the intended load and use. Switch designations are as follows: single pole is SP, double pole is DP, three-way is 3W, four-way is 4W, brown is BR, ivory is IV, white is W, red is R, gray is GY, locking is L.

Material	Craft@Hrs	Unit	Material Cost	Labor Cost	Installed Cost
Rotary 60 Hz medium voltage incandescent dimmers					
Single pole, 600W	1E@0.25	Ea	12.80	6.72	19.52
Three-way, 600W	1E@0.30	Ea	15.90	8.06	23.96
Rotary 60 Hz commercial grade incandescent dimmers					
Single pole, 600W	1E@0.25	Ea	23.40	6.72	30.12
Single pole, 1000W	1E@0.25	Ea	46.30	6.72	53.02
Single pole, 1500W	1E@0.25	Ea	85.00	6.72	91.72
Single pole, 2000W	1E@0.25	Ea	123.00	6.72	129.72
Slide control 120 volt 60 Hz incandescent dimming switches					
Single pole, 600W	1E@0.30	Ea	30.20	8.06	38.26
Single pole, 1000W	1E@0.30	Ea	72.40	8.06	80.46
Single pole, 1500W	1E@0.30	Ea	135.00	8.06	143.06
Single pole, 2000W	1E@0.30	Ea	191.00	8.06	199.06
Slide control 120 volt 60 Hz fluorescent dimming switches					
1 to 10 lamps	1E@0.35	Ea	112.00	9.41	121.41
6 to 20 lamps	1E@0.35	Ea	146.00	9.41	155.41
14 to 30 lamps	1E@0.35	Ea	263.00	9.41	272.41
Switchplate mounted 1000-watt tungsten photoelectric switches					
120 volt, AC	1E@0.30	Ea	20.70	8.06	28.76
208 volt, AC	1E@0.30	Ea	30.50	8.06	38.56
277 volt, AC	1E@0.30	Ea	30.50	8.06	38.56
15 amp, 1000-watt photoelectric switches with weatherproof enclosure					
120 volt, AC	1E@0.35	Ea	33.90	9.41	43.31
120 to 277 volt, AC	1E@0.35	Ea	46.30	9.41	55.71
10 amp 125 volt, 3 amp 250 volt AC door switches with flush enclosure					
Single pole	1E@0.50	Ea	16.30	13.40	29.70

Use these figures to estimate the cost of lighting switches installed in outlet boxes under the conditions described on pages 5 and 6. Costs listed are for each switch installed. The crew is one electrician working at a labor cost of $26.88 per manhour. These costs include mounting screws, layout, material handling, and normal waste. Add for the outlet box, trim plate, plaster ring, sales tax, delivery, supervision, mobilization, demobilization, cleanup, overhead and profit. Labor costs will be slightly lower when switches are ganged or when many switches are mounted on the same wall. Note: Be careful to select the switch size and type appropriate for the intended load and use. Switch designations are as follows: single pole is SP, double pole is DP, three-way is 3W, four-way is 4W, brown is BR, ivory is IV, white is W, red is R, gray is GY, locking is L.

15 Amp, 125 Volt AC Duplex Receptacles

Material	Craft@Hrs	Unit	Material Cost	Labor Cost	Installed Cost

15 amp 125v AC duplex receptacles, residential grade, push-in

Material	Craft@Hrs	Unit	Material Cost	Labor Cost	Installed Cost
Brown	1E@0.20	Ea	.53	5.38	5.91
Ivory	1E@0.20	Ea	.53	5.38	5.91
White	1E@0.20	Ea	.53	5.38	5.91

15 amp 125v AC duplex receptacles, residential grade, push-in and side-wire

Material	Craft@Hrs	Unit	Material Cost	Labor Cost	Installed Cost
Brown	1E@0.20	Ea	.53	5.38	5.91
Ivory	1E@0.20	Ea	.53	5.38	5.91
White	1E@0.20	Ea	.53	5.38	5.91

15 amp 125v decorator AC duplex receptacles, residential grade, push-in and side-wire

Material	Craft@Hrs	Unit	Material Cost	Labor Cost	Installed Cost
Brown	1E@0.20	Ea	2.31	5.38	7.69
Ivory	1E@0.20	Ea	2.31	5.38	7.69
White	1E@0.20	Ea	2.31	5.38	7.69
Gray	1E@0.20	Ea	2.31	5.38	7.69
Almond	1E@0.20	Ea	2.31	5.38	7.69
Black	1E@0.20	Ea	2.71	5.38	8.09

15 amp 125v AC duplex receptacles, commercial grade, side-wire

Material	Craft@Hrs	Unit	Material Cost	Labor Cost	Installed Cost
Brown	1E@0.20	Ea	1.28	5.38	6.66
Ivory	1E@0.20	Ea	1.28	5.38	6.66
White	1E@0.20	Ea	1.28	5.38	6.66

15 amp 125v AC duplex receptacles, commercial grade, smooth face, side-wire

Material	Craft@Hrs	Unit	Material Cost	Labor Cost	Installed Cost
Brown	1E@0.20	Ea	1.28	5.38	6.66
Ivory	1E@0.20	Ea	1.28	5.38	6.66
White	1E@0.20	Ea	1.28	5.38	6.66
Gray	1E@0.20	Ea	1.28	5.38	6.66

15 amp 125v AC duplex receptacles, commercial grade, back and side-wire

Material	Craft@Hrs	Unit	Material Cost	Labor Cost	Installed Cost
Brown	1E@0.20	Ea	2.24	5.38	7.62
Ivory	1E@0.20	Ea	2.24	5.38	7.62
White	1E@0.20	Ea	2.24	5.38	7.62
Gray	1E@0.20	Ea	2.24	5.38	7.62

15 amp 125v AC duplex receptacles, industrial grade, back and side-wire

Material	Craft@Hrs	Unit	Material Cost	Labor Cost	Installed Cost
Brown	1E@0.20	Ea	3.84	5.38	9.22
Ivory	1E@0.20	Ea	3.84	5.38	9.22
White	1E@0.20	Ea	3.84	5.38	9.22
Gray	1E@0.20	Ea	3.84	5.38	9.22

Use these figures to estimate the cost of power receptacles installed in outlet boxes under the conditions described on pages 5 and 6. Costs listed are for each receptacle installed. The crew is one electrician working at a labor cost of $26.88 per manhour. These costs include mounting screws, layout, material handling, and normal waste. Add for the outlet box, trim plate, plaster ring, sales tax, delivery, supervision, mobilization, demobilization, cleanup, overhead and profit. Note: Be careful to select the receptacle size and type appropriate for the intended load and voltage. Receptacles with special contact materials designed for durability under heavy use can be considerably more expensive than standard grade receptacles.

20 Amp, 125 Volt AC Duplex Receptacles

Material	Craft@Hrs	Unit	Material Cost	Labor Cost	Installed Cost

20 amp 125v AC duplex receptacles commercial grade, side-wire

Material	Craft@Hrs	Unit	Material Cost	Labor Cost	Installed Cost
Brown	1E@0.20	Ea	1.92	5.38	7.30
Ivory	1E@0.20	Ea	1.92	5.38	7.30
White	1E@0.20	Ea	1.92	5.38	7.30
Gray	1E@0.20	Ea	1.92	5.38	7.30

20 amp 125v AC duplex receptacles, commercial grade, smooth face, side-wire

Material	Craft@Hrs	Unit	Material Cost	Labor Cost	Installed Cost
Brown	1E@0.20	Ea	1.92	5.38	7.30
Ivory	1E@0.20	Ea	1.92	5.38	7.30
White	1E@0.20	Ea	1.92	5.38	7.30
Gray	1E@0.20	Ea	1.92	5.38	7.30

20 amp 125v AC duplex receptacles, industrial grade, side-wire

Material	Craft@Hrs	Unit	Material Cost	Labor Cost	Installed Cost
Brown	1E@0.20	Ea	5.71	5.38	11.09
Ivory	1E@0.20	Ea	5.71	5.38	11.09
White	1E@0.20	Ea	5.71	5.38	11.09
Gray	1E@0.20	Ea	5.71	5.38	11.09
Red	1E@0.20	Ea	5.71	5.38	11.09

20 amp 125v AC duplex receptacles, industrial grade, back and side-wire

Material	Craft@Hrs	Unit	Material Cost	Labor Cost	Installed Cost
Brown	1E@0.20	Ea	11.00	5.38	16.38
Ivory	1E@0.20	Ea	11.00	5.38	16.38
White	1E@0.20	Ea	11.00	5.38	16.38
Gray	1E@0.20	Ea	11.00	5.38	16.38
Red	1E@0.20	Ea	11.00	5.38	16.38

20 amp 125v AC duplex receptacles, hospital grade, side-wire

Material	Craft@Hrs	Unit	Material Cost	Labor Cost	Installed Cost
Brown	1E@0.20	Ea	16.20	5.38	21.58
Ivory	1E@0.20	Ea	16.20	5.38	21.58
Red	1E@0.20	Ea	16.20	5.38	21.58

20 amp 125v AC duplex receptacles, hospital grade, back and side-wire

Material	Craft@Hrs	Unit	Material Cost	Labor Cost	Installed Cost
Brown	1E@0.20	Ea	18.60	5.38	23.98
Ivory	1E@0.20	Ea	18.60	5.38	23.98
Red	1E@0.20	Ea	18.60	5.38	23.98

Use these figures to estimate the cost of power receptacles installed in outlet boxes under the conditions described on pages 5 and 6. Costs listed are for each receptacle installed. The crew is one electrician working at a labor cost of $26.88 per manhour. These costs include mounting screws, layout, material handling, and normal waste. Add for the outlet box, trim plate, plaster ring, sales tax, delivery, supervision, mobilization, demobilization, cleanup, overhead and profit. Note: Be careful to select the receptacle size and type appropriate for the intended load and voltage. Receptacles with special contact materials designed for durability under heavy use can be considerably more expensive than standard grade receptacles.

15 Amp, 125 Volt AC Decorator Duplex Receptacles

Material	Craft@Hrs	Unit	Material Cost	Labor Cost	Installed Cost

15 amp 125v AC duplex receptacles, commercial grade, side-wire, self grounding

Material	Craft@Hrs	Unit	Material Cost	Labor Cost	Installed Cost
Brown	1E@0.20	Ea	4.04	5.38	9.42
Ivory	1E@0.20	Ea	4.04	5.38	9.42
White	1E@0.20	Ea	4.04	5.38	9.42
Gray	1E@0.20	Ea	4.04	5.38	9.42
Black	1E@0.20	Ea	4.84	5.38	10.22

15 amp 125v AC duplex receptacles, commercial grade, back and side-wire, self grounding

Material	Craft@Hrs	Unit	Material Cost	Labor Cost	Installed Cost
Brown	1E@0.20	Ea	5.82	5.38	11.20
Ivory	1E@0.20	Ea	5.82	5.38	11.20
White	1E@0.20	Ea	5.82	5.38	11.20
Gray	1E@0.20	Ea	5.82	5.38	11.20
Almond	1E@0.20	Ea	6.99	5.38	12.37
Black	1E@0.20	Ea	6.99	5.38	12.37

15 amp 125v AC duplex receptacles, industrial grade, back & side-wire

Material	Craft@Hrs	Unit	Material Cost	Labor Cost	Installed Cost
Brown	1E@0.20	Ea	13.80	5.38	19.18
Ivory	1E@0.20	Ea	13.80	5.38	19.18
White	1E@0.20	Ea	13.80	5.38	19.18
Gray	1E@0.20	Ea	13.80	5.38	19.18

15 amp 125v AC duplex receptacles, hospital grade, back & side-wire

Material	Craft@Hrs	Unit	Material Cost	Labor Cost	Installed Cost
Brown	1E@0.20	Ea	16.70	5.38	22.08
Ivory	1E@0.20	Ea	16.70	5.38	22.08
White	1E@0.20	Ea	16.70	5.38	22.08
Gray	1E@0.20	Ea	16.70	5.38	22.08
Red	1E@0.20	Ea	16.70	5.38	22.08

15 amp 125v AC duplex receptacles, hospital grade, back and side-wire, isolated ground

Material	Craft@Hrs	Unit	Material Cost	Labor Cost	Installed Cost
Brown	1E@0.20	Ea	17.20	5.38	22.58
Ivory	1E@0.20	Ea	17.20	5.38	22.58
White	1E@0.20	Ea	17.20	5.38	22.58
Gray	1E@0.20	Ea	17.20	5.38	22.58
Red	1E@0.20	Ea	17.20	5.38	22.58
Orange	1E@0.20	Ea	17.20	5.38	22.58

15 amp 125v AC duplex receptacles, hospital grade, back and side-wire, tamper resistant

Material	Craft@Hrs	Unit	Material Cost	Labor Cost	Installed Cost
Brown	1E@0.25	Ea	21.40	6.72	28.12
Ivory	1E@0.25	Ea	21.40	6.72	28.12
White	1E@0.25	Ea	21.40	6.72	28.12
Gray	1E@0.25	Ea	21.40	6.72	28.12
Red	1E@0.25	Ea	21.40	6.72	28.12

Use these figures to estimate the cost of power receptacles installed in outlet boxes under the conditions described on pages 5 and 6. Costs listed are for each receptacle installed. The crew is one electrician working at a labor cost of $26.88 per manhour. These costs include mounting screws, layout, material handling, and normal waste. Add for the outlet box, trim plate, plaster ring, sales tax, delivery, supervision, mobilization, demobilization, cleanup, overhead and profit. Note: Be careful to select the receptacle size and type appropriate for the intended load and voltage. Receptacles with special contact materials designed for durability under heavy use can be considerably more expensive than standard grade receptacles.

20 Amp, 125 Volt AC Decorator Duplex Receptacles

Material	Craft@Hrs	Unit	Material Cost	Labor Cost	Installed Cost

20 amp 125v AC duplex receptacles, commercial grade, side-wire, self grounding

Material	Craft@Hrs	Unit	Material Cost	Labor Cost	Installed Cost
Brown	1E@0.20	Ea	6.61	5.38	11.99
Ivory	1E@0.20	Ea	6.61	5.38	11.99
White	1E@0.20	Ea	6.61	5.38	11.99
Gray	1E@0.20	Ea	6.61	5.38	11.99
Black	1E@0.20	Ea	6.61	5.38	11.99

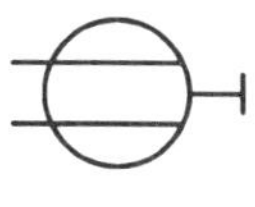

20 amp 125v AC duplex receptacles, commercial grade, back and side-wire, self grounding

Material	Craft@Hrs	Unit	Material Cost	Labor Cost	Installed Cost
Brown	1E@0.20	Ea	9.98	5.38	15.36
Ivory	1E@0.20	Ea	9.98	5.38	15.36
White	1E@0.20	Ea	9.98	5.38	15.36
Gray	1E@0.20	Ea	9.98	5.38	15.36
Almond	1E@0.20	Ea	9.98	5.38	15.36
Black	1E@0.20	Ea	9.98	5.38	15.36

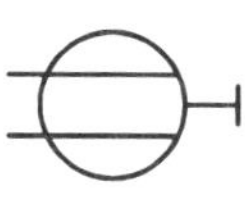

20 amp 125v AC decorator duplex receptacles, industrial grade, back and side-wire

Material	Craft@Hrs	Unit	Material Cost	Labor Cost	Installed Cost
Brown	1E@0.20	Ea	16.90	5.38	22.28
Ivory	1E@0.20	Ea	16.90	5.38	22.28
White	1E@0.20	Ea	16.90	5.38	22.28
Gray	1E@0.20	Ea	16.90	5.38	22.28

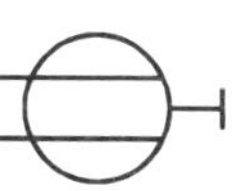

20 amp 125v AC decorator duplex receptacles, hospital grade, back and side-wire

Material	Craft@Hrs	Unit	Material Cost	Labor Cost	Installed Cost
Brown	1E@0.20	Ea	19.60	5.38	24.98
Ivory	1E@0.20	Ea	19.60	5.38	24.98
White	1E@0.20	Ea	19.60	5.38	24.98
Gray	1E@0.20	Ea	19.60	5.38	24.98
Red	1E@0.20	Ea	19.60	5.38	24.98

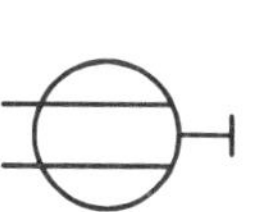

20 amp 125v AC duplex receptacles, hospital grade, back and side-wire, isolated ground

Material	Craft@Hrs	Unit	Material Cost	Labor Cost	Installed Cost
Brown	1E@0.20	Ea	29.30	5.38	34.68
Ivory	1E@0.20	Ea	20.30	5.38	25.68
White	1E@0.20	Ea	20.30	5.38	25.68
Gray	1E@0.20	Ea	20.30	5.38	25.68
Red	1E@0.20	Ea	20.30	5.38	25.68
Orange	1E@0.20	Ea	14.60	5.38	19.98

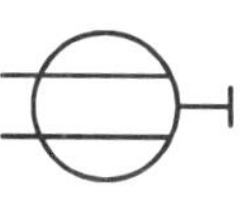

Use these figures to estimate the cost of power receptacles installed in outlet boxes under the conditions described on pages 5 and 6. Costs listed are for each receptacle installed. The crew is one electrician working at a labor cost of $26.88 per manhour. These costs include mounting screws, layout, material handling, and normal waste. Add for the outlet box, trim plate, plaster ring, sales tax, delivery, supervision, mobilization, demobilization, cleanup, overhead and profit. Note: Be careful to select the receptacle size and type appropriate for the intended load and voltage. Receptacles with special contact materials designed for durability under heavy use can be considerably more expensive than standard grade receptacles.

Special Purpose Receptacles

Material	Craft@Hrs	Unit	Material Cost	Labor Cost	Installed Cost
15 amp, 125 volt single receptacles					
Brown	1E@0.20	Ea	2.43	5.38	7.81
Ivory	1E@0.20	Ea	2.48	5.38	7.86
White	1E@0.20	Ea	2.75	5.38	8.13
15 amp, 250 volt single receptacles					
Brown	1E@0.20	Ea	2.69	5.38	8.07
Ivory	1E@0.20	Ea	2.73	5.38	8.11
White	1E@0.20	Ea	2.92	5.38	8.30
20 amp, 125 volt single receptacles					
Brown	1E@0.20	Ea	3.13	5.38	8.51
Ivory	1E@0.20	Ea	3.19	5.38	8.57
White	1E@0.20	Ea	3.23	5.38	8.61
20 amp, 250 volt single receptacles					
Brown	1E@0.20	Ea	3.96	5.38	9.34
Ivory	1E@0.20	Ea	4.00	5.38	9.38
White	1E@0.20	Ea	4.95	5.38	10.33
15 amp, 125 volt recessed U-ground clock hanger receptacles					
Brown	1E@0.25	Ea	4.58	6.72	11.30
Ivory	1E@0.25	Ea	4.70	6.72	11.42
Brass	1E@0.25	Ea	4.68	6.72	11.40
Stainless steel	1E@0.25	Ea	11.10	6.72	17.82
Ivory Bakelite	1E@0.25	Ea	4.75	6.72	11.47
15 amp, 125 volt isolated ground duplex receptacles, orange face					
Locking type	1E@0.20	Ea	19.10	5.38	24.48
Straight blade	1E@0.20	Ea	17.20	5.38	22.58
Hospital grade	1E@0.20	Ea	19.10	5.38	24.48
15 amp, 125 volt isolated ground single receptacles, orange face					
Locking type	1E@0.20	Ea	18.50	5.38	23.88
Straight blade	1E@0.20	Ea	13.60	5.38	18.98
Hospital grade	1E@0.20	Ea	18.50	5.38	23.88

Use these figures to estimate the cost of power receptacles installed in outlet boxes under the conditions described on pages 5 and 6. Costs listed are for each receptacle installed. The crew is one electrician working at a labor cost of $26.88 per manhour. These costs include mounting screws, layout, material handling, and normal waste. Add for the outlet box, trim plate, plaster ring, sales tax, delivery, supervision, mobilization, demobilization, cleanup, overhead and profit. Note: Orange receptacles have an isolated ground and are intended for use where the electrical system ground and the ground for the receptacle must be separate. For example, receptacles used by data processing equipment may require an isolated ground. Be careful to select the receptacle size and type appropriate for the load and voltage. Receptacles with special contact materials designed for durability under heavy use can be considerably more expensive than standard grade receptacles. 15 amp, 125 volt 3-wire grounding and locking duplex receptacles

Material	Craft@Hrs	Unit	Material Cost	Labor Cost	Installed Cost
15 amp,125 volt 3-wire grounding and locking duplex receptacles					
Brown	1E@0.25	Ea	18.20	6.72	24.92
15 amp, 3-wire locking duplex receptacles					
Brown, 125 volt	1E@0.25	Ea	14.80	6.72	21.52
Ivory, 125 volt	1E@0.25	Ea	17.50	6.72	24.22
Orange, 125 volt	1E@0.25	Ea	20.20	6.72	26.92
Yellow, 125 volt	1E@0.25	Ea	24.70	6.72	31.42
Brown, 250 volt	1E@0.25	Ea	14.80	6.72	21.52
Orange, 250 volt	1E@0.25	Ea	22.20	6.72	28.92
15 amp, 3-wire locking single receptacles					
Brown, 125 volt	1E@0.25	Ea	9.72	6.72	16.44
Orange, 125 volt	1E@0.25	Ea	19.10	6.72	25.82
Brown, 250 volt	1E@0.25	Ea	9.99	6.72	16.71
Orange, 250 volt	1E@0.25	Ea	19.30	6.72	26.02
Brown, 277 volt	1E@0.25	Ea	10.90	6.72	17.62
20 amp, 3-wire locking single receptacles					
Black, 125 volt	1E@0.25	Ea	11.50	6.72	18.22
Orange, 125 volt	1E@0.25	Ea	14.30	6.72	21.02
Yellow, 125 volt	1E@0.25	Ea	15.40	6.72	22.12
Black, 250 volt	1E@0.25	Ea	11.30	6.72	18.02
Orange, 250 volt	1E@0.25	Ea	14.60	6.72	21.32
Black, 277 volt	1E@0.25	Ea	11.30	6.72	18.02
Orange, 277 volt	1E@0.25	Ea	16.60	6.72	23.32
Black, 600 volt	1E@0.25	Ea	18.40	6.72	25.12
30 amp, 3-wire locking single receptacles					
Black, 125 volt	1E@0.30	Ea	16.40	8.06	24.46
Orange, 125 volt	1E@0.30	Ea	19.90	8.06	27.96
Yellow, 125 volt	1E@0.30	Ea	20.00	8.06	28.06
Black, 250 volt	1E@0.30	Ea	17.50	8.06	25.56
Orange, 250 volt	1E@0.30	Ea	23.50	8.06	31.56
Yellow, 250 volt	1E@0.30	Ea	17.90	8.06	25.96
Black, 277 volt	1E@0.30	Ea	17.50	8.06	25.56
Black, 480 volt	1E@0.30	Ea	19.90	8.06	27.96
Black, 600 volt	1E@0.30	Ea	19.90	8.06	27.96

Use these figures to estimate the cost of power receptacles installed in outlet boxes under the conditions described on pages 5 and 6. Costs listed are for each receptacle installed. The crew is one electrician working at a labor cost of $26.88 per manhour. These costs include mounting screws, layout, material handling, and normal waste. Add for the outlet box, trim plate, plaster ring, sales tax, delivery, supervision, mobilization, demobilization, cleanup, overhead and profit. Note: Locking receptacles are sometimes called "Twist Lock" receptacles. They're generally used on ceiling drops and also for construction extension cords. Wall mounted locking receptacles are used in shops to connect portable equipment. Be careful to select the receptacle size and type appropriate for the load and voltage.

Single Receptacles

Material	Craft@Hrs	Unit	Material Cost	Labor Cost	Installed Cost
50 amp, 3-wire locking single receptacles					
Black, 125 volt	1E@0.30	Ea	28.90	8.06	36.96
Black, 250 volt	1E@0.30	Ea	29.40	8.06	37.46
Black, 480 volt	1E@0.30	Ea	29.50	8.06	37.56
20 amp, 4-wire locking single receptacles					
Black, 125/250 volt	1E@0.35	Ea	13.00	9.41	22.41
Orange, 125/250 volt	1E@0.35	Ea	17.50	9.41	26.91
Black, 250 volt	1E@0.35	Ea	13.00	9.41	22.41
Orange, 250 volt	1E@0.35	Ea	17.50	9.41	26.91
Black, 480 volt	1E@0.35	Ea	14.90	9.41	24.31
Orange, 480 volt	1E@0.35	Ea	17.80	9.41	27.21
30 amp, 4-wire locking single receptacles					
Black, 125/250 volt	1E@0.35	Ea	19.70	9.41	29.11
Orange, 125/250 volt	1E@0.35	Ea	26.60	9.41	36.01
Black, 250 volt	1E@0.35	Ea	19.70	9.41	29.11
Orange, 250 volt	1E@0.35	Ea	30.40	9.41	39.81
Black, 480 volt	1E@0.35	Ea	22.60	9.41	32.01
50 amp, 4-wire locking single receptacles					
Black, 125/250 volt	1E@0.40	Ea	31.60	10.80	42.40
Black, 250 volt	1E@0.40	Ea	32.20	10.80	43.00
20 amp, 5-wire locking single receptacles					
Black, 120/208 volt	1E@0.35	Ea	15.90	9.41	25.31
Orange, 120/208 volt	1E@0.35	Ea	20.90	9.41	30.31
Black, 277/480 volt	1E@0.35	Ea	18.20	9.41	27.61
Orange, 277/480 volt	1E@0.35	Ea	21.70	9.41	31.11
Black, 600 volt	1E@0.35	Ea	18.20	9.41	27.61
Orange, 600 volt	1E@0.35	Ea	21.70	9.41	31.11
30 amp, 5-wire locking single receptacles					
Black, 120/208 volt	1E@0.40	Ea	22.00	10.80	32.80
Orange, 120/208 volt	1E@0.40	Ea	29.80	10.80	40.60
Black, 277/480 volt	1E@0.40	Ea	25.40	10.80	36.20
Black, 600 volt	1E@0.40	Ea	29.00	10.80	39.80

Use these figures to estimate the cost of power receptacles installed in outlet boxes under the conditions described on pages 5 and 6. Costs listed are for each receptacle installed. The crew is one electrician working at a labor cost of $26.88 per manhour. These costs include mounting screws, layout, material handling, and normal waste. Add for the outlet box, switch plaster ring, sales tax, delivery, supervision, mobilization, demobilization, cleanup, overhead and profit. Note: Special receptacles designed to meet OSHA requirements are available from some manufacturers. There's a special configuration for each voltage. Be careful to select the receptacle size and type appropriate for the load and voltage. Receptacles with special contact materials designed for durability under heavy use can be considerably more expensive than standard grade receptacles.

Material	Craft@Hrs	Unit	Material Cost	Labor Cost	Installed Cost
Switch cover plates					
1 gang brown	1E@0.05	Ea	.27	1.34	1.61
1 gang ivory	1E@0.05	Ea	.27	1.34	1.61
1 gang gray	1E@0.05	Ea	.30	1.34	1.64
1 gang white	1E@0.05	Ea	.30	1.34	1.64
1 gang red	1E@0.05	Ea	.30	1.34	1.64
1 gang black	1E@0.05	Ea	.30	1.34	1.64
1 gang brass	1E@0.05	Ea	3.28	1.34	4.62
1 gang aluminum	1E@0.05	Ea	1.66	1.34	3.00
1 gang stainless steel	1E@0.05	Ea	1.32	1.34	2.66
2 gang brown	1E@0.10	Ea	.56	2.69	3.25
2 gang ivory	1E@0.10	Ea	.56	2.69	3.25
2 gang gray	1E@0.10	Ea	.61	2.69	3.30
2 gang white	1E@0.10	Ea	.61	2.69	3.30
2 gang red	1E@0.10	Ea	.61	2.69	3.30
2 gang black	1E@0.10	Ea	.61	2.69	3.30
2 gang brass	1E@0.10	Ea	7.30	2.69	9.99
2 gang aluminum	1E@0.10	Ea	3.28	2.69	5.97
2 gang stainless steel	1E@0.10	Ea	2.51	2.69	5.20
3 gang brown	1E@0.15	Ea	.82	4.03	4.85
3 gang ivory	1E@0.15	Ea	.82	4.03	4.85
3 gang gray	1E@0.15	Ea	.91	4.03	4.94
3 gang white	1E@0.15	Ea	.91	4.03	4.94
3 gang red	1E@0.15	Ea	.91	4.03	4.94
3 gang black	1E@0.15	Ea	.91	4.03	4.94
3 gang brass	1E@0.15	Ea	11.30	4.03	15.33
3 gang aluminum	1E@0.15	Ea	5.62	4.03	9.65
3 gang stainless steel	1E@0.15	Ea	3.78	4.03	7.81
4 gang brown	1E@0.20	Ea	1.24	5.38	6.62
4 gang ivory	1E@0.20	Ea	1.24	5.38	6.62
4 gang gray	1E@0.20	Ea	1.40	5.38	6.78
4 gang white	1E@0.20	Ea	1.40	5.38	6.78
4 gang red	1E@0.20	Ea	1.40	5.38	6.78
4 gang black	1E@0.20	Ea	1.40	5.38	6.78
4 gang brass	1E@0.20	Ea	16.90	5.38	22.28
4 gang aluminum	1E@0.20	Ea	8.50	5.38	13.88
4 gang stainless steel	1E@0.20	Ea	6.43	5.38	11.81

Use these figures to estimate the cost of trim plates installed on outlet boxes under the conditions described on pages 5 and 6. Costs listed are for each plate installed. The crew is one electrician working at a labor cost of $26.88 per manhour. These costs include layout, material handling, and normal waste. Add for the outlet box, switch, plaster ring, sales tax, delivery, supervision, mobilization, demobilization, cleanup, overhead and profit. Note: Be careful to select the right plate color and material. Special plates with non-standard configurations can be ordered for custom applications.

Wiring Device Plates

Material	Craft@Hrs	Unit	Material Cost	Labor Cost	Installed Cost
Switch cover plates					
5 gang brown	1E@0.25	Ea	2.29	6.72	9.01
5 gang ivory	1E@0.25	Ea	2.29	6.72	9.01
5 gang gray	1E@0.25	Ea	2.59	6.72	9.31
5 gang white	1E@0.25	Ea	2.59	6.72	9.31
5 gang red	1E@0.25	Ea	2.59	6.72	9.31
5 gang black	1E@0.25	Ea	2.59	6.72	9.31
5 gang brass	1E@0.25	Ea	21.00	6.72	27.72
5 gang aluminum	1E@0.25	Ea	10.60	6.72	17.32
5 gang stainless steel	1E@0.25	Ea	8.03	6.72	14.75
6 gang brown	1E@0.30	Ea	2.68	8.06	10.74
6 gang ivory	1E@0.30	Ea	2.68	8.06	10.74
6 gang gray	1E@0.30	Ea	3.05	8.06	11.11
6 gang white	1E@0.30	Ea	3.05	8.06	11.11
6 gang red	1E@0.30	Ea	3.05	8.06	11.11
6 gang black	1E@0.30	Ea	3.05	8.06	11.11
6 gang brass	1E@0.30	Ea	24.60	8.06	32.66
6 gang aluminum	1E@0.30	Ea	14.90	8.06	22.96
6 gang stainless steel	1E@0.30	Ea	9.76	8.06	17.82
Duplex receptacle cover plates					
1 gang brown	1E@0.05	Ea	.27	1.34	1.61
1 gang ivory	1E@0.05	Ea	.27	1.34	1.61
1 gang gray	1E@0.05	Ea	.30	1.34	1.64
1 gang white	1E@0.05	Ea	.30	1.34	1.64
1 gang red	1E@0.05	Ea	.30	1.34	1.64
1 gang black	1E@0.05	Ea	.30	1.34	1.64
1 gang brass	1E@0.05	Ea	3.28	1.34	4.62
1 gang aluminum	1E@0.05	Ea	1.66	1.34	3.00
1 gang stainless steel	1E@0.05	Ea	1.32	1.34	2.66
2 gang brown	1E@0.10	Ea	.61	2.69	3.30
2 gang ivory	1E@0.10	Ea	.61	2.69	3.30
2 gang gray	1E@0.10	Ea	.71	2.69	3.40
2 gang white	1E@0.10	Ea	.71	2.69	3.40
2 gang red	1E@0.10	Ea	.71	2.69	3.40
2 gang black	1E@0.10	Ea	.71	2.69	3.40
2 gang brass	1E@0.10	Ea	8.19	2.69	10.88
2 gang aluminum	1E@0.10	Ea	7.81	2.69	10.50
2 gang stainless steel	1E@0.10	Ea	3.11	2.69	5.80

Use these figures to estimate the cost of trim plates installed on outlet boxes under the conditions described on pages 5 and 6. Costs listed are for each plate installed. The crew is one electrician working at a labor cost of $26.88 per manhour. These costs include layout, material handling, and normal waste. Add for the outlet box, switch, plaster ring, sales tax, delivery, supervision, mobilization, demobilization, cleanup, overhead and profit. Note: Be careful to select the right plate color and material. Special plates with non-standard configurations can be ordered for custom applications.

Material	Craft@Hrs	Unit	Material Cost	Labor Cost	Installed Cost

Combination switch and duplex receptacle cover plates

Material	Craft@Hrs	Unit	Material Cost	Labor Cost	Installed Cost
2 gang brown	1E@0.10	Ea	.56	2.69	3.25
2 gang ivory	1E@0.10	Ea	.56	2.69	3.25
2 gang gray	1E@0.10	Ea	.61	2.69	3.30
2 gang white	1E@0.10	Ea	.61	2.69	3.30
2 gang red	1E@0.10	Ea	.61	2.69	3.30
2 gang black	1E@0.10	Ea	.61	2.69	3.30
2 gang brass	1E@0.10	Ea	7.56	2.69	10.25
2 gang aluminum	1E@0.10	Ea	3.77	2.69	6.46
2 gang stainless steel	1E@0.10	Ea	2.74	2.69	5.43
3 gang brown	1E@0.15	Ea	.98	4.03	5.01
3 gang ivory	1E@0.15	Ea	.98	4.03	5.01
3 gang gray	1E@0.15	Ea	1.09	4.03	5.12
3 gang white	1E@0.15	Ea	1.09	4.03	5.12
3 gang red	1E@0.15	Ea	1.09	4.03	5.12
3 gang black	1E@0.15	Ea	1.09	4.03	5.12
3 gang brass	1E@0.15	Ea	11.30	4.03	15.33
3 gang aluminum	1E@0.15	Ea	6.36	4.03	10.39
3 gang stainless steel	1E@0.15	Ea	4.21	4.03	8.24

Single receptacle cover plates

Material	Craft@Hrs	Unit	Material Cost	Labor Cost	Installed Cost
1 gang brown	1E@0.05	Ea	.28	1.34	1.62
1 gang ivory	1E@0.05	Ea	.28	1.34	1.62
1 gang gray	1E@0.05	Ea	.32	1.34	1.66
1 gang white	1E@0.05	Ea	.32	1.34	1.66
1 gang red	1E@0.05	Ea	.32	1.34	1.66
1 gang black	1E@0.05	Ea	.32	1.34	1.66
1 gang brass	1E@0.05	Ea	3.66	1.34	5.00
1 gang aluminum	1E@0.05	Ea	1.70	1.34	3.04
1 gang stainless steel	1E@0.05	Ea	1.42	1.34	2.76

Blank wiring device cover plates

Material	Craft@Hrs	Unit	Material Cost	Labor Cost	Installed Cost
1 gang brown	1E@0.05	Ea	.35	1.34	1.69
1 gang ivory	1E@0.05	Ea	.35	1.34	1.69
1 gang gray	1E@0.05	Ea	.42	1.34	1.76
1 gang white	1E@0.05	Ea	.42	1.34	1.76
1 gang red	1E@0.05	Ea	.42	1.34	1.76
1 gang black	1E@0.05	Ea	.42	1.34	1.76
1 gang brass	1E@0.05	Ea	3.96	1.34	5.30
1 gang aluminum	1E@0.05	Ea	1.73	1.34	3.07
1 gang stainless steel	1E@0.05	Ea	1.42	1.34	2.76
2 gang brown	1E@0.10	Ea	1.25	2.69	3.94
2 gang ivory	1E@0.10	Ea	1.25	2.69	3.94
2 gang gray	1E@0.10	Ea	1.42	2.69	4.11
2 gang white	1E@0.10	Ea	1.42	2.69	4.11
2 gang red	1E@0.10	Ea	1.42	2.69	4.11
2 gang black	1E@0.10	Ea	1.42	2.69	4.11

Use these figures to estimate the cost of trim plates installed on outlet boxes under the conditions described on pages 5 and 6. Costs listed are for each plate installed. The crew is one electrician working at a labor cost of $26.88 per manhour. These costs include layout, material handling, and normal waste. Add for the outlet box, receptacle and switch, plaster ring, sales tax, delivery, supervision, mobilization, demobilization, cleanup, overhead and profit. Note: Be careful to select the right plate color and material. Special plates with non-standard configurations can be ordered for custom applications.

Wiring Device Plates

Material	Craft@Hrs	Unit	Material Cost	Labor Cost	Installed Cost
Blank wiring device plates					
2 gang brass	1E@0.10	Ea	9.38	2.69	12.07
2 gang aluminum	1E@0.10	Ea	2.98	2.69	5.67
2 gang stainless steel	1E@0.10	Ea	1.42	2.69	4.11
Telephone wiring device plates					
1 gang brown	1E@0.05	Ea	.35	1.34	1.69
1 gang ivory	1E@0.05	Ea	.35	1.34	1.69
1 gang gray	1E@0.05	Ea	.42	1.34	1.76
1 gang white	1E@0.05	Ea	.42	1.34	1.76
1 gang red	1E@0.05	Ea	.42	1.34	1.76
1 gang brass	1E@0.05	Ea	4.00	1.34	5.34
1 gang aluminum	1E@0.05	Ea	1.89	1.34	3.23
1 gang stainless steel	1E@0.05	Ea	1.43	1.34	2.77
Decorator wiring device plates					
1 gang brown	1E@0.05	Ea	.53	1.34	1.87
1 gang ivory	1E@0.05	Ea	.49	1.34	1.83
1 gang gray	1E@0.05	Ea	.65	1.34	1.99
1 gang white	1E@0.05	Ea	.49	1.34	1.83
1 gang red	1E@0.05	Ea	.67	1.34	2.01
1 gang black	1E@0.05	Ea	.65	1.34	1.99
1 gang brass	1E@0.05	Ea	3.40	1.34	4.74
1 gang aluminum	1E@0.05	Ea	1.96	1.34	3.30
1 gang stainless steel	1E@0.05	Ea	1.48	1.34	2.82
2 gang brown	1E@0.10	Ea	1.21	2.69	3.90
2 gang ivory	1E@0.10	Ea	1.09	2.69	3.78
2 gang gray	1E@0.10	Ea	1.30	2.69	3.99
2 gang white	1E@0.10	Ea	.98	2.69	3.67
2 gang red	1E@0.10	Ea	1.30	2.69	3.99
2 gang black	1E@0.10	Ea	1.30	2.69	3.99
2 gang brass	1E@0.10	Ea	6.87	2.69	9.56
2 gang aluminum	1E@0.10	Ea	3.91	2.69	6.60
2 gang stainless steel	1E@0.10	Ea	3.74	2.69	6.43
3 gang brown	1E@0.15	Ea	2.86	4.03	6.89
3 gang ivory	1E@0.15	Ea	2.86	4.03	6.89
3 gang gray	1E@0.15	Ea	2.86	4.03	6.89
3 gang white	1E@0.15	Ea	2.86	4.03	6.89
3 gang red	1E@0.15	Ea	2.86	4.03	6.89
3 gang black	1E@0.15	Ea	2.86	4.03	6.89
3 gang brass	1E@0.15	Ea	14.70	4.03	18.73
3 gang stainless steel	1E@0.15	Ea	5.73	4.03	9.76

Use these figures to estimate the cost of trim plates installed on outlet boxes under the conditions described on pages 5 and 6. Costs listed are for each plate installed. The crew is one electrician working at a labor cost of $26.88 per manhour. These costs include layout, material handling, and normal waste. Add for the outlet box, plaster ring, sales tax, delivery, supervision, mobilization, demobilization, cleanup, overhead and profit. Note: Be careful to select the right plate color and material. Special plates with non-standard configurations can be ordered for custom applications.

Material	Craft@Hrs	Unit	Material Cost	Labor Cost	Installed Cost
Decorator wiring device plates					
4 gang brown	1E@0.20	Ea	5.12	5.38	10.50
4 gang ivory	1E@0.20	Ea	5.12	5.38	10.50
4 gang gray	1E@0.20	Ea	5.12	5.38	10.50
4 gang white	1E@0.20	Ea	5.12	5.38	10.50
4 gang black	1E@0.20	Ea	5.12	5.38	10.50
4 gang brass	1E@0.20	Ea	18.20	5.38	23.58
4 gang stainless steel	1E@0.20	Ea	7.61	5.38	12.99
5 gang brown	1E@0.25	Ea	7.17	6.72	13.89
5 gang ivory	1E@0.25	Ea	7.17	6.72	13.89
5 gang gray	1E@0.25	Ea	7.17	6.72	13.89
5 gang white	1E@0.25	Ea	7.17	6.72	13.89
5 gang red	1E@0.25	Ea	7.17	6.72	13.89
5 gang black	1E@0.25	Ea	7.17	6.72	13.89
5 gang brass	1E@0.25	Ea	24.70	6.72	31.42
5 gang stainless steel	1E@0.25	Ea	22.90	6.72	29.62

Combination decorator and standard switch plates

Material	Craft@Hrs	Unit	Material Cost	Labor Cost	Installed Cost
2 gang brown	1E@0.10	Ea	1.25	2.69	3.94
2 gang ivory	1E@0.10	Ea	1.25	2.69	3.94
2 gang gray	1E@0.10	Ea	1.30	2.69	3.99
2 gang white	1E@0.10	Ea	1.30	2.69	3.99
2 gang red	1E@0.10	Ea	1.30	2.69	3.99
2 gang black	1E@0.10	Ea	1.30	2.69	3.99
2 gang brass	1E@0.10	Ea	7.42	2.69	10.11
2 gang aluminum	1E@0.10	Ea	3.91	2.69	6.60
2 gang stainless steel	1E@0.10	Ea	3.74	2.69	6.43

Combination decorator and two standard switch plates

Material	Craft@Hrs	Unit	Material Cost	Labor Cost	Installed Cost
3 gang brown	1E@0.15	Ea	2.52	4.03	6.55
3 gang ivory	1E@0.15	Ea	2.52	4.03	6.55
3 gang gray	1E@0.15	Ea	2.52	4.03	6.55
3 gang white	1E@0.15	Ea	2.52	4.03	6.55
3 gang brass	1E@0.15	Ea	11.10	4.03	15.13
3 gang aluminum	1E@0.15	Ea	5.95	4.03	9.98
3 gang stainless steel	1E@0.15	Ea	5.69	4.03	9.72

Use these figures to estimate the cost of switch trim plates installed on outlet boxes under the conditions described on pages 5 and 6. Costs listed are for each plate installed. The crew is one electrician working at a labor cost of $26.88 per manhour. These costs include layout, material handling, and normal waste. Add for the outlet box, switch, plaster ring, sales tax, delivery, supervision, mobilization, demobilization, cleanup, overhead and profit. Note: Be careful to select the right plate color and material. Special plates with non-standard configurations can be ordered for custom applications.

Wiring Device Plates

Material	Craft@Hrs	Unit	Material Cost	Labor Cost	Installed Cost
Combination decorator and three standard switch plates					
4 gang brown	1E@0.20	Ea	5.62	5.38	11.00
4 gang ivory	1E@0.20	Ea	5.62	5.38	11.00
4 gang white	1E@0.20	Ea	5.62	5.38	11.00
Semi-jumbo switch plates					
1 gang brown	1E@0.05	Ea	.80	1.34	2.14
1 gang ivory	1E@0.05	Ea	.80	1.34	2.14
1 gang white	1E@0.05	Ea	.87	1.34	2.21
1 gang gray	1E@0.05	Ea	.87	1.34	2.21
2 gang brown	1E@0.10	Ea	1.74	2.69	4.43
2 gang ivory	1E@0.10	Ea	1.74	2.69	4.43
2 gang white	1E@0.10	Ea	1.74	2.69	4.43
2 gang gray	1E@0.10	Ea	1.74	2.69	4.43
3 gang brown	1E@0.15	Ea	2.60	4.03	6.63
3 gang ivory	1E@0.15	Ea	2.60	4.03	6.63
3 gang white	1E@0.15	Ea	2.60	4.03	6.63
3 gang gray	1E@0.15	Ea	2.60	4.03	6.63
Semi-jumbo duplex wiring device plates					
1 gang brown	1E@0.05	Ea	.72	1.34	2.06
1 gang ivory	1E@0.05	Ea	.72	1.34	2.06
1 gang white	1E@0.05	Ea	.87	1.34	2.21
1 gang gray	1E@0.05	Ea	.87	1.34	2.21
Semi-jumbo single receptacle plates					
1 gang brown	1E@0.05	Ea	.97	1.34	2.31
1 gang ivory	1E@0.05	Ea	.97	1.34	2.31
1 gang white	1E@0.05	Ea	.97	1.34	2.31
1 gang gray	1E@0.05	Ea	.97	1.34	2.31
Semi-jumbo double duplex wiring device plates					
2 gang brown	1E@0.10	Ea	1.75	2.69	4.44
2 gang ivory	1E@0.10	Ea	1.75	2.69	4.44
2 gang white	1E@0.10	Ea	1.75	2.69	4.44
2 gang gray	1E@0.10	Ea	1.75	2.69	4.44

Use these figures to estimate the cost of switch and receptacle trim plates installed on outlet boxes under the conditions described on pages 5 and 6. Costs listed are for each plate installed. The crew is one electrician working at a labor cost of $26.88 per manhour. These costs include layout, material handling, and normal waste. Add for the outlet box, switch and receptacle, plaster ring, sales tax, delivery, supervision, mobilization, demobilization, cleanup, overhead and profit. Note: Be careful to select the right plate color and material. Special plates with non-standard configurations can be ordered for custom applications.

Material	Craft@Hrs	Unit	Material Cost	Labor Cost	Installed Cost
Semi-jumbo decorator wiring device plates					
1 gang brown	1E@0.05	Ea	.87	1.34	2.21
1 gang ivory	1E@0.05	Ea	.87	1.34	2.21
1 gang white	1E@0.05	Ea	.87	1.34	2.21
2 gang brown	1E@0.10	Ea	2.56	2.69	5.25
2 gang ivory	1E@0.10	Ea	2.56	2.69	5.25
2 gang white	1E@0.10	Ea	2.56	2.69	5.25
3 gang brown	1E@0.15	Ea	5.02	4.03	9.05
3 gang ivory	1E@0.15	Ea	5.02	4.03	9.05
3 gang white	1E@0.15	Ea	5.02	4.03	9.05
Jumbo switch plates					
1 gang brown	1E@0.05	Ea	1.15	1.34	2.49
1 gang ivory	1E@0.05	Ea	1.03	1.34	2.37
1 gang white	1E@0.05	Ea	1.15	1.34	2.49
2 gang brown	1E@0.10	Ea	2.55	2.69	5.24
2 gang ivory	1E@0.10	Ea	2.30	2.69	4.99
Jumbo duplex wiring device plates					
1 gang brown	1E@0.05	Ea	1.15	1.34	2.49
1 gang ivory	1E@0.05	Ea	1.03	1.34	2.37
1 gang white	1E@0.05	Ea	1.15	1.34	2.49
2 gang brown	1E@0.10	Ea	2.55	2.69	5.24
2 gang ivory	1E@0.10	Ea	2.30	2.69	4.99
Jumbo combination switch and duplex receptacle plates					
2 gang brown	1E@0.10	Ea	2.55	2.69	5.24
2 gang ivory	1E@0.10	Ea	2.30	2.69	4.99
Jumbo combination switch and blank wiring device plates					
2 gang brown	1E@0.10	Ea	3.04	2.69	5.73
2 gang ivory	1E@0.10	Ea	3.04	2.69	5.73

Use these figures to estimate the cost of switch and receptacle trim plates installed on outlet boxes under the conditions described on pages 5 and 6. Costs listed are for each plate installed. The crew is one electrician working at a labor cost of $26.88 per manhour. These costs include layout, material handling, and normal waste. Add for the outlet box, switch and receptacle, plaster ring, sales tax, delivery, supervision, mobilization, demobilization, cleanup, overhead and profit. Note: Be careful to select the right plate color and material. Special plates with non-standard configurations can be ordered for custom applications.

Wiring Device Plates

Material	Craft@Hrs	Unit	Material Cost	Labor Cost	Installed Cost
Jumbo combination duplex and blank wiring device plates					
2 gang brown	1E@0.10	Ea	2.67	2.69	5.36
2 gang ivory	1E@0.10	Ea	2.67	2.69	5.36
Jumbo blank wiring device plates					
1 gang brown	1E@0.05	Ea	1.53	1.34	2.87
1 gang ivory	1E@0.05	Ea	1.40	1.34	2.74
2 gang brown	1E@0.10	Ea	3.09	2.69	5.78
2 gang ivory	1E@0.10	Ea	3.09	2.69	5.78
Jumbo decorator wiring device plates					
1 gang brown	1E@0.05	Ea	1.27	1.34	2.61
1 gang ivory	1E@0.05	Ea	1.15	1.34	2.49
1 gang white	1E@0.05	Ea	1.27	1.34	2.61
2 gang brown	1E@0.10	Ea	2.56	2.69	5.25
2 gang ivory	1E@0.10	Ea	2.56	2.69	5.25
2 gang white	1E@0.10	Ea	2.56	2.69	5.25
Jumbo combination decorator and blank wiring device plates					
2 gang brown	1E@0.10	Ea	3.04	2.69	5.73
2 gang ivory	1E@0.10	Ea	3.04	2.69	5.73
2 gang white	1E@0.10	Ea	3.04	2.69	5.73
Deep switch plates					
1 gang brown	1E@0.05	Ea	.57	1.34	1.91
1 gang ivory	1E@0.05	Ea	.52	1.34	1.86
2 gang brown	1E@0.10	Ea	1.11	2.69	3.80
2 gang ivory	1E@0.10	Ea	1.01	2.69	3.70

Use these figures to estimate the cost of switch and receptacle trim plates installed on outlet boxes under the conditions described on pages 5 and 6. Costs listed are for each plate installed. The crew is one electrician working at a labor cost of $26.88 per manhour. These costs include layout, material handling, and normal waste. Add for the outlet box, switch and receptacle, plaster ring, sales tax, delivery, supervision, mobilization, demobilization, cleanup, overhead and profit. Note: Use deep plates when the wall has been built out away from the outlet box and the screws from a standard plate won't reach the box. This is common in remodeling work when a wall has been covered with new paneling.

Material	Craft@Hrs	Unit	Material Cost	Labor Cost	Installed Cost
Deep duplex receptacle plates					
1 gang brown	1E@0.05	Ea	.52	1.34	1.86
1 gang ivory	1E@0.05	Ea	.52	1.34	1.86
2 gang brown	1E@0.10	Ea	1.24	2.69	3.93
2 gang ivory	1E@0.10	Ea	1.01	2.69	3.70
Deep combination switch and duplex receptacle plates					
2 gang brown	1E@0.10	Ea	1.35	2.69	4.04
2 gang ivory	1E@0.10	Ea	1.22	2.69	3.91
Deep combination switch and blank wiring device plates					
2 gang brown	1E@0.10	Ea	1.35	2.69	4.04
2 gang ivory	1E@0.10	Ea	1.22	2.69	3.91
Deep combination duplex receptacle and blank wiring device plates					
2 gang brown	1E@0.10	Ea	1.35	2.69	4.04
2 gang ivory	1E@0.10	Ea	1.22	2.69	3.91
15 amp size single receptacle plates					
1 gang brass	1E@0.05	Ea	8.63	1.34	9.97
1 gang aluminum	1E@0.05	Ea	2.32	1.34	3.66
1 gang stainless 430	1E@0.05	Ea	1.83	1.34	3.17
1 gang stainless 302	1E@0.05	Ea	2.87	1.34	4.21
20 amp size single receptacle plates					
1 gang brass	1E@0.05	Ea	8.64	1.34	9.98
1 gang stainless 430	1E@0.05	Ea	2.09	1.34	3.43
1 gang stainless 302	1E@0.05	Ea	2.09	1.34	3.43

Use these figures to estimate the cost of switch and receptacle trim plates installed on outlet boxes under the conditions described on pages 5 and 6. Costs listed are for each plate installed. The crew is one electrician working at a labor cost of $26.88 per manhour. These costs include layout, material handling, and normal waste. Add for the outlet box, switch and receptacle, plaster ring, sales tax, delivery, supervision, mobilization, demobilization, cleanup, overhead and profit. Note: Be careful to select the right plate color and material. Special plates with non-standard configurations can be ordered for custom applications.

Wiring Device Plates

Material	Craft@Hrs	Unit	Material Cost	Labor Cost	Installed Cost
30 amp size single receptacle plates					
1 gang brass	1E@0.05	Ea	8.63	1.34	9.97
1 gang stainless 430	1E@0.05	Ea	2.35	1.34	3.69
1 gang stainless 302	1E@0.05	Ea	2.43	1.34	3.77
50 amp size single receptacle plates					
2 gang brass	1E@0.10	Ea	10.50	2.69	13.19
2 gang stainless 430	1E@0.10	Ea	3.98	2.69	6.67
Weatherproof single receptacle plates					
15A cast metal	1E@0.10	Ea	21.60	2.69	24.29
20A cast metal	1E@0.10	Ea	22.90	2.69	25.59
Weatherproof horizontal duplex receptacle plates with twin covers					
1 gang cast metal	1E@0.10	Ea	19.30	2.69	21.99

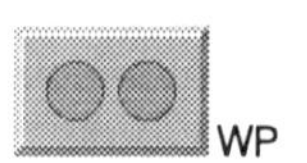

Use these figures to estimate the cost of receptacle trim plates installed on outlet boxes under the conditions described on pages 5 and 6. Costs listed are for each plate installed. The crew is one electrician working at a labor cost of $26.88 per manhour. These costs include layout, material handling, and normal waste. Add for the outlet box, receptacle, plaster ring, sales tax, delivery, supervision, mobilization, demobilization, cleanup, overhead and profit. Note: Be careful to select the right plate color and material. Special plates with non-standard configurations can be ordered for custom applications.

Section 6:
Service Entrance Equipment

This section deals with the equipment that's located at the point where electrical service enters the building — hence the name service entrance equipment. It includes safety switches, circuit breakers, a meter socket for mounting the utility company's meter, perhaps a transformer, at least one panelboard for distribution to the various loads, and wireway. This gear may be housed in a single steel cabinet or it may consist of several components, each with a separate enclosure.

Service entrance gear is custom-designed for larger commercial and industrial buildings. The type, size and ampacity of the equipment depend on the power that's needed, of course, and on requirements imposed by the local inspection authority and the electrical utility. Both the inspector and the electric company have standards that must be met. Be sure the equipment you're pricing meets those standards.

Description of
Service Entrance Equipment

Figure 6-1 shows service entrance equipment for a small home. Figures 6-2 and 6-3 show service entrance gear that might be used in commercial and larger industrial buildings.

Notice in Figures 6-1, 6-2 and 6-3 that all three installations have two main parts: the service section and the distribution section.

The **service section** connects to the utility company power feed and outputs power to the distribution section. It includes either a cable pull section where the feed is pulled into the service entrance cabinet from underground power lines, or conduit to a service entrance cap which receives lines from an overhead distribution system.

The main disconnect or circuit breaker is also located in the service section. It provides a quick and convenient way to cut off power to all circuits when necessary, or when an overload endangers the system. The *NEC* requires that each conductor entering the service entrance section have a readily accessible disconnect. Exceptions are made for very small installations.

Also in the service section will be the meter socket which receives the utility company's meter. The service side may include instrumentation such as a voltmeter or ammeter, a fire alarm breaker, and perhaps a transformer if needed to reduce distribution voltage to the voltage required by loads in the building.

The **distribution section** is usually mounted right beside the service section and is connected by a set of metal bars called **bus bars**. These bars carry power from the service section to the distribution section.

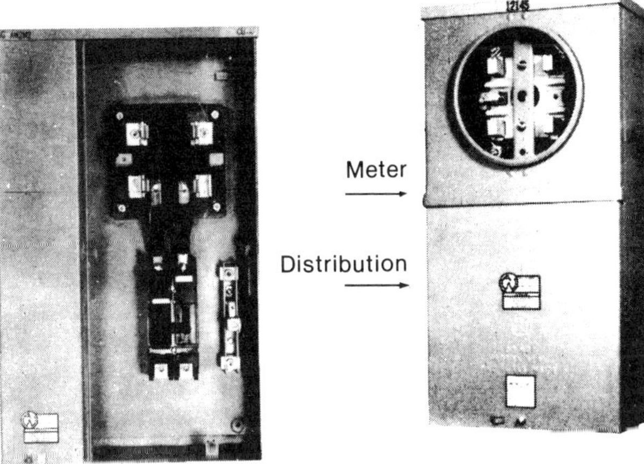

Figure 6-1
Residential Service Entrance Equipment

Figure 6-2
Multi-metering Center

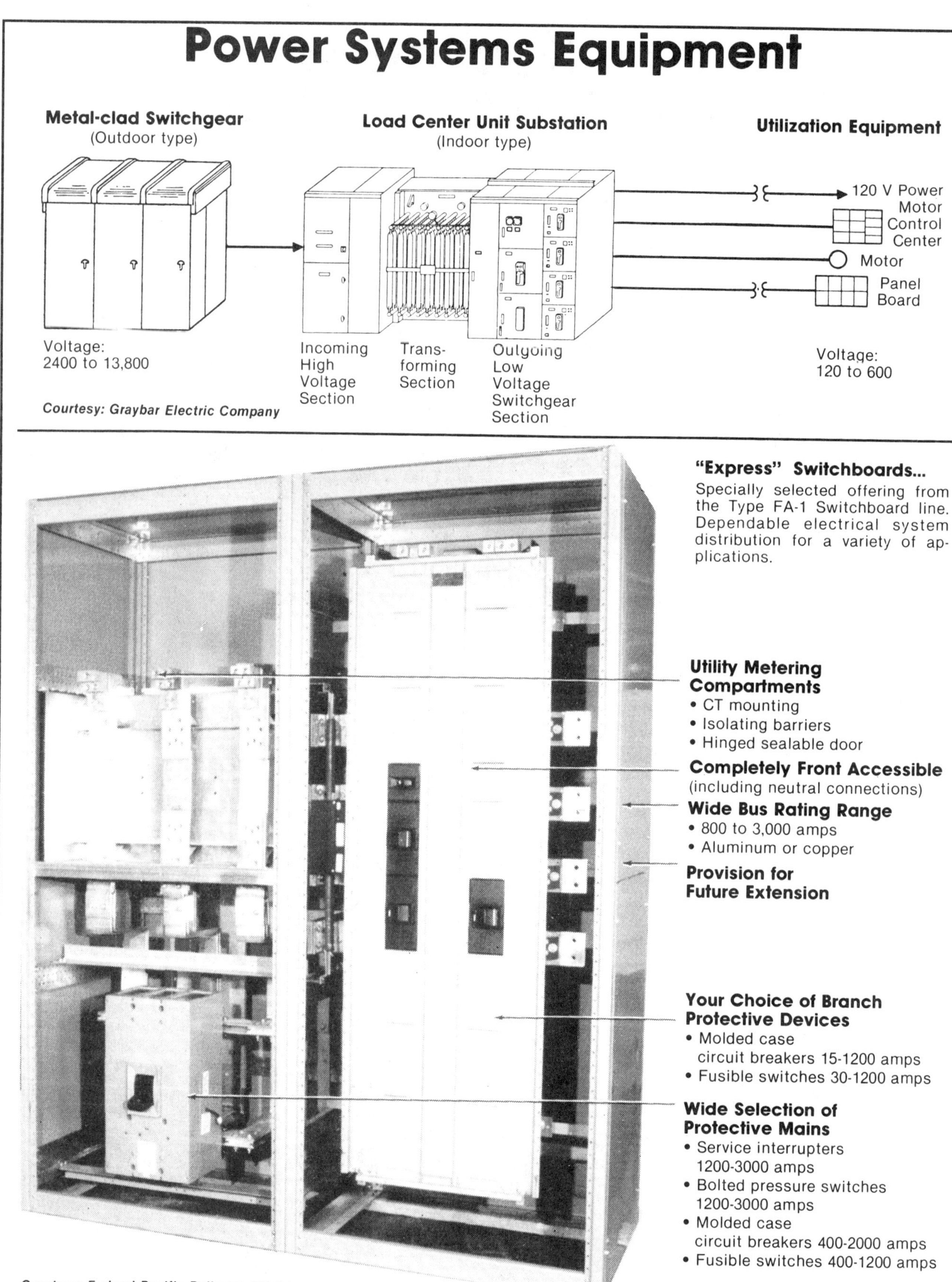

Figure 6-3
Power Systems Equipment

The purpose of the distribution section is to subdivide electrical power among the individual circuits, each of which is protected by a circuit breaker mounted in the distribution panel. The distribution section in a larger commercial building will include feeder breakers in addition to individual breakers for every circuit in the building's electrical system. If the building covers a large area or has electrical equipment like large motors or air conditioning units, the distribution section may include circuits that feed subpanels located in other parts of the building.

Service entrance equipment listed in this section is only representative of the many types of equipment that are available. But from what's listed here you can see the range of material prices and typical installation costs. Prices quoted for larger pieces of equipment are for the most competitively-priced units. Prices can vary widely. It's good practice to get several quotes.

Much of this equipment is custom-assembled to order. There will be only a few competitive suppliers in your area. But note that multiple units ordered at the same time can reduce costs 20 percent or more.

Safety Switches

Safety switches are also known as **disconnect switches and externally operated (EXO) devices**. A safety switch is a convenient way of shutting off electrical equipment in an emergency. For example, the *NEC* requires that a safety switch be installed at or near heavy machinery. In some cases the switch will be protected with a padlock so the power can't be shut off accidentally.

There are many types of safety switches. Some have specific applications. Others are appropriate for a wide range of uses. Ratings for safety switches are by amperage (30 amp, 60 amp, 100 amp, etc.), and by voltage (120 volt, 240 volt, 480 volt, etc.). Safety switches intended for use with electric motors are rated by motor horsepower. Safety switches are also classified as general duty, heavy duty, and by NEMA (National Electrical Manufacturer's Association) class. The most common classes are 1 (indoor), 3R (weatherproof), 4 and 4X (waterproof and dustproof), and 7 through 9 (for hazardous locations). Explosion-proof enclosures are also available.

Safety switches can be either fusible or non-fusible. The fusible type has a renewable fuse that can be changed to alter the overload protection characteristics of the circuit. A safety switch that isn't fused has no overload protection.

The engineer who designs the electrical system should identify the type and rating of each safety switch on the plans. But note that it's acceptable practice (but an unnecessary expense) to use over-size safety switches. For example, a 100A, 480V, 3-phase NEMA class 3R (weatherproof) safety switch can be installed inside a building to serve a machine with a 20A, 240V load. But it's much less expensive to use a switch that just meets rather than exceeds circuit requirements.

The fusible switches listed in this section are for cartridge fuses. Screw-in fuses can be used in some safety switches, but the cartridge type is much more common. Most suppliers carry cartridge fuses in these ranges. The fuse amperage ratings are as follows for the switch ratings indicated:

30A	fusible switches use fractional amp fuses to 30 amps
60A	fusible switches use 35 amp to 60 amp fuses
100A	fusible switches use 70 amp to 100 amp fuses
200A	fusible switches use 110 amp to 200 amp fuses
400A	fusible switches use 225 amp to 400 amp fuses
600A	fusible switches use 450 amp to 600 amp fuses
800A	fusible switches use 601 amp to 800 amp fuses
1200A	fusible switches use 1000 amp to 1200 amp fuses

The *National Electrical Code* lists several installation rules for safety switches. Some job specifications will include additional requirements for safety switches. For example, some specs require switches with interlocks that prevent opening the switch door or cover when the switch is in the "on" position. The specs might require that the switch have a block for connecting a neutral wire.

Circuit Breakers

Like safety switches, circuit breakers can be used to shut off the power to any circuit. Generally, a circuit breaker can be used anywhere a safety switch is required. But an automatic circuit breaker's most important function is to protect the system by opening when an overload is detected. Unlike a fuse, a breaker doesn't have to be replaced after tripping. Resetting the breaker restores power to the circuit.

Although non-automatic circuit breakers are available, automatic breakers with permanent trip settings are most common. They're rated by amperage capacity and voltage. They're also rated for maximum short-circuit interrupting capacity. The *NEC* dictates the type of circuit breaker to use for each application.

The most common use for circuit breakers is in loadcenters and panelboards. They can be either the plug-in or bolt-on type. Loadcenters use plug-in breakers. Panelboards use bolt-on breakers that bolt directly to the panelboard bus and to the panel frame for support. The vast majority of our jobs use bolt-on. Most well-engineered jobs require bolt-on breakers. Use bolt-on for high A.I.C. interrupting capacities.

Circuit breakers are available with more than one pole. A single handle opens and closes contact between two or more conductors. But single-pole breakers can be ganged to protect each line. When that's done, the operating handles for all poles on a circuit should be tied together with a bar device. Usually these breakers are made from single-pole breakers riveted together to form a two- or three-pole device.

Circuit breakers can be either thermal, magnetic or a combination of the two. Thermal breakers react to changes in temperature, opening the circuit in response to extra heat from a short. Magnetic breakers react to changes in current flow. A sudden increase in current flow creates enough magnetic force to activate an armature, opening the circuit.

Thermal-magnetic breakers combine the best features of both types of breakers and are the type commonly used in panelboards and loadcenters. Magnetic breakers are generally used in special applications where the breaker must be unaffected by air temperature.

Meter Sockets

Meter sockets are usually installed by the contractor. The power company will usually install the actual kilowatt hour meter.

Be sure to install the right socket. Each type of socket has a certain number of jaws that make contact with the meter. The socket must match the meter the power company intends to use. That depends on the size of the service feed and the service voltage provided.

In some cases the meter will require a current transformer in the service section. The current transformer is connected to the meter base. Test blocks may also be required. Check with your local utility company before buying or installing the meter socket.

Multi-socket assemblies are required in multi-family residences. Usually all the meters will be located in the same panel for convenience. Your supplier will help you select the right meter panel. But it's your responsibility to get approval from the utility company for the panel selected.

Loadcenters and Panelboards

Loadcenters are electrical distribution panels that use *plug-in* circuit breakers. They're usually used in residential and light commercial buildings where the electrical loads are not heavy.

Loadcenters are enclosed in a sheet metal box which can be ordered for surface, flush or semi-flush mounting in a wall. The cover has knockouts where breakers are mounted. Most loadcenters come with a door to cover the breaker handles. Both indoor- and outdoor-rated loadcenters are available. Units can have as few as two single-pole spaces or as many as 42 single-pole spaces. Two- and three-pole plug-in breakers can also be inserted. Individual breakers can be sized up to 100 amps.

Panelboards are electrical distribution panels that use *bolt-on* circuit breakers. They're usually used in heavy commercial and industrial buildings where higher amperages are needed.

Panelboards are enclosed in a heavy gauge sheet metal box which can be ordered for surface or flush mounting on a wall. Panelboards have interiors like similar loadcenters except that the circuit breakers must be bolted to the interior bus system. An interior cover is placed over the circuit breakers and extends to the panel cover. The panel cover is made of heavy gauge sheet metal and usually has a door which covers the circuit-breaker handles. Two- and three-pole bolt-on breakers can be installed. The breakers can be sized up to 100 amps for lighting panels, and higher for power panels.

Panelboards are rated for total load, such as 100A, 225A, 400A, etc. The circuit breakers installed cannot exceed the panel rating. Single-width panelboards are limited to 42 circuits. If additional circuits are required, a double-width panelboard can be used.

Panelboards are usually custom-assembled at the factory to meet specific job requirements. Loadcenters are usually stock items. Your supplier probably offers a good selection of loadcenter housings and large quantities of the more common plug-in breakers that are inserted into the housing. When you order a loadcenter, they can pull the loadcenter housing from stock and furnish whatever breakers you will insert into the housing.

Wireway

Wireway is an enclosed metal channel with one hinged or removable side so wire can be laid, rather than pulled, into place. Indoor wireway may have either a screw cover or a hinged cover. Both interior and exterior wireways are available in several sizes and lengths.

Wireway is useful in many situations. For example, it's used in service entrance equipment to enclose wire running between the various components. It can be installed in long rows for circuits, control or communications systems. Fittings are available for making turns up, down, right or left. Flange fittings allow connection to pull boxes and cabinets.

Wireway usually has knockouts every few inches on each side wall. These knockouts are spaced every few inches so a junction is possible at nearly all points.

Transformers

The most common transformer for buildings is the dry air-cooled type. Both indoor and outdoor dry transformers are available. Other types of transformers are cooled with oil or silicon compounds. The larger units are generally used at primary service facilities or substations and can be installed on poles, pads or submerged.

Most transformers are available with adjustable taps for making small changes in the output voltage. When the connected load voltage is too high or too low, the adjustable taps are changed to bring the voltage closer to the desired range. Usually the transformers must be de-energized before the taps are changed. The taps are connected inside the transformer. Each tap allows a 2½ percent adjustment. Most transformers have two taps above normal and two taps below the normal setting.

Estimating Service Equipment

Service entrance equipment is sized to meet the needs of the electrical loads served. The design engineer will supply the load calculations. Use these figures to select the right loadcenters or panelboards. The main service equipment must be based on the total building load, plus some capacity for future expansion. Since every building can be different, all service entrance equipment can be different. But you'll notice that most manufacturers offer only certain load capacities, such as 100 amp, 225 amp, 400 amp, etc.

Your estimate for service entrance equipment will always include several components: panels, breakers, and metering. Larger and more specialized installations will include specially-engineered gear.

Making the material estimate is usually a simple matter of getting a supplier to make a take-off and furnish a quote. The supplier will pass the take-off on to the manufacturer's representative for pricing major components. The supplier will prepare a quote on smaller stock items. You'll usually get a single lump sum price. That makes your job easy. But watch out for exceptions in the quote. And be sure the quote you get covers all service equipment and panels.

Labor for Service Equipment

Finding the installation cost for service entrance equipment is much harder than finding the material cost. First, study the plans and specs carefully. Then list all the major components on a pricing sheet.

Suppose the specs show a meter socket and main combination as the first line item. The next line shows a 12-circuit, flush-mounted loadcenter with 12 single-pole plug-in circuit breakers. First, find the labor cost for the meter socket and main combination. Then figure the labor for the loadcenter. The loadcenter is assembled by inserting plug-in circuit breakers into the panel. My usual practice is to have a single labor cost for the panel, housing and cover. The cost of installing breakers is a separate item. Circuit breaker labor in the tables that follow includes connection to the circuit conductor.

Items to Watch For

Most mistakes in estimating service entrance equipment are the result of omissions and oversights. Run through the checklist that follows when your estimate is complete. These questions should help you spot an error before it becomes an expensive mistake.

1) Are the service voltage and size correct?

2) Is it single- or three-phase?

3) Is this an indoor or outdoor installation?

4) Is special grounding needed?

5) Are the interrupt capacities correct?

6) Is it flush or surface mounted?

7) Is this for overhead or underground service?

8) Does it require a special nameplate?

9) What are the ground fault provisions?

10) Is a special corrosion-resisting finish needed?

11) Did you include the fuses?

12) Are there shunt trip breakers?

13) Is the undervoltage protection correct?

14) Are there surge arrestors?

15) What are the load shedding provisions?

16) Are there energy-management provisions?

17) Is any standby power equipment included?

18) Are transfer switches included?

19) Is there any special metering equipment?

20) Do physical clearances comply with the code?

21) Is it accessible for installation?

22) What are the utility company hookup charges?

23) Are there housekeeping pads?

24) Are special equipment pads needed?

25) Will it fit in the space provided?

Installation

Installation times for service equipment in commercial and industrial buildings will vary because the equipment is usually custom designed. Only on housing tracts will you install the same service equipment over and over again.

First, determine where the service equipment and panels are to be installed. Good access to the installation areas is important. If there is a substation in the service equipment, where will it be located? Lifting equipment may be needed to install heavy pieces of service equipment in vaults, basements, and mezzanines. Special rigging may be required. Moving heavy items into awkward places can be expensive.

If you suspect that a crane will be needed, get a quote from a local crane company. There will usually be a minimum charge. They'll probably charge for the entire time the crane is out of the yard. To get a quote, you'll need to describe the weight, maximum height and all obstructions that are likely to affect the lift. The angle of the boom will also be a factor. The longer the horizontal reach, the bigger the crane has to be.

Include an allowance for unloading when installing any heavy service equipment. The manufacturer's quote probably includes delivery. But it's your responsibility to unload it promptly. Always include a note on your purchase order requiring that you be notified at least 24 hours before heavy deliveries are made. You need some time to arrange for unloading equipment.

The labor units in these tables include time required to connect wires or conductors, clean the unit and put it into service. If any special testing is required, add that time separately. The contract documents will outline the testing procedure, if any.

The building inspector will probably have to approve your work before the utility company will install the meter. Usually the inspector is the one who calls for the meter. Before the utility company sets the meter, an application for service may be necessary. They need to know who's going to pay the electric bill.

In some cases the inspection authority will permit installation of a temporary meter while the building is under construction. The meter is later assigned to the owner after final inspection.

Estimates of service entrance equipment require a close study of the plans. It's a good idea to estimate the service gear last. That way you're fully familiar with the project when estimating the installation times.

Put some extra thought into estimates for service equipment placement and hookup. Use the service entrance checklist to spot an error or omission. This is one of the more difficult areas for most electrical estimators.

Material		Craft@Hrs	Unit	Material Cost	Labor Cost	Installed Cost

NEMA 1 general duty non-fused 240 volt safety switches

Material		Craft@Hrs	Unit	Material Cost	Labor Cost	Installed Cost
3P	30A	1E@0.50	Ea	40.80	13.40	54.20
3P	60A	1E@0.70	Ea	55.10	18.80	73.90
3P	100A	1E@1.00	Ea	127.00	26.90	153.90
3P	200A	1E@1.50	Ea	234.00	40.30	274.30
3P	400A	1E@2.00	Ea	581.00	53.80	634.80
3P	600A	1E@4.00	Ea	1,110.00	108.00	1,218.00

NEMA 3R general duty non-fused 240 volt safety switches

Material		Craft@Hrs	Unit	Material Cost	Labor Cost	Installed Cost
3P	30A	1E@0.50	Ea	77.20	13.40	90.60
3P	60A	1E@0.70	Ea	117.00	18.80	135.80
3P	100A	1E@1.00	Ea	216.00	26.90	242.90
3P	200A	1E@1.50	Ea	387.00	40.30	427.30

NEMA 1 general duty fusible 240 volt safety switches

Material		Craft@Hrs	Unit	Material Cost	Labor Cost	Installed Cost
2P	30A	1E@0.50	Ea	32.00	13.40	45.40
2P	60A	1E@0.70	Ea	55.10	18.80	73.90
2P	100A	1E@1.00	Ea	112.00	26.90	138.90
2P	200A	1E@1.50	Ea	234.00	40.30	274.30
2P	400A	1E@4.00	Ea	676.00	108.00	784.00
2P	600A	1E@4.00	Ea	1,350.00	108.00	1,458.00
3P	30A	1E@0.50	Ea	49.70	13.40	63.10
3P	60A	1E@0.70	Ea	87.20	18.80	106.00
3P	100A	1E@1.00	Ea	149.00	26.90	175.90
3P	200A	1E@1.50	Ea	317.00	40.30	357.30
3P	400A	1E@2.00	Ea	823.00	53.80	876.80
3P	600A	1E@4.00	Ea	1,540.00	108.00	1,648.00

NEMA 3R general duty fusible 240 volt safety switches

Material		Craft@Hrs	Unit	Material Cost	Labor Cost	Installed Cost
3P	30A	1E@0.50	Ea	77.20	13.40	90.60
3P	60A	1E@0.70	Ea	117.00	18.80	135.80
3P	100A	1E@1.00	Ea	216.00	26.90	242.90
3P	200A	1E@1.50	Ea	387.00	40.30	427.30
3P	400A	1E@2.00	Ea	1,030.00	53.80	1,083.80
3P	600A	1E@4.00	Ea	2,090.00	108.00	2,198.00

Use these figures to estimate the cost of safety switches installed in buildings under the conditions described on pages 5 and 6. Costs listed are for each switch installed. The crew is one electrician working at a labor cost of $26.88 per manhour. These costs include layout, material handling, and normal waste. Add for the fuses, hubs, supports, sales tax, delivery, supervision, mobilization, demobilization, cleanup, overhead and profit. Note: Safety switches can be purchased with provision for neutral connection and are used for many purposes besides installation as part of the service entrance gear. NEMA class designations are as follows: 1 is for indoor use, 3R is for outdoor use, 4 is dusttight or watertight, 12 is watertight. Class 3R safety switches usually come with blank top hubs. Be sure the conduit size is appropriate for the switch hubs. Many safety switches are available with door interlocks or with eyes for attaching a padlock.

Fuses are on pages 255 to 277.

240 Volt Heavy Duty Safety Switches

Material		Craft@Hrs	Unit	Material Cost	Labor Cost	Installed Cost

NEMA 1 heavy duty non-fused 240 volt safety switches

Material		Craft@Hrs	Unit	Material Cost	Labor Cost	Installed Cost
2P	30A	1E@0.50	Ea	77.20	13.40	90.60
2P	60A	1E@0.70	Ea	135.00	18.80	153.80
2P	100A	1E@1.00	Ea	217.00	26.90	243.90
2P	200A	1E@1.50	Ea	336.00	40.30	376.30
2P	400A	1E@2.00	Ea	779.00	53.80	832.80
2P	600A	1E@4.00	Ea	1,390.00	108.00	1,498.00
2P	800A	1E@5.00	Ea	2,770.00	134.00	2,904.00
2P	1200A	1E@6.00	Ea	3,730.00	161.00	3,891.00
3P	30A	1E@0.50	Ea	77.20	13.40	90.60
3P	60A	1E@0.70	Ea	135.00	18.80	153.80
3P	100A	1E@1.00	Ea	217.00	26.90	243.90
3P	200A	1E@1.50	Ea	336.00	40.30	376.30
3P	400A	1E@2.00	Ea	779.00	53.80	832.80
3P	600A	1E@4.00	Ea	1,390.00	108.00	1,498.00
3P	800A	1E@5.00	Ea	2,770.00	134.00	2,904.00
3P	1200A	1E@6.00	Ea	3,730.00	161.00	3,891.00

NEMA 3R heavy duty non-fused 240 volt safety switches

Material		Craft@Hrs	Unit	Material Cost	Labor Cost	Installed Cost
2P	30A	1E@0.50	Ea	135.00	13.40	148.40
2P	60A	1E@0.70	Ea	244.00	18.80	262.80
2P	100A	1E@1.00	Ea	340.00	26.90	366.90
2P	200A	1E@1.50	Ea	413.00	40.30	453.30
2P	400A	1E@2.00	Ea	1,070.00	53.80	1,123.80
2P	600A	1E@4.00	Ea	2,130.00	108.00	2,238.00
2P	800A	1E@5.00	Ea	3,620.00	134.00	3,754.00
2P	1200A	1E@6.00	Ea	4,970.00	161.00	5,131.00
3P	30A	1E@0.50	Ea	135.00	13.40	148.40
3P	60A	1E@0.70	Ea	244.00	18.80	262.80
3P	100A	1E@1.00	Ea	340.00	26.90	366.90
3P	200A	1E@1.50	Ea	413.00	40.30	453.30
3P	400A	1E@2.00	Ea	1,070.00	53.80	1,123.80
3P	600A	1E@4.00	Ea	2,130.00	108.00	2,238.00
3P	800A	1E@5.00	Ea	3,620.00	134.00	3,754.00
3P	1200A	1E@6.00	Ea	4,970.00	161.00	5,131.00

Use these figures to estimate the cost of safety switches installed in buildings under the conditions described on pages 5 and 6. Costs listed are for each switch installed. The crew is one electrician working at a labor cost of $26.88 per manhour. These costs include layout, material handling, and normal waste. Add for the fuses, hubs, supports, sales tax, delivery, supervision, mobilization, demobilization, cleanup, overhead and profit. Note: Safety switches can be purchased with provision for neutral connection and are used for many purposes besides installation as part of the service entrance gear. NEMA class designations are as follows: 1 is for indoor use, 3R is for outdoor use, 4 is dusttight or watertight, 12 is watertight. Class 3R safety switches usually come with blank top hubs. Be sure the conduit size is appropriate for the switch hubs. Many safety switches are available with door interlocks or with eyes for attaching a padlock.

240 Volt Heavy Duty Safety Switches

Material		Craft@Hrs	Unit	Material Cost	Labor Cost	Installed Cost

NEMA 4 heavy duty fusible 240 volt safety switches

Material		Craft@Hrs	Unit	Material Cost	Labor Cost	Installed Cost
2P	30A	1E@0.60	Ea	561.00	16.10	577.10
2P	60A	1E@0.80	Ea	667.00	21.50	688.50
2P	100A	1E@1.25	Ea	1,350.00	33.60	1,383.60
2P	200A	1E@1.70	Ea	1,840.00	45.70	1,885.70
3P	30A	1E@0.60	Ea	561.00	16.10	577.10
3P	60A	1E@0.80	Ea	667.00	21.50	688.50
3P	100A	1E@1.25	Ea	1,350.00	33.60	1,383.60
3P	200A	1E@1.70	Ea	1,840.00	45.70	1,885.70

NEMA 12 heavy duty fusible 240 volt safety switches

Material		Craft@Hrs	Unit	Material Cost	Labor Cost	Installed Cost
2P	30A	1E@0.50	Ea	171.00	13.40	184.40
2P	60A	1E@0.70	Ea	221.00	18.80	239.80
2P	100A	1E@1.00	Ea	316.00	26.90	342.90
2P	200A	1E@1.50	Ea	424.00	40.30	464.30
2P	400A	1E@2.00	Ea	851.00	53.80	904.80
2P	600A	1E@4.00	Ea	1,430.00	108.00	1,538.00
3P	30A	1E@0.50	Ea	171.00	13.40	184.40
3P	60A	1E@0.70	Ea	221.00	18.80	239.80
3P	100A	1E@1.00	Ea	316.00	26.90	342.90
3P	200A	1E@1.50	Ea	424.00	40.30	464.30
3P	400A	1E@2.00	Ea	1,060.00	53.80	1,113.80
3P	600A	1E@4.00	Ea	1,770.00	108.00	1,878.00
4P	30A	1E@0.60	Ea	243.00	16.10	259.10
4P	60A	1E@0.80	Ea	267.00	21.50	288.50
4P	100A	1E@1.25	Ea	475.00	33.60	508.60
4P	200A	1E@1.70	Ea	750.00	45.70	795.70
4P	400A	1E@2.25	Ea	1,500.00	60.50	1,560.50

Use these figures to estimate the cost of safety switches installed in buildings under the conditions described on pages 5 and 6. Costs listed are for each switch installed. The crew is one electrician working at a labor cost of $26.80 per manhour. These costs include layout, material handling, and normal waste. Add for the fuses, hubs, supports, sales tax, delivery, supervision, mobilization, demobilization, cleanup, overhead and profit. Note: Safety switches can be purchased with provision for neutral connection and are used for many purposes besides installation as part of the service entrance gear. NEMA class designations are as follows: 1 is for indoor use, 3R is for outdoor use, 4 is dusttight or watertight, 12 is watertight. Class 3R safety switches usually come with blank top hubs. Be sure the conduit size is appropriate for the switch hubs. Many safety switches are available with door interlocks or with eyes for attaching a padlock.

Fuses are on pages 255 to 277.

600 Volt Heavy Duty Safety Switches

NEMA 1 heavy duty non-fused 600 volt safety switches

Material		Craft@Hrs	Unit	Material Cost	Labor Cost	Installed Cost
2P	30A	1E@0.50	Ea	77.20	13.40	90.60
2P	60A	1E@0.70	Ea	135.00	18.80	153.80
2P	100A	1E@1.00	Ea	217.00	26.90	243.90
2P	200A	1E@1.50	Ea	336.00	40.30	376.30
2P	400A	1E@2.00	Ea	779.00	53.80	832.80
2P	600A	1E@4.00	Ea	1,390.00	108.00	1,498.00
2P	800A	1E@5.00	Ea	2,770.00	134.00	2,904.00
2P	1200A	1E@6.00	Ea	3,730.00	161.00	3,891.00
3P	30A	1E@0.50	Ea	77.20	13.40	90.60
3P	60A	1E@0.70	Ea	135.00	18.80	153.80
3P	100A	1E@1.00	Ea	217.00	26.90	243.90
3P	200A	1E@1.50	Ea	336.00	40.30	376.30
3P	400A	1E@2.00	Ea	779.00	53.80	832.80
3P	600A	1E@4.00	Ea	1,390.00	108.00	1,498.00
3P	800A	1E@5.00	Ea	2,770.00	134.00	2,904.00
3P	1200A	1E@6.00	Ea	3,730.00	161.00	3,891.00
4P	30A	1E@0.60	Ea	218.00	16.10	234.10
4P	60A	1E@0.80	Ea	242.00	21.50	263.50
4P	100A	1E@1.25	Ea	437.00	33.60	470.60
4P	200A	1E@1.70	Ea	635.00	45.70	680.70
4P	400A	1E@2.25	Ea	1,380.00	60.50	1,440.50
4P	600A	1E@4.50	Ea	2,400.00	121.00	2,521.00

NEMA 3R heavy duty non-fused 600 volt safety switches

Material		Craft@Hrs	Unit	Material Cost	Labor Cost	Installed Cost
2P	30A	1E@0.50	Ea	135.00	13.40	148.40
2P	60A	1E@0.70	Ea	244.00	18.80	262.80
2P	100A	1E@1.00	Ea	340.00	26.90	366.90
2P	200A	1E@1.50	Ea	413.00	40.30	453.30
2P	400A	1E@2.00	Ea	1,070.00	53.80	1,123.80
2P	600A	1E@4.00	Ea	2,130.00	108.00	2,238.00
2P	800A	1E@5.00	Ea	3,620.00	134.00	3,754.00
2P	1200A	1E@6.00	Ea	4,970.00	161.00	5,131.00
3P	30A	1E@0.50	Ea	135.00	13.40	148.40
3P	60A	1E@0.70	Ea	244.00	18.80	262.80
3P	100A	1E@1.00	Ea	340.00	26.90	366.90
3P	200A	1E@1.50	Ea	413.00	40.30	453.30
3P	400A	1E@2.00	Ea	1,070.00	53.80	1,123.80
3P	600A	1E@4.00	Ea	2,130.00	108.00	2,238.00
3P	800A	1E@5.00	Ea	3,620.00	134.00	3,754.00
3P	1200A	1E@6.00	Ea	4,970.00	161.00	5,131.00

Use these figures to estimate the cost of safety switches installed in buildings under the conditions described on pages 5 and 6. Costs listed are for each switch installed. The crew is one electrician working at a labor cost of $26.88 per manhour. These costs include layout, material handling, and normal waste. Add for the fuses, hubs, supports, sales tax, delivery, supervision, mobilization, demobilization, cleanup, overhead and profit. Note: Safety switches can be purchased with provision for neutral connection and are used for many purposes besides installation as part of the service entrance gear. NEMA class designations are as follows: 1 is for indoor use, 3R is for outdoor use, 4 is dusttight or watertight, 12 is watertight. Class 3R safety switches usually come with blank top hubs. Be sure the conduit size is appropriate for the switch hubs. Many safety switches are available with door interlocks or with eyes for attaching a padlock.

600 Volt Heavy Duty Safety Switches

Material		Craft@Hrs	Unit	Material Cost	Labor Cost	Installed Cost

NEMA 4 heavy duty non-fused 600 volt safety switches

Material		Craft@Hrs	Unit	Material Cost	Labor Cost	Installed Cost
2P	30A	1E@0.60	Ea	561.00	16.10	577.10
2P	60A	1E@0.80	Ea	667.00	21.50	688.50
2P	100A	1E@1.25	Ea	1,350.00	33.60	1,383.60
2P	200A	1E@1.70	Ea	1,840.00	45.70	1,885.70
3P	30A	1E@0.60	Ea	561.00	16.10	577.10
3P	60A	1E@0.80	Ea	667.00	21.50	688.50
3P	100A	1E@1.25	Ea	1,350.00	33.60	1,383.60
3P	200A	1E@1.70	Ea	1,840.00	45.70	1,885.70

NEMA 12 heavy duty non-fused 600 volt safety switches

Material		Craft@Hrs	Unit	Material Cost	Labor Cost	Installed Cost
2P	30A	1E@0.50	Ea	171.00	13.40	184.40
2P	60A	1E@0.70	Ea	221.00	18.80	239.80
2P	100A	1E@1.00	Ea	316.00	26.90	342.90
2P	200A	1E@1.50	Ea	424.00	40.30	464.30
2P	400A	1E@2.00	Ea	851.00	53.80	904.80
2P	600A	1E@4.00	Ea	1,430.00	108.00	1,538.00
3P	30A	1E@0.50	Ea	171.00	13.40	184.40
3P	60A	1E@0.70	Ea	221.00	18.80	239.80
3P	100A	1E@1.00	Ea	316.00	26.90	342.90
3P	200A	1E@1.50	Ea	424.00	40.30	464.30
3P	400A	1E@2.00	Ea	1,060.00	53.80	1,113.80
3P	600A	1E@4.00	Ea	1,770.00	108.00	1,878.00
4P	30A	1E@0.60	Ea	243.00	16.10	259.10
4P	60A	1E@0.80	Ea	267.00	21.50	288.50
4P	100A	1E@1.25	Ea	475.00	33.60	508.60
4P	200A	1E@1.70	Ea	750.00	45.70	795.70
4P	400A	1E@2.25	Ea	1,500.00	60.50	1,560.50

NEMA 1 heavy duty fusible 600 volt safety switches

Material		Craft@Hrs	Unit	Material Cost	Labor Cost	Installed Cost
2P	30A	1E@0.50	Ea	147.00	13.40	160.40
2P	60A	1E@0.70	Ea	177.00	18.80	195.80
2P	100A	1E@1.00	Ea	330.00	26.90	356.90
2P	200A	1E@1.50	Ea	476.00	40.30	516.30
2P	400A	1E@2.00	Ea	1,170.00	53.80	1,223.80

Use these figures to estimate the cost of safety switches installed in buildings under the conditions described on pages 5 and 6. Costs listed are for each switch installed. The crew is one electrician working at a labor cost of $26.88 per manhour. These costs include layout, material handling, and normal waste. Add for the fuses, hubs, supports, sales tax, delivery, supervision, mobilization, demobilization, cleanup, overhead and profit. Note: Safety switches can be purchased with provision for neutral connection and are used for many purposes besides installation as part of the service entrance gear. NEMA class designations are as follows: 1 is for indoor use, 3R is for outdoor use, 4 is dusttight or watertight, 12 is watertight. Class 3R safety switches usually come with blank top hubs. Be sure the conduit size is appropriate for the switch hubs. Many safety switches are available with door interlocks or with eyes for attaching a padlock.

Fuses are on pages 255 to 277.

600 Volt Heavy Duty Safety Switches

Material		Craft@Hrs	Unit	Material Cost	Labor Cost	Installed Cost

NEMA 1 heavy duty fusible 600 volt safety switches

Material		Craft@Hrs	Unit	Material Cost	Labor Cost	Installed Cost
2P	600A	1E@4.00	Ea	1,850.00	108.00	1,958.00
2P	800A	1E@5.00	Ea	2,880.00	134.00	3,014.00
2P	1200A	1E@6.00	Ea	4,050.00	161.00	4,211.00
3P	30A	1E@0.50	Ea	147.00	13.40	160.40
3P	60A	1E@0.70	Ea	177.00	18.80	195.80
3P	100A	1E@1.00	Ea	330.00	26.90	356.90
3P	200A	1E@1.50	Ea	476.00	40.30	516.30
3P	400A	1E@2.00	Ea	1,270.00	53.80	1,323.80
3P	600A	1E@4.00	Ea	2,120.00	108.00	2,228.00
3P	800A	1E@5.00	Ea	3,910.00	134.00	4,044.00
3P	1200A	1E@6.00	Ea	4,860.00	161.00	5,021.00
4P	30A	1E@0.60	Ea	242.00	16.10	258.10
4P	60A	1E@0.80	Ea	282.00	21.50	303.50
4P	100A	1E@1.25	Ea	470.00	33.60	503.60
4P	200A	1E@1.70	Ea	783.00	45.70	828.70
4P	400A	1E@2.25	Ea	1,650.00	60.50	1,710.50
4P	600A	1E@4.25	Ea	2,680.00	114.00	2,794.00

NEMA 3R heavy duty fusible 600 volt safety switches

Material		Craft@Hrs	Unit	Material Cost	Labor Cost	Installed Cost
2P	30A	1E@0.50	Ea	249.00	13.40	262.40
2P	60A	1E@0.70	Ea	293.00	18.80	311.80
2P	100A	1E@1.00	Ea	457.00	26.90	483.90
2P	200A	1E@1.50	Ea	629.00	40.30	669.30
2P	400A	1E@2.00	Ea	1,510.00	53.80	1,563.80
2P	600A	1E@4.00	Ea	3,030.00	108.00	3,138.00
2P	800A	1E@5.00	Ea	4,560.00	134.00	4,694.00
2P	1200A	1E@6.00	Ea	5,000.00	161.00	5,161.00
3P	30A	1E@0.50	Ea	249.00	13.40	262.40
3P	60A	1E@0.70	Ea	293.00	18.80	311.80
3P	100A	1E@1.00	Ea	457.00	26.90	483.90
3P	200A	1E@1.50	Ea	629.00	40.30	669.30
3P	400A	1E@2.00	Ea	1,530.00	53.80	1,583.80
3P	600A	1E@4.00	Ea	3,030.00	108.00	3,138.00
3P	800A	1E@5.00	Ea	4,590.00	134.00	4,724.00
3P	1200A	1E@6.00	Ea	5,570.00	161.00	5,731.00

Use these figures to estimate the cost of safety switches installed in buildings under the conditions described on pages 5 and 6. Costs listed are for each switch installed. The crew is one electrician working at a labor cost of $26.88 per manhour. These costs include layout, material handling, and normal waste. Add for the fuses, hubs, supports, sales tax, delivery, supervision, mobilization, demobilization, cleanup, overhead and profit. Note: Safety switches can be purchased with provision for neutral connection and are used for many purposes besides installation as part of the service entrance gear. NEMA class designations are as follows: 1 is for indoor use, 3R is for outdoor use, 4 is dusttight or watertight, 12 is watertight. Class 3R safety switches usually come with blank top hubs. Be sure the conduit size is appropriate for the switch hubs. Many safety switches are available with door interlocks or with eyes for attaching a padlock.

Fuses are on pages 255 to 277.

600 Volt Heavy Duty Safety Switches

Material		Craft@Hrs	Unit	Material Cost	Labor Cost	Installed Cost

NEMA 4 heavy duty fusible 600 volt safety switches

Material		Craft@Hrs	Unit	Material Cost	Labor Cost	Installed Cost
2P	30A	1E@0.60	Ea	667.00	16.10	683.10
2P	60A	1E@0.80	Ea	733.00	21.50	754.50
2P	100A	1E@1.25	Ea	1,450.00	33.60	1,483.60
2P	200A	1E@1.75	Ea	2,040.00	47.00	2,087.00
2P	400A	1E@2.25	Ea	4,040.00	60.50	4,100.50
2P	600A	1E@4.25	Ea	5,790.00	114.00	5,904.00
3P	30A	1E@0.60	Ea	668.00	16.10	684.10
3P	60A	1E@0.80	Ea	733.00	21.50	754.50
3P	100A	1E@1.25	Ea	1,450.00	33.60	1,483.60
3P	200A	1E@1.75	Ea	2,040.00	47.00	2,087.00
3P	400A	1E@2.25	Ea	4,050.00	60.50	4,110.50
3P	600A	1E@4.25	Ea	5,880.00	114.00	5,994.00

NEMA 12 heavy duty fusible 600 volt safety switches

Material		Craft@Hrs	Unit	Material Cost	Labor Cost	Installed Cost
2P	30A	1E@0.50	Ea	260.00	13.40	273.40
2P	60A	1E@0.70	Ea	261.00	18.80	279.80
2P	100A	1E@1.00	Ea	406.00	26.90	432.90
2P	200A	1E@1.50	Ea	635.00	40.30	675.30
2P	400A	1E@2.00	Ea	1,350.00	53.80	1,403.80
2P	600A	1E@4.00	Ea	2,030.00	108.00	2,138.00
3P	30A	1E@0.50	Ea	260.00	13.40	273.40
3P	60A	1E@0.70	Ea	261.00	18.80	279.80
3P	100A	1E@1.00	Ea	406.00	26.90	432.90
3P	200A	1E@1.50	Ea	635.00	40.30	675.30
3P	400A	1E@2.00	Ea	1,450.00	53.80	1,503.80
3P	600A	1E@4.00	Ea	2,440.00	108.00	2,548.00
4P	30A	1E@0.60	Ea	295.00	16.10	311.10
4P	60A	1E@0.80	Ea	333.00	21.50	354.50
4P	100A	1E@1.25	Ea	511.00	33.60	544.60
4P	200A	1E@1.75	Ea	873.00	47.00	920.00
4P	400A	1E@2.25	Ea	1,810.00	60.50	1,870.50

Use these figures to estimate the cost of safety switches installed in buildings under the conditions described on pages 5 and 6. Costs listed are for each switch installed. The crew is one electrician working at a labor cost of $26.88 per manhour. These costs include layout, material handling, and normal waste. Add for the fuses, hubs, supports, sales tax, delivery, supervision, mobilization, demobilization, cleanup, overhead and profit. Note: Safety switches can be purchased with provision for neutral connection and are used for many purposes besides installation as part of the service entrance gear. NEMA class designations are as follows: 1 is for indoor use, 3R is for outdoor use, 4 is dusttight or watertight, 12 is watertight. Class 3R safety switches usually come with blank top hubs. Be sure the conduit size is appropriate for the switch hubs. Many safety switches are available with door interlocks or with eyes for attaching a padlock.

Fuses are on pages 255 to 277.

Non-fused Safety Switches

Material		Craft@Hrs	Unit	Material Cost	Labor Cost	Installed Cost

NEMA 1 heavy duty non-fused 600 volt safety switches

Material		Craft@Hrs	Unit	Material Cost	Labor Cost	Installed Cost
6P	30A	1E@0.70	Ea	910.00	18.80	928.80
6P	60A	1E@1.00	Ea	1,060.00	26.90	1,086.90
6P	100A	1E@1.50	Ea	1,290.00	40.30	1,330.30
6P	200A	1E@2.00	Ea	2,860.00	53.80	2,913.80

NEMA 7 non-fused 600 volt safety switches

Material		Craft@Hrs	Unit	Material Cost	Labor Cost	Installed Cost
3P	60A	1E@0.90	Ea	691.00	24.20	715.20
3P	100A	1E@1.25	Ea	779.00	33.60	812.60
3P	200A	1E@1.50	Ea	1,710.00	40.30	1,750.30

NEMA 1 non-fused 240 volt double throw safety switches

Material		Craft@Hrs	Unit	Material Cost	Labor Cost	Installed Cost
2P	30A	1E@0.75	Ea	154.00	20.20	174.20
2P	60A	1E@1.25	Ea	246.00	33.60	279.60
2P	100A	1E@1.75	Ea	436.00	47.00	483.00
2P	200A	1E@2.25	Ea	680.00	60.50	740.50
2P	400A	1E@4.50	Ea	1,410.00	121.00	1,531.00
2P	600A	1E@6.00	Ea	2,630.00	161.00	2,791.00
3P	30A	1E@0.80	Ea	182.00	21.50	203.50
3P	60A	1E@1.30	Ea	266.00	34.90	300.90
3P	100A	1E@1.80	Ea	436.00	48.40	484.40
3P	200A	1E@2.30	Ea	680.00	61.80	741.80
3P	400A	1E@4.60	Ea	1,930.00	124.00	2,054.00
3P	600A	1E@6.25	Ea	2,630.00	168.00	2,798.00
4P	30A	1E@1.00	Ea	252.00	26.90	278.90
4P	60A	1E@1.50	Ea	393.00	40.30	433.30
4P	100A	1E@2.00	Ea	800.00	53.80	853.80
4P	200A	1E@2.50	Ea	1,110.00	67.20	1,177.20
4P	400A	1E@5.00	Ea	2,540.00	134.00	2,674.00
4P	600A	1E@6.50	Ea	3,060.00	175.00	3,235.00

NEMA 3R non-fused 240 volt double throw safety switches

Material		Craft@Hrs	Unit	Material Cost	Labor Cost	Installed Cost
2P	100A	1E@1.75	Ea	218.00	47.00	265.00
2P	200A	1E@2.00	Ea	576.00	53.80	629.80
2P	400A	1E@2.50	Ea	3,450.00	67.20	3,517.20
3P	100A	1E@2.00	Ea	797.00	53.80	850.80
3P	200A	1E@2.50	Ea	1,550.00	67.20	1,617.20
3P	400A	1E@2.75	Ea	3,450.00	73.90	3,523.90
3P	600A	1E@5.00	Ea	4,140.00	134.00	4,274.00

Use these figures to estimate the cost of safety switches installed in buildings under the conditions described on pages 5 and 6. Costs listed are for each switch installed. The crew is one electrician working at a labor cost of $26.88 per manhour. These costs include layout, material handling, and normal waste. Add for the fuses, hubs, supports, sales tax, delivery, supervision, mobilization, demobilization, cleanup, overhead and profit. Note: Safety switches can be purchased with provision for neutral connection and are used for many purposes besides installation as part of the service entrance gear. NEMA class designations are as follows: 1 is for indoor use, 3R is for outdoor use, 4 is dusttight or watertight, 12 is watertight. Class 3R safety switches usually come with blank top hubs. Be sure the conduit size is appropriate for the switch hubs. Many safety switches are available with door interlocks or with eyes for attaching a padlock.

Fast acting single element current limiting plug fuses, 120 volt, class T

Material	Craft@Hrs	Unit	Material Cost	Labor Cost	Installed Cost
1 amp	1E@0.05	Ea	1.04	1.34	2.38
2 amp	1E@0.05	Ea	1.10	1.34	2.44
3 amp	1E@0.05	Ea	1.10	1.34	2.44
5 amp	1E@0.05	Ea	1.10	1.34	2.44
6 amp	1E@0.05	Ea	1.10	1.34	2.44
8 amp	1E@0.05	Ea	1.06	1.34	2.40
10 amp	1E@0.05	Ea	1.10	1.34	2.44
15 amp	1E@0.05	Ea	.60	1.34	1.94
20 amp	1E@0.05	Ea	.51	1.34	1.85
25 amp	1E@0.05	Ea	.65	1.34	1.99
30 amp	1E@0.05	Ea	.53	1.34	1.87

Fast acting dual element current limiting plug fuses, 120 volt, class T

Material	Craft@Hrs	Unit	Material Cost	Labor Cost	Installed Cost
3/10 amp	1E@0.05	Ea	4.01	1.34	5.35
4/10 amp	1E@0.05	Ea	4.01	1.34	5.35
1/2 amp	1E@0.05	Ea	4.01	1.34	5.35
6/10 amp	1E@0.05	Ea	4.01	1.34	5.35
8/10 amp	1E@0.05	Ea	4.01	1.34	5.35
1 amp	1E@0.05	Ea	4.01	1.34	5.35
1-1/8 amp	1E@0.05	Ea	4.01	1.34	5.35
1-1/4 amp	1E@0.05	Ea	4.01	1.34	5.35
1-4/10 amp	1E@0.05	Ea	4.01	1.34	5.35
1-6/10 amp	1E@0.05	Ea	4.01	1.34	5.35
1-8/10 amp	1E@0.05	Ea	4.01	1.34	5.35
2 amp	1E@0.05	Ea	4.01	1.34	5.35
2-1/4 amp	1E@0.05	Ea	4.01	1.34	5.35
2-8/10 amp	1E@0.05	Ea	4.01	1.34	5.35
3-2/10 amp	1E@0.05	Ea	4.01	1.34	5.35
3-1/2 amp	1E@0.05	Ea	4.05	1.34	5.39
4 amp	1E@0.05	Ea	3.60	1.34	4.94
4-1/2 amp	1E@0.05	Ea	4.01	1.34	5.35
5 amp	1E@0.05	Ea	3.60	1.34	4.94
5-6/10 amp	1E@0.05	Ea	4.01	1.34	5.35
6-1/4 amp	1E@0.05	Ea	3.15	1.34	4.49
7 amp	1E@0.05	Ea	4.01	1.34	5.35
8 amp	1E@0.05	Ea	3.15	1.34	4.49
9 amp	1E@0.05	Ea	4.05	1.34	5.39
10 amp	1E@0.05	Ea	2.79	1.34	4.13
12 amp	1E@0.05	Ea	3.60	1.34	4.94
14 amp	1E@0.05	Ea	4.01	1.34	5.35
15 amp	1E@0.05	Ea	1.14	1.34	2.48
20 amp	1E@0.05	Ea	1.14	1.34	2.48
25 amp	1E@0.05	Ea	1.76	1.34	3.10
30 amp	1E@0.05	Ea	1.14	1.34	2.48

Use this figures to estimate the cost of fuses installed in safety switches and combination starters under the conditions described on pages 5 and 6. Costs listed are for each fuse installed. The crew size is one electrician working at a labor cost of $26.88 per manhour. These costs include layout, material handling, and normal waste. Add for sales tax, delivery supervision, mobilization, demobilization, cleanup, overhead and profit.

Plug Fuses, 120 Volt

Material	Craft@Hrs	Unit	Material Cost	Labor Cost	Installed Cost
Dual element type "S" 120 volt plug fuses					
3/10 amp	1E@0.05	Ea	4.05	1.34	5.39
4/10 amp	1E@0.05	Ea	4.05	1.34	5.39
1/2 amp	1E@0.05	Ea	4.05	1.34	5.39
6/10 amp	1E@0.05	Ea	4.05	1.34	5.39
8/10 amp	1E@0.05	Ea	4.05	1.34	5.39
1 amp	1E@0.05	Ea	4.05	1.34	5.39
1-1/8 amp	1E@0.05	Ea	4.05	1.34	5.39
1-1/4 amp	1E@0.05	Ea	3.39	1.34	4.73
1-4/10 amp	1E@0.05	Ea	4.05	1.34	5.39
1-6/10 amp	1E@0.05	Ea	4.05	1.34	5.39
1-8/10 amp	1E@0.05	Ea	4.05	1.34	5.39
2 amp	1E@0.05	Ea	4.05	1.34	5.39
2-1/4 amp	1E@0.05	Ea	4.05	1.34	5.39
2-8/10 amp	1E@0.05	Ea	4.05	1.34	5.39
3 amp	1E@0.05	Ea	4.05	1.34	5.39
3-2/10 amp	1E@0.05	Ea	3.17	1.34	4.51
3-1/2 amp	1E@0.05	Ea	4.05	1.34	5.39
4 amp	1E@0.05	Ea	3.88	1.34	5.22
4-1/2 amp	1E@0.05	Ea	4.05	1.34	5.39
5 amp	1E@0.05	Ea	3.62	1.34	4.96
5-6/10 amp	1E@0.05	Ea	4.05	1.34	5.39
6 amp	1E@0.05	Ea	4.05	1.34	5.39
6-1/4 amp	1E@0.05	Ea	3.17	1.34	4.51
7 amp	1E@0.05	Ea	4.05	1.34	5.39
8 amp	1E@0.05	Ea	3.62	1.34	4.96
9 amp	1E@0.05	Ea	4.05	1.34	5.39
10 amp	1E@0.05	Ea	3.17	1.34	4.51
12 amp	1E@0.05	Ea	3.88	1.34	5.22
14 amp	1E@0.05	Ea	4.05	1.34	5.39
15 amp	1E@0.05	Ea	1.18	1.34	2.52
20 amp	1E@0.05	Ea	1.18	1.34	2.52
25 amp	1E@0.05	Ea	1.60	1.34	2.94
30 amp	1E@0.05	Ea	1.60	1.34	2.94
Time delay type "SL" 120 volt plug fuses					
15 amp	1E@0.05	Ea	.58	1.34	1.92
20 amp	1E@0.05	Ea	.70	1.34	2.04
25 amp	1E@0.05	Ea	.98	1.34	2.32
30 amp	1E@0.05	Ea	.79	1.34	2.13

Use this figures to estimate the cost of fuses installed in safety switches and combination starters under the conditions described on pages 5 and 6. Costs listed are for each fuse installed. The crew size is one electrician working at a labor cost of $26.88 per manhour. These costs include layout, material handling, and normal waste. Add for sales tax, delivery supervision, mobilization, demobilization, cleanup, overhead and profit.

Cartridge Fuses, Non-Renewable, 250 Volt, Current Limiting, Class H

Material	Craft@Hrs	Unit	Material Cost	Labor Cost	Installed Cost
Ferrule type non-renewable 250 volt class H current limiting cartridge fuses					
1 amp	1E@0.06	Ea	1.06	1.61	2.67
2 amp	1E@0.06	Ea	1.30	1.61	2.91
3 amp	1E@0.06	Ea	1.06	1.61	2.67
4 amp	1E@0.06	Ea	1.30	1.61	2.91
5 amp	1E@0.06	Ea	1.06	1.61	2.67
6 amp	1E@0.06	Ea	1.06	1.61	2.67
7 amp	1E@0.06	Ea	1.30	1.61	2.91
8 amp	1E@0.06	Ea	1.30	1.61	2.91
10 amp	1E@0.06	Ea	0.91	1.61	2.52
12 amp	1E@0.06	Ea	1.30	1.61	2.91
15 amp	1E@0.06	Ea	0.82	1.61	2.43
20 amp	1E@0.06	Ea	0.82	1.61	2.43
25 amp	1E@0.06	Ea	0.82	1.61	2.43
30 amp	1E@0.06	Ea	0.82	1.61	2.43
35 amp	1E@0.06	Ea	1.55	1.61	3.16
40 amp	1E@0.06	Ea	1.31	1.61	2.92
45 amp	1E@0.06	Ea	1.55	1.61	3.16
50 amp	1E@0.06	Ea	1.55	1.61	3.16
60 amp	1E@0.06	Ea	1.31	1.61	2.92

Material	Craft@Hrs	Unit	Material Cost	Labor Cost	Installed Cost
Blade type non-renewable 250 volt class H current limiting cartridge fuses					
70 amp	1E@0.08	Ea	6.39	2.15	8.54
80 amp	1E@0.08	Ea	7.52	2.15	9.67
90 amp	1E@0.08	Ea	7.52	2.15	9.67
100 amp	1E@0.08	Ea	5.53	2.15	7.68
110 amp	1E@0.08	Ea	19.00	2.15	21.15
125 amp	1E@0.08	Ea	15.60	2.15	17.75
150 amp	1E@0.08	Ea	15.60	2.15	17.75
175 amp	1E@0.08	Ea	19.00	2.15	21.15
200 amp	1E@0.08	Ea	13.40	2.15	15.55
225 amp	1E@0.10	Ea	33.90	2.69	36.59
250 amp	1E@0.10	Ea	33.90	2.69	36.59
300 amp	1E@0.10	Ea	33.90	2.69	36.59
350 amp	1E@0.10	Ea	33.90	2.69	36.59
400 amp	1E@0.15	Ea	24.00	4.03	28.03
450 amp	1E@0.15	Ea	53.10	4.03	57.13
500 amp	1E@0.15	Ea	53.10	4.03	57.13
600 amp	1E@0.15	Ea	46.20	4.03	50.23

Use this figures to estimate the cost of fuses installed in safety switches and combination starters under the conditions described on pages 5 and 6. Costs listed are for each fuse installed. The crew size is one electrician working at a labor cost of $26.88 per manhour. These costs include layout, material handling, and normal waste. Add for sales tax, delivery supervision, mobilization, demobilization, cleanup, overhead and profit.

Cartridge Fuses, Non-Renewable, 600 Volt, Class H

Material	Craft@Hrs	Unit	Material Cost	Labor Cost	Installed Cost
Ferrule type non-renewable 600 volt class H cartridge fuses					
1 amp	1E@0.06	Ea	5.56	1.61	7.17
2 amp	1E@0.06	Ea	5.56	1.61	7.17
3 amp	1E@0.06	Ea	5.34	1.61	6.95
4 amp	1E@0.06	Ea	5.56	1.61	7.17
5 amp	1E@0.06	Ea	5.34	1.61	6.95
6 amp	1E@0.06	Ea	4.53	1.61	6.14
8 amp	1E@0.06	Ea	5.56	1.61	7.17
10 amp	1E@0.06	Ea	4.53	1.61	6.14
12 amp	1E@0.06	Ea	5.56	1.61	7.17
15 amp	1E@0.06	Ea	3.75	1.61	5.36
20 amp	1E@0.06	Ea	3.75	1.61	5.36
25 amp	1E@0.06	Ea	4.43	1.61	6.04
30 amp	1E@0.06	Ea	3.75	1.61	5.36
35 amp	1E@0.06	Ea	7.86	1.61	9.47
40 amp	1E@0.06	Ea	6.69	1.61	8.30
45 amp	1E@0.06	Ea	7.86	1.61	9.47
50 amp	1E@0.06	Ea	6.69	1.61	8.30
60 amp	1E@0.06	Ea	5.72	1.61	7.33
Blade type non-renewable 600 volt class H cartridge fuses					
70 amp	1E@0.08	Ea	16.30	2.15	18.45
80 amp	1E@0.08	Ea	16.30	2.15	18.45
90 amp	1E@0.08	Ea	16.30	2.15	18.45
100 amp	1E@0.08	Ea	11.90	2.15	14.05
110 amp	1E@0.08	Ea	33.10	2.15	35.25
125 amp	1E@0.08	Ea	33.10	2.15	35.25
150 amp	1E@0.08	Ea	33.10	2.15	35.25
175 amp	1E@0.08	Ea	33.10	2.15	35.25
200 amp	1E@0.08	Ea	28.30	2.15	30.45
225 amp	1E@0.10	Ea	66.10	2.69	68.79
250 amp	1E@0.10	Ea	66.10	2.69	68.79
300 amp	1E@0.10	Ea	66.10	2.69	68.79
350 amp	1E@0.10	Ea	66.10	2.69	68.79
400 amp	1E@0.15	Ea	95.20	4.03	99.23
450 amp	1E@0.15	Ea	95.20	4.03	99.23
500 amp	1E@0.15	Ea	95.20	4.03	99.23
600 amp	1E@0.15	Ea	95.20	4.03	99.23

Use this figures to estimate the cost of fuses installed in safety switches and combination starters under the conditions described on pages 5 and 6. Costs listed are for each fuse installed. The crew size is one electrician working at a labor cost of $26.88 per manhour. These costs include layout, material handling, and normal waste. Add for sales tax, delivery supervision, mobilization, demobilization, cleanup, overhead and profit.

Material	Craft@Hrs	Unit	Material Cost	Labor Cost	Installed Cost

Ferrule type renewable 250 volt current limiting class H cartridge fuses with links

Material	Craft@Hrs	Unit	Material Cost	Labor Cost	Installed Cost
1 amp	1E@0.06	Ea	6.24	1.61	7.85
2 amp	1E@0.06	Ea	6.24	1.61	7.85
3 amp	1E@0.06	Ea	5.36	1.61	6.97
4 amp	1E@0.06	Ea	6.24	1.61	7.85
5 amp	1E@0.06	Ea	5.36	1.61	6.97
6 amp	1E@0.06	Ea	5.36	1.61	6.97
8 amp	1E@0.06	Ea	6.24	1.61	7.85
10 amp	1E@0.06	Ea	5.76	1.61	7.37
12 amp	1E@0.06	Ea	5.76	1.61	7.37
15 amp	1E@0.06	Ea	4.15	1.61	5.76
20 amp	1E@0.06	Ea	4.15	1.61	5.76
25 amp	1E@0.06	Ea	5.76	1.61	7.37
30 amp	1E@0.06	Ea	4.15	1.61	5.76
35 amp	1E@0.06	Ea	11.30	1.61	12.91
40 amp	1E@0.06	Ea	9.67	1.61	11.28
45 amp	1E@0.06	Ea	11.30	1.61	12.91
50 amp	1E@0.06	Ea	9.67	1.61	11.28
60 amp	1E@0.06	Ea	8.09	1.61	9.70

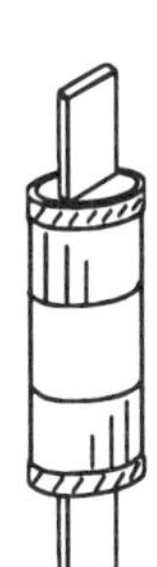

Blade type renewable 250 volt current limiting class H cartridge fuses with links

Material	Craft@Hrs	Unit	Material Cost	Labor Cost	Installed Cost
70 amp	1E@0.08	Ea	21.70	2.15	23.85
80 amp	1E@0.08	Ea	25.30	2.15	27.45
90 amp	1E@0.08	Ea	25.30	2.15	27.45
100 amp	1E@0.08	Ea	18.20	2.15	20.35
110 amp	1E@0.08	Ea	57.00	2.15	59.15
125 amp	1E@0.08	Ea	57.00	2.15	59.15
150 amp	1E@0.08	Ea	41.00	2.15	43.15
175 amp	1E@0.08	Ea	57.00	2.15	59.15
200 amp	1E@0.08	Ea	41.00	2.15	43.15
225 amp	1E@0.10	Ea	103.00	2.69	105.69
250 amp	1E@0.10	Ea	103.00	2.69	105.69
300 amp	1E@0.10	Ea	103.00	2.69	105.69
350 amp	1E@0.10	Ea	112.00	2.69	114.69
400 amp	1E@0.15	Ea	74.10	4.03	78.13
450 amp	1E@0.15	Ea	171.00	4.03	175.03
500 amp	1E@0.15	Ea	171.00	4.03	175.03
600 amp	1E@0.15	Ea	135.00	4.03	139.03

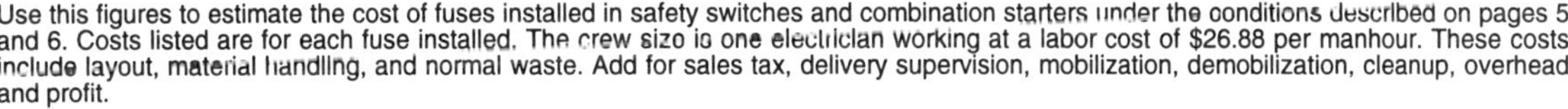

Use this figures to estimate the cost of fuses installed in safety switches and combination starters under the conditions described on pages 5 and 6. Costs listed are for each fuse installed. The crew size is one electrician working at a labor cost of $26.88 per manhour. These costs include layout, material handling, and normal waste. Add for sales tax, delivery supervision, mobilization, demobilization, cleanup, overhead and profit.

Cartridge Fuses, Links, 250 Volt, Class H

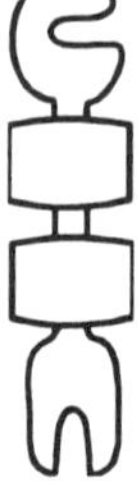

Material	Craft@Hrs	Unit	Material Cost	Labor Cost	Installed Cost
Links only for 250 volt class H cartridge fuses					
1 amp	1E@0.06	Ea	1.27	1.61	2.88
2 amp	1E@0.06	Ea	1.27	1.61	2.88
3 amp	1E@0.06	Ea	1.46	1.61	3.07
4 amp	1E@0.06	Ea	1.27	1.61	2.88
5 amp	1E@0.06	Ea	1.09	1.61	2.70
6 amp	1E@0.06	Ea	1.09	1.61	2.70
8 amp	1E@0.06	Ea	1.27	1.61	2.88
10 amp	1E@0.06	Ea	.92	1.61	2.53
12 amp	1E@0.06	Ea	.41	1.61	2.02
15 amp	1E@0.06	Ea	.41	1.61	2.02
20 amp	1E@0.06	Ea	.41	1.61	2.02
25 amp	1E@0.06	Ea	.41	1.61	2.02
30 amp	1E@0.06	Ea	.31	1.61	1.92
35 amp	1E@0.06	Ea	.78	1.61	2.39
40 amp	1E@0.06	Ea	.78	1.61	2.39
45 amp	1E@0.06	Ea	.78	1.61	2.39
50 amp	1E@0.06	Ea	.78	1.61	2.39
60 amp	1E@0.06	Ea	.78	1.61	2.39
70 amp	1E@0.08	Ea	1.52	2.15	3.67
80 amp	1E@0.08	Ea	1.52	2.15	3.67
90 amp	1E@0.08	Ea	1.52	2.15	3.67
100 amp	1E@0.08	Ea	1.12	2.15	3.27
110 amp	1E@0.08	Ea	3.19	2.15	5.34
125 amp	1E@0.08	Ea	3.19	2.15	5.34
150 amp	1E@0.08	Ea	3.19	2.15	5.34
175 amp	1E@0.08	Ea	3.19	2.15	5.34
200 amp	1E@0.08	Ea	2.53	2.15	4.68
225 amp	1E@0.10	Ea	6.17	2.69	8.86
250 amp	1E@0.10	Ea	5.30	2.69	7.99
300 amp	1E@0.10	Ea	5.30	2.69	7.99
350 amp	1E@0.10	Ea	6.77	2.69	9.46
400 amp	1E@0.15	Ea	6.32	4.03	10.35
450 amp	1E@0.15	Ea	9.38	4.03	13.41
500 amp	1E@0.15	Ea	9.38	4.03	13.41
600 amp	1E@0.15	Ea	8.01	4.03	12.04

Use this figures to estimate the cost of fuses installed in safety switches and combination starters under the conditions described on pages 5 and 6. Costs listed are for each fuse installed. The crew size is one electrician working at a labor cost of $26.88 per manhour. These costs include layout, material handling, and normal waste. Add for sales tax, delivery supervision, mobilization, demobilization, cleanup, overhead and profit.

Cartridge Fuses, Renewable, 600 Volt, Current Limiting, Class H

Material	Craft@Hrs	Unit	Material Cost	Labor Cost	Installed Cost

Ferrule type renewable 600 volt current limiting class H cartridge fuses with links

Material	Craft@Hrs	Unit	Material Cost	Labor Cost	Installed Cost
1 amp	1E@0.06	Ea	15.80	1.61	17.41
2 amp	1E@0.06	Ea	15.80	1.61	17.41
3 amp	1E@0.06	Ea	15.80	1.61	17.41
4 amp	1E@0.06	Ea	15.80	1.61	17.41
5 amp	1E@0.06	Ea	15.80	1.61	17.41
6 amp	1E@0.06	Ea	13.50	1.61	15.11
8 amp	1E@0.06	Ea	15.80	1.61	17.41
10 amp	1E@0.06	Ea	11.30	1.61	12.91
12 amp	1E@0.06	Ea	14.30	1.61	15.91
15 amp	1E@0.06	Ea	10.30	1.61	11.91
20 amp	1E@0.06	Ea	10.30	1.61	11.91
25 amp	1E@0.06	Ea	12.20	1.61	13.81
30 amp	1E@0.06	Ea	10.30	1.61	11.91
35 amp	1E@0.06	Ea	19.20	1.61	20.81
40 amp	1E@0.06	Ea	19.20	1.61	20.81
45 amp	1E@0.06	Ea	22.40	1.61	24.01
50 amp	1E@0.06	Ea	19.20	1.61	20.81
60 amp	1E@0.06	Ea	16.10	1.61	17.71

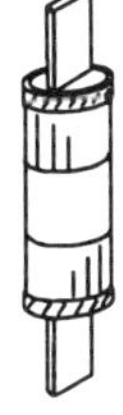

Blade type renewable 600 volt current limiting class H cartridge fuses with links

Material	Craft@Hrs	Unit	Material Cost	Labor Cost	Installed Cost
70 amp	1E@0.08	Ea	43.70	2.15	45.85
80 amp	1E@0.08	Ea	43.70	2.15	45.85
90 amp	1E@0.08	Ea	51.00	2.15	53.15
100 amp	1E@0.08	Ea	36.70	2.15	38.85
110 amp	1E@0.08	Ea	98.90	2.15	101.05
125 amp	1E@0.08	Ea	84.80	2.15	86.95
150 amp	1E@0.08	Ea	84.80	2.15	86.95
175 amp	1E@0.08	Ea	98.90	2.15	101.05
200 amp	1E@0.08	Ea	71.10	2.15	73.25
225 amp	1E@0.10	Ea	199.00	2.69	201.69
250 amp	1E@0.10	Ea	199.00	2.69	201.69
300 amp	1E@0.10	Ea	199.00	2.69	201.69
350 amp	1E@0.10	Ea	199.00	2.69	201.69
400 amp	1E@0.15	Ea	141.00	4.03	145.03
450 amp	1E@0.15	Ea	285.00	4.03	289.03
500 amp	1E@0.15	Ea	285.00	4.03	289.03
600 amp	1E@0.15	Ea	205.00	4.03	209.03

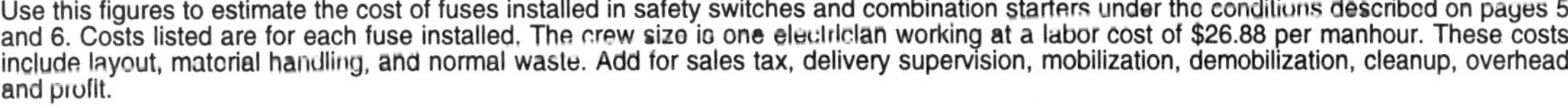

Use this figures to estimate the cost of fuses installed in safety switches and combination starters under the conditions described on pages 5 and 6. Costs listed are for each fuse installed. The crew size is one electrician working at a labor cost of $26.88 per manhour. These costs include layout, material handling, and normal waste. Add for sales tax, delivery supervision, mobilization, demobilization, cleanup, overhead and profit.

Cartridge Fuses Links, 600 Volt, Class H

Material	Craft@Hrs	Unit	Material Cost	Labor Cost	Installed Cost
Links only for 600 volt class H cartridge fuses					
1 amp	1E@0.06	Ea	2.06	1.61	3.67
3 amp	1E@0.06	Ea	2.06	1.61	3.67
4 amp	1E@0.06	Ea	2.06	1.61	3.67
5 amp	1E@0.06	Ea	2.06	1.61	3.67
6 amp	1E@0.06	Ea	1.80	1.61	3.41
8 amp	1E@0.06	Ea	2.06	1.61	3.67
10 amp	1E@0.06	Ea	1.49	1.61	3.10
12 amp	1E@0.06	Ea	.83	1.61	2.44
15 amp	1E@0.06	Ea	.62	1.61	2.23
20 amp	1E@0.06	Ea	.62	1.61	2.23
25 amp	1E@0.06	Ea	.83	1.61	2.44
30 amp	1E@0.06	Ea	.62	1.61	2.23
35 amp	1E@0.06	Ea	1.52	1.61	3.13
40 amp	1E@0.06	Ea	1.52	1.61	3.13
45 amp	1E@0.06	Ea	1.52	1.61	3.13
50 amp	1E@0.06	Ea	1.52	1.61	3.13
60 amp	1E@0.06	Ea	1.12	1.61	2.73
70 amp	1E@0.08	Ea	3.17	2.15	5.32
80 amp	1E@0.08	Ea	3.17	2.15	5.32
90 amp	1E@0.08	Ea	3.17	2.15	5.32
100 amp	1E@0.08	Ea	2.29	2.15	4.44
110 amp	1E@0.08	Ea	5.81	2.15	7.96
125 amp	1E@0.08	Ea	5.81	2.15	7.96
150 amp	1E@0.08	Ea	5.81	2.15	7.96
175 amp	1E@0.08	Ea	5.81	2.15	7.96
200 amp	1E@0.08	Ea	4.18	2.15	6.33
225 amp	1E@0.10	Ea	11.00	2.69	13.69
250 amp	1E@0.10	Ea	11.00	2.69	13.69
300 amp	1E@0.10	Ea	9.40	2.69	12.09
350 amp	1E@0.10	Ea	11.00	2.69	13.69
400 amp	1E@0.15	Ea	9.40	4.03	13.43
450 amp	1E@0.15	Ea	15.10	4.03	19.13
500 amp	1E@0.15	Ea	15.10	4.03	19.13
600 amp	1E@0.15	Ea	15.10	4.03	19.13

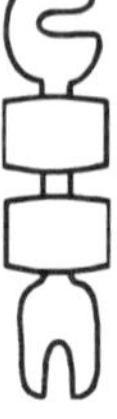

Use this figures to estimate the cost of fuses installed in safety switches and combination starters under the conditions described on pages 5 and 6. Costs listed are for each fuse installed. The crew size is one electrician working at a labor cost of $26.88 per manhour. These costs include layout, material handling, and normal waste. Add for sales tax, delivery supervision, mobilization, demobilization, cleanup, overhead and profit.

Material	Craft@Hrs	Unit	Material Cost	Labor Cost	Installed Cost
Ferrule type non-time delay 600 volt current limiting class J cartridge fuses					
1 amp	1E@0.06	Ea	13.50	1.61	15.11
3 amp	1E@0.06	Ea	13.00	1.61	14.61
4 amp	1E@0.06	Ea	13.50	1.61	15.11
5 amp	1E@0.06	Ea	13.50	1.61	15.11
6 amp	1E@0.06	Ea	11.40	1.61	13.01
10 amp	1E@0.06	Ea	10.60	1.61	12.21
15 amp	1E@0.06	Ea	10.60	1.61	12.21
20 amp	1E@0.06	Ea	10.60	1.61	12.21
25 amp	1E@0.06	Ea	13.00	1.61	14.61
30 amp	1E@0.06	Ea	9.18	1.61	10.79
35 amp	1E@0.06	Ea	21.10	1.61	22.71
40 amp	1E@0.06	Ea	17.30	1.61	18.91
45 amp	1E@0.06	Ea	21.10	1.61	22.71
50 amp	1E@0.06	Ea	15.00	1.61	16.61
60 amp	1E@0.06	Ea	15.00	1.61	16.61

Material	Craft@Hrs	Unit	Material Cost	Labor Cost	Installed Cost
Blade type non-time delay 600 volt current limiting class J cartridge fuses					
70 amp	1E@0.08	Ea	28.10	2.15	30.25
80 amp	1E@0.08	Ea	32.00	2.15	34.15
90 amp	1E@0.08	Ea	33.40	2.15	35.55
100 amp	1E@0.08	Ea	22.70	2.15	24.85
110 amp	1E@0.08	Ea	61.60	2.15	63.75
125 amp	1E@0.08	Ea	59.10	2.15	61.25
150 amp	1E@0.08	Ea	51.90	2.15	54.05
175 amp	1E@0.08	Ea	59.10	2.15	61.25
200 amp	1E@0.08	Ea	42.70	2.15	44.85
225 amp	1E@0.10	Ea	130.00	2.69	132.69
250 amp	1E@0.10	Ea	130.00	2.69	132.69
300 amp	1E@0.10	Ea	106.00	2.69	108.69
350 amp	1E@0.10	Ea	130.00	2.69	132.69
400 amp	1E@0.15	Ea	94.10	4.03	98.13
450 amp	1E@0.15	Ea	192.00	4.03	196.03
500 amp	1E@0.15	Ea	184.00	4.03	188.03
600 amp	1E@0.15	Ea	133.00	4.03	137.03

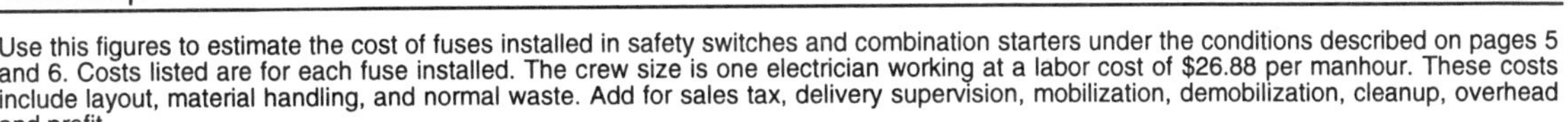

Use this figures to estimate the cost of fuses installed in safety switches and combination starters under the conditions described on pages 5 and 6. Costs listed are for each fuse installed. The crew size is one electrician working at a labor cost of $26.88 per manhour. These costs include layout, material handling, and normal waste. Add for sales tax, delivery supervision, mobilization, demobilization, cleanup, overhead and profit.

Cartridge Fuses, Time Delay, 600 Volt, Current Limiting, Class J

Material	Craft@Hrs	Unit	Material Cost	Labor Cost	Installed Cost
Ferrule type time delay 600 volt current limiting class J cartridge fuses					
1 amp	1E@0.06	Ea	7.62	1.61	9.23
3 amp	1E@0.06	Ea	7.62	1.61	9.23
6 amp	1E@0.06	Ea	7.62	1.61	9.23
10 amp	1E@0.06	Ea	7.62	1.61	9.23
15 amp	1E@0.06	Ea	7.62	1.61	9.23
20 amp	1E@0.06	Ea	7.62	1.61	9.23
25 amp	1E@0.06	Ea	7.62	1.61	9.23
30 amp	1E@0.06	Ea	7.62	1.61	9.23
35 amp	1E@0.06	Ea	13.00	1.61	14.61
40 amp	1E@0.06	Ea	13.00	1.61	14.61
45 amp	1E@0.06	Ea	13.00	1.61	14.61
50 amp	1E@0.06	Ea	13.00	1.61	14.61
60 amp	1E@0.06	Ea	13.00	1.61	14.61
Blade type time delay 600 volt current limiting class J cartridge fuses					
70 amp	1E@0.08	Ea	26.90	2.15	29.05
80 amp	1E@0.08	Ea	26.90	2.15	29.05
90 amp	1E@0.08	Ea	32.10	2.15	34.25
100 amp	1E@0.08	Ea	26.90	2.15	29.05
110 amp	1E@0.08	Ea	64.20	2.15	66.35
125 amp	1E@0.08	Ea	53.80	2.15	55.95
150 amp	1E@0.08	Ea	53.80	2.15	55.95
175 amp	1E@0.08	Ea	64.20	2.15	66.35
200 amp	1E@0.08	Ea	53.80	2.15	55.95
225 amp	1E@0.10	Ea	129.00	2.69	131.69
250 amp	1E@0.10	Ea	108.00	2.69	110.69
300 amp	1E@0.10	Ea	108.00	2.69	110.69
350 amp	1E@0.10	Ea	129.00	2.69	131.69
400 amp	1E@0.15	Ea	108.00	4.03	112.03
450 amp	1E@0.15	Ea	208.00	4.03	212.03
500 amp	1E@0.15	Ea	208.00	4.03	212.03
600 amp	1E@0.15	Ea	184.00	4.03	188.03

Use this figures to estimate the cost of fuses installed in safety switches and combination starters under the conditions described on pages 5 and 6. Costs listed are for each fuse installed. The crew size is one electrician working at a labor cost of $26.88 per manhour. These costs include layout, material handling, and normal waste. Add for sales tax, delivery supervision, mobilization, demobilization, cleanup, overhead and profit.

Bolt-on Type Cartridge Fuses, Non-Time Delay, 600 Volt, Current Limiting, Class L

Material	Craft@Hrs	Unit	Material Cost	Labor Cost	Installed Cost

Non-time delay 600 volt current limiting class L cartridge fuses, bolt-on type

Material	Craft@Hrs	Unit	Material Cost	Labor Cost	Installed Cost
601 amp	1E@0.40	Ea	380.00	10.80	390.80
650 amp	1E@0.40	Ea	396.00	10.80	406.80
700 amp	1E@0.40	Ea	379.00	10.80	389.80
800 amp	1E@0.40	Ea	239.00	10.80	249.80
1000 amp	1E@0.40	Ea	316.00	10.80	326.80
1200 amp	1E@0.50	Ea	316.00	13.40	329.40
1350 amp	1E@0.50	Ea	534.00	13.40	547.40
1500 amp	1E@0.50	Ea	534.00	13.40	547.40
1600 amp	1E@0.50	Ea	425.00	13.40	438.40
1800 amp	1E@0.50	Ea	704.00	13.40	717.40
2000 amp	1E@0.60	Ea	628.00	16.10	644.10
2500 amp	1E@0.60	Ea	848.00	16.10	864.10
3000 amp	1E@0.60	Ea	969.00	16.10	985.10
3500 amp	1E@0.60	Ea	1,490.00	16.10	1,506.10
4000 amp	1E@0.75	Ea	1,280.00	20.20	1,300.20
4500 amp	1E@0.75	Ea	1,810.00	20.20	1,830.20
5000 amp	1E@0.75	Ea	1,810.00	20.20	1,830.20
6000 amp	1E@0.75	Ea	2,270.00	20.20	2,290.20

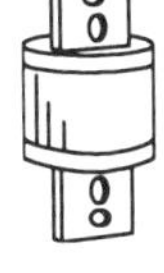

Time delay 600 volt current limiting class L cartridge fuses, bolt-on type

Material	Craft@Hrs	Unit	Material Cost	Labor Cost	Installed Cost
601 amp	1E@0.40	Ea	304.00	10.80	314.80
650 amp	1E@0.40	Ea	304.00	10.80	314.80
700 amp	1E@0.40	Ea	304.00	10.80	314.80
750 amp	1E@0.40	Ea	304.00	10.80	314.80
800 amp	1E@0.40	Ea	236.00	10.80	246.80
900 amp	1E@0.50	Ea	305.00	13.40	318.40
1000 amp	1E@0.50	Ea	276.00	13.40	289.40
1200 amp	1E@0.50	Ea	276.00	13.40	289.40
1350 amp	1E@0.50	Ea	406.00	13.40	419.40
1400 amp	1E@0.50	Ea	406.00	13.40	419.40
1500 amp	1E@0.60	Ea	406.00	16.10	422.10
1600 amp	1E@0.60	Ea	368.00	16.10	384.10
1800 amp	1E@0.60	Ea	536.00	16.10	552.10
2000 amp	1E@0.60	Ea	490.00	16.10	506.10
2500 amp	1E@0.60	Ea	679.00	16.10	695.10
3000 amp	1E@0.60	Ea	746.00	16.10	762.10
3500 amp	1E@0.75	Ea	1,200.00	20.20	1,220.20
4000 amp	1E@0.75	Ea	1,020.00	20.20	1,040.20
4500 amp	1E@0.75	Ea	1,480.00	20.20	1,500.20
5000 amp	1E@0.75	Ea	1,480.00	20.20	1,500.20
6000 amp	1E@0.55	Ea	1,860.00	14.80	1,874.80

Use this figures to estimate the cost of fuses installed in safety switches and combination starters under the conditions described on pages 5 and 6. Costs listed are for each fuse installed. The crew size is one electrician working at a labor cost of $26.88 per manhour. These costs include layout, material handling, and normal waste. Add for sales tax, delivery supervision, mobilization, demobilization, cleanup, overhead and profit.

Cartridge Fuses, Non-Time Delay, 250 Volt, Current Limiting, Class RK1

Material	Craft@Hrs	Unit	Material Cost	Labor Cost	Installed Cost
Ferrule type non-time delay 250 volt current limiting class RK1 cartridge fuses					
1 amp	1E@0.06	Ea	8.92	1.61	10.53
3 amp	1E@0.06	Ea	8.92	1.61	10.53
5 amp	1E@0.06	Ea	8.92	1.61	10.53
6 amp	1E@0.06	Ea	8.51	1.61	10.12
10 amp	1E@0.06	Ea	8.92	1.61	10.53
15 amp	1E@0.06	Ea	8.51	1.61	10.12
20 amp	1E@0.06	Ea	8.51	1.61	10.12
25 amp	1E@0.06	Ea	8.51	1.61	10.12
30 amp	1E@0.06	Ea	8.51	1.61	10.12
35 amp	1E@0.06	Ea	20.40	1.61	22.01
40 amp	1E@0.06	Ea	20.40	1.61	22.01
45 amp	1E@0.06	Ea	20.40	1.61	22.01
50 amp	1E@0.06	Ea	22.60	1.61	24.21
60 amp	1E@0.06	Ea	16.20	1.61	17.81
Blade type non-time delay 250 volt current limiting class RK1 cartridge fuses					
70 amp	1E@0.08	Ea	41.40	2.15	43.55
80 amp	1E@0.08	Ea	43.20	2.15	45.35
90 amp	1E@0.08	Ea	43.20	2.15	45.35
100 amp	1E@0.08	Ea	35.90	2.15	38.05
110 amp	1E@0.08	Ea	83.70	2.15	85.85
125 amp	1E@0.08	Ea	83.70	2.15	85.85
150 amp	1E@0.08	Ea	83.70	2.15	85.85
175 amp	1E@0.08	Ea	83.70	2.15	85.85
200 amp	1E@0.08	Ea	71.50	2.15	73.65
225 amp	1E@0.10	Ea	170.00	2.69	172.69
250 amp	1E@0.10	Ea	172.00	2.69	174.69
300 amp	1E@0.10	Ea	172.00	2.69	174.69
350 amp	1E@0.10	Ea	172.00	2.69	174.69
400 amp	1E@0.15	Ea	146.00	4.03	150.03
450 amp	1E@0.15	Ea	236.00	4.03	240.03
500 amp	1E@0.15	Ea	236.00	4.03	240.03
600 amp	1E@0.15	Ea	188.00	4.03	192.03

Use this figures to estimate the cost of fuses installed in safety switches and combination starters under the conditions described on pages 5 and 6. Costs listed are for each fuse installed. The crew size is one electrician working at a labor cost of $26.88 per manhour. These costs include layout, material handling, and normal waste. Add for sales tax, delivery supervision, mobilization, demobilization, cleanup, overhead and profit.

Cartridge Fuses, Time Delay, 250 Volt, Current Limiting, Class RK1

Material	Craft@Hrs	Unit	Material Cost	Labor Cost	Installed Cost
Ferrule type time delay 250 volt current limiting class RK1 cartridge fuses					
1/10 amp	1E@0.06	Ea	8.72	1.61	10.33
15/100 amp	1E@0.06	Ea	8.72	1.61	10.33
2/10 amp	1E@0.06	Ea	8.72	1.61	10.33
3/10 amp	1E@0.06	Ea	8.72	1.61	10.33
4/10 amp	1E@0.06	Ea	8.72	1.61	10.33
1/2 amp	1E@0.06	Ea	6.99	1.61	8.60
6/10 amp	1E@0.06	Ea	6.99	1.61	8.60
8/10 amp	1E@0.06	Ea	8.72	1.61	10.33
1 amp	1E@0.06	Ea	5.83	1.61	7.44
1-1/8 amp	1E@0.06	Ea	8.72	1.61	10.33
1-1/4 amp	1E@0.06	Ea	8.72	1.61	10.33
1-4/10 amp	1E@0.06	Ea	8.72	1.61	10.33
1-6/10 amp	1E@0.06	Ea	6.99	1.61	8.60
2 amp	1E@0.06	Ea	5.83	1.61	7.44
2-1/4 amp	1E@0.06	Ea	8.72	1.61	10.33
2-1/2 amp	1E@0.06	Ea	5.83	1.61	7.44
2-8/10 amp	1E@0.06	Ea	8.72	1.61	10.33
3 amp	1E@0.06	Ea	5.83	1.61	7.44
3-2/10 amp	1E@0.06	Ea	5.83	1.61	7.44
3-1/2 amp	1E@0.06	Ea	6.99	1.61	8.60
4 amp	1E@0.06	Ea	5.83	1.61	7.44
4-1/2 amp	1E@0.06	Ea	6.99	1.61	8.60
5 amp	1E@0.06	Ea	5.83	1.61	7.44
5-6/10 amp	1E@0.06	Ea	8.72	1.61	10.33
6 amp	1E@0.06	Ea	5.83	1.61	7.44
6-1/4 amp	1E@0.06	Ea	5.83	1.61	7.44
7 amp	1E@0.06	Ea	6.99	1.61	8.60
8 amp	1E@0.06	Ea	5.83	1.61	7.44
9 amp	1E@0.06	Ea	6.99	1.61	8.60
10 amp	1E@0.06	Ea	5.83	1.61	7.44
12 amp	1E@0.06	Ea	5.83	1.61	7.44
15 amp	1E@0.06	Ea	5.36	1.61	6.97
17-1/2 amp	1E@0.06	Ea	6.32	1.61	7.93

Use this figures to estimate the cost of fuses installed in safety switches and combination starters under the conditions described on pages 5 and 6. Costs listed are for each fuse installed. The crew size is one electrician working at a labor cost of $26.88 per manhour. These costs include layout, material handling, and normal waste. Add for sales tax, delivery supervision, mobilization, demobilization, cleanup, overhead and profit.

Cartridge Fuses, Time Delay, 250 Volt, Current Limiting, Class RK1

Material	Craft@Hrs	Unit	Material Cost	Labor Cost	Installed Cost
Ferrule type time delay 250 volt current limiting class RK1 cartridge fuses					
20 amp	1E@0.06	Ea	5.36	1.61	6.97
25 amp	1E@0.06	Ea	5.36	1.61	6.97
30 amp	1E@0.06	Ea	5.36	1.61	6.97
35 amp	1E@0.06	Ea	12.50	1.61	14.11
40 amp	1E@0.06	Ea	12.50	1.61	14.11
45 amp	1E@0.06	Ea	12.50	1.61	14.11
50 amp	1E@0.06	Ea	12.50	1.61	14.11
60 amp	1E@0.06	Ea	12.50	1.61	14.11
Blade type time delay 250 volt current limiting class RK1 cartridge fuses					
70 amp	1E@0.08	Ea	29.30	2.15	31.45
80 amp	1E@0.08	Ea	29.30	2.15	31.45
90 amp	1E@0.08	Ea	29.30	2.15	31.45
100 amp	1E@0.08	Ea	29.30	2.15	31.45
110 amp	1E@0.08	Ea	71.30	2.15	73.45
125 amp	1E@0.08	Ea	60.90	2.15	63.05
150 amp	1E@0.08	Ea	60.90	2.15	63.05
175 amp	1E@0.08	Ea	60.90	2.15	63.05
200 amp	1E@0.08	Ea	60.90	2.15	63.05
225 amp	1E@0.10	Ea	123.00	2.69	125.69
250 amp	1E@0.10	Ea	123.00	2.69	125.69
300 amp	1E@0.10	Ea	123.00	2.69	125.69
350 amp	1E@0.10	Ea	144.00	2.69	146.69
400 amp	1E@0.15	Ea	121.00	4.03	125.03
450 amp	1E@0.15	Ea	210.00	4.03	214.03
500 amp	1E@0.15	Ea	177.00	4.03	181.03
600 amp	1E@0.15	Ea	150.00	4.03	154.03

Use this figures to estimate the cost of fuses installed in safety switches and combination starters under the conditions described on pages 5 and 6. Costs listed are for each fuse installed. The crew size is one electrician working at a labor cost of $26.88 per manhour. These costs include layout, material handling, and normal waste. Add for sales tax, delivery supervision, mobilization, demobilization, cleanup, overhead and profit.

Ferrule type non-time delay 600 volt current limiting class RK1 cartridge fuses

Material	Craft@Hrs	Unit	Material Cost	Labor Cost	Installed Cost
1 amp	1E@0.06	Ea	11.40	1.61	13.01
3 amp	1E@0.06	Ea	11.40	1.61	13.01
4 amp	1E@0.06	Ea	11.40	1.61	13.01
5 amp	1E@0.06	Ea	11.40	1.61	13.01
6 amp	1E@0.06	Ea	11.40	1.61	13.01
10 amp	1E@0.06	Ea	11.40	1.61	13.01
15 amp	1E@0.06	Ea	10.80	1.61	12.41
20 amp	1E@0.06	Ea	10.80	1.61	12.41
25 amp	1E@0.06	Ea	10.80	1.61	12.41
30 amp	1E@0.06	Ea	10.80	1.61	12.41
35 amp	1E@0.06	Ea	23.80	1.61	25.41
40 amp	1E@0.06	Ea	23.80	1.61	25.41
45 amp	1E@0.06	Ea	23.80	1.61	25.41
50 amp	1E@0.06	Ea	23.80	1.61	25.41
60 amp	1E@0.06	Ea	23.80	1.61	25.41

Blade type non-time delay 600 volt current limiting class RK1 cartridge fuses

Material	Craft@Hrs	Unit	Material Cost	Labor Cost	Installed Cost
70 amp	1E@0.08	Ea	51.00	2.15	53.15
80 amp	1E@0.08	Ea	51.00	2.15	53.15
90 amp	1E@0.08	Ea	51.00	2.15	53.15
100 amp	1E@0.08	Ea	42.90	2.15	45.05
110 amp	1E@0.08	Ea	93.20	2.15	95.35
125 amp	1E@0.08	Ea	93.20	2.15	95.35
150 amp	1E@0.08	Ea	80.60	2.15	82.75
175 amp	1E@0.08	Ea	93.20	2.15	95.35
200 amp	1E@0.08	Ea	80.60	2.15	82.75
225 amp	1E@0.10	Ea	186.00	2.69	188.69
250 amp	1E@0.10	Ea	186.00	2.69	188.69
300 amp	1E@0.10	Ea	164.00	2.69	166.69
350 amp	1E@0.10	Ea	186.00	2.69	188.69
400 amp	1E@0.15	Ea	164.00	4.03	168.03
450 amp	1E@0.15	Ea	253.00	4.03	257.03
500 amp	1E@0.15	Ea	253.00	4.03	257.03
600 amp	1E@0.15	Ea	219.00	4.03	223.03

Use this figures to estimate the cost of fuses installed in safety switches and combination starters under the conditions described on pages 5 and 6. Costs listed are for each fuse installed. The crew size is one electrician working at a labor cost of $26.88 per manhour. These costs include layout, material handling, and normal waste. Add for sales tax, delivery supervision, mobilization, demobilization, cleanup, overhead and profit.

Cartridge Fuses, Time Delay, 600 Volt, Current Limiting, Class RK1

Material	Craft@Hrs	Unit	Material Cost	Labor Cost	Installed Cost
Ferrule type time delay 600 volt current limiting class RK1 cartridge fuses					
1/10 amp	1E@0.06	Ea	11.80	1.61	13.41
15/100 amp	1E@0.06	Ea	11.80	1.61	13.41
2/10 amp	1E@0.06	Ea	11.80	1.61	13.41
3/10 amp	1E@0.06	Ea	11.80	1.61	13.41
4/10 amp	1E@0.06	Ea	11.80	1.61	13.41
1/2 amp	1E@0.06	Ea	12.00	1.61	13.61
6/10 amp	1E@0.06	Ea	12.00	1.61	13.61
8/10 amp	1E@0.06	Ea	11.80	1.61	13.41
1 amp	1E@0.06	Ea	8.04	1.61	9.65
1-1/8 amp	1E@0.06	Ea	11.80	1.61	13.41
1-1/4 amp	1E@0.06	Ea	9.41	1.61	11.02
1-4/10 amp	1E@0.06	Ea	11.80	1.61	13.41
1-6/10 amp	1E@0.06	Ea	11.80	1.61	13.41
1-8/10 amp	1E@0.06	Ea	11.80	1.61	13.41
2 amp	1E@0.06	Ea	11.80	1.61	13.41
2-1/4 amp	1E@0.06	Ea	11.80	1.61	13.41
2-1/2 amp	1E@0.06	Ea	8.04	1.61	9.65
2-8/10 amp	1E@0.06	Ea	8.04	1.61	9.65
3 amp	1E@0.06	Ea	8.04	1.61	9.65
3-1/2 amp	1E@0.06	Ea	8.04	1.61	9.65
4 amp	1E@0.06	Ea	8.04	1.61	9.65
4-1/2 amp	1E@0.06	Ea	9.41	1.61	11.02
5 amp	1E@0.06	Ea	7.98	1.61	9.59
5-6/10 amp	1E@0.06	Ea	9.41	1.61	11.02
6 amp	1E@0.06	Ea	7.98	1.61	9.59
6-1/4 amp	1E@0.06	Ea	7.98	1.61	9.59
7 amp	1E@0.06	Ea	7.98	1.61	9.59
8 amp	1E@0.06	Ea	7.98	1.61	9.59
9 amp	1E@0.06	Ea	7.98	1.61	9.59
10 amp	1E@0.06	Ea	7.98	1.61	9.59
12 amp	1E@0.06	Ea	7.98	1.61	9.59
15 amp	1E@0.06	Ea	7.52	1.61	9.13
17-1/2 amp	1E@0.06	Ea	7.52	1.61	9.13

Use this figures to estimate the cost of fuses installed in safety switches and combination starters under the conditions described on pages 5 and 6. Costs listed are for each fuse installed. The crew size is one electrician working at a labor cost of $26.88 per manhour. These costs include layout, material handling, and normal waste. Add for sales tax, delivery supervision, mobilization, demobilization, cleanup, overhead and profit.

Material	Craft@Hrs	Unit	Material Cost	Labor Cost	Installed Cost
Ferrule type time delay 600 volt current limiting class RK1 cartridge fuses					
20 amp	1E@0.06	Ea	7.52	1.61	9.13
25 amp	1E@0.06	Ea	7.52	1.61	9.13
30 amp	1E@0.06	Ea	7.52	1.61	9.13
35 amp	1E@0.06	Ea	16.00	1.61	17.61
40 amp	1E@0.06	Ea	16.00	1.61	17.61
45 amp	1E@0.06	Ea	16.00	1.61	17.61
50 amp	1E@0.06	Ea	16.00	1.61	17.61
60 amp	1E@0.06	Ea	16.00	1.61	17.61

Material	Craft@Hrs	Unit	Material Cost	Labor Cost	Installed Cost
Blade type time delay 600 volt current limiting class RK1 cartridge fuses					
70 amp	1E@0.08	Ea	36.20	2.15	38.35
80 amp	1E@0.08	Ea	36.20	2.15	38.35
90 amp	1E@0.08	Ea	36.20	2.15	38.35
100 amp	1E@0.08	Ea	36.20	2.15	38.35
110 amp	1E@0.08	Ea	72.50	2.15	74.65
125 amp	1E@0.08	Ea	72.50	2.15	74.65
150 amp	1E@0.08	Ea	72.50	2.15	74.65
200 amp	1E@0.08	Ea	60.90	2.15	63.05
225 amp	1E@0.10	Ea	149.00	2.69	151.69
250 amp	1E@0.10	Ea	149.00	2.69	151.69
300 amp	1E@0.10	Ea	149.00	2.69	151.69
350 amp	1E@0.10	Ea	149.00	2.69	151.69
400 amp	1E@0.15	Ea	149.00	4.03	153.03
450 amp	1E@0.15	Ea	216.00	4.03	220.03
500 amp	1E@0.15	Ea	185.00	4.03	189.03
600 amp	1E@0.15	Ea	185.00	4.03	189.03

Use this figures to estimate the cost of fuses installed in safety switches and combination starters under the conditions described on pages 5 and 6. Costs listed are for each fuse installed. The crew size is one electrician working at a labor cost of $26.88 per manhour. These costs include layout, material handling, and normal waste. Add for sales tax, delivery supervision, mobilization, demobilization, cleanup, overhead and profit.

Cartridge Fuses, Time Delay, 250 Volt, Current Limiting, Class RK5

Material	Craft@Hrs	Unit	Material Cost	Labor Cost	Installed Cost
Ferrule type time delay 250 volt current limiting class RK5 cartridge fuses					
1/10 amp	1E@0.06	Ea	4.68	1.61	6.29
15/100 amp	1E@0.06	Ea	4.68	1.61	6.29
2/10 amp	1E@0.06	Ea	4.68	1.61	6.29
1/4 amp	1E@0.06	Ea	5.75	1.61	7.36
3/10 amp	1E@0.06	Ea	4.68	1.61	6.29
4/10 amp	1E@0.06	Ea	4.68	1.61	6.29
1/2 amp	1E@0.06	Ea	3.85	1.61	5.46
6/10 amp	1E@0.06	Ea	3.85	1.61	5.46
8/10 amp	1E@0.06	Ea	4.50	1.61	6.11
1 amp	1E@0.06	Ea	3.28	1.61	4.89
1-1/8 amp	1E@0.06	Ea	4.68	1.61	6.29
1-1/4 amp	1E@0.06	Ea	4.68	1.61	6.29
1-4/10 amp	1E@0.06	Ea	4.68	1.61	6.29
1-6/10 amp	1E@0.06	Ea	3.85	1.61	5.46
1-8/10 amp	1E@0.06	Ea	4.68	1.61	6.29
2 amp	1E@0.06	Ea	3.28	1.61	4.89
2-1/4 amp	1E@0.06	Ea	4.68	1.61	6.29
2-1/2 amp	1E@0.06	Ea	3.28	1.61	4.89
2-8/10 amp	1E@0.06	Ea	4.68	1.61	6.29
3 amp	1E@0.06	Ea	3.02	1.61	4.63
3-2/10 amp	1E@0.06	Ea	3.02	1.61	4.63
3-1/2 amp	1E@0.06	Ea	3.51	1.61	5.12
4 amp	1E@0.06	Ea	3.02	1.61	4.63
4-1/2 amp	1E@0.06	Ea	3.51	1.61	5.12
5 amp	1E@0.06	Ea	3.02	1.61	4.63
5-6/10 amp	1E@0.06	Ea	4.10	1.61	5.71
6 amp	1E@0.06	Ea	3.02	1.61	4.63
6-1/4 amp	1E@0.06	Ea	3.02	1.61	4.63
7 amp	1E@0.06	Ea	3.51	1.61	5.12
8 amp	1E@0.06	Ea	3.02	1.61	4.63
9 amp	1E@0.06	Ea	3.51	1.61	5.12
10 amp	1E@0.06	Ea	3.02	1.61	4.63
12 amp	1E@0.06	Ea	3.02	1.61	4.63
15 amp	1E@0.06	Ea	2.47	1.61	4.08
17-1/2 amp	1E@0.06	Ea	2.90	1.61	4.51

Use this figures to estimate the cost of fuses installed in safety switches and combination starters under the conditions described on pages 5 and 6. Costs listed are for each fuse installed. The crew size is one electrician working at a labor cost of $26.88 per manhour. These costs include layout, material handling, and normal waste. Add for sales tax, delivery supervision, mobilization, demobilization, cleanup, overhead and profit.

Material	Craft@Hrs	Unit	Material Cost	Labor Cost	Installed Cost
Ferrule type time delay 250 volt current limiting class RK5 cartridge fuses					
20 amp	1E@0.06	Ea	2.47	1.61	4.08
25 amp	1E@0.06	Ea	2.47	1.61	4.08
30 amp	1E@0.06	Ea	2.47	1.61	4.08
35 amp	1E@0.06	Ea	4.49	1.61	6.10
40 amp	1E@0.06	Ea	4.49	1.61	6.10
45 amp	1E@0.06	Ea	4.49	1.61	6.10
50 amp	1E@0.06	Ea	4.49	1.61	6.10
60 amp	1E@0.06	Ea	4.49	1.61	6.10
Blade type time delay 250 volt current limiting class RK5 cartridge fuses					
70 amp	1E@0.08	Ea	9.90	2.15	12.05
75 amp	1E@0.08	Ea	14.40	2.15	16.55
80 amp	1E@0.08	Ea	10.10	2.15	12.25
90 amp	1E@0.08	Ea	10.10	2.15	12.25
100 amp	1E@0.08	Ea	10.10	2.15	12.25
110 amp	1E@0.08	Ea	25.90	2.15	28.05
125 amp	1E@0.08	Ea	22.10	2.15	24.25
150 amp	1E@0.08	Ea	22.10	2.15	24.25
175 amp	1E@0.08	Ea	22.10	2.15	24.25
200 amp	1E@0.08	Ea	22.10	2.15	24.25
225 amp	1E@0.10	Ea	39.90	2.69	42.59
250 amp	1E@0.10	Ea	39.90	2.69	42.59
300 amp	1E@0.10	Ea	39.90	2.69	42.59
350 amp	1E@0.10	Ea	46.70	2.69	49.39
400 amp	1E@0.15	Ea	39.90	4.03	43.93
450 amp	1E@0.15	Ea	83.90	4.03	87.93
500 amp	1E@0.15	Ea	71.50	4.03	75.53
600 amp	1E@0.15	Ea	61.10	4.03	65.13

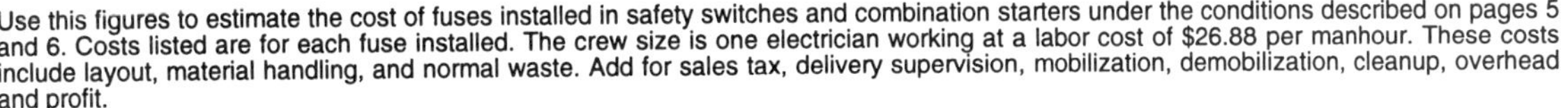

Use this figures to estimate the cost of fuses installed in safety switches and combination starters under the conditions described on pages 5 and 6. Costs listed are for each fuse installed. The crew size is one electrician working at a labor cost of $26.88 per manhour. These costs include layout, material handling, and normal waste. Add for sales tax, delivery supervision, mobilization, demobilization, cleanup, overhead and profit.

Cartridge Fuses, Time Delay, 600 Volt, Current Limiting, Class RK5

Material	Craft@Hrs	Unit	Material Cost	Labor Cost	Installed Cost
Ferrule type time delay 600 volt current limiting class RK5 cartridge fuses					
1/10 amp	1E@0.06	Ea	9.50	1.61	11.11
15/100 amp	1E@0.06	Ea	9.50	1.61	11.11
2/10 amp	1E@0.06	Ea	9.50	1.61	11.11
3/10 amp	1E@0.06	Ea	9.50	1.61	11.11
4/10 amp	1E@0.06	Ea	9.50	1.61	11.11
1/2 amp	1E@0.06	Ea	9.04	1.61	10.65
6/10 amp	1E@0.06	Ea	9.04	1.61	10.65
8/10 amp	1E@0.06	Ea	9.04	1.61	10.65
1 amp	1E@0.06	Ea	6.59	1.61	8.20
1-1/8 amp	1E@0.06	Ea	9.04	1.61	10.65
1-1/4 amp	1E@0.06	Ea	7.70	1.61	9.31
1-4/10 amp	1E@0.06	Ea	9.04	1.61	10.65
1-6/10 amp	1E@0.06	Ea	7.70	1.61	9.31
1-8/10 amp	1E@0.06	Ea	7.70	1.61	9.31
2 amp	1E@0.06	Ea	6.59	1.61	8.20
2-1/4 amp	1E@0.06	Ea	7.70	1.61	9.31
2-1/2 amp	1E@0.06	Ea	6.59	1.61	8.20
2-8/10 amp	1E@0.06	Ea	6.59	1.61	8.20
3 amp	1E@0.06	Ea	6.12	1.61	7.73
3-2/10 amp	1E@0.06	Ea	6.12	1.61	7.73
3-1/2 amp	1E@0.06	Ea	6.12	1.61	7.73
4 amp	1E@0.06	Ea	6.12	1.61	7.73
4-1/2 amp	1E@0.06	Ea	7.17	1.61	8.78
5 amp	1E@0.06	Ea	6.12	1.61	7.73
5-6/10 amp	1E@0.06	Ea	7.17	1.61	8.78
6 amp	1E@0.06	Ea	6.12	1.61	7.73
6-1/4 amp	1E@0.06	Ea	6.12	1.61	7.73
7 amp	1E@0.06	Ea	6.12	1.61	7.73
8 amp	1E@0.06	Ea	6.12	1.61	7.73
9 amp	1E@0.06	Ea	6.12	1.61	7.73
10 amp	1E@0.06	Ea	6.12	1.61	7.73
12 amp	1E@0.06	Ea	6.12	1.61	7.73
15 amp	1E@0.06	Ea	5.49	1.61	7.10
17-1/2 amp	1E@0.06	Ea	5.49	1.61	7.10

Use this figures to estimate the cost of fuses installed in safety switches and combination starters under the conditions described on pages 5 and 6. Costs listed are for each fuse installed. The crew size is one electrician working at a labor cost of $26.88 per manhour. These costs include layout, material handling, and normal waste. Add for sales tax, delivery supervision, mobilization, demobilization, cleanup, overhead and profit.

Material	Craft@Hrs	Unit	Material Cost	Labor Cost	Installed Cost
Ferrule type time delay 600 volt current limiting class RK5 cartridge fuses					
20 amp	1E@0.06	Ea	5.49	1.61	7.10
25 amp	1E@0.06	Ea	5.49	1.61	7.10
30 amp	1E@0.06	Ea	5.49	1.61	7.10
35 amp	1E@0.06	Ea	9.34	1.61	10.95
40 amp	1E@0.06	Ea	9.34	1.61	10.95
45 amp	1E@0.06	Ea	9.34	1.61	10.95
50 amp	1E@0.06	Ea	9.34	1.61	10.95
60 amp	1E@0.06	Ea	9.34	1.61	10.95

Material	Craft@Hrs	Unit	Material Cost	Labor Cost	Installed Cost
Blade type time delay 600 volt current limiting class RK5 cartridge fuses					
70 amp	1E@0.08	Ea	19.30	2.15	21.45
75 amp	1E@0.08	Ea	26.60	2.15	28.75
80 amp	1E@0.08	Ea	19.30	2.15	21.45
90 amp	1E@0.08	Ea	19.30	2.15	21.45
100 amp	1E@0.08	Ea	19.30	2.15	21.45
110 amp	1E@0.08	Ea	38.70	2.15	40.85
125 amp	1E@0.08	Ea	38.70	2.15	40.85
150 amp	1E@0.08	Ea	38.70	2.15	40.85
175 amp	1E@0.08	Ea	38.70	2.15	40.85
200 amp	1E@0.08	Ea	38.70	2.15	40.85
225 amp	1E@0.10	Ea	77.30	2.69	79.99
250 amp	1E@0.10	Ea	77.30	2.69	79.99
300 amp	1E@0.10	Ea	77.30	2.69	79.99
350 amp	1E@0.10	Ea	77.30	2.69	79.99
400 amp	1E@0.15	Ea	77.30	4.03	81.33
450 amp	1E@0.15	Ea	129.00	4.03	133.03
500 amp	1E@0.15	Ea	111.00	4.03	115.03
600 amp	1E@0.15	Ea	111.00	4.03	115.03

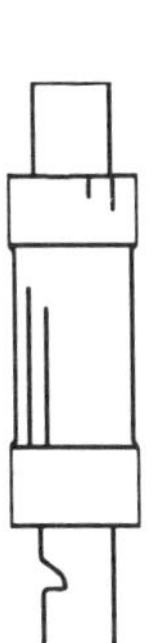

Use this figures to estimate the cost of fuses installed in safety switches and combination starters under the conditions described on pages 5 and 6. Costs listed are for each fuse installed. The crew size is one electrician working at a labor cost of $26.88 per manhour. These costs include layout, material handling, and normal waste. Add for sales tax, delivery supervision, mobilization, demobilization, cleanup, overhead and profit.

Cartridge Fuses, Fast Acting, 300 Volt, Current Limiting, Class T

Material	Craft@Hrs	Unit	Material Cost	Labor Cost	Installed Cost
Blade type fast acting 300 volt current limiting class T cartridge fuses					
1 amp	1E@0.06	Ea	8.44	1.61	10.05
2 amp	1E@0.06	Ea	8.44	1.61	10.05
3 amp	1E@0.06	Ea	8.44	1.61	10.05
6 amp	1E@0.06	Ea	8.44	1.61	10.05
10 amp	1E@0.06	Ea	8.44	1.61	10.05
15 amp	1E@0.06	Ea	8.44	1.61	10.05
20 amp	1E@0.06	Ea	8.44	1.61	10.05
25 amp	1E@0.06	Ea	8.44	1.61	10.05
30 amp	1E@0.06	Ea	8.13	1.61	9.74
35 amp	1E@0.06	Ea	8.99	1.61	10.60
40 amp	1E@0.06	Ea	8.58	1.61	10.19
45 amp	1E@0.06	Ea	8.99	1.61	10.60
50 amp	1E@0.06	Ea	8.99	1.61	10.60
60 amp	1E@0.06	Ea	6.67	1.61	8.28
70 amp	1E@0.08	Ea	10.70	2.15	12.85
80 amp	1E@0.08	Ea	12.80	2.15	14.95
90 amp	1E@0.08	Ea	12.80	2.15	14.95
100 amp	1E@0.08	Ea	9.98	2.15	12.13
110 amp	1E@0.08	Ea	19.30	2.15	21.45
125 amp	1E@0.08	Ea	18.50	2.15	20.65
150 amp	1E@0.08	Ea	16.20	2.15	18.35
175 amp	1E@0.08	Ea	19.30	2.15	21.45
200 amp	1E@0.08	Ea	11.00	2.15	13.15
225 amp	1E@0.10	Ea	42.60	2.69	45.29
250 amp	1E@0.10	Ea	42.60	2.69	45.29
300 amp	1E@0.10	Ea	40.90	2.69	43.59
350 amp	1E@0.10	Ea	33.40	2.69	36.09
400 amp	1E@0.15	Ea	24.30	4.03	28.33
450 amp	1E@0.15	Ea	43.90	4.03	47.93
500 amp	1E@0.15	Ea	56.00	4.03	60.03
600 amp	1E@0.15	Ea	31.90	4.03	35.93

Use this figures to estimate the cost of fuses installed in safety switches and combination starters under the conditions described on pages 5 and 6. Costs listed are for each fuse installed. The crew size is one electrician working at a labor cost of $26.88 per manhour. These costs include layout, material handling, and normal waste. Add for sales tax, delivery supervision, mobilization, demobilization, cleanup, overhead and profit.

Material	Craft@Hrs	Unit	Material Cost	Labor Cost	Installed Cost
Blade type fast acting 600 volt current limiting class T cartridge fuses					
1 amp	1E@0.06	Ea	8.60	1.61	10.21
2 amp	1E@0.06	Ea	8.99	1.61	10.60
3 amp	1E@0.06	Ea	8.60	1.61	10.21
6 amp	1E@0.06	Ea	7.02	1.61	8.63
10 amp	1E@0.06	Ea	7.02	1.61	8.63
15 amp	1E@0.06	Ea	7.02	1.61	8.63
20 amp	1E@0.06	Ea	7.55	1.61	9.16
25 amp	1E@0.06	Ea	7.02	1.61	8.63
30 amp	1E@0.06	Ea	5.12	1.61	6.73
35 amp	1E@0.06	Ea	10.60	1.61	12.21
40 amp	1E@0.06	Ea	11.40	1.61	13.01
45 amp	1E@0.06	Ea	13.00	1.61	14.61
50 amp	1E@0.06	Ea	10.60	1.61	12.21
60 amp	1E@0.06	Ea	7.70	1.61	9.31
70 amp	1E@0.08	Ea	14.10	2.15	16.25
80 amp	1E@0.08	Ea	17.20	2.15	19.35
90 amp	1E@0.08	Ea	20.40	2.15	22.55
100 amp	1E@0.08	Ea	17.20	2.15	19.35
110 amp	1E@0.08	Ea	24.80	2.15	26.95
125 amp	1E@0.08	Ea	23.70	2.15	25.85
150 amp	1E@0.08	Ea	20.80	2.15	22.95
175 amp	1E@0.08	Ea	24.80	2.15	26.95
200 amp	1E@0.08	Ea	23.70	2.15	25.85
225 amp	1E@0.10	Ea	114.00	2.69	116.69
250 amp	1E@0.10	Ea	100.00	2.69	102.69
300 amp	1E@0.10	Ea	105.00	2.69	107.69
350 amp	1E@0.10	Ea	119.00	2.69	121.69
400 amp	1E@0.15	Ea	100.00	4.03	104.03
450 amp	1E@0.15	Ea	159.00	4.03	163.03
500 amp	1E@0.15	Ea	159.00	4.03	163.03
600 amp	1E@0.15	Ea	153.00	4.03	157.03

Use this figures to estimate the cost of fuses installed in safety switches and combination starters under the conditions described on pages 5 and 6. Costs listed are for each fuse installed. The crew size is one electrician working at a labor cost of $26.88 per manhour. These costs include layout, material handling, and normal waste. Add for sales tax, delivery supervision, mobilization, demobilization, cleanup, overhead and profit.

Plug-in Circuit Breakers

Material		Craft@Hrs	Unit	Material Cost	Labor Cost	Installed Cost
120/240 volt AC plug-in circuit breakers with 10,000 amp interrupt capacity						
1 pole	10A	1E@0.10	Ea	8.49	2.69	11.18
1 pole	15A	1E@0.10	Ea	8.49	2.69	11.18
1 pole	20A	1E@0.10	Ea	8.49	2.69	11.18
1 pole	25A	1E@0.10	Ea	8.49	2.69	11.18
1 pole	30A	1E@0.10	Ea	8.49	2.69	11.18
1 pole	35A	1E@0.10	Ea	8.49	2.69	11.18
1 pole	40A	1E@0.10	Ea	8.49	2.69	11.18
1 pole	45A	1E@0.10	Ea	8.49	2.69	11.18
1 pole	50A	1E@0.10	Ea	8.49	2.69	11.18
1 pole	60A	1E@0.10	Ea	8.49	2.69	11.18
1 pole	70A	1E@0.10	Ea	19.00	2.69	21.69
2 pole	15A	1E@0.15	Ea	19.00	4.03	23.03
2 pole	20A	1E@0.15	Ea	19.00	4.03	23.03
2 pole	25A	1E@0.15	Ea	19.00	4.03	23.03
2 pole	30A	1E@0.15	Ea	19.00	4.03	23.03
2 pole	35A	1E@0.15	Ea	19.00	4.03	23.03
2 pole	40A	1E@0.15	Ea	19.00	4.03	23.03
2 pole	45A	1E@0.15	Ea	19.00	4.03	23.03
2 pole	50A	1E@0.15	Ea	19.00	4.03	23.03
2 pole	60A	1E@0.15	Ea	19.00	4.03	23.03
2 pole	70A	1E@0.15	Ea	39.70	4.03	43.73
2 pole	90A	1E@0.15	Ea	55.10	4.03	59.13
2 pole	100A	1E@0.15	Ea	55.10	4.03	59.13
3 pole	15A	1E@0.20	Ea	68.40	5.38	73.78
3 pole	20A	1E@0.20	Ea	68.40	5.38	73.78
3 pole	25A	1E@0.20	Ea	68.40	5.38	73.78
3 pole	30A	1E@0.20	Ea	68.40	5.38	73.78
3 pole	35A	1E@0.20	Ea	68.40	5.38	73.78
3 pole	40A	1E@0.20	Ea	68.40	5.38	73.78
3 pole	45A	1E@0.20	Ea	68.40	5.38	73.78
3 pole	50A	1E@0.20	Ea	68.40	5.38	73.78
3 pole	60A	1E@0.20	Ea	68.40	5.38	73.78
3 pole	70A	1E@0.20	Ea	88.20	5.38	93.58
3 pole	90A	1E@0.20	Ea	103.00	5.38	108.38
3 pole	100A	1E@0.20	Ea	103.00	5.38	108.38

Use these figures to estimate the cost of circuit breakers installed in enclosures under the conditions described on pages 5 and 6. Costs listed are for each circuit breaker installed. The crew is one electrician working at a labor cost of $26.88 per manhour. These costs include layout, material handling, and normal waste. Add for the enclosure, support, sales tax, delivery, supervision, mobilization, demobilization, cleanup, overhead and profit. The labor required for wire termination is not included in these figures. Add the labor cost for making terminations according to the wire size and the size of the enclosure. Note: Be sure to select the right mounting style and amp interrupt capacity (A.I.C.). Two and three pole circuit breakers have a common operating handle. Some multi-pole breakers have a common tie bar that holds several operating handles together. Most engineered systems require one operating handle on each circuit breaker.

Material	Craft@Hrs	Unit	Material Cost	Labor Cost	Installed Cost

Tandem 120/240 volt circuit breakers in single pole space

Material		Craft@Hrs	Unit	Material Cost	Labor Cost	Installed Cost
15A - 15A		1E@0.20	Ea	24.30	5.38	29.68
15A - 20A		1E@0.20	Ea	24.30	5.38	29.68
20A - 20A		1E@0.20	Ea	24.30	5.38	29.68

120/240 volt bolt-on circuit breakers, 10,000 A.I.C.

Material		Craft@Hrs	Unit	Material Cost	Labor Cost	Installed Cost
1 pole	15A	1E@0.15	Ea	10.90	4.03	14.93
1 pole	20A	1E@0.15	Ea	10.90	4.03	14.93
1 pole	25A	1E@0.15	Ea	10.90	4.03	14.93
1 pole	30A	1E@0.15	Ea	10.90	4.03	14.93
1 pole	40A	1E@0.15	Ea	10.90	4.03	14.93
1 pole	50A	1E@0.15	Ea	10.90	4.03	14.93
2 pole	15A	1E@0.20	Ea	24.30	5.38	29.68
2 pole	20A	1E@0.20	Ea	24.30	5.38	29.68
2 pole	30A	1E@0.20	Ea	24.30	5.38	29.68
2 pole	40A	1E@0.20	Ea	24.30	5.38	29.68
2 pole	50A	1E@0.20	Ea	24.30	5.38	29.68
2 pole	60A	1E@0.20	Ea	24.30	5.38	29.68
2 pole	70A	1E@0.20	Ea	24.30	5.38	29.68
3 pole	15A	1E@0.25	Ea	79.40	6.72	86.12
3 pole	20A	1E@0.25	Ea	79.40	6.72	86.12
3 pole	30A	1E@0.25	Ea	79.40	6.72	86.12
3 pole	40A	1E@0.25	Ea	79.40	6.72	86.12
3 pole	50A	1E@0.25	Ea	79.40	6.72	86.12
3 pole	60A	1E@0.25	Ea	79.40	6.72	86.12
3 pole	70A	1E@0.25	Ea	79.40	6.72	86.12
3 pole	80A	1E@0.25	Ea	101.00	6.72	107.72
3 pole	90A	1E@0.25	Ea	114.00	6.72	120.72
3 pole	100A	1E@0.25	Ea	114.00	6.72	120.72

Use these figures to estimate the cost of circuit breakers installed in enclosures under the conditions described on pages 5 and 6. Costs listed are for each circuit breaker installed. The crew is one electrician working at a labor cost of $26.88 per manhour. These costs include layout, material handling, and normal waste. Add for the enclosure, support, sales tax, delivery, supervision, mobilization, demobilization, cleanup, overhead and profit. The labor required for wire termination is not included in these figures. Add the labor cost for making terminations according to the wire size and the size of the enclosure. Note: Be sure to select the right mounting style and amp interrupt capacity (A.I.C.). Two and three pole circuit breakers have a common operating handle. Some multi-pole breakers have a common tie bar that holds several operating handles together. Most engineered systems require one operating handle on each circuit breaker.

120/240 Volt AC Thermal Magnetic Breakers

Material		Craft@Hrs	Unit	Material Cost	Labor Cost	Installed Cost
120/240 volt AC thermal magnetic bolt-on breakers with 100 amp frame						
1 pole	15A	1E@0.20	Ea	54.10	5.38	59.48
1 pole	20A	1E@0.20	Ea	54.10	5.38	59.48
1 pole	30A	1E@0.20	Ea	54.10	5.38	59.48
1 pole	40A	1E@0.20	Ea	54.10	5.38	59.48
1 pole	50A	1E@0.20	Ea	54.10	5.38	59.48
1 pole	60A	1E@0.20	Ea	54.10	5.38	59.48
1 pole	70A	1E@0.20	Ea	72.80	5.38	78.18
1 pole	90A	1E@0.20	Ea	72.80	5.38	78.18
1 pole	100A	1E@0.20	Ea	72.80	5.38	78.18
2 pole	15A	1E@0.25	Ea	91.60	6.72	98.32
2 pole	20A	1E@0.25	Ea	91.60	6.72	98.32
2 pole	30A	1E@0.25	Ea	91.60	6.72	98.32
2 pole	40A	1E@0.25	Ea	91.60	6.72	98.32
2 pole	50A	1E@0.25	Ea	91.60	6.72	98.32
2 pole	60A	1E@0.25	Ea	91.60	6.72	98.32
2 pole	70A	1E@0.25	Ea	149.00	6.72	155.72
2 pole	90A	1E@0.25	Ea	149.00	6.72	155.72
2 pole	100A	1E@0.25	Ea	149.00	6.72	155.72
3 pole	15A	1E@0.30	Ea	134.00	8.06	142.06
3 pole	20A	1E@0.30	Ea	134.00	8.06	142.06
3 pole	30A	1E@0.30	Ea	134.00	8.06	142.06
3 pole	40A	1E@0.30	Ea	134.00	8.06	142.06
3 pole	50A	1E@0.30	Ea	134.00	8.06	142.06
3 pole	60A	1E@0.30	Ea	134.00	8.06	142.06
3 pole	70A	1E@0.30	Ea	193.00	8.06	201.06
3 pole	90A	1E@0.30	Ea	193.00	8.06	201.06
3 pole	100A	1E@0.30	Ea	193.00	8.06	201.06

Use these figures to estimate the cost of circuit breakers installed in enclosures under the conditions described on pages 5 and 6. Costs listed are for each circuit breaker installed. The crew is one electrician working at a labor cost of $26.88 per manhour. These costs include layout, material handling, and normal waste. Add for the enclosure, support, sales tax, delivery, supervision, mobilization, demobilization, cleanup, overhead and profit. The labor required for wire termination is not included in these figures. Add the labor cost for making terminations according to the wire size and the size of the enclosure. Note: Be sure to select the right mounting style and amp interrupt capacity (A.I.C.). Two and three pole circuit breakers have a common operating handle. Some multi-pole breakers have a common tie bar that holds several operating handles together. Most engineered systems require one operating handle on each circuit breaker.

Thermal Magnetic Breakers

Material		Craft@Hrs	Unit	Material Cost	Labor Cost	Installed Cost
277/480 volt AC thermal magnetic breakers with 100 amp frame						
1 pole	15A	1E@0.30	Ea	68.40	8.06	76.46
1 pole	20A	1E@0.30	Ea	68.40	8.06	76.46
1 pole	30A	1E@0.30	Ea	68.40	8.06	76.46
1 pole	40A	1E@0.30	Ea	68.40	8.06	76.46
1 pole	50A	1E@0.30	Ea	68.40	8.06	76.46
1 pole	60A	1E@0.30	Ea	68.40	8.06	76.46
1 pole	70A	1E@0.30	Ea	86.00	8.06	94.06
1 pole	90A	1E@0.30	Ea	86.00	8.06	94.06
1 pole	100A	1E@0.30	Ea	86.00	8.06	94.06
2 pole	15A	1E@0.40	Ea	167.00	10.80	177.80
2 pole	20A	1E@0.40	Ea	167.00	10.80	177.80
2 pole	30A	1E@0.40	Ea	167.00	10.80	177.80
2 pole	40A	1E@0.40	Ea	167.00	10.80	177.80
2 pole	50A	1E@0.40	Ea	167.00	10.80	177.80
2 pole	60A	1E@0.40	Ea	167.00	10.80	177.80
2 pole	70A	1E@0.40	Ea	215.00	10.80	225.80
2 pole	90A	1E@0.40	Ea	215.00	10.80	225.80
2 pole	100A	1E@0.40	Ea	215.00	10.80	225.80
3 pole	15A	1E@0.50	Ea	214.00	13.40	227.40
3 pole	20A	1E@0.50	Ea	214.00	13.40	227.40
3 pole	30A	1E@0.50	Ea	214.00	13.40	227.40
3 pole	40A	1E@0.50	Ea	214.00	13.40	227.40
3 pole	50A	1E@0.50	Ea	214.00	13.40	227.40
3 pole	60A	1E@0.50	Ea	214.00	13.40	227.40
3 pole	70A	1E@0.50	Ea	254.00	13.40	267.40
3 pole	90A	1E@0.50	Ea	254.00	13.40	267.40
3 pole	100A	1E@0.50	Ea	254.00	13.40	267.40

Material		Craft@Hrs	Unit	Material Cost	Labor Cost	Installed Cost
600 volt AC, 250 volt DC thermal magnetic breakers with 100 amp frame						
2 pole	15A	1E@0.70	Ea	193.00	18.80	211.80
2 pole	20A	1E@0.70	Ea	193.00	18.80	211.80
2 pole	30A	1E@0.70	Ea	193.00	18.80	211.80
2 pole	40A	1E@0.70	Ea	193.00	18.80	211.80
2 pole	50A	1E@0.70	Ea	193.00	18.80	211.80
2 pole	60A	1E@0.70	Ea	193.00	18.80	211.80
2 pole	70A	1E@0.70	Ea	244.00	18.80	262.80
2 pole	90A	1E@0.70	Ea	244.00	18.80	262.80
2 pole	100A	1E@0.70	Ea	244.00	18.80	262.80

Use these figures to estimate the cost of circuit breakers installed in enclosures under the conditions described on pages 5 and 6. Costs listed are for each circuit breaker installed. The crew is one electrician working at a labor cost of $26.88 per manhour. These costs include layout, material handling, and normal waste. Add for the enclosure, supports, sales tax, delivery, supervision, mobilization, demobilization, cleanup, overhead and profit. The labor required for wire termination is not included in these figures. Add the labor cost for making terminations according to the wire size and the size of the enclosure. Note: Be sure to select the right mounting style and amp interrupt capacity (A.I.C.). Two and three pole circuit breakers have a common operating handle. Some multi-pole breakers have a common tie bar that holds several operating handles together. Most engineered systems require one operating handle on each circuit breaker.

Thermal Magnetic Breakers

Material		Craft@Hrs	Unit	Material Cost	Labor Cost	Installed Cost

600 volt AC, 250 volt DC thermal magnetic breakers with 100 amp frame

Material		Craft@Hrs	Unit	Material Cost	Labor Cost	Installed Cost
3 pole	15A	1E@0.75	Ea	248.00	20.20	268.20
3 pole	20A	1E@0.75	Ea	248.00	20.20	268.20
3 pole	30A	1E@0.75	Ea	248.00	20.20	268.20
3 pole	40A	1E@0.75	Ea	248.00	20.20	268.20
3 pole	50A	1E@0.75	Ea	248.00	20.20	268.20
3 pole	60A	1E@0.75	Ea	248.00	20.20	268.20
3 pole	70A	1E@0.75	Ea	305.00	20.20	325.20
3 pole	90A	1E@0.75	Ea	305.00	20.20	325.20
3 pole	100A	1E@0.75	Ea	305.00	20.20	325.20

600 volt AC, 250 volt DC thermal magnetic breakers with 250 amp frame

Material		Craft@Hrs	Unit	Material Cost	Labor Cost	Installed Cost
2 pole	70A	1E@1.00	Ea	568.00	26.90	594.90
2 pole	80A	1E@1.00	Ea	568.00	26.90	594.90
2 pole	90A	1E@1.00	Ea	568.00	26.90	594.90
2 pole	100A	1E@1.00	Ea	568.00	26.90	594.90
2 pole	110A	1E@1.00	Ea	568.00	26.90	594.90
2 pole	125A	1E@1.00	Ea	568.00	26.90	594.90
2 pole	150A	1E@1.00	Ea	568.00	26.90	594.90
2 pole	175A	1E@1.00	Ea	568.00	26.90	594.90
2 pole	200A	1E@1.00	Ea	568.00	26.90	594.90
2 pole	225A	1E@1.00	Ea	568.00	26.90	594.90
3 pole	70A	1E@1.25	Ea	711.00	33.60	744.60
3 pole	80A	1E@1.25	Ea	711.00	33.60	744.60
3 pole	90A	1E@1.25	Ea	711.00	33.60	744.60
3 pole	100A	1E@1.25	Ea	711.00	33.60	744.60
3 pole	110A	1E@1.25	Ea	711.00	33.60	744.60
3 pole	125A	1E@1.25	Ea	711.00	33.60	744.60
3 pole	150A	1E@1.25	Ea	711.00	33.60	744.60
3 pole	175A	1E@1.25	Ea	711.00	33.60	744.60
3 pole	200A	1E@1.25	Ea	711.00	33.60	744.60
3 pole	225A	1E@1.25	Ea	711.00	33.60	744.60

600 volt AC, 250 volt DC thermal magnetic breakers with 400 amp frame

Material		Craft@Hrs	Unit	Material Cost	Labor Cost	Installed Cost
2 pole	125A	1E@1.50	Ea	1,040.00	40.30	1,080.30
2 pole	150A	1E@1.50	Ea	1,040.00	40.30	1,080.30
2 pole	175A	1E@1.50	Ea	1,040.00	40.30	1,080.30
2 pole	200A	1E@1.50	Ea	1,040.00	40.30	1,080.30
2 pole	225A	1E@1.50	Ea	1,040.00	40.30	1,080.30
2 pole	250A	1E@1.50	Ea	1,040.00	40.30	1,080.30

Use these figures to estimate the cost of circuit breakers installed in enclosures under the conditions described on pages 5 and 6. Costs listed are for each circuit breaker installed. The crew is one electrician working at a labor cost of $26.88 per manhour. These costs include layout, material handling, and normal waste. Add for the enclosure, support, sales tax, delivery, supervision, mobilization, demobilization, cleanup, overhead and profit. The labor required for wire termination is not included in these figures. Add the labor cost for making terminations according to the wire size and the size of the enclosure. Note: Be sure to select the right mounting style and amp interrupt capacity (A.I.C.). Two and three pole circuit breakers have a common operating handle. Some multi-pole breakers have a common tie bar that holds several operating handles together. Most engineered systems require one operating handle on each circuit breaker.

Material		Craft@Hrs	Unit	Material Cost	Labor Cost	Installed Cost

600 volt AC, 250 volt DC thermal magnetic breakers with 400 amp frame

Material		Craft@Hrs	Unit	Material Cost	Labor Cost	Installed Cost
2 pole	300A	1E@1.50	Ea	1,040.00	40.30	1,080.30
2 pole	350A	1E@1.50	Ea	1,040.00	40.30	1,080.30
2 pole	400A	1E@1.50	Ea	1,040.00	40.30	1,080.30
3 pole	125A	1E@1.75	Ea	1,260.00	47.00	1,307.00
3 pole	150A	1E@1.75	Ea	1,260.00	47.00	1,307.00
3 pole	175A	1E@1.75	Ea	1,260.00	47.00	1,307.00
3 pole	200A	1E@1.75	Ea	1,260.00	47.00	1,307.00
3 pole	225A	1E@1.75	Ea	1,260.00	47.00	1,307.00
3 pole	250A	1E@1.75	Ea	1,260.00	47.00	1,307.00
3 pole	300A	1E@1.75	Ea	1,260.00	47.00	1,307.00
3 pole	350A	1E@1.75	Ea	1,260.00	47.00	1,307.00
3 pole	400A	1E@1.75	Ea	1,260.00	47.00	1,307.00

600 volt AC, 250 volt DC thermal magnetic breakers with 1,000 amp frame

Material		Craft@Hrs	Unit	Material Cost	Labor Cost	Installed Cost
2 pole	125A	1E@1.75	Ea	1,640.00	47.00	1,687.00
2 pole	150A	1E@1.75	Ea	1,640.00	47.00	1,687.00
2 pole	175A	1E@1.75	Ea	1,640.00	47.00	1,687.00
2 pole	200A	1E@1.75	Ea	1,640.00	47.00	1,687.00
2 pole	225A	1E@1.75	Ea	1,640.00	47.00	1,687.00
2 pole	250A	1E@1.75	Ea	1,640.00	47.00	1,687.00
2 pole	300A	1E@1.75	Ea	1,640.00	47.00	1,687.00
2 pole	350A	1E@1.75	Ea	1,640.00	47.00	1,687.00
2 pole	400A	1E@1.75	Ea	1,640.00	47.00	1,687.00
2 pole	450A	1E@1.75	Ea	1,640.00	47.00	1,687.00
2 pole	500A	1E@1.75	Ea	1,640.00	47.00	1,687.00
2 pole	600A	1E@1.75	Ea	1,640.00	47.00	1,687.00
2 pole	700A	1E@1.75	Ea	2,110.00	47.00	2,157.00
2 pole	800A	1E@1.75	Ea	2,110.00	47.00	2,157.00
2 pole	900A	1E@1.75	Ea	3,010.00	47.00	3,057.00
2 pole	1000A	1E@1.75	Ea	3,010.00	47.00	3,057.00
3 pole	125A	1E@2.00	Ea	2,070.00	53.80	2,123.80
3 pole	150A	1E@2.00	Ea	2,070.00	53.80	2,123.80
3 pole	175A	1E@2.00	Ea	2,070.00	53.80	2,123.80
3 pole	200A	1E@2.00	Ea	2,070.00	53.80	2,123.80
3 pole	225A	1E@2.00	Ea	2,070.00	53.80	2,123.80
3 pole	250A	1E@2.00	Ea	2,070.00	53.80	2,123.80
3 pole	300A	1E@2.00	Ea	2,070.00	53.80	2,123.80
3 pole	350A	1E@2.00	Ea	2,070.00	53.80	2,123.80
3 pole	400A	1E@2.00	Ea	2,070.00	53.80	2,123.80
3 pole	450A	1E@2.00	Ea	2,070.00	53.80	2,123.80

Use these figures to estimate the cost of circuit breakers installed in enclosures under the conditions described on pages 5 and 6. Costs listed are for each circuit breaker installed. The crew is one electrician working at a labor cost of $26.88 per manhour. These costs include layout, material handling, and normal waste. Add for the enclosure, support, sales tax, delivery, supervision, mobilization, demobilization, cleanup, overhead and profit. The labor required for wire termination is not included in these figures. Add the labor cost for making terminations according to the wire size and the size of the enclosure. Note: Be sure to select the right mounting style and amp interrupt capacity (A.I.C.). Two and three pole circuit breakers have a common operating handle. Some multi-pole breakers have a common tie bar that holds several operating handles together. Most engineered systems require one operating handle on each circuit breaker.

Thermal Magnetic Breakers

Material		Craft@Hrs	Unit	Material Cost	Labor Cost	Installed Cost

600 volt AC, 250 volt DC thermal magnetic breakers with 1,000 amp frame

Material		Craft@Hrs	Unit	Material Cost	Labor Cost	Installed Cost
3 pole	500A	1E@2.00	Ea	2,070.00	53.80	2,123.80
3 pole	600A	1E@2.00	Ea	2,070.00	53.80	2,123.80
3 pole	700A	1E@2.00	Ea	2,720.00	53.80	2,773.80
3 pole	800A	1E@2.00	Ea	2,720.00	53.80	2,773.80
3 pole	900A	1E@2.00	Ea	3,480.00	53.80	3,533.80
3 pole	1000A	1E@2.00	Ea	3,480.00	53.80	3,533.80

600 volt AC, 250 volt DC thermal magnetic breakers with 1,200 amp frame

Material		Craft@Hrs	Unit	Material Cost	Labor Cost	Installed Cost
2 pole	600A	1E@2.00	Ea	4,760.00	53.80	4,813.80
2 pole	700A	1E@2.00	Ea	4,760.00	53.80	4,813.80
2 pole	800A	1E@2.00	Ea	4,760.00	53.80	4,813.80
2 pole	1000A	1E@2.00	Ea	4,760.00	53.80	4,813.80
2 pole	1200A	1E@2.00	Ea	4,760.00	53.80	4,813.80
3 pole	600A	1E@2.50	Ea	5,220.00	67.20	5,287.20
3 pole	700A	1E@2.50	Ea	5,220.00	67.20	5,287.20
3 pole	800A	1E@2.50	Ea	5,220.00	67.20	5,287.20
3 pole	1000A	1E@2.50	Ea	5,220.00	67.20	5,287.20
3 pole	1200A	1E@2.50	Ea	5,220.00	67.20	5,287.20

277 volt AC plug-on breakers with 100 amp frame, 14,000 A.I.C.

Material		Craft@Hrs	Unit	Material Cost	Labor Cost	Installed Cost
1 pole	15A	1E@0.25	Ea	36.40	6.72	43.12
1 pole	20A	1E@0.25	Ea	36.40	6.72	43.12
1 pole	25A	1E@0.25	Ea	36.40	6.72	43.12
1 pole	30A	1E@0.25	Ea	36.40	6.72	43.12
1 pole	40A	1E@0.25	Ea	36.40	6.72	43.12
1 pole	50A	1E@0.25	Ea	36.40	6.72	43.12
1 pole	60A	1E@0.25	Ea	36.40	6.72	43.12

277 volt AC bolt-on breakers with 100 amp frame, 14,000 A.I.C.

Material		Craft@Hrs	Unit	Material Cost	Labor Cost	Installed Cost
1 pole	15A	1E@0.30	Ea	41.90	8.06	49.96
1 pole	20A	1E@0.30	Ea	41.90	8.06	49.96
1 pole	25A	1E@0.30	Ea	41.90	8.06	49.96
1 pole	30A	1E@0.30	Ea	41.90	8.06	49.96
1 pole	40A	1E@0.30	Ea	41.90	8.06	49.96
1 pole	50A	1E@0.30	Ea	41.90	8.06	49.96
1 pole	60A	1E@0.30	Ea	41.90	8.06	49.96

Use these figures to estimate the cost of circuit breakers installed in enclosures under the conditions described on pages 5 and 6. Costs listed are for each circuit breaker installed. The crew is one electrician working at a labor cost of $26.88 per manhour. These costs include layout, material handling, and normal waste. Add for the enclosure, support, sales tax, delivery, supervision, mobilization, demobilization, cleanup, overhead and profit. The labor required for wire termination is not included in these figures. Add the labor cost for making terminations according to the wire size and the size of the enclosure. Note: Be sure to select the right mounting style and amp interrupt capacity (A.I.C.). Two and three pole circuit breakers have a common operating handle. Some multi-pole breakers have a common tie bar that holds several operating handles together. Most engineered systems require one operating handle on each circuit breaker.

Thermal Magnetic Breakers, 14,000 A.I.C.

Material		Craft@Hrs	Unit	Material Cost	Labor Cost	Installed Cost

277 volt AC plug-on breakers with 100 amp frame, 14,000 A.I.C.

Material		Craft@Hrs	Unit	Material Cost	Labor Cost	Installed Cost
2 pole	15A	1E@0.40	Ea	105.00	10.80	115.80
2 pole	20A	1E@0.40	Ea	105.00	10.80	115.80
2 pole	25A	1E@0.40	Ea	105.00	10.80	115.80
2 pole	30A	1E@0.40	Ea	105.00	10.80	115.80
2 pole	40A	1E@0.40	Ea	105.00	10.80	115.80
2 pole	50A	1E@0.40	Ea	105.00	10.80	115.80
2 pole	60A	1E@0.40	Ea	105.00	10.80	115.80

277 volt AC bolt-on breakers with 100 amp frame, 14,000 A.I.C.

Material		Craft@Hrs	Unit	Material Cost	Labor Cost	Installed Cost
2 pole	15A	1E@0.50	Ea	112.00	13.40	125.40
2 pole	20A	1E@0.50	Ea	112.00	13.40	125.40
2 pole	25A	1E@0.50	Ea	112.00	13.40	125.40
2 pole	30A	1E@0.50	Ea	112.00	13.40	125.40
2 pole	40A	1E@0.50	Ea	112.00	13.40	125.40
2 pole	50A	1E@0.50	Ea	112.00	13.40	125.40
2 pole	60A	1E@0.50	Ea	112.00	13.40	125.40

277 volt AC plug-on breakers with 100 amp frame, 14,000 A.I.C.

Material		Craft@Hrs	Unit	Material Cost	Labor Cost	Installed Cost
3 pole	15A	1E@0.60	Ea	213.00	16.10	229.10
3 pole	20A	1E@0.60	Ea	213.00	16.10	229.10
3 pole	25A	1E@0.60	Ea	213.00	16.10	229.10
3 pole	30A	1E@0.60	Ea	213.00	16.10	229.10
3 pole	40A	1E@0.60	Ea	213.00	16.10	229.10
3 pole	50A	1E@0.60	Ea	213.00	16.10	229.10
3 pole	60A	1E@0.60	Ea	213.00	16.10	229.10

277 volt AC bolt-on breakers with 100 amp frame, 14,000 A.I.C.

Material		Craft@Hrs	Unit	Material Cost	Labor Cost	Installed Cost
3 pole	15A	1E@0.70	Ea	182.00	18.80	200.80
3 pole	20A	1E@0.70	Ea	182.00	18.80	200.80
3 pole	25A	1E@0.70	Ea	182.00	18.80	200.80
3 pole	30A	1E@0.70	Ea	182.00	18.80	200.80
3 pole	40A	1E@0.70	Ea	182.00	18.80	200.80
3 pole	50A	1E@0.70	Ea	182.00	18.80	200.80
3 pole	60A	1E@0.70	Ea	182.00	18.80	200.80

Use these figures to estimate the cost of circuit breakers installed in enclosures under the conditions described on pages 5 and 6. Costs listed are for each circuit breaker installed. The crew is one electrician working at a labor cost of $26.88 per manhour. These costs include layout, material handling, and normal waste. Add for the enclosure, support, sales tax, delivery, supervision, mobilization, demobilization, cleanup, overhead and profit. The labor required for wire termination is not included in these figures. Add the labor cost for making terminations according to the wire size and the size of the enclosure. Note: Be sure to select the right mounting style and amp interrupt capacity (A.I.C.). Two and three pole circuit breakers have a common operating handle. Some multi-pole breakers have a common tie bar that holds several operating handles together. Most engineered systems require one operating handle on each circuit breaker.

Thermal Magnetic Breakers

Material		Craft@Hrs	Unit	Material Cost	Labor Cost	Installed Cost

600 volt AC, 250 volt DC breakers with 225 amp frame, 10,000 A.I.C.

Material		Craft@Hrs	Unit	Material Cost	Labor Cost	Installed Cost
3 pole	125A	1E@1.50	Ea	600.00	40.30	640.30
3 pole	150A	1E@1.50	Ea	600.00	40.30	640.30
3 pole	175A	1E@1.50	Ea	600.00	40.30	640.30
3 pole	200A	1E@1.50	Ea	600.00	40.30	640.30
3 pole	225A	1E@1.50	Ea	600.00	40.30	640.30

600 volt AC, 250 volt DC breakers with 400 amp frame, 22,000 A.I.C.

Material		Craft@Hrs	Unit	Material Cost	Labor Cost	Installed Cost
2 pole	125A	1E@1.30	Ea	1,110.00	34.90	1,144.90
2 pole	150A	1E@1.30	Ea	1,110.00	34.90	1,144.90
2 pole	175A	1E@1.30	Ea	1,110.00	34.90	1,144.90
2 pole	200A	1E@1.30	Ea	1,110.00	34.90	1,144.90
2 pole	225A	1E@1.30	Ea	1,110.00	34.90	1,144.90
2 pole	250A	1E@1.30	Ea	1,110.00	34.90	1,144.90
2 pole	300A	1E@1.30	Ea	1,110.00	34.90	1,144.90
2 pole	350A	1E@1.30	Ea	1,110.00	34.90	1,144.90
2 pole	400A	1E@1.30	Ea	1,110.00	34.90	1,144.90
3 pole	125A	1E@1.70	Ea	1,350.00	45.70	1,395.70
3 pole	150A	1E@1.70	Ea	1,350.00	45.70	1,395.70
3 pole	175A	1E@1.70	Ea	1,350.00	45.70	1,395.70
3 pole	200A	1E@1.70	Ea	1,350.00	45.70	1,395.70
3 pole	225A	1E@1.70	Ea	1,350.00	45.70	1,395.70
3 pole	250A	1E@1.70	Ea	1,350.00	45.70	1,395.70
3 pole	300A	1E@1.70	Ea	1,350.00	45.70	1,395.70
3 pole	350A	1E@1.70	Ea	1,350.00	45.70	1,395.70
3 pole	400A	1E@1.70	Ea	1,350.00	45.70	1,395.70

600 volt AC, 250 volt DC breakers with 1,000 amp frame, 30,000 A.I.C.

Material		Craft@Hrs	Unit	Material Cost	Labor Cost	Installed Cost
2 pole	125A	1E@1.50	Ea	1,640.00	40.30	1,680.30
2 pole	150A	1E@1.50	Ea	1,640.00	40.30	1,680.30
2 pole	175A	1E@1.50	Ea	1,640.00	40.30	1,680.30
2 pole	200A	1E@1.50	Ea	1,640.00	40.30	1,680.30
2 pole	225A	1E@1.50	Ea	1,640.00	40.30	1,680.30
2 pole	250A	1E@1.50	Ea	1,640.00	40.30	1,680.30
2 pole	300A	1E@1.50	Ea	1,640.00	40.30	1,680.30
2 pole	350A	1E@1.50	Ea	1,640.00	40.30	1,680.30
2 pole	400A	1E@1.50	Ea	1,640.00	40.30	1,680.30

Use these figures to estimate the cost of circuit breakers installed in enclosures under the conditions described on pages 5 and 6. Costs listed are for each circuit breaker installed. The crew is one electrician working at a labor cost of $26.88 per manhour. These costs include layout, material handling, and normal waste. Add for the enclosure, support, sales tax, delivery, supervision, mobilization, demobilization, cleanup, overhead and profit. The labor required for wire termination is not included in these figures. Add the labor cost for making terminations according to the wire size and the size of the enclosure. Note: Be sure to select the right mounting style and amp interrupt capacity (A.I.C.). Two and three pole circuit breakers have a common operating handle. Some multi-pole breakers have a common tie bar that holds several operating handles together. Most engineered systems require one operating handle on each circuit breaker.

Thermal Magnetic Circuit Breakers and Enclosures

Material		Craft@Hrs	Unit	Material Cost	Labor Cost	Installed Cost

600 volt AC, 250 volt DC breakers with 1,000 amp frame, 30,000 A.I.C.

Material		Craft@Hrs	Unit	Material Cost	Labor Cost	Installed Cost
2 pole	450A	1E@1.50	Ea	1,640.00	40.30	1,680.30
2 pole	500A	1E@1.50	Ea	1,640.00	40.30	1,680.30
2 pole	600A	1E@1.50	Ea	1,640.00	40.30	1,680.30
2 pole	700A	1E@1.50	Ea	2,110.00	40.30	2,150.30
2 pole	800A	1E@1.50	Ea	2,110.00	40.30	2,150.30
2 pole	900A	1E@1.50	Ea	3,010.00	40.30	3,050.30
2 pole	1000A	1E@1.50	Ea	3,010.00	40.30	3,050.30
3 pole	125A	1E@2.00	Ea	2,070.00	53.80	2,123.80
3 pole	150A	1E@2.00	Ea	2,070.00	53.80	2,123.80
3 pole	175A	1E@2.00	Ea	2,070.00	53.80	2,123.80
3 pole	200A	1E@2.00	Ea	2,070.00	53.80	2,123.80
3 pole	225A	1E@2.00	Ea	2,070.00	53.80	2,123.80
3 pole	250A	1E@2.00	Ea	2,070.00	53.80	2,123.80
3 pole	300A	1E@2.00	Ea	2,070.00	53.80	2,123.80
3 pole	350A	1E@2.00	Ea	2,070.00	53.80	2,123.80
3 pole	400A	1E@2.00	Ea	2,070.00	53.80	2,123.80
3 pole	450A	1E@2.00	Ea	2,070.00	53.80	2,123.80
3 pole	500A	1E@2.00	Ea	2,070.00	53.80	2,123.80
3 pole	600A	1E@2.00	Ea	2,070.00	53.80	2,123.80
3 pole	700A	1E@2.00	Ea	2,720.00	53.80	2,773.80
3 pole	800A	1E@2.00	Ea	2,720.00	53.80	2,773.80
3 pole	900A	1E@2.00	Ea	3,480.00	53.80	3,533.80
3 pole	1000A	1E@2.00	Ea	3,480.00	53.80	3,533.80

NEMA 1 flush mounted circuit breaker enclosures

Material		Craft@Hrs	Unit	Material Cost	Labor Cost	Installed Cost
100A	120/240V	1E@0.30	Ea	72.70	8.06	80.76
100A	277/480V	1E@0.30	Ea	93.10	8.06	101.16
225A	120/240V	1E@0.35	Ea	86.70	9.41	96.11
225A	600V	1E@0.35	Ea	69.90	9.41	79.31
400A	600V	1E@0.50	Ea	127.00	13.40	140.40
1000A	600V	1E@0.70	Ea	227.00	18.80	245.80

NEMA 1 surface mounted circuit breaker enclosures

Material		Craft@Hrs	Unit	Material Cost	Labor Cost	Installed Cost
100A	120/240V	1E@0.30	Ea	72.70	8.06	80.76
100A	277/480V	1E@0.30	Ea	93.10	8.06	101.16
225A	120/240V	1E@0.35	Ea	86.70	9.41	96.11
225A	277/480V	1E@0.35	Ea	69.90	9.41	79.31

Use these figures to estimate the cost of circuit breakers installed in enclosures and circuit breaker enclosures installed in buildings under the conditions described on pages 5 and 6. Costs listed are for each circuit breaker or circuit breaker enclosure installed. The crew is one electrician working at a labor cost of $26.88 per manhour. These costs include layout, material handling, and normal waste. Add for the enclosure (except as noted above), support, sales tax, delivery, supervision, mobilization, demobilization, cleanup, overhead and profit. The labor required for wire termination is not included in these figures. Add the labor cost for making terminations according to the wire size and the size of the enclosure. Note: Be sure to select the right mounting style and amp interrupt capacity (A.I.C.). Two and three pole circuit breakers have a common operating handle. Some multi-pole breakers have a common tie bar that holds several operating handles together. Most engineered systems require one operating handle on each circuit breaker.

Surface Mounted Circuit Breaker Enclosures

Material		Craft@Hrs	Unit	Material Cost	Labor Cost	Installed Cost
NEMA 1 surface mounted circuit breaker enclosures						
225A	600V	1E@0.35	Ea	69.50	9.41	78.91
400A	600V	1E@0.50	Ea	127.00	13.40	140.40
1000A	600V	1E@0.70	Ea	227.00	18.80	245.80
1200A	600V	1E@1.00	Ea	403.00	26.90	429.90
NEMA 3R surface mounted circuit breaker enclosures						
100A	120/240	1E@0.30	Ea	195.00	8.06	203.06
100A	277/480V	1E@0.30	Ea	218.00	8.06	226.06
225A	120/240V	1E@0.35	Ea	150.00	9.41	159.41
225A	277/480V	1E@0.35	Ea	168.00	9.41	177.41
225A	600V	1E@0.35	Ea	287.00	9.41	296.41
400A	600V	1E@0.50	Ea	646.00	13.40	659.40
1000A	600V	1E@0.70	Ea	843.00	18.80	861.80
1200A	600V	1E@1.00	Ea	1,040.00	26.90	1,066.90
NEMA 4 & 5 surface mounted circuit breaker enclosures						
100A	277/480V	1E@0.50	Ea	560.00	13.40	573.40
225A	600V	1E@0.70	Ea	1,140.00	18.80	1,158.80
400A	600V	1E@1.00	Ea	2,220.00	26.90	2,246.90
1000A	600V	1E@1.50	Ea	3,960.00	40.30	4,000.30
NEMA 12 surface mounted circuit breaker enclosures						
100A	277/480V	1E@0.35	Ea	120.00	9.41	129.41
225A	600V	1E@0.50	Ea	202.00	13.40	215.40
400A	600V	1E@0.70	Ea	342.00	18.80	360.80
1000A	600V	1E@1.00	Ea	646.00	26.90	672.90
NEMA 7 aluminum surface mounted circuit breaker enclosures						
100A	277/480V	1E@0.50	Ea	784.00	13.40	797.40
225A	600V	1E@0.70	Ea	1,600.00	18.80	1,618.80
400A	600V	1E@1.00	Ea	3,740.00	26.90	3,766.90
800A	600V	1E@1.50	Ea	5,450.00	40.30	5,490.30

Use these figures to estimate the cost of circuit breaker enclosures installed in buildings under the conditions described on pages 5 and 6. Costs listed are for each enclosure installed. The crew is one electrician working at a labor cost of $26.88 per manhour. These costs include installation screws, layout, material handling, and normal waste. Add for circuit breakers, supports, sales tax, delivery, supervision, mobilization, demobilization, cleanup, overhead and profit. Note: Be sure to select the right enclosure for the style of circuit breaker required and for the installation location.

NEMA 3R 120/240 Volt Meter Sockets, 10,000 A.I.C.

Material	Craft@Hrs	Unit	Material Cost	Labor Cost	Installed Cost
NEMA 3R 120/240 volt surface mounted meter sockets for overhead service					
100A ring	1E@0.30	Ea	28.50	8.06	36.56
100A ringless	1E@0.30	Ea	28.50	8.06	36.56
150A ring	1E@0.40	Ea	43.10	10.80	53.90
150A ringless	1E@0.40	Ea	43.10	10.80	53.90
200A ring	1E@0.50	Ea	43.10	13.40	56.50
200A ringless	1E@0.50	Ea	43.10	13.40	56.50
NEMA 3R 120/240 volt surface mounted meter sockets for underground service					
200A ring	1E@0.40	Ea	43.10	10.80	53.90
200A ringless	1E@0.40	Ea	43.10	10.80	53.90
NEMA 3R 120/240 volt surface mounted meter socket & main breaker for overhead service					
100A main	1E@0.60	Ea	204.00	16.10	220.10
125A main	1E@0.60	Ea	315.00	16.10	331.10
150A main	1E@0.70	Ea	570.00	18.80	588.80
200A main	1E@0.70	Ea	570.00	18.80	588.80
NEMA 3R 120/240v semi-flush meter socket & main, overhead service					
100A main	1E@0.70	Ea	382.00	18.80	400.80
125A main	1E@0.70	Ea	780.00	18.80	798.80
150A main	1E@0.90	Ea	780.00	24.20	804.20
200A main	1E@0.90	Ea	780.00	24.20	804.20
NEMA 3R 120/240 volt surface meter socket & main, underground service					
100A main	1E@0.60	Ea	348.00	16.10	364.10
125A main	1E@0.60	Ea	465.00	16.10	481.10
150A main	1E@0.70	Ea	641.00	18.80	659.80
200A main	1E@0.70	Ea	641.00	18.80	659.80
NEMA 3R 120/240v semi-flush meter socket & main, underground service					
100A main	1E@0.70	Ea	348.00	18.80	366.80
125A main	1E@0.70	Ea	465.00	18.80	483.80
150A main	1E@0.90	Ea	641.00	24.20	665.20
200A main	1E@0.90	Ea	641.00	24.20	665.20

Use these figures to estimate the cost of meter sockets installed in buildings under the conditions described on pages 5 and 6. Costs listed are for each meter socket installed. The crew is one electrician working at a labor cost of $26.88 per manhour. These costs include sealing rings, layout, material handling, and normal waste. Add for supports, top conduit hubs, sales tax, delivery, supervision, mobilization, demobilization, cleanup, overhead and profit. Note: Top conduit hubs cost extra. Order the size appropriate for the size of conduit to be used. Blank caps are also available.

Safety Sockets and Meter Centers

Material	Craft@Hrs	Unit	Material Cost	Labor Cost	Installed Cost

NEMA 3R surface mounted safety sockets with test blocks

Material	Craft@Hrs	Unit	Material Cost	Labor Cost	Installed Cost
100A 5 jaw 240V	1E@0.50	Ea	237.00	13.40	250.40
100A 5 jaw 208V	1E@0.50	Ea	237.00	13.40	250.40
100A 5 jaw 480V	1E@0.50	Ea	296.00	13.40	309.40
100A 7 jaw 240V	1E@0.50	Ea	318.00	13.40	331.40
100A 7 jaw 480V	1E@0.50	Ea	318.00	13.40	331.40
200A 4 jaw 240V	1E@0.60	Ea	424.00	16.10	440.10
200A 5 jaw 208V	1E@0.60	Ea	424.00	16.10	440.10
200A 5 jaw 480V	1E@0.60	Ea	497.00	16.10	513.10
200A 7 jaw 240V	1E@0.60	Ea	550.00	16.10	566.10
200A 7 jaw 480V	1E@0.60	Ea	550.00	16.10	566.10

120/240v meter center & main, 800A bus, 4 jaw sockets, 125A max., 2 pole branch, indoor

Material	Craft@Hrs	Unit	Material Cost	Labor Cost	Installed Cost
3 meters & mains	1E@1.25	Ea	536.00	33.60	569.60
4 meters & mains	1E@1.50	Ea	715.00	40.30	755.30
5 meters & mains	1E@1.75	Ea	894.00	47.00	941.00
6 meters & mains	1E@2.00	Ea	1,040.00	53.80	1,093.80
7 meters & mains	1E@2.25	Ea	1,380.00	60.50	1,440.50
8 meters & mains	1E@2.50	Ea	1,380.00	67.20	1,447.20
10 meters & mains	1E@3.00	Ea	1,720.00	80.60	1,800.60

120/240v meter center, 1,200A bus, 4 jaw sockets, 200A max., 2 pole branch, indoor

Material	Craft@Hrs	Unit	Material Cost	Labor Cost	Installed Cost
3 meters & mains	1E@1.25	Ea	1,060.00	33.60	1,093.60
4 meters & mains	1E@1.50	Ea	1,410.00	40.30	1,450.30
6 meters & mains	1E@2.00	Ea	2,050.00	53.80	2,103.80
7 meters & mains	1E@2.25	Ea	2,050.00	60.50	2,110.50
8 meters & mains	1E@2.50	Ea	2,730.00	67.20	2,797.20

120/208v meter center, 800A bus, 5 jaw sockets, 125A max., 2 pole branch, indoor

Material	Craft@Hrs	Unit	Material Cost	Labor Cost	Installed Cost
3 meters & mains	1E@1.50	Ea	573.00	40.30	613.30
4 meters & mains	1E@1.75	Ea	764.00	47.00	811.00
5 meters & mains	1E@2.00	Ea	954.00	53.80	1,007.80
6 meters & mains	1E@2.25	Ea	1,110.00	60.50	1,170.50
7 meters & mains	1E@2.50	Ea	1,490.00	67.20	1,557.20
8 meters & mains	1E@2.75	Ea	1,490.00	73.90	1,563.90
10 meters & mains	1E@3.25	Ea	1,850.00	87.40	1,937.40

Use these figures to estimate the cost of meter centers installed in buildings under the conditions described on pages 5 and 6. Costs listed are for each meter socket installed. The crew is one electrician working at a labor cost of $26.88 per manhour. These costs include sealing rings, layout, material handling, and normal waste. Add for supports, top conduit hubs, sales tax, delivery, supervision, mobilization, demobilization, cleanup, overhead and profit. Note: Top conduit hubs cost extra. Order the size appropriate for the size of conduit to be used. Blank caps are also available.

Material	Craft@Hrs	Unit	Material Cost	Labor Cost	Installed Cost	

120/208v meter centers, 1,200A bus, 5 jaw sockets, 10,000 A.I.C., 2 pole, indoor

Material	Craft@Hrs	Unit	Material Cost	Labor Cost	Installed Cost	
3 meters & mains	1E@1.50	Ea	1,100.00	40.30	1,140.30	
4 meters & mains	1E@1.75	Ea	1,470.00	47.00	1,517.00	
6 meters & mains	1E@2.25	Ea	2,140.00	60.50	2,200.50	
7 meters & mains	1E@2.50	Ea	2,140.00	67.20	2,207.20	
8 meters & mains	1E@2.75	Ea	2,870.00	73.90	2,943.90	

120/240v meter centers, 800A bus, 4 jaw sockets, 42,000 A.I.C., 2 pole, indoor

Material	Craft@Hrs	Unit	Material Cost	Labor Cost	Installed Cost	
3 meters & mains	1E@1.75	Ea	832.00	47.00	879.00	
4 meters & mains	1E@2.00	Ea	1,110.00	53.80	1,163.80	
5 meters & mains	1E@2.25	Ea	1,380.00	60.50	1,440.50	
6 meters & mains	1E@2.50	Ea	1,630.00	67.20	1,697.20	
7 meters & mains	1E@2.75	Ea	2,080.00	73.90	2,153.90	
8 meters & mains	1E@3.00	Ea	2,170.00	80.60	2,250.60	
10 meters & mains	1E@3.50	Ea	2,710.00	94.10	2,804.10	

120/240v meter centers, 1,200A bus, 4 jaw sockets, 42,000 A.I.C., 2 pole, indoor

Material	Craft@Hrs	Unit	Material Cost	Labor Cost	Installed Cost	
3 meters & mains	1E@1.75	Ea	1,350.00	47.00	1,397.00	
4 meters & mains	1E@2.00	Ea	1,800.00	53.80	1,853.80	
6 meters & mains	1E@2.50	Ea	2,550.00	67.20	2,617.20	
7 meters & mains	1E@2.75	Ea	2,730.00	73.90	2,803.90	
8 meters & mains	1E@3.00	Ea	3,530.00	80.60	3,610.60	

120/208v meter centers, 800A bus, 5 jaw sockets, 42,000 A.I.C., 2 pole, indoor

Material	Craft@Hrs	Unit	Material Cost	Labor Cost	Installed Cost	
3 meters & mains	1E@2.00	Ea	869.00	53.80	922.80	
4 meters & mains	1E@2.25	Ea	1,160.00	60.50	1,220.50	
5 meters & mains	1E@2.50	Ea	1,450.00	67.20	1,517.20	
6 meters & mains	1E@2.75	Ea	1,700.00	73.90	1,773.90	
7 meters & mains	1E@3.00	Ea	2,170.00	80.60	2,250.60	
8 meters & mains	1E@3.25	Ea	2,270.00	87.40	2,357.40	
10 meters & mains	1E@3.75	Ea	2,850.00	101.00	2,951.00	

Use these figures to estimate the cost of meter centers installed in buildings under the conditions described on pages 5 and 6. Costs listed are for each meter socket installed. The crew is one electrician working at a labor cost of $26.88 per manhour. These costs include sealing rings, layout, material handling, and normal waste. Add for supports, top conduit hubs, sales tax, delivery, supervision, mobilization, demobilization, cleanup, overhead and profit. Note: Top conduit hubs cost extra. Order the size appropriate for the size of conduit to be used. Blank caps are also available.

Meter Centers

Material	Craft@Hrs	Unit	Material Cost	Labor Cost	Installed Cost

120/208v meter centers, 1,200A bus, 5 jaw sockets, 42,000 A.I.C., 2 pole, indoor

Material	Craft@Hrs	Unit	Material Cost	Labor Cost	Installed Cost
3 meters & mains	1E@2.00	Ea	1,410.00	53.80	1,463.80
4 meters & mains	1E@2.25	Ea	1,880.00	60.50	1,940.50
6 meters & mains	1E@2.75	Ea	2,730.00	73.90	2,803.90
7 meters & mains	1E@3.00	Ea	2,850.00	80.60	2,930.60
8 meters & mains	1E@3.25	Ea	3,650.00	87.40	3,737.40

120/240 volt meter centers, 800A bus, 4 jaw sockets, 10,000 A.I.C., 2 pole, NEMA 3R

Material	Craft@Hrs	Unit	Material Cost	Labor Cost	Installed Cost
3 meters & mains	1E@1.25	Ea	536.00	33.60	569.60
4 meters & mains	1E@1.50	Ea	715.00	40.30	755.30
6 meters & mains	1E@2.00	Ea	1,040.00	53.80	1,093.80
7 meters & mains	1E@2.25	Ea	1,380.00	60.50	1,440.50
8 meters & mains	1E@2.50	Ea	1,720.00	67.20	1,787.20

120/240v meter centers, 1,200A bus, 4 jaw sockets, 10,000 A.I.C., 2 pole, NEMA 3R

Material	Craft@Hrs	Unit	Material Cost	Labor Cost	Installed Cost
3 meters & mains	1E@1.25	Ea	1,060.00	33.60	1,093.60
4 meters & mains	1E@1.50	Ea	1,410.00	40.30	1,450.30
6 meters & mains	1E@2.00	Ea	2,050.00	53.80	2,103.80
7 meters & mains	1E@2.25	Ea	2,050.00	60.50	2,110.50
8 meters & mains	1E@2.50	Ea	2,730.00	67.20	2,797.20

120/208v meter centers, 800A bus, 5 jaw sockets, 10,000 A.I.C., 2 pole, NEMA 3R

Material	Craft@Hrs	Unit	Material Cost	Labor Cost	Installed Cost
3 meters & mains	1E@1.50	Ea	573.00	40.30	613.30
4 meters & mains	1E@1.75	Ea	764.00	47.00	811.00
6 meters & mains	1E@2.25	Ea	1,110.00	60.50	1,170.50
7 meters & mains	1E@2.50	Ea	1,490.00	67.20	1,557.20
8 meters & mains	1E@2.75	Ea	1,490.00	73.90	1,563.90

Use these figures to estimate the cost of meter centers installed in buildings under the conditions described on pages 5 and 6. Costs listed are for each meter socket installed. The crew is one electrician working at a labor cost of $26.88 per manhour. These costs include sealing rings, layout, material handling, and normal waste. Add for supports, top conduit hubs, sales tax, delivery, supervision, mobilization, demobilization, cleanup, overhead and profit. Note: Main circuit breakers are included in these meter centers. But meters are not included. They are usually furnished by the electrical utility. Top conduit hubs cost extra. Order the size appropriate for the size of conduit to be used. Blank caps are also available.

Material	Craft@Hrs	Unit	Material Cost	Labor Cost	Installed Cost

120/208v raintight meter centers, 1,200A bus, 5 jaw sockets, 10,000 A.I.C., 2 pole, NEMA 3R

Material	Craft@Hrs	Unit	Material Cost	Labor Cost	Installed Cost
3 meters & mains	1E@1.50	Ea	1,100.00	40.30	1,140.30
4 meters & mains	1E@1.75	Ea	1,470.00	47.00	1,517.00
6 meters & mains	1E@2.25	Ea	2,140.00	60.50	2,200.50
7 meters & mains	1E@2.50	Ea	2,140.00	67.20	2,207.20
8 meters & mains	1E@2.75	Ea	2,870.00	73.90	2,943.90

120/240 volt raintight meter centers, 800A bus, 4 jaw sockets, 42,000 A.I.C., 2 pole, NEMA 3R

Material	Craft@Hrs	Unit	Material Cost	Labor Cost	Installed Cost
3 meters & mains	1E@1.75	Ea	832.00	47.00	879.00
4 meters & mains	1E@2.00	Ea	1,110.00	53.80	1,163.80
6 meters & mains	1E@2.50	Ea	1,630.00	67.20	1,697.20
7 meters & mains	1E@2.75	Ea	2,080.00	73.90	2,153.90
8 meters & mains	1E@3.00	Ea	2,170.00	80.60	2,250.60

120/240 volt raintight meter centers, 1,200A bus, 4 jaw sockets, 42,000 A.I.C., 2 pole, NEMA 3R

Material	Craft@Hrs	Unit	Material Cost	Labor Cost	Installed Cost
3 meters & mains	1E@1.75	Ea	1,350.00	47.00	1,397.00
4 meters & mains	1E@2.00	Ea	1,800.00	53.80	1,853.80
6 meters & mains	1E@2.50	Ea	2,550.00	67.20	2,617.20
7 meters & mains	1E@2.75	Ea	2,730.00	73.90	2,803.90
8 meters & mains	1E@3.00	Ea	3,530.00	80.60	3,610.60

120/208 volt raintight meter centers, 800A bus, 5 jaw sockets, 42,000 A.I.C., 2 pole, NEMA 3R

Material	Craft@Hrs	Unit	Material Cost	Labor Cost	Installed Cost
3 meters & mains	1E@2.00	Ea	869.00	53.80	922.80
4 meters & mains	1E@2.25	Ea	1,160.00	60.50	1,220.50
6 meters & mains	1E@2.75	Ea	1,700.00	73.90	1,773.90
7 meters & mains	1E@3.00	Ea	2,170.00	80.60	2,250.60
8 meters & mains	1E@3.25	Ea	2,270.00	87.40	2,357.40

120/208 volt raintight meter centers, 1,200A bus, 5 jaw sockets, 42,000 A.I.C., 2 pole, NEMA 3R

Material	Craft@Hrs	Unit	Material Cost	Labor Cost	Installed Cost
3 meters & mains	1E@2.00	Ea	1,410.00	53.80	1,463.80
4 meters & mains	1E@2.25	Ea	1,880.00	60.50	1,940.50
6 meters & mains	1E@2.75	Ea	2,730.00	73.90	2,803.90
7 meters & mains	1E@3.00	Ea	2,850.00	80.60	2,930.60
8 meters & mains	1E@3.25	Ea	3,650.00	87.40	3,737.40

Use these figures to estimate the cost of meter centers installed in buildings under the conditions described on pages 5 and 6. Costs listed are for each meter socket installed. The crew is one electrician working at a labor cost of $26.88 per manhour. These costs include sealing rings, layout, material handling, and normal waste. Add for supports, top conduit hubs, sales tax, delivery, supervision, mobilization, demobilization, cleanup, overhead and profit. Note: Main circuit breakers are included in these meter centers. But meters are not included. They are usually furnished by the electrical utility. Top conduit hubs cost extra. Order the size appropriate for the size of conduit to be used. Blank caps are also available.

120/240 Volt Circuit Breaker Loadcenters

Material		Craft@Hrs	Unit	Material Cost	Labor Cost	Installed Cost

Indoor 120/240 volt loadcenters with main lugs only, no breakers

Material		Craft@Hrs	Unit	Material Cost	Labor Cost	Installed Cost
30A	2 spaces	1E@0.35	Ea	11.00	9.41	20.41
40A	2 spaces	1E@0.35	Ea	32.00	9.41	41.41
70A	4 spaces	1E@0.40	Ea	36.40	10.80	47.20
100A	6 spaces	1E@0.50	Ea	22.10	13.40	35.50
100A	8 spaces	1E@0.60	Ea	36.40	16.10	52.50
100A	12 spaces	1E@0.70	Ea	65.10	18.80	83.90
125A	16 spaces	1E@0.75	Ea	50.70	20.20	70.90
125A	20 spaces	1E@0.80	Ea	63.90	21.50	85.40
125A	24 spaces	1E@0.90	Ea	92.60	24.20	116.80
150A	12 spaces	1E@0.75	Ea	82.70	20.20	102.90
150A	16 spaces	1E@0.80	Ea	82.70	21.50	104.20
150A	24 spaces	1E@1.00	Ea	93.80	26.90	120.70
150A	30 spaces	1E@1.20	Ea	93.80	32.30	126.10
200A	8 spaces	1E@0.70	Ea	108.00	18.80	126.80
200A	12 spaces	1E@0.80	Ea	108.00	21.50	129.50
200A	16 spaces	1E@0.95	Ea	108.00	25.50	133.50
200A	24 spaces	1E@1.10	Ea	108.00	29.60	137.60
200A	30 spaces	1E@1.30	Ea	108.00	34.90	142.90
225A	42 spaces	1E@1.50	Ea	194.00	40.30	234.30
400A	30 spaces	1E@1.25	Ea	410.00	33.60	443.60
400A	42 spaces	1E@1.75	Ea	410.00	47.00	457.00

Raintight 120/240 volt loadcenters with main lugs only, no breakers

Material		Craft@Hrs	Unit	Material Cost	Labor Cost	Installed Cost
40A	2 spaces	1E@0.35	Ea	36.40	9.41	45.81
70A	4 spaces	1E@0.40	Ea	38.60	10.80	49.40
100A	6 spaces	1E@0.50	Ea	51.90	13.40	65.30
100A	8 spaces	1E@0.60	Ea	51.90	16.10	68.00
100A	12 spaces	1E@0.70	Ea	51.90	18.80	70.70
125A	12 spaces	1E@0.70	Ea	84.90	18.80	103.70
125A	16 spaces	1E@0.75	Ea	103.00	20.20	123.20
125A	20 spaces	1E@0.80	Ea	120.00	21.50	141.50
125A	24 spaces	1E@0.90	Ea	120.00	24.20	144.20
150A	12 spaces	1E@0.75	Ea	159.00	20.20	179.20
150A	16 spaces	1E@0.80	Ea	159.00	21.50	180.50
150A	24 spaces	1E@1.00	Ea	159.00	26.90	185.90
150A	30 spaces	1E@1.25	Ea	298.00	33.60	331.60
200A	8 spaces	1E@0.70	Ea	159.00	18.80	177.80
200A	12 spaces	1E@0.80	Ea	159.00	21.50	180.50
200A	16 spaces	1E@0.95	Ea	159.00	25.50	184.50
200A	24 spaces	1E@1.10	Ea	159.00	29.60	188.60
200A	30 spaces	1E@1.30	Ea	298.00	34.90	332.90
225A	42 spaces	1E@1.50	Ea	427.00	40.30	467.30

Use these figures to estimate the cost of loadcenters installed in buildings under the conditions described on pages 5 and 6. Costs listed are for each loadcenter installed. The crew is one electrician working at a labor cost of $26.88 per manhour. These costs include the cover, bus, neutral bar, layout, material handling, and normal waste. Add for circuit breakers, supports, sales tax, delivery, supervision, mobilization, demobilization, cleanup, overhead and profit. Note: Be careful to select the right type of enclosure and voltage.

Material		Craft@Hrs	Unit	Material Cost	Labor Cost	Installed Cost

Indoor 120/208 volt loadcenters with main lugs only, no breakers

Material		Craft@Hrs	Unit	Material Cost	Labor Cost	Installed Cost
100A	12 spaces	1E@0.70	Ea	137.00	18.80	155.80
100A	16 spaces	1E@0.75	Ea	137.00	20.20	157.20
100A	20 spaces	1E@0.80	Ea	137.00	21.50	158.50
125A	24 spaces	1E@0.90	Ea	175.00	24.20	199.20
150A	16 spaces	1E@0.80	Ea	175.00	21.50	196.50
150A	20 spaces	1E@1.00	Ea	175.00	26.90	201.90
150A	24 spaces	1E@1.20	Ea	175.00	32.30	207.30
150A	30 spaces	1E@1.25	Ea	175.00	33.60	208.60
200A	16 spaces	1E@0.95	Ea	173.00	25.50	198.50
200A	20 spaces	1E@1.00	Ea	173.00	26.90	199.90
200A	24 spaces	1E@1.10	Ea	173.00	29.60	202.60
200A	30 spaces	1E@1.20	Ea	173.00	32.30	205.30
200A	40 spaces	1E@1.30	Ea	235.00	34.90	269.90
300A	42 spaces	1E@1.50	Ea	741.00	40.30	781.30
400A	42 spaces	1E@1.75	Ea	741.00	47.00	788.00

Raintight 120/208 volt loadcenters with main lugs only, no breakers

Material		Craft@Hrs	Unit	Material Cost	Labor Cost	Installed Cost
100A	12 spaces	1E@0.70	Ea	123.00	18.80	141.80
100A	16 spaces	1E@0.75	Ea	159.00	20.20	179.20
100A	20 spaces	1E@0.80	Ea	202.00	21.50	223.50
125A	24 spaces	1E@0.90	Ea	202.00	24.20	226.20
150A	16 spaces	1E@0.80	Ea	159.00	21.50	180.50
150A	20 spaces	1E@1.00	Ea	202.00	26.90	228.90
150A	24 spaces	1E@1.20	Ea	202.00	32.30	234.30
150A	30 spaces	1E@1.25	Ea	202.00	33.60	235.60
200A	16 spaces	1E@0.95	Ea	184.00	25.50	209.50
200A	20 spaces	1E@1.00	Ea	177.00	26.90	203.90

Use these figures to estimate the cost of loadcenters installed in buildings under the conditions described on pages 5 and 6. Costs listed are for each loadcenter installed. The crew is one electrician working at a labor cost of $26.88 per manhour. These costs include the cover, bus, neutral bar, layout, material handling, and normal waste. Add for circuit breakers, supports, sales tax, delivery, supervision, mobilization, demobilization, cleanup, overhead and profit. Note: Be careful to select the right type of enclosure and voltage.

Loadcenters and Panelboards

Material		Craft@Hrs	Unit	Material Cost	Labor Cost	Installed Cost

Raintight 120/208 volt C.B. loadcenters with main lugs but no other breakers

Material		Craft@Hrs	Unit	Material Cost	Labor Cost	Installed Cost
200A	24 spaces	1E@1.10	Ea	239.00	29.60	268.60
200A	30 spaces	1E@1.20	Ea	239.00	32.30	271.30
200A	40 spaces	1E@1.30	Ea	239.00	34.90	273.90
225A	30 spaces	1E@1.40	Ea	432.00	37.60	469.60
225A	40 spaces	1E@1.50	Ea	432.00	40.30	472.30

Indoor 120/240 volt C.B. panelboards (NQO) with plug-in breakers to 60 amps, main lugs only

Material		Craft@Hrs	Unit	Material Cost	Labor Cost	Installed Cost
100A	8 poles	1E@3.40	Ea	415.00	91.40	506.40
100A	10 poles	1E@4.00	Ea	437.00	108.00	545.00
100A	12 poles	1E@4.60	Ea	458.00	124.00	582.00
100A	14 poles	1E@5.70	Ea	532.00	153.00	685.00
100A	16 poles	1E@6.30	Ea	554.00	169.00	723.00
100A	18 poles	1E@6.90	Ea	575.00	185.00	760.00
100A	20 poles	1E@7.50	Ea	596.00	202.00	798.00
225A	22 poles	1E@8.10	Ea	685.00	218.00	903.00
225A	24 poles	1E@8.70	Ea	706.00	234.00	940.00
225A	26 poles	1E@9.80	Ea	728.00	263.00	991.00
225A	28 poles	1E@10.4	Ea	750.00	280.00	1,030.00
225A	30 poles	1E@11.0	Ea	772.00	296.00	1,068.00
225A	32 poles	1E@11.6	Ea	849.00	312.00	1,161.00
225A	34 poles	1E@12.2	Ea	880.00	328.00	1,208.00
225A	36 poles	1E@12.8	Ea	903.00	344.00	1,247.00
225A	38 poles	1E@13.9	Ea	926.00	374.00	1,300.00
225A	40 poles	1E@14.5	Ea	949.00	390.00	1,339.00
225A	42 poles	1E@15.1	Ea	965.00	406.00	1,371.00
400A	38 poles	1E@14.4	Ea	1,090.00	387.00	1,477.00
400A	40 poles	1E@15.0	Ea	1,110.00	403.00	1,513.00
400A	42 poles	1E@15.6	Ea	1,130.00	419.00	1,549.00

Indoor 120/208 volt C.B. panelboards (NQO) with plug-in breakers to 60 amps and main breaker

Material		Craft@Hrs	Unit	Material Cost	Labor Cost	Installed Cost
100A	8 poles	1E@3.40	Ea	579.00	91.40	670.40
100A	12 poles	1E@4.00	Ea	622.00	108.00	730.00
100A	16 poles	1E@4.60	Ea	696.00	124.00	820.00

Use these figures to estimate the cost of loadcenters and panelboards installed in buildings under the conditions described on pages 5 and 6. Costs listed are for each panelboard installed. The crew is one electrician working at a labor cost of $26.88 per manhour. These costs include the cover, bus, neutral bar, circuit breakers, layout, material handling, and normal waste. Labor costs include mounting the panel can, installing the interior section, making the circuit connections for each breaker and the main breaker, and hanging the door or cover. Reduce the labor cost for each spare breaker not connected to a circuit. Add for supports, sales tax, delivery, supervision, mobilization, demobilization, cleanup, overhead and profit. Note: Be sure to select the loadcenter or panelboard with the right voltage and type of enclosure.

Circuit Breaker Panelboards Including Breakers to 60 Amps

Material	Craft@Hrs	Unit	Material Cost	Labor Cost	Installed Cost

Indoor 120/208 volt C.B. panelboards (NQO) with plug-in breakers to 60 amps and main breaker

Material		Craft@Hrs	Unit	Material Cost	Labor Cost	Installed Cost
100A	18 poles	1E@6.90	Ea	718.00	185.00	903.00
100A	20 poles	1E@7.50	Ea	740.00	202.00	942.00
225A	22 poles	1E@8.10	Ea	1,050.00	218.00	1,268.00
225A	24 poles	1E@8.70	Ea	1,070.00	234.00	1,304.00
225A	26 poles	1E@9.80	Ea	1,090.00	263.00	1,353.00
225A	28 poles	1E@10.4	Ea	1,120.00	280.00	1,400.00
225A	30 poles	1E@11.0	Ea	1,130.00	296.00	1,426.00
225A	32 poles	1E@11.6	Ea	1,230.00	312.00	1,542.00
225A	34 poles	1E@12.2	Ea	1,250.00	328.00	1,578.00
225A	36 poles	1E@12.8	Ea	1,270.00	344.00	1,614.00
225A	38 poles	1E@13.9	Ea	1,300.00	374.00	1,674.00
225A	40 poles	1E@14.5	Ea	1,310.00	390.00	1,700.00
225A	42 poles	1E@15.1	Ea	1,320.00	406.00	1,726.00

Indoor 120/240 volt C.B. panelboards (NQOB) with breakers to 60 amps, with main lugs only

Material		Craft@Hrs	Unit	Material Cost	Labor Cost	Installed Cost
100A	8 poles	1E@3.40	Ea	415.00	91.40	506.40
100A	10 poles	1E@4.00	Ea	437.00	108.00	545.00
100A	12 poles	1E@4.60	Ea	458.00	124.00	582.00
100A	14 poles	1E@5.70	Ea	532.00	153.00	685.00
100A	16 poles	1E@6.30	Ea	554.00	169.00	723.00
100A	18 poles	1E@6.90	Ea	575.00	185.00	760.00
100A	20 poles	1E@7.50	Ea	597.00	202.00	799.00
225A	22 poles	1E@8.10	Ea	685.00	218.00	903.00
225A	24 poles	1E@8.70	Ea	706.00	234.00	940.00
225A	26 poles	1E@9.80	Ea	728.00	263.00	991.00
225A	28 poles	1E@10.4	Ea	750.00	280.00	1,030.00
225A	30 poles	1E@11.0	Ea	771.00	296.00	1,067.00
225A	32 poles	1E@11.6	Ea	857.00	312.00	1,169.00
225A	34 poles	1E@12.2	Ea	880.00	328.00	1,208.00
225A	36 poles	1E@12.8	Ea	903.00	344.00	1,247.00
225A	38 poles	1E@13.9	Ea	926.00	374.00	1,300.00
225A	40 poles	1E@14.5	Ea	949.00	390.00	1,339.00
225A	42 poles	1E@15.1	Ea	965.00	406.00	1,371.00

Use these figures to estimate the cost of panelboards installed in buildings under the conditions described on pages 5 and 6. Costs listed are for each panelboard installed. The crew is one electrician working at a labor cost of $26.88 per manhour. These costs include the cover, bus, neutral bar, circuit breakers, layout, material handling, and normal waste. Labor costs include mounting the panel can, installing the interior section, making the circuit connections for each breaker and the main breaker, and hanging the door or cover. Reduce the labor cost for each spare breaker not connected to a circuit. Add for supports, sales tax, delivery, supervision, mobilization, demobilization, cleanup, overhead and profit. Note: Be sure to select the panelboard with the right voltage and type of enclosure.

Circuit Breaker Panelboards Including Bolt-on Breakers to 60 Amps

Material		Craft@Hrs	Unit	Material Cost	Labor Cost	Installed Cost

Indoor 120/208 volt C.B. panelboards with bolt-on breakers to 60 amps, with main lugs only

Material		Craft@Hrs	Unit	Material Cost	Labor Cost	Installed Cost
100A	8 poles	1E@3.40	Ea	420.00	91.40	511.40
100A	10 poles	1E@4.00	Ea	442.00	108.00	550.00
100A	12 poles	1E@4.60	Ea	463.00	124.00	587.00
100A	14 poles	1E@5.70	Ea	532.00	153.00	685.00
100A	16 poles	1E@6.30	Ea	554.00	169.00	723.00
100A	18 poles	1E@6.90	Ea	575.00	185.00	760.00
100A	20 poles	1E@7.50	Ea	597.00	202.00	799.00
100A	22 poles	1E@8.10	Ea	685.00	218.00	903.00
100A	24 poles	1E@8.70	Ea	706.00	234.00	940.00
225A	26 poles	1E@9.80	Ea	728.00	263.00	991.00
225A	28 poles	1E@10.4	Ea	750.00	280.00	1,030.00
225A	30 poles	1E@11.0	Ea	771.00	296.00	1,067.00
225A	32 poles	1E@11.6	Ea	1,120.00	312.00	1,432.00
225A	34 poles	1E@12.2	Ea	1,140.00	328.00	1,468.00
225A	36 poles	1E@12.8	Ea	1,170.00	344.00	1,514.00
225A	38 poles	1E@13.9	Ea	1,200.00	374.00	1,574.00
225A	40 poles	1E@14.5	Ea	1,230.00	390.00	1,620.00
225A	42 poles	1E@15.1	Ea	1,940.00	406.00	2,346.00

Indoor 120/240 volt C.B. panelboards with bolt-on breakers to 60 amps and main breaker

Material		Craft@Hrs	Unit	Material Cost	Labor Cost	Installed Cost
100A	8 poles	1E@3.40	Ea	579.00	91.40	670.40
100A	10 poles	1E@4.00	Ea	601.00	108.00	709.00
100A	12 poles	1E@4.60	Ea	622.00	124.00	746.00
100A	14 poles	1E@5.70	Ea	675.00	153.00	828.00
100A	16 poles	1E@6.30	Ea	696.00	169.00	865.00
100A	18 poles	1E@6.90	Ea	718.00	185.00	903.00
100A	20 poles	1E@7.50	Ea	740.00	202.00	942.00
225A	22 poles	1E@8.10	Ea	1,050.00	218.00	1,268.00
225A	24 poles	1E@8.70	Ea	1,070.00	234.00	1,304.00
225A	26 poles	1E@9.80	Ea	1,090.00	263.00	1,353.00
225A	28 poles	1E@10.4	Ea	1,120.00	280.00	1,400.00
225A	30 poles	1E@11.0	Ea	1,130.00	296.00	1,426.00
225A	32 poles	1E@11.6	Ea	1,230.00	312.00	1,542.00
225A	34 poles	1E@12.2	Ea	1,250.00	328.00	1,578.00
225A	36 poles	1E@12.8	Ea	1,270.00	344.00	1,614.00
225A	38 poles	1E@13.9	Ea	1,300.00	374.00	1,674.00
225A	40 poles	1E@14.5	Ea	1,310.00	390.00	1,700.00
225A	42 poles	1E@15.1	Ea	1,320.00	406.00	1,726.00

Use these figures to estimate the cost of panelboards installed in buildings under the conditions described on pages 5 and 6. Costs listed are for each panelboard installed. The crew is one electrician working at a labor cost of $26.88 per manhour. These costs include the cover, bus, neutral bar, circuit breakers, layout, material handling, and normal waste. Labor costs include mounting the panel can, installing the interior section, making the circuit connections for each breaker and the main breaker, and hanging the door or cover. Reduce the labor cost for each spare breaker not connected to a circuit. Add for supports, sales tax, delivery, supervision, mobilization, demobilization, cleanup, overhead and profit. Note: Be sure to select the panelboard with the right voltage and type of enclosure.

Material		Craft@Hrs	Unit	Material Cost	Labor Cost	Installed Cost

120/240 volt 14" wide C.B. panelboards with bolt-on breakers up to 60 amps and main lugs only

Material		Craft@Hrs	Unit	Material Cost	Labor Cost	Installed Cost
100A	8 poles	1E@3.40	Ea	539.00	91.40	630.40
100A	10 poles	1E@4.00	Ea	561.00	108.00	669.00
100A	12 poles	1E@4.60	Ea	583.00	124.00	707.00
100A	14 poles	1E@5.70	Ea	604.00	153.00	757.00
100A	16 poles	1E@6.30	Ea	626.00	169.00	795.00
100A	18 poles	1E@6.90	Ea	679.00	185.00	864.00
100A	20 poles	1E@7.50	Ea	704.00	202.00	906.00
225A	22 poles	1E@8.10	Ea	726.00	218.00	944.00
225A	24 poles	1E@8.70	Ea	750.00	234.00	984.00
225A	26 poles	1E@9.80	Ea	771.00	263.00	1,034.00
225A	28 poles	1E@10.4	Ea	795.00	280.00	1,075.00
225A	30 poles	1E@11.0	Ea	818.00	296.00	1,114.00
225A	32 poles	1E@11.6	Ea	917.00	312.00	1,229.00
225A	34 poles	1E@12.2	Ea	941.00	328.00	1,269.00
225A	36 poles	1E@12.8	Ea	965.00	344.00	1,309.00
225A	38 poles	1E@13.9	Ea	989.00	374.00	1,363.00
225A	40 poles	1E@14.5	Ea	1,000.00	390.00	1,390.00
225A	42 poles	1E@15.1	Ea	1,040.00	406.00	1,446.00

NEMA 1 telephone & signal terminal cabinets with keyed door lock, wood backing

Material	Craft@Hrs	Unit	Material Cost	Labor Cost	Installed Cost
12"W x 12"H x 4"D	1E@0.45	Ea	78.30	12.10	90.40
12"W x 16"H x 4"D	1E@0.60	Ea	78.30	16.10	94.40
12"W x 16"H x 6"D	1E@0.70	Ea	109.00	18.80	127.80
12"W x 18"H x 4"D	1E@0.90	Ea	94.10	24.20	118.30
12"W x 18"H x 6"D	1E@1.00	Ea	114.00	26.90	140.90
12"W x 24"H x 4"D	1E@1.20	Ea	102.00	32.30	134.30
12"W x 24"H x 6"D	1E@1.30	Ea	160.00	34.90	194.90
18"W x 18"H x 4"D	1E@1.20	Ea	105.00	32.30	137.30
18"W x 18"H x 6"D	1E@1.30	Ea	123.00	34.90	157.90
18"W x 24"H x 4"D	1E@1.50	Ea	142.00	40.30	182.30
18"W x 24"H x 6"D	1E@1.60	Ea	175.00	43.00	218.00
18"W x 30"H x 4"D	1E@1.80	Ea	160.00	48.40	208.40
18"W x 30"H x 6"D	1E@1.90	Ea	185.00	51.10	236.10
24"W x 24"H x 4"D	1E@2.10	Ea	170.00	56.40	226.40
24"W x 24"H x 6"D	1E@2.20	Ea	197.00	59.10	256.10
24"W x 30"H x 4"D	1E@2.50	Ea	215.00	67.20	282.20
24"W x 30"H x 6"D	1E@2.60	Ea	236.00	69.90	305.90
24"W x 36"H x 4"D	1E@3.00	Ea	250.00	80.60	330.60
24"W x 36"H x 6"D	1E@3.10	Ea	274.00	83.30	357.30

Use these figures to estimate the cost of panelboards and terminal cabinets installed in buildings under the conditions described on pages 5 and 6. Costs listed are for each panelboard or terminal cabinet installed. The crew is one electrician working at a labor cost of $26.88 per manhour. These costs include the cover, bus, neutral bar, circuit breakers, layout, material handling, and normal waste. Labor costs include mounting the panel can, installing the interior section, making the circuit connections for each breaker and the main breaker, and hanging the door or cover. Reduce the labor cost for each spare breaker not connected to a circuit. Add for supports, sales tax, delivery, supervision, mobilization, demobilization, cleanup, overhead and profit. Note: Be sure to select the panelboard with the right voltage and type of enclosure.

Signal Cabinets and Wireway

Material	Craft@Hrs	Unit	Material Cost	Labor Cost	Installed Cost
NEMA 1 telephone & signal terminal cabinets with keyed door lock, wood backing					
30"W x 30"H x 4"D	1E@3.00	Ea	258.00	80.60	338.60
30"W x 30"H x 6"D	1E@3.10	Ea	278.00	83.30	361.30
30"W x 36"H x 4"D	1E@3.30	Ea	316.00	88.70	404.70
30"W x 36"H x 6"D	1E@3.40	Ea	343.00	91.40	434.40
36"W x 48"H x 4"D	1E@3.60	Ea	580.00	96.80	676.80
36"W x 48"H x 6"D	1E@3.70	Ea	657.00	99.50	756.50
NEMA 1 screw cover wireway					
3" x 3" x 12"	1E@0.40	Ea	6.64	10.80	17.44
3" x 3" x 18"	1E@0.45	Ea	9.95	12.10	22.05
3" x 3" x 24"	1E@0.50	Ea	11.00	13.40	24.40
3" x 3" x 36"	1E@0.60	Ea	16.80	16.10	32.90
3" x 3" x 48"	1E@0.70	Ea	22.70	18.80	41.50
3" x 3" x 60"	1E@0.80	Ea	25.60	21.50	47.10
3" x 3" x 72"	1E@1.00	Ea	31.80	26.90	58.70
3" x 3" x 120"	1E@1.15	Ea	52.70	30.90	83.60
4" x 4" x 12"	1E@0.50	Ea	7.19	13.40	20.59
4" x 4" x 18"	1E@0.60	Ea	11.00	16.10	27.10
4" x 4" x 24"	1E@0.70	Ea	11.30	18.80	30.10
4" x 4" x 36"	1E@0.80	Ea	17.60	21.50	39.10
4" x 4" x 48"	1E@1.00	Ea	23.00	26.90	49.90
4" x 4" x 60"	1E@1.20	Ea	25.10	32.30	57.40
4" x 4" x 72"	1E@1.50	Ea	35.00	40.30	75.30
4" x 4" x 120"	1E@1.75	Ea	53.70	47.00	100.70
4" x 6" x 12"	1E@0.55	Ea	13.10	14.80	27.90
4" x 6" x 18"	1E@0.65	Ea	16.00	17.50	33.50
4" x 6" x 24"	1E@0.75	Ea	18.40	20.20	38.60
4" x 6" x 36"	1E@0.95	Ea	25.90	25.50	51.40
4" x 6" x 48"	1E@1.05	Ea	33.10	28.20	61.30
4" x 6" x 60"	1E@1.25	Ea	41.00	33.60	74.60
4" x 6" x 72"	1E@1.55	Ea	43.70	41.70	85.40
6" x 6" x 12"	1E@0.60	Ea	14.50	16.10	30.60
6" x 6" x 18"	1E@0.70	Ea	19.50	18.80	38.30
6" x 6" x 24"	1E@0.80	Ea	19.70	21.50	41.20
6" x 6" x 36"	1E@1.00	Ea	26.10	26.90	53.00
6" x 6" x 48"	1E@1.20	Ea	35.50	32.30	67.80
6" x 6" x 60"	1E@1.30	Ea	38.50	34.90	73.40
6" x 6" x 72"	1E@1.60	Ea	49.50	43.00	92.50
6" x 6" x 120"	1E@1.75	Ea	91.60	47.00	138.60

Use these figures to estimate the cost of terminal cabinets and wireway installed in buildings under the conditions described on pages 5 and 6. Costs listed are for each cabinet and wireway installed. The crew is one electrician working at a labor cost of $26.88 per manhour. These costs include the cover, layout, material handling, and normal waste. Add for wireway ends, couplings, wireway fittings, supports, sales tax, delivery, supervision, mobilization, demobilization, cleanup, overhead and profit. Note: Wireway size depends on the number of wires and the equipment that will be attached.

Material	Craft@Hrs	Unit	Material Cost	Labor Cost	Installed Cost
NEMA 1 screw cover wireway					
8" x 8" x 12"	1E@0.65	Ea	24.40	17.50	41.90
8" x 8" x 24"	1E@0.85	Ea	37.20	22.80	60.00
8" x 8" x 36"	1E@1.05	Ea	55.80	28.20	84.00
8" x 8" x 48"	1E@1.25	Ea	66.50	33.60	100.10
8" x 8" x 60"	1E@1.35	Ea	77.20	36.30	113.50
8" x 8" x 72"	1E@1.65	Ea	106.00	44.40	150.40
8" x 8" x 120"	1E@1.80	Ea	160.00	48.40	208.40
10" x 10" x 24"	1E@0.70	Ea	57.20	18.80	76.00
10" x 10" x 36"	1E@1.15	Ea	79.60	30.90	110.50
10" x 10" x 48"	1E@1.35	Ea	95.90	36.30	132.20
10" x 10" x 60"	1E@1.45	Ea	116.00	39.00	155.00
10" x 10" x 72"	1E@1.75	Ea	152.00	47.00	199.00
12" x 12" x 24"	1E@0.90	Ea	66.20	24.20	90.40
12" x 12" x 36"	1E@1.25	Ea	95.90	33.60	129.50
12" x 12" x 48"	1E@1.45	Ea	117.00	39.00	156.00
12" x 12" x 60"	1E@1.55	Ea	133.00	41.70	174.70
12" x 12" x 72"	1E@1.85	Ea	170.00	49.70	219.70
NEMA 1 hinged cover wireway					
4" x 4" x 12"	1E@0.50	Ea	8.63	13.40	22.03
4" x 4" x 24"	1E@0.70	Ea	12.70	18.80	31.50
4" x 4" x 36"	1E@0.80	Ea	18.60	21.50	40.10
4" x 4" x 48"	1E@1.00	Ea	25.40	26.90	52.30
4" x 4" x 60"	1E@1.20	Ea	27.90	32.30	60.20
4" x 4" x 72"	1E@1.50	Ea	34.90	40.30	75.20
4" x 4" x 120"	1E@1.75	Ea	57.20	47.00	104.20
6" x 6" x 12"	1E@0.60	Ea	16.60	16.10	32.70
6" x 6" x 18"	1E@0.70	Ea	19.40	18.80	38.20
6" x 6" x 24"	1E@0.80	Ea	20.70	21.50	42.20
6" x 6" x 36"	1E@1.00	Ea	28.80	26.90	55.70
6" x 6" x 48"	1E@1.20	Ea	38.20	32.30	70.50
6" x 6" x 60"	1E@1.30	Ea	41.30	34.90	76.20
6" x 6" x 72"	1E@1.60	Ea	49.40	43.00	92.40
6" x 6" x 120"	1E@1.75	Ea	97.70	47.00	144.70
8" x 8" x 12"	1E@0.65	Ea	26.30	17.50	43.80
8" x 8" x 24"	1E@0.85	Ea	40.50	22.80	63.30
8" x 8" x 36"	1E@1.05	Ea	61.10	28.20	89.30
8" x 8" x 48"	1E@1.25	Ea	74.10	33.60	107.70
8" x 8" x 60"	1E@1.35	Ea	83.30	36.30	119.60
8" x 8" x 72"	1E@1.65	Ea	106.00	44.40	150.40
8" x 8" x 120"	1E@1.80	Ea	160.00	48.40	208.40

Use these figures to estimate the cost of wireway installed in buildings under the conditions described on pages 5 and 6. Costs listed are for each wireway installed. The crew is one electrician working at a labor cost of $26.88 per manhour. These costs include the cover, layout, material handling, and normal waste. Add for supports, wireway ends, couplings, wireway fittings, sales tax, delivery, supervision, mobilization, demobilization, cleanup, overhead and profit. Note: Wireway size depends on the number of wires and the equipment that will be attached.

Wireway Fittings

Material	Craft@Hrs	Unit	Material Cost	Labor Cost	Installed Cost
NEMA 1 wireway fittings					
3" x 3" couplings	1E@0.10	Ea	2.57	2.69	5.26
4" x 4" couplings	1E@0.10	Ea	2.57	2.69	5.26
4" x 6" couplings	1E@0.10	Ea	3.35	2.69	6.04
6" x 6" couplings	1E@0.10	Ea	3.35	2.69	6.04
8" x 8" couplings	1E@0.15	Ea	4.87	4.03	8.90
10" x 10" couplings	1E@0.20	Ea	6.63	5.38	12.01
12" x 12" couplings	1E@0.25	Ea	10.30	6.72	17.02
3" x 3" elbows 45	1E@0.20	Ea	20.00	5.38	25.38
4" x 4" elbows 45	1E@0.20	Ea	20.60	5.38	25.98
4" x 6" elbows 45	1E@0.20	Ea	26.00	5.38	31.38
6" x 6" elbows 45	1E@0.20	Ea	23.80	5.38	29.18
8" x 8" elbows 45	1E@0.30	Ea	36.30	8.06	44.36
3" x 3" elbows 90	1E@0.20	Ea	19.00	5.38	24.38
4" x 4" elbows 90	1E@0.20	Ea	19.70	5.38	25.08
4" x 6" elbows 90	1E@0.20	Ea	23.60	5.38	28.98
6" x 6" elbows 90	1E@0.20	Ea	24.50	5.38	29.88
8" x 8" elbows 90	1E@0.30	Ea	36.50	8.06	44.56
10" x 10" elbows 90	1E@0.40	Ea	47.70	10.80	58.50
12" x 12" elbows 90	1E@0.50	Ea	66.40	13.40	79.80
3" x 3" ends	1E@0.10	Ea	2.57	2.69	5.26
4" x 4" ends	1E@0.10	Ea	2.57	2.69	5.26
4" x 6" ends	1E@0.10	Ea	3.68	2.69	6.37
6" x 6" ends	1E@0.10	Ea	3.68	2.69	6.37
8" x 8" ends	1E@0.15	Ea	4.87	4.03	8.90
10" x 10" ends	1E@0.20	Ea	6.63	5.38	12.01
12" x 12" ends	1E@0.25	Ea	10.30	6.72	17.02
3" x 3" flanges	1E@0.20	Ea	7.23	5.38	12.61
4" x 4" flanges	1E@0.20	Ea	7.23	5.38	12.61
4" x 6" flanges	1E@0.20	Ea	7.68	5.38	13.06
6" x 6" flanges	1E@0.30	Ea	10.30	8.06	18.36
8" x 8" flanges	1E@0.40	Ea	15.10	10.80	25.90
10" x 10" flanges	1E@0.50	Ea	17.10	13.40	30.50
12" x 12" flanges	1E@0.70	Ea	31.00	18.80	49.80
3" x 3" hangers	1E@0.25	Ea	6.24	6.72	12.96
4" x 4" hangers	1E@0.25	Ea	6.24	6.72	12.96
6" x 6" hangers	1E@0.30	Ea	10.10	8.06	18.16
8" x 8" hangers	1E@0.40	Ea	13.80	10.80	24.60

Use these figures to estimate the cost of wireway fittings installed in buildings under the conditions described on pages 5 and 6. Costs listed are for each fitting installed. The crew is one electrician working at a labor cost of $26.88 per manhour. These costs include connecting screws, layout, material handling, and normal waste. Add for wireway, supports, sales tax, delivery, supervision, mobilization, demobilization, cleanup, overhead and profit.

Material	Craft@Hrs	Unit	Material Cost	Labor Cost	Installed Cost
NEMA 1 wireway fittings					
4" - 3" reducers	1E@0.15	Ea	23.20	4.03	27.23
6" - 4" reducers	1E@0.20	Ea	23.20	5.38	28.58
8" - 6" reducers	1E@0.25	Ea	26.70	6.72	33.42
3" x 3" tees	1E@0.25	Ea	21.50	6.72	28.22
4" x 4" tees	1E@0.25	Ea	22.80	6.72	29.52
4" x 6" tees	1E@0.30	Ea	27.90	8.06	35.96
6" x 6" tees	1E@0.30	Ea	32.20	8.06	40.26
8" x 8" tees	1E@0.40	Ea	38.20	10.80	49.00
10" x 10" tees	1E@0.50	Ea	91.40	13.40	104.80
12" x 12" tees	1E@0.70	Ea	108.00	18.80	126.80
3" x 3" X's	1E@0.30	Ea	23.80	8.06	31.86
4" x 4" X's	1E@0.30	Ea	20.40	8.06	28.46
6" x 6" X's	1E@0.40	Ea	24.50	10.80	35.30
8" x 8" X's	1E@0.50	Ea	37.60	13.40	51.00
NEMA 3R wireway					
4" x 4" x 12"	1E@0.50	Ea	23.40	13.40	36.80
4" x 4" x 48"	1E@1.00	Ea	53.10	26.90	80.00
4" x 4" x 72"	1E@1.50	Ea	78.50	40.30	118.80
4" x 6" x 12"	1E@0.55	Ea	27.80	14.80	42.60
4" x 6" x 48"	1E@1.05	Ea	65.00	28.20	93.20
6" x 6" x 12"	1E@0.60	Ea	32.80	16.10	48.90
6" x 6" x 48"	1E@1.20	Ea	75.20	32.30	107.50
6" x 6" x 72"	1E@1.60	Ea	104.00	43.00	147.00
12" x 12" x 36"	1E@1.25	Ea	252.00	33.60	285.60
12" x 12" x 60"	1E@1.55	Ea	374.00	41.70	415.70
12" x 12" x 72"	1E@1.85	Ea	448.00	49.70	497.70

Use these figures to estimate the cost of wireway fittings installed in buildings under the conditions described on pages 5 and 6. See the footnote on the previous page for notes on wireway fittings. The crew is one electrician. The cost per manhour is $26.88. These costs include layout, material handling, and normal waste. Add for sales tax, delivery, supervision, mobilization, demobilization, cleanup, overhead and profit.

Dry Type Transformers, Indoor/Outdoor

Material		Craft@Hrs	Unit	Material Cost	Labor Cost	Installed Cost

Dry type indoor/outdoor transformers, 240/480 volt primary, 120/240 volt secondary

Material		Craft@Hrs	Unit	Material Cost	Labor Cost	Installed Cost
.050	KVA	1E@0.25	Ea	38.20	6.72	44.92
.075	KVA	1E@0.25	Ea	48.10	6.72	54.82
.100	KVA	1E@0.25	Ea	48.10	6.72	54.82
.150	KVA	1E@0.30	Ea	54.50	8.06	62.56
.250	KVA	1E@0.30	Ea	66.40	8.06	74.46
.500	KVA	1E@0.35	Ea	90.60	9.41	100.01
.750	KVA	1E@0.40	Ea	116.00	10.80	126.80
1	KVA	1E@0.50	Ea	138.00	13.40	151.40
1.5	KVA	1E@0.55	Ea	170.00	14.80	184.80
2	KVA	1E@0.60	Ea	207.00	16.10	223.10
3	KVA	1E@0.70	Ea	257.00	18.80	275.80

Dry type indoor/outdoor transformers, 120/240 volt primary and secondary

Material		Craft@Hrs	Unit	Material Cost	Labor Cost	Installed Cost
.050	KVA	1E@0.25	Ea	54.50	6.72	61.22
.100	KVA	1E@0.25	Ea	64.50	6.72	71.22
.150	KVA	1E@0.25	Ea	79.70	6.72	86.42
.250	KVA	1E@0.30	Ea	91.70	8.06	99.76
.500	KVA	1E@0.30	Ea	129.00	8.06	137.06
.750	KVA	1E@0.35	Ea	166.00	9.41	175.41

Dry type indoor/outdoor transformers, 120/240 volt primary and secondary

Material		Craft@Hrs	Unit	Material Cost	Labor Cost	Installed Cost
1	KVA	1E@0.50	Ea	207.00	13.40	220.40
1.5	KVA	1E@0.55	Ea	255.00	14.80	269.80
2	KVA	1E@0.60	Ea	311.00	16.10	327.10
3	KVA	1E@0.70	Ea	376.00	18.80	394.80
5	KVA	1E@1.00	Ea	563.00	26.90	589.90
7.5	KVA	1E@1.25	Ea	783.00	33.60	816.60
10	KVA	2E@1.50	Ea	996.00	40.30	1,036.30
15	KVA	2E@1.75	Ea	1,320.00	47.00	1,367.00
25	KVA	2E@2.00	Ea	1,760.00	53.80	1,813.80

Dry type indoor/outdoor transformers, 120/240 volt primary, 12/24 volt secondary

Material		Craft@Hrs	Unit	Material Cost	Labor Cost	Installed Cost
.050	KVA	1E@0.25	Ea	50.60	6.72	57.32
.075	KVA	1E@0.25	Ea	56.30	6.72	63.02
.100	KVA	1E@0.25	Ea	60.80	6.72	67.52
.150	KVA	1E@0.30	Ea	74.00	8.06	82.06
.250	KVA	1E@0.30	Ea	89.00	8.06	97.06
.500	KVA	1E@0.35	Ea	122.00	9.41	131.41
.750	KVA	1E@0.40	Ea	156.00	10.80	166.80
1	KVA	1E@0.50	Ea	195.00	13.40	208.40
1.5	KVA	1E@0.55	Ea	239.00	14.80	253.80
2	KVA	1E@0.60	Ea	289.00	16.10	305.10
3	KVA	1E@0.70	Ea	389.00	18.80	407.80

Use these figures to estimate the cost of transformers installed in buildings under the conditions described on pages 5 and 6. Costs listed are for each transformer installed. The crew is one electrician for transformers to 7.5 KVA and two electricians for transformers over 7.5 KVA. The cost per manhour is $26.88. These costs include layout, material handling, and normal waste. Add for protective devices, supports, lifting equipment (if needed), sales tax, delivery, supervision, mobilization, demobilization, cleanup, overhead and profit. Note: Transformers with a 120/240 volt secondary are single phase. Transformers with a 208/120 volt secondary are three phase. Taps on transformers usually permit 2-1/2% adjustments in the output voltage. Generally two taps are available for raising the voltage and two are available for lowering the voltage.

Material		Craft@Hrs	Unit	Material Cost	Labor Cost	Installed Cost

Dry type indoor/outdoor transformers, 120/240 volt primary, 16/32 volt secondary

.050	KVA	1E@0.25	Ea	50.60	6.72	57.32
.075	KVA	1E@0.25	Ea	56.30	6.72	63.02
.100	KVA	1E@0.25	Ea	60.80	6.72	67.52
.150	KVA	1E@0.30	Ea	74.00	8.06	82.06
.250	KVA	1E@0.30	Ea	89.00	8.06	97.06
.500	KVA	1E@0.35	Ea	122.00	9.41	131.41
.750	KVA	1E@0.40	Ea	156.00	10.80	166.80
1	KVA	1E@0.50	Ea	195.00	13.40	208.40
1.5	KVA	1E@0.55	Ea	239.00	14.80	253.80
2	KVA	1E@0.60	Ea	289.00	16.10	305.10
3	KVA	1E@0.70	Ea	389.00	18.80	407.80

Dry type indoor/outdoor transformers, 240/480 volt primary, 120/240 volt secondary

5	KVA	1E@1.00	Ea	443.00	26.90	469.90
7.5	KVA	1E@1.25	Ea	615.00	33.60	648.60
10	KVA	2E@1.50	Ea	790.00	40.30	830.30
15	KVA	2E@1.75	Ea	1,070.00	47.00	1,117.00
25	KVA	2E@2.00	Ea	1,600.00	53.80	1,653.80

Dry type indoor/outdoor transformers, 240/480 volt primary, 120/240 volt secondary, with taps

37.5	KVA	2E@2.50	Ea	1,440.00	67.20	1,507.20
50	KVA	2E@3.00	Ea	1,740.00	80.60	1,820.60
75	KVA	2E@3.50	Ea	2,150.00	94.10	2,244.10
100	KVA	2E@4.00	Ea	2,520.00	108.00	2,628.00
167	KVA	2E@4.50	Ea	4,890.00	121.00	5,011.00

Dry type indoor/outdoor transformers, 480 volt primary, 120/240 volt secondary, with taps

5	KVA	1E@1.00	Ea	455.00	26.90	481.90
7.5	KVA	1E@1.25	Ea	623.00	33.60	656.60
10	KVA	2E@1.50	Ea	800.00	40.30	840.30
15	KVA	2E@1.75	Ea	1,080.00	47.00	1,127.00
25	KVA	2E@2.00	Ea	1,610.00	53.80	1,663.80

Use these figures to estimate the cost of transformers installed in buildings under the conditions described on pages 5 and 6. See the footnote on page 276 for notes on wireway fittings. Costs listed are for each transformer installed. The crew is one electrician for transformers to 7.5 KVA and two electricians for transformers over 7.5 KVA. The cost per manhour is $26.88. These costs include layout, material handling, and normal waste. Add for protective devices, supports, lifting equipment (if needed), sales tax, delivery, supervision, mobilization, demobilization, cleanup, overhead and profit. Note: Transformers with a 120/240 volt secondary are single phase. Transformers with a 208/120 volt secondary are three phase. Taps on transformers usually permit 2-1/2% adjustments in the output voltage. Generally two taps are available for raising the voltage and two are available for lowering the voltage.

Dry Type Transformers, Indoor/Outdoor

Material		Craft@Hrs	Unit	Material Cost	Labor Cost	Installed Cost
Dry type indoor/outdoor transformers, 600 volt primary, 120/240 volt secondary						
5	KVA	1E@1.00	Ea	485.00	26.90	511.90
7.5	KVA	1E@1.25	Ea	656.00	33.60	689.60
10	KVA	2E@1.50	Ea	836.00	40.30	876.30
15	KVA	2E@1.75	Ea	1,140.00	47.00	1,187.00
25	KVA	2E@2.00	Ea	1,700.00	53.80	1,753.80
Dry type indoor/outdoor transformers, 600 volt primary, 120/240 volt secondary, with taps						
5	KVA	1E@1.00	Ea	543.00	26.90	569.90
7.5	KVA	1E@1.25	Ea	746.00	33.60	779.60
10	KVA	2E@1.50	Ea	863.00	40.30	903.30
15	KVA	2E@1.75	Ea	1,300.00	47.00	1,347.00
25	KVA	2E@2.00	Ea	1,940.00	53.80	1,993.80
Dry type indoor/outdoor transformers, 480 volt primary, 208/120 volt secondary						
3	KVA	2E@0.75	Ea	670.00	20.20	690.20
6	KVA	2E@1.00	Ea	840.00	26.90	866.90
9	KVA	2E@1.40	Ea	1,050.00	37.60	1,087.60
30	KVA	2E@2.25	Ea	2,000.00	60.50	2,060.50
45	KVA	2E@2.75	Ea	2,250.00	73.90	2,323.90
50	KVA	2E@3.00	Ea	2,890.00	80.60	2,970.60
75	KVA	2E@3.50	Ea	3,600.00	94.10	3,694.10
112	KVA	2E@4.25	Ea	4,790.00	114.00	4,904.00
150	KVA	2E@4.25	Ea	6,240.00	114.00	6,354.00
225	KVA	2E@5.00	Ea	8,530.00	134.00	8,664.00
300	KVA	2E@5.50	Ea	10,700.00	148.00	10,848.00
400	KVA	2E@6.00	Ea	14,100.00	161.00	14,261.00
500	KVA	2E@6.50	Ea	17,700.00	175.00	17,875.00

Use these figures to estimate the cost of transformers installed in buildings under the conditions described on pages 5 and 6. Costs listed are for each transformer installed. The crew is one electrician for transformers to 7.5 KVA and two electricians for transformers over 7.5 KVA. The cost per manhour is $26.88. These costs include layout, material handling, and normal waste. Add for protective devices, supports, lifting equipment (if needed), sales tax, delivery, supervision, mobilization, demobilization, cleanup, overhead and profit. Note: Transformers with a 120/240 volt secondary are single phase. Transformers with a 208/120 volt secondary are three phase. Taps on transformers usually permit 2-1/2% adjustments in the output voltage. Generally two taps are available for raising the voltage and two are available for lowering the voltage.

Section 7:
Underfloor Raceway

Most residential buildings have electrical requirements that can be expected to change very little during the normal life of the building. In most cases, you can wire it and forget it. Larger office buildings are different. They need electrical systems that offer more flexibility. They need a wiring system that adapts easily each time a new tenant moves in, and that is easy to expand as the needs of each occupant change. That's why owners of many office and industrial buildings install underfloor raceway systems.

Underfloor raceway permits major changes in the size and number of conductors or signal cables during the life of the building, without compromise in fire resistance — install a major piece of new equipment; add a new row of desks down the middle of a room; put computer terminals on each desk and connect them with the computer located in another room. Changes like this can be expected in an office building. Each may require major electrical work — unless the building was equipped with an underfloor raceway system.

Underfloor raceway is used principally in fire-resistant construction. The raceway provides an open channel in or under the floor slab for power and communications lines. Wiring devices can be installed under every work space in an office or plant. Figure 7-1 shows a typical underfloor duct system.

Duct is usually set in the floor before the concrete deck is poured. Concrete surrounds the duct and finishes flush with each duct insert. Outlet fixtures are installed in the inserts wherever electrical and communications access will be needed. This provides power and signal lines to each desk and work space, even to areas located away from wall outlets.

The network of duct lines are joined by junction boxes that make the system more rigid and keep raceway laterals in place and at right angles. Junction boxes don't provide outlets for the system. They're used only when pulling wire through the duct. Junction boxes should be placed close enough together to make it easy to fish wire or cable through the cells. It's good practice to provide a regular pattern of junction boxes that can be rewired easily later when additional circuits or cables are needed.

Junction boxes are used at right angle turns, at tee connections, at cross connections and at feed-in positions. The junction box finish ring should be set at screed level so the finished concrete floor is level with the junction box lid.

Minor adjustment is possible in junction box cover rings. The ring can be raised or lowered slightly after floor tile is laid. When the floor is carpeted, a special cover is used to bring the junction box cover up to the level of the carpet. This cover is called a **carpet pan**. When the finished floor is terrazzo, a similar ring can be used.

Underfloor Duct Materials

The underfloor duct is made in two common sizes. The Type 1 duct is usually used for power circuits. Cell size is $1\frac{3}{8}$" x $3\frac{1}{8}$" x 10'. The Type 2 duct is commonly used for communications cables such as telephone, alarm, and computer systems. Type 2 duct measures $1\frac{3}{8}$" x $7\frac{1}{4}$" x 10'. Both types of duct are made in blank sections for feeder lines and with insert access holes placed 24 inches on center along one face of the duct. The first insert can be set 6 inches from one end and the last is 18 inches from the other end. Other spacing is available from some manufacturers on special order.

Cell insert height can vary from $\frac{7}{8}$" to $3\frac{3}{8}$". But note that inserts aren't usually interchangeable. Each must be ordered to fit the selected type of duct. When the junction box is set level with the screed line, the duct inserts must also be at the screed line.

Courtesy: Walkerduct

Figure 7-1
Underfloor Duct System

Duct Supports

You'll hear duct supports called *chairs*. Supports hold the duct in place and at the correct height while the concrete floor is being placed. Once the concrete has hardened, supports serve no useful purpose. A special type of support serves both as a support and as a coupling at duct joints. It replaces the common duct coupling.

Install enough supports to hold the duct in place as the deck is being poured. For planning purposes, allow one support every 5 feet and at the end of each duct run.

Some duct systems use more than one level of duct in the same floor. For example, the system may combine Type 1 duct for electrical lines and Type 2 duct for communication lines. When that's the case, use duct supports made for combination duct runs and combination junction boxes. Use standard supports and duct couplings at each duct joint.

All duct supports have leveling screws so you can level access rings at the screed line between junction boxes. Leveling legs are available in any length to meet any slab thicknesses.

Duct Elbows

Duct elbows are used to make vertical or horizontal turns. Vertical elbows turn the duct above the level of the finish floor. That's necessary when duct has to be connected directly to a terminal cabinet in a wall, for example. Use blank feeder duct in an exposed vertical section above an elbow. Vertical elbows form right angles, but have a fairly long radius to make pulling wire or cable easier.

Horizontal elbows are used to change direction without leaving the concrete deck. They come in 90-degree bends, 45-degree bends and adjustable 15- to 30-degree bends. All of the bends have a radius long enough to make pulling wire relatively easy.

A special offset elbow is made to simplify running duct under obstacles or other underfloor duct. The offset elbow is available for both Type 1 and Type 2 duct and is a one-piece unit. Notice that it takes two offset elbows to go under an obstacle and return to the original level.

Fittings

The fittings made for underfloor duct will meet any need. For the estimator, the only hard part is counting all the fittings you'll need. Here are the most common fittings:

Wyes are used when one cell divides into two runs. Both 30- and 45-degree wyes are made for both sizes of duct.

Plugs are used to fill any unused openings in junction boxes so liquid concrete doesn't flow into the box. The plugs come in both duct sizes. Corners in junction boxes have entrances for 2 inch conduit. You'll also need plugs to fill any openings that aren't used for conduit.

There are special duct plugs for capping the ends of duct runs that terminate without a junction box. The caps go into the end of the duct to keep concrete out of the cell.

Adapters are sometimes needed when conduit joins an underfloor duct system. When the electrical feed to a duct system comes from conduit in the slab, the conduit usually enters at a corner of a junction box. Entrance hubs at the junction box corner are usually for 2 inch conduit. If the conduit feed isn't 2 inches, an adapter will be needed to make a concrete-tight connection. Adapter sizes range from ½" to 1¼".

Adapters are also available for joining the end of both Type 1 and Type 2 duct to conduit. Insert the right size adapter into the end of the duct and connect the conduit.

Insert caps are installed in the duct inserts at the factory. Leave the caps in place until it's time to install the service fittings. If you need extra insert caps, they're available separately. There are two common types of inserts. The most common has a 2 inch diameter. Another has an ellipsoid shape. Ellipsoid insert caps are also available.

Cabinet connectors are used to attach the duct to a service panel or cabinet. The connector provides ground continuity within the metallic raceway system. It's attached with machine screws and has a protective bushing to keep wire insulation from getting stripped off as it's pulled into the duct.

Electrical outlet receptacles are installed in the service fittings. Note that the fittings listed in this section include the cost of labor and material for the basic housing only — not the receptacle itself. Add the cost of buying and installing receptacles separately. Job specs will identify the type and grade of receptacles to use.

Abandon plugs are used when a service fitting has to be relocated. The opening that remains is plugged with a special abandon plug. They're made either of brass or aluminum. Install the plug immediately after the service fitting has been removed to keep debris out of the duct. The plug has a toggle-type bar that extends into the duct through the insert. When the screw is tightened, the plug seals the hole.

The abandon plug can be removed later so the service fitting can be reinstalled.

When a service fitting is abandoned, the wire or cable can either be removed or taped off. If it's left in place, tape the end carefully and tuck it back into the insert.

Finding Caps and Installing Fittings

Sometimes underfloor raceway gets buried a little too deep in the concrete deck. That makes finding the inserts more difficult. To locate "lost" inserts, measure from the last junction box or the last service fitting previously installed. Remember, standard insert spacing is 24 inches on center and the first insert is 6 inches from the box. Be careful to avoid damaging the floor when searching for the insert. Don't start chipping concrete until you know where the insert is.

The further away from the junction box, the harder it is to find inserts. To simplify locating inserts, put a marking screw in the last insert in every duct run. The screw should extend ½ inch above the surface of the finished floor. Snap a chalk line between the marking screw and the junction box. Inserts will be on that line. Find the first insert from the junction box and measure in 2 foot increments along the line to the next service fitting. Mark each insert location and then remove the marking screw.

When each insert is located in the floor, use a chipping gun or chisel to remove the small amount of concrete cover. Then vacuum up the concrete dust and chips before opening the insert cap. That keeps debris out of the duct. Then fish wire or cable to the new service fitting location and install the service fitting.

Service fittings go on over the duct insert and are mechanically connected with a chase nipple-type fitting. Tighten the nipple to hold the service fitting in place and ground the fitting to the duct.

Labor for Underfloor Duct

The labor units listed in this section assume no unusual conditions, and good access. Be alert for other situations. Start by visualizing what the site will look like on installation day. A congested site will add to installation time and increase the labor cost. A tight installation schedule will increase conflicts with other trades. A deeper slab will usually extend the installation time. If the floor carries several other duct runs, expect some problems. Several extra offset fittings may be needed to avoid other lines.

If it looks like the system will be hard to lay out, hard to get to, will share the slab with grade beams or structural beams, if there's too much conduit or piping in the area, and if the job involves less than 100 linear feet of duct, expect labor time to be 25 to 50 percent higher than listed in this section.

Installations over 100 feet may take less time, especially if the duct runs are long. If you're installing over 200 linear feet of duct, labor time can be 10 to 20 percent less. Even an inexperienced crew will have picked up enough about duct installation to increase their productivity after running the first 100 feet.

Doing the Take-off

Your take-off of underfloor duct begins with a lined worksheet just like the other estimating worksheets you've used. Draw a vertical line down the left side about 1½ inches from the left edge. At the top of the page, on the right side of the vertical line, head the column with the plan page number. In the left column, list the material descriptions.

First, list the junction boxes for single duct runs. Start the junction boxes for Type 1 duct. On the next line list junction boxes for Type 2 duct, if there are any. Then do the same for combination duct runs.

Next measure the lengths of Type 1 and Type 2 duct. List all elbows and offsets as you find them on the plan. After measuring all of the duct, calculate the number of duct supports of each type.

Next, locate all conduit that feeds the duct system. Determine the conduit size and list the adapter size needed for each connection. Watch for duct couplings or combination coupling and support pieces. There won't be too many couplings. Underfloor duct comes in 10-foot lengths.

Watch for blank duct. Blank duct doesn't have inserts. It will usually be installed in the floor of rooms or corridors that don't require service fittings. The plans should show where blank duct is to be used.

Blank duct is also used in risers from the underfloor system to panels or cabinets. It can be exposed or concealed in the wall. When a riser is to be connected to a panel or cabinet, list an adapter for that duct size.

Next, identify and count all the service fittings that will be required. Check the job specifications and see if additional service fittings are to be delivered to the owner for future installation. These extra service fittings won't show up on the plans.

Using Duct Systems

Most underfloor duct jobs require many cut pieces of duct. The runs between junction boxes are seldom spaced the full length of a duct section. Usually you'll lay two or three full-duct sections and have to cut the next section to complete the run to the next junction box.

Cutting duct by hand with a hacksaw is slow work. A power hacksaw will both save time and produce straighter cuts. Be sure to ream the inside edge with a file to remove the cutting burr.

Learn From Your Mistakes

My most memorable underfloor duct job was a large office building for a state Department of Motor Vehicles. The plans called for both Type 1 and Type 2 duct in one large room.

The construction schedule was tight. Close coordination with the general contractor was essential. Notice to proceed with construction was issued before the subcontracts were completed. The contractor wanted to start the job right away, but full descriptions of all key materials, including the underfloor duct, had to be submitted for approval before we could place orders.

We worked quickly with suppliers to prepare shop drawings and catalog cuts for the submittals. It took about two weeks to gather all the information that was required. We prepared a material list and packaged it with the catalog cuts for the submittal. The shop drawings for switchboards, motor control centers and panels would be late, so we planned to submit them separately at a later date.

The final submittal package included ten submittal brochures. We rushed these to the general contractor and he passed them immediately to the owner. Along with the brochures went a request for prompt review so we could stay on the construction schedule.

The approved brochures were returned in three weeks — not exactly record time, but pretty good for a state agency. By then the contractor had graded the site and excavated for the footings. The plans called for wire mesh in the floor slabs and the mesh was already on the job site. The contractor was ready to form the floor slab. But the underfloor duct had to go in that slab! Construction would have to wait until we could set the duct.

We notified the supplier that the formwork was about to start. Our supplier put pressure on the manufacturer to rush the underfloor duct and fittings. This was probably a mistake, but we didn't realize it at the time. The duct arrived about a week later. The forms were in place for the slab and concrete delivery had been scheduled. Our underfloor duct was right on the critical path, but everything was looking good at this point. We were already congratulating ourselves on a job well done. As it turned out, that was premature.

Our tradesmen began to set duct at one end of the building. The slab crew went to work beside them setting screed pins for leveling the concrete to a close tolerance. The junction boxes were spotted and set in small batches of bagged concrete mix. The concrete would anchor the junction box firmly in place until the slab was poured. The top of each box was set at screed level.

As the slab excavation crew would screed the underfloor material, fill dirt was moved back across the floor area. Screeding was done in patterns because the floor was going to be poured in a checkerboard pattern. A 2 x 4 header marked the boundary of each pour.

With the junction boxes in place, we began to unload the duct sections. It was at this time that one of our electricians noticed a problem. A seam on one length of duct wasn't welded closed. We quickly started checking other duct. Out of the first few dozen sections checked, about one-third had the same flaw! Now we were in trouble.

We immediately notified the contractor of the problem. Next, we phoned our supplier. Based on our guess that 30 percent was going to be rejected, the manufacturer immediately shipped 30 percent more duct. The project waited until the duct arrived, of course. But we were able to finish on time by compressing other parts of the job.

It's always embarrassing to be the cause of a delay on a project — especially when everyone on the job is working against a tight deadline. But in truth, there wasn't much we could have done to avoid the problem. And it could have been worse. Think how embarrassed we would have been if that defective duct had been in our warehouse for two weeks before we started unloading it on the job.

We learned one other lesson on this job. Most of the duct was staked in place right on the ground while the slab crew poured concrete around it. This can lead to trouble. When the laborers placed, screeded and tamped the concrete, they walked on and stumbled over the underfloor duct. That was unavoidable. The stakes tended to hold the duct in place, of course. But they also tended to hold it wherever it was moved during the pour. We found that some of the inserts ended up deeper than others.

The duct that wasn't staked survived the pour better. The laborers walked on the unstaked duct too, of course. And the unstaked duct would move when it was disturbed. But it would also return closer to its intended position after being disturbed because no stakes were holding it down.

On the whole, this job was both successful and profitable, both for the general contractor and my company. And the state got the facility it paid for. But there were more than a few anxious moments for the electrical crew that day when we were setting seamless underfloor duct.

Material	Craft@Hrs	Unit	Material Cost	Labor Cost	Installed Cost

Underfloor raceway 14 gauge power duct, 3-1/4" wide, 10' long, Type 1 duct

Material	Craft@Hrs	Unit	Material Cost	Labor Cost	Installed Cost
Blank duct	2E@0.30	Ea	56.30	8.06	64.36
7/8" insert height	2E@0.30	Ea	56.30	8.06	64.36
1-3/8" insert height	2E@0.30	Ea	64.90	8.06	72.96
1-7/8" insert height	2E@0.30	Ea	64.90	8.06	72.96
2-3/8" insert height	2E@0.30	Ea	64.90	8.06	72.96
3-3/8" insert height	2E@0.30	Ea	64.90	8.06	72.96

Underfloor raceway 14 gauge communications duct, 7-1/4" wide, 10' long, Type 2 duct

Material	Craft@Hrs	Unit	Material Cost	Labor Cost	Installed Cost
Blank duct	2E@0.40	Ea	104.00	10.80	114.80
7/8" insert height	2E@0.40	Ea	104.00	10.80	114.80
1-3/8" insert height	2E@0.40	Ea	112.00	10.80	122.80
1-7/8" insert height	2E@0.40	Ea	112.00	10.80	122.80
2-3/8" insert height	2E@0.40	Ea	112.00	10.80	122.80
3-3/8" insert height	2E@0.40	Ea	112.00	10.80	122.80

Underfloor raceway one level junction boxes, 7/8" insert height

Material	Craft@Hrs	Unit	Material Cost	Labor Cost	Installed Cost
4 #1 ducts	2E@0.50	Ea	216.00	13.40	229.40
8 #1 ducts	2E@0.60	Ea	216.00	16.10	232.10
12 #1 ducts	2E@0.75	Ea	431.00	20.20	451.20
4 #2 ducts	2E@0.60	Ea	224.00	16.10	240.10
8 #2 ducts	2E@1.00	Ea	557.00	26.90	583.90
4 #1 & 4 #2 ducts	2E@0.80	Ea	431.00	21.50	452.50

Underfloor raceway one level junction boxes, 1-3/8" insert height

Material	Craft@Hrs	Unit	Material Cost	Labor Cost	Installed Cost
4 #1 ducts	2E@0.50	Ea	216.00	13.40	229.40
8 #1 ducts	2E@0.60	Ea	216.00	16.10	232.10
12 #1 ducts	2E@0.75	Ea	431.00	20.20	451.20
4 #2 ducts	2E@0.60	Ea	224.00	16.10	240.10
8 #2 ducts	2E@1.00	Ea	557.00	26.90	583.90
4 #1 & 4 #2 ducts	2E@0.80	Ea	429.00	21.50	450.50

Use these figures to estimate the cost of underfloor raceway installed in building floors under the conditions described on pages 5 and 6. Costs listed are for each length installed. The crew is two electricians working at a labor cost of $26.88 per manhour. These costs include leveling, layout, material handling, and normal waste. Add for supports, sales tax, delivery, supervision, mobilization, demobilization, cleanup, overhead and profit. Labor costs will be higher when installing duct in slabs on grade. Extra care is needed to anchor the duct properly and prevent it from floating as the concrete is poured. Some installers place all the concrete that's poured around duct to be sure the duct isn't disturbed. When duct is installed on wood deck forms, the junction boxes and duct supports can be fastened to the form to prevent floating. No matter where the duct is placed, be sure to seal all openings in the duct before concrete is poured. In the table above, No. 1 duct is power duct and No. 2 duct is communications duct. Junction boxes have one, two or three duct entrances in each side. When a dead end, straight through, or tee junction is needed, fill the unused duct outlets with blank plugs.

Underfloor Raceway

Material	Craft@Hrs	Unit	Material Cost	Labor Cost	Installed Cost

Underfloor raceway one level junction boxes, 1-7/8" insert height

Material	Craft@Hrs	Unit	Material Cost	Labor Cost	Installed Cost
4 #1 ducts	2E@0.50	Ea	305.00	13.40	318.40
8 #1 ducts	2E@0.60	Ea	305.00	16.10	321.10
12 #1 ducts	2E@0.75	Ea	466.00	20.20	486.20
4 #2 ducts	2E@0.60	Ea	305.00	16.10	321.10
8 #2 ducts	2E@1.00	Ea	573.00	26.90	599.90
4 #1 & 4 #2 ducts	2E@0.80	Ea	466.00	21.50	487.50

Underfloor raceway one level junction boxes, 2-3/8" insert height

Material	Craft@Hrs	Unit	Material Cost	Labor Cost	Installed Cost
4 #1 ducts	2E@0.50	Ea	323.00	13.40	336.40
8 #1 ducts	2E@0.60	Ea	323.00	16.10	339.10
12 #1 ducts	2E@0.75	Ea	494.00	20.20	514.20
4 #2 ducts	2E@0.60	Ea	323.00	16.10	339.10
8 #2 ducts	2E@1.00	Ea	638.00	26.90	664.90
4 #1 & 4 #2 ducts	2E@0.80	Ea	494.00	21.50	515.50

Underfloor raceway one level junction boxes, 3-3/8" insert height

Material	Craft@Hrs	Unit	Material Cost	Labor Cost	Installed Cost
4 #1 ducts	2E@0.50	Ea	358.00	13.40	371.40
8 #1 ducts	2E@0.60	Ea	358.00	16.10	374.10
12 #1 ducts	2E@0.75	Ea	557.00	20.20	577.20
4 #2 ducts	2E@0.60	Ea	358.00	16.10	374.10
8 #2 ducts	2E@1.00	Ea	708.00	26.90	734.90
4 #1 & 4 #2 ducts	2E@0.80	Ea	557.00	21.50	578.50

Underfloor raceway duct supports with leveling screws

Material	Craft@Hrs	Unit	Material Cost	Labor Cost	Installed Cost
1 #1 duct	2E@0.15	Ea	12.10	4.03	16.13
2 #1 ducts	2E@0.20	Ea	14.70	5.38	20.08
3 #1 ducts	2E@0.25	Ea	21.60	6.72	28.32
1 #2 ducts	2E@0.20	Ea	12.10	5.38	17.48
2 #2 ducts	2E@0.25	Ea	14.70	6.72	21.42
1 #1 & 1 #2 ducts	2E@0.25	Ea	12.10	6.72	18.82
2 #1 & 1 #2 ducts	2E@0.25	Ea	23.40	6.72	30.12
1 #1 & 2 #2 ducts	2E@0.30	Ea	25.30	8.06	33.36

Use these figures to estimate the cost of underfloor raceway installed in building floors under the conditions described on pages 5 and 6. Costs listed are for each length installed. The crew is two electricians working at a labor cost of $26.88 per manhour. These costs include leveling, layout, material handling, and normal waste. Add for supports, sales tax, delivery, supervision, mobilization, demobilization, cleanup, overhead and profit. Labor costs will be higher when installing duct in slabs on grade. Extra care is needed to anchor the duct properly and prevent it from floating as the concrete is poured. Some installers place all the concrete that's poured around duct to be sure the duct isn't disturbed. When duct is installed on wood deck forms, the junction boxes and duct supports can be fastened to the form to prevent floating. No matter where the duct is placed, be sure to seal all openings in the duct before concrete is poured. In the table above, No. 1 duct is power duct and No. 2 duct is communications duct. Junction boxes have one, two or three duct entrances in each side. When a dead end, straight through, or tee junction is needed, fill the unused duct outlets with blank plugs.

Material	Craft@Hrs	Unit	Material Cost	Labor Cost	Installed Cost
Fittings for underfloor raceway					
#1 couplings	1E@0.05	Ea	6.08	1.34	7.42
#2 couplings	1E@0.06	Ea	7.79	1.61	9.40
#1 vertical elbows	1E@0.20	Ea	29.40	5.38	34.78
#2 vertical elbows	1E@0.25	Ea	36.20	6.72	42.92
#1 horizontal elbows	1E@0.20	Ea	56.00	5.38	61.38
#2 horizontal elbows	1E@0.25	Ea	73.60	6.72	80.32
#1 horz. elbows, 30	1E@0.20	Ea	10.40	5.38	15.78
#2 horz. elbows, 30	1E@0.25	Ea	21.60	6.72	28.32
#1 horz. elbows, 45	1E@0.20	Ea	17.30	5.38	22.68
#2 horz. elbows, 45	1E@0.25	Ea	43.20	6.72	49.92
#1 offset elbows	1E@0.20	Ea	23.30	5.38	28.68
#2 offset elbows	1E@0.25	Ea	38.00	6.72	44.72
#1 box plugs	1E@0.05	Ea	2.18	1.34	3.52
#2 box plugs	1E@0.06	Ea	2.54	1.61	4.15
#1 duct plugs	1E@0.05	Ea	2.18	1.34	3.52
#2 duct plugs	1E@0.06	Ea	2.54	1.61	4.15
1/2" pipe adapters	1E@0.05	Ea	9.03	1.34	10.37
3/4" pipe adapters	1E@0.06	Ea	10.40	1.61	12.01
1" pipe adapters	1E@0.08	Ea	10.40	2.15	12.55
1-1/4" pipe adapters	1E@0.10	Ea	10.40	2.69	13.09
2" pipe adapters	1E@0.15	Ea	16.40	4.03	20.43
2"-1-1/4" pipe adapters	1E@0.20	Ea	16.40	5.38	21.78
3"-1-1/4" pipe adapters	1E@0.25	Ea	16.40	6.72	23.12
1/2" pipe inserts	1E@0.05	Ea	9.89	1.34	11.23
3/4" pipe inserts	1E@0.06	Ea	9.89	1.61	11.50
1" pipe inserts	1E@0.08	Ea	9.89	2.15	12.04
1-1/4" pipe inserts	1E@0.10	Ea	9.89	2.69	12.58
1-1/2" pipe inserts	1E@0.10	Ea	9.89	2.69	12.58
1-1/4"-3/4" re bush	1E@0.10	Ea	9.89	2.69	12.58
1-1/4"-1" re bush	1E@0.10	Ea	9.89	2.69	12.58
2"-1-1/2" re bush	1E@0.15	Ea	16.40	4.03	20.43
Insert cap	1E@0.05	Ea	.87	1.34	2.21
Marker screw	1E@0.25	Ea	2.43	6.72	9.15
#1 cabinet connectors	1E@0.30	Ea	8.68	8.06	16.74
#2 cabinet connectors	1E@0.50	Ea	16.40	13.40	29.80

Use these figures to estimate the cost of underfloor raceway fittings installed in building floors under the conditions described on pages 5 and 6. Costs listed are for each fitting installed. The crew is one electrician working at a labor cost of $26.88 per manhour. These costs include leveling, mounting accessories, layout, material handling, and normal waste. Add for wiring devices, sales tax, delivery, supervision, mobilization, demobilization, cleanup, overhead and profit.

Underfloor Raceway Fittings

Material	Craft@Hrs	Unit	Material Cost	Labor Cost	Installed Cost
Fittings for underfloor raceway					
#1 wye coupling	1E@0.25	Ea	24.70	6.72	31.42
#2 wye connectors	1E@0.30	Ea	43.20	8.06	51.26
2" leveling legs	1E@0.05	Ea	1.84	1.34	3.18
3" leveling legs	1E@0.06	Ea	1.84	1.61	3.45
4" leveling legs	1E@0.08	Ea	2.18	2.15	4.33
6" leveling legs	1E@0.10	Ea	2.18	2.69	4.87
8" leveling legs	1E@0.12	Ea	2.68	3.23	5.91
10" leveling legs	1E@0.15	Ea	2.91	4.03	6.94
12" leveling legs	1E@0.20	Ea	2.91	5.38	8.29
14" leveling legs	1E@0.25	Ea	2.91	6.72	9.63
Service fittings without receptacles for underfloor raceway					
1 single recept. 20A	1E@0.25	Ea	36.90	6.72	43.62
2 single recept. 20A	1E@0.25	Ea	36.90	6.72	43.62
1 single recept. 30A	1E@0.25	Ea	36.90	6.72	43.62
1 single recept. 50A	1E@0.25	Ea	36.90	6.72	43.62
1 duplex receptacle	1E@0.25	Ea	36.90	6.72	43.62
2 duplex receptacles	1E@0.25	Ea	36.90	6.72	43.62
1 1" insulated bushing	1E@0.25	Ea	36.90	6.72	43.62
2 1" insulated bushing	1E@0.25	Ea	36.90	6.72	43.62
3/4" x 3" standpipes	1E@0.20	Ea	60.50	5.38	65.88
1" x 3" standpipes	1E@0.20	Ea	74.10	5.38	79.48
2" x 3" standpipes	1E@0.25	Ea	82.30	6.72	89.02
Brass abandon plugs	1E@0.15	Ea	20.40	4.03	24.43
Alum. abandon plugs	1E@0.15	Ea	20.40	4.03	24.43

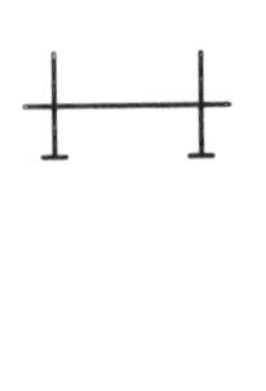

Use these figures to estimate the cost of service fittings installed in building floors under the conditions described on pages 5 and 6. Costs listed are for each fitting installed. The crew is one electrician working at a labor cost of $26.88 per manhour. These costs include mounting accessories, layout, material handling, and normal waste. Add for wiring devices, sales tax, delivery, supervision, mobilization, demobilization, cleanup, overhead and profit. Note: Many other underfloor raceway fittings are available. Supplier catalogs list other service fittings, and fittings for special applications.

Section 8: Bus Duct

Bus duct (sometimes called busway) is used in place of wire to distribute high amperage current within a building. It's made of copper or aluminum bus bars mounted in a protective sheet metal enclosure. Bus duct comes in prefabricated sections from one to 10 feet long and can't be cut to length or modified on the job.

Bus duct is common in high-rise commercial and apartment buildings. It reduces installation cost when long runs of high-amperage feeder or branch circuit are required.

Several types of bus duct are made. Each serves a different purpose. *Feeder bus duct* is used for large feeder circuits. *Plug-in bus duct* is used where multiple-power take-offs are needed.

Feeder Bus Duct

Feeder bus duct is rated as either indoor (NEMA 1) or raintight (NEMA 3R). The capacity of the bus duct usually begins at 225 amps and ranges up to several thousands of amps. Standard voltage ratings run from 250 volts to 600.

The feeder duct is used primarily for carrying large circuit loads from primary distribution centers or switchboards to other distribution centers or switchboards. Feeder bus duct is also used in high-rise buildings to carry the power feed vertically to the power center on each floor. It replaces large multiple conduit and wire feeders.

Many fittings are available for feeder duct: 90-degree flat elbows, 45-degree flat elbows, special bend flat elbows, tees, crosses or X's, flange adapters for switchboard connections, hangers, supports, and expansion joints.

A special transition section is needed to connect a run of outdoor raintight NEMA class 3R bus duct to an extension of indoor NEMA class 1 bus duct. Another type of transition section may be needed when extending bus duct from an area with one fire or service rating to an area with a different fire or service rating.

Feeder bus duct is available with either aluminum or copper bus bars. Job specs usually dictate which is to be used. Aluminum bus duct is lighter than copper duct. Most inspection authorities will not permit mixing the two types together in the same system.

Plug-In Bus Duct

Many machine shops uses plug-in bus duct to supply power to heavy electrical equipment. The duct is run overhead with individual feeds plugged into the bus where needed. Each service location is equipped with a plug-in circuit breaker.

The bus duct has ports placed along the face of the housing. Wherever power is needed, the port cover is moved aside so a plug-in device can be installed. After the device is secured in place, conduit and wire are extended to the machine's disconnect switch.

The plug-in device is operated from the floor with a *hook stick*. The hook stick switches power off and on.

Fittings for plug-in bus duct include elbows, tees, crosses, hangers, ends, feed-in cable tap boxes and expansion joints. Bus duct reducers are used down the line when amperage requirements have dropped and smaller duct is appropriate.

Plug-In Devices

Plug-in devices are made to fit every need. When selecting the appropriate device, be sure the voltage rating and conductor configuration matches the bus duct. For example, bus duct might be 3-phase, 3-wire service or 3-phase, 4-wire service. The device should be the same.

The plug-in device is secured directly to the bus duct housing with screw-type clamps. Installation is usually very simple. But be careful when inserting the plug-in device. Line up the device squarely with the bus duct. Insert the contacts through the installation ports on the face of the bus duct.

Hangers

All bus duct manufacturers provide instructions for installing their duct. Be especially careful to follow the instructions for duct supports. Placement and spacing of hangers are important. Be sure the structure the bus duct is attached to is strong enough to support the duct weight.

Generally, you'll place hangers at least every 10 feet. The size of the hanger rod should be a minimum of ⅜ inch for the small size duct and increased proportionately for heavier duct. Usually two hangers are used, one on each side of the duct. A trapeze bar or channel is used in some cases to seat the duct. Rods carry the trapeze bar.

Plug-in Bus Duct

Conventional Type
225 Amp Through 1000 Amp

Materials For Run # 1

Quantity	Cat. No.	Description
1	LF1003	1000A 3 wire left flanged end with bus extension
1	DE1003	1000A 3 wire downward elbow
2	FST1003-2	1000A 3 wire feeder 2 ft. straight lengths
1	EC1003	1000A 3 wire edgewise cross
5	FST1003	1000A 3 wire feeder 10 ft. straight lengths
1	FST1003-8	1000A 3 wire feeder 8 ft. straight length
2	FE1003	1000A 3 wire forward elbows
2	RR1043	1000/400A 3 wire right unfused reducer adapters
1	LR1043	1000/400A 3 wire left unfused reducer adapter
12	ST403	400A 3 wire plug-in 10 ft. straight lengths
3	ST403-2'-6"	400A 3 wire plug-in 2'-6" straight lengths
2	RN403	400A 3 wire right end closers
1	LN403	400A 3 wire left end closer

Materials For Run # 2

Quantity	Cat. No.	Description
1	RF1003	1000A 3 wire right flanged end with bus extension
1	DE1003	1000A 3 wire downward elbow
1	FST1003-2	1000A 3 wire feeder 2 ft. straight length
10	FST1003	1000A 3 wire feeder 10 ft. straight lengths
2	FST1003-9	1000A 3 wire feeder 9 ft. straight lengths
2	FT1003	1000A 3 wire forward tees
1	FE1003	1000A 3 wire forward elbow
2	RR1043	1000/400A 3 wire right unfused reducer adapters
1	LR1043	1000/400A 3 wire left unfused reducer adapter
12	ST403	400A 3 wire plug-in 10 ft. straight lengths
3	ST403-7'-6"	400A 3 wire plug-in 7'-6" straight lengths
2	RN403	400A 3 wire right end closers
1	LN403	400A 3 wire left end closer

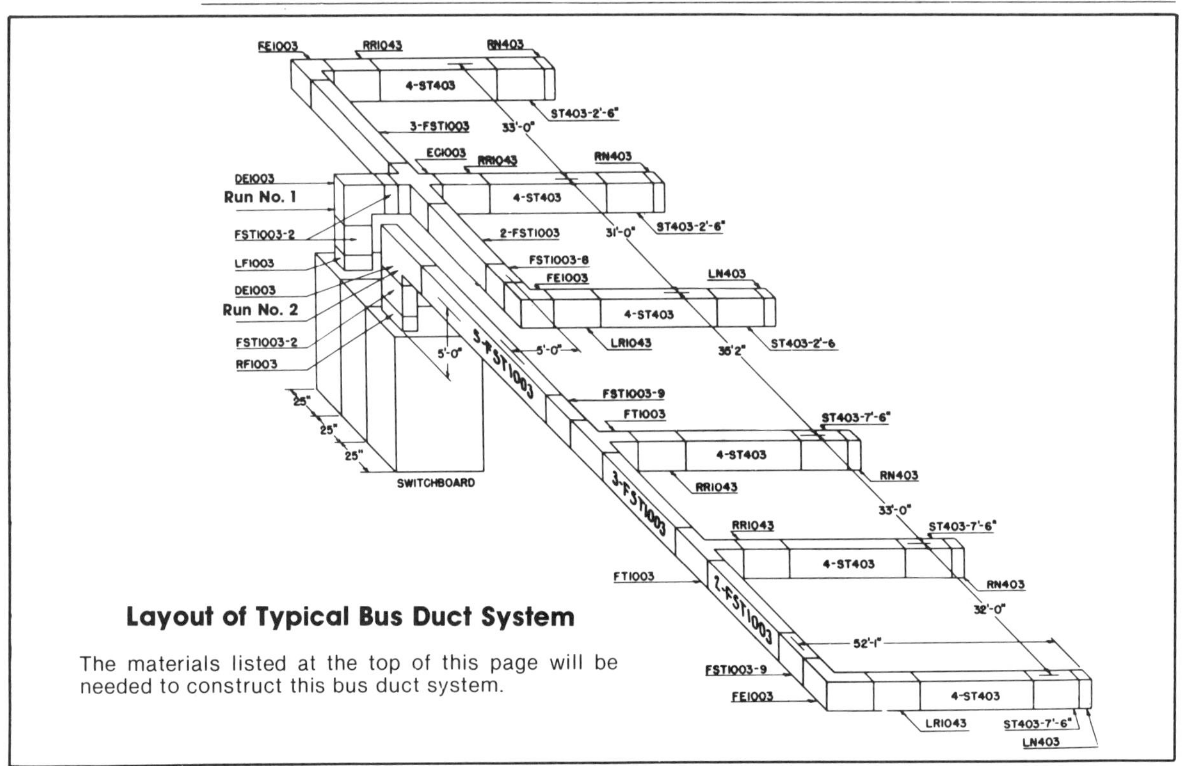

Layout of Typical Bus Duct System

The materials listed at the top of this page will be needed to construct this bus duct system.

FEDERAL PACIFIC
RELIANCE ELECTRIC

Figure 8-1
Plug-in Bus Duct

When the bus duct is run vertically through a floor, a special collar-like support is used at the floor line to carry the weight of the duct.

When laying out a bus duct system, be careful to avoid obstructions. Routing duct around a column or overhead crane can add four elbows and several hundred dollars to the cost. It's hard to see some of these potential conflicts on the plans. Consider what an elevation view would look like, even if you have only a plan view of the electrical and mechanical system.

Before beginning installation, be sure all the right fittings and devices are on hand. If you haven't estimated or installed the particular type of bus duct before, get help from a manufacturer's representative.

Estimating Bus Duct

Start your estimate by finding the plan sheet details and sections in the specs that cover bus duct. Determine the amperage and voltage required. Locate the run or runs of bus duct on the plans. All bus duct on most jobs will be the same size and be rated at the same voltage. But this isn't always the case.

Study the floor plan and details. Figure out what purpose the bus duct serves and try to visualize the way it should be installed. Then scale off the length of duct as shown on the floor plan. Calculate the length of vertical risers. List these measurements on your worksheet. Then look for the fittings you'll need. List the elbows as either vertical or horizontal and by the degree of bend, either 15, 30, 45, 60 or 90 degrees. List the tees, crosses and offsets. List

flanges needed for switchboard connections. When duct penetrates a fire wall, there will be a fire stop fitting.

After all fittings have been listed, check for hangers. The hangers will probably need beam clamps, nuts, rods, and trapeze racks. Bracing to eliminate sway movement may be required. List all block-outs or chases that must be made in structural walls or floors.

It's easiest to install bus duct in open spaces where overhead structural support is about 12 feet above the floor. This permits easier handling and hoisting. Power bus duct made with copper bus bars is both heavy and difficult to handle.

Some types of bus duct have simple through-bolt connections; others require bolting each bus bar together. Bolts have to be tightened with a torque wrench. A cover assembly is then placed over the joint. Connections are needed between each duct section and at each fitting.

Aluminum bus duct is much lighter and is easier to handle. Reduce the labor cost by about 30 percent when installing aluminum bus duct and fittings. But note that installing the hangers will take about the same time as copper bus duct.

If the installation area is cluttered with obstructions and if runs are short, increase the labor units by 15 or 20 percent.

You may need lifting equipment to move bus duct from the delivery truck to the storage area. Be sure to protect the duct sections from moisture and dust while they're in storage. Keep the ends covered until it's time to install the system.

Bus Duct, Aluminum

Material		Craft@Hrs	Unit	Material Cost	Labor Cost	Installed Cost
Aluminum feeder bus duct, 600 volt, 3-wire						
800A	12"	4E@0.30	Ea	109.00	8.06	117.06
1000A	12"	4E@0.40	Ea	122.00	10.80	132.80
1200A	12"	4E@0.50	Ea	163.00	13.40	176.40
1350A	12"	4E@0.60	Ea	188.00	16.10	204.10
1600A	12"	4E@0.70	Ea	229.00	18.80	247.80
2000A	12"	4E@0.80	Ea	277.00	21.50	298.50
2500A	12"	4E@0.90	Ea	336.00	24.20	360.20
3000A	12"	4E@1.00	Ea	383.00	26.90	409.90
4000A	12"	4E@1.25	Ea	525.00	33.60	558.60
800A	18"	4E@0.50	Ea	217.00	13.40	230.40
1000A	18"	4E@0.60	Ea	244.00	16.10	260.10
1200A	18"	4E@0.70	Ea	327.00	18.80	345.80
1600A	18"	4E@0.90	Ea	460.00	24.20	484.20
2000A	18"	4E@1.00	Ea	551.00	26.90	577.90
800A	24"	4E@0.60	Ea	217.00	16.10	233.10
1000A	24"	4E@0.70	Ea	244.00	18.80	262.80
1200A	24"	4E@0.80	Ea	327.00	21.50	348.50
1600A	24"	4E@1.00	Ea	460.00	26.90	486.90
2000A	24"	4E@1.10	Ea	551.00	29.60	580.60
800A	72"	4E@1.50	Ea	652.00	40.30	692.30
1000A	72"	4E@1.75	Ea	729.00	47.00	776.00
1200A	72"	4E@2.00	Ea	978.00	53.80	1,031.80
1600A	72"	4E@2.50	Ea	1,380.00	67.20	1,447.20
2000A	72"	4E@2.75	Ea	1,660.00	73.90	1,733.90
800A	84"	4E@1.75	Ea	761.00	47.00	808.00
1000A	84"	4E@2.00	Ea	849.00	53.80	902.80
1200A	84"	4E@2.25	Ea	1,130.00	60.50	1,190.50
1600A	84"	4E@2.75	Ea	1,610.00	73.90	1,683.90
2000A	84"	4E@3.00	Ea	1,930.00	80.60	2,010.60
800A	120"	4E@2.00	Ea	1,090.00	53.80	1,143.80
1000A	120"	4E@2.25	Ea	1,220.00	60.50	1,280.50
1200A	120"	4E@2.50	Ea	1,630.00	67.20	1,697.20
1350A	120"	4E@2.75	Ea	1,880.00	73.90	1,953.90
1600A	120"	4E@3.00	Ea	2,290.00	80.60	2,370.60
2000A	120"	4E@3.25	Ea	2,770.00	87.40	2,857.40
2500A	120"	4E@3.50	Ea	3,360.00	94.10	3,454.10
3000A	120"	4E@3.75	Ea	3,830.00	101.00	3,931.00
4000A	120"	4E@4.00	Ea	5,250.00	108.00	5,358.00

Use these figures to estimate the cost of indoor bus duct installed in buildings under the conditions described on pages 5 and 6. Costs listed are for each length of duct installed. The crew is four electricians working at a labor cost of $26.88 per manhour. These costs include connecting hardware, layout, material handling, and normal waste. Add for supports, service devices, sales tax, delivery, supervision, mobilization, demobilization, cleanup, overhead and profit. Note: Never mix copper and aluminum duct or fittings in a single system. Dissimilar metals will lead to a failure in a very short time. Using contact compounds won't overcome this problem. Some bus duct and fittings are made with quick connects. The duct slips together and is held in place with a tightening stud. Use a torque wrench to apply the specified pressure to the stud. All bus duct must be supported according to the manufacturer's recommendations.

Material		Craft@Hrs	Unit	Material Cost	Labor Cost	Installed Cost

Aluminum feeder bus duct, 600 volt, 4-wire

Material		Craft@Hrs	Unit	Material Cost	Labor Cost	Installed Cost
800A	12"	4E@0.40	Ea	127.00	10.80	137.80
1000A	12"	4E@0.50	Ea	159.00	13.40	172.40
1200A	12"	4E@0.60	Ea	197.00	16.10	213.10
1350A	12"	4E@0.70	Ea	227.00	18.80	245.80
1600A	12"	4E@0.80	Ea	276.00	21.50	297.50
2000A	12"	4E@0.90	Ea	336.00	24.20	360.20
2500A	12"	4E@1.00	Ea	413.00	26.90	439.90
3000A	12"	4E@1.10	Ea	480.00	29.60	509.60
4000A	12"	4E@1.20	Ea	643.00	32.30	675.30
800A	18"	4E@0.50	Ea	254.00	13.40	267.40
1000A	18"	4E@0.60	Ea	316.00	16.10	332.10
1200A	18"	4E@0.70	Ea	396.00	18.80	414.80
1600A	18"	4E@0.90	Ea	549.00	24.20	573.20
2000A	18"	4E@1.00	Ea	670.00	26.90	696.90
800A	24"	4E@0.60	Ea	254.00	16.10	270.10
1000A	24"	4E@0.70	Ea	316.00	18.80	334.80
1200A	24"	4E@0.80	Ea	396.00	21.50	417.50
1600A	24"	4E@1.00	Ea	549.00	26.90	575.90
2000A	24"	4E@1.10	Ea	670.00	29.60	699.60
800A	72"	4E@1.75	Ea	761.00	47.00	808.00
1000A	72"	4E@2.00	Ea	952.00	53.80	1,005.80
1200A	72"	4E@2.25	Ea	1,190.00	60.50	1,250.50
1600A	72"	4E@2.75	Ea	1,640.00	73.90	1,713.90
2000A	72"	4E@3.00	Ea	2,020.00	80.60	2,100.60
800A	84"	4E@2.00	Ea	887.00	53.80	940.80
1000A	84"	4E@2.25	Ea	1,110.00	60.50	1,170.50
1200A	84"	4E@2.50	Ea	1,380.00	67.20	1,447.20
1600A	84"	4E@3.00	Ea	1,930.00	80.60	2,010.60
2000A	84"	4E@3.25	Ea	2,350.00	87.40	2,437.40
800A	120"	4E@2.25	Ea	1,270.00	60.50	1,330.50
1000A	120"	4E@2.50	Ea	1,590.00	67.20	1,657.20
1200A	120"	4E@2.75	Ea	2,080.00	73.90	2,153.90
1350A	120"	4E@3.00	Ea	2,270.00	80.60	2,350.60
1600A	120"	4E@3.25	Ea	2,760.00	87.40	2,847.40
2000A	120"	4E@3.50	Ea	3,360.00	94.10	3,454.10
2500A	120"	4E@3.75	Ea	4,130.00	101.00	4,231.00
3000A	120"	4E@4.00	Ea	4,800.00	108.00	4,908.00
4000A	120"	4E@4.25	Ea	6,430.00	114.00	6,544.00

Use these figures to estimate the cost of indoor bus duct installed in buildings under the conditions described on pages 5 and 6. Costs listed are for each length of duct installed. The crew is four electricians working at a labor cost of $26.88 per manhour. These costs include connecting hardware, layout, material handling, and normal waste. Add for supports, service devices, sales tax, delivery, supervision, mobilization, demobilization, cleanup, overhead and profit. Note: Never mix copper and aluminum duct or fittings in a single system. Dissimilar metals will lead to a failure in a very short time. Using contact compounds won't overcome this problem. Some bus duct and fittings are made with quick connects. The duct slips together and is held in place with a tightening stud. Use a torque wrench to apply the specified pressure to the stud. All bus duct must be supported according to the manufacturer's recommendations.

Bus Duct, Aluminum

Material		Craft@Hrs	Unit	Material Cost	Labor Cost	Installed Cost
Aluminum plug-in bus duct, 600 volt, 3-wire						
800A	12"	4E@0.30	Ea	109.00	8.06	117.06
1000A	12"	4E@0.40	Ea	122.00	10.80	132.80
1200A	12"	4E@0.50	Ea	163.00	13.40	176.40
1350A	12"	4E@0.60	Ea	188.00	16.10	204.10
1600A	12"	4E@0.70	Ea	229.00	18.80	247.80
2000A	12"	4E@0.80	Ea	277.00	21.50	298.50
2500A	12"	4E@0.90	Ea	336.00	24.20	360.20
3000A	12"	4E@1.00	Ea	383.00	26.90	409.90
4000A	12"	4E@1.25	Ea	525.00	33.60	558.60
800A	18"	4E@0.50	Ea	217.00	13.40	230.40
1000A	18"	4E@0.60	Ea	244.00	16.10	260.10
1200A	18"	4E@0.70	Ea	327.00	18.80	345.80
1600A	18"	4E@0.90	Ea	460.00	24.20	484.20
2000A	18"	4E@1.00	Ea	551.00	26.90	577.90
800A	24"	4E@0.60	Ea	217.00	16.10	233.10
1000A	24"	4E@0.70	Ea	244.00	18.80	262.80
1200A	24"	4E@0.80	Ea	327.00	21.50	348.50
1600A	24"	4E@1.00	Ea	460.00	26.90	486.90
2000A	24"	4E@1.10	Ea	551.00	29.60	580.60
800A	72"	4E@1.50	Ea	652.00	40.30	692.30
1000A	72"	4E@1.75	Ea	729.00	47.00	776.00
1200A	72"	4E@2.00	Ea	978.00	53.80	1,031.80
1600A	72"	4E@2.50	Ea	1,380.00	67.20	1,447.20
2000A	72"	4E@2.75	Ea	1,670.00	73.90	1,743.90
800A	84"	4E@1.75	Ea	761.00	47.00	808.00
1000A	84"	4E@2.00	Ea	849.00	53.80	902.80
1200A	84"	4E@2.25	Ea	1,130.00	60.50	1,190.50
1600A	84"	4E@2.75	Ea	1,610.00	73.90	1,683.90
2000A	84"	4E@3.00	Ea	1,930.00	80.60	2,010.60
800A	120"	4E@2.00	Ea	1,090.00	53.80	1,143.80
1000A	120"	4E@2.25	Ea	1,220.00	60.50	1,280.50
1200A	120"	4E@2.50	Ea	1,630.00	67.20	1,697.20
1350A	120"	4E@2.75	Ea	1,880.00	73.90	1,953.90
1600A	120"	4E@3.00	Ea	2,290.00	80.60	2,370.60
2000A	120"	4E@3.25	Ea	2,770.00	87.40	2,857.40
2500A	120"	4E@3.50	Ea	3,360.00	94.10	3,454.10
3000A	120"	4E@3.75	Ea	3,830.00	101.00	3,931.00
4000A	120"	4E@4.00	Ea	5,250.00	108.00	5,358.00

Use these figures to estimate the cost of indoor bus duct installed in buildings under the conditions described on pages 5 and 6. Costs listed are for each length of duct installed. The crew is four electricians working at a labor cost of $26.88 per manhour. These costs include connecting hardware, layout, material handling, and normal waste. Add for supports, service devices, sales tax, delivery, supervision, mobilization, demobilization, cleanup, overhead and profit. Note: Never mix copper and aluminum duct or fittings in a single system. Dissimilar metals will lead to a failure in a very short time. Using contact compounds won't overcome this problem. Some bus duct and fittings are made with quick connects. The duct slips together and is held in place with a tightening stud. Use a torque wrench to apply the specified pressure to the stud. All bus duct must be supported according to the manufacturer's recommendations.

Material		Craft@Hrs	Unit	Material Cost	Labor Cost	Installed Cost
Aluminum plug-in bus duct, 600 volt, 4-wire						
800A	12"	4E@0.40	Ea	127.00	10.80	137.80
1000A	12"	4E@0.50	Ea	159.00	13.40	172.40
1200A	12"	4E@0.60	Ea	197.00	16.10	213.10
1350A	12"	4E@0.70	Ea	227.00	18.80	245.80
1600A	12"	4E@0.80	Ea	276.00	21.50	297.50
2000A	12"	4E@0.90	Ea	336.00	24.20	360.20
2500A	12"	4E@1.00	Ea	413.00	26.90	439.90
3000A	12"	4E@1.10	Ea	480.00	29.60	509.60
4000A	12"	4E@1.20	Ea	643.00	32.30	675.30
800A	18"	4E@0.50	Ea	254.00	13.40	267.40
1000A	18"	4E@0.60	Ea	316.00	16.10	332.10
1200A	18"	4E@0.70	Ea	396.00	18.80	414.80
1600A	18"	4E@0.90	Ea	549.00	24.20	573.20
2000A	18"	4E@1.00	Ea	670.00	26.90	696.90
800A	24"	4E@0.60	Ea	254.00	16.10	270.10
1000A	24"	4E@0.70	Ea	316.00	18.80	334.80
1200A	24"	4E@0.80	Ea	396.00	21.50	417.50
1600A	24"	4E@1.00	Ea	549.00	26.90	575.90
2000A	24"	4E@1.10	Ea	670.00	29.60	699.60
800A	72"	4E@1.75	Ea	761.00	47.00	808.00
1000A	72"	4E@2.00	Ea	952.00	53.80	1,005.80
1200A	72"	4E@2.25	Ea	1,190.00	60.50	1,250.50
1600A	72"	4E@2.75	Ea	1,640.00	73.90	1,713.90
2000A	72"	4E@3.00	Ea	2,100.00	80.60	2,180.60
800A	84"	4E@2.00	Ea	887.00	53.80	940.80
1000A	84"	4E@2.25	Ea	1,110.00	60.50	1,170.50
1200A	84"	4E@2.50	Ea	1,380.00	67.20	1,447.20
1600A	84"	4E@3.00	Ea	1,930.00	80.60	2,010.60
2000A	84"	4E@3.25	Ea	2,350.00	87.40	2,437.40
800A	120"	4E@2.25	Ea	1,270.00	60.50	1,330.50
1000A	120"	4E@2.50	Ea	1,590.00	67.20	1,657.20
1200A	120"	4E@2.75	Ea	2,080.00	73.90	2,153.90
1350A	120"	4E@3.00	Ea	2,270.00	80.60	2,350.60
1600A	120"	4E@3.25	Ea	2,760.00	87.40	2,847.40
2000A	120"	4E@3.50	Ea	3,360.00	94.10	3,454.10
2500A	120"	4E@3.75	Ea	4,130.00	101.00	4,231.00
3000A	120"	4E@4.00	Ea	4,800.00	108.00	4,908.00
4000A	120"	4E@4.25	Ea	6,430.00	114.00	6,544.00

Use these figures to estimate the cost of indoor bus duct installed in buildings under the conditions described on pages 5 and 6. Costs listed are for each length of duct installed. The crew is four electricians working at a labor cost of $26.88 per manhour. These costs include connecting hardware, layout, material handling, and normal waste. Add for supports, service devices, sales tax, delivery, supervision, mobilization, demobilization, cleanup, overhead and profit. Note: Never mix copper and aluminum duct or fittings in a single system. Dissimilar metals will lead to a failure in a very short time. Using contact compounds won't overcome this problem. Some bus duct and fittings are made with quick connects. The duct slips together and is held in place with a tightening stud. Use a torque wrench to apply the specified pressure to the stud. All bus duct must be supported according to the manufacturer's recommendations.

Bus Duct, Copper

Material		Craft@Hrs	Unit	Material Cost	Labor Cost	Installed Cost
Copper feeder bus duct, 600 volt, 3-wire						
800A	12"	4E@0.50	Ea	176.00	13.40	189.40
1000A	12"	4E@0.60	Ea	188.00	16.10	204.10
1200A	12"	4E@0.70	Ea	244.00	18.80	262.80
1350A	12"	4E@0.80	Ea	282.00	21.50	303.50
1600A	12"	4E@0.90	Ea	327.00	24.20	351.20
2000A	12"	4E@1.00	Ea	416.00	26.90	442.90
2500A	12"	4E@1.10	Ea	529.00	29.60	558.60
3000A	12"	4E@1.20	Ea	629.00	32.30	661.30
4000A	12"	4E@1.30	Ea	818.00	34.90	852.90
800A	18"	4E@0.60	Ea	354.00	16.10	370.10
1000A	18"	4E@0.70	Ea	375.00	18.80	393.80
1200A	18"	4E@0.80	Ea	485.00	21.50	506.50
1600A	18"	4E@1.00	Ea	652.00	26.90	678.90
2000A	18"	4E@1.10	Ea	834.00	29.60	863.60
800A	24"	4E@0.80	Ea	354.00	21.50	375.50
1000A	24"	4E@0.90	Ea	375.00	24.20	399.20
1200A	24"	4E@1.00	Ea	485.00	26.90	511.90
1600A	24"	4E@1.20	Ea	652.00	32.30	684.30
2000A	24"	4E@1.30	Ea	834.00	34.90	868.90
800A	72"	4E@3.00	Ea	1,060.00	80.60	1,140.60
1000A	72"	4E@3.50	Ea	1,120.00	94.10	1,214.10
1200A	72"	4E@3.70	Ea	1,460.00	99.50	1,559.50
1600A	72"	4E@4.10	Ea	1,950.00	110.00	2,060.00
2000A	72"	4E@4.25	Ea	2,500.00	114.00	2,614.00
800A	84"	4E@3.30	Ea	1,240.00	88.70	1,328.70
1000A	84"	4E@3.85	Ea	1,310.00	103.00	1,413.00
1200A	84"	4E@4.00	Ea	1,700.00	108.00	1,808.00
1600A	84"	4E@4.40	Ea	2,280.00	118.00	2,398.00
2000A	84"	4E@4.60	Ea	2,910.00	124.00	3,034.00
800A	120"	4E@3.75	Ea	1,760.00	101.00	1,861.00
1000A	120"	4E@4.25	Ea	1,880.00	114.00	1,994.00
1200A	120"	4E@4.40	Ea	2,440.00	118.00	2,558.00
1350A	120"	4E@4.60	Ea	2,820.00	124.00	2,944.00
1600A	120"	4E@4.80	Ea	3,270.00	129.00	3,399.00
2000A	120"	4E@5.00	Ea	4,160.00	134.00	4,294.00
2500A	120"	4E@5.20	Ea	5,290.00	140.00	5,430.00
3000A	120"	4E@5.40	Ea	6,290.00	145.00	6,435.00
4000A	120"	4E@5.60	Ea	8,180.00	151.00	8,331.00

Use these figures to estimate the cost of indoor bus duct installed in buildings under the conditions described on pages 5 and 6. Costs listed are for each length of duct installed. The crew is four electricians working at a labor cost of $26.88 per manhour. These costs include connecting hardware, layout, material handling, and normal waste. Add for supports, service devices, sales tax, delivery, supervision, mobilization, demobilization, cleanup, overhead and profit. Note: Never mix copper and aluminum duct or fittings in a single system. Dissimilar metals will lead to a failure in a very short time. Using contact compounds won't overcome this problem. Some bus duct and fittings are made with quick connects. The duct slips together and is held in place with a tightening stud. Use a torque wrench to apply the specified pressure to the stud. All bus duct must be supported according to the manufacturer's recommendations.

Material		Craft@Hrs	Unit	Material Cost	Labor Cost	Installed Cost

Copper feeder bus duct, 600 volt, 4-wire

Material		Craft@Hrs	Unit	Material Cost	Labor Cost	Installed Cost
800A	12"	4E@0.60	Ea	195.00	16.10	211.10
1000A	12"	4E@0.70	Ea	203.00	18.80	221.80
1200A	12"	4E@0.80	Ea	259.00	21.50	280.50
1350A	12"	4E@0.90	Ea	299.00	24.20	323.20
1600A	12"	4E@1.00	Ea	353.00	26.90	379.90
2000A	12"	4E@1.10	Ea	438.00	29.60	467.60
2500A	12"	4E@1.20	Ea	543.00	32.30	575.30
3000A	12"	4E@1.30	Ea	649.00	34.90	683.90
4000A	12"	4E@1.40	Ea	843.00	37.60	880.60
800A	18"	4E@0.70	Ea	390.00	18.80	408.80
1000A	18"	4E@0.80	Ea	404.00	21.50	425.50
1200A	18"	4E@0.90	Ea	516.00	24.20	540.20
1600A	18"	4E@1.10	Ea	705.00	29.60	734.60
2000A	18"	4E@1.20	Ea	875.00	32.30	907.30
800A	24"	4E@0.90	Ea	390.00	24.20	414.20
1000A	24"	4E@1.00	Ea	404.00	26.90	430.90
1200A	24"	4E@1.10	Ea	516.00	29.60	545.60
1600A	24"	4E@1.30	Ea	705.00	34.90	739.90
2000A	24"	4E@1.40	Ea	875.00	37.60	912.60
800A	72"	4E@3.10	Ea	1,170.00	83.30	1,253.30
1000A	72"	4E@3.60	Ea	1,220.00	96.80	1,316.80
1200A	72"	4E@3.80	Ea	1,540.00	102.00	1,642.00
1600A	72"	4E@4.20	Ea	2,120.00	113.00	2,233.00
2000A	72"	4E@4.40	Ea	2,360.00	118.00	2,478.00
800A	84"	4E@3.40	Ea	1,370.00	91.40	1,461.40
1000A	84"	4E@3.95	Ea	1,420.00	106.00	1,526.00
1200A	84"	4E@4.10	Ea	1,810.00	110.00	1,920.00
1600A	84"	4E@4.50	Ea	2,470.00	121.00	2,591.00
2000A	84"	4E@4.70	Ea	3,070.00	126.00	3,196.00
800A	120"	4E@4.00	Ea	1,950.00	108.00	2,058.00
1000A	120"	4E@4.50	Ea	2,030.00	121.00	2,151.00
1200A	120"	4E@4.75	Ea	2,590.00	128.00	2,718.00
1350A	120"	4E@5.00	Ea	2,990.00	134.00	3,124.00
1600A	120"	4E@5.25	Ea	3,530.00	141.00	3,671.00
2000A	120"	4E@5.50	Ea	4,380.00	148.00	4,528.00
2500A	120"	4E@5.75	Ea	5,430.00	155.00	5,585.00
3000A	120"	4E@6.00	Ea	6,490.00	161.00	6,651.00
4000A	120"	4E@6.25	Ea	8,430.00	168.00	8,598.00

Use these figures to estimate the cost of indoor bus duct installed in buildings under the conditions described on pages 5 and 6. Costs listed are for each length of duct installed. The crew is four electricians working at a labor cost of $26.88 per manhour. These costs include connecting hardware, layout, material handling, and normal waste. Add for supports, service devices, sales tax, delivery, supervision, mobilization, demobilization, cleanup, overhead and profit. Note: Never mix copper and aluminum duct or fittings in a single system. Dissimilar metals will lead to a failure in a very short time. Using contact compounds won't overcome this problem. Some bus duct and fittings are made with quick connects. The duct slips together and is held in place with a tightening stud. Use a torque wrench to apply the specified pressure to the stud. All bus duct must be supported according to the manufacturer's recommendations.

Bus Duct, Copper

Material		Craft@Hrs	Unit	Material Cost	Labor Cost	Installed Cost
Copper plug-in bus duct, 600 volt, 3-wire						
800A	12"	4E@0.50	Ea	176.00	13.40	189.40
1000A	12"	4E@0.60	Ea	188.00	16.10	204.10
1200A	12"	4E@0.70	Ea	244.00	18.80	262.80
1350A	12"	4E@0.80	Ea	282.00	21.50	303.50
1600A	12"	4E@0.90	Ea	327.00	24.20	351.20
2000A	12"	4E@1.00	Ea	416.00	26.90	442.90
2500A	12"	4E@1.10	Ea	529.00	29.60	558.60
3000A	12"	4E@1.20	Ea	629.00	32.30	661.30
4000A	12"	4E@1.30	Ea	820.00	34.90	854.90
800A	18"	4E@0.60	Ea	354.00	16.10	370.10
1000A	18"	4E@0.70	Ea	375.00	18.80	393.80
1200A	18"	4E@0.80	Ea	485.00	21.50	506.50
1600A	18"	4E@1.00	Ea	652.00	26.90	678.90
2000A	18"	4E@1.10	Ea	834.00	29.60	863.60
800A	24"	4E@0.80	Ea	354.00	21.50	375.50
1000A	24"	4E@0.90	Ea	375.00	24.20	399.20
1200A	24"	4E@1.00	Ea	485.00	26.90	511.90
1600A	24"	4E@1.20	Ea	652.00	32.30	684.30
2000A	24"	4E@1.30	Ea	834.00	34.90	868.90
800A	72"	4E@3.00	Ea	1,060.00	80.60	1,140.60
1000A	72"	4E@3.50	Ea	1,120.00	94.10	1,214.10
1200A	72"	4E@3.70	Ea	1,460.00	99.50	1,559.50
1600A	72"	4E@4.10	Ea	1,950.00	110.00	2,060.00
2000A	72"	4E@4.25	Ea	2,500.00	114.00	2,614.00
800A	84"	4E@3.30	Ea	1,240.00	88.70	1,328.70
1000A	84"	4E@3.85	Ea	1,310.00	103.00	1,413.00
1200A	84"	4E@4.00	Ea	1,700.00	108.00	1,808.00
1600A	84"	4E@4.40	Ea	2,280.00	118.00	2,398.00
2000A	84"	4E@4.60	Ea	2,910.00	124.00	3,034.00
800A	120"	4E@3.75	Ea	1,760.00	101.00	1,861.00
1000A	120"	4E@4.25	Ea	1,880.00	114.00	1,994.00
1200A	120"	4E@4.40	Ea	2,440.00	118.00	2,558.00
1350A	120"	4E@4.60	Ea	2,820.00	124.00	2,944.00
1600A	120"	4E@4.80	Ea	3,270.00	129.00	3,399.00
2000A	120"	4E@5.00	Ea	4,160.00	134.00	4,294.00
2500A	120"	4E@5.20	Ea	5,290.00	140.00	5,430.00
3000A	120"	4E@5.40	Ea	6,290.00	145.00	6,435.00
4000A	120"	4E@5.60	Ea	8,180.00	151.00	8,331.00

Use these figures to estimate the cost of indoor bus duct installed in buildings under the conditions described on pages 5 and 6. Costs listed are for each length of duct installed. The crew is four electricians working at a labor cost of $26.88 per manhour. These costs include connecting hardware, layout, material handling, and normal waste. Add for supports, service devices, sales tax, delivery, supervision, mobilization, demobilization, cleanup, overhead and profit. Note: Never mix copper and aluminum duct or fittings in a single system. Dissimilar metals will lead to a failure in a very short time. Using contact compounds won't overcome this problem. Some bus duct and fittings are made with quick connects. The duct slips together and is held in place with a tightening stud. Use a torque wrench to apply the specified pressure to the stud. All bus duct must be supported according to the manufacturer's recommendations.

Material		Craft@Hrs	Unit	Material Cost	Labor Cost	Installed Cost

Copper plug-in bus duct, 600 volt, 4-wire

Material		Craft@Hrs	Unit	Material Cost	Labor Cost	Installed Cost
800A	12"	4E@0.60	Ea	195.00	16.10	211.10
1000A	12"	4E@0.70	Ea	203.00	18.80	221.80
1200A	12"	4E@0.80	Ea	259.00	21.50	280.50
1350A	12"	4E@0.90	Ea	299.00	24.20	323.20
1600A	12"	4E@1.00	Ea	353.00	26.90	379.90
2000A	12"	4E@1.10	Ea	438.00	29.60	467.60
2500A	12"	4E@1.20	Ea	543.00	32.30	575.30
3000A	12"	4E@1.30	Ea	649.00	34.90	683.90
4000A	12"	4E@1.40	Ea	843.00	37.60	880.60
800A	18"	4E@0.70	Ea	390.00	18.80	408.80
1000A	18"	4E@0.80	Ea	404.00	21.50	425.50
1200A	18"	4E@0.90	Ea	516.00	24.20	540.20
1600A	18"	4E@1.10	Ea	705.00	29.60	734.60
2000A	18"	4E@1.20	Ea	875.00	32.30	907.30
800A	24"	4E@0.90	Ea	390.00	24.20	414.20
1000A	24"	4E@1.00	Ea	404.00	26.90	430.90
1200A	24"	4E@1.10	Ea	516.00	29.60	545.60
1600A	24"	4E@1.30	Ea	705.00	34.90	739.90
2000A	24"	4E@1.40	Ea	875.00	37.60	912.60
800A	72"	4E@3.10	Ea	1,170.00	83.30	1,253.30
1000A	72"	4E@3.60	Ea	1,220.00	96.80	1,316.80
1200A	72"	4E@3.80	Ea	1,540.00	102.00	1,642.00
1600A	72"	4E@4.20	Ea	2,120.00	113.00	2,233.00
2000A	72"	4E@4.40	Ea	2,360.00	118.00	2,478.00
800A	84"	4E@3.40	Ea	1,370.00	91.40	1,461.40
1000A	84"	4E@3.95	Ea	1,420.00	106.00	1,526.00
1200A	84"	4E@4.10	Ea	1,810.00	110.00	1,920.00
1600A	84"	4E@4.50	Ea	2,470.00	121.00	2,591.00
2000A	84"	4E@4.70	Ea	3,070.00	126.00	3,196.00
800A	120"	4E@4.00	Ea	1,950.00	108.00	2,058.00
1000A	120"	4E@4.50	Ea	2,030.00	121.00	2,151.00
1200A	120"	4E@4.75	Ea	2,590.00	128.00	2,718.00
1350A	120"	4E@5.00	Ea	2,990.00	134.00	3,124.00
1600A	120"	4E@5.25	Ea	3,530.00	141.00	3,671.00
2000A	120"	4E@5.50	Ea	4,380.00	148.00	4,528.00
2500A	120"	4E@5.75	Ea	5,430.00	155.00	5,585.00
3000A	120"	4E@6.00	Ea	6,490.00	161.00	6,651.00
4000A	120"	4E@6.25	Ea	8,430.00	168.00	8,598.00

Use these figures to estimate the cost of indoor bus duct installed in buildings under the conditions described on pages 5 and 6. Costs listed are for each length of duct installed. The crew is four electricians working at a labor cost of $26.88 per manhour. These costs include connecting hardware, layout, material handling, and normal waste. Add for supports, service devices, sales tax, delivery, supervision, mobilization, demobilization, cleanup, overhead and profit. Note: Never mix copper and aluminum duct or fittings in a single system. Dissimilar metals will lead to a failure in a very short time. Using contact compounds won't overcome this problem. Some bus duct and fittings are made with quick connects. The duct slips together and is held in place with a tightening stud. Use a torque wrench to apply the specified pressure to the stud. All bus duct must be supported according to the manufacturer's recommendations.

Bus Duct Fittings

Material	Craft@Hrs	Unit	Material Cost	Labor Cost	Installed Cost
Flat elbows for bus duct, 600 volt, 3 or 4-wire					
800A	4E@1.50	Ea	491.00	40.30	531.30
1000A	4E@1.75	Ea	491.00	47.00	538.00
1200A	4E@2.00	Ea	681.00	53.80	734.80
1350A	4E@2.25	Ea	681.00	60.50	741.50
1600A	4E@2.50	Ea	681.00	67.20	748.20
2000A	4E@2.75	Ea	681.00	73.90	754.90
2500A	4E@3.00	Ea	681.00	80.60	761.60
3000A	4E@3.25	Ea	681.00	87.40	768.40
4000A	4E@3.50	Ea	802.00	94.10	896.10
Tee for bus duct, 600 volt, 3 or 4-wire					
800A	4E@2.25	Ea	573.00	60.50	633.50
1000A	4E@2.50	Ea	573.00	67.20	640.20
1200A	4E@2.75	Ea	802.00	73.90	875.90
1350A	4E@3.00	Ea	802.00	80.60	882.60
1600A	4E@3.25	Ea	802.00	87.40	889.40
2000A	4E@3.50	Ea	802.00	94.10	896.10
2500A	4E@3.75	Ea	802.00	101.00	903.00
3000A	4E@4.00	Ea	802.00	108.00	910.00
4000A	4E@4.25	Ea	938.00	114.00	1,052.00
Cross for bus duct, 600 volt, 3 or 4-wire					
800A	4E@2.50	Ea	711.00	67.20	778.20
1000A	4E@2.75	Ea	711.00	73.90	784.90
1200A	4E@3.00	Ea	1,000.00	80.60	1,080.60
1350A	4E@3.25	Ea	1,000.00	87.40	1,087.40
1600A	4E@3.50	Ea	1,000.00	94.10	1,094.10
2000A	4E@3.75	Ea	1,000.00	101.00	1,101.00
2500A	4E@4.00	Ea	1,000.00	108.00	1,108.00
3000A	4E@4.25	Ea	1,000.00	114.00	1,114.00
4000A	4E@4.50	Ea	1,140.00	121.00	1,261.00
Flanged end for bus duct, 600 volt, 3 or 4-wire					
800A	4E@1.50	Ea	439.00	40.30	479.30
1000A	4E@1.75	Ea	524.00	47.00	571.00
1200A	4E@2.00	Ea	564.00	53.80	617.80
1350A	4E@2.25	Ea	590.00	60.50	650.50
1600A	4E@2.50	Ea	683.00	67.20	750.20
2000A	4E@2.75	Ea	792.00	73.90	865.90
2500A	4E@3.00	Ea	1,130.00	80.60	1,210.60
3000A	4E@3.25	Ea	1,410.00	87.40	1,497.40
4000A	4E@3.50	Ea	1,760.00	94.10	1,854.10

Use these figures to estimate the cost of indoor bus duct fittings installed in buildings under the conditions described on pages 5 and 6. Costs listed are for each fitting installed. The crew is four electricians working at a labor cost of $26.88 per manhour. These costs include connecting hardware, layout, material handling, and normal waste. Add for supports, service devices, sales tax, delivery, supervision, mobilization, demobilization, cleanup, overhead and profit. Note: Never mix copper and aluminum duct or fittings in a single system. Dissimilar metals will lead to a failure in a very short time. Using contact compounds won't overcome this problem. Material and labor costs for the elbows listed above will be about the same whether the elbow is vertical or horizontal. But horizontal and vertical elbows are not interchangeable. Order the type needed on your job. Some bus duct and fittings are made with quick connects. The duct slips together and is held in place with a tightening stud. Use a torque wrench to apply the specified pressure to the stud. All bus duct must be supported according to the manufacturer's recommendations.

Material	Craft@Hrs	Unit	Material Cost	Labor Cost	Installed Cost

Tap box for bus duct, 600 volt, 3 or 4-wire

Material	Craft@Hrs	Unit	Material Cost	Labor Cost	Installed Cost
800A	4E@1.50	Ea	1,380.00	40.30	1,420.30
1000A	4E@1.75	Ea	1,440.00	47.00	1,487.00
1200A	4E@2.00	Ea	1,500.00	53.80	1,553.80
1350A	4E@2.25	Ea	1,530.00	60.50	1,590.50
1600A	4E@2.50	Ea	1,610.00	67.20	1,677.20
2000A	4E@2.75	Ea	1,760.00	73.90	1,833.90
2500A	4E@3.00	Ea	2,130.00	80.60	2,210.60
3000A	4E@3.25	Ea	2,320.00	87.40	2,407.40
4000A	4E@3.50	Ea	2,660.00	94.10	2,754.10

Reducer for bus duct, 600 volt, 3 or 4-wire

Material	Craft@Hrs	Unit	Material Cost	Labor Cost	Installed Cost
800A	4E@1.75	Ea	473.00	47.00	520.00
1000A	4E@2.00	Ea	555.00	53.80	608.80
1200A	4E@2.25	Ea	948.00	60.50	1,008.50
1350A	4E@2.50	Ea	1,220.00	67.20	1,287.20
1600A	4E@2.75	Ea	1,310.00	73.90	1,383.90
2000A	4E@3.00	Ea	1,760.00	80.60	1,840.60
2500A	4E@3.25	Ea	2,180.00	87.40	2,267.40
3000A	4E@3.50	Ea	2,630.00	94.10	2,724.10
4000A	4E@3.75	Ea	3,500.00	101.00	3,601.00

Expansion joint for bus duct, 600 volt, 3 or 4-wire

Material	Craft@Hrs	Unit	Material Cost	Labor Cost	Installed Cost
800A	4E@1.75	Ea	1,200.00	47.00	1,247.00
1000A	4E@2.00	Ea	1,350.00	53.80	1,403.80
1200A	4E@2.25	Ea	1,620.00	60.50	1,680.50
1350A	4E@2.50	Ea	1,800.00	67.20	1,867.20
1600A	4E@2.75	Ea	2,180.00	73.90	2,253.90
2000A	4E@3.00	Ea	2,320.00	80.60	2,400.60
2500A	4E@3.25	Ea	2,510.00	87.40	2,597.40
3000A	4E@3.50	Ea	3,300.00	94.10	3,394.10
4000A	4E@3.75	Ea	3,780.00	101.00	3,881.00

End closures for bus duct, 3 or 4-wire

Material	Craft@Hrs	Unit	Material Cost	Labor Cost	Installed Cost
800A	4E@0.25	Ea	123.00	6.72	129.72
1000A	4E@0.30	Ea	123.00	8.06	131.06
1200A	4E@0.35	Ea	123.00	9.41	132.41
1350A	4E@0.40	Ea	123.00	10.80	133.80
1600A	4E@0.45	Ea	123.00	12.10	135.10
2000A	4E@0.50	Ea	159.00	13.40	172.40
2500A	4E@0.55	Ea	159.00	14.80	173.80
3000A	4E@0.60	Ea	159.00	16.10	175.10
4000A	4E@0.70	Ea	202.00	18.80	220.80

Use these figures to estimate the cost of indoor bus duct fittings installed in buildings under the conditions described on pages 5 and 6. Costs listed are for each fitting installed. The crew is four electricians working at a labor cost of $26.88 per manhour. These costs include connecting hardware, layout, material handling, and normal waste. Add for supports, service devices, sales tax, delivery, supervision, mobilization, demobilization, cleanup, overhead and profit. Note: Never mix copper and aluminum duct or fittings in a single system. Dissimilar metals will lead to a failure in a very short time. Using contact compounds won't overcome this problem. Some bus duct and fittings are made with quick connects. The duct slips together and is held in place with a tightening stud. Use a torque wrench to apply the specified pressure to the stud. All bus duct must be supported according to the manufacturer's recommendations.

Bus Duct Plug-in Units

Material	Craft@Hrs	Unit	Material Cost	Labor Cost	Installed Cost
Fusible plug-in switches for bus duct					
30A, 3P, 240V	2E@0.40	Ea	310.00	10.80	320.80
60A, 3P, 240V	2E@0.50	Ea	332.00	13.40	345.40
100A, 3P, 240V	2E@0.75	Ea	470.00	20.20	490.20
200A, 3P, 240V	2E@1.00	Ea	785.00	26.90	811.90
400A, 3P, 240V	2E@1.50	Ea	2,050.00	40.30	2,090.30
600A, 3P, 240V	2E@2.00	Ea	2,940.00	53.80	2,993.80
30A, 4P, 240V	2E@0.50	Ea	362.00	13.40	375.40
60A, 4P, 240V	2E@0.75	Ea	377.00	20.20	397.20
100A, 4P, 240V	2E@1.00	Ea	510.00	26.90	536.90
200A, 4P, 240V	2E@1.50	Ea	869.00	40.30	909.30
400A, 4P, 240V	2E@2.00	Ea	2,230.00	53.80	2,283.80
600A, 3P, 240V	2E@2.50	Ea	3,220.00	67.20	3,287.20
30A, 3P, 480V	2E@0.50	Ea	377.00	13.40	390.40
60A, 3P, 480V	2E@0.75	Ea	392.00	20.20	412.20
100A, 3P, 480V	2E@1.00	Ea	552.00	26.90	578.90
200A, 3P, 480V	2E@1.50	Ea	914.00	40.30	954.30
400A, 3P, 480V	2E@2.00	Ea	2,230.00	53.80	2,283.80
600A, 3P, 480V	2E@2.50	Ea	3,220.00	67.20	3,287.20
30A, 3P, 600V	2E@0.50	Ea	332.00	13.40	345.40
60A, 3P, 600V	2E@0.75	Ea	354.00	20.20	374.20
100A, 3P, 600V	2E@1.00	Ea	483.00	26.90	509.90
200A, 3P, 600V	2E@1.50	Ea	818.00	40.30	858.30
400A, 3P, 600V	2E@2.00	Ea	2,050.00	53.80	2,103.80
600A, 3P, 600V	2E@2.50	Ea	2,940.00	67.20	3,007.20
Circuit breaker plug-in switches for bus duct					
15A, 3P, 240V	2E@0.50	Ea	464.00	13.40	477.40
20A, 3P, 240V	2E@0.50	Ea	464.00	13.40	477.40
30A, 3P, 240V	2E@0.50	Ea	464.00	13.40	477.40
40A, 3P, 240V	2E@0.50	Ea	464.00	13.40	477.40
50A, 3P, 240V	2E@0.50	Ea	464.00	13.40	477.40
60A, 3P, 240V	2E@0.50	Ea	464.00	13.40	477.40
70A, 3P, 240V	2E@0.75	Ea	544.00	20.20	564.20
90A, 3P, 240V	2E@0.75	Ea	544.00	20.20	564.20
100A, 3P, 240V	2E@0.75	Ea	544.00	20.20	564.20
15A, 3P, 240V, S/N	2E@0.60	Ea	523.00	16.10	539.10
20A, 3P, 240V, S/N	2E@0.60	Ea	523.00	16.10	539.10
30A, 3P, 240V, S/N	2E@0.60	Ea	523.00	16.10	539.10
40A, 3P, 240V, S/N	2E@0.60	Ea	523.00	16.10	539.10
50A, 3P, 240V, S/N	2E@0.60	Ea	523.00	16.10	539.10
60A, 3P, 240V, S/N	2E@0.60	Ea	523.00	16.10	539.10
70A, 3P, 240V, S/N	2E@0.90	Ea	603.00	24.20	627.20

Use these figures to estimate the cost of plug-in switches installed in bus duct under the conditions described on pages 5 and 6. Costs listed are for each switch installed. The crew is two electricians working at a labor cost of $26.88 per manhour. These costs include connecting hardware, layout, material handling, and normal waste. Add for supports, service devices, sales tax, delivery, supervision, mobilization, demobilization, cleanup, overhead and profit. Note: Bus duct made as feeder duct only will not accept plug-in breakers.

Material	Craft@Hrs	Unit	Material Cost	Labor Cost	Installed Cost
Circuit breaker plug-in switches for bus duct					
90A, 3P, 240V, S/N	2E@1.00	Ea	603.00	26.90	629.90
100A, 3P, 240V, S/N	2E@1.25	Ea	603.00	33.60	636.60
15A, 3P, 480V	2E@0.50	Ea	567.00	13.40	580.40
20A, 3P, 480V	2E@0.50	Ea	567.00	13.40	580.40
30A, 3P, 480V	2E@0.50	Ea	567.00	13.40	580.40
40A, 3P, 480V	2E@0.50	Ea	567.00	13.40	580.40
50A, 3P, 480V	2E@0.50	Ea	567.00	13.40	580.40
60A, 3P, 480V	2E@0.50	Ea	567.00	13.40	580.40
70A, 3P, 480V	2E@0.75	Ea	618.00	20.20	638.20
90A, 3P, 480V	2E@0.75	Ea	618.00	20.20	638.20
100A, 3P, 480V	2E@0.75	Ea	618.00	20.20	638.20
15A, 3P, 480V, S/N	2E@0.60	Ea	631.00	16.10	647.10
20A, 3P, 480V, S/N	2E@0.60	Ea	631.00	16.10	647.10
30A, 3P, 480V, S/N	2E@0.60	Ea	631.00	16.10	647.10
40A, 3P, 480V, S/N	2E@0.60	Ea	631.00	16.10	647.10
50A, 3P, 480V, S/N	2E@0.60	Ea	631.00	16.10	647.10
60A, 3P, 480V, S/N	2E@0.60	Ea	631.00	16.10	647.10
70A, 3P, 480V, S/N	2E@0.90	Ea	677.00	24.20	701.20
90A, 3P, 480V, S/N	2E@0.90	Ea	677.00	24.20	701.20
100A, 3P, 480V, S/N	2E@0.90	Ea	677.00	24.20	701.20
15A, 3P, 600V	2E@0.50	Ea	616.00	13.40	629.40
20A, 3P, 600V	2E@0.50	Ea	616.00	13.40	629.40
30A, 3P, 600V	2E@0.50	Ea	616.00	13.40	629.40
40A, 3P, 600V	2E@0.50	Ea	616.00	13.40	629.40
50A, 3P, 600V	2E@0.50	Ea	616.00	13.40	629.40
60A, 3P, 600V	2E@0.50	Ea	616.00	13.40	629.40
70A, 3P, 600V	2E@0.75	Ea	679.00	20.20	699.20
90A, 3P, 600V	2E@0.75	Ea	679.00	20.20	699.20
100A, 3P, 600V	2E@0.75	Ea	679.00	20.20	699.20
Combination starter/fusible switch plug-in unit for bus duct					
Size 0, 3P, 240V	2E@1.50	Ea	1,070.00	40.30	1,110.30
Size 1, 3P, 240V	2E@1.75	Ea	1,130.00	47.00	1,177.00
Size 2, 3P, 240V	2E@2.00	Ea	1,440.00	53.80	1,493.80
Size 3, 3P, 240V	2E@3.00	Ea	2,290.00	80.60	2,370.60
Combination starter/circuit breaker plug-in unit for bus duct					
Size 0, 3P, 240V	2E@1.50	Ea	1,100.00	40.30	1,140.30
Size 1, 3P, 240V	2E@1.75	Ea	1,140.00	47.00	1,187.00
Size 2, 3P, 240V	2E@2.00	Ea	1,630.00	53.80	1,683.80
Size 3, 3P, 240V	2E@3.00	Ea	2,120.00	80.60	2,200.60

Use these figures to estimate the cost of plug-in switches or breakers installed in bus duct under the conditions described on pages 5 and 6. Costs listed are for each switch or breaker installed. The crew is two electricians working at a labor cost of $26.88 per manhour. These costs include connecting hardware, layout, material handling, and normal waste. Add for supports, service devices, sales tax, delivery, supervision, mobilization, demobilization, cleanup, overhead and profit. Note: Bus duct made as feeder duct only will not accept plug-in breakers.

Section 9: Cable Tray

Cable tray has been installed in industrial and commercial buildings for many years. In fact, the first cable tray was developed when large buildings were first wired for electric power. Since then, there have been many improvements in design, carrying capacities and materials.

Cable tray in its simplest form is just a continuous tray that carries cable. The types of cable tray available include the following:

Ladder tray has two side rails connected by uniformly-spaced rungs.

Trough tray has a ventilated bottom and closely-spaced supports that are carried by side rails.

Channel tray has a one-piece bottom and side rails.

No matter what type of tray is used, cable is laid in the tray bottom. Sides of the tray keep the cable from shifting out of place. Figure 9-1 shows a typical installation.

This section covers the most common sizes and types of cable tray systems. The sizes listed are the sizes used most commonly today. Some manufacturers will fabricate special sizes to meet unusual job requirements. Your suppliers can provide information on custom orders.

Most cable tray is open to permit free air circulation. But some cable tray comes with a cover to enclose the system. A cover helps protect cable run in areas where it could be damaged, and reduces the chance of developing dust problems in a dusty environment.

Angle fittings are made for routing the cable tray over and under obstructions or to provide a lateral offset.

Most cable tray catalogs don't list prices. Pricing is usually done by a factory representative from a material takeoff done by the electrical contractor. Some factory reps will do the takeoff from your plans and then route the price quotation to your supplier.

Hanger spacing is critical with cable tray. Be sure that the hangers are spaced to carry the weight of the tray and the weight of all cables that can be installed. Some catalogs offer guidelines for calculating hanger spacing. Cable tray class is a guide to hanger spacing. The National Electrical Manufacturer's Association has established four classes of cable tray. Class 1 can span 12 feet between hangers, and supports up to 35 pounds of cable per linear foot. Class 2 can also span 12 feet, but supports up to 50 pounds per linear foot. Class 3 spans 20 feet and supports up to 45 pounds. Class 4 spans 20 feet and supports up to 75 pounds.

Be careful when attaching hangers to structural members. What you're attaching to has to be strong enough to support a lot of weight. In some cases you'll need bracing to spread the load over a wider area.

The most common support for cable tray is threaded hanger rod attached to overhead beams with beam clamps or angle brackets. Rod is usually attached to a metal channel that forms a trapeze support for the tray. The size of the rod depends on the gross weight. The smallest rod to use is 3/8" in diameter. It should be installed on each side of the tray. If you use larger diameter rod, a single rod hanger running through the middle of the tray may be enough.

When centered single hangers are used, the tray may not hang level if the cable load isn't distributed evenly. One easy way to remedy this problem is to install a piece of conduit over nearly the full length of rod. Be sure to cut the ends of the conduit square. When the rod nuts are tightened, the conduit becomes a spacer that stiffens the hanger. With a conduit spacer, even a moderately unbalanced load won't affect the tray.

It's common to install cable tray but no cable. Tray is installed when the building is constructed. Cable will be run later when the building is occupied by a tenant who needs power for high amperage loads or when the phone company installs a phone system.

Planning the layout is a very important part of every cable tray job. Do detailed planning before ordering materials. Study the drawings carefully. Ask yourself some questions: What is the elevation of the cable tray? What obstruction will also be at that level? What's the best way to avoid potential obstructions? Can you improve the layout to eliminate any unnecessary turns or offsets? Special fittings are expensive, both to buy from the manufacturer and to install. Discovering that you need a special fitting after installation has begun can delay completion for days or weeks. Try to plan the simplest layout possible and anticipate all of the potential problems.

Most tray material is lightweight and easy to handle. But the 12-foot sections are hard to install in an existing drop ceiling. Allow additional time for this

P-W Cable Tray Systems

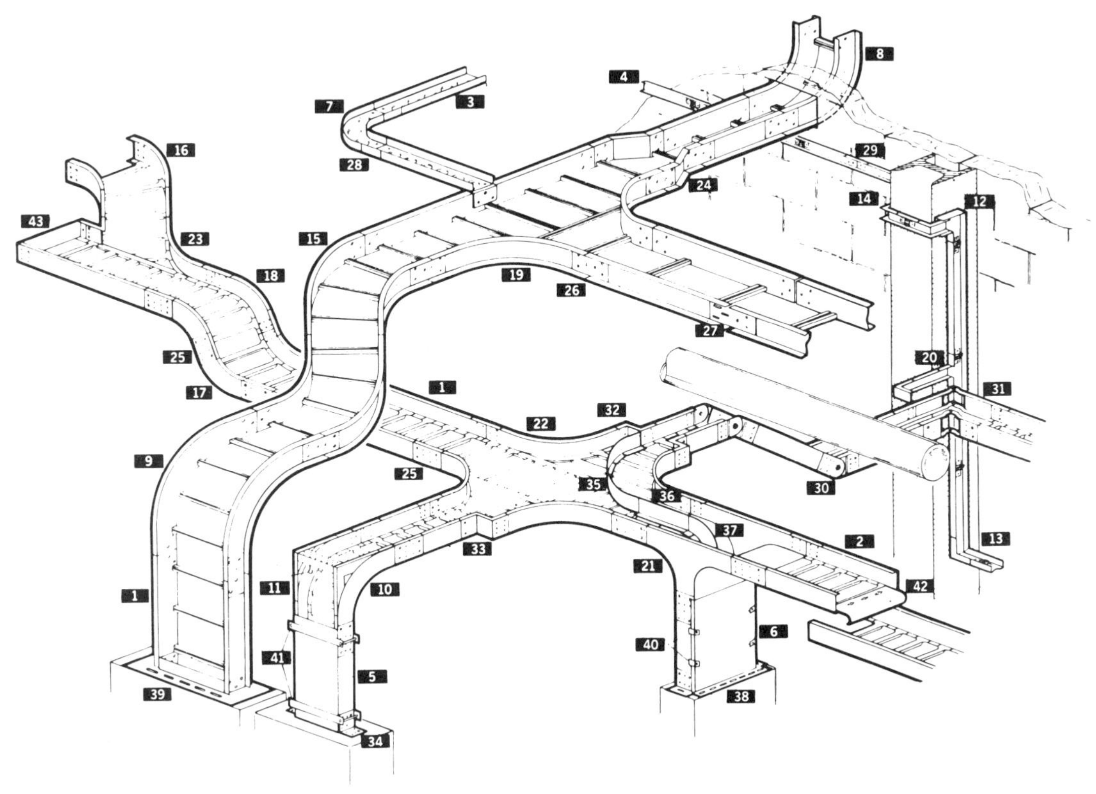

1. Straight Length
2. Straight Length
3. Straight Length
4. Straight Length
5. Flat Covers
6. Flanged Covers
7. Horizontal Fittings
8. Inside 90° Vertical Elbow
9. Outside 90° Verticle Elbow
10. Cable Support Elbow
11. "J" Hooks for hanging Cable
12. 90° Raceway Connectors
13. 90°Inside Elbow Raceway
14. 90° Outside Elbow, Raceway
15. 45° Horizontal Elbow

16. 45°Horizontal Elbow
17. 45° Inside Vertical Elbow
18. 45° Outside Vertical Elbow
19. Horizontal Tee
20. Tee Connector Raceway
21. Veritcal Tee
22. Horizontal Cross
23. 45° Horizontal Wye Branch
24. Reducers
25. Channel Connector
26. Plate Connector
27. Expansion Connector
28. 4" Channel Connector
29. Straight Raceway Connector

30. Adj. Vertical Connector
31. Adj. Horizontal Connector
32. Straight Reducer
33. Offset Reducer
34. Angle Connector
35. Adjustable Horizontal Divider
36. Straight Divider
37. Vertical Elbow Divider
38. Trough-to-Box Connector
39. Trough-to-Box Connector
40. Cover Connector Clip
41. Cover Connector Strap
42. Dropout
43. Blind End

Figure 9-1
Cable Tray

type of work and for any retrofit job that involves cable tray.

Installation in an existing building usually requires working over and around machinery, office furniture, drop ceilings or in confined areas. There may even be a problem with storage space on site. Sometimes only small bundles of material can be stockpiled on site at one time. Allow extra labor if you anticipate conditions like this. With cable tray, even an apparently inconsequential discrepancy can turn normal installation into a nightmare. That turns a good estimate into a major loss.

The labor units listed in this section apply on standard overhead installations at heights up to 12 feet above the floor. Higher installations will take longer.

Estimating Cable Tray

Cable tray take-off is very similar to bus duct take-off. First, study the plans and specifications. Find the type and size of cable tray to be used. Study the area where the system will be installed.

The ideal location is in an open area about 12 feet above the floor without any significant obstructions. That's an easy job, especially if the plans indicate very clearly all bends and tray drops required. Most cable tray jobs aren't that easy. The hardest cable tray job would be in an existing building, where your electricians have to work around furniture or machinery. If you have to run tray in a drop ceiling that's already filled with air conditioning duct, fire sprinkler piping, electrical conduit and recessed lighting fixtures, installation time will be much longer than the tables in this section indicate.

In some cases it may be nearly impossible to install cable tray as specified. On any job over a dropped ceiling, remember that cable tray comes in 12-foot lengths. Most tee-bar ceiling grid has 2 foot by 4 foot openings. It will take more than grease to pass tray through that grid. Some of the ceiling will have to be removed to install the tray. Your electricians shouldn't have to remove and reset the ceiling grid. If removal and replacement aren't covered separately in the specs, get a quote on the gridwork from a ceiling contractor.

The labor cost for installing cable tray in some buildings may be double the figures listed in the tables that follow. Don't be reluctant to increase the labor cost when you anticipate access problems or note a lot of possible obstructions. Remember that a greater installation height will increase your labor cost. Figure that cable tray installed 18 feet above the floor may require 15 percent more labor than the same tray installed 12 feet above the floor.

On many jobs there will be conflicts with fire sprinkler piping, air ducts and light fixtures. This is the most common problem. A slight change in elevation might help. But remember that tray that isn't accessible for installation of cable has little value.

Material	Craft@Hrs	Unit	Material Cost	Labor Cost	Installed Cost
16 gauge galvanized cable tray with louvered openings					
12" wide	2E@1.25	Ea	139.00	33.60	172.60
18" wide	2E@1.50	Ea	148.00	40.30	188.30
24" wide	2E@1.75	Ea	170.00	47.00	217.00
12" radius elbows for louvered cable tray					
12" 45 degree, flat	2E@0.50	Ea	95.60	13.40	109.00
18" 45 degree, flat	2E@0.60	Ea	113.00	16.10	129.10
24" 45 degree, flat	2E@0.75	Ea	130.00	20.20	150.20
12" 90 degree, flat	2E@0.60	Ea	110.00	16.10	126.10
18" 90 degree, flat	2E@0.75	Ea	129.00	20.20	149.20
24" 90 degree, flat	2E@1.00	Ea	145.00	26.90	171.90
12" 90 degree, vert.	2E@0.60	Ea	103.00	16.10	119.10
18" 90 degree, vert.	2E@0.75	Ea	125.00	20.20	145.20
24" 90 degree, vert.	2E@1.00	Ea	137.00	26.90	163.90
12" radius tees for louvered cable tray					
12" flat	2E@0.75	Ea	205.00	20.20	225.20
18" flat	2E@1.00	Ea	233.00	26.90	259.90
24" flat	2E@1.25	Ea	260.00	33.60	293.60
Panel or box adapter for louvered cable tray					
12"	2E@0.30	Ea	10.00	8.06	18.06
18"	2E@0.30	Ea	12.60	8.06	20.66
24"	2E@0.30	Ea	15.10	8.06	23.16
End caps for louvered cable tray					
12"	2E@0.10	Ea	14.60	2.69	17.29
18"	2E@0.15	Ea	16.90	4.03	20.93
24"	2E@0.20	Ea	19.10	5.38	24.48

Use these figures to estimate the cost of cable tray and fittings installed in buildings under the conditions described on pages 5 and 6. Costs listed are for each tray section or fitting installed. The crew is two electricians working at a labor cost of $26.88 per manhour. These costs include layout, material handling, and normal waste. Add for supports, sales tax, delivery, supervision, mobilization, demobilization, cleanup, overhead and profit. Note: Covers are available for cable tray to keep dust away from the conductors. Order special extra-wide sections or reducing sections when changing from one width to another. Be sure to use the right size supports. Cable tray can be suspended from single support hangers. The hanger attaches to the center of the tray. A stiffener bar goes under the tray and the rod hanger extends up through the tray to the building structure above. If you use single hangers, be sure the hanger is sized for the tray and cable load. Spacing of hangers can't exceed 10' and must be closer for higher carrying capacities. The manufacturer's catalog will list approved spacing and other fittings that may be needed for special applications.

Fittings for 16 Gauge Louvered Cable Tray

Material	Craft@Hrs	Unit	Material Cost	Labor Cost	Installed Cost
Hangers and joiners for 16 gauge louvered cable tray					
Hangers, pair	2E@0.50	Ea	10.20	13.40	23.60
Joiners, standard, pair	2E@0.30	Ea	10.00	8.06	18.06
Joiners, adj., flat, pair	2E@0.40	Ea	35.30	10.80	46.10
Joiners, adj., vert, pair	2E@0.30	Ea	23.50	8.06	31.56
Dropouts for 16 gauge louvered cable tray					
12"	2E@0.25	Ea	10.00	6.72	16.72
18"	2E@0.30	Ea	11.00	8.06	19.06
24"	2E@0.35	Ea	12.20	9.41	21.61

Use these figures to estimate the cost of cable tray and fittings installed in bus duct under the conditions described on pages 5 and 6. Costs listed are for each tray section or fitting installed. The crew is two electricians working at a labor cost of $26.88 per manhour. These costs include layout, material handling, and normal waste. Add for supports, sales tax, delivery, supervision, mobilization, demobilization, cleanup, overhead and profit. Note: Covers are available for cable tray to keep dust away from the conductors. Order special extra-wide sections or reducing sections when changing form one width to another. Be sure to use the right size supports. Cable tray can be suspended from single support hangers. The hanger attaches to the center of the tray. A stiffener bar goes under the tray and the rod hanger extends up through the tray to the building structure above. If you use single hangers, be sure the hanger is sized for the tray and cable load. Spacing of hangers can't exceed 10' and must be closer for higher carrying capacities. The manufacturer's catalog will list approved spacing and other fittings that may be needed for special applications.

Aluminum Ladder Type Cable Tray

Material	Craft@Hrs	Unit	Material Cost	Labor Cost	Installed Cost
12' long aluminum ladder cable tray					
12" wide	2E@1.00	Ea	177.00	26.90	203.90
18" wide	2E@1.25	Ea	187.00	33.60	220.60
24" wide	2E@1.50	Ea	203.00	40.30	243.30
12" radius elbows for aluminum ladder cable tray					
12" 45 degree, flat	2E@0.40	Ea	107.00	10.80	117.80
18" 45 degree, flat	2E@0.50	Ea	113.00	13.40	126.40
24" 45 degree, flat	2E@0.60	Ea	120.00	16.10	136.10
12" 90 degree, flat	2E@0.50	Ea	111.00	13.40	124.40
18" 90 degree, flat	2E@0.60	Ea	131.00	16.10	147.10
24" 90 degree, flat	2E@0.75	Ea	167.00	20.20	187.20
12" 90 degree, vert.	2E@0.50	Ea	103.00	13.40	116.40
18" 90 degree, vert.	2E@0.60	Ea	106.00	16.10	122.10
24" 90 degree, vert.	2E@0.75	Ea	108.00	20.20	128.20
12" radius tees for aluminum ladder cable tray					
12"	2E@0.70	Ea	224.00	18.80	242.80
18"	2E@0.90	Ea	239.00	24.20	263.20
24"	2E@1.20	Ea	243.00	32.30	275.30
Panel or box adapters for aluminum ladder cable tray					
12"	2E@0.30	Ea	14.00	8.06	22.06
18"	2E@0.30	Ea	14.00	8.06	22.06
24"	2E@0.30	Ea	14.00	8.06	22.06
End caps for aluminum ladder cable tray					
12"	2E@0.10	Ea	20.00	2.69	22.69
18"	2E@0.15	Ea	20.00	4.03	24.03
24"	2E@0.20	Ea	20.00	5.38	25.38

Use these figures to estimate the cost of cable tray and fittings installed in bus duct under the conditions described on pages 5 and 6. Costs listed are for each tray section or fitting installed. The crew is two electricians working at a labor cost of $26.88 per manhour. These costs include layout, material handling, and normal waste. Add for supports, sales tax, delivery, supervision, mobilization, demobilization, cleanup, overhead and profit. Note: Covers are available for cable tray to keep dust away from the conductors. Order special extra-wide sections or reducing sections when changing form one width to another. Be sure to use the right size supports. Cable tray can be suspended from single support hangers. The hanger attaches to the center of the tray. A stiffener bar goes under the tray and the rod hanger extends up through the tray to the building structure above. If you use single hangers, be sure the hanger is sized for the tray and cable load. Spacing of hangers can't exceed 10' and must be closer for higher carrying capacities. The manufacturer's catalog will list approved spacing and other fittings that may be needed for special applications.

Aluminum Ladder Cable Tray Fittings

Material	Craft@Hrs	Unit	Material Cost	Labor Cost	Installed Cost
Dropouts for aluminum ladder cable tray					
12"	2E@0.25	Ea	14.40	6.72	21.12
18"	2E@0.30	Ea	19.80	8.06	27.86
24"	2E@0.35	Ea	22.10	9.41	31.51
Joiners for aluminum ladder cable tray					
Standard, pairs	2E@0.30	Ea	15.40	8.06	23.46
Adjustable, flat	2E@0.40	Ea	57.30	10.80	68.10
Adjustable, vert.	2E@0.30	Ea	27.60	8.06	35.66
Hangers in pairs for aluminum ladder cable tray					
Hangers, pairs	2E@0.50	Ea	11.00	13.40	24.40

Use these figures to estimate the cost of cable tray and fittings installed in bus duct under the conditions described on pages 5 and 6. Costs listed are for each tray section or fitting installed. The crew is two electricians working at a labor cost of $26.88 per manhour. These costs include layout, material handling, and normal waste. Add for supports, sales tax, delivery, supervision, mobilization, demobilization, cleanup, overhead and profit. Note: Covers are available for cable tray to keep dust away from the conductors. Order special extra-wide sections or reducing sections when changing form one width to another. Be sure to use the right size supports. Cable tray can be suspended from single support hangers. The hanger attaches to the center of the tray. A stiffener bar goes under the tray and the rod hanger extends up through the tray to the building structure above. If you use single hangers, be sure the hanger is sized for the tray and cable load. Spacing of hangers can't exceed 10' and must be closer for higher carrying capacities. The manufacturer's catalog will list approved spacing and other fittings that may be needed for special applications.

Section 10:
Signal Systems

Most buildings have some type of signal system. It could be anything from a simple bell circuit to a complex building management and control system. Many of the more sophisticated systems are installed by specialists. If you don't want to tackle the work yourself, get a quote from a company that handles the work required. If the project includes a sophisticated signal system, the designers probably wrote a separate specification section that covers signal work. Detach that portion of the plans and specs and send it out for bid.

Buildings with high-tech management systems are sometimes called "smart" buildings. Building management systems are designed to control all energy consumption: heating, cooling, ventilation, and lighting. Some management systems can detect when a room isn't being used and will shut off the lights and adjust the room temperature to conserve energy. Devices in the ceiling or in the lighting fixtures can sense the amount of lighting needed throughout the work day. When daylight is available, artificial lighting is reduced. Other sensors control the demand for electrical service during peak consumption hours by shedding nonessential loads.

Building management systems usually have security circuits and sensors to detect unauthorized entry. Security devices may include video cameras, door and window alarms, coded entry stations, automatic parking entry control, building emergency evacuation systems, standby electrical generator units, standby battery-powered centers, and uninterruptable power supplies, to name a few.

Bells

Bells are the most common signal device. In fact, door bells have been in common use longer than electric lighting. School jobs nearly always include bell systems to signal the beginning and ending of class periods.

Bells are intended for either indoor or outdoor use and may be either single stroke or the vibrating type. Low-voltage bells are powered by a small transformer that's activated by pressing a push button. Low-voltage bell wiring uses twisted pair conductors. No conduit is needed unless the conductors go through masonry or concrete.

Buzzers

Buzzer wiring is like bell wiring. Because buzzers are usually smaller than bells, the entire buzzer may be mounted in a flush single-gang box. Buzzers are made in a variety of sizes and voltages. Some are adjustable so the sound can be controlled.

Sirens

Sirens are warning devices and emergency alarms. For example, a siren would be required in a noisy engine room if an automatic halon or carbon dioxide fire-extinguishing system is used.

Sirens are made in many sizes and voltages and are made either for indoor or outdoor use.

Horns

Horns are speakers used for amplification, usually of the human voice. For example, a horn might be required as part of a fire-protection system to signal occupants to evacuate a building. Horns are also used in paging and public address systems.

Horns are designed for either indoor or outdoor use. Some have grilles. Others have double projection cones for use in corridors where the sound has to be aimed in two directions.

Beacons

Beacons provide a visual signal. The light can be steady, flashing, or rotating to give the effect of a flashing light. Beacons are required where it's important to draw attention to some hazard or to mark the boundary of a hazard.

Beacons are designed either for indoor or outdoor use. Navigation hazard beacons usually flash at intervals that are set by rules of navigation.

Chimes

Some chimes can produce several bars of music. The most common application is in place of a door bell.

Chimes are designed either for indoor or outdoor use. They're made in several sizes and voltages and may have an adjustable volume level.

Push Buttons

Push buttons operate signal devices. They can be either surface or flush mounted with either a plain or decorative exterior.

Note that push buttons are rated by voltage, from simple residential types to heavy-duty industrial models. Some push buttons have lighted buttons. Even color-coded buttons are available. Some heavy-duty push buttons are rated as oiltight.

Signal Transformers

Every low-voltage signal device needs a small transformer to reduce the line voltage. Signal transformers are usually the dry type and are intended for continuous service.

Signal transformers are made in many sizes and voltages. Some have multiple taps so you can select the voltage that's needed.

Clocks

Commercial wall clocks are available for use with most of the standard voltages. Some are centrally operated. A main program control panel transmits impulses that advance all clocks in the system, keeping all clocks synchronized. All clocks in the system have to be connected to the central control panel.

Where precise control of many clocks is needed, a frequency generator can be used to superimpose a precise frequency on a building's electrical supply. That way every clock in the building is synchronized with the control center's impulse frequency — without any special wiring.

Smoke Detectors

Some fire alarm systems use both smoke detectors and manual pull stations. In an office, detectors mount in the drop ceiling cavity. The code also requires smoke detectors in some larger air conditioning units. Detectors can be wired so they both sound alarms and shut down the equipment that they monitor.

There are ionization detectors, fixed-temperature detectors and rate-of-rise detectors. The local fire marshal will explain which type is needed for your building. You'll probably have to submit plans for approval by the fire-protection authority if smoke or fire detectors are required.

Intrusion Detectors

There's an intrusion detection device made to meet every need. Some are very simple. Others have silent alarms, video filming, general alarms, visual or audible alarms and extensive communication ability.

Some of the more common intrusion detecting devices are motion detectors, magnetic window or door switches, floor mats with pressure switches built in, door releases, door openers, door closers and cameras.

Many larger apartment buildings have entry control systems. Each apartment has a microphone and speaker that permits communication with selected other stations. Occupants can respond to someone at the entry without leaving their living unit and without permitting entry.

Material		Craft@Hrs	Unit	Material Cost	Labor Cost	Installed Cost
Bells						
4"	12 VAC	1E@0.25	Ea	56.80	6.72	63.52
4"	16 VAC	1E@0.25	Ea	56.80	6.72	63.52
4"	18 VAC	1E@0.30	Ea	56.80	8.06	64.86
4	24 VAC	1E@0.30	Ea	56.80	8.06	64.86
4"	120 VAC	1E@0.30	Ea	56.80	8.06	64.86
4"	240 VAC	1E@0.30	Ea	56.80	8.06	64.86
6"	12 VAC	1E@0.30	Ea	67.70	8.06	75.76
6"	16 VAC	1E@0.35	Ea	67.70	9.41	77.11
6"	18 VAC	1E@0.35	Ea	67.70	9.41	77.11
6"	24 VAC	1E@0.35	Ea	67.70	9.41	77.11
6"	120 VAC	1E@0.35	Ea	67.70	9.41	77.11
6"	240 VAC	1E@0.45	Ea	67.70	12.10	79.80
10"	120 VAC	1E@0.45	Ea	79.80	12.10	91.90
10"	240 VAC	1E@0.45	Ea	79.80	12.10	91.90
Single stroke bells						
2-1/2"	24 VAC	1E@0.30	Ea	14.00	8.06	22.06
2-1/2"	120 VAC	1E@0.30	Ea	15.10	8.06	23.16
4"	24 VAC	1E@0.35	Ea	59.90	9.41	69.31
4"	120 VAC	1E@0.35	Ea	59.90	9.41	69.31
6"	24 VAC	1E@0.40	Ea	71.80	10.80	82.60
6"	120 VAC	1E@0.40	Ea	71.80	10.80	82.60
10"	24 VAC	1E@0.45	Ea	77.90	12.10	90.00
10"	120 VAC	1E@0.45	Ea	77.90	12.10	90.00
Surface mounted buzzers						
8 VAC	71.5 db	1E@0.30	Ea	12.80	8.06	20.86
24 VAC	71.5 db	1E@0.30	Ea	12.80	8.06	20.86
8 VAC	74 db	1E@0.30	Ea	12.80	8.06	20.86
24 VAC	74 db	1E@0.30	Ea	12.80	8.06	20.86
8 VAC	75 db	1E@0.30	Ea	13.10	8.06	21.16
24 VAC	75 db	1E@0.30	Ea	13.10	8.06	21.16
24 VAC	80.5 db	1E@0.30	Ea	14.00	8.06	22.06
Sirens						
12 VAC		1E@0.50	Ea	279.00	13.40	292.40
24 VAC		1E@0.50	Ea	279.00	13.40	292.40
120 VAC		1E@0.50	Ea	279.00	13.40	292.40
240 VAC		1E@0.50	Ea	296.00	13.40	309.40
12 VDC		1E@0.50	Ea	279.00	13.40	292.40
24 VDC		1E@0.50	Ea	279.00	13.40	292.40
125 VDC		1E@0.50	Ea	279.00	13.40	292.40
250 VDC		1E@0.50	Ea	296.00	13.40	309.40

Use these figures to estimate the cost of signal devices installed in buildings under the conditions described on pages 5 and 6. Costs listed are for each device installed. The crew is one electrician working at a labor cost of $26.88 per manhour. These costs include mounting screws, layout, material handling, and normal waste. Add for the outlet box, supports, sales tax, delivery, supervision, mobilization, demobilization, cleanup, overhead and profit.

Sirens and Horns

Material	Craft@Hrs	Unit	Material Cost	Labor Cost	Installed Cost
Horn sirens, explosion proof					
12 VAC or DC	1E@0.50	Ea	771.00	13.40	784.40
24 VAC or DC	1E@0.50	Ea	771.00	13.40	784.40
120 VAC or 125 VDC	1E@0.50	Ea	771.00	13.40	784.40
240 VAC	1E@0.50	Ea	771.00	13.40	784.40
250 VDC	1E@0.50	Ea	771.00	13.40	784.40
Horns, AC					
24 VAC, flush	1E@0.35	Ea	55.20	9.41	64.61
120 VAC, flush	1E@0.35	Ea	55.20	9.41	64.61
240 VAC, flush	1E@0.35	Ea	56.80	9.41	66.21
24 VAC, projector	1E@0.40	Ea	56.80	10.80	67.60
120 VAC, projector	1E@0.40	Ea	56.80	10.80	67.60
240 VAC, projector	1E@0.40	Ea	58.30	10.80	69.10
24 VAC, grille	1E@0.40	Ea	47.80	10.80	58.60
120 VAC, grille	1E@0.40	Ea	47.80	10.80	58.60
240 VAC, grille	1E@0.40	Ea	49.10	10.80	59.90
24 VAC, weatherproof	1E@0.40	Ea	60.90	10.80	71.70
120 VAC, weatherproof	1E@0.40	Ea	60.90	10.80	71.70
240 VAC, weatherproof	1E@0.40	Ea	62.50	10.80	73.30
Horns, DC					
24 VDC, flush	1E@0.40	Ea	77.60	10.80	88.40
125 VDC, flush	1E@0.40	Ea	77.60	10.80	88.40
24 VDC, projector	1E@0.40	Ea	79.20	10.80	90.00
125 VDC, projector	1E@0.40	Ea	79.20	10.80	90.00
12 VDC, grille	1E@0.40	Ea	68.10	10.80	78.90
24 VDC, grille	1E@0.40	Ea	68.10	10.80	78.90
32 VDC, grille	1E@0.40	Ea	68.10	10.80	78.90
125 VDC, grille	1E@0.40	Ea	68.10	10.80	78.90
250 VDC, grille	1E@0.40	Ea	69.80	10.80	80.60

Use these figures to estimate the cost of signal devices installed in buildings under the conditions described on pages 5 and 6. Costs listed are for each device installed. The crew is one electrician working at a labor cost of $26.88 per manhour. These costs include mounting screws, layout, material handling, and normal waste. Add for the outlet box, supports, sales tax, delivery, supervision, mobilization, demobilization, cleanup, overhead and profit.

Material	Craft@Hrs	Unit	Material Cost	Labor Cost	Installed Cost
Beacons					
Amber, 12VDC	1E@0.50	Ea	87.70	13.40	101.10
Blue, 12VDC	1E@0.50	Ea	87.60	13.40	101.00
Red, 12VDC	1E@0.50	Ea	87.70	13.40	101.10
Clear, 12VDC	1E@0.50	Ea	87.60	13.40	101.00
Flashing beacons					
Amber, 120VAC	1E@0.50	Ea	103.00	13.40	116.40
Blue, 120VAC	1E@0.50	Ea	103.00	13.40	116.40
Red, 120VAC	1E@0.50	Ea	103.00	13.40	116.40
Clear, 120VAC	1E@0.50	Ea	103.00	13.40	116.40
Rotating beacons					
Amber, 120VAC	1E@0.50	Ea	116.00	13.40	129.40
Blue, 120VAC	1E@0.50	Ea	116.00	13.40	129.40
Red, 120VAC	1E@0.50	Ea	116.00	13.40	129.40
Clear, 120VAC	1E@0.50	Ea	116.00	13.40	129.40
Amber, 12VDC	1E@0.50	Ea	86.80	13.40	100.20
Blue, 12VDC	1E@0.50	Ea	86.80	13.40	100.20
Red, 12VDC	1E@0.50	Ea	86.80	13.40	100.20
Clear, 12VDC	1E@0.50	Ea	86.80	13.40	100.20
Chimes					
One entrance, white	1E@0.40	Ea	12.10	10.80	22.90
One entrance, beige	1E@0.40	Ea	12.10	10.80	22.90
Two entrance, white	1E@0.40	Ea	13.40	10.80	24.20
Two entrance, beige	1E@0.40	Ea	13.40	10.80	24.20
Single stroke chimes					
24 VAC	1E@0.40	Ea	105.00	10.80	115.80
120 VAC	1E@0.40	Ea	105.00	10.80	115.80
24 VDC	1E@0.40	Ea	98.00	10.80	108.80
125 VDC	1E@0.40	Ea	98.00	10.80	108.80

Use these figures to estimate the cost of signal devices installed in buildings under the conditions described on pages 5 and 6. Costs listed are for each device installed. The crew is one electrician working at a labor cost of $26.88 per manhour. These costs include mounting screws, layout, material handling, and normal waste. Add for the outlet box, supports, sales tax, delivery, supervision, mobilization, demobilization, cleanup, overhead and profit.

Signal Systems

Material	Craft@Hrs	Unit	Material Cost	Labor Cost	Installed Cost
Corridor dome lights for signal systems					
24 VAC, 1-lamp	1E@0.40	Ea	25.00	10.80	35.80
120 VAC, 1-lamp	1E@0.40	Ea	26.80	10.80	37.60
24 VAC, 2-lamp	1E@0.40	Ea	32.10	10.80	42.90
120 VAC, 2-lamp	1E@0.40	Ea	32.10	10.80	42.90
Surface mounted push buttons for signal systems					
Round, plain	1E@0.20	Ea	2.08	5.38	7.46
Round, gold	1E@0.20	Ea	2.08	5.38	7.46
Oblong, brown	1E@0.20	Ea	1.94	5.38	7.32
Oblong, ivory	1E@0.20	Ea	1.94	5.38	7.32
Flush mounted push buttons for signal systems					
5/8" chrome	1E@0.25	Ea	1.88	6.72	8.60
5/8" brass	1E@0.25	Ea	1.88	6.72	8.60
5/8" lighted, plain	1E@0.25	Ea	3.96	6.72	10.68
5/8" lighted, brass	1E@0.25	Ea	3.96	6.72	10.68
Signal transformers					
5W 120/10V	1E@0.30	Ea	8.43	8.06	16.49
5W 240/10V	1E@0.30	Ea	11.80	8.06	19.86
10W 120/16V	1E@0.30	Ea	9.38	8.06	17.44
15W 120/6-12-18V	1E@0.30	Ea	10.80	8.06	18.86
20W 120/8-16-24V	1E@0.35	Ea	12.10	9.41	21.51
20W 240/8-16-24V	1E@0.35	Ea	13.80	9.41	23.21
30W 120/8-16-24V	1E@0.35	Ea	14.00	9.41	23.41
30W 240/8-16-24V	1E@0.35	Ea	16.90	9.41	26.31
40W 120/24V	1E@0.35	Ea	14.90	9.41	24.31
Adapter plate	1E@0.10	Ea	4.49	2.69	7.18
Commercial grade clocks					
8" dia. silver	1E@0.50	Ea	33.80	13.40	47.20
8" dia. brown	1E@0.50	Ea	33.80	13.40	47.20
12" dia. silver	1E@0.50	Ea	35.60	13.40	49.00
12" dia. brown	1E@0.50	Ea	35.60	13.40	49.00
15" dia. silver	1E@0.50	Ea	79.30	13.40	92.70
15" dia. brown	1E@0.50	Ea	79.30	13.40	92.70

Use these figures to estimate the cost of signal devices installed in buildings under the conditions described on pages 5 and 6. Costs listed are for each device installed. The crew is one electrician working at a labor cost of $26.88 per manhour. These costs include mounting screws, layout, material handling, and normal waste. Add for the outlet box, supports, sales tax, delivery, supervision, mobilization, demobilization, cleanup, overhead and profit.

Material	Craft@Hrs	Unit	Material Cost	Labor Cost	Installed Cost
Battery type smoke detectors					
No LED indicator	1E@0.30	Ea	13.40	8.06	21.46
Blinking red LED	1E@0.30	Ea	13.80	8.06	21.86
Flashing yellow light	1E@0.30	Ea	28.00	8.06	36.06
120 volt smoke detectors					
Continuous red LED	1E@0.50	Ea	13.00	13.40	26.40
Multi or single station, 120 volt, smoke detectors					
Up to 6 units	1E@0.50	Ea	13.10	13.40	26.50
Up to 12 units	1E@0.50	Ea	14.10	13.40	27.50
Temperature or rate-of-rise heat detectors					
Rise heat, 135 degree	1E@0.50	Ea	10.60	13.40	24.00
Rise heat, 200 degree	1E@0.50	Ea	10.60	13.40	24.00
Fixed temp, 135 degree	1E@0.50	Ea	6.87	13.40	20.27
Fixed temp, 190 degree	1E@0.50	Ea	6.87	13.40	20.27
Intrusion detectors					
Magnetic detector	1E@0.30	Ea	4.02	8.06	12.08
Floor mat, 17" x 23"	1E@0.25	Ea	32.90	6.72	39.62
Roller/plunger det	1E@0.25	Ea	7.13	6.72	13.85
Door switch	1E@0.25	Ea	8.03	6.72	14.75
Door opener, 3-6VDC	1E@0.50	Ea	30.20	13.40	43.60
Door opener, 16VAC	1E@0.50	Ea	30.20	13.40	43.60
Door opener, 24VAC	1E@0.50	Ea	37.40	13.40	50.80
Door opener, 24VDC	1E@0.50	Ea	40.10	13.40	53.50
Door trip	1E@0.25	Ea	10.20	6.72	16.92
Desk push button	1E@0.25	Ea	16.90	6.72	23.62
Door switch, closed cir.	1E@0.40	Ea	8.14	10.80	18.94

Use these figures to estimate the cost of signal devices installed in buildings under the conditions described on pages 5 and 6. Costs listed are for each device installed. The crew is one electrician working at a labor cost of $26.88 per manhour. These costs include mounting screws, layout, material handling, and normal waste. Add for the outlet box, supports, sales tax, delivery, supervision, mobilization, demobilization, cleanup, overhead and profit.

Apartment Entry Control

Material	Craft@Hrs	Unit	Material Cost	Labor Cost	Installed Cost
Apartment entry control with speakers and flush mounted push buttons					
4 buttons, brass	1E@0.75	Ea	71.70	20.20	91.90
4 buttons, silver	1E@0.75	Ea	71.70	20.20	91.90
6 buttons, brass	1E@1.00	Ea	78.90	26.90	105.80
6 buttons, silver	1E@1.00	Ea	78.90	26.90	105.80
8 buttons, brass	1E@1.25	Ea	86.20	33.60	119.80
8 buttons, silver	1E@1.25	Ea	86.20	33.60	119.80
12 buttons, brass	1E@1.50	Ea	100.00	40.30	140.30
12 buttons, silver	1E@1.50	Ea	100.00	40.30	140.30
16 buttons, brass	1E@1.75	Ea	114.00	47.00	161.00
16 buttons, silver	1E@1.75	Ea	114.00	47.00	161.00
20 buttons, brass	1E@2.00	Ea	129.00	53.80	182.80
20 buttons, silver	1E@2.00	Ea	129.00	53.80	182.80
Apartment speakers	1E@0.30	Ea	49.40	8.06	57.46
Entry release switch	1E@0.20	Ea	47.50	5.38	52.88
Timed release switch	1E@0.20	Ea	40.10	5.38	45.48

Use these figures to estimate the cost of signal devices installed in buildings under the conditions described on pages 5 and 6. Costs listed are for each device installed. The crew is one electrician working at a labor cost of $26.88 per manhour. These costs include mounting screws, layout, material handling, and normal waste. Add for outlet box, supports, sales tax, delivery, supervision, mobilization, demobilization, cleanup, overhead and profit.

Section 11:
Precast Concrete Access Boxes

This section covers precast concrete utility boxes, handholes, pull boxes and manholes. These boxes provide access to underground electrical cable during construction and after installation is complete. Of course, it's possible to form and pour your own access boxes. But it's nearly always cheaper to buy a precast box if a box of the right size and shape is available.

Note that the boxes listed in this section are typical sizes. Your local electric company probably dictates the exact size required. Be sure the box you price will meet job specs.

Access boxes usually (but not always) come with the cover that's required. Covers may be precast concrete, flat metal or cast iron. The location of the precast box dictates the type of cover to use. A concrete cover is adequate if it's placed away from vehicle traffic. Cast iron and flat steel covers are used in roadways and parking areas. Traffic covers are rated by the anticipated wheel load. Watch for these requirements in the job specs.

The steel covers are usually galvanized and have a checker-plate surface. Larger sizes intended for traffic areas have structural reinforcing under the cover.

Handholes

Handholes comes in several standard sizes. The size to use depends on the wire diameter and intended service. Most handholes are composed of a base section, a body section and a cover. Some have open bottoms. Others have a cast concrete bottom. Handholes have knockouts that are removed where conduit is to enter the box.

The cover will have a word that identifies the type of service. Standard markings are *Electric, Telephone, CATV, Street Lighting, Cathodic Protection* and *Signal*. Many suppliers can provide other wording on request.

Precast Pull Boxes and Manholes

Precast pull boxes and manholes usually come in sections. The larger sizes have a bottom section, midsection and top section. Large tunnel vaults are usually made in vertical sections that can be joined to extend the tunnel to any length.

Some specs require mastic in the joints between sections. This mastic comes in long, thin lengths that are laid in the tongue and groove of each joint. Mastic keeps moisture out of the vault.

Some specs also require an exterior waterseal. Most suppliers can provide precast sections with the waterseal already on the outer surface.

Most boxes, manholes and vaults have knockouts in the ends and sides. Placement of knockouts and the number of knockouts depends on the size of the box. Remove knockouts with a hammer.

If a duct system is installed by direct burial, attach an end bell to the first piece of duct. Insert the bell into the knockout opening and grout it into place with masonry grout.

The procedure for concrete-encased duct is different. Again, attach an end bell to the first length of duct and insert the bell in a knockout. Use a short length of wood placed inside the box as a brace to hold the bell in place. When encasing the duct, be sure to *rod* the concrete around the end bell. That makes the concrete flow completely around the end bell and creates a good seal. Sometimes a little grout is needed to completely seal the end bell in place.

Larger pull boxes, manholes and tunnels have steel ladders installed from the cover to the bottom. The ladder must be installed in inserts when the box is placed. Inserts are cast into the box by the manufacturer.

Some boxes, manholes and tunnels require cable racks for holding cable that's to be installed later on. If cable racks are needed, be sure to order boxes with inserts or flush channels for cable racks. The specs should indicate spacing of racks in the box.

Most precast boxes and vaults have sumps in the bottom. The sump may be a simple block-out for a sump pump suction hose. Some specifications require that a drain line be installed from the sump so any accumulation of water is drained away. Other specifications require a sump pump in the box.

Excavation for the box usually has to be several feet larger than the box itself. That allows room to compact backfill around the box. The amount of overexcavation depends on the compaction equipment to be used. Minimum overexcavation is usually 2 feet on each side.

The depth of the excavation is determined by the overall height of the box, plus the neck and cover. The finish ring should be at least ½ inch above the finish grade or surface. Place sand or other fine soil at the bottom of the pit so it's easy to make small final adjustments in the box position.

Most underground installers set the precast box first and then place the duct. Usually the box is completely installed and backfilled before duct trenching even begins. Duct is then run from point to point or box to box.

Precast Pad Mount Transformer Slabs

Pad mount transformers are installed on concrete slabs, either poured and finished or precast. Precast slabs are often cheaper because there is no forming, pouring or finishing required. And precast slabs come with a block-out where the transformer will be placed. That makes it easy to stub conduit up into the transformer entrance section. Pads come with inserts already installed for anchoring the transformer in place.

Many sizes of slabs are available. Some come with underground pull boxes for cable entrance to the transformer. The size of the pull box has to conform with utility company requirements.

Note that pad mounted transformers must be grounded. Grounding has to be installed before the slab is set. Your power company will describe grounding requirements for the transformer you're placing.

Delivery Costs

Delivery cost is usually a major part of the installed cost of precast boxes, handholes and slabs.

Don't settle for a price quote F.O.B. the manufacturer's yard. Get a price that includes delivery to the site.

Most precast concrete companies charge extra for setting a box or pad in place. They'll quote setting prices, usually based upon a flat rate per item. Some will only quote hourly costs.

Some vaults are too large to be set with the precast company's equipment. In that case, you'll need hoisting equipment. Some large manholes require a large motor crane to move sections from the delivery truck to the pit or trench.

Minimize setting costs with careful planning and scheduling. Be sure the pit is ready before the delivery truck arrives. Have someone meet the delivery truck and direct it to the point of installation. The area around the pit should be clear so unloading isn't obstructed.

Pricing Precast Concrete Products

Get at least three quotes for major items like large boxes and vaults. Call for quotes early enough so the yard can submit a written quotation. Some precast companies will include all of the hardware, cable racks, hooks and insulators in their quote.

Most manufacturers of precast boxes and vaults offer standard products that meet the requirements of the local power company. But box and vault sizes can vary slightly from one manufacturer to another. Protect yourself. Get written quotes for each job.

Material	Craft@Hrs	Unit	Material Cost	Labor Cost	Installed Cost
Precast concrete handholes for underground systems					
14" x 27"	2E@0.50	Ea	89.00	13.40	102.40
18" x 30"	2E@0.70	Ea	119.00	18.80	137.80
Precast concrete pull boxes					
24" x 36" x 24"	2E@1.00	Ea	522.00	26.90	548.90
24" x 36" x 36"	2E@1.25	Ea	670.00	33.60	703.60
30" x 48" x 36"	2E@1.50	Ea	816.00	40.30	856.30
36" x 72" x 24"	2E@2.25	Ea	1,490.00	60.50	1,550.50
36" x 60" x 36"	2E@2.50	Ea	1,640.00	67.20	1,707.20
48" x 48" x 36"	2E@2.50	Ea	1,800.00	67.20	1,867.20
48" x 60" x 48"	2E@2.75	Ea	2,240.00	73.90	2,313.90
Precast concrete manholes					
48" x 90" x 84"	2E@3.50	Ea	1,800.00	94.10	1,894.10
48" x 48" x 72"	2E@3.00	Ea	1,930.00	80.60	2,010.60
48" x 48" x 84"	2E@3.00	Ea	2,080.00	80.60	2,160.60
48" x 78" x 86"	2E@3.50	Ea	1,930.00	94.10	2,024.10
54" x 102" x 78"	2E@4.00	Ea	2,380.00	108.00	2,488.00
72" x 72" x 72"	2E@4.00	Ea	2,540.00	108.00	2,648.00
72" x 72" x 60"	2E@3.75	Ea	2,240.00	101.00	2,341.00
72" x 120" x 84"	2E@4.50	Ea	3,420.00	121.00	3,541.00
96" x 120" x 72"	2E@5.00	Ea	3,280.00	134.00	3,414.00
60" x 120" x 72"	2E@4.50	Ea	2,980.00	121.00	3,101.00
60" x 120" x 86"	2E@4.50	Ea	3,130.00	121.00	3,251.00
72" x 180" x 108"	2E@6.00	Ea	4,460.00	161.00	4,621.00
72" x 180" x 144"	2E@6.50	Ea	4,460.00	175.00	4,635.00
96" x 168" x 100"	2E@8.00	Ea	9,230.00	215.00	9,445.00
72" x 228" x 108"	2E@9.00	Ea	7,440.00	242.00	7,682.00
96" x 168" x 112"	2E@10.0	Ea	5,950.00	269.00	6,219.00
96" x 240" x 112"	2E@12.0	Ea	8,190.00	323.00	8,513.00
96" x 312" x 112"	2E@14.0	Ea	10,400.00	376.00	10,776.00
84" x 264" x 96"	2E@12.0	Ea	7,440.00	323.00	7,763.00
120" x 180" x 120"	2E@16.0	Ea	14,900.00	430.00	15,330.00

Use these figures to estimate the cost of precast concrete products installed under the conditions described on pages 5 and 6. Costs listed are for each item installed. The crew is two electricians working at a labor cost of $26.88 per manhour. These costs include unloading and setting, layout, material handling, and normal waste. Add for excavation, backfill, hoisting equipment (when needed), surface patching, mastic water sealing around the item set, interior hardware, grounding, sales tax, delivery, supervision, mobilization, demobilization, cleanup, overhead and profit.

Precast Concrete Products

Material	Craft@Hrs	Unit	Material Cost	Labor Cost	Installed Cost
Precast concrete manhole necking					
Concrete cone 27" dia	2E@0.30	Ea	74.30	8.06	82.36
Concrete grade ring	2E@0.30	Ea	60.90	8.06	68.96
Iron ring & cover	2E@1.00	Ea	373.00	26.90	399.90
48" square cover set	2E@1.00	Ea	1,190.00	26.90	1,216.90
Precast concrete pad mount transformer slabs					
31" x 46"	2E@0.50	Ea	97.10	13.40	110.50
32" x 38"	2E@0.50	Ea	74.30	13.40	87.70
42" x 49"	2E@0.70	Ea	74.30	18.80	93.10
44" x 46"	2E@0.70	Ea	112.00	18.80	130.80
48" x 54"	2E@1.00	Ea	149.00	26.90	175.90
57" x 76"	2E@1.25	Ea	274.00	33.60	307.60
72" x 66"	2E@1.50	Ea	373.00	40.30	413.30
73" x 75"	2E@1.50	Ea	373.00	40.30	413.30
94" x 72"	2E@1.50	Ea	408.00	40.30	448.30

Use these figures to estimate the cost of precast concrete products installed under the conditions described on pages 5 and 6. Costs listed are for each item installed. The crew is two electricians working at a labor cost of $26.88 per manhour. These costs include unloading and setting, layout, material handling, and normal waste. Add for excavation, backfill, hoisting equipment (when needed), surface patching, mastic water sealing around the item set, interior hardware, grounding, sales tax, delivery, supervision, mobilization, demobilization, cleanup, overhead and profit.

Section 12:
Equipment Hookup

This section covers the simple equipment hookups that are common on small commercial jobs. If you're handling more complex work or the hookup of heavy industrial equipment, there won't be much information here to help you. But what's found in this section covers nearly all the hookups most electrical contractors have to handle.

Equipment hookup isn't as easy as it may seem to the novice. My advice is to read every word in the specs and examine the plans very carefully any time hookup is required. Even small details can make a big difference in the labor and material cost. Study the type of equipment, connection details and the power requirements. Be sure to include a labor and material cost for each accessory. Make an allowance for difficult conditions when you anticipate problems.

Motors

The labor costs in this section for electric motor hookup assume that the motor is easily accessible, ready for wire connection and that there are no complications. Labor and material costs will be higher for any other type of work. Prices for hookup materials are assumed to be standard commodity grade. The motors are assumed to be standard frame, open or drip proof, and either single or three phase up to 480 volts.

Before connecting any electric motor, check for wiring instructions on or near the dataplate. Sometimes you'll find wiring instructions in the motor terminal box or pasted to the inside of the terminal box cover.

To make the connection, first determine what direction the motor must turn. If you're in doubt, the equipment supplier should be able to supply the answer. Look for rotation arrows around the outside of the motor case. Some types of motors are damaged if the motor turns the wrong direction. Most pumps are damaged if the rotation is wrong. The impeller inside the pump may even unscrew from the shaft.

To reverse the rotation of a standard single phase motor, open the terminal box and disconnect the line wiring. Look for two internal wires attached to the terminals where the line wires make contact. Reverse those two internal lead wires and reconnect the line wiring. That should reverse the direction of rotation.

If a single phase motor has four wires connected to the terminal posts, two on each side, reverse the top two internal lead wires and reconnect the line wiring. Single phase motors have two separate windings inside the stator. One starts the motor; the other is the power source after the motor is started. Both are energized when the motor is first turned on.

On larger single phase motors, there's a centrifugal switch built onto the rotor that energizes the starting coil until the rotation reaches a higher R.P.M. When the motor is at rest, the centrifugal switch is at rest and presses against a stationary contact device. The starting winding is connected to this stationary switch device. When the motor starts and gains speed, the centrifugal switch opens the starting circuit. The motor then runs completely on the main coil. To reverse rotation, swap the two starting leads. That starts rotation in the opposite direction. The running leads don't need to be changed.

A three phase motor is even simpler. Three line wires are connected to the motor. To reverse rotation, swap any two of the three line wires at either the motor, the disconnect switch or at the motor starter. The line wires are usually identified as phase A, B or C. Simply change A and B or B and C or A and C. Any swap will change the direction of rotation. There's no starting device in a three phase motor and there is no starting winding in the stator.

Some motors have thermal overload switches or built-in stator heaters that have to be connected also. Heaters are connected to a separate circuit that's turned on before the motor is started. Usually these connections are made at the motor terminal box.

All motors need overload protection to prevent motor damage. The overload protection is usually built in on smaller motors. Motor starters provide this protection on larger motors. The largest electric motors operate from motor control centers. That's the subject of the next section.

Manual motor starters are built so that overload heating elements can be inserted. When the current flow to the motor exceeds the design maximum, the heater element opens a bi-metallic switch to disconnect the line wiring.

Magnetic starters can also have overload heater elements. Many specifications require that all three phases be protected from overload. In that case all three line wires have to be connected to an overload device. When any one of the three overload heaters opens, the starter circuit opens and stops the motor.

Overload heater elements are sold in sizes that cover a small range of amperages. Each manufacturer offers heaters of a distinct style and shape. Generally heaters of different manufacturers are not interchangeable. To size the overload heater element, check the motor nameplate for the full load amperage rating (FLA). Then check the chart inside the starter for the corresponding heater catalog number for that FLA.

Don't install the wrong heater. If you use a heater with an FLA rating that's too low, the normal running load will activate the heater switch, dropping the starter coil out of the circuit and stopping the motor. If you install a heater with an FLA rating that's too high, an overload will probably damage the motor eventually.

There are many types of electric motors: two speed, variable speed, wound rotor, synchronous, and a variety of D.C. and special wound motors. Be sure to check the wiring diagrams before making any line hookup.

Mechanical Equipment Hookup

The mechanical equipment covered in this section is the type of equipment that you'll find in small and moderate size commercial buildings. The figures assume no special conditions or complicated control schemes. High capacity mechanical equipment will cost considerably more.

Equipment hookup can be very complex. Some mechanical equipment is controlled with electronic programmers or microprocessors. Other equipment has multiple sections that have to be interconnected either electrically or mechanically. Study the control diagrams for the equipment. Have a clear picture of what work is covered in your bid.

Most mechanical equipment controls are supplied by the mechanical contractor and installed or wired by the electrical contractor. You'll see these controls on the installation drawings and note them in the specs. Solenoid valves, stats, pressure switches and control switches are usually easy to hook up. Assume that hookup of these controls is your work until you are advised otherwise.

Kitchen Equipment

Commercial food handling equipment comes in many sizes. Smaller units are easy to connect to the power source. Larger pieces of equipment take more time, material, skill and planning. Consultants who specialize in food handling facilities usually do the complete kitchen layout. The electrical system will be specially designed to meet the needs of the equipment.

Only the more common commercial grade kitchen equipment is listed in this section. Note that costs will be higher if the plan calls for a special control system that interconnects several pieces of equipment.

Standby Engine-Generator Units

Standby engine-generators provide backup power in case of a general power failure. Larger generators are usually diesel powered. The fuel used can have an effect on the type of hookup you make. Special features and accessories on the generator can increase hookup costs considerably.

Emergency engine-generators have precise start-up controls that bring the system on line in the shortest time possible. Some engine-generators have block heaters to keep the engine oil heated — and sometimes circulating. They also have expensive transfer switches that cut in emergency power when that power is needed and available. Expect connecting the power transfer system to be fairly complex.

Engine-generators need special long-life and high capacity batteries for startup. The voltage needed determines the number of cells required. Some specifications require dual batteries. And, of course, each battery must have a charger — perhaps one of the more expensive solid state chargers.

Look for several auxiliary items in any emergency electrical system: fuel transfer pumps, cooling system heat exchanging units, oil purifiers and monitoring devices. Each will add to the cost of electrical hookup.

It's a good idea to discuss installation with the engine-generator supplier before bidding the job. Go over system requirements in detail. Make your contact as early as possible during the bidding period so your supplier can get answers from the manufacturer if necessary.

Other Electrical Hookup Items

Most commercial jobs include several miscellaneous hookup items. Some of this equipment may have been designed and built specially for the job at hand. If you don't know what type of hookup is required, get help. Experienced electrical estimators have learned not to be bashful about calling the design engineer. If he can't supply an answer, get a suggestion on where more information can be found. Usually the design engineer is both well-informed and anxious to be helpful.

Material		Craft@Hrs	Unit	Material Cost	Labor Cost	Installed Cost

Hook up single or three-phase motors to 600 volts

Material		Craft@Hrs	Unit	Material Cost	Labor Cost	Installed Cost
Fractional	HP	1E@0.50	Ea	2.16	13.40	15.56
1	HP	1E@0.75	Ea	2.78	20.20	22.98
1-1/2	HP	1E@0.75	Ea	2.78	20.20	22.98
2	HP	1E@0.75	Ea	2.78	20.20	22.98
3	HP	1E@0.75	Ea	2.78	20.20	22.98
5	HP	1E@1.00	Ea	4.09	26.90	30.99
7-1/2	HP	1E@1.00	Ea	4.09	26.90	30.99
10	HP	1E@1.25	Ea	8.04	33.60	41.64
15	HP	1E@1.25	Ea	8.04	33.60	41.64
20	HP	1E@1.50	Ea	13.70	40.30	54.00
25	HP	1E@1.50	Ea	13.70	40.30	54.00
30	HP	1E@2.00	Ea	27.20	53.80	81.00
50	HP	1E@3.00	Ea	40.90	80.60	121.50
75	HP	1E@4.00	Ea	68.10	108.00	176.10
100	HP	1E@5.00	Ea	95.20	134.00	229.20

Hook up mechanical equipment

Material	Craft@Hrs	Unit	Material Cost	Labor Cost	Installed Cost
Air compressor	1E@1.50	Ea	20.30	40.30	60.60
Air alternator	1E@1.25	Ea	13.70	33.60	47.30
Air handlers	1E@1.50	Ea	20.30	40.30	60.60
Aquastats	1E@0.50	Ea	68.10	13.40	81.50
Boilers	1E@8.00	Ea	137.00	215.00	352.00
Boiler control panels	1E@8.00	Ea	68.10	215.00	283.10
Bridge cranes	1E@16.0	Ea	171.00	430.00	601.00
Budget hoists	1E@2.00	Ea	54.40	53.80	108.20
Chiller, water	1E@24.0	Ea	409.00	645.00	1,054.00
Chiller control panels	1E@16.0	Ea	68.10	430.00	498.10
Conveyors, single dr	1E@16.0	Ea	171.00	430.00	601.00
Conveyor control panels	1E@16.0	Ea	68.10	430.00	498.10
Cooling tower, single	1E@10.0	Ea	171.00	269.00	440.00
Cooling tower, twin	1E@16.0	Ea	238.00	430.00	668.00
Duct heaters, 1-ph	1E@2.00	Ea	6.81	53.80	60.61
Duct heaters, 3-ph	1E@2.25	Ea	13.70	60.50	74.20
Elevators, small	1E@24.0	Ea	340.00	645.00	985.00
Evap coolers, small	1E@2.00	Ea	103.00	53.80	156.80
Exhaust fans, small	1E@1.50	Ea	20.30	40.30	60.60
Fan coil units	1E@2.00	Ea	20.30	53.80	74.10
Flow switches	1E@1.00	Ea	6.81	26.90	33.71
Furnaces, small	1E@2.50	Ea	34.60	67.20	101.80
Humidistat	1E@1.00	Ea	6.81	26.90	33.71

Use these figures to estimate the cost of equipment hookup installed in buildings under the conditions described on pages 5 and 6. Costs listed are for each piece of equipment connected. The crew is one electrician working at a labor cost of $26.88 per manhour. These costs include flexible connections, connectors, wire, phase and rotation identification, testing, layout, material handling, and normal waste. Add for junction boxes, safety switches, control devices, electric motors, disconnect switches, grounding, sales tax, delivery, supervision, mobilization, demobilization, cleanup, overhead and profit. Note: Some equipment items require sequence planning.

Equipment Hookup

Material	Craft@Hrs	Unit	Material Cost	Labor Cost	Installed Cost
Hook up mechanical equipment					
Mech control panel, small	1E@4.00	Ea	40.90	108.00	148.90
Modulating valves	1E@1.50	Ea	6.81	40.30	47.11
Monorail trolley, small	1E@8.00	Ea	68.10	215.00	283.10
Motorized valves	1E@1.50	Ea	6.81	40.30	47.11
Overhead door, res.	1E@2.50	Ea	40.90	67.20	108.10
Overhead door, comm.	1E@8.00	Ea	203.00	215.00	418.00
Pneumatic switches	1E@1.00	Ea	6.81	26.90	33.71
Pressure switches	1E@1.00	Ea	6.81	26.90	33.71
Pump control panels	1E@2.50	Ea	34.60	67.20	101.80
Refrigeration, small	1E@4.00	Ea	68.10	108.00	176.10
Sail switches	1E@1.25	Ea	6.81	33.60	40.41
Solenoid valves	1E@1.50	Ea	6.81	40.30	47.11
Stats, outside air	1E@1.25	Ea	6.81	33.60	40.41
Sump pumps, small	1E@2.00	Ea	13.70	53.80	67.50
Sump control panels	1E@2.50	Ea	13.70	67.20	80.90
Sump alternators	1E@1.25	Ea	6.81	33.60	40.41
Unit heaters, small	1E@2.00	Ea	35.40	53.80	89.20
Valves, 3-way	1E@1.50	Ea	6.81	40.30	47.11
Hook up kitchen equipment					
Booster water heater	1E@2.50	Ea	47.00	67.20	114.20
Dishwasher, small	1E@5.00	Ea	95.20	134.00	229.20
Elec hot tables	1E@2.00	Ea	20.30	53.80	74.10
Elec hot water tables	1E@1.50	Ea	20.30	40.30	60.60
Elec steam tables	1E@2.00	Ea	20.30	53.80	74.10
Garbage disposals	1E@1.25	Ea	13.70	33.60	47.30
Grill	1E@1.50	Ea	13.70	40.30	54.00
Grill hood fan	1E@2.00	Ea	13.70	53.80	67.50
Portable food warmer	1E@1.00	Ea	20.30	26.90	47.20
Ovens	1E@1.50	Ea	13.70	40.30	54.00
Stationary coffee urn	1E@1.25	Ea	6.81	33.60	40.41
Stationary food mixer	1E@1.25	Ea	6.81	33.60	40.41
Stationary refrig.	1E@1.50	Ea	6.81	40.30	47.11
Walk-in freezer	1E@8.00	Ea	103.00	215.00	318.00
Walk-in refrigerator	1E@8.00	Ea	103.00	215.00	318.00

Use these figures to estimate the cost of equipment hookup installed in buildings under the conditions described on pages 5 and 6. Costs listed are for each piece of equipment connected. The crew is one electrician working at a labor cost of $26.88 per manhour. These costs include flexible connections, connectors, wire, phase and rotation identification, testing, layout, material handling, and normal waste. Add for junction boxes, safety switches, control devices, electric motors, disconnect switches, grounding, sales tax, delivery, supervision, mobilization, demobilization, cleanup, overhead and profit. Note: Some equipment items require sequence planning.

Material	Craft@Hrs	Unit	Material Cost	Labor Cost	Installed Cost

Hook up standby engine-generators to 600 volts, no accessories included

Material	Craft@Hrs	Unit	Material Cost	Labor Cost	Installed Cost
10KW, single phase	1E@1.50	Ea	49.10	40.30	89.40
15KW, single phase	1E@2.00	Ea	71.20	53.80	125.00
25KW, single phase	1E@4.00	Ea	142.00	108.00	250.00
50KW, single phase	1E@6.00	Ea	213.00	161.00	374.00
75KW, single phase	1E@8.00	Ea	249.00	215.00	464.00
100KW, single phase	1E@14.0	Ea	355.00	376.00	731.00
10KW, three phase	1E@2.00	Ea	71.20	53.80	125.00
15KW, three phase	1E@2.50	Ea	107.00	67.20	174.20
25KW, three phase	1E@5.00	Ea	213.00	134.00	347.00
50KW, three phase	1E@7.00	Ea	285.00	188.00	473.00
75KW, three phase	1E@10.0	Ea	355.00	269.00	624.00
100KW, three phase	1E@16.0	Ea	498.00	430.00	928.00

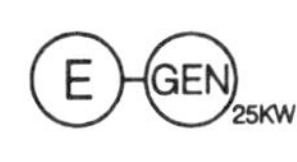

Use these figures to estimate the cost of equipment hookup installed in buildings under the conditions described on pages 5 and 6. Costs listed are for each piece of equipment connected. The crew is one electrician working at a labor cost of $26.88 per manhour. These costs include flexible connections, connectors, wire, phase and rotation identification, testing, layout, material handling, and normal waste. Add for junction boxes, safety switches, control devices, electric motors, disconnect switches, grounding, sales tax, delivery, supervision, mobilization, demobilization, cleanup, overhead and profit. Note: Some equipment items require sequence planning.

Section 13: Motor Control Equipment

Many commercial jobs include some kind of motor control equipment. It may be very simple or highly complex, with interlocking devices such as relays or contact blocks attached to magnetic starters for switching control circuits.

This section covers only some of the motor control equipment that's available. But each item listed here is representative of a larger group of motor control devices that your supplier sells.

The electrical designer will provide a schematic wiring diagram of the motor control system. It should show the control devices needed and how they're connected. It takes some experience to understand the sequence of control the designer intends. Sometimes it's hard enough just to follow a control circuit from its point of origin to its termination.

When doing a take-off of motor control equipment, start by checking the mechanical plans for mechanical equipment controls. Usually the designer will indicate which devices are furnished and installed by the electrical contractor and which are to be furnished by the mechanical contractor and installed by the electrical contractor. Some specs require that the mechanical contractor be responsible for all control systems that operate at less than 100 volts.

Some motor control systems aren't electric, although electric components may be included. For example, pneumatic controls are available for heating, ventilating and air conditioning systems. Pneumatic tubing controls thermostats, valves, pressure switches and other devices. No wiring is required. Air pressure is regulated by a pneumatic control panel.

Manual Motor Starters

Manual motor starters prevent motor overload damage. A single pole starter has a toggle switch handle that controls the **on** and **off** position. Included in the starter will be a bi-metallic switch with an overload heater element sized for the full motor load. If the current feeding the motor exceeds the value of the element, the element heats up and opens the switch. The toggle handle goes into a trip position. It has to be reset to **off** before it can be turned on again. Most overloads happen when the motor starts under stress. Some adjustment is needed to reduce that load.

Manual motor starters come with single pole, double pole and three pole contacts for various applications. Some have a pilot light which indicates when the circuit is on.

Manual motor starters are rated in the horsepower of the motor they're intended to control. Sizes range from fractional horsepower to 10 horsepower. Voltages vary from 115 to 600 volts. Your supplier's catalog will list the appropriate heater element size and the type of enclosure for each starter.

Magnetic Contactors

Magnetic contactors control high amperage loads but don't have overload relays for overcurrent protection. They're made in a wide range of sizes and are available with enclosures to meet any need. Contactors are rated by horsepower and are available in various voltages. The holding coil can be changed to accommodate any voltage needed.

Another type of contactor is the mechanically held contactor. It's operated with a momentary control switch. Once the contactor is closed, it stays closed until opened by a momentary control switch. The opening circuit is connected to another operating coil which causes the switch to jump open and release the contactor.

AC Magnetic Starters

The most common motor starter for single phase and poly-phase motors is the magnetic starter. The contacts are held in position electrically. The starter has overload relays that require overload heater elements. Size the relays to match the full load amp rating on the motor nameplate.

Enclosures for magnetic starters come in several classes: NEMA 1, NEMA 3R, NEMA 4, NEMA 4X, NEMA 7 and 9 and NEMA 12. They're made in all the common voltages and are rated by the horsepower of the motor controlled. Operating coils can be changed to match the control voltage. Most have a button on the face of the enclosure that resets the relay after it's been tripped by an overload.

Only the most common magnetic starters are listed in this section. Many more will be available from your supplier: reversing starters, reduced voltage starters, two speed starters, variable speed starters, special pump starters and multi-speed starters, among others. Each type of starter comes in a range of sizes and types appropriate to protect, control and operate the motor that's connected.

When magnetic starters are housed in a single enclosure, it's usually called a **motor control center** (MCC). These units are custom-made at the factory to meet individual job specs. The MCC's are assembled at the factory from standard modular components. The magnetic starter is usually mounted in a basket-type box which is inserted into an open module in the MCC. Power supply probes extend from the starter to the housing bus. A door assembly is then attached to enclose the starter. The door may have a reset push button, pilot lights, selector switches, operating handles and nameplates that indicate how the motor or equipment is to be controlled. Motor control centers usually stand directly on the floor.

Other devices can be installed in a MCC. Panelboards, step down transformers, circuit breakers, combination disconnect switch and magnetic starters, combination circuit breaker and magnetic starters, blank spaces and relay or control sections are all found in some motor control centers.

Control Devices

A few common control devices are listed in this section. These devices are usually assembled from standard components and installed in a single enclosure. They may use a common operating mechanism, standard contact blocks and standard enclosures. The assembly will include all of the switches needed for control purposes. Most can be ganged into a common enclosure or adapted to other enclosures, such as a motor control center or relay panel.

Manual Motor Starters Without Overload Relays

Material	Craft@Hrs	Unit	Material Cost	Labor Cost	Installed Cost
NEMA 1 manual motor starters without overload relays, toggle handle					
2P standard	1E@0.35	Ea	18.30	9.41	27.71
2P w/ pilot light, 115V	1E@0.35	Ea	37.80	9.41	47.21
2P w/ pilot light, 230V	1E@0.35	Ea	37.80	9.41	47.21
3P standard	1E@0.40	Ea	41.10	10.80	51.90
3P w/ pilot light, 240V	1E@0.40	Ea	60.80	10.80	71.60
3P w/ pilot light, 600V	1E@0.40	Ea	60.80	10.80	71.60
NEMA 1 flush mounted manual motor starters without overload relays					
2P standard	1E@0.35	Ea	16.30	9.41	25.71
2P w/ pilot light, 115V	1E@0.35	Ea	35.90	9.41	45.31
2P w/ pilot light, 230V	1E@0.35	Ea	35.90	9.41	45.31
3P standard	1E@0.40	Ea	39.20	10.80	50.00
3P w/ pilot light, 240V	1E@0.40	Ea	58.80	10.80	69.60
3P w/ pilot light, 600V	1E@0.40	Ea	58.80	10.80	69.60
NEMA 1 flush mounted manual motor starters, no overload relays, stainless cover					
2P standard	1E@0.35	Ea	18.30	9.41	27.71
2P w/ pilot light, 115V	1E@0.35	Ea	37.80	9.41	47.21
2P w/ pilot light, 230V	1E@0.35	Ea	37.80	9.41	47.21
3P standard	1E@0.40	Ea	41.10	10.80	51.90
3P w/ pilot light, 240V	1E@0.40	Ea	60.80	10.80	71.60
3P w/ pilot light, 600V	1E@0.40	Ea	60.80	10.80	71.60
NEMA 4 & 5 manual motor starters without overload relays					
2P standard	1E@0.50	Ea	86.40	13.40	99.80
2P w/ pilot light, 115V	1E@0.50	Ea	117.00	13.40	130.40
2P w/ pilot light, 230V	1E@0.50	Ea	117.00	13.40	130.40
3P standard	1E@0.60	Ea	106.00	16.10	122.10
3P w/ pilot light, 240V	1E@0.60	Ea	137.00	16.10	153.10
3P w/ pilot light, 600V	1E@0.60	Ea	137.00	16.10	153.10

Use these figures to estimate the cost of manual motor starters installed in buildings under the conditions described on pages 5 and 6. Costs listed are for each starter installed. The crew size is one electrician working at a labor cost of $26.88 per manhour. These costs include layout, material handling, and normal waste. Add for thermal overload devices, sales tax, delivery, supervision, mobilization, demobilization, cleanup, overhead and profit. Overload heating elements are not included in the cost of these manual motor starters. Size of the overload relay is based on the full motor load in amps. Check the manufacturer's overload heater chart for the correct heater size. These manual motor starters will fit into a standard single gang switch box. They are used as disconnect switches for small machinery, appliances, pumps and fans.

Material	Craft@Hrs	Unit	Material Cost	Labor Cost	Installed Cost
Key operated manual motor starters without overload relays					
2P standard	1E@0.35	Ea	26.10	9.41	35.51
2P w/ pilot light, 115V	1E@0.35	Ea	45.70	9.41	55.11
2P w/ pilot light, 230V	1E@0.35	Ea	45.70	9.41	55.11
3P standard	1E@0.40	Ea	48.90	10.80	59.70
3P w/ pilot light, 240V	1E@0.40	Ea	68.60	10.80	79.40
3P w/ pilot light, 600V	1E@0.40	Ea	68.60	10.80	79.40

STₖ

Material	Craft@Hrs	Unit	Material Cost	Labor Cost	Installed Cost
Open type manual motor starters without overload relays					
2P standard	1E@0.30	Ea	14.40	8.06	22.46
2P w/ pilot light, 115V	1E@0.30	Ea	33.80	8.06	41.86
2P w/ pilot light, 230V	1E@0.30	Ea	33.80	8.06	41.86
3P standard	1E@0.40	Ea	37.20	10.80	48.00
3P w/ pilot light, 240V	1E@0.40	Ea	56.80	10.80	67.60
3P w/ pilot light, 600V	1E@0.40	Ea	56.80	10.80	67.60

STO

Material	Craft@Hrs	Unit	Material Cost	Labor Cost	Installed Cost
Manual motor starters with overload relays					
1P standard	1E@0.30	Ea	23.60	8.06	31.66
1P w/ pilot light, 115V	1E@0.30	Ea	35.30	8.06	43.36
2P standard	1E@0.40	Ea	27.40	10.80	38.20
2P w/ pilot light, 115V	1E@0.40	Ea	39.20	10.80	50.00
2P w/ pilot light, 230V	1E@0.40	Ea	39.20	10.80	50.00

ST

Material	Craft@Hrs	Unit	Material Cost	Labor Cost	Installed Cost
Flush mounted manual motor starters with overload relays					
1P standard	1E@0.30	Ea	21.50	8.06	29.56
1P w/ pilot light, 115V	1E@0.30	Ea	33.30	8.06	41.36
2P standard	1E@0.40	Ea	25.50	10.80	36.30
2P w/ pilot light, 115V	1E@0.40	Ea	37.20	10.80	48.00
2P w/ pilot light, 230V	1E@0.40	Ea	37.20	10.80	48.00

ST

Use these figures to estimate the cost of manual motor starters installed in buildings under the conditions described on pages 5 and 6. Costs listed are for each starter installed. The crew size is one electrician working at a labor cost of $26.88 per manhour. These costs include layout, material handling, and normal waste. Add for thermal overload devices, sales tax, delivery, supervision, mobilization, demobilization, cleanup, overhead and profit. Overload heating elements are not included in the cost of these manual motor starters. Size of the overload relay is based on the full motor load in amps. Check the manufacturer's overload heater chart for the correct heater size. These manual motor starters will fit into a standard single gang switch box. They are used as disconnect switches for small machinery, appliances, pumps and fans.

Manual Motor Starters With Overload Relays

Material		Craft@Hrs	Unit	Material Cost	Labor Cost	Installed Cost
NEMA 4 & 5 horsepower rated manual motor starters with overload relays						
1P standard		1E@0.40	Ea	88.00	10.80	98.80
1P w/ pilot light, 115V		1E@0.40	Ea	120.00	10.80	130.80
2P standard		1E@0.50	Ea	91.90	13.40	105.30
2P w/ pilot light, 115V		1E@0.50	Ea	124.00	13.40	137.40
2P w/ pilot light, 230V		1E@0.50	Ea	124.00	13.40	137.40
NEMA 1 flush manual motor starters with overload relays, stainless steel cover						
1P standard		1E@0.30	Ea	22.80	8.06	30.86
1P w/ pilot light, 115V		1E@0.30	Ea	39.20	8.06	47.26
2P standard		1E@0.40	Ea	31.30	10.80	42.10
2P w/ pilot light, 115V		1E@0.40	Ea	43.20	10.80	54.00
2P w/ pilot light, 230V		1E@0.40	Ea	43.20	10.80	54.00
NEMA 7 & 9 horsepower rated manual motor starters with overload relays						
1P standard		1E@0.60	Ea	95.70	16.10	111.80
1P w/ pilot light, 115V		1E@0.60	Ea	104.00	16.10	120.10
2P standard		1E@0.70	Ea	99.50	18.80	118.30
2P w/ pilot light, 115V		1E@0.70	Ea	109.00	18.80	127.80
2P w/ pilot light, 230V		1E@0.70	Ea	109.00	18.80	127.80
Open type horsepower rated manual motor starters with overload relays						
2 pole	1 HP	1E@0.35	Ea	64.60	9.41	74.01
2 pole	2 HP	1E@0.35	Ea	84.10	9.41	93.51
2 pole	3 HP	1E@0.35	Ea	127.00	9.41	136.41
2 pole	5 HP	1E@0.35	Ea	127.00	9.41	136.41
3 pole	3 HP	1E@0.40	Ea	77.90	10.80	88.70
3 pole	5 HP	1E@0.40	Ea	77.90	10.80	88.70
3 pole	7.5 HP	1E@0.40	Ea	97.20	10.80	108.00
3 pole	10 HP	1E@0.40	Ea	97.20	10.80	108.00

ST_{4/5} — ST — ST_{7/9} — STO

Use these figures to estimate the cost of manual motor starters installed in buildings under the conditions described on pages 5 and 6. Costs listed are for each starter installed. The crew size is one electrician working at a labor cost of $26.88 per manhour. These costs include layout, material handling, and normal waste. Add for thermal overload devices, sales tax, delivery, supervision, mobilization, demobilization, cleanup, overhead and profit. Overload heating elements are not included in the cost of these manual motor starters. Size of the overload relay is based on the full motor load in amps. Check the manufacturer's overload heater chart for the correct heater size.

Manual Motor Starters With Overload Relays

Material	Craft@Hrs	Unit	Material Cost	Labor Cost	Installed Cost

NEMA 1 manual motor starters with overload relays

Material		Craft@Hrs	Unit	Material Cost	Labor Cost	Installed Cost
2 pole	1 HP	1E@0.40	Ea	72.50	10.80	83.30
2 pole	2 HP	1E@0.40	Ea	91.90	10.80	102.70
2 pole	3 HP	1E@0.40	Ea	135.00	10.80	145.80
2 pole	5 HP	1E@0.40	Ea	135.00	10.80	145.80
3 pole	3 HP	1E@0.45	Ea	86.40	12.10	98.50
3 pole	5 HP	1E@0.45	Ea	86.40	12.10	98.50
3 pole	7.5 HP	1E@0.45	Ea	106.00	12.10	118.10
3 pole	10 HP	1E@0.45	Ea	106.00	12.10	118.10

ST

NEMA 4 manual motor starters with overload relays

Material		Craft@Hrs	Unit	Material Cost	Labor Cost	Installed Cost
2 pole	1 HP	1E@0.35	Ea	197.00	9.41	206.41
2 pole	2 HP	1E@0.35	Ea	245.00	9.41	254.41
2 pole	3 HP	1E@0.35	Ea	300.00	9.41	309.41
2 pole	5 HP	1E@0.35	Ea	300.00	9.41	309.41
3 pole	3 HP	1E@0.40	Ea	212.00	10.80	222.80
3 pole	5 HP	1E@0.40	Ea	212.00	10.80	222.80
3 pole	7.5 HP	1E@0.40	Ea	259.00	10.80	269.80
3 pole	10 HP	1E@0.40	Ea	259.00	10.80	269.80

ST_4

NEMA 4X manual motor starters with overload relays

Material		Craft@Hrs	Unit	Material Cost	Labor Cost	Installed Cost
2 pole	1 HP	1E@0.40	Ea	197.00	10.80	207.80
2 pole	2 HP	1E@0.40	Ea	245.00	10.80	255.80
2 pole	3 HP	1E@0.40	Ea	245.00	10.80	255.80
2 pole	5 HP	1E@0.40	Ea	300.00	10.80	310.80
3 pole	3 HP	1E@0.45	Ea	300.00	12.10	312.10
3 pole	5 HP	1E@0.45	Ea	212.00	12.10	224.10
3 pole	7.5 HP	1E@0.45	Ea	259.00	12.10	271.10
3 pole	10 HP	1E@0.45	Ea	259.00	12.10	271.10

ST_{4X}

NEMA 7 & 9 manual motor starters with overload relays

Material		Craft@Hrs	Unit	Material Cost	Labor Cost	Installed Cost
2 pole	1 HP	1E@0.40	Ea	277.00	10.80	287.80
2 pole	2 HP	1E@0.40	Ea	329.00	10.80	339.80
2 pole	3 HP	1E@0.40	Ea	379.00	10.80	389.80
2 pole	5 HP	1E@0.40	Ea	379.00	10.80	389.80

$ST_{7/9}$

Use these figures to estimate the cost of manual motor starters installed in buildings under the conditions described on pages 5 and 6. Costs listed are for each starter installed. The crew size is one electrician working at a labor cost of $26.88 per manhour. These costs include layout, material handling, and normal waste. Add for thermal overload devices, sales tax, delivery, supervision, mobilization, demobilization, cleanup, overhead and profit. Overload heating elements are not included in the cost of these manual motor starters. Size of the overload relay is based on the full motor load in amps. Check the manufacturer's overload heater chart for the correct heater size. These manual motor starters will fit into a standard single gang switch box. They are used as disconnect switches for small machinery, appliances, pumps and fans.

HP Rated Manual Motor Starters with Overload Relays

Material	Craft@Hrs	Unit	Material Cost	Labor Cost	Installed Cost
NEMA 7 & 9 HP rated manual motor starters with overload relays					
3 pole 3 HP	1E@0.45	Ea	292.00	12.10	304.10
3 pole 5 HP	1E@0.45	Ea	292.00	12.10	304.10
3 pole 7.5 HP	1E@0.45	Ea	344.00	12.10	356.10
3 pole 10 HP	1E@0.45	Ea	344.00	12.10	356.10
NEMA 1 reversing HP rated manual motor starters with overload relays					
3 pole 3 HP	1E@1.25	Ea	270.00	33.60	303.60
3 pole 5 HP	1E@1.25	Ea	270.00	33.60	303.60
3 pole 7.5 HP	1E@1.50	Ea	329.00	40.30	369.30
3 pole 10 HP	1E@1.50	Ea	329.00	40.30	369.30
NEMA 1 two speed HP rated manual motor starters with overload relays					
3 pole 3 HP	1E@1.00	Ea	270.00	26.90	296.90
3 pole 5 HP	1E@1.00	Ea	270.00	26.90	296.90
3 pole 7.5 HP	1E@1.25	Ea	329.00	33.60	362.60
3 pole 10 HP	1E@1.25	Ea	329.00	33.60	362.60
NEMA 1 reversing HP rated manual starters, overload relays, low voltage protection					
3 pole 3 HP, 208V	1E@1.25	Ea	356.00	33.60	389.60
3 pole 3 HP, 240V	1E@1.25	Ea	356.00	33.60	389.60
3 pole 5 HP, 480V	1E@1.25	Ea	356.00	33.60	389.60
3 pole 5 HP, 600V	1E@1.25	Ea	356.00	33.60	389.60
3 pole 7.5 HP, 208V	1E@1.50	Ea	415.00	40.30	455.30
3 pole 7.5 HP, 240V	1E@1.50	Ea	415.00	40.30	455.30
3 pole 10 HP, 480V	1E@1.50	Ea	415.00	40.30	455.30
3 pole 10 HP, 600V	1E@1.50	Ea	415.00	40.30	455.30
NEMA 1 two speed HP rated manual starters, overload relays, low voltage protection					
3 pole 3 HP, 208V	1E@1.25	Ea	356.00	33.60	389.60
3 pole 3 HP, 240V	1E@1.25	Ea	356.00	33.60	389.60
3 pole 5 HP, 480V	1E@1.25	Ea	356.00	33.60	389.60
3 pole 5 HP, 600V	1E@1.25	Ea	356.00	33.60	389.60
3 pole 7.5 HP, 208V	1E@1.50	Ea	415.00	40.30	455.30
3 pole 7.5 HP, 240V	1E@1.50	Ea	415.00	40.30	455.30
3 pole 10 HP, 480V	1E@1.50	Ea	415.00	40.30	455.30
3 pole 10 HP, 600V	1E@1.50	Ea	415.00	40.30	455.30

Use these figures to estimate the cost of manual motor starters installed in buildings under the conditions described on pages 5 and 6. Costs listed are for each starter installed. The crew size is one electrician working at a labor cost of $26.88 per manhour. These costs include layout, material handling, and normal waste. Add for thermal overload devices, sales tax, delivery, supervision, mobilization, demobilization, cleanup, overhead and profit. Overload heating elements are not included in the cost of these manual motor starters. Size of the overload relay is based on the full motor load in amps. Check the manufacturer's overload heater chart for the correct heater size.

Material	Craft@Hrs	Unit	Material Cost	Labor Cost	Installed Cost

Open type HP rated reversing two-speed manual starters, overload relays, low voltage protection

Material	Craft@Hrs	Unit	Material Cost	Labor Cost	Installed Cost
3 pole 3 HP, 208V	1E@1.00	Ea	333.00	26.90	359.90
3 pole 3 HP, 240V	1E@1.00	Ea	333.00	26.90	359.90
3 pole 5 HP, 480V	1E@1.00	Ea	333.00	26.90	359.90
3 pole 5 HP, 600V	1E@1.00	Ea	333.00	26.90	359.90
3 pole 7.5 HP, 208V	1E@1.25	Ea	392.00	33.60	425.60
3 pole 7.5 HP, 240V	1E@1.25	Ea	392.00	33.60	425.60
3 pole 10 HP, 480V	1E@1.25	Ea	392.00	33.60	425.60
3 pole 10 HP, 600V	1E@1.25	Ea	392.00	33.60	425.60

STO

NEMA 1 two pole AC magnetic contactors without overload relays

Material	Craft@Hrs	Unit	Material Cost	Labor Cost	Installed Cost
Size 00 1/3 HP, 115V	1E@0.50	Ea	112.00	13.40	125.40
Size 00 1 HP, 230V	1E@0.50	Ea	112.00	13.40	125.40
Size 0 1 HP, 115V	1E@0.60	Ea	144.00	16.10	160.10
Size 0 2 HP, 230V	1E@0.60	Ea	144.00	16.10	160.10
Size 1 2 HP, 115V	1E@0.75	Ea	168.00	20.20	188.20
Size 1 3 HP, 230V	1E@0.75	Ea	168.00	20.20	188.20
Size 2 3 HP, 115V	1E@1.00	Ea	343.00	26.90	369.90
Size 2 7.5 HP, 230V	1E@1.00	Ea	343.00	26.90	369.90
Size 3 7.5 HP, 115V	1E@1.25	Ea	565.00	33.60	598.60
Size 3 15 HP, 230V	1E@1.25	Ea	565.00	33.60	598.60

0 C

NEMA 4 two pole AC magnetic contactors without overload relays

Material	Craft@Hrs	Unit	Material Cost	Labor Cost	Installed Cost
Size 0 1 HP, 115V	1E@1.00	Ea	318.00	26.90	344.90
Size 0 2 HP, 230V	1E@1.00	Ea	318.00	26.90	344.90
Size 1 2 HP, 115V	1E@1.25	Ea	350.00	33.60	383.60
Size 1 3 HP, 230V	1E@1.25	Ea	350.00	33.60	383.60
Size 2 3 HP, 115V	1E@1.50	Ea	707.00	40.30	747.30
Size 2 7.5 HP, 230V	1E@1.50	Ea	707.00	40.30	747.30
Size 3 7.5 HP, 115V	1E@1.75	Ea	1,080.00	47.00	1,127.00
Size 3 15 HP, 230V	1E@1.75	Ea	1,080.00	47.00	1,127.00

1 C 4

Use these figures to estimate the cost of manual motor starters and magnetic contactors installed in buildings under the conditions described on pages 5 and 6. Costs listed are for each starter or contactor installed. The crew size is one electrician working at a labor cost of $26.88 per manhour. These costs include layout, material handling, and normal waste. Add for supports, sales tax, delivery, supervision, mobilization, demobilization, cleanup, overhead and profit. Magnetic contactors do not provide overload protection. They are used as a relay device only.

Two Pole AC Magnetic Contactors Without Overload Relays

Material	Craft@Hrs	Unit	Material Cost	Labor Cost	Installed Cost
NEMA 4X two-pole AC magnetic contactors without overload relays					
Size 0 1 HP, 115V	1E@1.00	Ea	354.00	26.90	380.90
Size 0 2 HP, 230V	1E@1.00	Ea	354.00	26.90	380.90
Size 1 2 HP, 115V	1E@1.25	Ea	389.00	33.60	422.60
Size 1 3 HP, 230V	1E@1.25	Ea	389.00	33.60	422.60
Size 2 3 HP, 115V	1E@1.50	Ea	785.00	40.30	825.30
Size 2 7.5 HP, 230V	1E@1.50	Ea	785.00	40.30	825.30
NEMA 7 & 9 two-pole AC magnetic contactors without overload relays					
Size 0 1 HP, 115V	1E@1.25	Ea	793.00	33.60	826.60
Size 0 2 HP, 230V	1E@1.25	Ea	793.00	33.60	826.60
Size 1 2 HP, 115V	1E@1.50	Ea	823.00	40.30	863.30
Size 1 3 HP, 230V	1E@1.50	Ea	823.00	40.30	863.30
Size 2 3 HP, 115V	1E@1.75	Ea	1,330.00	47.00	1,377.00
Size 2 7.5 HP, 230V	1E@1.75	Ea	1,330.00	47.00	1,377.00
Size 3 7.5 HP, 115V	1E@2.00	Ea	2,000.00	53.80	2,053.80
Size 3 15 HP, 230V	1E@2.00	Ea	2,000.00	53.80	2,053.80
NEMA 12 two-pole AC magnetic contactors without overload relays					
Size 0 1 HP, 115V	1E@0.50	Ea	226.00	13.40	239.40
Size 0 2 HP, 230V	1E@0.50	Ea	226.00	13.40	239.40
Size 1 2 HP, 115V	1E@0.60	Ea	254.00	16.10	270.10
Size 1 3 HP, 230V	1E@0.60	Ea	254.00	16.10	270.10
Size 2 3 HP, 115V	1E@0.75	Ea	505.00	20.20	525.20
Size 2 7.5 HP, 230V	1E@0.75	Ea	505.00	20.20	525.20
Size 3 7.5 HP, 115V	1E@1.00	Ea	773.00	26.90	799.90
Size 3 15 HP, 230V	1E@1.00	Ea	773.00	26.90	799.90
Open type two-pole AC magnetic contactors without overload relays					
Size 00 1/3 HP, 115V	1E@0.40	Ea	111.00	10.80	121.80
Size 00 1 HP, 230V	1E@0.40	Ea	111.00	10.80	121.80
Size 0 1 HP, 115V	1E@0.50	Ea	146.00	13.40	159.40
Size 0 2 HP, 230V	1E@0.50	Ea	146.00	13.40	159.40

Use these figures to estimate the cost of magnetic contactors installed in buildings under the conditions described on pages 5 and 6. Costs listed are for each contactor installed. The crew size is one electrician working at a labor cost of $26.88 per manhour. These costs include layout, material handling, and normal waste. Add for supports, sales tax, delivery, supervision, mobilization, demobilization, cleanup, overhead and profit. These magnetic contactors do not provide overload protection. They are used as a relay device only. For motor overload protection, use magnetic starters with overload relays and heating elements. Other types of magnetic contactors are available for incandescent lamp loads and induction loads.

AC Magnetic Contactors Without Overload Relays

Material	Craft@Hrs	Unit	Material Cost	Labor Cost	Installed Cost

Two pole open type AC magnetic contactors without overload relays

Material	Craft@Hrs	Unit	Material Cost	Labor Cost	Installed Cost
Size 1 2 HP, 115V	1E@0.60	Ea	174.00	16.10	190.10
Size 1 3 HP, 230V	1E@0.60	Ea	174.00	16.10	190.10
Size 2 3 HP, 115V	1E@0.70	Ea	324.00	18.80	342.80
Size 2 7.5 HP, 230V	1E@0.70	Ea	324.00	18.80	342.80
Size 3 7.5 HP, 115V	1E@0.90	Ea	515.00	24.20	539.20
Size 3 15 HP, 230V	1E@0.90	Ea	515.00	24.20	539.20

1 C

NEMA 1 three pole AC magnetic contactors without overload relays

Material	Craft@Hrs	Unit	Material Cost	Labor Cost	Installed Cost
Size 00 1.5 HP, 208V	1E@0.60	Ea	142.00	16.10	158.10
Size 00 2 HP, 480V	1E@0.60	Ea	142.00	16.10	158.10
Size 0 3 HP, 208V	1E@0.60	Ea	175.00	16.10	191.10
Size 0 5 HP, 480V	1E@0.60	Ea	175.00	16.10	191.10
Size 1 7.5 HP, 208V	1E@0.70	Ea	203.00	18.80	221.80
Size 1 10 HP, 480V	1E@0.70	Ea	203.00	18.80	221.80
Size 2 10 HP, 208V	1E@1.00	Ea	404.00	26.90	430.90
Size 2 25 HP, 480V	1E@1.00	Ea	404.00	26.90	430.90
Size 3 25 HP, 208V	1E@1.25	Ea	674.00	33.60	707.60
Size 3 50 HP, 480V	1E@1.25	Ea	674.00	33.60	707.60
Size 4 40 HP, 208V	1E@2.00	Ea	1,590.00	53.80	1,643.80
Size 4 100 HP, 480V	1E@2.00	Ea	1,590.00	53.80	1,643.80
Size 5 75 HP, 208V	1E@4.00	Ea	3,360.00	108.00	3,468.00
Size 5 200 HP, 480V	1E@4.00	Ea	3,360.00	108.00	3,468.00
Size 6 150 HP, 208V	1E@6.00	Ea	9,880.00	161.00	10,041.00
Size 6 400 HP, 480V	1E@6.00	Ea	9,880.00	161.00	10,041.00
Size 7 300 HP, 208V	1E@8.00	Ea	13,300.00	215.00	13,515.00
Size 7 600 HP, 480V	1E@8.00	Ea	13,300.00	215.00	13,515.00

2 C

NEMA 4 three pole AC magnetic contactors without overload relays

Material	Craft@Hrs	Unit	Material Cost	Labor Cost	Installed Cost
Size 0 3 HP, 208V	1E@1.25	Ea	372.00	33.60	405.60
Size 0 5 HP, 480V	1E@1.25	Ea	372.00	33.60	405.60
Size 1 7.5 HP, 208V	1E@1.50	Ea	404.00	40.30	444.30
Size 1 10 HP, 480V	1E@1.50	Ea	404.00	40.30	444.30
Size 2 10 HP, 208V	1E@1.75	Ea	809.00	47.00	856.00
Size 2 25 HP, 480V	1E@1.75	Ea	809.00	47.00	856.00
Size 3 25 HP, 208V	1E@2.00	Ea	1,250.00	53.80	1,303.80
Size 3 50 HP, 480V	1E@2.00	Ea	1,250.00	53.80	1,303.80
Size 4 40 HP, 208V	1E@3.00	Ea	2,550.00	80.60	2,630.60
Size 4 100 HP, 480V	1E@3.00	Ea	2,550.00	80.60	2,630.60
Size 5 75 HP, 208V	1E@6.00	Ea	4,590.00	161.00	4,751.00
Size 5 200 HP, 480V	1E@6.00	Ea	4,590.00	161.00	4,751.00

3 C 4

Use these figures to estimate the cost of magnetic contactors installed in buildings under the conditions described on pages 5 and 6. Costs listed are for each contactor installed. The crew size is one electrician working at a labor cost of $26.88 per manhour. These costs include layout, material handling, and normal waste. Add for supports, sales tax, delivery, supervision, mobilization, demobilization, cleanup, overhead and profit. These magnetic contactors do not provide overload protection. They are used as a relay device only. For motor overload protection, use magnetic starters with overload relays and heating elements. Other types of magnetic contactors are available for incandescent lamp loads and induction loads.

Three Pole AC Magnetic Contactors Without Overload Relays

Material	Craft@Hrs	Unit	Material Cost	Labor Cost	Installed Cost
NEMA 4 three pole AC magnetic contactors without overload relays					
Size 6 150 HP, 208V	1E@8.00	Ea	13,300.00	215.00	13,515.00
Size 6 400 HP, 480V	1E@8.00	Ea	13,300.00	215.00	13,515.00
Size 7 300 HP, 208V	1E@12.0	Ea	16,100.00	323.00	16,423.00
Size 7 600 HP, 480V	1E@12.0	Ea	16,100.00	323.00	16,423.00
NEMA 4X three pole AC magnetic contactors without overload relays					
Size 0 3 HP, 208V	1E@1.25	Ea	372.00	33.60	405.60
Size 0 5 HP, 480V	1E@1.25	Ea	372.00	33.60	405.60
Size 1 7.5 HP, 208V	1E@1.50	Ea	404.00	40.30	444.30
Size 1 10 HP, 480V	1E@1.50	Ea	404.00	40.30	444.30
Size 2 10 HP, 208V	1E@1.75	Ea	809.00	47.00	856.00
Size 2 25 HP, 480V	1E@1.75	Ea	809.00	47.00	856.00
Size 3 25 HP, 208V	1E@2.00	Ea	1,550.00	53.80	1,603.80
Size 3 50 HP, 480V	1E@2.00	Ea	1,550.00	53.80	1,603.80
Size 4 40 HP, 208V	1E@3.00	Ea	3,200.00	80.60	3,280.60
Size 4 100 HP, 480V	1E@3.00	Ea	3,200.00	80.60	3,280.60
NEMA 7 & 9 three pole AC magnetic contactors without overload relays					
Size 0 3 HP, 208V	1E@1.50	Ea	813.00	40.30	853.30
Size 0 5 HP, 480V	1E@1.50	Ea	813.00	40.30	853.30
Size 1 7.5 HP, 208V	1E@1.75	Ea	849.00	47.00	896.00
Size 1 10 HP, 480V	1E@1.75	Ea	849.00	47.00	896.00
Size 2 10 HP, 208V	1E@2.00	Ea	1,370.00	53.80	1,423.80
Size 2 25 HP, 480V	1E@2.00	Ea	1,370.00	53.80	1,423.80
Size 3 25 HP, 208V	1E@3.00	Ea	2,040.00	80.60	2,120.60
Size 3 50 HP, 480V	1E@3.00	Ea	2,040.00	80.60	2,120.60
Size 4 40 HP, 208V	1E@4.00	Ea	3,310.00	108.00	3,418.00
Size 4 100 HP, 480V	1E@4.00	Ea	3,310.00	108.00	3,418.00
Size 5 75 HP, 208V	1E@8.00	Ea	7,280.00	215.00	7,495.00
Size 5 200 HP, 480V	1E@8.00	Ea	7,280.00	215.00	7,495.00
Size 6 150 HP, 208V	1E@12.0	Ea	20,800.00	323.00	21,123.00
Size 6 400 HP, 480V	1E@12.0	Ea	20,800.00	323.00	21,123.00

Use these figures to estimate the cost of magnetic contactors installed in buildings under the conditions described on pages 5 and 6. Costs listed are for each contactor installed. The crew size is one electrician working at a labor cost of $26.88 per manhour. These costs include layout, material handling, and normal waste. Add for supports, sales tax, delivery, supervision, mobilization, demobilization, cleanup, overhead and profit. These magnetic contactors do not provide overload protection. They are used as a relay device only. For motor overload protection, use magnetic starters with overload relays and heating elements. Other types of magnetic contactors are available for incandescent lamp loads and induction loads.

Three Pole AC Magnetic Contactors Without Overload Relays

Material	Craft@Hrs	Unit	Material Cost	Labor Cost	Installed Cost
NEMA 12 three pole AC magnetic contactors without overload relays					
Size 0 3 HP, 208V	1E@0.70	Ea	243.00	18.80	261.80
Size 0 5 HP, 480V	1E@0.70	Ea	243.00	18.80	261.80
Size 1 7.5 HP, 208V	1E@0.90	Ea	270.00	24.20	294.20
Size 1 10 HP, 480V	1E@0.90	Ea	270.00	24.20	294.20
Size 2 10 HP, 208V	1E@1.00	Ea	527.00	26.90	553.90
Size 2 25 HP, 480V	1E@1.00	Ea	527.00	26.90	553.90
Size 3 25 HP, 208V	1E@1.25	Ea	818.00	33.60	851.60
Size 3 50 HP, 480V	1E@1.25	Ea	818.00	33.60	851.60
Size 4 40 HP, 208V	1E@2.00	Ea	2,050.00	53.80	2,103.80
Size 4 100 HP, 480V	1E@2.00	Ea	2,050.00	53.80	2,103.80
Size 5 75 HP, 208V	1E@4.00	Ea	4,590.00	108.00	4,698.00
Size 5 200 HP, 480V	1E@4.00	Ea	4,590.00	108.00	4,698.00
Size 6 150 HP, 208V	1E@6.00	Ea	11,300.00	161.00	11,461.00
Size 6 400 HP, 480V	1E@6.00	Ea	11,300.00	161.00	11,461.00
Size 7 300 HP, 208V	1E@8.00	Ea	14,900.00	215.00	15,115.00
Size 7 600 HP, 480V	1E@8.00	Ea	14,900.00	215.00	15,115.00

0 C 12

Material	Craft@Hrs	Unit	Material Cost	Labor Cost	Installed Cost
Open type three pole AC magnetic contactors without overload relays					
Size 00 1.5 HP, 208V	1E@0.50	Ea	129.00	13.40	142.40
Size 00 2 HP, 480V	1E@0.50	Ea	129.00	13.40	142.40
Size 0 3 HP, 208V	1E@0.50	Ea	162.00	13.40	175.40
Size 0 5 HP, 480V	1E@0.50	Ea	162.00	13.40	175.40
Size 1 7.5 HP, 208V	1E@0.60	Ea	190.00	16.10	206.10
Size 1 10 HP, 480V	1E@0.60	Ea	190.00	16.10	206.10
Size 2 10 HP, 208V	1E@0.75	Ea	347.00	20.20	367.20
Size 2 25 HP, 480V	1E@0.75	Ea	347.00	20.20	367.20
Size 3 25 HP, 208V	1E@1.00	Ea	559.00	26.90	585.90
Size 3 50 HP, 480V	1E@1.00	Ea	559.00	26.90	585.90
Size 4 40 HP, 208V	1E@1.75	Ea	1,340.00	47.00	1,387.00
Size 4 100 HP, 480V	1E@1.75	Ea	1,340.00	47.00	1,387.00
Size 5 75 HP, 208V	1E@3.50	Ea	2,930.00	94.10	3,024.10
Size 5 200 HP, 480V	1E@3.50	Ea	2,930.00	94.10	3,024.10
Size 6 150 HP, 208V	1E@5.50	Ea	7,980.00	148.00	8,128.00
Size 6 400 HP, 480V	1E@5.50	Ea	7,980.00	148.00	8,128.00
Size 7 300 HP, 208V	1E@7.50	Ea	11,300.00	202.00	11,502.00
Size 7 600 HP, 480V	1E@7.50	Ea	11,300.00	202.00	11,502.00

I C

Use these figures to estimate the cost of magnetic contactors installed in buildings under the conditions described on pages 5 and 6. Costs listed are for each contactor installed. The crew size is one electrician working at a labor cost of $26.88 per manhour. These costs include layout, material handling, and normal waste. Add for supports, sales tax, delivery, supervision, mobilization, demobilization, cleanup, overhead and profit. These magnetic contactors do not provide overload protection. They are used as a relay device only. For motor overload protection, use magnetic starters with overload relays and heating elements. Other types of magnetic contactors are available for incandescent lamp loads and induction loads.

Four Pole AC Magnetic Contactors Without Overload Relays

Material		Craft@Hrs	Unit	Material Cost	Labor Cost	Installed Cost

NEMA 1 four pole AC magnetic contactors without overload relays

Material		Craft@Hrs	Unit	Material Cost	Labor Cost	Installed Cost
Size 0	3 HP, 208V	1E@0.70	Ea	219.00	18.80	237.80
Size 0	5 HP, 480V	1E@0.70	Ea	219.00	18.80	237.80
Size 1	7.5 HP, 208V	1E@0.90	Ea	248.00	24.20	272.20
Size 1	10 HP, 480V	1E@0.90	Ea	248.00	24.20	272.20
Size 2	10 HP, 208V	1E@1.25	Ea	505.00	33.60	538.60
Size 2	15 HP, 230V	1E@1.25	Ea	505.00	33.60	538.60
Size 2	25 HP, 480V	1E@1.25	Ea	505.00	33.60	538.60
Size 3	25 HP, 208V	1E@1.50	Ea	830.00	40.30	870.30
Size 3	30 HP, 230V	1E@1.50	Ea	830.00	40.30	870.30
Size 3	50 HP, 480V	1E@1.50	Ea	830.00	40.30	870.30
Size 4	40 HP, 208V	1E@1.75	Ea	2,110.00	47.00	2,157.00
Size 4	75 HP, 230V	1E@1.75	Ea	2,110.00	47.00	2,157.00
Size 4	100 HP, 480V	1E@1.75	Ea	2,110.00	47.00	2,157.00

NEMA 4 four pole AC magnetic contactors without overload relays

Material		Craft@Hrs	Unit	Material Cost	Labor Cost	Installed Cost
Size 0	3 HP, 208V	1E@1.50	Ea	421.00	40.30	461.30
Size 0	5 HP, 480V	1E@1.50	Ea	421.00	40.30	461.30
Size 1	7.5 HP, 208V	1E@1.75	Ea	448.00	47.00	495.00
Size 1	10 HP, 480V	1E@1.75	Ea	448.00	47.00	495.00
Size 2	10 HP, 208V	1E@2.00	Ea	1,060.00	53.80	1,113.80
Size 2	15 HP, 230V	1E@2.00	Ea	1,060.00	53.80	1,113.80
Size 2	25 HP, 480V	1E@2.00	Ea	1,060.00	53.80	1,113.80
Size 3	25 HP, 208V	1E@2.25	Ea	1,550.00	60.50	1,610.50
Size 3	30 HP, 230V	1E@2.25	Ea	1,550.00	60.50	1,610.50
Size 3	50 HP, 480V	1E@2.25	Ea	1,550.00	60.50	1,610.50
Size 4	40 HP, 208V	1E@2.75	Ea	3,480.00	73.90	3,553.90
Size 4	75 HP, 230V	1E@2.75	Ea	3,480.00	73.90	3,553.90
Size 4	100 HP, 480V	1E@2.75	Ea	3,480.00	73.90	3,553.90

NEMA 4x four pole AC magnetic contactors without overload relays

Material		Craft@Hrs	Unit	Material Cost	Labor Cost	Installed Cost
Size 0	3 HP, 208V	1E@1.50	Ea	421.00	40.30	461.30
Size 0	5 HP, 480V	1E@1.50	Ea	421.00	40.30	461.30
Size 1	7.5 HP, 208V	1E@1.75	Ea	448.00	47.00	495.00
Size 1	10 HP, 480V	1E@1.75	Ea	448.00	47.00	495.00
Size 2	10 HP, 208V	1E@2.00	Ea	1,060.00	53.80	1,113.80
Size 2	15 HP, 230V	1E@2.00	Ea	1,060.00	53.80	1,113.80
Size 2	25 HP, 480V	1E@2.00	Ea	1,060.00	53.80	1,113.80

Use these figures to estimate the cost of magnetic contactors installed in buildings under the conditions described on pages 5 and 6. Costs listed are for each contactor installed. The crew size is one electrician working at a labor cost of $26.88 per manhour. These costs include layout, material handling, and normal waste. Add for supports, sales tax, delivery, supervision, mobilization, demobilization, cleanup, overhead and profit. These magnetic contactors do not provide overload protection. They are used as a relay device only. For motor overload protection, use magnetic starters with overload relays and heating elements. Other types of magnetic contactors are available for incandescent lamp loads and induction loads.

Four Pole AC Magnetic Contactors Without Overload Relays

Material		Craft@Hrs	Unit	Material Cost	Labor Cost	Installed Cost

NEMA 7 & 9 four pole AC magnetic contactors without overload relays

Material		Craft@Hrs	Unit	Material Cost	Labor Cost	Installed Cost
Size 0	3 HP, 208V	1E@1.75	Ea	863.00	47.00	910.00
Size 0	5 HP, 480V	1E@1.75	Ea	863.00	47.00	910.00
Size 1	7.5 HP, 208V	1E@2.00	Ea	899.00	53.80	952.80
Size 1	10 HP, 480V	1E@2.00	Ea	899.00	53.80	952.80
Size 2	10 HP, 208V	1E@2.00	Ea	1,640.00	53.80	1,693.80
Size 2	15 HP, 230V	1E@2.25	Ea	1,640.00	60.50	1,700.50
Size 2	25 HP, 480V	1E@2.25	Ea	1,640.00	60.50	1,700.50
Size 3	25 HP, 208V	1E@2.50	Ea	2,380.00	67.20	2,447.20
Size 3	30 HP, 230V	1E@2.50	Ea	2,380.00	67.20	2,447.20
Size 3	50 HP, 480V	1E@2.50	Ea	2,380.00	67.20	2,447.20
Size 4	40 HP, 208V	1E@3.00	Ea	4,480.00	80.60	4,560.60
Size 4	75 HP, 230V	1E@3.00	Ea	4,480.00	80.60	4,560.60
Size 4	100 HP, 480V	1E@3.00	Ea	4,480.00	80.60	4,560.60

NEMA 12 four pole AC magnetic contactors without overload relays

Material		Craft@Hrs	Unit	Material Cost	Labor Cost	Installed Cost
Size 0	3 HP, 208V	1E@0.75	Ea	287.00	20.20	307.20
Size 0	5 HP, 480V	1E@0.75	Ea	287.00	20.20	307.20
Size 1	7.5 HP, 208V	1E@1.00	Ea	314.00	26.90	340.90
Size 1	10 HP, 480V	1E@1.00	Ea	314.00	26.90	340.90
Size 2	10 HP, 208V	1E@1.30	Ea	627.00	34.90	661.90
Size 2	15 HP, 230V	1E@1.30	Ea	627.00	34.90	661.90
Size 2	25 HP, 480V	1E@1.30	Ea	627.00	34.90	661.90
Size 3	25 HP, 208V	1E@1.60	Ea	974.00	43.00	1,017.00
Size 3	30 HP, 230V	1E@1.60	Ea	974.00	43.00	1,017.00
Size 3	50 HP, 480V	1E@1.60	Ea	974.00	43.00	1,017.00
Size 4	40 HP, 208V	1E@3.00	Ea	2,760.00	80.60	2,840.60
Size 4	75 HP, 230V	1E@3.00	Ea	2,760.00	80.60	2,840.60
Size 4	100 HP, 480V	1E@3.00	Ea	2,760.00	80.60	2,840.60

Use these figures to estimate the cost of magnetic contactors installed in buildings under the conditions described on pages 5 and 6. Costs listed are for each contactor installed. The crew size is one electrician working at a labor cost of $26.88 per manhour. These costs include layout, material handling, and normal waste. Add for supports, sales tax, delivery, supervision, mobilization, demobilization, cleanup, overhead and profit. These magnetic contactors do not provide overload protection. They are used as a relay device only. For motor overload protection, use magnetic starters with overload relays and heating elements. Other types of magnetic contactors are available for incandescent lamp loads and induction loads.

AC Magnetic Contactors Without Overload Relays

Material		Craft@Hrs	Unit	Material Cost	Labor Cost	Installed Cost

Open type four pole AC magnetic contactors without overload relays

Material		Craft@Hrs	Unit	Material Cost	Labor Cost	Installed Cost
Size 0	3 HP, 208V	1E@0.60	Ea	206.00	16.10	222.10
Size 0	5 HP, 480V	1E@0.60	Ea	206.00	16.10	222.10
Size 1	7.5 HP, 208V	1E@0.75	Ea	235.00	20.20	255.20
Size 1	10 HP, 480V	1E@0.75	Ea	235.00	20.20	255.20
Size 2	10 HP, 208V	1E@1.00	Ea	447.00	26.90	473.90
Size 2	15 HP, 230V	1E@1.00	Ea	447.00	26.90	473.90
Size 2	25 HP, 480V	1E@1.00	Ea	447.00	26.90	473.90
Size 3	25 HP, 208V	1E@1.25	Ea	715.00	33.60	748.60
Size 3	30 HP, 230V	1E@1.25	Ea	715.00	33.60	748.60
Size 3	50 HP, 480V	1E@1.25	Ea	715.00	33.60	748.60
Size 4	40 HP, 208V	1E@1.50	Ea	1,860.00	40.30	1,900.30
Size 4	75 HP, 230V	1E@1.50	Ea	1,860.00	40.30	1,900.30
Size 4	100 HP, 480V	1E@1.50	Ea	1,860.00	40.30	1,900.30

NEMA 1 five pole AC magnetic contactors without overload relays

Material		Craft@Hrs	Unit	Material Cost	Labor Cost	Installed Cost
Size 0	3 HP, 208V	1E@1.00	Ea	281.00	26.90	307.90
Size 0	5 HP, 480V	1E@1.00	Ea	281.00	26.90	307.90
Size 1	7.5 HP, 208V	1E@1.25	Ea	310.00	33.60	343.60
Size 1	10 HP, 480V	1E@1.25	Ea	310.00	33.60	343.60
Size 2	10 HP, 208V	1E@1.50	Ea	729.00	40.30	769.30
Size 2	15 HP, 230V	1E@1.50	Ea	729.00	40.30	769.30
Size 2	25 HP, 480V	1E@1.50	Ea	729.00	40.30	769.30
Size 3	25 HP, 208V	1E@1.75	Ea	1,190.00	47.00	1,237.00
Size 3	30 HP, 230V	1E@1.75	Ea	1,190.00	47.00	1,237.00
Size 3	50 HP, 480V	1E@1.75	Ea	1,190.00	47.00	1,237.00
Size 4	40 HP, 208V	1E@2.00	Ea	2,810.00	53.80	2,863.80
Size 4	75 HP, 230V	1E@2.00	Ea	2,810.00	53.80	2,863.80
Size 4	100 HP, 480V	1E@2.00	Ea	2,810.00	53.80	2,863.80

NEMA 4 five pole AC magnetic contactors without overload relays

Material		Craft@Hrs	Unit	Material Cost	Labor Cost	Installed Cost
Size 0	3 HP, 208V	1E@1.75	Ea	483.00	47.00	530.00
Size 0	5 HP, 480V	1E@1.75	Ea	483.00	47.00	530.00
Size 1	7.5 HP, 208V	1E@2.00	Ea	510.00	53.80	563.80
Size 1	10 HP, 480V	1E@2.00	Ea	510.00	53.80	563.80
Size 2	10 HP, 208V	1E@2.25	Ea	1,290.00	60.50	1,350.50
Size 2	15 HP, 230V	1E@2.25	Ea	1,290.00	60.50	1,350.50
Size 2	25 HP, 480V	1E@2.25	Ea	1,290.00	60.50	1,350.50
Size 3	25 HP, 208V	1E@2.50	Ea	1,920.00	67.20	1,987.20
Size 3	30 HP, 230V	1E@2.50	Ea	1,920.00	67.20	1,987.20
Size 3	50 HP, 480V	1E@2.50	Ea	1,920.00	67.20	1,987.20
Size 4	40 HP, 208V	1E@2.75	Ea	4,200.00	73.90	4,273.90
Size 4	75 HP, 230V	1E@2.75	Ea	4,200.00	73.90	4,273.90
Size 4	100 HP, 480V	1E@2.75	Ea	4,200.00	73.90	4,273.90

Use these figures to estimate the cost of magnetic contactors installed in buildings under the conditions described on pages 5 and 6. Costs listed are for each contactor installed. The crew size is one electrician working at a labor cost of $26.88 per manhour. These costs include layout, material handling, and normal waste. Add for supports, sales tax, delivery, supervision, mobilization, demobilization, cleanup, overhead and profit. These magnetic contactors do not provide overload protection. They are used as a relay device only. For motor overload protection, use magnetic starters with overload relays and heating elements. Other types of magnetic contactors are available for incandescent lamp loads and induction loads.

Five Pole AC Magnetic Contactors Without Overload Relays

Material		Craft@Hrs	Unit	Material Cost	Labor Cost	Installed Cost
Open type five pole AC magnetic contactors without overload relays						
Size 0	3 HP, 208V	1E@0.75	Ea	269.00	20.20	289.20
Size 0	5 HP, 480V	1E@0.75	Ea	269.00	20.20	289.20
Size 1	7.5 HP, 208V	1E@1.00	Ea	296.00	26.90	322.90
Size 1	10 HP, 480V	1E@1.00	Ea	296.00	26.90	322.90
Size 2	10 HP, 208V	1E@1.25	Ea	670.00	33.60	703.60
Size 2	15 HP, 230V	1E@1.25	Ea	670.00	33.60	703.60
Size 2	25 HP, 480V	1E@1.25	Ea	670.00	33.60	703.60
Size 3	25 HP, 208V	1E@1.50	Ea	1,070.00	40.30	1,110.30
Size 3	30 HP, 230V	1E@1.50	Ea	1,070.00	40.30	1,110.30
Size 3	50 HP, 480V	1E@1.50	Ea	1,070.00	40.30	1,110.30
Size 4	40 HP, 208V	1E@2.00	Ea	2,590.00	53.80	2,643.80
Size 4	75 HP, 230V	1E@2.00	Ea	2,590.00	53.80	2,643.80
Size 4	100 HP, 480V	1E@2.00	Ea	2,590.00	53.80	2,643.80

Use these figures to estimate the cost of magnetic contactors installed in buildings under the conditions described on pages 5 and 6. Costs listed are for each contactor installed. The crew size is one electrician working at a labor cost of $26.88 per manhour. These costs include layout, material handling, and normal waste. Add for supports, sales tax, delivery, supervision, mobilization, demobilization, cleanup, overhead and profit. These magnetic contactors do not provide overload protection. They are used as a relay device only. For motor overload protection, use magnetic starters with overload relays and heating elements. Other types of magnetic contactors are available for incandescent lamp loads and induction loads.

AC Magnetic Two Pole Starters with Overload Relays

Material		Craft@Hrs	Unit	Material Cost	Labor Cost	Installed Cost
NEMA 1 two pole AC magnetic starters with overload relays						
Size 00	1/3 HP, 115V	1E@0.60	Ea	147.00	16.10	163.10
Size 00	1 HP, 230V	1E@0.60	Ea	147.00	16.10	163.10
Size 0	1 HP, 115V	1E@0.70	Ea	168.00	18.80	186.80
Size 0	2 HP, 230V	1E@0.70	Ea	168.00	18.80	186.80
Size 1	2 HP, 115V	1E@0.80	Ea	191.00	21.50	212.50
Size 1	3 HP, 230V	1E@0.80	Ea	191.00	21.50	212.50
Size 2	3 HP, 115V	1E@1.10	Ea	245.00	29.60	274.60
Size 2	7.5 HP, 230V	1E@1.10	Ea	245.00	29.60	274.60
Size 3	7.5 HP, 115V	1E@1.30	Ea	376.00	34.90	410.90
Size 3	15 HP, 230V	1E@1.30	Ea	376.00	34.90	410.90
NEMA 3R two pole AC magnetic starters with overload relays						
Size 00	1/3 HP, 115V	1E@0.75	Ea	226.00	20.20	246.20
Size 00	1 HP, 230V	1E@0.75	Ea	226.00	20.20	246.20
Size 0	1 HP, 115V	1E@0.75	Ea	226.00	20.20	246.20
Size 0	2 HP, 230V	1E@0.75	Ea	226.00	20.20	246.20
Size 1	2 HP, 115V	1E@1.00	Ea	251.00	26.90	277.90
Size 1	3 HP, 230V	1E@1.00	Ea	251.00	26.90	277.90
Size 2	3 HP, 115V	1E@1.25	Ea	306.00	33.60	339.60
Size 2	7.5 HP, 230V	1E@1.25	Ea	306.00	33.60	339.60
Size 3	7.5 HP, 115V	1E@1.50	Ea	487.00	40.30	527.30
Size 3	15 HP, 230V	1E@1.50	Ea	487.00	40.30	527.30
NEMA 4 two pole AC magnetic starters with overload relays						
Size 00	1/3 HP, 115V	1E@1.00	Ea	342.00	26.90	368.90
Size 00	1 HP, 230V	1E@1.00	Ea	342.00	26.90	368.90
Size 0	1 HP, 115V	1E@1.00	Ea	342.00	26.90	368.90
Size 0	2 HP, 230V	1E@1.00	Ea	342.00	26.90	368.90
Size 1	2 HP, 115V	1E@1.25	Ea	373.00	33.60	406.60
Size 1	3 HP, 230V	1E@1.25	Ea	373.00	33.60	406.60
Size 2	3 HP, 115V	1E@1.50	Ea	426.00	40.30	466.30
Size 2	7.5 HP, 230V	1E@1.50	Ea	426.00	40.30	466.30
Size 3	7.5 HP, 115V	1E@1.75	Ea	740.00	47.00	787.00
Size 3	15 HP, 230V	1E@1.75	Ea	740.00	47.00	787.00

Use these figures to estimate the cost of magnetic starters installed in buildings under the conditions described on pages 5 and 6. Costs listed are for each starter installed. The crew size is one electrician working at a labor cost of $26.88 per manhour. These costs include layout, material handling, and normal waste. Add for thermal overload devices, supports, sales tax, delivery, supervision, mobilization, demobilization, cleanup, overhead and profit. Note: Select the right size for load and voltage. To select the right overload relay element, find the full load amps rating opposite the appropriate voltage on the motor nameplate. Then check the manufacturer's overload element chart for the correct element or heater. Heater elements are inserted into the overload relay. If the running current exceeds the full load amperage rating, the heater element heats a bi-metallic switch which opens the control circuit and interrupts power to the motor.

AC Magnetic Two Pole Starters with Overload Relays

Material		Craft@Hrs	Unit	Material Cost	Labor Cost	Installed Cost

NEMA 4X two pole AC magnetic starters with overload relays

Material		Craft@Hrs	Unit	Material Cost	Labor Cost	Installed Cost
Size 00	1/3 HP, 115V	1E@1.00	Ea	342.00	26.90	368.90
Size 00	1 HP, 230V	1E@1.00	Ea	342.00	26.90	368.90
Size 0	1 HP, 115V	1E@1.00	Ea	342.00	26.90	368.90
Size 0	2 HP, 230V	1E@1.00	Ea	342.00	26.90	368.90
Size 1	2 HP, 115V	1E@1.25	Ea	373.00	33.60	406.60
Size 1	3 HP, 230V	1E@1.25	Ea	373.00	33.60	406.60
Size 2	3 HP, 115V	1E@1.50	Ea	426.00	40.30	466.30
Size 2	7.5 HP, 230V	1E@1.50	Ea	426.00	40.30	466.30
Size 3	7.5 HP, 115V	1E@1.75	Ea	740.00	47.00	787.00
Size 3	15 HP, 230V	1E@1.75	Ea	740.00	47.00	787.00

NEMA 7 & 9 two pole AC magnetic starters with overload relays

Material		Craft@Hrs	Unit	Material Cost	Labor Cost	Installed Cost
Size 00	1/3 HP, 115V	1E@1.25	Ea	740.00	33.60	773.60
Size 00	1 HP, 230V	1E@1.25	Ea	740.00	33.60	773.60
Size 0	1 HP, 115V	1E@1.25	Ea	740.00	33.60	773.60
Size 0	2 HP, 230V	1E@1.25	Ea	740.00	33.60	773.60
Size 1	2 HP, 115V	1E@1.50	Ea	771.00	40.30	811.30
Size 1	3 HP, 230V	1E@1.50	Ea	771.00	40.30	811.30
Size 2	3 HP, 115V	1E@1.75	Ea	1,250.00	47.00	1,297.00
Size 2	7.5 HP, 230V	1E@1.75	Ea	1,250.00	47.00	1,297.00

Open type two pole AC magnetic starters with overload relays

Material		Craft@Hrs	Unit	Material Cost	Labor Cost	Installed Cost
Size 00	1/3 HP, 115V	1E@0.50	Ea	135.00	13.40	148.40
Size 00	1 HP, 230V	1E@0.50	Ea	135.00	13.40	148.40
Size 0	1 HP, 115V	1E@0.50	Ea	154.00	13.40	167.40
Size 0	2 HP, 230V	1E@0.50	Ea	154.00	13.40	167.40
Size 1	2 HP, 115V	1E@0.70	Ea	180.00	18.80	198.80
Size 1	3 HP, 230V	1E@0.70	Ea	180.00	18.80	198.80
Size 2	3 HP, 115V	1E@1.00	Ea	234.00	26.90	260.90
Size 2	7.5 HP, 230V	1E@1.00	Ea	234.00	26.90	260.90
Size 3	7.5 HP, 115V	1E@1.25	Ea	324.00	33.60	357.60
Size 3	15 HP, 230V	1E@1.25	Ea	324.00	33.60	357.60

Use these figures to estimate the cost of magnetic starters installed in buildings under the conditions described on pages 5 and 6. Costs listed are for each starter installed. The crew size is one electrician working at a labor cost of $26.88 per manhour. These costs include layout, material handling, and normal waste. Add for thermal overload devices, supports, sales tax, delivery, supervision, mobilization, demobilization, cleanup, overhead and profit. Note: Select the right size for load and voltage. To select the right overload relay element, find the full load amps rating opposite the appropriate voltage on the motor nameplate. Then check the manufacturer's overload element chart for the correct element or heater. Heater elements are inserted into the overload relay. If the running current exceeds the full load amperage rating, the heater element heats a bi-metallic switch which opens the control circuit and interrupts power to the motor.

AC Magnetic Three Pole Starters with Overload Relays

Material		Craft@Hrs	Unit	Material Cost	Labor Cost	Installed Cost

NEMA 1 three pole AC magnetic starters with overload relays

Material		Craft@Hrs	Unit	Material Cost	Labor Cost	Installed Cost
Size 00	1.5 HP, 208V	1E@0.70	Ea	147.00	18.80	165.80
Size 00	2 HP, 480V	1E@0.70	Ea	147.00	18.80	165.80
Size 0	3 HP, 208V	1E@0.80	Ea	183.00	21.50	204.50
Size 0	5 HP, 480V	1E@0.80	Ea	183.00	21.50	204.50
Size 1	7.5 HP, 208V	1E@1.00	Ea	207.00	26.90	233.90
Size 1	10 HP, 480V	1E@1.00	Ea	207.00	26.90	233.90
Size 2	10 HP, 208V	1E@1.25	Ea	411.00	33.60	444.60
Size 2	15 HP, 230V	1E@1.25	Ea	411.00	33.60	444.60
Size 2	25 HP, 480V	1E@1.25	Ea	411.00	33.60	444.60
Size 3	25 HP, 208V	1E@1.50	Ea	681.00	40.30	721.30
Size 3	30 HP, 230V	1E@1.50	Ea	681.00	40.30	721.30
Size 3	50 HP, 480V	1E@1.50	Ea	681.00	40.30	721.30
Size 4	40 HP, 208V	1E@2.25	Ea	1,540.00	60.50	1,600.50
Size 4	75 HP, 230V	1E@2.25	Ea	1,540.00	60.50	1,600.50
Size 4	100 HP, 480V	1E@2.25	Ea	1,540.00	60.50	1,600.50
Size 5	75 HP, 208V	2E@4.50	Ea	3,620.00	121.00	3,741.00
Size 5	100 HP, 230V	2E@4.50	Ea	3,620.00	121.00	3,741.00
Size 5	200 HP, 480V	2E@4.50	Ea	3,620.00	121.00	3,741.00
Size 6	150 HP, 208V	2E@6.50	Ea	10,200.00	175.00	10,375.00
Size 6	200 HP, 230V	2E@6.50	Ea	10,200.00	175.00	10,375.00
Size 6	400 HP, 480V	2E@6.50	Ea	10,200.00	175.00	10,375.00
Size 7	300 HP, 230V	2E@8.50	Ea	13,500.00	228.00	13,728.00
Size 7	600 HP, 480V	2E@8.50	Ea	13,500.00	228.00	13,728.00

NEMA 3R three pole AC magnetic starters with overload relays

Material		Craft@Hrs	Unit	Material Cost	Labor Cost	Installed Cost
Size 00	1.5 HP, 208V	1E@0.80	Ea	244.00	21.50	265.50
Size 00	2 HP, 480V	1E@0.80	Ea	244.00	21.50	265.50
Size 0	3 HP, 208V	1E@1.00	Ea	244.00	26.90	270.90
Size 0	5 HP, 480V	1E@1.00	Ea	244.00	26.90	270.90
Size 1	7.5 HP, 208V	1E@1.25	Ea	268.00	33.60	301.60
Size 1	10 HP, 480V	1E@1.25	Ea	268.00	33.60	301.60
Size 2	10 HP, 208V	1E@1.50	Ea	521.00	40.30	561.30
Size 2	15 HP, 230V	1E@1.50	Ea	521.00	40.30	561.30
Size 2	25 HP, 480V	1E@1.50	Ea	521.00	40.30	561.30
Size 3	25 HP, 208V	1E@1.75	Ea	812.00	47.00	859.00
Size 3	30 HP, 230V	1E@1.75	Ea	812.00	47.00	859.00
Size 3	50 HP, 480V	1E@1.75	Ea	812.00	47.00	859.00
Size 4	40 HP, 208V	1E@2.00	Ea	1,970.00	53.80	2,023.80
Size 4	75 HP, 230V	1E@2.00	Ea	1,970.00	53.80	2,023.80
Size 4	100 HP, 480V	1E@2.00	Ea	1,970.00	53.80	2,023.80

Use these figures to estimate the cost of magnetic starters installed in buildings under the conditions described on pages 5 and 6. Costs listed are for each starter installed. The crew size is one electrician for starters up to size 4 and two electricians for larger starters. The labor cost per manhour is $26.88. These costs include layout, material handling, and normal waste. Add for thermal overload devices, supports, sales tax, delivery, supervision, mobilization, demobilization, cleanup, overhead and profit. Note: Select the right size for load and voltage. To select the right overload relay element, find the full load amps rating opposite the appropriate voltage on the motor nameplate. Then check the manufacturer's overload element chart for the correct element or heater. Heater elements are inserted into the overload relay. If the running current exceeds the full load amperage rating, the heater element heats a bi-metallic switch which opens the control circuit and interrupts power to the motor.

AC Magnetic Three Pole Starters with Overload Relays

Material	Craft@Hrs	Unit	Material Cost	Labor Cost	Installed Cost

NEMA 4 three pole AC magnetic starters with overload relays

Material	Craft@Hrs	Unit	Material Cost	Labor Cost	Installed Cost
Size 00 1.5 HP, 208V	1E@1.25	Ea	359.00	33.60	392.60
Size 00 2 HP, 480V	1E@1.25	Ea	359.00	33.60	392.60
Size 0 3 HP, 208V	1E@1.25	Ea	359.00	33.60	392.60
Size 0 5 HP, 480V	1E@1.25	Ea	359.00	33.60	392.60
Size 1 7.5 HP, 208V	1E@1.50	Ea	390.00	40.30	430.30
Size 1 10 HP, 480V	1E@1.50	Ea	390.00	40.30	430.30
Size 2 10 HP, 208V	1E@1.75	Ea	771.00	47.00	818.00
Size 2 15 HP, 230V	1E@1.75	Ea	771.00	47.00	818.00
Size 2 25 HP, 480V	1E@1.75	Ea	771.00	47.00	818.00
Size 3 25 HP, 208V	1E@2.00	Ea	1,190.00	53.80	1,243.80
Size 3 30 HP, 230V	1E@2.00	Ea	1,190.00	53.80	1,243.80
Size 3 50 HP, 480V	1E@2.00	Ea	1,190.00	53.80	1,243.80
Size 4 40 HP, 208V	1E@2.50	Ea	2,420.00	67.20	2,487.20
Size 4 75 HP, 230V	1E@2.50	Ea	2,420.00	67.20	2,487.20
Size 4 100 HP, 480V	1E@2.50	Ea	2,420.00	67.20	2,487.20
Size 5 75 HP, 208V	2E@5.50	Ea	4,710.00	148.00	4,858.00
Size 5 100 HP, 230V	2E@5.50	Ea	4,710.00	148.00	4,858.00
Size 5 200 HP, 480V	2E@5.50	Ea	4,710.00	148.00	4,858.00
Size 6 150 HP, 208V	2E@7.00	Ea	12,700.00	188.00	12,888.00
Size 6 200 HP, 230V	2E@7.00	Ea	12,700.00	188.00	12,888.00
Size 6 400 HP, 480V	2E@7.00	Ea	12,700.00	188.00	12,888.00
Size 7 300 HP, 230V	2E@9.00	Ea	16,100.00	242.00	16,342.00
Size 7 600 HP, 480V	2E@9.00	Ea	16,100.00	242.00	16,342.00

NEMA 4X three pole AC magnetic starters with overload relays

Material	Craft@Hrs	Unit	Material Cost	Labor Cost	Installed Cost
Size 00 1.5 HP, 208V	1E@1.25	Ea	359.00	33.60	392.60
Size 00 2 HP, 480V	1E@1.25	Ea	359.00	33.60	392.60
Size 0 3 HP, 208V	1E@1.25	Ea	359.00	33.60	392.60
Size 0 5 HP, 480V	1E@1.25	Ea	359.00	33.60	392.60
Size 1 7.5 HP, 208V	1E@1.50	Ea	390.00	40.30	430.30
Size 1 10 HP, 480V	1E@1.50	Ea	390.00	40.30	430.30
Size 2 10 HP, 208V	1E@1.75	Ea	771.00	47.00	818.00
Size 2 15 HP. 230V	1E@1.75	Ea	771.00	47.00	818.00
Size 2 25 HP, 480V	1E@1.75	Ea	771.00	47.00	818.00
Size 3 25 HP, 208V	1E@2.00	Ea	1,500.00	53.80	1,553.80
Size 3 30 HP, 230V	1E@2.00	Ea	1,500.00	53.80	1,553.80
Size 3 50 HP, 480V	1E@2.00	Ea	1,500.00	53.80	1,553.80

Use these figures to estimate the cost of magnetic starters installed in buildings under the conditions described on pages 5 and 6. Costs listed are for each starter installed. The crew size is one electrician for starters up to size 4 and two electricians for larger starters. The labor cost per manhour is $26.88. These costs include layout, material handling, and normal waste. Add for thermal overload devices, supports, sales tax, delivery, supervision, mobilization, demobilization, cleanup, overhead and profit. Note: Select the right size for load and voltage. To select the right overload relay element, find the full load amps rating opposite the appropriate voltage on the motor nameplate. Then check the manufacturer's overload element chart for the correct element or heater. Heater elements are inserted into the overload relay. If the running current exceeds the full load amperage rating, the heater element heats a bi-metallic switch which opens the control circuit and interrupts power to the motor.

AC Magnetic Three Pole Starters with Overload Relays

Material	Craft@Hrs	Unit	Material Cost	Labor Cost	Installed Cost
NEMA 4X three pole AC magnetic starters with overload relays					
Size 4 40 HP, 208V	2E@2.50	Ea	3,020.00	67.20	3,087.20
Size 4 75 HP, 230V	2E@2.50	Ea	3,020.00	67.20	3,087.20
Size 4 100 HP, 480V	2E@2.50	Ea	3,020.00	67.20	3,087.20
NEMA 7 & 9 three pole AC magnetic starters with overload relays					
Size 00 1.5 HP, 208V	1E@1.50	Ea	759.00	40.30	799.30
Size 00 2 HP, 480V	1E@1.50	Ea	759.00	40.30	799.30
Size 0 3 HP, 208V	1E@1.50	Ea	759.00	40.30	799.30
Size 0 5 HP, 480V	1E@1.50	Ea	759.00	40.30	799.30
Size 1 7.5 HP, 208V	1E@1.75	Ea	792.00	47.00	839.00
Size 1 10 HP, 480V	1E@1.75	Ea	792.00	47.00	839.00
Size 2 10 HP, 208V	1E@2.00	Ea	1,280.00	53.80	1,333.80
Size 2 15 HP, 230V	1E@2.00	Ea	1,280.00	53.80	1,333.80
Size 2 25 HP, 480V	1E@2.00	Ea	1,280.00	53.80	1,333.80
Size 3 25 HP, 208V	1E@2.25	Ea	1,920.00	60.50	1,980.50
Size 3 30 HP, 230V	1E@2.25	Ea	1,920.00	60.50	1,980.50
Size 3 50 HP, 480V	1E@2.25	Ea	1,920.00	60.50	1,980.50
Size 4 40 HP, 208V	1E@3.00	Ea	3,110.00	80.60	3,190.60
Size 4 75 HP, 230V	1E@3.00	Ea	3,110.00	80.60	3,190.60
Size 4 100 HP, 480V	1E@3.00	Ea	3,110.00	80.60	3,190.60
Size 5 75 HP, 208V	2E@6.00	Ea	8,330.00	161.00	8,491.00
Size 5 100 HP, 230V	2E@6.00	Ea	8,330.00	161.00	8,491.00
Size 5 200 HP, 480V	2E@6.00	Ea	8,330.00	161.00	8,491.00
Size 6 150 HP, 208V	2E@8.00	Ea	19,300.00	215.00	19,515.00
Size 6 200 HP, 230V	2E@8.00	Ea	19,300.00	215.00	19,515.00
Size 6 400 HP, 480V	2E@8.00	Ea	19,300.00	215.00	19,515.00
NEMA 12 three pole AC magnetic starters with overload relays					
Size 00 1.5 HP, 208V	1E@0.80	Ea	244.00	21.50	265.50
Size 00 2 HP, 480V	1E@0.80	Ea	244.00	21.50	265.50
Size 0 3 HP, 208V	1E@1.00	Ea	244.00	26.90	270.90
Size 0 5 HP, 480V	1E@1.00	Ea	244.00	26.90	270.90
Size 1 7.5 HP, 208V	1E@1.25	Ea	268.00	33.60	301.60
Size 1 10 HP, 480V	1E@1.25	Ea	268.00	33.60	301.60

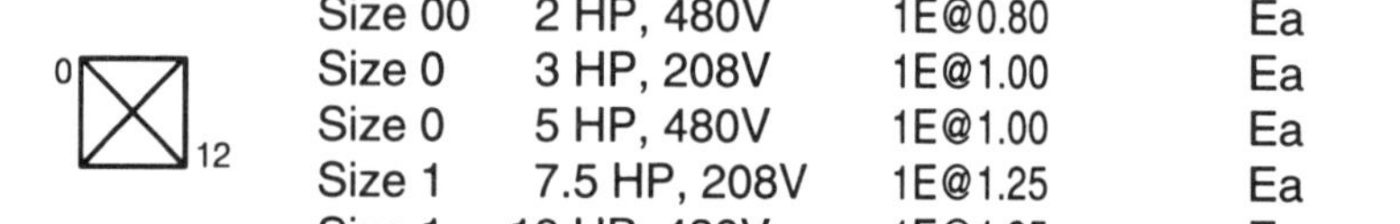

Use these figures to estimate the cost of magnetic starters installed in buildings under the conditions described on pages 5 and 6. Costs listed are for each starter installed. The crew size is one electrician for starters up to size 4 and two electricians for larger starters. The labor cost per manhour is $26.88. These costs include layout, material handling, and normal waste. Add for thermal overload devices, supports, sales tax, delivery, supervision, mobilization, demobilization, cleanup, overhead and profit. Note: Select the right size for load and voltage. To select the right overload relay element, find the full load amps rating opposite the appropriate voltage on the motor nameplate. Then check the manufacturer's overload element chart for the correct element or heater. Heater elements are inserted into the overload relay. If the running current exceeds the full load amperage rating, the heater element heats a bi-metallic switch which opens the control circuit and interrupts power to the motor.

AC Magnetic Three Pole Starters with Overload Relays

Material		Craft@Hrs	Unit	Material Cost	Labor Cost	Installed Cost
NEMA 12 three pole AC magnetic starters with overload relays						
Size 2	10 HP, 208V	1E@1.50	Ea	521.00	40.30	561.30
Size 2	15 HP, 230V	1E@1.50	Ea	521.00	40.30	561.30
Size 2	25 HP, 480V	1E@1.50	Ea	521.00	40.30	561.30
Size 3	25 HP, 208V	1E@1.75	Ea	812.00	47.00	859.00
Size 3	30 HP, 230V	1E@1.75	Ea	812.00	47.00	859.00
Size 3	50 HP, 480V	1E@1.75	Ea	812.00	47.00	859.00
Size 4	40 HP, 208V	1E@2.00	Ea	1,970.00	53.80	2,023.80
Size 4	75 HP, 230V	1E@2.00	Ea	1,970.00	53.80	2,023.80
Size 4	100 HP, 480V	1E@2.00	Ea	1,970.00	53.80	2,023.80
Size 5	75 HP, 208V	2E@5.00	Ea	4,710.00	134.00	4,844.00
Size 5	100 HP, 230V	2E@5.00	Ea	4,710.00	134.00	4,844.00
Size 5	200 HP, 480V	2E@5.00	Ea	4,710.00	134.00	4,844.00
Size 6	150 HP, 208V	2E@7.00	Ea	11,400.00	188.00	11,588.00
Size 6	200 HP, 230V	2E@7.00	Ea	11,400.00	188.00	11,588.00
Size 6	400 HP, 480V	2E@7.00	Ea	11,400.00	188.00	11,588.00
Size 7	300 HP, 230V	2E@9.00	Ea	15,000.00	242.00	15,242.00
Size 7	600 HP, 480V	2E@9.00	Ea	15,000.00	242.00	15,242.00

Material		Craft@Hrs	Unit	Material Cost	Labor Cost	Installed Cost
Open type three pole AC magnetic starters with overload relays						
Size 00	1.5 HP, 208V	1E@0.70	Ea	135.00	18.80	153.80
Size 00	2 HP, 480V	1E@0.70	Ea	135.00	18.80	153.80
Size 0	3 HP, 208V	1E@0.70	Ea	171.00	18.80	189.80
Size 0	5 HP, 480V	1E@0.70	Ea	171.00	18.80	189.80
Size 1	7.5 HP, 208V	1E@0.90	Ea	196.00	24.20	220.20
Size 1	10 HP, 480V	1E@0.90	Ea	196.00	24.20	220.20
Size 2	10 HP, 208V	1E@1.10	Ea	357.00	29.60	386.60
Size 2	15 HP, 230V	1E@1.10	Ea	357.00	29.60	386.60
Size 2	25 HP, 480V	1E@1.10	Ea	357.00	29.60	386.60
Size 3	25 HP, 208V	1E@1.20	Ea	579.00	32.30	611.30
Size 3	30 HP, 230V	1E@1.20	Ea	579.00	32.30	611.30
Size 3	50 HP, 480V	1E@1.20	Ea	579.00	32.30	611.30
Size 4	40 HP, 208V	1E@2.00	Ea	1,320.00	53.80	1,373.80
Size 4	75 HP, 230V	1E@2.00	Ea	1,320.00	53.80	1,373.80
Size 4	100 HP, 480V	1E@2.00	Ea	1,320.00	53.80	1,373.80
Size 5	75 HP, 208V	2E@4.25	Ea	3,230.00	114.00	3,344.00
Size 5	100 HP, 230V	2E@4.25	Ea	3,230.00	114.00	3,344.00
Size 5	200 HP, 480V	2E@4.25	Ea	3,230.00	114.00	3,344.00
Size 6	150 HP, 208V	2E@6.25	Ea	7,700.00	168.00	7,868.00
Size 6	200 HP, 230V	2E@6.25	Ea	7,700.00	168.00	7,868.00
Size 6	400 HP, 480V	2E@6.25	Ea	7,700.00	168.00	7,868.00
Size 7	300 HP, 230V	2E@8.25	Ea	11,000.00	222.00	11,222.00
Size 7	600 HP, 480V	2E@8.25	Ea	11,000.00	222.00	11,222.00

Use these figures to estimate the cost of magnetic starters installed in buildings under the conditions described on pages 5 and 6. Costs listed are for each starter installed. The crew size is one electrician for starters up to size 4 and two electricians for larger starters. The labor cost per manhour is $26.88. These costs include layout, material handling, and normal waste. Add for thermal overload devices, supports, sales tax, delivery, supervision, mobilization, demobilization, cleanup, overhead and profit. Note: Select the right size for load and voltage. To select the right overload relay element, find the full load amps rating opposite the appropriate voltage on the motor nameplate. Then check the manufacturer's overload element chart for the correct element or heater. Heater elements are inserted into the overload relay. If the running current exceeds the full load amperage rating, the heater element heats a bi-metallic switch which opens the control circuit and interrupts power to the motor.

Combination AC Magnetic Three Pole Starters with Non-fused Disconnect and Overload Relays

Material		Craft@Hrs	Unit	Material Cost	Labor Cost	Installed Cost

NEMA 1 combination three pole AC magnetic starters, non-fused disconnect

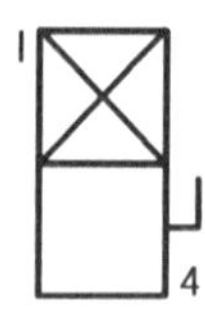

Material		Craft@Hrs	Unit	Material Cost	Labor Cost	Installed Cost
Size 0	3 HP, 208V	1E@1.25	Ea	460.00	33.60	493.60
Size 0	3 HP, 230V	1E@1.25	Ea	460.00	33.60	493.60
Size 0	5 HP, 480V	1E@1.25	Ea	460.00	33.60	493.60
Size 1	7.5 HP, 208V	1E@1.25	Ea	485.00	33.60	518.60
Size 1	7.5 HP, 230V	1E@1.25	Ea	485.00	33.60	518.60
Size 1	10 HP, 480V	1E@1.25	Ea	485.00	33.60	518.60
Size 2	10 HP, 208V	1E@1.50	Ea	768.00	40.30	808.30
Size 2	15 HP, 230V	1E@1.50	Ea	768.00	40.30	808.30
Size 2	25 HP, 480V	1E@1.50	Ea	768.00	40.30	808.30
Size 3	25 HP, 208V	1E@2.00	Ea	1,280.00	53.80	1,333.80
Size 3	30 HP, 230V	1E@2.00	Ea	1,280.00	53.80	1,333.80
Size 3	50 HP, 480V	1E@2.00	Ea	1,280.00	53.80	1,333.80
Size 4	40 HP, 208V	1E@3.00	Ea	2,450.00	80.60	2,530.60
Size 4	50 HP, 230V	1E@3.00	Ea	2,450.00	80.60	2,530.60
Size 4	100 HP, 480V	1E@3.00	Ea	2,450.00	80.60	2,530.60
Size 5	75 HP, 208V	2E@6.00	Ea	5,510.00	161.00	5,671.00
Size 5	100 HP, 230V	2E@6.00	Ea	5,510.00	161.00	5,671.00
Size 5	200 HP, 480V	2E@6.00	Ea	5,510.00	161.00	5,671.00
Size 6	150 HP, 208V	2E@8.00	Ea	14,500.00	215.00	14,715.00
Size 6	200 HP, 230V	2E@8.00	Ea	14,500.00	215.00	14,715.00
Size 6	400 HP, 480V	2E@8.00	Ea	14,500.00	215.00	14,715.00

NEMA 4 combination three pole AC magnetic starters, non-fused disconnect

Material		Craft@Hrs	Unit	Material Cost	Labor Cost	Installed Cost
Size 0	3 HP, 208V	1E@1.50	Ea	943.00	40.30	983.30
Size 0	3 HP, 230V	1E@1.50	Ea	943.00	40.30	983.30
Size 0	5 HP, 480V	1E@1.50	Ea	943.00	40.30	983.30
Size 1	7.5 HP, 208V	1E@1.50	Ea	968.00	40.30	1,008.30
Size 1	7.5 HP, 230V	1E@1.50	Ea	968.00	40.30	1,008.30
Size 1	10 HP, 480V	1E@1.50	Ea	968.00	40.30	1,008.30
Size 2	10 HP, 208V	1E@2.00	Ea	1,510.00	53.80	1,563.80
Size 2	15 HP, 230V	1E@2.00	Ea	1,510.00	53.80	1,563.80
Size 2	25 HP, 480V	1E@2.00	Ea	1,510.00	53.80	1,563.80
Size 3	25 HP, 208V	1E@3.00	Ea	2,570.00	80.60	2,650.60
Size 3	30 HP, 230V	1E@3.00	Ea	2,570.00	80.60	2,650.60
Size 3	50 HP, 480V	1E@3.00	Ea	2,570.00	80.60	2,650.60
Size 4	40 HP, 208V	1E@4.00	Ea	4,130.00	108.00	4,238.00
Size 4	50 HP, 230V	1E@4.00	Ea	4,130.00	108.00	4,238.00
Size 4	100 HP, 480V	1E@4.00	Ea	4,130.00	108.00	4,238.00
Size 5	75 HP, 208V	2E@7.00	Ea	9,750.00	188.00	9,938.00
Size 5	100 HP, 230V	2E@7.00	Ea	9,750.00	188.00	9,938.00
Size 5	200 HP, 480V	2E@7.00	Ea	9,750.00	188.00	9,938.00
Size 6	150 HP, 208V	2E@10.0	Ea	18,500.00	269.00	18,769.00
Size 6	200 HP, 230V	2E@10.0	Ea	18,500.00	269.00	18,769.00
Size 6	400 HP, 480V	2E@10.0	Ea	18,500.00	269.00	18,769.00

Use these figures to estimate the cost of combination starters installed in buildings under the conditions described on pages 5 and 6. Costs listed are for each starter installed. The crew size is one electrician for starters up to size 4 and two electricians for starters over size 4. Cost per manhour is $26.88. These costs include layout, material handling, and normal waste. Add for overload relay heater elements, sales tax, delivery, supervision, mobilization, demobilization, cleanup, overhead and profit. Note: AC magnetic starters are non-reversing with three melting alloy overload relays for motor protection. The heating elements must be appropriate for the full load amp rating of the connected motor.

Combination AC Magnetic Three Pole Starters with Non-fused Disconnect and Overload Relays

Material	Craft@Hrs	Unit	Material Cost	Labor Cost	Installed Cost

NEMA 4X combination three pole AC magnetic starters, non-fused disconnect

Material		Craft@Hrs	Unit	Material Cost	Labor Cost	Installed Cost
Size 0	3 HP, 208V	1E@1.75	Ea	1,080.00	47.00	1,127.00
Size 0	3 HP, 230V	1E@1.75	Ea	1,080.00	47.00	1,127.00
Size 0	5 HP, 480V	1E@1.75	Ea	1,080.00	47.00	1,127.00
Size 1	7.5 HP, 208V	1E@1.75	Ea	1,110.00	47.00	1,157.00
Size 1	7.5 HP, 230V	1E@1.75	Ea	1,110.00	47.00	1,157.00
Size 1	10 HP, 480V	1E@1.75	Ea	1,110.00	47.00	1,157.00
Size 2	10 HP, 208V	1E@2.00	Ea	1,670.00	53.80	1,723.80
Size 2	15 HP, 230V	1E@2.00	Ea	1,670.00	53.80	1,723.80
Size 2	25 HP, 480V	1E@2.00	Ea	1,670.00	53.80	1,723.80
Size 3	25 HP, 208V	1E@2.50	Ea	2,840.00	67.20	2,907.20
Size 3	30 HP, 230V	1E@2.50	Ea	2,840.00	67.20	2,907.20
Size 3	50 HP, 480V	1E@2.50	Ea	2,840.00	67.20	2,907.20

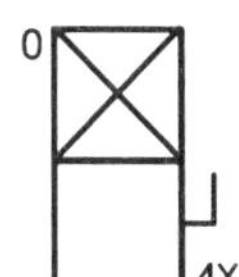

NEMA 12 combination three pole AC magnetic starters, non-fused disconnect

Material		Craft@Hrs	Unit	Material Cost	Labor Cost	Installed Cost
Size 0	3 HP, 208V	1E@1.50	Ea	581.00	40.30	621.30
Size 0	3 HP, 230V	1E@1.50	Ea	581.00	40.30	621.30
Size 0	5 HP, 480V	1E@1.50	Ea	581.00	40.30	621.30
Size 1	7.5 HP, 208V	1E@1.50	Ea	607.00	40.30	647.30
Size 1	7.5 HP, 230V	1E@1.50	Ea	607.00	40.30	647.30
Size 1	10 HP, 480V	1E@1.50	Ea	607.00	40.30	647.30
Size 2	10 HP, 208V	1E@2.00	Ea	939.00	53.80	992.80
Size 2	15 HP, 230V	1E@2.00	Ea	939.00	53.80	992.80
Size 2	25 HP, 480V	1E@2.00	Ea	939.00	53.80	992.80
Size 3	25 HP, 208V	1E@3.00	Ea	1,500.00	80.60	1,580.60
Size 3	30 HP, 230V	1E@3.00	Ea	1,500.00	80.60	1,580.60
Size 3	50 HP, 480V	1E@3.00	Ea	1,500.00	80.60	1,580.60
Size 4	40 HP, 208V	1E@4.00	Ea	3,070.00	108.00	3,178.00
Size 4	50 HP, 230V	1E@4.00	Ea	3,070.00	108.00	3,178.00
Size 4	100 HP, 480V	1E@4.00	Ea	3,070.00	108.00	3,178.00
Size 5	75 HP, 208V	2E@7.00	Ea	7,000.00	188.00	7,188.00
Size 5	100 HP, 230V	2E@7.00	Ea	7,000.00	188.00	7,188.00
Size 5	200 HP, 480V	2E@7.00	Ea	7,000.00	188.00	7,188.00
Size 6	150 HP, 208V	2E@10.0	Ea	16,200.00	269.00	16,469.00
Size 6	200 HP, 230V	2E@10.0	Ea	16,200.00	269.00	16,469.00
Size 6	400 HP, 480V	2E@10.0	Ea	16,200.00	269.00	16,469.00

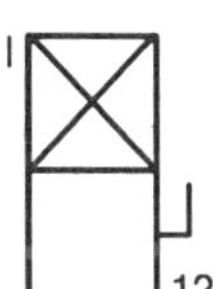

Use these figures to estimate the cost of combination starters installed in buildings under the conditions described on pages 5 and 6. Costs listed are for each starter installed. The crew size is one electrician for starters up to size 4 and two electricians for starters over size 4. Cost per manhour is $26.88. These costs include layout, material handling, and normal waste. Add for overload relay heater elements, sales tax, delivery, supervision, mobilization, demobilization, cleanup, overhead and profit. Note: AC magnetic starters are non-reversing with three melting alloy overload relays for motor protection. The heating elements must be appropriate for the full load amp rating of the connected motor.

Combination AC Magnetic Three Pole Starters with Fusible Disconnect and Overload Relays

Material		Craft@Hrs	Unit	Material Cost	Labor Cost	Installed Cost

NEMA 1 combination three pole AC magnetic starters, fusible disconnect

Material		Craft@Hrs	Unit	Material Cost	Labor Cost	Installed Cost
Size 0	3 HP, 208V	1E@1.25	Ea	475.00	33.60	508.60
Size 0	3 HP, 230V	1E@1.50	Ea	475.00	40.30	515.30
Size 0	5 HP, 480V	1E@1.50	Ea	475.00	40.30	515.30
Size 1	5 HP, 208V	1E@1.25	Ea	501.00	33.60	534.60
Size 1	7.5 HP, 208V	1E@1.25	Ea	510.00	33.60	543.60
Size 1	5 HP, 230V	1E@1.25	Ea	501.00	33.60	534.60
Size 1	7.5 HP, 230V	1E@1.25	Ea	510.00	33.60	543.60
Size 1	10 HP, 480V	1E@1.25	Ea	510.00	33.60	543.60
Size 2	10 HP, 208V	1E@1.50	Ea	786.00	40.30	826.30
Size 2	15 HP, 230V	1E@1.50	Ea	786.00	40.30	826.30
Size 2	25 HP, 480V	1E@1.50	Ea	801.00	40.30	841.30
Size 3	20 HP, 208V	1E@2.00	Ea	1,320.00	53.80	1,373.80
Size 3	25 HP, 208V	1E@2.00	Ea	1,450.00	53.80	1,503.80
Size 3	25 HP, 230V	1E@2.00	Ea	1,320.00	53.80	1,373.80
Size 3	50 HP, 480V	1E@2.00	Ea	1,340.00	53.80	1,393.80
Size 4	40 HP, 208V	1E@3.00	Ea	2,540.00	80.60	2,620.60
Size 4	50 HP, 230V	1E@3.00	Ea	2,540.00	80.60	2,620.60
Size 4	100 HP, 480V	1E@3.00	Ea	2,570.00	80.60	2,650.60
Size 5	75 HP, 208V	2E@4.00	Ea	5,700.00	108.00	5,808.00
Size 5	100 HP, 230V	2E@4.00	Ea	5,700.00	108.00	5,808.00
Size 5	200 HP, 480V	2E@4.00	Ea	5,700.00	108.00	5,808.00
Size 6	150 HP, 208V	2E@8.00	Ea	15,000.00	215.00	15,215.00
Size 6	200 HP, 230V	2E@8.00	Ea	15,000.00	215.00	15,215.00
Size 6	400 HP, 480V	2E@8.00	Ea	15,000.00	215.00	15,215.00

NEMA 4 combination three pole AC magnetic starters, fusible disconnect

Material		Craft@Hrs	Unit	Material Cost	Labor Cost	Installed Cost
Size 0	3 HP, 208V	1E@1.50	Ea	958.00	40.30	998.30
Size 0	3 HP, 230V	1E@1.50	Ea	958.00	40.30	998.30
Size 0	5 HP, 480V	1E@1.50	Ea	958.00	40.30	998.30
Size 1	5 HP, 208V	1E@1.50	Ea	983.00	40.30	1,023.30
Size 1	7.5 HP, 208V	1E@1.50	Ea	992.00	40.30	1,032.30
Size 1	5 HP, 230V	1E@1.50	Ea	983.00	40.30	1,023.30
Size 1	7.5 HP, 230V	1E@1.50	Ea	992.00	40.30	1,032.30
Size 1	10 HP, 480V	1E@1.50	Ea	992.00	40.30	1,032.30
Size 2	10 HP, 208V	1E@2.00	Ea	1,530.00	53.80	1,583.80
Size 2	15 HP, 230V	1E@2.00	Ea	1,530.00	53.80	1,583.80
Size 2	25 HP, 480V	1E@2.00	Ea	1,540.00	53.80	1,593.80

Use these figures to estimate the cost of combination starters installed in buildings under the conditions described on pages 5 and 6. Costs listed are for each starter installed. The crew size is one electrician for starters up to size 4 and two electricians for starters over size 4. Cost per manhour is $26.88. These costs include layout, material handling, and normal waste. Add for overload relay heater elements, sales tax, delivery, supervision, mobilization, demobilization, cleanup, overhead and profit. Note: AC magnetic starters are non-reversing with three melting alloy overload relays for motor protection. The heating elements must be appropriate for the full load amp rating of the connected motor.

Fuses are on pages 255 to 277.

Combination AC Magnetic Three Pole Starters with Fusible Disconnect and Overload Relays

Material		Craft@Hrs	Unit	Material Cost	Labor Cost	Installed Cost

NEMA 4 combination three pole AC magnetic starters, fusible disconnect

Material		Craft@Hrs	Unit	Material Cost	Labor Cost	Installed Cost
Size 3	20 HP, 208V	1E@3.00	Ea	2,630.00	80.60	2,710.60
Size 3	25 HP, 208V	1E@3.00	Ea	2,730.00	80.60	2,810.60
Size 3	25 HP, 230V	1E@3.00	Ea	2,630.00	80.60	2,710.60
Size 3	30 HP, 230V	1E@3.00	Ea	2,730.00	80.60	2,810.60
Size 3	50 HP, 480V	1E@3.00	Ea	2,650.00	80.60	2,730.60
Size 4	40 HP, 208V	1E@4.00	Ea	4,200.00	108.00	4,308.00
Size 4	50 HP, 230V	1E@4.00	Ea	4,200.00	108.00	4,308.00
Size 4	100 HP, 480V	1E@4.00	Ea	4,220.00	108.00	4,328.00
Size 5	75 HP, 208V	2E@7.00	Ea	9,920.00	188.00	10,108.00
Size 5	100 HP, 230V	2E@7.00	Ea	9,920.00	188.00	10,108.00
Size 5	200 HP, 480V	2E@7.00	Ea	9,920.00	188.00	10,108.00
Size 6	150 HP, 208V	2E@10.0	Ea	19,100.00	269.00	19,369.00
Size 6	200 HP, 230V	2E@10.0	Ea	19,100.00	269.00	19,369.00
Size 6	400 HP, 480V	2E@10.0	Ea	19,100.00	269.00	19,369.00

NEMA 4X combination three pole AC magnetic starters, fusible disconnect

Material		Craft@Hrs	Unit	Material Cost	Labor Cost	Installed Cost
Size 0	3 HP, 208V	1E@1.75	Ea	1,100.00	47.00	1,147.00
Size 0	3 HP, 230V	1E@1.75	Ea	1,100.00	47.00	1,147.00
Size 0	5 HP, 480V	1E@1.75	Ea	1,100.00	47.00	1,147.00
Size 1	5 HP, 208V	1E@1.75	Ea	1,130.00	47.00	1,177.00
Size 1	7.5 HP, 208V	1E@1.75	Ea	1,140.00	47.00	1,187.00
Size 1	5 HP, 230V	1E@1.75	Ea	1,130.00	47.00	1,177.00
Size 1	7.5 HP, 230V	1E@1.75	Ea	1,140.00	47.00	1,187.00
Size 1	10 HP, 480V	1E@1.75	Ea	1,140.00	47.00	1,187.00
Size 2	10 HP, 208V	1E@2.00	Ea	1,690.00	53.80	1,743.80
Size 2	15 HP, 230V	1E@2.00	Ea	1,690.00	53.80	1,743.80
Size 2	15 HP, 480V	1E@2.00	Ea	1,690.00	53.80	1,743.80
Size 2	25 HP, 480V	1E@2.00	Ea	1,700.00	53.80	1,753.80
Size 3	20 HP, 208V	1E@3.00	Ea	2,890.00	80.60	2,970.60
Size 3	25 HP, 230V	1E@3.00	Ea	2,890.00	80.60	2,970.60
Size 3	50 HP, 480V	1E@3.00	Ea	2,910.00	80.60	2,990.60

NEMA 12 combination three pole AC magnetic starters, fusible disconnect

Material		Craft@Hrs	Unit	Material Cost	Labor Cost	Installed Cost
Size 0	3 HP, 208V	1E@1.50	Ea	595.00	40.30	635.30
Size 0	3 HP, 230V	1E@1.50	Ea	595.00	40.30	635.30
Size 0	5 HP, 480V	1E@1.50	Ea	595.00	40.30	635.30

Use these figures to estimate the cost of combination starters installed in buildings under the conditions described on pages 5 and 6. Costs listed are for each starter installed. The crew size is one electrician for starters up to size 4 and two electricians for starters over size 4. Cost per manhour is $26.88. These costs include layout, material handling, and normal waste. Add for overload relay heater elements, sales tax, delivery, supervision, mobilization, demobilization, cleanup, overhead and profit. Note: AC magnetic starters are non-reversing with three melting alloy overload relays for motor protection. The heating elements must be appropriate for the full load amp rating of the connected motor.

Fuses are on pages 255 to 277.

Combination AC Magnetic Three Pole Starters with Fusible Disconnect and Overload Relays

Material	Craft@Hrs	Unit	Material Cost	Labor Cost	Installed Cost
NEMA 12 combination three pole AC magnetic starters with fusible disconnect					
Size 1 5 HP, 208V	1E@1.50	Ea	621.00	40.30	661.30
Size 1 7.5 HP, 208V	1E@1.50	Ea	631.00	40.30	671.30
Size 1 5 HP, 230V	1E@1.50	Ea	621.00	40.30	661.30
Size 1 7.5 HP, 230V	1E@1.50	Ea	631.00	40.30	671.30
Size 1 10 HP, 480V	1E@1.50	Ea	631.00	40.30	671.30
Size 2 10 HP, 208V	1E@2.00	Ea	958.00	53.80	1,011.80
Size 2 15 HP, 230V	1E@2.00	Ea	958.00	53.80	1,011.80
Size 2 15 HP, 480V	1E@2.00	Ea	958.00	53.80	1,011.80
Size 2 25 HP, 480V	1E@2.00	Ea	973.00	53.80	1,026.80
Size 3 20 HP, 208V	1E@3.00	Ea	1,540.00	80.60	1,620.60
Size 3 25 HP, 208V	1E@3.00	Ea	1,660.00	80.60	1,740.60
Size 3 25 HP, 230V	1E@3.00	Ea	1,540.00	80.60	1,620.60
Size 3 30 HP, 230V	1E@3.00	Ea	1,660.00	80.60	1,740.60
Size 3 50 HP, 480V	1E@3.00	Ea	1,580.00	80.60	1,660.60
Size 4 40 HP, 208V	1E@4.00	Ea	3,160.00	108.00	3,268.00
Size 4 50 HP, 230V	1E@4.00	Ea	3,160.00	108.00	3,268.00
Size 4 100 HP, 480V	1E@4.00	Ea	3,170.00	108.00	3,278.00
Size 5 75 HP, 208V	2E@7.00	Ea	7,190.00	188.00	7,378.00
Size 5 100 HP, 230V	2E@7.00	Ea	7,190.00	188.00	7,378.00
Size 5 200 HP, 480V	2E@7.00	Ea	7,190.00	188.00	7,378.00
Size 6 150 HP, 208V	2E@10.0	Ea	16,700.00	269.00	16,969.00
Size 6 200 HP, 230V	2E@10.0	Ea	16,700.00	269.00	16,969.00
Size 6 400 HP, 480V	2E@10.0	Ea	16,700.00	269.00	16,969.00
NEMA 1 combination three pole AC magnetic starters with class R fuse disconnect					
Size 0 3 HP, 208V	1E@1.25	Ea	482.00	33.60	515.60
Size 0 3 HP, 230V	1E@1.25	Ea	482.00	33.60	515.60
Size 0 5 HP, 480V	1E@1.25	Ea	491.00	33.60	524.60
Size 1 5 HP, 208V	1E@1.25	Ea	508.00	33.60	541.60
Size 1 7.5 HP, 208V	1E@1.25	Ea	517.00	33.60	550.60
Size 1 5 HP, 230V	1E@1.25	Ea	508.00	33.60	541.60
Size 1 7.5 HP, 230V	1E@1.25	Ea	517.00	33.60	550.60
Size 1 10 HP, 480V	1E@1.25	Ea	517.00	33.60	550.60
Size 2 10 HP, 208V	1E@1.50	Ea	794.00	40.30	834.30
Size 2 15 HP, 230V	1E@1.50	Ea	794.00	40.30	834.30
Size 2 15 HP, 480V	1E@1.50	Ea	799.00	40.30	839.30
Size 2 25 HP, 480V	1E@1.50	Ea	810.00	40.30	850.30
Size 3 20 HP, 208V	1E@2.00	Ea	1,340.00	53.80	1,393.80
Size 3 25 HP, 208V	1E@2.00	Ea	1,450.00	53.80	1,503.80

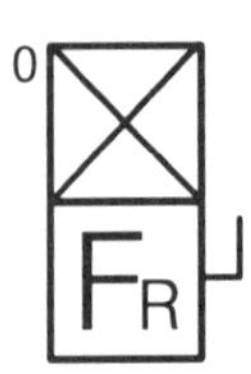

Use these figures to estimate the cost of combination starters installed in buildings under the conditions described on pages 5 and 6. Costs listed are for each starter installed. The crew size is one electrician for starters up to size 4 and two electricians for over size 4. Cost per manhour is $26.88. These costs include layout, material handling, and normal waste. Add for overload relay heater elements, sales tax, delivery, supervision, mobilization, demobilization, cleanup, overhead and profit. Note: AC magnetic starters are non-reversing with three melting alloy overload relays for motor protection. The heating elements must be appropriate for the full load amp rating of the connected motor.

Fuses are on pages 255 to 277.

Combination AC Magnetic Three Pole Starters with Fusible Disconnect and Overload Relays for Class R Fuses

Material	Craft@Hrs	Unit	Material Cost	Labor Cost	Installed Cost

NEMA 1 combination three pole AC magnetic starters with class R fuse disconnect

Material	Craft@Hrs	Unit	Material Cost	Labor Cost	Installed Cost
Size 3 25 HP, 230V	1E@2.00	Ea	1,340.00	53.80	1,393.80
Size 3 30 HP, 230V	1E@2.00	Ea	1,450.00	53.80	1,503.80
Size 3 50 HP, 480V	1E@2.00	Ea	1,370.00	53.80	1,423.80
Size 4 40 HP, 208V	1E@3.00	Ea	2,560.00	80.60	2,640.60
Size 4 50 HP, 230V	1E@3.00	Ea	2,560.00	80.60	2,640.60
Size 4 100 HP, 480V	1E@3.00	Ea	2,580.00	80.60	2,660.60
Size 5 75 HP, 208V	2E@4.00	Ea	5,720.00	108.00	5,828.00
Size 5 100 HP, 230V	2E@4.00	Ea	5,720.00	108.00	5,828.00
Size 5 200 HP, 480V	2E@4.00	Ea	5,720.00	108.00	5,828.00
Size 6 150 HP, 208V	2E@8.00	Ea	15,100.00	215.00	15,315.00
Size 6 200 HP, 230V	2E@8.00	Ea	15,100.00	215.00	15,315.00
Size 6 400 HP, 480V	2E@8.00	Ea	15,100.00	215.00	15,315.00

NEMA 4 combination three pole AC magnetic starters with class R fuse disconnect

Material	Craft@Hrs	Unit	Material Cost	Labor Cost	Installed Cost
Size 0 3 HP, 208V	1E@1.50	Ea	966.00	40.30	1,006.30
Size 0 3 HP, 230V	1E@1.50	Ea	966.00	40.30	1,006.30
Size 0 5 HP, 480V	1E@1.50	Ea	975.00	40.30	1,015.30
Size 1 5 HP, 208V	1E@1.50	Ea	990.00	40.30	1,030.30
Size 1 7.5 HP, 208V	1E@1.50	Ea	990.00	40.30	1,030.30
Size 1 5 HP, 230V	1E@1.50	Ea	1,000.00	40.30	1,040.30
Size 1 7.5 HP, 230V	1E@1.50	Ea	1,000.00	40.30	1,040.30
Size 1 10 HP, 480V	1E@1.50	Ea	1,000.00	40.30	1,040.30
Size 2 10 HP, 208V	1E@2.00	Ea	1,540.00	53.80	1,593.80
Size 2 15 HP, 230V	1E@2.00	Ea	1,540.00	53.80	1,593.80
Size 2 15 HP, 480V	1E@2.00	Ea	1,540.00	53.80	1,593.80
Size 2 25 HP, 480V	1E@2.00	Ea	1,540.00	53.80	1,593.80
Size 3 20 HP, 208V	1E@3.00	Ea	2,640.00	80.60	2,720.60
Size 3 25 HP, 208V	1E@3.00	Ea	2,740.00	80.60	2,820.60
Size 3 25 HP, 230V	1E@3.00	Ea	2,640.00	80.60	2,720.60
Size 3 30 HP, 480V	1E@3.00	Ea	2,660.00	80.60	2,740.60
Size 4 40 HP, 208V	1E@4.00	Ea	4,220.00	108.00	4,328.00
Size 4 50 HP, 230V	1E@4.00	Ea	4,220.00	108.00	4,328.00
Size 4 100 HP, 480V	1E@4.00	Ea	4,230.00	108.00	4,338.00
Size 5 75 HP, 208V	2E@7.00	Ea	10,000.00	188.00	10,188.00
Size 5 100 HP, 230V	2E@7.00	Ea	10,000.00	188.00	10,188.00
Size 5 200 HP, 480V	2E@7.00	Ea	10,000.00	188.00	10,188.00
Size 6 150 HP, 208V	2E@10.0	Ea	19,200.00	269.00	19,469.00
Size 6 200 HP, 230V	2E@10.0	Ea	19,200.00	269.00	19,469.00
Size 6 400 HP, 480V	2E@10.0	Ea	19,200.00	269.00	19,469.00

Use these figures to estimate the cost of combination starters installed in buildings under the conditions described on pages 5 and 6. Costs listed are for each starter installed. The crew size is one electrician for starters up to size 4 and two electricians for over size 4. Cost per manhour is $26.88. These costs include layout, material handling, and normal waste. Add for overload relay heater elements, sales tax, delivery, supervision, mobilization, demobilization, cleanup, overhead and profit. Note: AC magnetic starters are non-reversing with three melting alloy overload relays for motor protection. The heating elements must be appropriate for the full load amp rating of the connected motor.

Fuses are on pages 255 to 277.

Combination AC Magnetic Three Pole Starters with Fusible Disconnect and Overload Relays for Class R Fuses

Material	Craft@Hrs	Unit	Material Cost	Labor Cost	Installed Cost
NEMA 4X combination three pole AC magnetic starters with class R fuse disconnect					
Size 0 3 HP, 208V	1E@1.75	Ea	1,110.00	47.00	1,157.00
Size 0 3 HP, 230V	1E@1.75	Ea	1,110.00	47.00	1,157.00
Size 0 5 HP, 480V	1E@1.75	Ea	1,120.00	47.00	1,167.00
Size 1 5 HP, 208V	1E@1.75	Ea	1,140.00	47.00	1,187.00
Size 1 7.5 HP, 230V	1E@1.75	Ea	1,140.00	47.00	1,187.00
Size 1 10 HP, 480V	1E@1.75	Ea	1,140.00	47.00	1,187.00
Size 2 10 HP, 208V	1E@2.00	Ea	1,690.00	53.80	1,743.80
Size 2 15 HP, 230V	1E@2.00	Ea	1,690.00	53.80	1,743.80
Size 2 15 HP, 480V	1E@2.00	Ea	1,700.00	53.80	1,753.80
Size 2 25 HP, 480V	1E@2.00	Ea	1,710.00	53.80	1,763.80
Size 3 20 HP, 208V	1E@3.00	Ea	2,900.00	80.60	2,980.60
Size 3 25 HP, 230V	1E@3.00	Ea	2,900.00	80.60	2,980.60
Size 3 50 HP, 480V	1E@3.00	Ea	2,930.00	80.60	3,010.60
NEMA 12 combination three pole AC magnetic starters with class R fuse disconnect					
Size 0 3 HP, 208V	1E@1.50	Ea	602.00	40.30	642.30
Size 0 3 HP, 230V	1E@1.50	Ea	602.00	40.30	642.30
Size 0 5 HP, 480V	1E@1.50	Ea	613.00	40.30	653.30
Size 1 5 HP, 208V	1E@1.50	Ea	630.00	40.30	670.30
Size 1 7.5 HP, 208V	1E@1.50	Ea	638.00	40.30	678.30
Size 1 5 HP, 230V	1E@1.50	Ea	630.00	40.30	670.30
Size 1 7.5 HP, 230V	1E@1.50	Ea	638.00	40.30	678.30
Size 1 10 HP, 480V	1E@1.50	Ea	638.00	40.30	678.30
Size 2 10 HP, 208V	1E@2.00	Ea	966.00	53.80	1,019.80
Size 2 15 HP, 230V	1E@2.00	Ea	966.00	53.80	1,019.80
Size 2 15 HP, 480V	1E@2.00	Ea	969.00	53.80	1,022.80
Size 2 25 HP, 480V	1E@2.00	Ea	980.00	53.80	1,033.80
Size 3 20 HP, 208V	1E@3.00	Ea	1,560.00	80.60	1,640.60
Size 3 25 HP, 208V	1E@3.00	Ea	1,680.00	80.60	1,760.60
Size 3 25 HP, 230V	1E@3.00	Ea	1,560.00	80.60	1,640.60
Size 3 50 HP, 480V	1E@3.00	Ea	1,590.00	80.60	1,670.60
Size 4 40 HP, 208V	1E@4.00	Ea	3,170.00	108.00	3,278.00
Size 4 50 HP, 230V	1E@4.00	Ea	3,170.00	108.00	3,278.00
Size 4 100 HP, 480V	1E@4.00	Ea	3,200.00	108.00	3,308.00

Use these figures to estimate the cost of combination starters installed in buildings under the conditions described on pages 5 and 6. Costs listed are for each starter installed. The crew size is one electrician working at a labor cost of $26.88. These costs include layout, material handling, and normal waste. Add for overload relay heater elements, sales tax, delivery, supervision, mobilization, demobilization, cleanup, overhead and profit. Note: AC magnetic starters are non-reversing with three melting alloy overload relays for motor protection. The heating elements must be appropriate for the full load amp rating of the connected motor.

Fuses are on pages 255 to 277.

Combination AC Magnetic Three Pole Starters with Fusible & Non-fusible Disconnect and Overload Relays

Material	Craft@Hrs	Unit	Material Cost	Labor Cost	Installed Cost

NEMA 12 combination three pole AC magnetic starters with class R fuse disconnect

Material	Craft@Hrs	Unit	Material Cost	Labor Cost	Installed Cost
Size 5 75 HP, 208V	2E@7.00	Ea	7,220.00	188.00	7,408.00
Size 5 100 HP, 230V	2E@7.00	Ea	7,220.00	188.00	7,408.00
Size 5 200 HP, 480V	2E@7.00	Ea	7,220.00	188.00	7,408.00
Size 6 150 HP, 208V	2E@10.0	Ea	16,900.00	269.00	17,169.00
Size 6 200 HP, 230V	2E@10.0	Ea	16,900.00	269.00	17,169.00
Size 6 400 HP, 480V	2E@10.0	Ea	16,900.00	269.00	17,169.00

NEMA 1 combination three pole AC magnetic starters, oversized enclosure, non-fused

Material	Craft@Hrs	Unit	Material Cost	Labor Cost	Installed Cost
Size 0 3 HP, 208V	1E@1.30	Ea	586.00	34.90	620.90
Size 0 3 HP, 230V	1E@1.30	Ea	586.00	34.90	620.90
Size 0 5 HP, 480V	1E@1.30	Ea	586.00	34.90	620.90
Size 1 7.5 HP, 208V	1E@1.30	Ea	611.00	34.90	645.90
Size 1 7.5 HP, 230V	1E@1.30	Ea	611.00	34.90	645.90
Size 1 10 HP, 480V	1E@1.30	Ea	611.00	34.90	645.90
Size 2 10 HP, 208V	1E@1.60	Ea	891.00	43.00	934.00
Size 2 15 HP, 230V	1E@1.60	Ea	891.00	43.00	934.00
Size 2 25 HP, 480V	1E@1.60	Ea	891.00	43.00	934.00

NEMA 4 combination three pole AC magnetic starters, oversized enclosure, non-fused

Material	Craft@Hrs	Unit	Material Cost	Labor Cost	Installed Cost
Size 0 3 HP, 208V	1E@1.60	Ea	1,320.00	43.00	1,363.00
Size 0 3 HP, 230V	1E@1.60	Ea	1,320.00	43.00	1,363.00
Size 0 5 HP, 480V	1E@1.60	Ea	1,320.00	43.00	1,363.00
Size 1 7.5 HP, 208V	1E@1.60	Ea	1,340.00	43.00	1,383.00
Size 1 7.5 HP, 230V	1E@1.60	Ea	1,340.00	43.00	1,383.00
Size 1 10 HP, 480V	1E@1.60	Ea	1,340.00	43.00	1,383.00
Size 2 10 HP, 208V	1E@2.10	Ea	1,900.00	56.40	1,956.40
Size 2 15 HP, 230V	1E@2.10	Ea	1,900.00	56.40	1,956.40
Size 2 25 HP, 480V	1E@2.10	Ea	1,900.00	56.40	1,956.40

NEMA 12 combination three pole AC magnetic starters, oversized enclosure, non-fused

Material	Craft@Hrs	Unit	Material Cost	Labor Cost	Installed Cost
Size 0 3 HP, 208V	1E@1.60	Ea	807.00	43.00	850.00
Size 0 3 HP, 230V	1E@1.60	Ea	807.00	43.00	850.00
Size 0 5 HP, 480V	1E@1.60	Ea	807.00	43.00	850.00
Size 1 7.5 HP, 208V	1E@1.60	Ea	821.00	43.00	864.00
Size 1 7.5 HP, 230V	1E@1.60	Ea	821.00	43.00	864.00
Size 1 10 HP, 480V	1E@1.60	Ea	821.00	43.00	864.00
Size 2 10 HP, 208V	1E@2.10	Ea	1,140.00	56.40	1,196.40
Size 2 15 HP, 230V	1E@2.10	Ea	1,140.00	56.40	1,196.40
Size 2 25 HP, 480V	1E@2.10	Ea	1,140.00	56.40	1,196.40

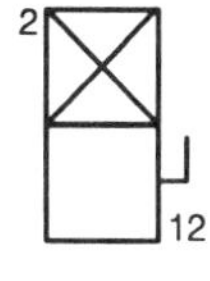

Use these figures to estimate the cost of combination starters installed in buildings under the conditions described on pages 5 and 6. Costs listed are for each starter installed. The crew size is one electrician for starters up to size 4 and two electricians for over size 4. Cost per manhour is $26.88. These costs include layout, material handling, and normal waste. Add for overload relay heater elements, sales tax, delivery, supervision, mobilization, demobilization, cleanup, overhead and profit. Note: AC magnetic starters are non-reversing with three melting alloy overload relays for motor protection. The heating elements must be appropriate for the full load amp rating of the connected motor.

Fuses are on pages 255 to 277.

Combination AC Magnetic Three Pole Starters in Oversized Enclosure with Fusible Disconnect and Overload Relays

Material	Craft@Hrs	Unit	Material Cost	Labor Cost	Installed Cost

NEMA 1 combination three pole AC magnetic starters, oversized enclosure, fusible disconnect

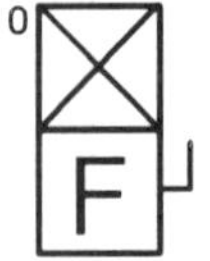

Material	Craft@Hrs	Unit	Material Cost	Labor Cost	Installed Cost
Size 0 3 HP, 208V	1E@1.30	Ea	601.00	34.90	635.90
Size 0 3 HP, 230V	1E@1.30	Ea	601.00	34.90	635.90
Size 0 5 HP, 480V	1E@1.30	Ea	611.00	34.90	645.90
Size 1 5 HP, 208V	1E@1.30	Ea	625.00	34.90	659.90
Size 1 7.5 HP, 208V	1E@1.30	Ea	636.00	34.90	670.90
Size 1 5 HP, 230V	1E@1.30	Ea	625.00	34.90	659.90
Size 1 7.5 HP, 230V	1E@1.30	Ea	636.00	34.90	670.90
Size 1 10 HP, 480V	1E@1.30	Ea	636.00	34.90	670.90
Size 2 10 HP, 208V	1E@1.60	Ea	912.00	43.00	955.00
Size 2 15 HP, 230V	1E@1.60	Ea	912.00	43.00	955.00
Size 2 15 HP, 480V	1E@1.60	Ea	919.00	43.00	962.00
Size 2 25 HP, 480V	1E@1.60	Ea	928.00	43.00	971.00

NEMA 4 combination three pole AC magnetic starters, oversized enclosure, fusible disconnect

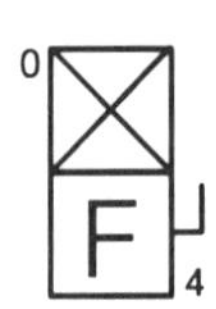

Material	Craft@Hrs	Unit	Material Cost	Labor Cost	Installed Cost
Size 0 3 HP, 208V	1E@1.60	Ea	1,340.00	43.00	1,383.00
Size 0 3 HP, 230V	1E@1.60	Ea	1,340.00	43.00	1,383.00
Size 0 5 HP, 480V	1E@1.60	Ea	1,340.00	43.00	1,383.00
Size 1 5 HP, 208V	1E@1.60	Ea	1,350.00	43.00	1,393.00
Size 1 7.5 HP, 208V	1E@1.60	Ea	1,370.00	43.00	1,413.00
Size 1 5 HP, 230V	1E@1.60	Ea	1,350.00	43.00	1,393.00
Size 1 7.5 HP, 230V	1E@1.60	Ea	1,370.00	43.00	1,413.00
Size 1 10 HP, 480V	1E@1.60	Ea	1,370.00	43.00	1,413.00
Size 2 10 HP, 208V	1E@2.10	Ea	1,910.00	56.40	1,966.40
Size 2 15 HP, 230V	1E@2.10	Ea	1,910.00	56.40	1,966.40
Size 2 15 HP, 480V	1E@2.10	Ea	1,920.00	56.40	1,976.40
Size 2 25 HP, 480V	1E@2.10	Ea	1,930.00	56.40	1,986.40

NEMA 12 combination three pole AC magnetic starters, oversized enclosure, fusible disconnect

Material	Craft@Hrs	Unit	Material Cost	Labor Cost	Installed Cost
Size 0 3 HP, 208V	1E@1.60	Ea	821.00	43.00	864.00
Size 0 3 HP, 230V	1E@1.60	Ea	821.00	43.00	864.00
Size 0 5 HP, 480V	1E@1.60	Ea	832.00	43.00	875.00
Size 1 5 HP, 208V	1E@1.60	Ea	838.00	43.00	881.00
Size 1 7.5 HP, 208V	1E@1.60	Ea	846.00	43.00	889.00
Size 1 5 HP, 230V	1E@1.60	Ea	838.00	43.00	881.00
Size 1 7.5 HP, 230V	1E@1.60	Ea	846.00	43.00	889.00
Size 1 10 HP, 480V	1E@1.60	Ea	846.00	43.00	889.00

Use these figures to estimate the cost of combination starters installed in buildings under the conditions described on pages 5 and 6. Costs listed are for each starter installed. The crew size is one electrician for starters up to size 4 and two electricians for over size 4. Cost per manhour is $26.88. These costs include layout, material handling, and normal waste. Add for overload relay heater elements, sales tax, delivery, supervision, mobilization, demobilization, cleanup, overhead and profit. Note: AC magnetic starters are non-reversing with three melting alloy overload relays for motor protection. The heating elements must be appropriate for the full load amp rating of the connected motor.

Fuses are on pages 255 to 277.

Combination AC Magnetic Three Pole Starters with Fusible & Circuit Breaker Disconnect and Overload Relays

Material	Craft@Hrs	Unit	Material Cost	Labor Cost	Installed Cost

NEMA 12 combination three pole AC magnetic starters, oversized enclosure, fusible disconnect

Material	Craft@Hrs	Unit	Material Cost	Labor Cost	Installed Cost
Size 2 10 HP, 208V	1E@2.10	Ea	1,170.00	56.40	1,226.40
Size 2 15 HP, 230V	1E@2.10	Ea	1,170.00	56.40	1,226.40
Size 2 25 HP, 480V	1E@2.10	Ea	1,190.00	56.40	1,246.40

NEMA 1 combination three pole AC magnetic starters, circuit breaker disconnect

Material	Craft@Hrs	Unit	Material Cost	Labor Cost	Installed Cost
Size 0 3 HP, 208V	1E@1.25	Ea	641.00	33.60	674.60
Size 0 3 HP, 230V	1E@1.25	Ea	641.00	33.60	674.60
Size 0 5 HP, 480V	1E@1.25	Ea	641.00	33.60	674.60
Size 0 5 HP, 600V	1E@1.25	Ea	641.00	33.60	674.60
Size 1 7.5 HP, 208V	1E@1.25	Ea	667.00	33.60	700.60
Size 1 7.5 HP, 230V	1E@1.25	Ea	667.00	33.60	700.60
Size 1 10 HP, 480V	1E@1.25	Ea	667.00	33.60	700.60
Size 1 10 HP, 600V	1E@1.25	Ea	667.00	33.60	700.60
Size 2 10 HP, 208V	1E@1.50	Ea	943.00	40.30	983.30
Size 2 15 HP, 230V	1E@1.50	Ea	943.00	40.30	983.30
Size 2 25 HP, 480V	1E@1.50	Ea	943.00	40.30	983.30
Size 2 25 HP, 600V	1E@1.50	Ea	943.00	40.30	983.30
Size 3 25 HP, 208V	1E@2.00	Ea	1,370.00	53.80	1,423.80
Size 3 30 HP, 230V	1E@2.00	Ea	1,370.00	53.80	1,423.80
Size 3 50 HP, 480V	1E@2.00	Ea	1,370.00	53.80	1,423.80
Size 3 50 HP, 600V	1E@2.00	Ea	1,370.00	53.80	1,423.80
Size 4 40 HP, 208V	1E@3.00	Ea	3,000.00	80.60	3,080.60
Size 4 50 HP, 230V	1E@3.00	Ea	3,000.00	80.60	3,080.60
Size 4 100 HP, 480V	1E@3.00	Ea	3,000.00	80.60	3,080.60
Size 4 100 HP, 600V	1E@3.00	Ea	3,000.00	80.60	3,080.60
Size 5 75 HP, 208V	2E@4.00	Ea	6,980.00	108.00	7,088.00
Size 5 100 HP, 230V	2E@4.00	Ea	6,980.00	108.00	7,088.00
Size 5 200 HP, 480V	2E@4.00	Ea	6,980.00	108.00	7,088.00
Size 5 200 HP, 600V	2E@4.00	Ea	6,980.00	108.00	7,088.00
Size 6 150 HP, 208V	2E@8.00	Ea	15,000.00	215.00	15,215.00
Size 6 200 HP, 230V	2E@8.00	Ea	15,200.00	215.00	15,415.00
Size 6 400 HP, 480V	2E@8.00	Ea	15,200.00	215.00	15,415.00
Size 6 400 HP, 600V	2E@8.00	Ea	15,200.00	215.00	15,415.00

Use these figures to estimate the cost of combination starters installed in buildings under the conditions described on pages 5 and 6. Costs listed are for each starter installed. The crew size is one electrician for starters up to size 4 and two electricians for over size 4. Cost per manhour is $26.88. These costs include layout, material handling, and normal waste. Add for overload relay heater elements, sales tax, delivery, supervision, mobilization, demobilization, cleanup, overhead and profit. Note: AC magnetic starters are non-reversing with three melting alloy overload relays for motor protection. The heating elements must be appropriate for the full load amp rating of the connected motor. Circuit breakers are either 22,000 or 100,000 amp interrupt capacity.

Fuses are on pages 255 to 277.

Combination AC Magnetic Three Pole Starters with Circuit Breaker Disconnect and Overload Relays

Material		Craft@Hrs	Unit	Material Cost	Labor Cost	Installed Cost
NEMA 4 combination three pole AC magnetic starters, C.B. disconnect and overload relays						
Size 0	3 HP, 208V	1E@1.50	Ea	1,120.00	40.30	1,160.30
Size 0	3 HP, 230V	1E@1.50	Ea	1,120.00	40.30	1,160.30
Size 0	5 HP, 480V	1E@1.50	Ea	1,120.00	40.30	1,160.30
Size 0	5 HP, 600V	1E@1.50	Ea	1,120.00	40.30	1,160.30
Size 1	7.5 HP, 208V	1E@1.50	Ea	1,140.00	40.30	1,180.30
Size 1	7.5 HP, 230V	1E@1.50	Ea	1,140.00	40.30	1,180.30
Size 1	10 HP, 480V	1E@1.50	Ea	1,140.00	40.30	1,180.30
Size 1	10 HP, 600V	1E@1.50	Ea	1,140.00	40.30	1,180.30
Size 2	10 HP, 208V	1E@2.00	Ea	1,690.00	53.80	1,743.80
Size 2	15 HP, 230V	1E@2.00	Ea	1,690.00	53.80	1,743.80
Size 2	25 HP, 480V	1E@2.00	Ea	1,690.00	53.80	1,743.80
Size 2	25 HP, 600V	1E@2.00	Ea	1,690.00	53.80	1,743.80
Size 3	25 HP, 208V	1E@3.00	Ea	2,670.00	80.60	2,750.60
Size 3	30 HP, 230V	1E@3.00	Ea	2,670.00	80.60	2,750.60
Size 3	50 HP, 480V	1E@3.00	Ea	2,670.00	80.60	2,750.60
Size 3	50 HP, 600V	1E@3.00	Ea	2,670.00	80.60	2,750.60
Size 4	40 HP, 208V	1E@4.00	Ea	4,660.00	108.00	4,768.00
Size 4	50 HP, 230V	1E@4.00	Ea	4,660.00	108.00	4,768.00
Size 4	100 HP, 480V	1E@4.00	Ea	4,660.00	108.00	4,768.00
Size 4	100 HP, 600V	1E@4.00	Ea	4,660.00	108.00	4,768.00
Size 5	75 HP, 208V	2E@7.00	Ea	11,200.00	188.00	11,388.00
Size 5	100 HP, 230V	2E@7.00	Ea	11,200.00	188.00	11,388.00
Size 5	200 HP, 480V	2E@7.00	Ea	11,200.00	188.00	11,388.00
Size 5	200 HP, 600V	2E@7.00	Ea	11,200.00	188.00	11,388.00
Size 6	150 HP, 208V	2E@10.0	Ea	17,600.00	269.00	17,869.00
Size 6	200 HP, 230V	2E@10.0	Ea	17,600.00	269.00	17,869.00
Size 6	400 HP, 480V	2E@10.0	Ea	17,600.00	269.00	17,869.00
Size 6	400 HP, 600V	2E@10.0	Ea	17,600.00	269.00	17,869.00

Material		Craft@Hrs	Unit	Material Cost	Labor Cost	Installed Cost
NEMA 4X combination three pole AC magnetic starters, C.B. disconnect and overload relays						
Size 0	3 HP, 208V	1E@1.75	Ea	1,290.00	47.00	1,337.00
Size 0	3 HP, 230V	1E@1.75	Ea	1,290.00	47.00	1,337.00
Size 0	5 HP, 480V	1E@1.75	Ea	1,290.00	47.00	1,337.00
Size 0	5 HP, 600V	1E@1.75	Ea	1,290.00	47.00	1,337.00
Size 1	7.5 HP, 208V	1E@1.75	Ea	1,320.00	47.00	1,367.00
Size 1	7.5 HP, 230V	1E@1.75	Ea	1,320.00	47.00	1,367.00
Size 1	10 HP, 480V	1E@1.75	Ea	1,320.00	47.00	1,367.00
Size 1	10 HP, 600V	1E@1.75	Ea	1,320.00	47.00	1,367.00

Use these figures to estimate the cost of combination starters installed in buildings under the conditions described on pages 5 and 6. Costs listed are for each starter installed. The crew size is one electrician for starters up to size 4 and two electricians for over size 4. Cost per manhour is $26.88. These costs include layout, material handling, and normal waste. Add for overload relay heater elements, sales tax, delivery, supervision, mobilization, demobilization, cleanup, overhead and profit. Note: AC magnetic starters are non-reversing with three melting alloy overload relays for motor protection. The heating elements must be appropriate for the full load amp rating of the connected motor. Circuit breakers are either 22,000 or 100,000 amp interrupt capacity.

Combination AC Magnetic Three Pole Starters with Circuit Breaker Disconnect and Overload Relays

Material		Craft@Hrs	Unit	Material Cost	Labor Cost	Installed Cost

NEMA 4X combination three pole AC magnetic starters, C.B. disconnect and overload relays

Material		Craft@Hrs	Unit	Material Cost	Labor Cost	Installed Cost
Size 2	10 HP, 208V	1E@2.00	Ea	1,850.00	53.80	1,903.80
Size 2	15 HP, 230V	1E@2.00	Ea	1,850.00	53.80	1,903.80
Size 2	25 HP, 480V	1E@2.00	Ea	1,850.00	53.80	1,903.80
Size 2	25 HP, 600V	1E@2.00	Ea	1,850.00	53.80	1,903.80
Size 3	25 HP, 208V	1E@3.00	Ea	2,930.00	80.60	3,010.60
Size 3	30 HP, 230V	1E@3.00	Ea	2,930.00	80.60	3,010.60
Size 3	50 HP, 480V	1E@3.00	Ea	2,930.00	80.60	3,010.60
Size 3	50 HP, 600V	1E@3.00	Ea	2,930.00	80.60	3,010.60
Size 4	40 HP, 208V	1E@4.00	Ea	5,120.00	108.00	5,228.00
Size 4	50 HP, 230V	1E@4.00	Ea	5,120.00	108.00	5,228.00
Size 4	100 HP, 480V	1E@4.00	Ea	5,120.00	108.00	5,228.00
Size 4	100 HP, 600V	1E@4.00	Ea	5,120.00	108.00	5,228.00

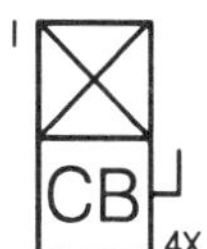

NEMA 7 & 9 combination three pole AC magnetic starters, C.B. disconnect and overload relays

Material		Craft@Hrs	Unit	Material Cost	Labor Cost	Installed Cost
Size 0	3 HP, 208V	1E@1.75	Ea	1,350.00	47.00	1,397.00
Size 0	3 HP, 230V	1E@1.75	Ea	1,350.00	47.00	1,397.00
Size 0	5 HP, 480V	1E@1.75	Ea	1,350.00	47.00	1,397.00
Size 0	5 HP, 600V	1E@1.75	Ea	1,350.00	47.00	1,397.00
Size 1	7.5 HP, 208V	1E@1.75	Ea	1,390.00	47.00	1,437.00
Size 1	7.5 HP, 230V	1E@1.75	Ea	1,390.00	47.00	1,437.00
Size 1	10 HP, 480V	1E@1.75	Ea	1,390.00	47.00	1,437.00
Size 1	10 HP, 600V	1E@1.75	Ea	1,390.00	47.00	1,437.00
Size 2	10 HP, 208V	1E@2.25	Ea	1,850.00	60.50	1,910.50
Size 2	15 HP, 230V	1E@2.25	Ea	1,850.00	60.50	1,910.50
Size 2	25 HP, 480V	1E@2.25	Ea	1,850.00	60.50	1,910.50
Size 2	25 HP, 600V	1E@2.25	Ea	1,850.00	60.50	1,910.50
Size 3	25 HP, 208V	1E@3.50	Ea	3,070.00	94.10	3,164.10
Size 3	30 HP, 230V	1E@3.50	Ea	3,070.00	94.10	3,164.10
Size 3	50 HP, 480V	1E@3.50	Ea	3,070.00	94.10	3,164.10
Size 3	50 HP, 600V	1E@3.50	Ea	3,070.00	94.10	3,164.10
Size 4	40 HP, 208V	1E@4.50	Ea	4,790.00	121.00	4,911.00
Size 4	50 HP, 230V	1E@4.50	Ea	4,790.00	121.00	4,911.00
Size 4	100 HP, 480V	1E@4.50	Ea	4,790.00	121.00	4,911.00
Size 4	100 HP, 600V	1E@4.50	Ea	4,790.00	121.00	4,911.00
Size 5	75 HP, 208V	2E@8.00	Ea	10,400.00	215.00	10,615.00
Size 5	100 HP, 230V	2E@8.00	Ea	10,400.00	215.00	10,615.00
Size 5	200 HP, 480V	2E@8.00	Ea	10,400.00	215.00	10,615.00
Size 5	200 HP, 600V	2E@8.00	Ea	10,400.00	215.00	10,615.00

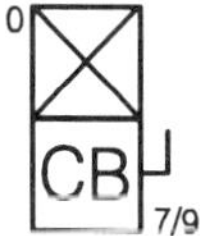

Use these figures to estimate the cost of combination starters installed in buildings under the conditions described on pages 5 and 6. Costs listed are for each starter installed. The crew size is one electrician for starters up to size 4 and two electricians for over size 4. Cost per manhour is $26.88. These costs include layout, material handling, and normal waste. Add for overload relay heater elements, sales tax, delivery, supervision, mobilization, demobilization, cleanup, overhead and profit. Note: AC magnetic starters are non-reversing with three melting alloy overload relays for motor protection. The heating elements must be appropriate for the full load amp rating of the connected motor. Circuit breakers are either 22,000 or 100,000 amp interrupt capacity.

Combination AC Magnetic Three Pole Starters with Circuit Breaker Disconnect and Overload Relays

Material		Craft@Hrs	Unit	Material Cost	Labor Cost	Installed Cost

NEMA 12 combination three pole AC magnetic starters, C.B. disconnect and overload relays

Material		Craft@Hrs	Unit	Material Cost	Labor Cost	Installed Cost
Size 0	3 HP, 208V	1E@1.60	Ea	761.00	43.00	804.00
Size 0	3 HP, 230V	1E@1.60	Ea	761.00	43.00	804.00
Size 0	5 HP, 480V	1E@1.60	Ea	761.00	43.00	804.00
Size 0	5 HP, 600V	1E@1.60	Ea	761.00	43.00	804.00
Size 1	7.5 HP, 208V	1E@1.60	Ea	786.00	43.00	829.00
Size 1	7.5 HP, 230V	1E@1.60	Ea	786.00	43.00	829.00
Size 1	10 HP, 480V	1E@1.60	Ea	786.00	43.00	829.00
Size 1	10 HP, 600V	1E@1.60	Ea	786.00	43.00	829.00
Size 2	10 HP, 208V	1E@2.10	Ea	1,110.00	56.40	1,166.40
Size 2	15 HP, 230V	1E@2.10	Ea	1,110.00	56.40	1,166.40
Size 2	25 HP, 480V	1E@2.10	Ea	1,110.00	56.40	1,166.40
Size 2	25 HP, 600V	1E@2.10	Ea	1,110.00	56.40	1,166.40
Size 3	25 HP, 208V	1E@3.25	Ea	1,600.00	87.40	1,687.40
Size 3	30 HP, 230V	1E@3.25	Ea	1,600.00	87.40	1,687.40
Size 3	50 HP, 480V	1E@3.25	Ea	1,600.00	87.40	1,687.40
Size 3	50 HP, 600V	1E@3.25	Ea	1,600.00	87.40	1,687.40
Size 4	40 HP, 208V	1E@4.50	Ea	3,620.00	121.00	3,741.00
Size 4	50 HP, 230V	1E@4.50	Ea	3,620.00	121.00	3,741.00
Size 4	100 HP, 480V	1E@4.50	Ea	3,620.00	121.00	3,741.00
Size 4	100 HP, 600V	1E@4.50	Ea	3,620.00	121.00	3,741.00
Size 5	75 HP, 208V	2E@8.50	Ea	8,090.00	228.00	8,318.00
Size 5	100 HP, 230V	2E@8.50	Ea	8,090.00	228.00	8,318.00
Size 5	200 HP, 480V	2E@8.50	Ea	8,090.00	228.00	8,318.00
Size 5	200 HP, 600V	2E@8.50	Ea	8,090.00	228.00	8,318.00
Size 6	150 HP, 208V	2E@12.5	Ea	16,600.00	336.00	16,936.00
Size 6	200 HP, 230V	2E@12.5	Ea	16,600.00	336.00	16,936.00
Size 6	400 HP, 480V	2E@12.5	Ea	16,600.00	336.00	16,936.00
Size 6	400 HP, 600V	2E@12.5	Ea	16,600.00	336.00	16,936.00

Use these figures to estimate the cost of combination starters installed in buildings under the conditions described on pages 5 and 6. Costs listed are for each starter installed. The crew size is one electrician for starters up to size 4 and two electricians for over size 4. Cost per manhour is $26.88. These costs include layout, material handling, and normal waste. Add for overload relay heater elements, sales tax, delivery, supervision, mobilization, demobilization, cleanup, overhead and profit. Note: AC magnetic starters are non-reversing with three melting alloy overload relays for motor protection. The heating elements must be appropriate for the full load amp rating of the connected motor. Circuit breakers are either 22,000 or 100,000 amp interrupt capacity.

Combination AC Magnetic Three Pole Starters in Oversized Enclosure with Circuit Breaker Disconnect and Overload Relays

Material	Craft@Hrs	Unit	Material Cost	Labor Cost	Installed Cost

NEMA 1 combination three pole AC magnetic starters, oversized enclosure, C.B. disconnect

Material	Craft@Hrs	Unit	Material Cost	Labor Cost	Installed Cost
Size 0 3 HP, 208V	1E@1.30	Ea	768.00	34.90	802.90
Size 0 3 HP, 230V	1E@1.30	Ea	768.00	34.90	802.90
Size 0 5 HP, 480V	1E@1.30	Ea	768.00	34.90	802.90
Size 0 5 HP, 600V	1E@1.30	Ea	768.00	34.90	802.90
Size 1 7.5 HP, 208V	1E@1.30	Ea	792.00	34.90	826.90
Size 1 7.5 HP, 230V	1E@1.30	Ea	792.00	34.90	826.90
Size 1 10 HP, 480V	1E@1.30	Ea	792.00	34.90	826.90
Size 1 10 HP, 600V	1E@1.30	Ea	792.00	34.90	826.90
Size 2 10 HP, 208V	1E@1.60	Ea	1,070.00	43.00	1,113.00
Size 2 15 HP, 230V	1E@1.60	Ea	1,070.00	43.00	1,113.00
Size 2 25 HP, 480V	1E@1.60	Ea	1,070.00	43.00	1,113.00
Size 2 25 HP, 600V	1E@1.60	Ea	1,070.00	43.00	1,113.00

NEMA 4 combination three pole AC magnetic starters, oversized enclosure, C.B. disconnect

Material	Craft@Hrs	Unit	Material Cost	Labor Cost	Installed Cost
Size 0 3 HP, 208V	1E@1.60	Ea	1,510.00	43.00	1,553.00
Size 0 3 HP, 230V	1E@1.60	Ea	1,510.00	43.00	1,553.00
Size 0 5 HP, 480V	1E@1.60	Ea	1,510.00	43.00	1,553.00
Size 0 5 HP, 600V	1E@1.60	Ea	1,510.00	43.00	1,553.00
Size 1 7.5 HP, 208V	1E@1.60	Ea	1,530.00	43.00	1,573.00
Size 1 7.5 HP, 230V	1E@1.60	Ea	1,530.00	43.00	1,573.00
Size 1 10 HP, 480V	1E@1.60	Ea	1,530.00	43.00	1,573.00
Size 1 10 HP, 600V	1E@1.60	Ea	1,530.00	43.00	1,573.00
Size 2 10 HP, 208V	1E@2.10	Ea	2,060.00	56.40	2,116.40
Size 2 15 HP, 230V	1E@2.10	Ea	2,060.00	56.40	2,116.40
Size 2 25 HP, 480V	1E@2.10	Ea	2,060.00	56.40	2,116.40
Size 2 25 HP, 600V	1E@2.10	Ea	2,060.00	56.40	2,116.40

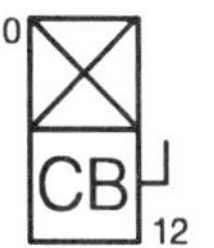

NEMA 12 combination three pole AC magnetic starters, oversized enclosure, C.B. disconnect

Material	Craft@Hrs	Unit	Material Cost	Labor Cost	Installed Cost
Size 0 3 HP, 208V	1E@1.60	Ea	988.00	43.00	1,031.00
Size 0 3 HP, 230V	1E@1.60	Ea	988.00	43.00	1,031.00
Size 0 5 HP, 480V	1E@1.60	Ea	988.00	43.00	1,031.00
Size 0 5 HP, 600V	1E@1.60	Ea	988.00	43.00	1,031.00

Use these figures to estimate the cost of combination starters installed in buildings under the conditions described on pages 5 and 6. Costs listed are for each starter installed. The crew size is one electrician for starters up to size 4 and two electricians for over size 4. Cost per manhour is $26.88. These costs include layout, material handling, and normal waste. Add for overload relay heater elements, sales tax, delivery, supervision, mobilization, demobilization, cleanup, overhead and profit. Note: AC magnetic starters are non-reversing with three melting alloy overload relays for motor protection. The heating elements must be appropriate for the full load amp rating of the connected motor. Circuit breakers are either 22,000 or 100,000 amp interrupt capacity.

Motor Control Equipment

Material	Craft@Hrs	Unit	Material Cost	Labor Cost	Installed Cost
NEMA 12 combination three pole AC magnetic starters in oversize enclosure with C.B. disconnect					
Size 1 7.5 HP, 208V	1E@1.60	Ea	1,010.00	43.00	1,053.00
Size 1 7.5 HP, 230V	1E@1.60	Ea	1,010.00	43.00	1,053.00
Size 1 10 HP, 480V	1E@1.60	Ea	1,010.00	43.00	1,053.00
Size 1 10 HP, 600V	1E@1.60	Ea	1,010.00	43.00	1,053.00
Size 2 10 HP, 208V	1E@2.10	Ea	1,340.00	56.40	1,396.40
Size 2 15 HP, 230V	1E@2.10	Ea	1,340.00	56.40	1,396.40
Size 2 25 HP, 480V	1E@2.10	Ea	1,340.00	56.40	1,396.40
Size 2 25 HP, 600V	1E@2.10	Ea	1,340.00	56.40	1,396.40
NEMA 1 two & three unit general purpose motor control stations					
Stop-start	1E@0.40	Ea	33.30	10.80	44.10
Stop-start lockout stop	1E@0.40	Ea	53.20	10.80	64.00
Stop-start main. contact	1E@0.40	Ea	50.00	10.80	60.80
Forward-reverse	1E@0.40	Ea	47.80	10.80	58.60
Open-close	1E@0.40	Ea	47.80	10.80	58.60
Up-down	1E@0.40	Ea	44.30	10.80	55.10
On-off, main. contact	1E@0.40	Ea	57.20	10.80	68.00
Hand-auto main. contact	1E@0.40	Ea	57.20	10.80	68.00
Hand-off-auto	1E@0.50	Ea	139.00	13.40	152.40
NEMA 1 two & three unit heavy duty motor control stations					
Stop-start	1E@0.50	Ea	105.00	13.40	118.40
Stop-start lockout stop	1E@0.50	Ea	134.00	13.40	147.40
Stop-start main. contact	1E@0.50	Ea	139.00	13.40	152.40
Forward-reverse	1E@0.50	Ea	122.00	13.40	135.40
Open-close	1E@0.50	Ea	122.00	13.40	135.40
Up-down	1E@0.50	Ea	122.00	13.40	135.40
On-off, main. contact	1E@0.50	Ea	139.00	13.40	152.40
Hand-auto main. contact	1E@0.50	Ea	197.00	13.40	210.40
Hand-off-auto	1E@0.60	Ea	152.00	16.10	168.10

Use these figures to estimate the cost of combination starters and control stations installed in buildings under the conditions described on pages 5 and 6. Costs listed for control stations are for each station installed. The crew size is one electrician working at a labor cost of $26.88 per manhour. These costs include the enclosure, layout, material handling, and normal waste. Add for sales tax, delivery, supervision, mobilization, demobilization, cleanup, overhead and profit. Note: Be sure to select the right switch for the control sequence. Many other general purpose and heavy duty control stations are available when other combinations and features are required. Most units can be assembled into special configurations to meet specific needs.

Material	Craft@Hrs	Unit	Material Cost	Labor Cost	Installed Cost

NEMA 1 general purpose flush mounted two & three unit motor control stations

Material	Craft@Hrs	Unit	Material Cost	Labor Cost	Installed Cost
Stop-start	1E@0.40	Ea	50.90	10.80	61.70
Stop-start lockout stop	1E@0.40	Ea	70.70	10.80	81.50
Stop-start main. contact	1E@0.40	Ea	70.70	10.80	81.50
Forward-reverse	1E@0.40	Ea	60.40	10.80	71.20
Open-close	1E@0.40	Ea	60.40	10.80	71.20
Up-down	1E@0.40	Ea	60.40	10.80	71.20
On-off	1E@0.40	Ea	60.40	10.80	71.20
On-off, main. contact	1E@0.40	Ea	70.70	10.80	81.50
Hand-auto, main. contact	1E@0.40	Ea	70.70	10.80	81.50
Hand-off auto	1E@0.60	Ea	152.00	16.10	168.10

NEMA 4 general purpose two & three unit motor control stations

Material	Craft@Hrs	Unit	Material Cost	Labor Cost	Installed Cost
Stop-start	1E@0.50	Ea	105.00	13.40	118.40
Stop-start lockout stop	1E@0.50	Ea	105.00	13.40	118.40
Stop-start main. contact	1E@0.50	Ea	139.00	13.40	152.40
Forward-reverse	1E@0.50	Ea	122.00	13.40	135.40
Open-close	1E@0.50	Ea	122.00	13.40	135.40
Up-down	1E@0.50	Ea	122.00	13.40	135.40
On-off	1E@0.50	Ea	122.00	13.40	135.40
On-off, main. contact	1E@0.50	Ea	139.00	13.40	152.40
Hand-auto main. contact	1E@0.50	Ea	139.00	13.40	152.40
Stop-start main. contact	1E@0.60	Ea	139.00	16.10	155.10

NEMA 4 heavy duty two & three unit motor control stations

Material	Craft@Hrs	Unit	Material Cost	Labor Cost	Installed Cost
Stop-start	1E@0.60	Ea	105.00	16.10	121.10
Stop-start lockout stop	1E@0.60	Ea	105.00	16.10	121.10
Stop-start main. contact	1E@0.60	Ea	139.00	16.10	155.10
Forward-reverse	1E@0.60	Ea	122.00	16.10	138.10
Open-close	1E@0.60	Ea	122.00	16.10	138.10
Up-down	1E@0.60	Ea	122.00	16.10	138.10
On-off	1E@0.60	Ea	122.00	16.10	138.10
Jog	1E@0.60	Ea	122.00	16.10	138.10
Manual-auto	1E@0.60	Ea	122.00	16.10	138.10
Hand-off-auto	1E@0.75	Ea	139.00	20.20	159.20

Use these figures to estimate the cost of control stations installed in buildings under the conditions described on pages 5 and 6. Costs listed are for each station installed. The crew size is one electrician working at a labor cost of $26.88 per manhour. These costs include the enclosure, layout, material handling, and normal waste. Add for sales tax, delivery, supervision, mobilization, demobilization, cleanup, overhead and profit. Note: Be sure to select the right switch for the control sequence. Many other general purpose and heavy duty control stations are available when other combinations and features are required. Most units can be assembled into special configurations to meet specific needs.

Section 14:
Trenching and Excavation

Many electrical estimators feel like a fish out of water when estimating trenching and excavation. There are too many unknowns. Every soil type has unique characteristics ready to trap the unwary and unprepared. What was supposed to be a simple trench can become a financial disaster when shoring or dewatering become necessary, when you discover an outcropping of rock at mid-trench, when you cut into an unrecorded water line or unmarked septic system or when a downpour fills a newly dug trench with silt.

If you feel uncomfortable about the unknowns when estimating trenching and excavation, take heart. Many seasoned excavation estimators feel exactly the same way. It goes with the territory. And as long as trenching and excavation are included with the electrical portion of the work, electrical estimators will have to work up bids for earthwork.

Of course, the easiest way to handle trenching and excavation is to get a firm quote from a dirt contractor. The more bids you get, the better the chances of getting an attractive price. But no matter how many bids you get, my advice is to protect yourself. Know enough about estimating trenching and excavation to read and evaluate each bid received. Be able to recognize a bid that doesn't include everything that's required or includes prices for more work than is necessary.

There are ways to reduce your risk and produce consistently reliable estimates for excavation and trenching — even if you're not a licensed, card-carrying earthwork estimator. With a little care and practice, you'll feel as confident about trench estimates as any other part of an electrical estimator's work.

Electrical contractors use many types of trenchers. Wheel trenchers dig a lot of trench in a short time and adapt well to most soil types. But they're impractical on the smaller jobs most electrical contractors handle regularly. Backhoes are the industry standard for deeper and wider trenches. But chain trenchers are a better choice for narrower and shallower trench work. The best choice in equipment depends on soil conditions, size of the trench, and the amount of work necessary. There's no single piece of equipment that fits every job.

Many electrical contractors have their own trenching equipment and routinely do all or nearly all of their own excavation. Others rent what's needed on a job by job basis. Of course, there are advantages to doing your own trenching. It's done when you want and the way you want, not when your trenching sub has the time. Many smaller electrical contractors can profit by doing their own trenching *if* they have an experienced operator available and *if* they estimate trench costs accurately.

My advice is to avoid buying excavation equipment if good rental equipment is available locally and if your need for a particular piece of equipment is less than two weeks out of most months. After that, buying or leasing on a long-term basis may lower equipment costs.

No matter whether you own or rent equipment, keep track of equipment productivity rates and operating costs. There should be two profits on every job: One goes into your wallet. The other is what you learn that will make future estimates more accurate. If you're not keeping good cost records, you're missing half the profit each job offers. This is especially true in earthwork where costs can multiply very quickly.

The manhour tables in this section will help if you have no other reliable data. But no manhour figure you find in a book can be as reliable as your own cost data on work done by your crews, with your equipment and under conditions you know and understand. What's listed in this section will be either too high or too low for most contractors on most jobs. But the figures I've selected will be a good starting point for most trenching work with most types of equipment and in most types of soil. Refining these figures for the work you do is your job.

Doing the Take-Off

The first step on every trench estimate is to look for boring data logs on the plans. They're your best clue about subgrade conditions. If there are no logs, make an on-site inspection — with a shovel. You'll seldom have to trench more than 3 or 4 feet below the surface. Scrape off the top 12 inches to remove any topsoil. What you see about a foot down is probably the same soil that's 3 feet below the surface.

Sandy soil is easy trenching. But sand tends to dry out and cave in. When digging in sand, schedule work so that duct lines go in immediately after the trenching. Backfill as soon as possible.

In very hard or rocky soil, a tracked trencher is best. A backhoe probably won't be adequate. No matter what type of equipment is used, make an allowance if it isn't in good condition, if it's prone to breakdowns or if the operator isn't experienced.

Location of the trench can make a big difference in the manhours and equipment time required. Poor access always slows production and can make it impossible to use the most productive equipment.

Always consider what you're going to do with the spoil that's removed, even when it's all going to go back in the trench eventually. On some sites there won't be room to stockpile spoil beside the trench. Even if there's plenty of room, some soil or debris will have to be hauled away on many jobs. If you're breaking out pavement, hauling broken pavement to a legal dump and paying the dump fee may be major cost items.

Don't automatically think of trenching equipment every time you see a trench on the plans. Sometimes it's cheaper to put a laborer to work with a shovel for a day or two. Consider all the costs involved in renting a trencher, moving it to the site, hiring an operator to do a few hours of work, and then taking the trencher back to the yard. The fastest way to dig a trench isn't always the cheapest.

Of course, there are limits to hand digging. Trenches over about 4 feet deep are usually machine work, especially in rocky soil or hardpan. And hot or cold weather will slow hand production more than machine work.

Consider jacking or boring conduit under streets and walkways rather than breaking pavement and trenching. Pipe jacking is expensive. But it's easy to get a bid on this type of work. And that bid may cut your costs when pavement sawing, excavation, backfill, hauling debris, dump fees, and patching pavement or concrete are considered. Jacking conduit is routine for street light and signal contractors.

Examine the cost of pavement removal and patching very carefully on any job where a line has to pass under an existing pavement or concrete surface. Most specs require that pavement be saw cut before being broken out. Saw cutting can be expensive, especially on a slope or where access is restricted. Backfill usually has to include a specific sub-base material between the compacted soil and the asphalt or concrete patch. Getting a small delivery of exactly the right sub-base may be expensive.

Occasionally some earthwork other than trenching will be included in the electrical portion of the plans. Foundations for light standards are probably the most common example. Estimating work like this is similar to estimating any trench. Study the plans. Figure the volume of soil to move. Make an educated guess of labor and equipment productivity per hour. Then multiply the cost per hour by the number of hours required.

You'll find that it's cheaper to drill light pole foundations on larger jobs. Consider renting a drill rig with an experienced operator if you don't have someone who meets that description on your payroll. In many areas you can get a quote on drilling per hole or for the entire job.

If you need less than a few hour's work with a drilling rig, consider hand digging. Even if the work's done with a drill rig, there's plenty of hand work. The pit may have to be shaped by hand. You'll have to spread the piles of spoil or load them for disposal. And some backfilling may be needed.

Trenching and Excavation

Material	Craft@Hrs	Unit	Material Cost	Labor Cost	Installed Cost
Trenching per linear foot of trench					
10" x 12"	1E@0.02	LF	2.66	.54	3.20
12" x 12"	1E@0.03	LF	2.83	.81	3.64
18" x 12"	1E@0.04	LF	3.10	1.08	4.18
24" x 12"	1E@0.05	LF	3.28	1.34	4.62
10" x 18"	1E@0.03	LF	3.00	.81	3.81
12" x 18"	1E@0.04	LF	3.17	1.08	4.25
18" x 18"	1E@0.05	LF	3.43	1.34	4.77
24" x 18"	1E@0.06	LF	3.61	1.61	5.22
10" x 24"	1E@0.04	LF	3.07	1.08	4.15
12" x 24"	1E@0.05	LF	3.25	1.34	4.59
18" x 24"	1E@0.06	LF	3.51	1.61	5.12
24" x 24"	1E@0.07	LF	3.70	1.88	5.58
10" x 30"	1E@0.05	LF	3.17	1.34	4.51
12" x 30"	1E@0.06	LF	3.34	1.61	4.95
18" x 30"	1E@0.07	LF	3.70	1.88	5.58
24" x 30"	1E@0.08	LF	3.86	2.15	6.01
10" x 36"	1E@0.06	LF	3.17	1.61	4.78
12" x 36"	1E@0.07	LF	3.51	1.88	5.39
18" x 36"	1E@0.08	LF	3.86	2.15	6.01
24" x 36"	1E@0.09	LF	4.04	2.42	6.46
12" x 40"	1E@0.07	LF	3.51	1.88	5.39
18" x 40"	1E@0.08	LF	3.86	2.15	6.01
24" x 40"	1E@0.09	LF	4.22	2.42	6.64
30" x 40"	1E@0.10	LF	4.57	2.69	7.26
12" x 48"	1E@0.08	LF	3.86	2.15	6.01
18" x 48"	1E@0.09	LF	4.22	2.42	6.64
24" x 48"	1E@0.10	LF	4.57	2.69	7.26
30" x 48"	1E@0.11	LF	5.61	2.96	8.57
18" x 54"	1E@0.10	LF	4.39	2.69	7.08
24" x 54"	1E@0.11	LF	4.74	2.96	7.70
30" x 54"	1E@0.12	LF	5.61	3.23	8.84
36" x 54"	1E@0.13	LF	6.15	3.49	9.64
18" x 60"	1E@0.11	LF	6.15	2.96	9.11
24" x 60"	1E@0.12	LF	7.03	3.23	10.26
30" x 60"	1E@0.13	LF	7.90	3.49	11.39
36" x 60"	1E@0.14	LF	9.67	3.76	13.43
Pit excavation by pit size					
24" x 24" x 24"	1E@0.14	Ea	8.79	3.76	12.55
24" x 36" x 24"	1E@0.18	Ea	7.03	4.84	11.87
24" x 48" x 24"	1E@0.20	Ea	14.10	5.38	19.48
24" x 24" x 36"	1E@0.21	Ea	18.50	5.64	24.14
24" x 36" x 36"	1E@0.27	Ea	21.00	7.26	28.26
24" x 48" x 36"	1E@0.30	Ea	24.60	8.06	32.66
24" x 24" x 48"	1E@0.28	Ea	35.10	7.53	42.63
24" x 36" x 48"	1E@0.36	Ea	70.30	9.68	79.98
24" x 48" x 48"	1E@0.40	Ea	105.00	10.80	115.80

Use these figures to estimate the cost of trenching and excavation done under the conditions described on pages 5 and 6. Costs listed are for each linear foot of trench excavated or for each pit excavated. The crew is one electrician working at a labor cost of $26.88 manhour. These costs include trenching or excavation equipment and layout. Add for surface cutting, patching, encasement, warning tape, barricades, shoring, clean-up, compaction, delivery, supervision, mobilization, demobilization, overhead and profit. Note: Costs will be higher when obstructions or poor digging conditions delay production. It's important to check ground conditions before estimating excavation work.

Section 15:
Surface Raceways

Surface raceways have been used since the early 1900's to cover and protect wiring that's not concealed in a wall. Originally it was used to add circuits in buildings that were erected before electrical power was widely available. Today, surface raceway can be used any time circuits have to be added or altered to meet changing needs of the occupants in the interior of offices, hospitals, stores, residences, and manufacturing plants.

Raceway and fittings are made from steel, aluminum and plastic to meet a wide variety of surface wiring requirements. Smaller sizes are made for situations where only a few conductors are needed to carry lower amperage. Larger sizes are available to carry more conductors and supply heavier electrical loads. Most manufacturers sell fittings that can be used to route raceway around nearly any obstacle on the wall surface. But it's important to use fittings appropriate for the type and size of raceway being installed.

Surface raceway can be installed on the ceiling, on the wall, in shelving, in counters or in any other dry interior location. It can be installed on drywall, plaster, wood, paneling, unit masonry or concrete surfaces. Surface raceway comes with a factory finish but can be painted after installation.

Use a hacksaw to cut surface raceway. Before assembling the cut piece and pulling wire, be sure to deburr the inside and outside of the cut end. This can be done with a small flat file or an electrician's pocket knife. The hacksaw blade should be shatterproof, high-speed steel with 40 teeth per inch. A shear can be used in place of a hacksaw when cutting the base or cover of $1\frac{9}{32}$" by ¾" raceway.

Many types of companion outlet boxes are available for each size of raceway. These boxes are designed for surface mounting. Most come in several depths, rectangular or round for junction boxes or fixtures, single gang for junction boxes or wiring devices, and multi-ganged for grouping wiring devices.

Fittings are available for connecting surface raceway to conduit, to other concealed wiring, or for connecting surface raceway of one size to surface raceway of another size.

Pulling wire through surface raceway is easier if you use a special wire pulley. The pulley is clamped to the raceway at internal elbows and guides wire around the fitting. Use a fish tape leader to pull several conductors through the raceway.

One-Piece Raceway

One-piece surface raceway comes in three sizes: ½" by $\frac{11}{32}$", ¾" by $\frac{17}{32}$", and ¾" by $\frac{21}{32}$". This raceway is shaped to look like wood molding so it isn't so obvious when installed. Fittings are made to look like the raceway. Wire size and number of conductors determine the size of raceway to use.

All surface raceway and fittings should carry the Underwriter's Laboratory label. The *National Electrical Code*, Articles 352 and 353 are your primary guide when installing surface raceway. But many other sections of the *NEC* will apply too.

Surface raceway is manufactured in 10 foot lengths with 10 pieces to the bundle. Each length comes with a coupling.

A hand bender is available for bending ¾" surface raceway. It's used like an EMT hand bender. The bender can make simple offsets or short bends in a flat position where the raceway is installed against a ceiling or wall. An adapter can be inserted into the bender for bending ½" surface raceway.

Two-Piece Raceway

Two-piece raceway comes without couplings and in the following sizes:

$1\frac{9}{32}$" by ¾" raceway is made of .040" galvanized steel and packed in bundles of 10. Bundles of 10-foot lengths have 100 feet to the bundle. Bundles of 5-foot lengths have 50 feet to the bundle. The cover is scored every 3 inches for easy breakoff.

$1\frac{9}{32}$" by ¾" type 304 stainless steel raceway is available in 5-foot lengths.

1¼" by ⅞" .040" steel raceway comes in 10-foot lengths, 100 feet to a carton. The cover comes in 5-foot lengths, 100 feet to a carton, and is scored every 3 inches. It can be ordered unscored if necessary.

$2\frac{3}{8}$" by ¾" .040" galvanized steel raceway comes in 5- or 10-foot lengths. Base and cover are packed separately.

2¾" by $1\frac{17}{32}$" .040" steel raceway comes 50 feet to a carton. Bases are 10 feet long and covers are 5 feet long.

Bases for 4¾" by 1¾" are made of .050" galvanized steel. Covers and dividers are .040" galvanized steel. Base comes in 10-foot lengths, 50 feet per carton. Cover and divider come in 5-foot lengths. Base, cover and divider are packed separately.

Bases for 4¾" by 3⁹⁄₁₆" surface raceway are .060" galvanized steel. Covers are .040" galvanized steel. Base and cover are packed 20 feet per carton. Base comes in 5- or 10-foot lengths, cover comes in 5-foot lengths.

Larger surface raceway is used in laboratories, TV studios, communication centers, computer rooms, control rooms, plants, and indoor areas that are dry and exposed. Base for two-piece raceway can be installed on any of the surfaces mentioned above for one-piece raceways.

Overfloor Raceway

Overfloor raceway comes in two sizes. The smaller size is 1⁹⁄₁₆" by ¹¹⁄₃₂", .040" galvanized steel and comes in 5- or 10-foot lengths with 50 feet per carton. The raceway base is scored every 3 inches for easy breakoff. It can be cut with the same hacksaw as other surface raceway. It's recommended that a reinforcing member be inserted in overfloor raceway where heavy traffic is likely. The larger overfloor raceway size is 2⁷⁄₃₂" by ²³⁄₃₂". The base is .040" galvanized steel and the cover is .050" steel. The raceway comes in 5- or 10-foot lengths with 5 pieces per carton. The base is scored every 3 inches for easy breakoff.

Many fittings can be used with overfloor raceway. Adapters are available for joining overfloor raceway to wall-mounted surface raceway. Of course, junction boxes and outlet boxes are made for distribution of power on the floor surface. Overfloor raceway is used to protect power, telephone, computer or signal wiring. It's tapered at the top edges so it's less of an obstruction when laid on the floor. The raceway comes with a factory finish and can be field painted.

Multi-Outlet Systems

Multi-outlet strips are a good choice in laboratories, office work stations, shops, stores, hobby shops, over work benches, in schools, hospitals, and clean rooms where many electrical appliances may be in use in a small area. These outlet strips can be used indoors in any non-hazardous location not subject to mechanical injury or abuse.

Multi-outlet strip is specified by the outlet spacing, the type of outlet, color and number of circuits. A wide range of fittings are available. The base and covers are cut to fit and deburred the same as other surface raceway. The covers are factory punched for outlet spacing. The outlets are prewired at the desired spacing as a single or multiple circuit.

Factory preassembled multi-outlet strip is available complete with cord and plug. The units come in a strip length of 2, 3, 5 and 6 feet, with a duplex receptacle or 6 single receptacles. The cords are usually 2 to 6 feet long. These units are handy when temporary displays are being installed. Colors and finishes available include antique bronze, ivory, brown, and stainless steel. Strips are available with a button-reset mini-breaker or lighted on-off switch.

Telepower Poles

A simple way to provide electrical, telephone, or computer wiring to a work station or desk located away from a wall is to use a telepower pole. The pole is attached to the drop ceiling grid system and secured to the work station or to the floor next to the desk. A junction box on the top of the pole above the grid system provides the connection point for electrical, telephone and computer system wiring. The telepower pole has divided raceway for each of the conductors. Receptacles for electrical service are factory installed in the assembled pole.

A range of colors and styles is available. The pole can be attached to the floor or merely set on a carpeted floor. A universal base plate is made for either application.

The telepower poles are about 10'5" inches long. When used in a room with a ceiling grid higher than that, use a section of two-piece raceway to extend the telepower pole through the ceiling grid. First remove the top junction box. Cut at least 1 foot from one side of the telepower pole, leaving one side longer than the other. Cut the two-piece raceway to the desired length. Tie the sections together with stove bolts. The section added should be painted to match the telepower pole.

Channel Wire Systems

Channel wire systems are popular in manufacturing plants, warehouses, storage areas, parts rooms, assembly areas and other large work areas. The channel is a simple 1⅝" by 1⅝" steel strut through which a flat conductor assembly has been drawn. The flat conductor is rated for 30 amps at 277/480 volts.

A special junction box is attached to the strut at the feed-in end. At the other end of the strut an insulated end cap is installed. Splice plates are used where the strut is coupled together.

Color coded caps can be inserted anywhere along the strut to tap into the flat conductor assembly. This offers access to power along the entire length of the channel. The color-coded caps provide connections to the phase conductors. For 277 volt power, the black colored cap connects phase A and neutral. The red cap connects phase B and neutral, and the blue cap connects phase C and neutral. For

480 volt power, the black over red cap connects to phase A and phase B, the red over blue cap connects to phase B and phase C, the blue over black cap connects to phase B and phase A.

A hanger assembly is used with each color coded cap to complete the tap to the flat conductors. Later the tap assembly can be relocated because the flat conductor is self healing.

For a high bay lighting job, the strut is simply installed in rows where the lighting will be placed. The feed-in box is installed on each run of strut, and an insulated end cap is installed on the end of each strut run. The strut is supported with standard strut hanger assemblies spaced close enough to provide adequate support for the lighting fixtures planned. After the strut is installed, the flat wire is drawn into the strut and connected to the feed-in junction boxes, and wired back through conduit to the panel. The color coded cap is installed with the fixture feed hanger. The fixture is then attached to the hanger assembly and lamped.

When fluorescent fixtures are installed, one color coded cap with hanger and one fixture hanger is used to secure each 4 foot or 8 foot fixture to the channel. Only one feed-in assembly is required for each fluorescent fixture.

Dual feed-in junction boxes are available. Two sets of lighting loads can be connected to the electrical system from the same junction box. On longer struts more dual feed-in junction boxes may be used.

System Engineering

Leading manufacturers of surface raceway and channel wiring can usually assist you with design and layout of a channel wiring system. An indoor lighting system can be designed for fluorescent, mercury vapor, metal halide, high pressure sodium or low pressure sodium lighting, depending on the light density needed. Your material supplier will probably be able to suggest the name of a manufacturer that can provide this design service.

Surface Raceway, Steel

Material	Craft@Hrs	Unit	Material Cost	Labor Cost	Installed Cost
One-piece steel surface raceway					
1/2" x 11/32"	1E@3.00	CLF	57.10	80.60	137.70
3/4" x 17/32"	1E@3.50	CLF	59.50	94.10	153.60
3/4" x 21/32"	1E@3.75	CLF	67.50	101.00	168.50
Two-piece steel surface raceway, base only					
1-9/32" x 3/4"	1E@3.50	CLF	62.60	94.10	156.70
1-1/4" x 7/8"	1E@3.75	CLF	93.80	101.00	194.80
2-3/8" x 3/4"	1E@4.00	CLF	93.10	108.00	201.10
2-3/4" x 1-17/32"	1E@5.00	CLF	155.00	134.00	289.00
4-3/4" x 1-3/4"	1E@6.00	CLF	261.00	161.00	422.00
4-3/4" x 3-9/16"	1E@6.50	CLF	486.00	175.00	661.00
Two-piece steel surface raceway, cover only					
1-1/4" x 7/8"	1E@2.00	CLF	59.50	53.80	113.30
2-3/8" x 3/4"	1E@2.25	CLF	56.30	60.50	116.80
2-3/4" x 1-17/32"	1E@2.50	CLF	80.50	67.20	147.70
4-3/4" x 1-3/4"	1E@3.00	CLF	151.00	80.60	231.60
4-3/4" x 3-9/16"	1E@3.25	CLF	166.00	87.40	253.40
Overfloor steel surface raceway					
1-9/16" x 11/32"	1E@4.50	CLF	125.00	121.00	246.00
2-7/32" x 23/32"	1E@6.00	CLF	193.00	161.00	354.00
Overfloor steel surface raceway fittings, 1-9/16" x 11/32"					
Reinforcing member	1E@2.75	CLF	295.00	73.90	368.90
Wire clip	1E@0.05	Ea	.18	1.34	1.52
Fiber bushing	1E@0.05	Ea	.39	1.34	1.73
Two-hole strap	1E@0.05	Ea	.34	1.34	1.68
Station tubing bracket	1E@0.10	Ea	.92	2.69	3.61
Flat elbow, 90 degree	1E@0.20	Ea	3.12	5.38	8.50
Internal elbow, 90 degree	1E@0.25	Ea	3.12	6.72	9.84
External elbow, 90 degree	1E@0.25	Ea	3.91	6.72	10.63
Adapter fitting	1E@0.15	Ea	4.01	4.03	8.04
Telephone outlet	1E@0.25	Ea	6.62	6.72	13.34
Narrow tel outlet	1E@0.25	Ea	6.31	6.72	13.03
Utility box	1E@0.25	Ea	5.80	6.72	12.52
Narrow junction box	1E@0.25	Ea	8.06	6.72	14.78
Combination connector	1E@0.20	Ea	3.93	5.38	9.31

Use these figures to estimate the cost of surface metal raceway installed in a building under the conditions described in the chapter text. Costs listed are for each 100 linear feet installed. The crew size is one electrician working at a labor cost of $26.88 per manhour. These costs include typical surface installation and layout for walls, ceiling, laboratory counters, work stations and overfloor, material handling and normal waste. Add for various fittings, boxes, sales tax, delivery, supervision, mobilization, demobilization, cleanup, overhead and profit. Note: Raceway runs are 100' long. Shorter runs will take more labor and longer runs will take less labor per linear foot.

Material	Craft@Hrs	Unit	Material Cost	Labor Cost	Installed Cost
Fittings for one-piece steel 1/2" x 11/32" surface raceway					
Coupling	1E@0.05	Ea	.26	1.34	1.60
Bushing	1E@0.05	Ea	.19	1.34	1.53
Supporting clip	1E@0.05	Ea	.23	1.34	1.57
One hole strap	1E@0.05	Ea	.17	1.34	1.51
Connection cover	1E@0.05	Ea	.28	1.34	1.62
Flat elbow, 90 degree	1E@0.15	Ea	1.06	4.03	5.09
Internal elbow, 90 degree	1E@0.15	Ea	2.02	4.03	6.05
Internal twisted 90 degree	1E@0.15	Ea	2.68	4.03	6.71
External elbow, 90 degree	1E@0.15	Ea	1.71	4.03	5.74
Adjustable J-box	1E@0.20	Ea	5.15	5.38	10.53
Extension adapter	1E@0.15	Ea	3.82	4.03	7.85
Reducing adapter	1E@0.15	Ea	2.10	4.03	6.13
Fittings for one-piece steel 3/4" x 17/32" surface raceway					
Bushing	1E@0.05	Ea	.15	1.34	1.49
One-hole strap	1E@0.05	Ea	.17	1.34	1.51
Connector cover	1E@0.05	Ea	.22	1.34	1.56
Flat elbow, 90 degree	1E@0.15	Ea	.88	4.03	4.91
Flat elbow, 45 degree	1E@0.15	Ea	2.58	4.03	6.61
Internal elbow, 90 degree	1E@0.15	Ea	1.10	4.03	5.13
External elbow, 90 degree	1E@0.15	Ea	1.04	4.03	5.07
Internal corner coupling	1E@0.15	Ea	1.24	4.03	5.27
Fittings for one-piece steel 3/4" x 17/32" or 3/4" x 21/32" surface raceway					
Flexible section, 18"	1E@0.20	Ea	10.00	5.38	15.38
Coupling	1E@0.05	Ea	.22	1.34	1.56
Supporting clip	1E@0.05	Ea	.25	1.34	1.59
Ground clamp	1E@0.05	Ea	2.69	1.34	4.03
Internal twist. elbow	1E@0.15	Ea	2.56	4.03	6.59
Tee	1E@0.20	Ea	2.13	5.38	7.51
Internal pull elbow	1E@0.20	Ea	7.07	5.38	12.45
Internal elbow, 90 degree	1E@0.15	Ea	1.72	4.03	5.75
Corner box	1E@0.20	Ea	7.30	5.38	12.68
Overfloor steel boxes for 1-9/16" x 11/32" surface raceway					
4" dia junction box	1E@0.25	Ea	5.13	6.72	11.85
Duplex recept & box	1E@0.35	Ea	9.61	9.41	19.02
Box for single recept	1E@0.25	Ea	6.13	6.72	12.85
Box for duplex recept	1E@0.25	Ea	7.72	6.72	14.44
Telephone outlet	1E@0.25	Ea	7.58	6.72	14.30

Use these figures to estimate the cost of surface metal raceway and fittings installed in a building under the conditions described in the chapter text. The crew size is one electrician working at a labor cost of $26.88 per manhour. These costs include layout, material handling and normal waste. Add for sales tax, delivery, supervision, mobilization, demobilization, cleanup, overhead and profit.

Steel Surface Raceway Fittings

Material	Craft@Hrs	Unit	Material Cost	Labor Cost	Installed Cost
Steel boxes for 1/2" x 11/32" or 3/4" x 17/32" surface raceway					

Material	Craft@Hrs	Unit	Material Cost	Labor Cost	Installed Cost
Single pole sw & box	1E@0.35	Ea	6.87	9.41	16.28
Duplex recept. & box	1E@0.25	Ea	8.12	6.72	14.84
Utility box	1E@0.20	Ea	4.60	5.38	9.98

Steel boxes for 3/4" x 17/32" or 3/4" x 21/32" surface raceway

Material	Craft@Hrs	Unit	Material Cost	Labor Cost	Installed Cost
Corner box, small	1E@0.25	Ea	5.48	6.72	12.20
Corner box, large	1E@0.25	Ea	7.72	6.72	14.44
Utility box, 3" dia.	1E@0.25	Ea	4.60	6.72	11.32
Blank cover	1E@0.05	Ea	2.45	1.34	3.79
Extension box, 3"	1E@0.20	Ea	5.57	5.38	10.95
Utility box, 4"	1E@0.25	Ea	4.94	6.72	11.66
Distribution box, 4"	1E@0.25	Ea	5.57	6.72	12.29
Blank cover, 4"	1E@0.05	Ea	2.46	1.34	3.80
Extension box, 4"	1E@0.20	Ea	6.01	5.38	11.39
Extension box, 5-1/2"	1E@0.25	Ea	6.14	6.72	12.86
Fixture box, 4-3/4"	1E@0.25	Ea	5.60	6.72	12.32
Fixture box, 5-1/2"	1E@0.30	Ea	5.69	8.06	13.75
Fixture box, 6-3/8"	1E@0.30	Ea	6.74	8.06	14.80
Device box, 1-3/8" deep	1E@0.25	Ea	3.77	6.72	10.49
Device box, 2 gang, 2-3/4"	1E@0.30	Ea	6.79	8.06	14.85
Device box, 3 gang, 2-3/4"	1E@0.40	Ea	16.80	10.80	27.60
Device box, 4 gang, 2-3/4"	1E@0.50	Ea	27.40	13.40	40.80
Device box, 5 gang, 2-3/4"	1E@0.65	Ea	46.80	17.50	64.30
Device box, 6 gang, 2-3/4"	1E@0.75	Ea	49.20	20.20	69.40

Steel fittings for 3/4" x 17/32" or 3/4" x 21/32" surface raceway

Material	Craft@Hrs	Unit	Material Cost	Labor Cost	Installed Cost
1/2" male box conn.	1E@0.05	Ea	1.62	1.34	2.96
3/4" male box conn.	1E@0.06	Ea	4.58	1.61	6.19
1/2" female box conn.	1E@0.05	Ea	1.67	1.34	3.01
3/4" female box conn.	1E@0.06	Ea	5.29	1.61	6.90
1/2" male elbow box	1E@0.20	Ea	5.27	5.38	10.65
1/2" female elbow box	1E@0.20	Ea	5.27	5.38	10.65
Combination connector	1E@0.10	Ea	1.88	2.69	4.57
Adjust. offset conn.	1E@0.10	Ea	5.11	2.69	7.80
Kick plate	1E@0.10	Ea	7.55	2.69	10.24
Armored cable conn.	1E@0.05	Ea	1.72	1.34	3.06

Use these figures to estimate the cost of surface metal raceway and fittings installed in a building under the conditions described in the chapter text. The crew size is one electrician working at a labor cost of $26.88 per manhour. These costs include layout, material handling and normal waste. Add for sales tax, delivery, supervision, mobilization, demobilization, cleanup, overhead and profit.

Surface Metal Raceway, Fittings, Tools and Assemblies

Material	Craft@Hrs	Unit	Material Cost	Labor Cost	Installed Cost
Overfloor steel fittings for 2-7/32" x 23/32" surface raceway					
Wire clip	1E@0.05	Ea	.38	1.34	1.72
Fiber bushing	1E@0.05	Ea	.60	1.34	1.94
Flat elbow, 90 degree	1E@0.25	Ea	4.92	6.72	11.64
Telephone elbow	1E@0.25	Ea	10.80	6.72	17.52
Tel service fit small	1E@0.25	Ea	9.03	6.72	15.75
Telephone service fit	1E@0.25	Ea	10.80	6.72	17.52
Junction box	1E@0.25	Ea	7.56	6.72	14.28
Tools for 1/2" and 3/4" steel surface raceway					
Bender	--	Ea	49.80	--	--
Bender adapter	--	Ea	9.68	--	--
Cutting box	--	Ea	52.70	--	--
Wire pulley	--	Ea	19.00	--	--
Cover removal tool	--	Ea	10.20	--	--
Canopy cutter	--	Ea	261.00	--	--
Shear for 2-3/8"	--	Ea	173.00	--	--
Enamel spray paint for steel surface raceway					
Buff	--	Ea	9.16	--	--
Gray	--	Ea	9.16	--	--
Putty	--	Ea	9.16	--	--
Two-piece 1-9/32" x 3/4" surface metal raceway assembly, with cord					
24" with 3 outlets	1E@0.15	Ea	42.50	4.03	46.53
36" with 6 outlets	1E@0.20	Ea	47.60	5.38	52.98
60" with 6 outlets	1E@0.25	Ea	53.10	6.72	59.82
72" with 6 outlets	1E@0.30	Ea	55.90	8.06	63.96
Two-piece surface metal raceway, 1-9/32" x 3/4", prewired, single circuit, grounding					
12" long, 2 outlets	1E@0.30	Ea	19.40	8.06	27.46
36" long, 6 outlets	1E@0.40	Ea	18.50	10.80	29.30
60" long, 5 outlets	1E@0.50	Ea	21.40	13.40	34.80
60" long, 10 outlets	1E@0.60	Ea	28.70	16.10	44.80
72" long, 4 outlets	1E@0.50	Ea	20.00	13.40	33.40
72" long, 6 outlets	1E@0.60	Ea	22.90	16.10	39.00
72" long, 8 outlets	1E@0.70	Ea	27.20	18.80	46.00
72" long, 12 outlets	1E@0.80	Ea	29.40	21.50	50.90

Use these figures to estimate the cost of surface metal raceway, assemblies and fittings installed in a building under the conditions described in the chapter text. The crew size is one electrician working at a labor cost of $26.88 per manhour. These costs include layout, material handling and normal waste. Add for sales tax, delivery, supervision, mobilization, demobilization, cleanup, overhead and profit.

Surface Metal Raceway Assemblies and Fittings

Material	Craft@Hrs	Unit	Material Cost	Labor Cost	Installed Cost

Two-piece surface metal raceway, 1-9/32" x 3/4", prewired, 1 circuit, 3-wire

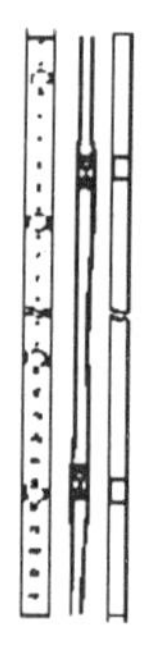

Material	Craft@Hrs	Unit	Material Cost	Labor Cost	Installed Cost
12" long, 2 outlets	1E@0.35	Ea	20.70	9.41	30.11
36" long, 6 outlets	1E@0.45	Ea	43.20	12.10	55.30
60" long, 5 outlets	1E@0.55	Ea	22.60	14.80	37.40
60" long, 10 outlets	1E@0.65	Ea	57.90	17.50	75.40
72" long, 4 outlets	1E@0.55	Ea	22.70	14.80	37.50
72" long, 6 outlets	1E@0.65	Ea	26.80	17.50	44.30
72" long, 8 outlets	1E@0.75	Ea	29.00	20.20	49.20
72" long, 12 outlets	1E@0.85	Ea	35.50	22.80	58.30

Two-piece surface metal raceway, 1-9/32" x 3/4" prewired, 2 circuit, 3-wire

Material	Craft@Hrs	Unit	Material Cost	Labor Cost	Installed Cost
60" long, 5 outlets	1E@0.60	Ea	25.00	16.10	41.10
72" long, 4 outlets	1E@0.60	Ea	27.80	16.10	43.90
72" long, 6 outlets	1E@0.70	Ea	33.40	18.80	52.20
72" long, 8 outlets	1E@0.80	Ea	43.30	21.50	64.80

Surface metal raceway, 2 circuit, 4-wire, grounding, outlets alternately wired

Material	Craft@Hrs	Unit	Material Cost	Labor Cost	Installed Cost
60" long, 5 outlets	1E@0.65	Ea	26.80	17.50	44.30
72" long, 4 outlets	1E@0.65	Ea	27.40	17.50	44.90
72" long, 6 outlets	1E@0.75	Ea	31.70	20.20	51.90
72" long, 8 outlets	1E@0.85	Ea	33.40	22.80	56.20

Two-piece 1-9/32" x 3/4" surface metal raceway assembly fittings

Material	Craft@Hrs	Unit	Material Cost	Labor Cost	Installed Cost
Flat elbow, 90 degree	1E@0.30	Ea	2.52	8.06	10.58
Splice cover	1E@0.05	Ea	1.60	1.34	2.94
Tee	1E@0.35	Ea	6.62	9.41	16.03
Internal corner coupling	1E@0.10	Ea	.82	2.69	3.51
External corner coupling	1E@0.10	Ea	1.88	2.69	4.57
Ivory receptacle cap	1E@0.05	Ea	.29	1.34	1.63
Single pole switch	1E@0.35	Ea	9.19	9.41	18.60
Flush plate adapter	1E@0.20	Ea	5.00	5.38	10.38
Side reducing conn.	1E@0.10	Ea	6.12	2.69	8.81
End reducing conn.	1E@0.05	Ea	1.97	1.34	3.31
Wire clip	1E@0.05	Ea	.24	1.34	1.58
Coupling	1E@0.05	Ea	.30	1.34	1.64
Supporting clip	1E@0.05	Ea	.70	1.34	2.04
Cover clip	1E@0.05	Ea	.59	1.34	1.93
T-bar clip	1E@0.10	Ea	2.58	2.69	5.27
Ground clip	1E@0.05	Ea	2.36	1.34	3.70
End entrance fitting	1E@0.15	Ea	2.78	4.03	6.81
End entr. fit. long	1E@0.20	Ea	9.27	5.38	14.65
Blank end fitting	1E@0.05	Ea	.44	1.34	1.78

Use these figures to estimate the cost of surface metal raceway assemblies and fittings installed in a building under the conditions described in the chapter text. The crew size is one electrician working at a labor cost of $26.88 per manhour. These costs include typical layout material handling and normal waste. Add for sales tax, delivery, supervision, mobilization, demobilization, cleanup, overhead and profit.

Surface Metal Raceway Assemblies and Fittings

Material	Craft@Hrs	Unit	Material Cost	Labor Cost	Installed Cost

Stainless steel surface raceway, 1-9/32" x 3/4", prewired, 1 circuit, 3-wire, with ends & couplings

Material	Craft@Hrs	Unit	Material Cost	Labor Cost	Installed Cost
36" long, 6 outlets	1E@0.50	Ea	43.20	13.40	56.60
60" long, 5 outlets	1E@0.60	Ea	52.00	16.10	68.10
60" long, 10 outlets	1E@0.70	Ea	57.90	18.80	76.70
72" long, 6 outlets	1E@0.70	Ea	65.10	18.80	83.90

Two-piece steel surface raceway, 2-3/8" x 3/4", prewired receptacles, single circuit, 3-wire

Material	Craft@Hrs	Unit	Material Cost	Labor Cost	Installed Cost
60" long, 1 outlet	1E@0.55	Ea	18.90	14.80	33.70
60" long, 2 outlets	1E@0.60	Ea	19.10	16.10	35.20
72" long, 4 outlets	1E@0.65	Ea	21.40	17.50	38.90

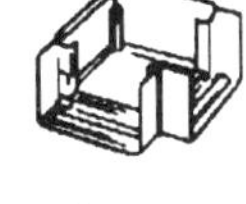

Fittings for 2-3/8" x 3/4" steel surface raceway with prewired receptacles

Material	Craft@Hrs	Unit	Material Cost	Labor Cost	Installed Cost
Coupling	1E@0.05	Ea	1.40	1.34	2.74
Wire clip	1E@0.05	Ea	.54	1.34	1.88
Cover clip	1E@0.05	Ea	.80	1.34	2.14
End blank fitting	1E@0.05	Ea	.86	1.34	2.20
End trim fitting	1E@0.05	Ea	7.31	1.34	8.65
End entrance fitting	1E@0.30	Ea	8.68	8.06	16.74
End reducing fitting	1E@0.10	Ea	4.07	2.69	6.76
Internal coupling	1E@0.25	Ea	4.02	6.72	10.74
External coupling	1E@0.25	Ea	8.46	6.72	15.18
Flat elbow, 90 degree	1E@0.30	Ea	17.40	8.06	25.46
Wall box connector	1E@0.30	Ea	6.04	8.06	14.10
Single pole switch 15A	1E@0.25	Ea	10.30	6.72	17.02
Three-way switch 15A	1E@0.30	Ea	14.20	8.06	22.26
Single recept. cover	1E@0.15	Ea	6.17	4.03	10.20

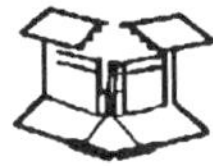

Use these figures to estimate the cost of surface metal raceway assemblies and fittings installed in a building under the conditions described in the chapter text. The crew size is one electrician working at a labor cost of $26.88 per manhour. These costs include typical layout material handling and normal waste. Add for sales tax, delivery, supervision, mobilization, demobilization, cleanup, overhead and profit.

Overhead Steel Channel Wiring System and Fittings

Material	Craft@Hrs	Unit	Material Cost	Labor Cost	Installed Cost
Overhead steel channel system, 1-5/8" x 1-5/8"					
10' lengths	1E@4.00	CLF	359.00	108.00	467.00
20' lengths	1E@3.75	CLF	359.00	101.00	460.00
Overhead steel channel flat wire, 600 volt					
#10 stranded	1E@8.00	MLF	303.00	215.00	518.00
Overhead steel channel system fittings					
Fixture hanger	1E@0.15	Ea	5.34	4.03	9.37
End cap, insulated	1E@0.15	Ea	7.56	4.03	11.59
Coupling	1E@0.15	Ea	7.37	4.03	11.40
Channel hanger	1E@0.10	Ea	2.64	2.69	5.33
Splice connector	1E@0.25	Ea	3.55	6.72	10.27
Junction box	1E@0.30	Ea	20.80	8.06	28.86
Dual feed J-box	1E@0.30	Ea	32.60	8.06	40.66
Overhead steel channel system power taps and fixture hangers					
Phase A, black	1E@0.20	Ea	12.30	5.38	17.68
Phase A-B, bk & red	1E@0.20	Ea	12.30	5.38	17.68
Phase B, red	1E@0.20	Ea	12.30	5.38	17.68
Phase B-C, red & blue	1E@0.20	Ea	12.30	5.38	17.68
Phase C, blue	1E@0.20	Ea	12.30	5.38	17.68
Phase A-C, bk & blue	1E@0.20	Ea	12.30	5.38	17.68
Overhead steel channel system power taps with box & locking receptacle					
Phase A, black	1E@0.20	Ea	51.70	5.38	57.08
Phase B, red	1E@0.20	Ea	51.70	5.38	57.08
Phase C, blue	1E@0.20	Ea	51.70	5.38	57.08
Overhead steel channel system tools					
Cable pulling leader	--	Ea	32.60	--	--
Cable pull-in guide	--	Ea	51.70	--	--

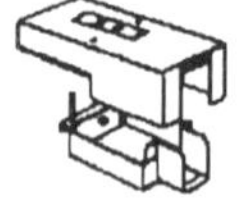

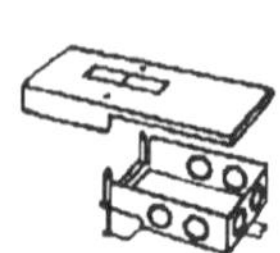

Use these figures to estimate the cost of surface metal raceway and fittings and installed in a building under the conditions described in the chapter text. The crew size is one electrician working at a labor cost of $26.88 per manhour. These costs include layout, material handling and normal waste. Add sales tax, delivery, supervision, mobilization, demobilization, cleanup, overhead and profit.

Material	Craft@Hrs	Unit	Material Cost	Labor Cost	Installed Cost
Two-piece overhead steel raceway lateral 1-9/32" x 3/4", wired with one 125 volt circuit					
24" outlet spacing	1E@6.00	CLF	382.00	161.00	543.00
30" outlet spacing	1E@6.00	CLF	331.00	161.00	492.00
Two-piece overhead steel raceway lateral 1-9/32" x 3/4", wired with one 277 volt circuit					
24" outlet spacing	1E@6.00	CLF	382.00	161.00	543.00
30" outlet spacing	1E@6.00	CLF	331.00	161.00	492.00
Two-piece overhead steel raceway lateral 1-9/32" x 3/4", wired with two 125 volt circuits					
24" outlet spacing	1E@6.00	CLF	400.00	161.00	561.00
30" outlet spacing	1E@6.00	CLF	349.00	161.00	510.00
Two-piece lateral 2-3/4" x 1-17/32", wired with one 125 volt circuit					
24" outlet spacing	1E@8.00	CLF	591.00	215.00	806.00
30" outlet spacing	1E@8.00	CLF	530.00	215.00	745.00
Two-piece lateral 2-3/4" x 1-17/32", wired with one 277 volt circuit					
24" outlet spacing	1E@8.00	CLF	591.00	215.00	806.00
30" outlet spacing	1E@8.00	CLF	530.00	215.00	745.00
Two-piece lateral 2-3/4" x 1-17/32", wired with two 125 volt circuits					
24" outlet spacing	1E@8.00	CLF	617.00	215.00	832.00
30" outlet spacing	1E@8.00	CLF	585.00	215.00	800.00
Two-piece lateral 2-3/4" x 1-17/32", wired with two 277 volt circuits					
24" outlet spacing	1E@8.00	CLF	617.00	215.00	832.00
Two-piece lateral 2-3/4" x 1-17/32", wired with two 125/277 volt circuits alternating					
24" outlet spacing	1E@8.25	CLF	634.00	222.00	856.00
Two-piece lateral 2-3/4" x 1-17/32", wired with two 125 and two 277 volt circuits alternating					
24" outlet spacing	1E@8.50	CLF	864.00	228.00	1,092.00

Use these figures to estimate the cost of surface metal raceway and fittings installed in a building under the conditions described in the chapter text. The crew size is one electrician working at a labor cost of $26.88 per manhour. These costs include layout, material handling and normal waste. Add for sales tax, delivery, supervision, mobilization, demobilization, cleanup, overhead and profit.

Overhead Distribution Systems

Material	Craft@Hrs	Unit	Material Cost	Labor Cost	Installed Cost

Two-piece steel raceway lateral 4-3/4" x 3-9/16", wired with four 277 volt circuits

Material	Craft@Hrs	Unit	Material Cost	Labor Cost	Installed Cost
24" outlet spacing	1E@10.0	CLF	698.00	269.00	967.00

Two-piece lateral 4-3/4" x 3-9/16", wired with three 125 volt and four 277 volt circuits

Material	Craft@Hrs	Unit	Material Cost	Labor Cost	Installed Cost
24" outlet spacing	1E@10.3	CLF	831.00	277.00	1,108.00

Two-piece lateral 4-3/4" x 3-9/16", wired with five 125 volt and six 277 volt circuits

Material	Craft@Hrs	Unit	Material Cost	Labor Cost	Installed Cost
24" outlet spacing	1E@10.5	CLF	911.00	282.00	1,193.00

Fixture whips for overhead steel raceway

Material	Craft@Hrs	Unit	Material Cost	Labor Cost	Installed Cost
15A, 125V, 6' long	1E@0.20	Ea	39.70	5.38	45.08
15A, 125V, 15' long	1E@0.25	Ea	58.60	6.72	65.32
15A, 277V, 6' long	1E@0.20	Ea	37.20	5.38	42.58
15A, 277V, 15' long	1E@0.25	Ea	67.70	6.72	74.42
20A, 125V, 6' long	1E@0.20	Ea	39.70	5.38	45.08
20A, 125V, 15' long	1E@0.25	Ea	58.60	6.72	65.32

Jumper whips for overhead steel raceway

Material	Craft@Hrs	Unit	Material Cost	Labor Cost	Installed Cost
20A, 125V, 18' long	1E@0.25	Ea	82.00	6.72	88.72
Flat elbow, 90 degree	1E@0.25	Ea	65.60	6.72	72.32

Switching whips for overhead steel raceway

Material	Craft@Hrs	Unit	Material Cost	Labor Cost	Installed Cost
20A, 125V, 18' long	1E@0.25	Ea	111.00	6.72	117.72
20A, 277V, 18' long	1E@0.50	Ea	83.20	13.40	96.60

Cable adapter assemblies for overhead steel raceway

Material	Craft@Hrs	Unit	Material Cost	Labor Cost	Installed Cost
20A, 125V, 6' long	1E@0.15	Ea	39.70	4.03	43.73
20A, 125V, 12' long	1E@0.20	Ea	58.60	5.38	63.98
20A, 125V, 18' long	1E@0.25	Ea	83.20	6.72	89.92
20A, 277V, 6' long	1E@0.15	Ea	37.10	4.03	41.13

Two-piece overhead steel raceway lateral extension with wiring harness, 4-3/4" x 3-9/16"

Material	Craft@Hrs	Unit	Material Cost	Labor Cost	Installed Cost
15A, 277V, 14" long	1E@0.15	Ea	106.00	4.03	110.03
20A, 277V, 22" long	1E@0.20	Ea	110.00	5.38	115.38
20A, 277V, 36" long	1E@0.30	Ea	172.00	8.06	180.06

Use these figures to estimate the cost of surface metal raceway and fittings installed in a building under the conditions described in the chapter text. The crew size is one electrician working at a labor cost of $26.88 per manhour. These costs include layout, material handling and normal waste. Add sales tax, delivery, supervision, mobilization, demobilization, cleanup, overhead and profit.

Overhead Distribution Systems and Fittings

Material	Craft@Hrs	Unit	Material Cost	Labor Cost	Installed Cost
Fittings for overhead steel raceway 2-3/4" x 1-17/32"					
Switch cover	1E@0.10	Ea	3.89	2.69	6.58
Sign recept cover	1E@0.10	Ea	3.89	2.69	6.58
Fixture/drop cover	1E@0.10	Ea	3.89	2.69	6.58
Tap-off fitting	1E@0.10	Ea	4.06	2.69	6.75
Single recept cover	1E@0.10	Ea	3.89	2.69	6.58
Circuit breaker box	1E@0.20	Ea	16.80	5.38	22.18
Extension adapter	1E@0.15	Ea	5.30	4.03	9.33
Take-off connector	1E@0.10	Ea	14.10	2.69	16.79
Conduit connector	1E@0.10	Ea	22.80	2.69	25.49
Panel connector	1E@0.30	Ea	5.29	8.06	13.35
End reducing conn.	1E@0.10	Ea	4.69	2.69	7.38
External corner coupl	1E@0.25	Ea	9.49	6.72	16.21
Sign recept. plate	1E@0.15	Ea	3.89	4.03	7.92
Single recept. plate	1E@0.15	Ea	3.89	4.03	7.92
Switch plate	1E@0.15	Ea	3.89	4.03	7.92
Duplex plate	1E@0.15	Ea	3.89	4.03	7.92
Utility box	1E@0.25	Ea	18.60	6.72	25.32
Deep sw box, 2-gang	1E@0.30	Ea	34.40	8.06	42.46
Fittings for overhead steel raceway 4-3/4" x 3-9/16"					
Coupling	1E@0.10	Ea	4.02	2.69	6.71
"C" hanger	1E@0.20	Ea	7.54	5.38	12.92
Blank end fitting	1E@0.15	Ea	5.15	4.03	9.18
Tap-off fitting	1E@0.25	Ea	5.32	6.72	12.04
Take-off fitting	1E@0.30	Ea	36.10	8.06	44.16
Tel conn housing	1E@0.30	Ea	49.00	8.06	57.06
Cable adapt assembly	1E@0.25	Ea	17.90	6.72	24.62
Receptacle cap	1E@0.05	Ea	.34	1.34	1.68

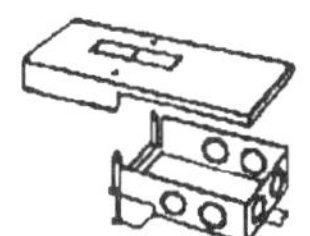

Use these figures to estimate the cost of surface metal raceway fittings installed in a building under the conditions described in the chapter text. The crew size is one electrician working at a labor cost of $26.88 per manhour. These costs include layout, material handling and normal waste. Add for sales tax, delivery, supervision, mobilization, demobilization, cleanup, overhead and profit.

Telephone-Power Poles

Material	Craft@Hrs	Unit	Material Cost	Labor Cost	Installed Cost
Telephone-power pole assemblies, 1-1/4" x 1-3/4", 2 power outlets					
10' pole, flush boot	1E@1.25	Ea	123.00	33.60	156.60
10'-4" pole, 2" boot	1E@1.25	Ea	114.00	33.60	147.60
12'-2" pole, 2" boot	1E@1.25	Ea	152.00	33.60	185.60
T-bar hanger clamp	1E@0.30	Ea	5.19	8.06	13.25
Telephone-power pole assemblies, 2-1/8" x 2-1/8", 4 power outlets					
10'-5" pole, flush boot	1E@1.30	Ea	100.00	34.90	134.90
12'-5" pole, flush boot	1E@1.35	Ea	130.00	36.30	166.30
15'-5" pole, flush boot	1E@1.40	Ea	167.00	37.60	204.60
10'-5" pole isol gr.	1E@1.35	Ea	118.00	36.30	154.30
Cabling network adapter	1E@0.50	Ea	15.50	13.40	28.90
Telephone-power pole assemblies, 2-3/4" x 1-7/16", 2 duplex outlets					
10'-5" pole, 2" foot	1E@1.30	Ea	157.00	34.90	191.90
12'-5" pole, 2" foot	1E@1.35	Ea	171.00	36.30	207.30
Cabling network adapter	1E@0.50	Ea	17.50	13.40	30.90
Telephone-power pole assemblies, 2-3/4" x 1-7/16", 4 power outlets					
10'-5" pole, 2" foot	1E@1.40	Ea	143.00	37.60	180.60
12'-5" pole, 2" foot	1E@1.45	Ea	171.00	39.00	210.00
Cabling network adapter	1E@0.50	Ea	17.50	13.40	30.90
Telephone-power pole assemblies, 2-3/4" x 2-7/8", 2 duplex outlets					
10'-5" pole, flush boot	1E@1.45	Ea	220.00	39.00	259.00
Telephone-power pole assemblies, 2" x 2", 4 power outlets					
10'-5" pole, adj foot	1E@1.30	Ea	225.00	34.90	259.90
10'-5" w/ lt woodgrain	1E@1.30	Ea	240.00	34.90	274.90
10'-5" w/ dk woodgrain	1E@1.30	Ea	240.00	34.90	274.90
12'-5" pole, adj foot	1E@1.40	Ea	204.00	37.60	241.60
15'-5" pole, adj foot	1E@1.50	Ea	260.00	40.30	300.30
Tel entrance fitting	1E@0.30	Ea	7.85	8.06	15.91
Cabling network adapter	1E@0.50	Ea	15.50	13.40	28.90

Use these figures to estimate the cost of telephone-power poles installed in a building under good conditions as described in the chapter text. Costs listed are for each telephone-power pole or listed fitting. The crew size is one electrician working at a labor cost of $26.88 per manhour. These costs include typical installation and layout for office furniture, counters and work stations, material handling and normal waste. Add for various fittings, boxes, sales tax, delivery, supervision, mobilization, demobilization, cleanup, overhead and profit.

Material	Craft@Hrs	Unit	Material Cost	Labor Cost	Installed Cost
Telephone-power pole assemblies, 1-5/16" x 1-5/8", 4 power outlets					
10'-4" pole, flush boot	1E@1.30	Ea	126.00	34.90	160.90
10'-4" 3 unit pole	1E@1.40	Ea	148.00	37.60	185.60
Tel entrance fitting	1E@0.30	Ea	7.85	8.06	15.91
Cabling network adapter	1E@0.50	Ea	15.50	13.40	28.90
Add-on compartment	1E@0.50	Ea	60.90	13.40	74.30
Telephone-power pole assemblies, 2" x 2", 4 power outlets					
10'-4" pole, flush boot	1E@1.40	Ea	174.00	37.60	211.60
10'-4" pole, w/lt woodgrain	1E@1.40	Ea	186.00	37.60	223.60
10'-4" pole, w/dk woodgrain	1E@1.40	Ea	186.00	37.60	223.60
12'-4" pole, flush boot	1E@1.50	Ea	204.00	40.30	244.30
15'-4" pole, flush boot	1E@1.70	Ea	260.00	45.70	305.70
Tel entrance fittings	1E@0.30	Ea	7.85	8.06	15.91
Telephone-power pole assemblies, 2" x 2", dedicated/isolated ground					
10'-4", dedicated	1E@1.40	Ea	194.00	37.60	231.60
10'-4", isolated	1E@1.40	Ea	188.00	37.60	225.60
Tel entrance fitting	1E@0.30	Ea	7.85	8.06	15.91
Telephone-communications pole, 2" x 2"					
10'-4" pole, flush boot	1E@1.00	Ea	88.70	26.90	115.60
Telephone-communications pole, 2-1/8" x 2-1/8"					
10'-4" pole, flush boot	1E@1.00	Ea	152.00	26.90	178.90

Use these figures to estimate the cost of telephone-power poles installed in a building under good conditions as described in the chapter text. Costs listed are for each telephone-power pole or listed fitting. The crew size is one electrician working at a labor cost of $26.88 per manhour. These costs include typical installation and layout for office furniture, counters and work stations, material handling and normal waste. Add for various fittings, boxes, sales tax, delivery, supervision, mobilization, demobilization, cleanup, overhead and profit.

Section 16: Grounding

Article 250 of the *National Electrical Code* sets the standard for grounding and bonding AC electrical systems from 50 volts to 1,000 volts when used to supply power to a premise's wiring systems.

Grounding is intended to provide a path of least resistance from the power source to the earth so voltage is drained away before it can do any damage to people or equipment. The *NEC* considers an electrical system grounded when it's "connected to earth or to some conducting body that serves in place of the earth."

The *NEC* describes the purpose of grounding as follows: "Systems and circuits conductors are grounded to limit voltages due to lightning, line surges or unintentional contact with higher voltage lines, and to stabilize the voltage to ground during normal operation. System and circuit conductors are solidly grounded to facilitate overcurrent device operation in case of ground faults."

Most building inspectors consider grounding to be an important part of every electrical system. They're likely to check the grounding system very carefully.

Seven Common Grounding Methods

There are seven common ways of grounding an electrical system. I'll describe each of these in the paragraphs that follow. It's important to understand that the grounding method you select has to be appropriate for the electrical system you're estimating. Your building inspector may approve some types of grounding and not others for any particular job. It's a good idea to check with the building official if you're not sure of the type of grounding system required.

Here are the seven common grounding methods:

1. Connect to the incoming water supply metallic piping.
2. Connect to a driven ground rod at the point of electrical service.
3. Connect to a driven ground pipe at the point of electrical service.
4. Connect to a Ufer ground system.
5. Connect to a buried copper ring.
6. Connect to the building frame.
7. Connect to a plate electrode.

Connect to Underground Water Supply

This is probably the most common type of grounding system. The underground metallic water pipe must be in direct contact with the earth for a distance of at least 10 feet. Connection to the water pipe is made with an approved water pipe clamp, usually above ground, and in an area protected from mechanical injury. If the grounding conductor is installed in conduit to prevent damage, the water pipe clamp should be equipped with a ground clamp hub sized for the conduit.

In most areas, the inspector will approve a ground conductor with an armor jacket installed from the service equipment to the water pipe clamp that's attached to the underground metallic water pipe.

Connect to Driven Electrode

A driven ground rod or electrode has to be installed so that at least 8 feet of electrode is in contact with the earth. The electrode has to be made of steel or iron and must be at least ⅝" in diameter. Nonferrous or stainless steel rods have to be at least ½" in diameter. The most common type of nonferrous electrode is made of steel and has a copper jacket. These are known as copper-clad ground rods. The exposed upper end of the ground rod must be protected against physical damage. If rock or hard soil below the surface makes it hard to drive the electrode, you can drive the rod at an angle not more than 45 degrees from the vertical. You can also bury it in a trench that's at least 2½ feet deep.

If you use metallic ground pipe or conduit for grounding, it has to be at least ¾" in diameter. The outer surface has to be galvanized or otherwise metal-coated for corrosion protection. In many areas the soil is highly corrosive. Even galvanized pipe will deteriorate in a few years. In any case, the pipe should be installed the same way as a ground rod would be installed.

It's not always easy to drive an 8-foot ground rod or pipe into the soil. If a sledge hammer won't drive a rod, a compressor or pneumatic drill may be required. A pneumatic drill can cut through hard earth and even some rock. When the hole has been cut to the required depth, insert the ground rod or pipe. Then wash backfill into the hole with water. The water helps increase the effectiveness of the grounding.

Connect to Ufer System

The Ufer ground system is installed in a building foundation when the foundation is poured. The Ufer electrode has to be surrounded by at least 2" of concrete poured near the bottom of a foundation or footing that's in direct contact with the earth. Two types of material are acceptable for a Ufer ground:

1. At least 20 feet of one or more ½" steel reinforcing bars or rods.

2. At least 20 feet of bare copper conductor not smaller than number 4 AWG.

Connect to a Bare Copper Ring

Use a ground ring that encircles the building in direct contact with the earth at a depth of not less than 2½ feet. The bare copper conductor has to be at least 20 feet long and not smaller than number 2 AWG.

Connect to Building Frame

The metal frame of the building has to be grounded to the earth. This connection should have a sufficiently low impedance and enough current-carrying capacity to prevent an accumulation of voltage within the frame.

Connect to Plate Electrode

Plate electrodes are acceptable for grounding if the electrode has at least 2 square feet of surface exposed to the earth. There are two types of plates:

1. Iron or steel plate electrodes have to be at least ¼" thick.

2. Nonferrous electrodes have to be at least .06" thick.

The grounding electrode conductor has to be one continuous piece. No splices are permitted. Use Table 16-1 (*NEC* Table 250-94) to size the grounding electrode conductor.

Other Grounding Requirements

Certain types of electrical equipment need a separate grounding system. Metallic raceway serving the equipment and the metallic enclosure may have to be grounded. Electrically connected equipment housings must also be grounded. Table 16-2 (*NEC* Table 250-95) identifies the size of these grounding conductors.

Here are some definitions the *NEC* uses to describe grounding requirements.

Grounded

Connected to earth or to some conducting body that serves in place of earth.

Grounded Conductor

A system of circuit conductors that is intentionally grounded.

Grounding Conductor

A conductor used to connect equipment or the grounded circuit of a wiring system to a grounding electrode or electrodes.

Table 250-94

Grounding Electrode Conductor for AC Systems

Size of Largest Service-Entrance Conductor or Equivalent Area for Parallel Conductors		Size of Grounding Electrode Conductor	
Copper	Aluminum or Copper-Clad Aluminum	Copper	Aluminum or Copper-Clad Aluminum*
2 or smaller	1/0 or smaller	8	6
1 or 1/0	2/0 or 3/0	6	4
2/0 or 3/0	4/0 or 250 kcmil	4	2
Over 3/0 kcmil through 350 kcmil	Over 250 kcmil through 500 kcmil	2	1/0
Over 350 kcmil through 600 kcmil	Over 500 kcmil through 900 kcmil	1/0	3/0
Over 600 kcmil through 1100 kcmil	Over 900 kcmil through 1750 kcmil	2/0	4/0
Over 1100 kcmil	Over 1750 kcmil	3/0	250 kcmil

Where multiple sets of service-entrance conductors are used as permitted in Section 230-40, Exception No. 2, the equivalent size of the largest service entrance conductor shall be determined by the largest sum of the areas of the corresponding conductors of each set.
Where there are no service-entrance conductors, the grounding electrode conductor size shall be determined by the equivalent size of the largest service-entrance conductor required for the load to be served.
*See installation restriction in Section 250-92(a).
(FPN): See Section 250-23(b) for size of alternating-current system grounded conductor brought to service equipment.
*From the **National Electric Code** ©1996, National Fire Protection Association.*

Table 16-1

Grounding Conductor, Equipment

The conductor used to connect the noncurrent-carrying metal parts of equipment, raceways, and other enclosures to the system grounded conductor and/or to the grounding electrode conductor at the service equipment or at the source of a separately derived system.

Grounding Electrode Conductor

The conductor used to connect the grounding electrode to the equipment grounding conductor and/or to the grounded conductor of the circuit at the service equipment or at the source of a separately derived system.

Ground Fault Circuit Interrupter

A device intended for the protection of personnel that functions to deenergize a circuit or portion thereof within an established period of time when a current to ground exceeds some predetermined value that is less than that required to operate the overcurrent protective device of the supply circuit.

Ground Fault Protection of Equipment

A system intended to provide protection of equipment from damaging line-to-ground fault currents by operating to cause a disconnecting means to open all ungrounded conductors of the faulted circuit. This protection is provided at current levels less than those required to protect conductors from damage through the operation of a supply circuit overcurrent device.

Ground Testing

Some contracts require that the grounding system be tested to see if it works as intended. Resistance is measured in ohms. The device used for this test is called a ***ground megger***. It checks the resistance of the earth.

To test resistance of soil, drive two ½" by 36" steel rods into the earth between 1 and 2 feet deep at a specified distance from the ground electrode. Then connect a #12 copper wire to each rod, to the ground electrode and to the megger. The megger indicates the resistance of the earth between electrodes.

Table 250-95
Minimum Size Equipment Grounding Conductors for Grounding Raceway and Equipment

Rating or Setting of Automatic Overcurrent Device in Circuit Ahead of Equipment, Conduit, etc., Not Exceeding (Amperes)	Size	Aluminum or Copper-Clad Aluminum Wire Number.*
	Copper Wire Number	
15	14	12
20	12	10
30	10	8
40	10	8
60	10	8
100	8	6
200	6	4
300	4	2
400	3	1
500	2	1/0
600	1	2/0
800	1/0	3/0
1000	2/0	4/0
1200	3/0	250 kcmil
1600	4/0	350 kcmil
2000	250 kcmil	400 kcmil
2500	350 kcmil	600 kcmil
3000	400 kcmil	600 kcmil
4000	500 kcmil	800 kcmil
5000	700 kcmil	1200 kcmil
6000	800 kcmil	1200 kcmil

*See installation restrictions in NEC Section 250-92(a)
Note: Equipment grounding conductors may need to be sized larger than specified in this table in order to comply with Section 250-51.
From the **National Electrical Code** ©1996, National Fire Protection Association.

Table 16-2

The *NEC* assumes ground resistance of 25 ohms or less. If resistance is higher than that, additional rod or rods will be required. Don't install these rods until a change order has been signed. When a megger reading is required, keep a record of those readings in your job file.

The earth's resistance will change as moisture levels in the soil change. Wetter soil is a better conductor. Dryer soil has higher resistance. When resistance is high and the climate is normally dry, you may have to increase the conducting capacity of the ground. Try drilling an oversized hole about 2" to 3" in diameter. Install a ½" to 1" ground rod. Mix copper sulfate with the backfill soil and wash it into the hole with water. After a few days, the copper sulfate should provide a much lower resistance reading.

Grounding Problems

Good grounding isn't always easy. Think about the type of grounding that will provide economical yet durable protection for the electric system. Some soils are very corrosive and limit the life of an electrode or even the underground water service pipe. As the metal is eaten away, the conducting capacity drops, reducing the margin of safety.

The Ufer ground system provides many advantages if the soil is corrosive. The main limitation is that a Ufer ground has to be installed in the building's foundation in contact with the earth when the foundation is poured.

Another choice would be dual grounding: Provide two or more grounds. That may help ensure an adequate grounding system for many years. For example, you might install a driven electrode and also connect to the underground water supply system. Any combination of two or more connections would be ideal. But be sure that both grounds meet code requirements.

Connecting the Ground

The connection to either a Ufer ground, a ground rod or a building frame should be made with an exothermic weld. The weld is made by fusing the conductor to the electrode.

Start by selecting a carbon mold the right size to accept the ground conductor and the electrode. The fusing process is quick and simple. Make sure both surfaces on the electrode and conductor are clean and bright. Then place the mold on the electrode and conductor. Select a powder charge that's the right size for the mold. Place the fusing slug and charge. Then set the powder off with a sparking device.

Let the weld cool before removing the mold. When the connection is cool, chip off the slag and check the weld for blistering. If the weld is puffy and weak, start over with a new weld.

Every conductor connection should be cleaned carefully. Remove any corrosion, insulation or dirt with a wire brush. A small piece of sandpaper is excellent for cleaning the connections.

Grounding Electrical Devices

Metallic outlet boxes have a threaded screw hole in the back of the box. Use this hole for attaching a ground conductor from the wiring device's ground terminal. Ground clips have been approved for clipping bonding conductors to the side of metallic boxes.

If you use nonmetallic conduit and boxes, install a ground or bonding conductor along with the circuit conductors. Splice grounding and bonding conductors and the device grounding jumpers in nonmetallic boxes.

Conduit will provide a continuous ground from each load all the way back to the service entrance equipment or distribution panel.

Wire comes with either a bare ground wire or with the ground wire insulated and colored green. Green insulation is used on wire sizes 8 and smaller. On heavier gauge wire, you may have to color the ground green or wrap green marking tape around the green conductor.

Metallic conduit that's stubbed up under electrical equipment should be bonded or grounded to the service equipment with a jumper. Install a threaded fitting called a ground bushing on the conduit or conduit fitting. The ground bushing has either a ground lug or a ground terminal for attaching the jumper wire.

No ground bushing is needed when nonmetallic conduit is used to connect electrical equipment. But a ground wire has to be attached to the electrical equipment ground. Ream the nonmetallic conduit to removed burrs before attaching the ground wire.

Aluminum Conductors

The *NEC* doesn't permit the use of aluminum electrodes for grounding purposes. When aluminum wire is used in a grounding system, be sure the points of contact are thoroughly cleaned of oxidation before making the connection. Clean the contacts with a brush and solvent. Then apply an anti-oxidation preparation to the joint. Make the connection with compression connectors rated for aluminum.

A few days after making the connection, retighten the connector. An aluminum connector tends to loosen when first installed. As it loosens, resistance increases. That weakens the connection and may increase resistance even more. It's a good idea to recheck connections to aluminum conductors a year after the work is completed.

Grounding

Material	Craft@Hrs	Unit	Material Cost	Labor Cost	Installed Cost
Soft drawn solid copper ground wire					
#14	1E@4.00	KLF	52.80	108.00	160.80
#12	1E@4.25	KLF	77.70	114.00	191.70
#10	1E@4.50	KLF	116.00	121.00	237.00
#8	1E@4.75	KLF	218.00	128.00	346.00
#6	1E@5.00	KLF	291.00	134.00	425.00
#4	1E@6.00	KLF	440.00	161.00	601.00
#2	1E@7.00	KLF	728.00	188.00	916.00
Soft drawn stranded bare copper ground wire					
#8	1E@4.75	KLF	220.00	128.00	348.00
#6	1E@5.00	KLF	301.00	134.00	435.00
#4	1E@6.00	KLF	475.00	161.00	636.00
#2	1E@8.00	KLF	739.00	215.00	954.00
#1	1E@10.0	KLF	969.00	269.00	1,238.00
#1/0	1E@11.0	KLF	1,240.00	296.00	1,536.00
#2/0	1E@12.0	KLF	1,540.00	323.00	1,863.00
#3/0	1E@13.0	KLF	1,950.00	349.00	2,299.00
#4/0	1E@14.0	KLF	2,440.00	376.00	2,816.00
Wiring device grounding jumpers, insulated					
#12	1E@0.05	Ea	.40	1.34	1.74
Ground clip					
Box grounding clip	1E@0.05	Ea	.19	1.34	1.53
Insulated ground bushing					
1/2"	1E@0.10	Ea	2.00	2.69	4.69
3/4"	1E@0.10	Ea	2.22	2.69	4.91
1"	1E@0.10	Ea	2.41	2.69	5.10
1-1/4"	1E@0.15	Ea	2.98	4.03	7.01
1-1/2"	1E@0.15	Ea	3.29	4.03	7.32
2"	1E@0.20	Ea	4.45	5.38	9.83
2-1/2"	1E@0.20	Ea	6.79	5.38	12.17
3"	1E@0.25	Ea	9.28	6.72	16.00
3-1/2"	1E@0.25	Ea	10.00	6.72	16.72
4"	1E@0.30	Ea	13.10	8.06	21.16
5"	1E@0.40	Ea	42.60	10.80	53.40
6"	1E@0.50	Ea	66.80	13.40	80.20

Use these figures to estimate the cost of bare copper wire, grounding jumpers, ground clips and insulated ground bushings installed in buildings, under the conditions described on pages 5 and 6. Costs listed are for each 1000 linear feet installed and each part installed. The crew size is one electrician working at a labor cost of $26.88 per manhour. These costs include wire make-up, continuity testing, reel set-up, layout, material handling, and normal waste. Add for connectors, sales tax, delivery, supervision, mobilization, demobilization, cleanup, overhead and profit. Note: The *NEC* permits installation of bare copper wire in conduit with conductors. But bare copper wire installed outside conduit must be protected against damage. Some dealers sell bare copper wire by weight rather than length. The table on page 102 shows weights per 1000 linear feet.

Material	Craft@Hrs	Unit	Material Cost	Labor Cost	Installed Cost

One hole solder type grounding lugs

Material	Craft@Hrs	Unit	Material Cost	Labor Cost	Installed Cost
#10	1E@0.10	Ea	.38	2.69	3.07
#8	1E@0.15	Ea	.48	4.03	4.51
#6	1E@0.15	Ea	.55	4.03	4.58
#4	1E@0.20	Ea	.58	5.38	5.96
#2	1E@0.20	Ea	.77	5.38	6.15
#1/0	1E@0.25	Ea	1.03	6.72	7.75
#2/0	1E@0.25	Ea	1.26	6.72	7.98
#3/0	1E@0.30	Ea	1.72	8.06	9.78
#4/0	1E@0.30	Ea	2.21	8.06	10.27
# 250 KCMIL	1E@0.40	Ea	4.20	10.80	15.00
# 400 KCMIL	1E@0.50	Ea	6.26	13.40	19.66
# 500 KCMIL	1E@0.50	Ea	9.85	13.40	23.25
# 600 KCMIL	1E@0.50	Ea	10.50	13.40	23.90
# 800 KCMIL	1E@0.60	Ea	17.50	16.10	33.60
#1000 KCMIL	1E@0.70	Ea	19.10	18.80	37.90
#1500 KCMIL	1E@0.75	Ea	34.30	20.20	54.50

One hole solderless type grounding lugs

Material	Craft@Hrs	Unit	Material Cost	Labor Cost	Installed Cost
#10	1E@0.05	Ea	.57	1.34	1.91
#6	1E@0.08	Ea	.60	2.15	2.75
#4	1E@0.10	Ea	.84	2.69	3.53
#2	1E@0.10	Ea	.84	2.69	3.53
#1/0	1E@0.15	Ea	1.63	4.03	5.66
#3/0	1E@0.20	Ea	3.16	5.38	8.54
#4/0	1E@0.20	Ea	3.76	5.38	9.14
#350 KCMIL	1E@0.30	Ea	6.98	8.06	15.04
#500 KCMIL	1E@0.40	Ea	10.20	10.80	21.00
#1000 KCMIL	1E@0.60	Ea	25.40	16.10	41.50

Water pipe ground clamps

Material	Craft@Hrs	Unit	Material Cost	Labor Cost	Installed Cost
1"	1E@0.10	Ea	1.60	2.69	4.29
2"	1E@0.15	Ea	2.20	4.03	6.23
2-1/2"	1E@0.20	Ea	9.49	5.38	14.87
5"	1E@0.30	Ea	14.40	8.06	22.46
6"	1E@0.40	Ea	21.90	10.80	32.70

Ground clamp conduit hubs

Material	Craft@Hrs	Unit	Material Cost	Labor Cost	Installed Cost
Armored cable	1E@0.15	Ea	3.30	4.03	7.33
1/2"	1E@0.20	Ea	1.60	5.38	6.98
3/4"	1E@0.25	Ea	2.87	6.72	9.59
1"	1E@0.30	Ea	5.59	8.06	13.65
1-1/4"	1E@0.35	Ea	9.24	9.41	18.65

Use these figures to estimate the cost of conductor lugs, ground clamps and ground hubs installed under conditions described on pages 5 and 6. Costs are for each lug installed. The crew size is one electrician working at a labor cost of $26.88 per manhour. The costs include lug or ground fitting (and no other material) wire cutting, insulation stripping for termination, layout, material handling and normal waste. Add for insulating the lug (if needed), sales tax, delivery, supervision, mobilization, demobilization, cleanup, overhead and profit. Note: Always recheck solderless lugs for tightness.

Grounding

Material	Craft@Hrs	Unit	Material Cost	Labor Cost	Installed Cost
Copper-clad ground rods					
1/2" dia x 8' long	1E@0.70	Ea	4.96	18.80	23.76
5/8" dia x 8' long	1E@0.75	Ea	5.67	20.20	25.87
3/4" dia x 8' long	1E@0.90	Ea	9.22	24.20	33.42
1/2" dia x 10' long	1E@0.75	Ea	6.10	20.20	26.30
5/8" dia x 10' long	1E@0.90	Ea	7.42	24.20	31.62
3/4" dia x 10' long	1E@1.00	Ea	11.40	26.90	38.30
1" dia x 10' long	1E@1.25	Ea	26.00	33.60	59.60
1" dia x 20' long	1E@2.00	Ea	53.40	53.80	107.20
Copper-clad sectional ground rods					
5/8" dia x 8' long	1E@1.70	Ea	5.86	45.70	51.56
1/2" dia x 10' long	1E@1.70	Ea	6.57	45.70	52.27
5/8" dia x 10' long	1E@1.75	Ea	7.61	47.00	54.61
3/4" dia x 10' long	1E@2.00	Ea	11.80	53.80	65.60
1" dia x 10' long	1E@2.25	Ea	26.90	60.50	87.40
Ground rod couplings					
1/2"	1E@0.10	Ea	4.70	2.69	7.39
5/8"	1E@0.10	Ea	6.13	2.69	8.82
3/4"	1E@0.15	Ea	9.09	4.03	13.12
1"	1E@0.20	Ea	34.50	5.38	39.88
Ground rod clamps					
1/2"	1E@0.10	Ea	2.84	2.69	5.53
5/8"	1E@0.10	Ea	3.33	2.69	6.02
3/4"	1E@0.15	Ea	4.04	4.03	8.07
1"	1E@0.20	Ea	6.46	5.38	11.84
Ground rod stud bolts for sectional rod					
1/2"	1E@0.25	Ea	2.84	6.72	9.56
5/8"	1E@0.25	Ea	3.33	6.72	10.05
3/4"	1E@0.30	Ea	4.04	8.06	12.10
1"	1E@0.40	Ea	6.46	10.80	17.26

Use these figures to estimate the cost of ground rods installed under the conditions described on pages 5 and 6. Costs listed are for each piece installed. The crew size is one electrician working at a labor cost of $26.88 per manhour. These costs include the ground rod or clamp only (and no other material), wire cutting, stripping for termination, layout, material handling and normal waste. Add for sales tax, delivery, supervision, mobilization, demobilization, cleanup, overhead and profit. Note: Always recheck mechanical connections for tightness.

Material	Craft@Hrs	Unit	Material Cost	Labor Cost	Installed Cost
Grounding locknuts					
1/2"	1E@0.05	Ea	.80	1.34	2.14
3/4"	1E@0.06	Ea	1.02	1.61	2.63
1"	1E@0.08	Ea	1.38	2.15	3.53
1-1/4"	1E@0.10	Ea	1.85	2.69	4.54
1-1/2"	1E@0.10	Ea	2.48	2.69	5.17
2"	1E@0.15	Ea	3.38	4.03	7.41
2-1/2"	1E@0.20	Ea	6.78	5.38	12.16
3"	1E@0.20	Ea	8.59	5.38	13.97
3-1/2"	1E@0.25	Ea	13.90	6.72	20.62
4"	1E@0.30	Ea	18.00	8.06	26.06
Exothermic copper grounding connections					
2/0 - 1/2" rod	1E@0.40	Ea	5.24	10.80	16.04
2/0 - 5/8" rod	1E@0.50	Ea	5.24	13.40	18.64
2/0 - 3/4" rod	1E@0.50	Ea	5.24	13.40	18.64
2/0 - 1" rod	1E@0.60	Ea	5.24	16.10	21.34
4/0 - 1/2" rod	1E@0.50	Ea	5.79	13.40	19.19
4/0 - 5/8" rod	1E@0.50	Ea	5.79	13.40	19.19
4/0 - 3/4" rod	1E@0.60	Ea	5.79	16.10	21.89
4/0 - 1" rod	1E@0.75	Ea	5.79	20.20	25.99
2/0 - 2/0 splice	1E@0.40	Ea	4.96	10.80	15.76
2/0 - 4/0 splice	1E@0.50	Ea	4.96	13.40	18.36
4/0 - 4/0 splice	1E@0.50	Ea	5.24	13.40	18.64
2/0 - 2/0 x 2/0	1E@0.70	Ea	5.24	18.80	24.04
2/0 - 4/0 x 4/0	1E@0.80	Ea	5.79	21.50	27.29
4/0 - 4/0 x 4/0	1E@1.00	Ea	6.07	26.90	32.97
2/0 - lug	1E@0.40	Ea	4.68	10.80	15.48
4/0 - lug	1E@0.50	Ea	4.96	13.40	18.36
2/0 - steel beam	1E@0.40	Ea	4.96	10.80	15.76
4/0 - steel beam	1E@0.50	Ea	5.24	13.40	18.64
Exothermic mold	--	Ea	49.00	--	--

Use these figures to estimate the cost of grounding connections installed under the conditions described on pages 5 and 6. Costs listed are for each piece installed. The crew size is one electrician working at a labor cost of $26.88 per manhour. These costs include the grounding locknut or exothermic connection only (and no other material), wire cutting, stripping for termination, layout, material handling and normal waste. Add for sales tax, delivery, supervision, mobilization, demobilization, cleanup, overhead and profit. Note: Always recheck mechanical connections for tightness. Be sure exothermic powder charges are stored in a dry place. Exothermic molds are designed for the connection required.

Section 17: Assemblies

The first sixteen chapters in this manual assume that your take-off form will list every box, plate, strap and plaster ring needed to do the work. That's the only way to make perfectly accurate electrical estimates: Measure and count everything that goes into the job. But it's also slow, tedious, demanding work.

Many experienced electrical estimators use shortcuts that save time without sacrificing accuracy. They group associated materials into assemblies that can be priced as a unit. For example, why price wire and couplings and conduit and connectors and straps separately when they're used together and in about the same proportion on nearly every circuit? Grouping materials into assemblies saves counting, measuring and calculating time and can produce results nearly as accurate as more exhaustive detailed estimates.

The remainder of this section has assembly cost estimates for the most common electrical work. Labor and material prices are the same as in the prior sections. The only difference here is that materials are grouped to speed and simplify your estimates. If you decide to use costs from this section, all you have to do is count symbols on the E sheets and measure circuit lengths. List each assembly on your take-off sheet. Then use the figures in this section to find the estimated cost per assembly.

Unfortunately, there's one major disadvantage to estimating by assembly: The number of possible assemblies is almost infinite. I couldn't possibly list every electrical assembly you're likely to need on the next job. There are far too many combinations. Even if you prefer assembly pricing, at least occasionally you're going to have to figure the wire, conduit, couplings and straps separately.

The assemblies listed here are the most common combinations, using standard commercial grade materials that meet code requirements. But you won't be able to use these costs on jobs that require oversize outlet boxes, set minimum conduit sizes, specify unusual wire insulation or identify lighting fixtures by catalog number. None of the assemblies in this section cover situations like that. But I believe you'll find that most switch and receptacle circuits on most jobs can be priced from the pages that follow.

Conduit and Wire

Cost tables for conduit and wire begin with assemblies for empty conduit with two set screw connectors, nine set screw couplings and nine one-hole straps per 100 linear feet. Simply measure conduit length for each size and type and multiply the length (in hundreds of feet) by the cost per 100 feet. The next set of cost tables include conduit, connectors, couplings, straps, a pull line, plus insulated copper wire. Note that costs will be slightly higher on short runs because more than two set screw connectors are required for each 100 linear feet.

Outlet Boxes and Wiring Devices

Assemblies on these pages include an outlet box, a wiring device, a cover plate and a plaster ring (when needed). Some outlet boxes can be assembled into multi-ganged outlets when more than one device is required. When using costs in this section, begin by deciding what type of assembly is required. Then count on the E sheets the symbols that represent each assembly. Finally, multiply the cost per assembly by the number of assemblies.

Switches

Assemblies are included for 15 amp and 20 amp switches, whether single pole, double pole, three way or four way and for both 120 volts and 277 volt circuits.

Receptacles

Assemblies are included for both 15 amp and 20 amp circuits and for both single and duplex receptacles. The 15 amp receptacles are NEMA 5-15R. The 20 amp receptacles are NEMA 5-20R.

Plastic Plates

Switch and receptacle plates are smooth plastic and are listed by color. All plates are good quality as produced by the major manufacturers.

Lighting Fixtures

Costs for lighting fixtures assume factory-assembled units with lamps and a 5' flexible conduit whip. The whip has copper wire and flex connectors. Using assembled fluorescent fixtures like these can reduce installation costs, especially in a T-bar grid ceiling.

1/2" EMT Conduit Assemblies

Material	Craft@Hrs	Unit	Material Cost	Labor Cost	Installed Cost

100' 1/2" EMT conduit, 2 set screw connectors, 9 set screw couplings and 9 one-hole straps

Material	Craft@Hrs	Unit	Material Cost	Labor Cost	Installed Cost
Empty conduit	1E@4.07	CLF	27.50	109.00	136.50
1 plastic pull line	1E@4.38	CLF	27.60	118.00	145.60
1 #14THHN, solid	1E@4.67	CLF	32.70	126.00	158.70
2 #14THHN, solid	1E@5.27	CLF	37.80	142.00	179.80
3 #14THHN, solid	1E@5.87	CLF	43.00	158.00	201.00
4 #14THHN, solid	1E@6.47	CLF	48.00	174.00	222.00
5 #14THHN, solid	1E@7.07	CLF	53.10	190.00	243.10
6 #14THHN, solid	1E@7.67	CLF	58.30	206.00	264.30
1 #14THHN, stranded	1E@4.67	CLF	33.50	126.00	159.50
2 #14THHN, stranded	1E@5.27	CLF	39.40	142.00	181.40
3 #14THHN, stranded	1E@5.87	CLF	45.40	158.00	203.40
4 #14THHN, stranded	1E@6.47	CLF	51.30	174.00	225.30
5 #14THHN, stranded	1E@7.07	CLF	57.20	190.00	247.20
6 #14THHN, stranded	1E@7.67	CLF	63.20	206.00	269.20
1 #12THHN, solid	1E@4.77	CLF	35.30	128.00	163.30
2 #12THHN, solid	1E@5.47	CLF	43.10	147.00	190.10
3 #12THHN, solid	1E@6.17	CLF	50.80	166.00	216.80
4 #12THHN, solid	1E@6.87	CLF	58.60	185.00	243.60
5 #12THHN, solid	1E@7.57	CLF	66.30	203.00	269.30
6 #12THHN, solid	1E@8.27	CLF	74.10	222.00	296.10
1 #12THHN, stranded	1E@4.77	CLF	36.50	128.00	164.50
2 #12THHN, stranded	1E@5.47	CLF	45.40	147.00	192.40
3 #12THHN, stranded	1E@6.17	CLF	54.30	166.00	220.30
4 #12THHN, stranded	1E@6.87	CLF	63.20	185.00	248.20
5 #12THHN, stranded	1E@7.57	CLF	72.10	203.00	275.10
6 #12THHN, stranded	1E@8.27	CLF	81.00	222.00	303.00
1 #10THHN, solid	1E@4.87	CLF	39.80	131.00	170.80
2 #10THHN, solid	1E@5.67	CLF	52.00	152.00	204.00
3 #10THHN, solid	1E@6.47	CLF	64.20	174.00	238.20
4 #10THHN, solid	1E@7.27	CLF	76.40	195.00	271.40
5 #10THHN, solid	1E@8.07	CLF	88.60	217.00	305.60
6 #10THHN, solid	1E@8.87	CLF	101.00	238.00	339.00
1 #10THHN, stranded	1E@4.87	CLF	41.00	131.00	172.00
2 #10THHN, stranded	1E@5.67	CLF	54.40	152.00	206.40
3 #10THHN, stranded	1E@6.47	CLF	67.80	174.00	241.80
4 #10THHN, stranded	1E@7.27	CLF	81.20	195.00	276.20
5 #10THHN, stranded	1E@8.07	CLF	94.60	217.00	311.60
6 #10THHN, stranded	1E@8.87	CLF	108.00	238.00	346.00

Use these figures to estimate the cost of installing assemblies under the conditions described on pages 5 and 6. Costs listed are for each assembly installed. The crew is one electrician working at a labor cost of $26.88 per manhour. These costs include layout, material handling and normal waste. Add for the sales tax, delivery, supervision, mobilization, demobilization, cleanup, overhead and profit.

3/4" EMT Conduit Assemblies

Material	Craft@Hrs	Unit	Material Cost	Labor Cost	Installed Cost

100' 3/4" EMT conduit, 2 set screw connectors, 9 set screw couplings and 9 one-hole straps

Material	Craft@Hrs	Unit	Material Cost	Labor Cost	Installed Cost
Empty conduit	1E@4.52	CLF	45.80	121.00	166.80
1 plastic pull line	1E@4.82	CLF	45.90	130.00	175.90
1 #14THHN, solid	1E@5.12	CLF	50.90	138.00	188.90
2 #14THHN, solid	1E@5.72	CLF	56.00	154.00	210.00
3 #14THHN, solid	1E@6.32	CLF	61.10	170.00	231.10
4 #14THHN, solid	1E@6.92	CLF	66.30	186.00	252.30
5 #14THHN, solid	1E@7.52	CLF	71.40	202.00	273.40
6 #14THHN, solid	1E@8.12	CLF	76.50	218.00	294.50
1 #14THHN, stranded	1E@5.12	CLF	51.70	138.00	189.70
2 #14THHN, stranded	1E@5.72	CLF	57.70	154.00	211.70
3 #14THHN, stranded	1E@6.32	CLF	63.60	170.00	233.60
4 #14THHN, stranded	1E@6.92	CLF	69.50	186.00	255.50
5 #14THHN, stranded	1E@7.52	CLF	75.50	202.00	277.50
6 #14THHN, stranded	1E@8.12	CLF	81.40	218.00	299.40
1 #12THHN, solid	1E@5.22	CLF	53.50	140.00	193.50
2 #12THHN, solid	1E@5.92	CLF	61.30	159.00	220.30
3 #12THHN, solid	1E@6.62	CLF	69.10	178.00	247.10
4 #12THHN, solid	1E@7.32	CLF	76.80	197.00	273.80
5 #12THHN, solid	1E@8.02	CLF	84.60	216.00	300.60
6 #12THHN, solid	1E@8.72	CLF	92.30	234.00	326.30
1 #12THHN, stranded	1E@5.22	CLF	54.70	140.00	194.70
2 #12THHN, stranded	1E@5.92	CLF	63.60	159.00	222.60
3 #12THHN, stranded	1E@6.62	CLF	71.50	178.00	249.50
4 #12THHN, stranded	1E@7.32	CLF	81.40	197.00	278.40
5 #12THHN, stranded	1E@8.02	CLF	90.30	216.00	306.30
6 #12THHN, stranded	1E@8.72	CLF	99.20	234.00	333.20
1 #10THHN, solid	1E@5.32	CLF	58.00	143.00	201.00
2 #10THHN, solid	1E@6.12	CLF	70.20	165.00	235.20
3 #10THHN, solid	1E@6.92	CLF	82.40	186.00	268.40
4 #10THHN, solid	1E@7.72	CLF	94.60	208.00	302.60
5 #10THHN, solid	1E@8.52	CLF	107.00	229.00	336.00
6 #10THHN, solid	1E@9.32	CLF	119.00	251.00	370.00
1 #10THHN, stranded	1E@5.32	CLF	59.20	143.00	202.20
2 #10THHN, stranded	1E@6.12	CLF	72.60	165.00	237.60
3 #10THHN, stranded	1E@6.92	CLF	86.00	186.00	272.00
4 #10THHN, stranded	1E@7.72	CLF	99.50	208.00	307.50
5 #10THHN, stranded	1E@8.52	CLF	113.00	229.00	342.00
6 #10THHN, stranded	1E@9.32	CLF	126.00	251.00	377.00

Use these figures to estimate the cost of installing assemblies under the conditions described on pages 5 and 6. Costs listed are for each assembly installed. The crew is one electrician working at a labor cost of $26.88 per manhour. These costs include layout, material handling and normal waste. Add for the sales tax, delivery, supervision, mobilization, demobilization, cleanup, overhead and profit.

1" EMT Conduit Assemblies

Material	Craft@Hrs	Unit	Material Cost	Labor Cost	Installed Cost
100' 1" EMT conduit, 2 set screw connectors, 9 set screw couplings and 9 one-hole straps					
Empty conduit	1E@5.33	CLF	74.90	143.00	217.90
1 plastic pull line	1E@5.63	CLF	75.00	151.00	226.00
1 #14THHN, solid	1E@5.93	CLF	80.00	159.00	239.00
2 #14THHN, solid	1E@6.53	CLF	85.10	176.00	261.10
3 #14THHN, solid	1E@7.13	CLF	90.20	192.00	282.20
4 #14THHN, solid	1E@7.73	CLF	95.40	208.00	303.40
5 #14THHN, solid	1E@8.33	CLF	100.00	224.00	324.00
6 #14THHN, solid	1E@8.93	CLF	105.00	240.00	345.00
1 #14THHN, stranded	1E@5.93	CLF	80.80	159.00	239.80
2 #14THHN, stranded	1E@6.53	CLF	86.80	176.00	262.80
3 #14THHN, stranded	1E@7.13	CLF	92.70	192.00	284.70
4 #14THHN, stranded	1E@7.73	CLF	98.60	208.00	306.60
5 #14THHN, stranded	1E@8.33	CLF	105.00	224.00	329.00
6 #14THHN, stranded	1E@8.93	CLF	111.00	240.00	351.00
1 #12THHN, solid	1E@6.03	CLF	82.60	162.00	244.60
2 #12THHN, solid	1E@6.73	CLF	90.40	181.00	271.40
3 #12THHN, solid	1E@7.43	CLF	98.20	200.00	298.20
4 #12THHN, solid	1E@8.13	CLF	106.00	219.00	325.00
5 #12THHN, solid	1E@8.83	CLF	114.00	237.00	351.00
6 #12THHN, solid	1E@9.53	CLF	121.00	256.00	377.00
1 #12THHN, stranded	1E@6.03	CLF	83.80	162.00	245.80
2 #12THHN, stranded	1E@6.73	CLF	92.70	181.00	273.70
3 #12THHN, stranded	1E@7.43	CLF	102.00	200.00	302.00
4 #12THHN, stranded	1E@8.13	CLF	111.00	219.00	330.00
5 #12THHN, stranded	1E@8.83	CLF	119.00	237.00	356.00
6 #12THHN, stranded	1E@9.53	CLF	128.00	256.00	384.00
1 #10THHN, solid	1E@6.13	CLF	87.10	165.00	252.10
2 #10THHN, solid	1E@6.93	CLF	99.30	186.00	285.30
3 #10THHN, solid	1E@7.73	CLF	112.00	208.00	320.00
4 #10THHN, solid	1E@8.53	CLF	124.00	229.00	353.00
5 #10THHN, solid	1E@9.33	CLF	136.00	251.00	387.00
6 #10THHN, solid	1E@10.1	CLF	148.00	271.00	419.00
1 #10THHN, stranded	1E@6.13	CLF	88.30	165.00	253.30
2 #10THHN, stranded	1E@6.93	CLF	102.00	186.00	288.00
3 #10THHN, stranded	1E@7.73	CLF	115.00	208.00	323.00
4 #10THHN, stranded	1E@8.53	CLF	129.00	229.00	358.00
5 #10THHN, stranded	1E@9.33	CLF	142.00	251.00	393.00
6 #10THHN, stranded	1E@10.1	CLF	155.00	271.00	426.00

Use these figures to estimate the cost of installing assemblies under the conditions described on pages 5 and 6. Costs listed are for each assembly installed. The crew is one electrician working at a labor cost of $26.88 per manhour. These costs include layout, material handling and normal waste. Add for the sales tax, delivery, supervision, mobilization, demobilization, cleanup, overhead and profit.

1-1/4" EMT Conduit Assemblies

Material	Craft@Hrs	Unit	Material Cost	Labor Cost	Installed Cost

100' 1-1/4" EMT conduit, 2 set screw connectors, 9 set screw couplings and 9 one-hole straps

Material	Craft@Hrs	Unit	Material Cost	Labor Cost	Installed Cost
Empty conduit	1E@6.14	CLF	108.00	165.00	273.00
1 plastic pull line	1E@6.44	CLF	108.00	173.00	281.00
1 #14THHN, solid	1E@6.74	CLF	113.00	181.00	294.00
2 #14THHN, solid	1E@7.34	CLF	119.00	197.00	316.00
3 #14THHN, solid	1E@7.94	CLF	124.00	213.00	337.00
4 #14THHN, solid	1E@8.54	CLF	129.00	230.00	359.00
5 #14THHN, solid	1E@9.14	CLF	134.00	246.00	380.00
6 #14THHN, solid	1E@9.74	CLF	139.00	262.00	401.00
1 #14THHN, stranded	1E@6.74	CLF	114.00	181.00	295.00
2 #14THHN, stranded	1E@7.34	CLF	120.00	197.00	317.00
3 #14THHN, stranded	1E@7.94	CLF	126.00	213.00	339.00
4 #14THHN, stranded	1E@8.54	CLF	132.00	230.00	362.00
5 #14THHN, stranded	1E@9.14	CLF	138.00	246.00	384.00
6 #14THHN, stranded	1E@9.74	CLF	144.00	262.00	406.00
1 #12THHN, solid	1E@6.84	CLF	116.00	184.00	300.00
2 #12THHN, solid	1E@7.54	CLF	124.00	203.00	327.00
3 #12THHN, solid	1E@8.24	CLF	132.00	221.00	353.00
4 #12THHN, solid	1E@8.94	CLF	139.00	240.00	379.00
5 #12THHN, solid	1E@9.64	CLF	147.00	259.00	406.00
6 #12THHN, solid	1E@10.3	CLF	155.00	277.00	432.00
1 #12THHN, stranded	1E@6.84	CLF	117.00	184.00	301.00
2 #12THHN, stranded	1E@7.54	CLF	126.00	203.00	329.00
3 #12THHN, stranded	1E@8.24	CLF	135.00	221.00	356.00
4 #12THHN, stranded	1E@8.94	CLF	144.00	240.00	384.00
5 #12THHN, stranded	1E@9.64	CLF	153.00	259.00	412.00
6 #12THHN, stranded	1E@10.3	CLF	162.00	277.00	439.00
1 #10THHN, solid	1E@6.94	CLF	121.00	187.00	308.00
2 #10THHN, solid	1E@7.74	CLF	133.00	208.00	341.00
3 #10THHN, solid	1E@8.54	CLF	145.00	230.00	375.00
4 #10THHN, solid	1E@9.34	CLF	157.00	251.00	408.00
5 #10THHN, solid	1E@10.1	CLF	169.00	271.00	440.00
6 #10THHN, solid	1E@10.9	CLF	182.00	293.00	475.00
1 #10THHN, stranded	1E@6.94	CLF	122.00	187.00	309.00
2 #10THHN, stranded	1E@7.74	CLF	135.00	208.00	343.00
3 #10THHN, stranded	1E@8.54	CLF	149.00	230.00	379.00
4 #10THHN, stranded	1E@9.34	CLF	162.00	251.00	413.00
5 #10THHN, stranded	1E@10.1	CLF	175.00	271.00	446.00
6 #10THHN, stranded	1E@10.9	CLF	189.00	293.00	482.00

Use these figures to estimate the cost of installing assemblies under the conditions described on pages 5 and 6. Costs listed are for each assembly installed. The crew is one electrician working at a labor cost of $26.88 per manhour. These costs include layout, material handling and normal waste. Add for the sales tax, delivery, supervision, mobilization, demobilization, cleanup, overhead and profit.

3/8" Aluminum Flex Conduit Assemblies

Material	Craft@Hrs	Unit	Material Cost	Labor Cost	Installed Cost
100' 3/8" aluminum flexible conduit, and 2 screw-in connectors					
Empty conduit	1E@2.53	CLF	21.60	68.00	89.60
1 plastic pull line	1E@2.83	CLF	21.70	76.10	97.80
1 #14THHN, solid	1E@3.13	CLF	26.80	84.10	110.90
2 #14THHN, solid	1E@3.73	CLF	31.90	100.00	131.90
3 #14THHN, solid	1E@4.33	CLF	37.10	116.00	153.10
4 #14THHN, solid	1E@4.93	CLF	42.20	133.00	175.20
5 #14THHN, solid	1E@5.53	CLF	47.30	149.00	196.30
6 #14THHN, solid	1E@6.13	CLF	52.40	165.00	217.40
1 #14THHN, stranded	1E@3.13	CLF	27.50	84.10	111.60
2 #14THHN, stranded	1E@3.73	CLF	33.50	100.00	133.50
3 #14THHN, stranded	1E@4.33	CLF	39.40	116.00	155.40
4 #14THHN, stranded	1E@4.93	CLF	45.40	133.00	178.40
5 #14THHN, stranded	1E@5.53	CLF	51.30	149.00	200.30
6 #14THHN, stranded	1E@6.13	CLF	57.20	165.00	222.20
1 #12THHN, solid	1E@3.23	CLF	29.40	86.80	116.20
2 #12THHN, solid	1E@3.93	CLF	37.10	106.00	143.10
3 #12THHN, solid	1E@4.63	CLF	44.90	124.00	168.90
4 #12THHN, solid	1E@5.33	CLF	52.60	143.00	195.60
5 #12THHN, solid	1E@6.03	CLF	60.40	162.00	222.40
6 #12THHN, solid	1E@6.73	CLF	68.20	181.00	249.20
1 #12THHN, stranded	1E@3.23	CLF	30.50	86.80	117.30
2 #12THHN, stranded	1E@3.93	CLF	39.40	106.00	145.40
3 #12THHN, stranded	1E@4.63	CLF	48.30	124.00	172.30
4 #12THHN, stranded	1E@5.33	CLF	57.20	143.00	200.20
5 #12THHN, stranded	1E@6.03	CLF	66.20	162.00	228.20
6 #12THHN, stranded	1E@6.73	CLF	75.10	181.00	256.10
1 #10THHN, solid	1E@3.33	CLF	33.80	89.50	123.30
2 #10THHN, solid	1E@4.13	CLF	46.00	111.00	157.00
3 #10THHN, solid	1E@4.93	CLF	58.20	133.00	191.20
4 #10THHN, solid	1E@5.73	CLF	70.40	154.00	224.40
5 #10THHN, solid	1E@6.53	CLF	82.70	176.00	258.70
6 #10THHN, solid	1E@7.33	CLF	94.90	197.00	291.90
1 #10THHN, stranded	1E@3.33	CLF	35.00	89.50	124.50
2 #10THHN, stranded	1E@4.13	CLF	48.40	111.00	159.40
3 #10THHN, stranded	1E@4.93	CLF	61.90	133.00	194.90
4 #10THHN, stranded	1E@5.73	CLF	75.30	154.00	229.30
5 #10THHN, stranded	1E@6.53	CLF	88.70	176.00	264.70
6 #10THHN, stranded	1E@7.33	CLF	102.00	197.00	299.00

Use these figures to estimate the cost of installing assemblies under the conditions described on pages 5 and 6. Costs listed are for each assembly installed. The crew is one electrician working at a labor cost of $26.88 per manhour. These costs include layout, material handling and normal waste. Add for the sales tax, delivery, supervision, mobilization, demobilization, cleanup, overhead and profit.

1/2" Aluminum Flex Conduit Assemblies

Material	Craft@Hrs	Unit	Material Cost	Labor Cost	Installed Cost
100' 1/2" aluminum flexible conduit and 2 screw-in connectors					
Empty conduit	1E@2.73	CLF	30.60	73.40	104.00
1 plastic pull line	1E@3.03	CLF	30.70	81.40	112.10
1 #14THHN, solid	1E@3.33	CLF	35.70	89.50	125.20
2 #14THHN, solid	1E@3.93	CLF	40.80	106.00	146.80
3 #14THHN, solid	1E@4.53	CLF	46.00	122.00	168.00
4 #14THHN, solid	1E@5.13	CLF	51.10	138.00	189.10
5 #14THHN, solid	1E@5.73	CLF	56.20	154.00	210.20
6 #14THHN, solid	1E@6.33	CLF	61.30	170.00	231.30
1 #14THHN, stranded	1E@3.33	CLF	36.50	89.50	126.00
2 #14THHN, stranded	1E@3.93	CLF	42.50	106.00	148.50
3 #14THHN, stranded	1E@4.53	CLF	48.40	122.00	170.40
4 #14THHN, stranded	1E@5.13	CLF	54.40	138.00	192.40
5 #14THHN, stranded	1E@5.73	CLF	60.30	154.00	214.30
6 #14THHN, stranded	1E@6.33	CLF	66.20	170.00	236.20
1 #12THHN, solid	1E@3.43	CLF	38.40	92.20	130.60
2 #12THHN, solid	1E@4.13	CLF	46.10	111.00	157.10
3 #12THHN, solid	1E@4.83	CLF	53.90	130.00	183.90
4 #12THHN, solid	1E@5.53	CLF	61.60	149.00	210.60
5 #12THHN, solid	1E@6.23	CLF	69.40	167.00	236.40
6 #12THHN, solid	1E@6.93	CLF	77.20	186.00	263.20
1 #12THHN, stranded	1E@3.43	CLF	39.50	92.20	131.70
2 #12THHN, stranded	1E@4.13	CLF	48.40	111.00	159.40
3 #12THHN, stranded	1E@4.83	CLF	57.30	130.00	187.30
4 #12THHN, stranded	1E@5.53	CLF	68.20	149.00	217.20
5 #12THHN, stranded	1E@6.23	CLF	75.20	167.00	242.20
6 #12THHN, stranded	1E@6.93	CLF	84.10	186.00	270.10
1 #10THHN, solid	1E@3.53	CLF	42.80	94.90	137.70
2 #10THHN, solid	1E@4.33	CLF	55.00	116.00	171.00
3 #10THHN, solid	1E@5.13	CLF	67.20	138.00	205.20
4 #10THHN, solid	1E@5.93	CLF	79.40	159.00	238.40
5 #10THHN, solid	1E@6.73	CLF	91.70	181.00	272.70
6 #10THHN, solid	1E@7.53	CLF	104.00	202.00	306.00
1 #10THHN, stranded	1E@3.53	CLF	44.00	94.90	138.90
2 #10THHN, stranded	1E@4.33	CLF	57.40	116.00	173.40
3 #10THHN, stranded	1E@5.13	CLF	70.90	138.00	208.90
4 #10THHN, stranded	1E@5.93	CLF	84.30	159.00	243.30
5 #10THHN, stranded	1E@6.73	CLF	97.70	181.00	278.70
6 #10THHN, stranded	1E@7.53	CLF	111.00	202.00	313.00

Use these figures to estimate the cost of installing assemblies under the conditions described on pages 5 and 6. Cost listed are for each assembly installed. The crew is one electrician working at a labor cost of $26.88 per manhour. These costs include layout, material handling and normal waste. Add for the sales tax, delivery, supervision, mobilization, demobilization, cleanup, overhead and profit.

3/4" Aluminum Flex Conduit Assemblies

Material	Craft@Hrs	Unit	Material Cost	Labor Cost	Installed Cost
100' 3/4" aluminum flexible conduit and 2 screw-in connectors					
Empty conduit	1E@3.05	CLF	43.00	82.00	125.00
1 plastic pull line	1E@3.35	CLF	43.10	90.00	133.10
1 #14THHN, solid	1E@3.65	CLF	48.20	98.10	146.30
2 #14THHN, solid	1E@4.25	CLF	53.30	114.00	167.30
3 #14THHN, solid	1E@4.85	CLF	58.40	130.00	188.40
4 #14THHN, solid	1E@5.45	CLF	63.50	146.00	209.50
5 #14THHN, solid	1E@6.05	CLF	68.60	163.00	231.60
6 #14THHN, solid	1E@6.65	CLF	73.80	179.00	252.80
1 #14THHN, stranded	1E@3.65	CLF	49.00	98.10	147.10
2 #14THHN, stranded	1E@4.25	CLF	54.90	114.00	168.90
3 #14THHN, stranded	1E@4.85	CLF	60.90	130.00	190.90
4 #14THHN, stranded	1E@5.45	CLF	66.80	146.00	212.80
5 #14THHN, stranded	1E@6.05	CLF	72.70	163.00	235.70
6 #14THHN, stranded	1E@6.65	CLF	78.70	179.00	257.70
1 #12THHN, solid	1E@3.75	CLF	50.80	101.00	151.80
2 #12THHN, solid	1E@4.45	CLF	58.60	120.00	178.60
3 #12THHN, solid	1E@5.15	CLF	66.30	138.00	204.30
4 #12THHN, solid	1E@5.85	CLF	74.10	157.00	231.10
5 #12THHN, solid	1E@6.55	CLF	81.80	176.00	257.80
6 #12THHN, solid	1E@7.25	CLF	89.60	195.00	284.60
1 #12THHN, stranded	1E@3.75	CLF	52.00	101.00	153.00
2 #12THHN, stranded	1E@4.45	CLF	60.90	120.00	180.90
3 #12THHN, stranded	1E@5.15	CLF	69.80	138.00	207.80
4 #12THHN, stranded	1E@5.85	CLF	78.70	157.00	235.70
5 #12THHN, stranded	1E@6.55	CLF	87.60	176.00	263.60
6 #12THHN, stranded	1E@7.25	CLF	96.50	195.00	291.50
1 #10THHN, solid	1E@3.85	CLF	55.30	103.00	158.30
2 #10THHN, solid	1E@4.65	CLF	67.50	125.00	192.50
3 #10THHN, solid	1E@5.45	CLF	79.70	146.00	225.70
4 #10THHN, solid	1E@6.25	CLF	91.90	168.00	259.90
5 #10THHN, solid	1E@7.05	CLF	104.00	190.00	294.00
6 #10THHN, solid	1E@7.85	CLF	116.00	211.00	327.00
1 #10THHN, stranded	1E@3.85	CLF	56.50	103.00	159.50
2 #10THHN, stranded	1E@4.65	CLF	69.90	125.00	194.90
3 #10THHN, stranded	1E@5.45	CLF	83.30	146.00	229.30
4 #10THHN, stranded	1E@6.25	CLF	96.70	168.00	264.70
5 #10THHN, stranded	1E@7.05	CLF	110.00	190.00	300.00
6 #10THHN, stranded	1E@7.85	CLF	124.00	211.00	335.00

Use these figures to estimate the cost of installing assemblies under the conditions described on pages 5 and 6. Cost listed are for each assembly installed. The crew is one electrician working at a labor cost of $26.88 per manhour. These costs include layout, material handling and normal waste. Add for the sales tax, delivery, supervision, mobilization, demobilization, cleanup, overhead and profit.

1" Aluminum Flex Conduit Assemblies

Material	Craft@Hrs	Unit	Material Cost	Labor Cost	Installed Cost
100' 1" aluminum flexible conduit and 2 screw-in connectors					
Empty conduit	1E@3.36	CLF	88.50	90.30	178.80
1 plastic pull line	1E@3.66	CLF	88.60	98.40	187.00
1 #14THHN, solid	1E@3.96	CLF	93.60	106.00	199.60
2 #14THHN, solid	1E@4.56	CLF	98.70	123.00	221.70
3 #14THHN, solid	1E@5.16	CLF	104.00	139.00	243.00
4 #14THHN, solid	1E@5.76	CLF	109.00	155.00	264.00
5 #14THHN, solid	1E@6.36	CLF	114.00	171.00	285.00
6 #14THHN, solid	1E@6.96	CLF	119.00	187.00	306.00
1 #14THHN, stranded	1E@3.96	CLF	94.40	106.00	200.40
2 #14THHN, stranded	1E@4.56	CLF	100.00	123.00	223.00
3 #14THHN, stranded	1E@5.16	CLF	106.00	139.00	245.00
4 #14THHN, stranded	1E@5.76	CLF	112.00	155.00	267.00
5 #14THHN, stranded	1E@6.36	CLF	118.00	171.00	289.00
6 #14THHN, stranded	1E@6.96	CLF	124.00	187.00	311.00
1 #12THHN, solid	1E@4.06	CLF	96.20	109.00	205.20
2 #12THHN, solid	1E@4.76	CLF	104.00	128.00	232.00
3 #12THHN, solid	1E@5.46	CLF	112.00	147.00	259.00
4 #12THHN, solid	1E@6.16	CLF	120.00	166.00	286.00
5 #12THHN, solid	1E@6.86	CLF	127.00	184.00	311.00
6 #12THHN, solid	1E@7.56	CLF	135.00	203.00	338.00
1 #12THHN, stranded	1E@4.06	CLF	97.40	109.00	206.40
2 #12THHN, stranded	1E@4.76	CLF	106.00	128.00	234.00
3 #12THHN, stranded	1E@5.46	CLF	115.00	147.00	262.00
4 #12THHN, stranded	1E@6.16	CLF	124.00	166.00	290.00
5 #12THHN, stranded	1E@6.86	CLF	133.00	184.00	317.00
6 #12THHN, stranded	1E@7.56	CLF	142.00	203.00	345.00
1 #10THHN, solid	1E@4.16	CLF	101.00	112.00	213.00
2 #10THHN, solid	1E@4.96	CLF	113.00	133.00	246.00
3 #10THHN, solid	1E@5.76	CLF	125.00	155.00	280.00
4 #10THHN, solid	1E@6.56	CLF	137.00	176.00	313.00
5 #10THHN, solid	1E@7.36	CLF	150.00	198.00	348.00
6 #10THHN, solid	1E@8.16	CLF	162.00	219.00	381.00
1 #10THHN, stranded	1E@4.16	CLF	102.00	112.00	214.00
2 #10THHN, stranded	1E@4.96	CLF	115.00	133.00	248.00
3 #10THHN, stranded	1E@5.76	CLF	128.00	155.00	283.00
4 #10THHN, stranded	1E@6.56	CLF	142.00	176.00	318.00
5 #10THHN, stranded	1E@7.36	CLF	156.00	198.00	354.00
6 #10THHN, stranded	1E@8.16	CLF	170.00	219.00	389.00

Use these figures to estimate the cost of installing assemblies under the conditions described on pages 5 and 6. Costs listed are for each assembly installed. The crew is one electrician working at a labor cost of $26.88 per manhour. These costs include layout, material handling and normal waste. Add for the sales tax, delivery, supervision, mobilization, demobilization, cleanup, overhead and profit.

3/8" Steel Flex Conduit Assemblies

Material	Craft@Hrs	Unit	Material Cost	Labor Cost	Installed Cost
100' 3/8" steel flexible conduit and 2 screw-in connectors					
Empty conduit	1E@2.73	CLF	17.80	73.40	91.20
1 plastic pull line	1E@3.03	CLF	17.90	81.40	99.30
1 #14THHN, solid	1E@3.33	CLF	22.90	89.50	112.40
2 #14THHN, solid	1E@3.93	CLF	28.00	106.00	134.00
3 #14THHN, solid	1E@4.53	CLF	33.10	122.00	155.10
4 #14THHN, solid	1E@5.13	CLF	38.30	138.00	176.30
5 #14THHN, solid	1E@5.73	CLF	43.40	154.00	197.40
6 #14THHN, solid	1E@6.33	CLF	48.50	170.00	218.50
1 #14THHN, stranded	1E@3.33	CLF	23.70	89.50	113.20
2 #14THHN, stranded	1E@3.93	CLF	29.70	106.00	135.70
3 #14THHN, stranded	1E@4.53	CLF	35.60	122.00	157.60
4 #14THHN, stranded	1E@5.13	CLF	41.50	138.00	179.50
5 #14THHN, stranded	1E@5.73	CLF	47.50	154.00	201.50
6 #14THHN, stranded	1E@6.33	CLF	53.40	170.00	223.40
1 #12THHN, solid	1E@3.43	CLF	25.50	92.20	117.70
2 #12THHN, solid	1E@4.13	CLF	33.30	111.00	144.30
3 #12THHN, solid	1E@4.83	CLF	41.10	130.00	171.10
4 #12THHN, solid	1E@5.53	CLF	48.80	149.00	197.80
5 #12THHN, solid	1E@6.23	CLF	56.60	167.00	223.60
6 #12THHN, solid	1E@6.93	CLF	64.30	186.00	250.30
1 #12THHN, stranded	1E@3.43	CLF	26.70	92.20	118.90
2 #12THHN, stranded	1E@4.13	CLF	35.60	111.00	146.60
3 #12THHN, stranded	1E@4.83	CLF	44.50	130.00	174.50
4 #12THHN, stranded	1E@5.53	CLF	53.40	149.00	202.40
5 #12THHN, stranded	1E@6.23	CLF	62.30	167.00	229.30
6 #12THHN, stranded	1E@6.93	CLF	71.20	186.00	257.20
1 #10THHN, solid	1E@3.53	CLF	30.00	94.90	124.90
2 #10THHN, solid	1E@4.33	CLF	42.20	116.00	158.20
3 #10THHN, solid	1E@5.13	CLF	54.40	138.00	192.40
4 #10THHN, solid	1E@5.93	CLF	66.60	159.00	225.60
5 #10THHN, solid	1E@6.73	CLF	78.80	181.00	259.80
6 #10THHN, solid	1E@7.53	CLF	91.00	202.00	293.00
1 #10THHN, stranded	1E@3.53	CLF	31.20	94.90	126.10
2 #10THHN, stranded	1E@4.33	CLF	44.60	116.00	160.60
3 #10THHN, stranded	1E@5.13	CLF	58.00	138.00	196.00
4 #10THHN, stranded	1E@5.93	CLF	71.50	159.00	230.50
5 #10THHN, stranded	1E@6.73	CLF	84.90	181.00	265.90
6 #10THHN, stranded	1E@7.53	CLF	98.30	202.00	300.30

Use these figures to estimate the cost of installing assemblies under the conditions described on pages 5 and 6. Cost listed are for each assembly installed. The crew is one electrician working at a labor cost of $26.88 per manhour. These costs include layout, material handling and normal waste. Add for the sales tax, delivery, supervision, mobilization, demobilization, cleanup, overhead and profit.

1/2" Steel Flex Conduit Assemblies

Material	Craft@Hrs	Unit	Material Cost	Labor Cost	Installed Cost
100' 1/2" steel flexible conduit and 2 screw-in connectors					
Empty conduit	1E@3.03	CLF	25.10	81.40	106.50
1 plastic pull line	1E@3.33	CLF	25.20	89.50	114.70
1 #14THHN, solid	1E@3.63	CLF	30.20	97.60	127.80
2 #14THHN, solid	1E@4.23	CLF	35.30	114.00	149.30
3 #14THHN, solid	1E@4.83	CLF	40.50	130.00	170.50
4 #14THHN, solid	1E@5.43	CLF	45.60	146.00	191.60
5 #14THHN, solid	1E@6.03	CLF	50.70	162.00	212.70
6 #14THHN, solid	1E@6.63	CLF	55.80	178.00	233.80
1 #14THHN, stranded	1E@3.63	CLF	31.00	97.60	128.60
2 #14THHN, stranded	1E@4.23	CLF	37.00	114.00	151.00
3 #14THHN, stranded	1E@4.83	CLF	42.90	130.00	172.90
4 #14THHN, stranded	1E@5.43	CLF	48.90	146.00	194.90
5 #14THHN, stranded	1E@6.03	CLF	54.80	162.00	216.80
6 #14THHN, stranded	1E@6.63	CLF	60.70	178.00	238.70
1 #12THHN, solid	1E@3.73	CLF	32.90	100.00	132.90
2 #12THHN, solid	1E@4.43	CLF	40.60	119.00	159.60
3 #12THHN, solid	1E@5.13	CLF	48.40	138.00	186.40
4 #12THHN, solid	1E@5.83	CLF	56.10	157.00	213.10
5 #12THHN, solid	1E@6.53	CLF	63.90	176.00	239.90
6 #12THHN, solid	1E@7.23	CLF	71.70	194.00	265.70
1 #12THHN, stranded	1E@3.73	CLF	34.00	100.00	134.00
2 #12THHN, stranded	1E@4.43	CLF	42.90	119.00	161.90
3 #12THHN, stranded	1E@5.13	CLF	51.80	138.00	189.80
4 #12THHN, stranded	1E@5.83	CLF	60.70	157.00	217.70
5 #12THHN, stranded	1E@6.53	CLF	69.60	176.00	245.60
6 #12THHN, stranded	1E@7.23	CLF	78.60	194.00	272.60
1 #10THHN, solid	1E@3.83	CLF	37.30	103.00	140.30
2 #10THHN, solid	1E@4.63	CLF	49.50	124.00	173.50
3 #10THHN, solid	1E@5.43	CLF	61.70	146.00	207.70
4 #10THHN, solid	1E@6.23	CLF	73.90	167.00	240.90
5 #10THHN, solid	1E@7.03	CLF	86.10	189.00	275.10
6 #10THHN, solid	1E@7.83	CLF	98.40	210.00	308.40
1 #10THHN, stranded	1E@3.83	CLF	38.50	103.00	141.50
2 #10THHN, stranded	1E@4.63	CLF	51.90	124.00	175.90
3 #10THHN, stranded	1E@5.43	CLF	65.40	146.00	211.40
4 #10THHN, stranded	1E@6.23	CLF	78.30	167.00	245.30
5 #10THHN, stranded	1E@7.03	CLF	92.20	189.00	281.20
6 #10THHN, stranded	1E@7.83	CLF	106.00	210.00	316.00

Use these figures to estimate the cost of installing assemblies under the conditions described on pages 5 and 6. Costs listed are for each assembly installed. The crew is one electrician working at a labor cost of $26.88 per manhour. These costs include layout, material handling and normal waste. Add for the sales tax, delivery, supervision, mobilization, demobilization, cleanup, overhead and profit.

3/4" Steel Flex Conduit Assemblies

Material	Craft@Hrs	Unit	Material Cost	Labor Cost	Installed Cost
100' 3/4" steel flexible conduit and 2 screw-in connectors					
Empty conduit	1E@3.30	CLF	32.70	88.70	121.40
1 plastic pull line	1E@3.60	CLF	32.80	96.80	129.60
1 #14THHN, solid	1E@3.90	CLF	37.80	105.00	142.80
2 #14THHN, solid	1E@4.50	CLF	42.90	121.00	163.90
3 #14THHN, solid	1E@5.10	CLF	48.10	137.00	185.10
4 #14THHN, solid	1E@5.70	CLF	53.20	153.00	206.20
5 #14THHN, solid	1E@6.30	CLF	58.30	169.00	227.30
6 #14THHN, solid	1E@6.90	CLF	63.40	185.00	248.40
1 #14THHN, stranded	1E@3.90	CLF	38.60	105.00	143.60
2 #14THHN, stranded	1E@4.50	CLF	44.60	121.00	165.60
3 #14THHN, stranded	1E@5.10	CLF	50.50	137.00	187.50
4 #14THHN, stranded	1E@5.70	CLF	56.50	153.00	209.50
5 #14THHN, stranded	1E@6.30	CLF	62.40	169.00	231.40
6 #14THHN, stranded	1E@6.90	CLF	68.30	185.00	253.30
1 #12THHN, solid	1E@4.00	CLF	40.50	108.00	148.50
2 #12THHN, solid	1E@4.70	CLF	48.20	126.00	174.20
3 #12THHN, solid	1E@5.40	CLF	56.00	145.00	201.00
4 #12THHN, solid	1E@6.10	CLF	63.70	164.00	227.70
5 #12THHN, solid	1E@6.80	CLF	71.50	183.00	254.50
6 #12THHN, solid	1E@7.50	CLF	79.30	202.00	281.30
1 #12THHN, stranded	1E@4.00	CLF	41.60	108.00	149.60
2 #12THHN, stranded	1E@4.70	CLF	50.50	126.00	176.50
3 #12THHN, stranded	1E@5.40	CLF	59.40	145.00	204.40
4 #12THHN, stranded	1E@6.10	CLF	68.30	164.00	232.30
5 #12THHN, stranded	1E@6.80	CLF	77.30	183.00	260.30
6 #12THHN, stranded	1E@7.50	CLF	86.20	202.00	288.20
1 #10THHN, solid	1E@4.10	CLF	44.90	110.00	154.90
2 #10THHN, solid	1E@4.90	CLF	57.10	132.00	189.10
3 #10THHN, solid	1E@5.70	CLF	69.30	153.00	222.30
4 #10THHN, solid	1E@6.50	CLF	81.50	175.00	256.50
5 #10THHN, solid	1E@7.30	CLF	93.80	196.00	289.80
6 #10THHN, solid	1E@8.10	CLF	106.00	218.00	324.00
1 #10THHN, stranded	1E@4.10	CLF	46.10	110.00	156.10
2 #10THHN, stranded	1E@4.90	CLF	59.50	132.00	191.50
3 #10THHN, stranded	1E@5.70	CLF	73.00	153.00	226.00
4 #10THHN, stranded	1E@6.50	CLF	86.40	175.00	261.40
5 #10THHN, stranded	1E@7.30	CLF	99.80	196.00	295.80
6 #10THHN, stranded	1E@8.10	CLF	113.00	218.00	331.00

Use these figures to estimate the cost of installing assemblies under the conditions described on pages 5 and 6. Costs listed are for each assembly installed. The crew is one electrician working at a labor cost of $26.88 per manhour. These costs include layout, material handling and normal waste. Add for the sales tax, delivery, supervision, mobilization, demobilization, cleanup, overhead and profit.

1" Steel Flex Conduit Assemblies

Material	Craft@Hrs	Unit	Material Cost	Labor Cost	Installed Cost
100' 1" steel flexible conduit and 2 screw-in connectors					
Empty conduit	1E@3.56	CLF	67.90	95.70	163.60
1 plastic pull line	1E@3.86	CLF	68.00	104.00	172.00
1 #14THHN, solid	1E@4.16	CLF	73.10	112.00	185.10
2 #14THHN, solid	1E@4.76	CLF	78.20	128.00	206.20
3 #14THHN, solid	1E@5.36	CLF	83.30	144.00	227.30
4 #14THHN, solid	1E@5.96	CLF	88.40	160.00	248.40
5 #14THHN, solid	1E@6.56	CLF	93.50	176.00	269.50
6 #14THHN, solid	1E@7.16	CLF	98.70	192.00	290.70
1 #14THHN, stranded	1E@4.16	CLF	73.90	112.00	185.90
2 #14THHN, stranded	1E@4.76	CLF	79.80	128.00	207.80
3 #14THHN, stranded	1E@5.36	CLF	85.80	144.00	229.80
4 #14THHN, stranded	1E@5.96	CLF	91.70	160.00	251.70
5 #14THHN, stranded	1E@6.56	CLF	97.60	176.00	273.60
6 #14THHN, stranded	1E@7.16	CLF	104.00	192.00	296.00
1 #12THHN, solid	1E@4.26	CLF	75.70	115.00	190.70
2 #12THHN, solid	1E@4.96	CLF	83.50	133.00	216.50
3 #12THHN, solid	1E@5.66	CLF	91.20	152.00	243.20
4 #12THHN, solid	1E@6.36	CLF	99.00	171.00	270.00
5 #12THHN, solid	1E@7.06	CLF	107.00	190.00	297.00
6 #12THHN, solid	1E@7.76	CLF	114.00	209.00	323.00
1 #12THHN, stranded	1E@4.26	CLF	76.80	115.00	191.80
2 #12THHN, stranded	1E@4.96	CLF	85.80	133.00	218.80
3 #12THHN, stranded	1E@5.66	CLF	94.70	152.00	246.70
4 #12THHN, stranded	1E@6.36	CLF	104.00	171.00	275.00
5 #12THHN, stranded	1E@7.06	CLF	112.00	190.00	302.00
6 #12THHN, stranded	1E@7.76	CLF	121.00	209.00	330.00
1 #10THHN, solid	1E@4.36	CLF	80.10	117.00	197.10
2 #10THHN, solid	1E@5.16	CLF	93.40	139.00	232.40
3 #10THHN, solid	1E@5.96	CLF	105.00	160.00	265.00
4 #10THHN, solid	1E@6.76	CLF	117.00	182.00	299.00
5 #10THHN, solid	1E@7.56	CLF	129.00	203.00	332.00
6 #10THHN, solid	1E@8.36	CLF	141.00	225.00	366.00
1 #10THHN, stranded	1E@4.36	CLF	81.40	117.00	198.40
2 #10THHN, stranded	1E@5.16	CLF	94.80	139.00	233.80
3 #10THHN, stranded	1E@5.96	CLF	108.00	160.00	268.00
4 #10THHN, stranded	1E@6.76	CLF	122.00	182.00	304.00
5 #10THHN, stranded	1E@7.56	CLF	135.00	203.00	338.00
6 #10THHN, stranded	1E@8.36	CLF	148.00	225.00	373.00

Use these figures to estimate the cost of installing assemblies under the conditions described on pages 5 and 6. Costs listed are for each assembly installed. The crew is one electrician working at a labor cost of $26.88 per manhour. These costs include layout, material handling and normal waste. Add for the sales tax, delivery, supervision, mobilization, demobilization, cleanup, overhead and profit.

Material	Craft@Hrs	Unit	Material Cost	Labor Cost	Installed Cost
100' 1/2" PVC conduit, 2 terminal adapters and 2 elbows					
Empty conduit	1E@3.30	CLF	11.50	88.70	100.20
1 plastic pull line	1E@3.60	CLF	11.60	96.80	108.40
1 #14THHN, solid	1E@4.20	CLF	16.60	113.00	129.60
2 #14THHN, solid	1E@4.80	CLF	21.80	129.00	150.80
3 #14THHN, solid	1E@5.40	CLF	26.90	145.00	171.90
4 #14THHN, solid	1E@6.00	CLF	32.00	161.00	193.00
5 #14THHN, solid	1E@6.60	CLF	37.10	177.00	214.10
6 #14THHN, solid	1E@7.20	CLF	42.20	194.00	236.20
1 #14THHN, stranded	1E@4.20	CLF	17.50	113.00	130.50
2 #14THHN, stranded	1E@4.80	CLF	23.40	129.00	152.40
3 #14THHN, stranded	1E@5.40	CLF	29.30	145.00	174.30
4 #14THHN, stranded	1E@6.00	CLF	35.30	161.00	196.30
5 #14THHN, stranded	1E@6.60	CLF	41.20	177.00	218.20
6 #14THHN, stranded	1E@7.20	CLF	47.20	194.00	241.20
1 #12THHN, solid	1E@4.00	CLF	19.30	108.00	127.30
2 #12THHN, solid	1E@4.70	CLF	27.00	126.00	153.00
3 #12THHN, solid	1E@5.40	CLF	34.80	145.00	179.80
4 #12THHN, solid	1E@6.10	CLF	42.60	164.00	206.60
5 #12THHN, solid	1E@6.80	CLF	50.30	183.00	233.30
6 #12THHN, solid	1E@7.50	CLF	58.10	202.00	260.10
1 #12THHN, stranded	1E@4.00	CLF	20.40	108.00	128.40
2 #12THHN, stranded	1E@4.70	CLF	29.30	126.00	155.30
3 #12THHN, stranded	1E@5.40	CLF	38.30	145.00	183.30
4 #12THHN, stranded	1E@6.10	CLF	47.20	164.00	211.20
5 #12THHN, stranded	1E@6.80	CLF	56.10	183.00	239.10
6 #12THHN, stranded	1E@7.50	CLF	65.00	202.00	267.00
1 #10THHN, solid	1E@4.10	CLF	23.70	110.00	133.70
2 #10THHN, solid	1E@4.90	CLF	35.90	132.00	167.90
3 #10THHN, solid	1E@5.70	CLF	48.20	153.00	201.20
4 #10THHN, solid	1E@6.50	CLF	60.40	175.00	235.40
5 #10THHN, solid	1E@7.30	CLF	72.60	196.00	268.60
6 #10THHN, solid	1E@8.10	CLF	84.80	218.00	302.80
1 #10THHN, stranded	1E@4.10	CLF	24.90	110.00	134.90
2 #10THHN, stranded	1E@4.90	CLF	38.40	132.00	170.40
3 #10THHN, stranded	1E@5.70	CLF	51.80	153.00	204.80
4 #10THHN, stranded	1E@6.50	CLF	65.20	175.00	240.20
5 #10THHN, stranded	1E@7.30	CLF	78.60	196.00	274.60
6 #10THHN, stranded	1E@8.10	CLF	92.00	218.00	310.00

Use these figures to estimate the cost of installing assemblies under the conditions described on pages 5 and 6. Costs listed are for each assembly installed. The crew is one electrician working at a labor cost of $26.88 per manhour. These costs include layout, material handling and normal waste. Add for the sales tax, delivery, supervision, mobilization, demobilization, cleanup, overhead and profit.

3/4" PVC Conduit Assemblies

Material	Craft@Hrs	Unit	Material Cost	Labor Cost	Installed Cost
100' 3/4" PVC conduit, 2 terminal adapters and 2 elbows					
Empty conduit	1E@3.32	CLF	15.20	89.20	104.40
1 plastic pull line	1E@3.63	CLF	15.30	97.60	112.90
1 #14THHN, solid	1E@3.92	CLF	20.30	105.00	125.30
2 #14THHN, solid	1E@4.52	CLF	25.50	121.00	146.50
3 #14THHN, solid	1E@5.12	CLF	30.60	138.00	168.60
4 #14THHN, solid	1E@5.72	CLF	35.70	154.00	189.70
5 #14THHN, solid	1E@6.32	CLF	40.80	170.00	210.80
6 #14THHN, solid	1E@6.92	CLF	45.90	186.00	231.90
1 #14THHN, stranded	1E@3.92	CLF	21.20	105.00	126.20
2 #14THHN, stranded	1E@4.52	CLF	27.10	121.00	148.10
3 #14THHN, stranded	1E@5.12	CLF	33.00	138.00	171.00
4 #14THHN, stranded	1E@5.72	CLF	39.00	154.00	193.00
5 #14THHN, stranded	1E@6.32	CLF	44.90	170.00	214.90
6 #14THHN, stranded	1E@6.92	CLF	50.90	186.00	236.90
1 #12THHN, solid	1E@4.02	CLF	23.00	108.00	131.00
2 #12THHN, solid	1E@4.72	CLF	30.70	127.00	157.70
3 #12THHN, solid	1E@5.42	CLF	38.50	146.00	184.50
4 #12THHN, solid	1E@6.12	CLF	46.30	165.00	211.30
5 #12THHN, solid	1E@6.82	CLF	54.00	183.00	237.00
6 #12THHN, solid	1E@7.52	CLF	61.80	202.00	263.80
1 #12THHN, stranded	1E@4.02	CLF	24.10	108.00	132.10
2 #12THHN, stranded	1E@4.72	CLF	33.00	127.00	160.00
3 #12THHN, stranded	1E@5.42	CLF	42.00	146.00	188.00
4 #12THHN, stranded	1E@6.12	CLF	50.90	165.00	215.90
5 #12THHN, stranded	1E@6.82	CLF	59.80	183.00	242.80
6 #12THHN, stranded	1E@7.52	CLF	68.70	202.00	270.70
1 #10THHN, solid	1E@4.12	CLF	27.40	111.00	138.40
2 #10THHN, solid	1E@4.92	CLF	39.60	132.00	171.60
3 #10THHN, solid	1E@5.72	CLF	51.90	154.00	205.90
4 #10THHN, solid	1E@6.52	CLF	64.10	175.00	239.10
5 #10THHN, solid	1E@7.32	CLF	76.30	197.00	273.30
6 #10THHN, solid	1E@8.12	CLF	88.50	218.00	306.50
1 #10THHN, stranded	1E@4.12	CLF	28.60	111.00	139.60
2 #10THHN, stranded	1E@4.92	CLF	42.10	132.00	174.10
3 #10THHN, stranded	1E@5.72	CLF	55.50	154.00	209.50
4 #10THHN, stranded	1E@6.52	CLF	68.90	175.00	243.90
5 #10THHN, stranded	1E@7.32	CLF	82.30	197.00	279.30
6 #10THHN, stranded	1E@8.12	CLF	95.70	218.00	313.70

Use these figures to estimate the cost of installing assemblies under the conditions described on pages 5 and 6. Costs listed are for each assembly installed. The crew is one electrician working at a labor cost of $26.88 per manhour. These costs include layout, material handling and normal waste. Add for the sales tax, delivery, supervision, mobilization, demobilization, cleanup, overhead and profit.

Material	Craft@Hrs	Unit	Material Cost	Labor Cost	Installed Cost
100' 1" PVC conduit, 2 terminal adapters and 2 elbows					
Empty conduit	1E@3.46	CLF	22.60	93.00	115.60
1 plastic pull line	1E@3.76	CLF	22.70	101.00	123.70
1 #14THHN, solid	1E@4.06	CLF	27.80	109.00	136.80
2 #14THHN, solid	1E@4.66	CLF	32.90	125.00	157.90
3 #14THHN, solid	1E@5.26	CLF	38.00	141.00	179.00
4 #14THHN, solid	1E@5.86	CLF	43.10	158.00	201.10
5 #14THHN, solid	1E@6.46	CLF	48.20	174.00	222.20
6 #14THHN, solid	1E@7.06	CLF	53.40	190.00	243.40
1 #14THHN, stranded	1E@4.06	CLF	28.60	109.00	137.60
2 #14THHN, stranded	1E@4.66	CLF	34.50	125.00	159.50
3 #14THHN, stranded	1E@5.26	CLF	40.50	141.00	181.50
4 #14THHN, stranded	1E@5.86	CLF	46.40	158.00	204.40
5 #14THHN, stranded	1E@6.46	CLF	52.30	174.00	226.30
6 #14THHN, stranded	1E@7.06	CLF	58.30	190.00	248.30
1 #12THHN, solid	1E@4.16	CLF	30.40	112.00	142.40
2 #12THHN, solid	1E@4.86	CLF	38.20	131.00	169.20
3 #12THHN, solid	1E@5.56	CLF	45.90	149.00	194.90
4 #12THHN, solid	1E@6.26	CLF	53.70	168.00	221.70
5 #12THHN, solid	1E@6.96	CLF	61.40	187.00	248.40
6 #12THHN, solid	1E@7.66	CLF	69.20	206.00	275.20
1 #12THHN, stranded	1E@4.16	CLF	31.60	112.00	143.60
2 #12THHN, stranded	1E@4.86	CLF	40.50	131.00	171.50
3 #12THHN, stranded	1E@5.56	CLF	49.40	149.00	198.40
4 #12THHN, stranded	1E@6.26	CLF	58.30	168.00	226.30
5 #12THHN, stranded	1E@6.96	CLF	67.20	187.00	254.20
6 #12THHN, stranded	1E@7.66	CLF	76.10	206.00	282.10
1 #10THHN, solid	1E@4.26	CLF	34.90	115.00	149.90
2 #10THHN, solid	1E@5.06	CLF	47.10	136.00	183.10
3 #10THHN, solid	1E@5.86	CLF	59.30	158.00	217.30
4 #10THHN, solid	1E@6.66	CLF	71.50	179.00	250.50
5 #10THHN, solid	1E@7.46	CLF	83.70	201.00	284.70
6 #10THHN, solid	1E@8.26	CLF	95.90	222.00	317.90
1 #10THHN, stranded	1E@4.26	CLF	36.10	115.00	151.10
2 #10THHN, stranded	1E@5.06	CLF	49.50	136.00	185.50
3 #10THHN, stranded	1E@5.86	CLF	62.90	158.00	220.90
4 #10THHN, stranded	1E@6.66	CLF	76.30	179.00	255.30
5 #10THHN, stranded	1E@7.46	CLF	89.70	201.00	290.70
6 #10THHN, stranded	1E@8.26	CLF	103.00	222.00	325.00

Use these figures to estimate the cost of installing assemblies under the conditions described on pages 5 and 6. Costs listed are for each assembly installed. The crew is one electrician working at a labor cost of $26.88 per manhour. These costs include layout, material handling and normal waste. Add for the sales tax, delivery, supervision, mobilization, demobilization, cleanup, overhead and profit.

1-1/4" PVC Conduit Assemblies

Material	Craft@Hrs	Unit	Material Cost	Labor Cost	Installed Cost
100' 1-1/4" PVC conduit, 2 terminal adapters and 2 elbows					
Empty conduit	1E@3.60	CLF	30.40	96.80	127.20
1 plastic pull line	1E@3.90	CLF	30.50	105.00	135.50
1 #14THHN, solid	1E@4.20	CLF	35.50	113.00	148.50
2 #14THHN, solid	1E@4.80	CLF	40.60	129.00	169.60
3 #14THHN, solid	1E@5.40	CLF	45.80	145.00	190.80
4 #14THHN, solid	1E@6.00	CLF	50.90	161.00	211.90
5 #14THHN, solid	1E@6.60	CLF	56.00	177.00	233.00
6 #14THHN, solid	1E@7.20	CLF	61.10	194.00	255.10
1 #14THHN, stranded	1E@4.20	CLF	36.30	113.00	149.30
2 #14THHN, stranded	1E@4.80	CLF	42.30	129.00	171.30
3 #14THHN, stranded	1E@5.40	CLF	48.20	145.00	193.20
4 #14THHN, stranded	1E@6.00	CLF	54.20	161.00	215.20
5 #14THHN, stranded	1E@6.60	CLF	60.10	177.00	237.10
6 #14THHN, stranded	1E@7.20	CLF	66.00	194.00	260.00
1 #12THHN, solid	1E@4.30	CLF	38.20	116.00	154.20
2 #12THHN, solid	1E@5.00	CLF	45.90	134.00	179.90
3 #12THHN, solid	1E@5.70	CLF	53.70	153.00	206.70
4 #12THHN, solid	1E@6.40	CLF	61.40	172.00	233.40
5 #12THHN, solid	1E@7.10	CLF	69.20	191.00	260.20
6 #12THHN, solid	1E@7.80	CLF	77.00	210.00	287.00
1 #12THHN, stranded	1E@4.30	CLF	39.30	116.00	155.30
2 #12THHN, stranded	1E@5.00	CLF	48.20	134.00	182.20
3 #12THHN, stranded	1E@5.70	CLF	57.10	153.00	210.10
4 #12THHN, stranded	1E@6.40	CLF	66.00	172.00	238.00
5 #12THHN, stranded	1E@7.10	CLF	75.00	191.00	266.00
6 #12THHN, stranded	1E@7.80	CLF	83.90	210.00	293.90
1 #10THHN, solid	1E@4.40	CLF	42.60	118.00	160.60
2 #10THHN, solid	1E@5.20	CLF	54.80	140.00	194.80
3 #10THHN, solid	1E@6.00	CLF	67.00	161.00	228.00
4 #10THHN, solid	1E@6.80	CLF	79.20	183.00	262.20
5 #10THHN, solid	1E@7.60	CLF	91.50	204.00	295.50
6 #10THHN, solid	1E@8.40	CLF	104.00	226.00	330.00
1 #10THHN, stranded	1E@4.40	CLF	43.80	118.00	161.80
2 #10THHN, stranded	1E@5.20	CLF	57.20	140.00	197.20
3 #10THHN, stranded	1E@6.00	CLF	70.70	161.00	231.70
4 #10THHN, stranded	1E@6.80	CLF	84.10	183.00	267.10
5 #10THHN, stranded	1E@7.60	CLF	97.50	204.00	301.50
6 #10THHN, stranded	1E@8.40	CLF	111.00	226.00	337.00

Use these figures to estimate the cost of installing assemblies under the conditions described on pages 5 and 6. Costs listed are for each assembly installed. The crew is one electrician working at a labor cost of $26.88 per manhour. These costs include layout, material handling and normal waste. Add for the sales tax, delivery, supervision, mobilization, demobilization, cleanup, overhead and profit.

1/2" Galvanized Rigid Conduit Assemblies

Material	Craft@Hrs	Unit	Material Cost	Labor Cost	Installed Cost
100' 1/2" galvanized rigid conduit, 2 terminations and 9 one-hole straps					
Empty conduit	1E@4.55	CLF	100.00	122.00	222.00
1 plastic pull line	1E@4.85	CLF	100.00	130.00	230.00
1 #14THHN, solid	1E@5.15	CLF	105.00	138.00	243.00
2 #14THHN, solid	1E@5.75	CLF	110.00	155.00	265.00
3 #14THHN, solid	1E@6.35	CLF	115.00	171.00	286.00
4 #14THHN, solid	1E@6.95	CLF	121.00	187.00	308.00
5 #14THHN, solid	1E@7.55	CLF	126.00	203.00	329.00
6 #14THHN, solid	1E@8.15	CLF	131.00	219.00	350.00
1 #14THHN, stranded	1E@5.15	CLF	106.00	138.00	244.00
2 #14THHN, stranded	1E@5.75	CLF	112.00	155.00	267.00
3 #14THHN, stranded	1E@6.35	CLF	118.00	171.00	289.00
4 #14THHN, stranded	1E@6.95	CLF	124.00	187.00	311.00
5 #14THHN, stranded	1E@7.55	CLF	130.00	203.00	333.00
6 #14THHN, stranded	1E@8.15	CLF	136.00	219.00	355.00
1 #12THHN, solid	1E@5.25	CLF	108.00	141.00	249.00
2 #12THHN, solid	1E@5.95	CLF	116.00	160.00	276.00
3 #12THHN, solid	1E@6.65	CLF	123.00	179.00	302.00
4 #12THHN, solid	1E@7.35	CLF	131.00	198.00	329.00
5 #12THHN, solid	1E@8.05	CLF	139.00	216.00	355.00
6 #12THHN, solid	1E@8.75	CLF	147.00	235.00	382.00
1 #12THHN, stranded	1E@5.25	CLF	109.00	141.00	250.00
2 #12THHN, stranded	1E@5.95	CLF	118.00	160.00	278.00
3 #12THHN, stranded	1E@6.65	CLF	127.00	179.00	306.00
4 #12THHN, stranded	1E@7.35	CLF	136.00	198.00	334.00
5 #12THHN, stranded	1E@8.05	CLF	145.00	216.00	361.00
6 #12THHN, stranded	1E@8.75	CLF	154.00	235.00	389.00
1 #10THHN, solid	1E@5.35	CLF	112.00	144.00	256.00
2 #10THHN, solid	1E@6.15	CLF	124.00	165.00	289.00
3 #10THHN, solid	1E@6.95	CLF	137.00	187.00	324.00
4 #10THHN, solid	1E@7.75	CLF	149.00	208.00	357.00
5 #10THHN, solid	1E@8.55	CLF	161.00	230.00	391.00
6 #10THHN, solid	1E@9.35	CLF	173.00	251.00	424.00
1 #10THHN, stranded	1E@5.35	CLF	113.00	144.00	257.00
2 #10THHN, stranded	1E@6.15	CLF	127.00	165.00	292.00
3 #10THHN, stranded	1E@6.95	CLF	140.00	187.00	327.00
4 #10THHN, stranded	1E@7.75	CLF	154.00	208.00	362.00
5 #10THHN, stranded	1E@8.55	CLF	167.00	230.00	397.00
6 #10THHN, stranded	1E@9.35	CLF	181.00	251.00	432.00

Use these figures to estimate the cost of installing assemblies under the conditions described on pages 5 and 6. Costs listed are for each assembly installed. The crew is one electrician working at a labor cost of $26.88 per manhour. These costs include layout, material handling and normal waste. Add for the sales tax, delivery, supervision, mobilization, demobilization, cleanup, overhead and profit.

3/4" Galvanized Rigid Conduit Assemblies

Material	Craft@Hrs	Unit	Material Cost	Labor Cost	Installed Cost
100' 3/4" galvanized rigid conduit, 2 terminations and 9 one-hole straps					
Empty conduit	1E@5.16	CLF	125.00	139.00	264.00
1 plastic pull line	1E@5.46	CLF	125.00	147.00	272.00
1 #14THHN, solid	1E@5.76	CLF	130.00	155.00	285.00
2 #14THHN, solid	1E@6.36	CLF	136.00	171.00	307.00
3 #14THHN, solid	1E@6.96	CLF	141.00	187.00	328.00
4 #14THHN, solid	1E@7.56	CLF	146.00	203.00	349.00
5 #14THHN, solid	1E@8.16	CLF	151.00	219.00	370.00
6 #14THHN, solid	1E@8.76	CLF	156.00	235.00	391.00
1 #14THHN, stranded	1E@5.76	CLF	131.00	155.00	286.00
2 #14THHN, stranded	1E@6.36	CLF	137.00	171.00	308.00
3 #14THHN, stranded	1E@6.96	CLF	143.00	187.00	330.00
4 #14THHN, stranded	1E@7.56	CLF	149.00	203.00	352.00
5 #14THHN, stranded	1E@8.16	CLF	155.00	219.00	374.00
6 #14THHN, stranded	1E@8.76	CLF	161.00	235.00	396.00
1 #12THHN, solid	1E@5.86	CLF	133.00	158.00	291.00
2 #12THHN, solid	1E@6.56	CLF	141.00	176.00	317.00
3 #12THHN, solid	1E@7.26	CLF	149.00	195.00	344.00
4 #12THHN, solid	1E@7.96	CLF	156.00	214.00	370.00
5 #12THHN, solid	1E@8.66	CLF	164.00	233.00	397.00
6 #12THHN, solid	1E@9.36	CLF	172.00	252.00	424.00
1 #12THHN, stranded	1E@5.86	CLF	134.00	158.00	292.00
2 #12THHN, stranded	1E@6.56	CLF	143.00	176.00	319.00
3 #12THHN, stranded	1E@7.26	CLF	152.00	195.00	347.00
4 #12THHN, stranded	1E@7.96	CLF	161.00	214.00	375.00
5 #12THHN, stranded	1E@8.66	CLF	170.00	233.00	403.00
6 #12THHN, stranded	1E@9.36	CLF	179.00	252.00	431.00
1 #10THHN, solid	1E@5.96	CLF	138.00	160.00	298.00
2 #10THHN, solid	1E@6.76	CLF	150.00	182.00	332.00
3 #10THHN, solid	1E@7.56	CLF	162.00	203.00	365.00
4 #10THHN, solid	1E@8.36	CLF	174.00	225.00	399.00
5 #10THHN, solid	1E@9.16	CLF	186.00	246.00	432.00
6 #10THHN, solid	1E@9.96	CLF	199.00	268.00	467.00
1 #10THHN, stranded	1E@5.96	CLF	139.00	160.00	299.00
2 #10THHN, stranded	1E@6.76	CLF	152.00	182.00	334.00
3 #10THHN, stranded	1E@7.56	CLF	166.00	203.00	369.00
4 #10THHN, stranded	1E@8.36	CLF	179.00	225.00	404.00
5 #10THHN, stranded	1E@9.16	CLF	192.00	246.00	438.00
6 #10THHN, stranded	1E@9.96	CLF	206.00	268.00	474.00

Use these figures to estimate the cost of installing assemblies under the conditions described on pages 5 and 6. Costs listed are for each assembly installed. The crew is one electrician working at a labor cost of $26.88 per manhour. These costs include layout, material handling and normal waste. Add for the sales tax, delivery, supervision, mobilization, demobilization, cleanup, overhead and profit.

1" Galvanized Rigid Conduit Assemblies

Material	Craft@Hrs	Unit	Material Cost	Labor Cost	Installed Cost
100' 1" galvanized rigid conduit, 2 terminations and 9 one-hole straps					
Empty conduit	1E@5.88	CLF	183.00	158.00	341.00
1 plastic pull line	1E@6.18	CLF	183.00	166.00	349.00
1 #14THHN, solid	1E@6.48	CLF	188.00	174.00	362.00
2 #14THHN, solid	1E@7.08	CLF	193.00	190.00	383.00
3 #14THHN, solid	1E@7.68	CLF	198.00	206.00	404.00
4 #14THHN, solid	1E@8.28	CLF	203.00	223.00	426.00
5 #14THHN, solid	1E@8.88	CLF	208.00	239.00	447.00
6 #14THHN, solid	1E@9.48	CLF	213.00	255.00	468.00
1 #14THHN, stranded	1E@6.48	CLF	189.00	174.00	363.00
2 #14THHN, stranded	1E@7.08	CLF	195.00	190.00	385.00
3 #14THHN, stranded	1E@7.68	CLF	200.00	206.00	406.00
4 #14THHN, stranded	1E@8.28	CLF	206.00	223.00	429.00
5 #14THHN, stranded	1E@8.88	CLF	212.00	239.00	451.00
6 #14THHN, stranded	1E@9.48	CLF	218.00	255.00	473.00
1 #12THHN, solid	1E@6.58	CLF	190.00	177.00	367.00
2 #12THHN, solid	1E@7.28	CLF	198.00	196.00	394.00
3 #12THHN, solid	1E@7.98	CLF	206.00	215.00	421.00
4 #12THHN, solid	1E@8.68	CLF	214.00	233.00	447.00
5 #12THHN, solid	1E@9.38	CLF	221.00	252.00	473.00
6 #12THHN, solid	1E@10.1	CLF	229.00	271.00	500.00
1 #12THHN, stranded	1E@6.58	CLF	192.00	177.00	369.00
2 #12THHN, stranded	1E@7.28	CLF	200.00	196.00	396.00
3 #12THHN, stranded	1E@7.98	CLF	209.00	215.00	424.00
4 #12THHN, stranded	1E@8.68	CLF	218.00	233.00	451.00
5 #12THHN, stranded	1E@9.38	CLF	222.00	252.00	474.00
6 #12THHN, stranded	1E@10.1	CLF	236.00	271.00	507.00
1 #10THHN, solid	1E@6.68	CLF	195.00	180.00	375.00
2 #10THHN, solid	1E@7.48	CLF	207.00	201.00	408.00
3 #10THHN, solid	1E@8.28	CLF	219.00	223.00	442.00
4 #10THHN, solid	1E@9.08	CLF	231.00	244.00	475.00
5 #10THHN, solid	1E@9.88	CLF	244.00	266.00	510.00
6 #10THHN, solid	1E@10.7	CLF	256.00	288.00	544.00
1 #10THHN, stranded	1E@6.68	CLF	196.00	180.00	376.00
2 #10THHN, stranded	1E@7.48	CLF	209.00	201.00	410.00
3 #10THHN, stranded	1E@8.28	CLF	223.00	223.00	446.00
4 #10THHN, stranded	1E@9.08	CLF	236.00	244.00	480.00
5 #10THHN, stranded	1E@9.88	CLF	250.00	266.00	516.00
6 #10THHN, stranded	1E@10.7	CLF	263.00	288.00	551.00

Use these figures to estimate the cost of installing assemblies under the conditions described on pages 5 and 6. Costs listed are for each assembly installed. The crew is one electrician working at a labor cost of $26.88 per manhour. These costs include layout, material handling and normal waste. Add for the sales tax, delivery, supervision, mobilization, demobilization, cleanup, overhead and profit.

1-1/4" Galvanized Rigid Conduit Assemblies

Material	Craft@Hrs	Unit	Material Cost	Labor Cost	Installed Cost
100' 1-1/4" galvanized rigid conduit, 2 terminations and 9 one-hole straps					
Empty conduit	1E@8.10	CLF	233.00	218.00	451.00
1 plastic pull line	1E@8.40	CLF	233.00	226.00	459.00
1 #14THHN, solid	1E@8.70	CLF	238.00	234.00	472.00
2 #14THHN, solid	1E@9.30	CLF	243.00	250.00	493.00
3 #14THHN, solid	1E@9.90	CLF	248.00	266.00	514.00
4 #14THHN, solid	1E@10.5	CLF	253.00	282.00	535.00
5 #14THHN, solid	1E@11.1	CLF	258.00	298.00	556.00
6 #14THHN, solid	1E@11.7	CLF	263.00	314.00	577.00
1 #14THHN, stranded	1E@8.70	CLF	239.00	234.00	473.00
2 #14THHN, stranded	1E@9.30	CLF	244.00	250.00	494.00
3 #14THHN, stranded	1E@9.90	CLF	250.00	266.00	516.00
4 #14THHN, stranded	1E@10.5	CLF	256.00	282.00	538.00
5 #14THHN, stranded	1E@11.1	CLF	262.00	298.00	560.00
6 #14THHN, stranded	1E@11.7	CLF	268.00	314.00	582.00
1 #12THHN, solid	1E@8.80	CLF	240.00	237.00	477.00
2 #12THHN, solid	1E@9.50	CLF	248.00	255.00	503.00
3 #12THHN, solid	1E@10.2	CLF	256.00	274.00	530.00
4 #12THHN, solid	1E@10.9	CLF	264.00	293.00	557.00
5 #12THHN, solid	1E@11.6	CLF	271.00	312.00	583.00
6 #12THHN, solid	1E@12.3	CLF	279.00	331.00	610.00
1 #12THHN, stranded	1E@8.80	CLF	241.00	237.00	478.00
2 #12THHN, stranded	1E@9.50	CLF	250.00	255.00	505.00
3 #12THHN, stranded	1E@10.2	CLF	259.00	274.00	533.00
4 #12THHN, stranded	1E@10.9	CLF	268.00	293.00	561.00
5 #12THHN, stranded	1E@11.6	CLF	277.00	312.00	589.00
6 #12THHN, stranded	1E@12.3	CLF	286.00	331.00	617.00
1 #10THHN, solid	1E@8.90	CLF	245.00	239.00	484.00
2 #10THHN, solid	1E@9.70	CLF	257.00	261.00	518.00
3 #10THHN, solid	1E@10.5	CLF	269.00	282.00	551.00
4 #10THHN, solid	1E@11.3	CLF	281.00	304.00	585.00
5 #10THHN, solid	1E@12.1	CLF	294.00	325.00	619.00
6 #10THHN, solid	1E@12.9	CLF	306.00	347.00	653.00
1 #10THHN, stranded	1E@8.90	CLF	246.00	239.00	485.00
2 #10THHN, stranded	1E@9.70	CLF	259.00	261.00	520.00
3 #10THHN, stranded	1E@10.5	CLF	273.00	282.00	555.00
4 #10THHN, stranded	1E@11.3	CLF	286.00	304.00	590.00
5 #10THHN, stranded	1E@12.1	CLF	300.00	325.00	625.00
6 #10THHN, stranded	1E@12.9	CLF	313.00	347.00	660.00

Use these figures to estimate the cost of installing assemblies under the conditions described on pages 5 and 6. Costs listed are for each assembly installed. The crew is one electrician working at a labor cost of $26.88 per manhour. These costs include layout, material handling and normal waste. Add for the sales tax, delivery, supervision, mobilization, demobilization, cleanup, overhead and profit.

1-1/2" Deep Handy Box Switch Assemblies

Material	Craft@Hrs	Unit	Material Cost	Labor Cost	Installed Cost
Switches in 1-1/2" deep handy boxes with handy box covers					
1-pole 15 amp brown	1E@0.43	Ea	3.52	11.60	15.12
2-pole 15 amp brown	1E@0.48	Ea	7.16	12.90	20.06
3-way 15 amp brown	1E@0.48	Ea	4.93	12.90	17.83
4-way 15 amp brown	1E@0.58	Ea	9.74	15.60	25.34
1-pole 15 amp brown 277V	1E@0.43	Ea	3.52	11.60	15.12
2-pole 15 amp brown 277V	1E@0.48	Ea	7.16	12.90	20.06
3-way 15 amp brown 277V	1E@0.48	Ea	4.93	12.90	17.83
4-way 15 amp brown 277V	1E@0.58	Ea	9.74	15.60	25.34
1-pole 15 amp ivory	1E@0.43	Ea	3.52	11.60	15.12
2-pole 15 amp ivory	1E@0.48	Ea	7.16	12.90	20.06
3-way 15 amp ivory	1E@0.48	Ea	4.93	12.90	17.83
4-way 15 amp ivory	1E@0.58	Ea	9.74	15.60	25.34
1-pole 15 amp ivory 277V	1E@0.43	Ea	3.52	11.60	15.12
2-pole 15 amp ivory 277V	1E@0.48	Ea	7.16	12.90	20.06
3-way 15 amp ivory 277V	1E@0.48	Ea	4.93	12.90	17.83
4-way 15 amp ivory 277V	1E@0.58	Ea	9.74	15.60	25.34
1-pole 20 amp brown	1E@0.43	Ea	4.50	11.60	16.10
2-pole 20 amp brown	1E@0.48	Ea	10.10	12.90	23.00
3-way 20 amp brown	1E@0.48	Ea	5.74	12.90	18.64
4-way 20 amp brown	1E@0.58	Ea	17.40	15.60	33.00
1-pole 20 amp brown 277V	1E@0.43	Ea	4.50	11.60	16.10
2-pole 20 amp brown 277V	1E@0.48	Ea	10.10	12.90	23.00
3-way 20 amp brown 277V	1E@0.48	Ea	5.74	12.90	18.64
4-way 20 amp brown 277V	1E@0.58	Ea	17.40	15.60	33.00
1-pole 20 amp ivory	1E@0.43	Ea	4.50	11.60	16.10
2-pole 20 amp ivory	1E@0.48	Ea	10.10	12.90	23.00
3-way 20 amp ivory	1E@0.48	Ea	5.74	12.90	18.64
4-way 20 amp ivory	1E@0.58	Ea	17.40	15.60	33.00
1-pole 20 amp ivory 277V	1E@0.43	Ea	4.50	11.60	16.10
2-pole 20 amp ivory 277V	1E@0.48	Ea	10.10	12.90	23.00
3-way 20 amp ivory 277V	1E@0.48	Ea	5.74	12.90	18.64
4-way 20 amp ivory 277V	1E@0.58	Ea	17.40	15.60	33.00
Key switches in 1-1/2" deep handy boxes with handy box covers					
1-pole 15 amp	1E@0.43	Ea	9.94	11.60	21.54
2-pole 15 amp	1E@0.48	Ea	20.10	12.90	33.00
3-way 15 amp	1E@0.48	Ea	13.70	12.90	26.60
4-way 15 amp	1E@0.58	Ea	39.60	15.60	55.20
1-pole 20 amp	1E@0.43	Ea	19.70	11.60	31.30
2-pole 20 amp	1E@0.48	Ea	41.90	12.90	54.80
3-way 20 amp	1E@0.48	Ea	21.60	12.90	34.50
4-way 20 amp	1E@0.58	Ea	42.30	15.60	57.90

S

S_K

Use these figures to estimate the cost of installing assemblies under the conditions described on pages 5 and 6. Costs listed are for each assembly installed. The crew is one electrician working at a labor cost of $26.88 per manhour. These costs include layout, material handling and normal waste. Add for the sales tax, delivery, supervision, mobilization, demobilization, cleanup, overhead and profit.

1-7/8" Deep Handy Box Assemblies

Material	Craft@Hrs	Unit	Material Cost	Labor Cost	Installed Cost
Switches in 1-7/8" deep handy boxes with handy box covers					
1-pole 15 amp brown	1E@0.45	Ea	3.54	12.10	15.64
2-pole 15 amp brown	1E@0.50	Ea	7.18	13.40	20.58
3-way 15 amp brown	1E@0.50	Ea	4.95	13.40	18.35
4-way 15 amp brown	1E@0.60	Ea	9.76	16.10	25.86
1-pole 15 amp brown 277V	1E@0.45	Ea	3.54	12.10	15.64
2-pole 15 amp brown 277V	1E@0.50	Ea	7.18	13.40	20.58
3-way 15 amp brown 277V	1E@0.50	Ea	4.95	13.40	18.35
4-way 15 amp brown 277V	1E@0.60	Ea	9.76	16.10	25.86
1-pole 15 amp ivory	1E@0.45	Ea	3.54	12.10	15.64
2-pole 15 amp ivory	1E@0.50	Ea	7.18	13.40	20.58
3-way 15 amp ivory	1E@0.50	Ea	4.95	13.40	18.35
4-way 15 amp ivory	1E@0.60	Ea	9.76	16.10	25.86
1-pole 15 amp ivory 277V	1E@0.45	Ea	3.54	12.10	15.64
2-pole 15 amp ivory 277V	1E@0.50	Ea	7.18	13.40	20.58
3-way 15 amp ivory 277V	1E@0.50	Ea	4.95	13.40	18.35
4-way 15 amp ivory 277V	1E@0.60	Ea	9.76	16.10	25.86
1-pole 20 amp brown	1E@0.45	Ea	4.52	12.10	16.62
2-pole 20 amp brown	1E@0.50	Ea	10.10	13.40	23.50
3-way 20 amp brown	1E@0.50	Ea	5.76	13.40	19.16
4-way 20 amp brown	1E@0.60	Ea	17.40	16.10	33.50
1-pole 20 amp brown 277V	1E@0.45	Ea	4.52	12.10	16.62
2-pole 20 amp brown 277V	1E@0.50	Ea	10.10	13.40	23.50
3-way 20 amp brown 277V	1E@0.50	Ea	5.76	13.40	19.16
4-way 20 amp brown 277V	1E@0.60	Ea	17.40	16.10	33.50
1-pole 20 amp ivory	1E@0.45	Ea	4.52	12.10	16.62
2-pole 20 amp ivory	1E@0.50	Ea	10.10	13.40	23.50
3-way 20 amp ivory	1E@0.50	Ea	5.76	13.40	19.16
4-way 20 amp ivory	1E@0.60	Ea	17.40	16.10	33.50
1-pole 20 amp ivory 277V	1E@0.45	Ea	4.52	12.10	16.62
2-pole 20 amp ivory 277V	1E@0.50	Ea	10.10	13.40	23.50
3-way 20 amp ivory 277V	1E@0.50	Ea	5.76	13.40	19.16
4-way 20 amp ivory 277V	1E@0.60	Ea	17.40	16.10	33.50
Key switches in 1-7/8" deep handy boxes with handy box covers					
1-pole 15 amp	1E@0.45	Ea	9.96	12.10	22.06
2-pole 15 amp	1E@0.50	Ea	20.20	13.40	33.60
3-way 15 amp	1E@0.50	Ea	13.80	13.40	27.20
4-way 15 amp	1E@0.60	Ea	39.70	16.10	55.80
1-pole 20 amp	1E@0.45	Ea	19.80	12.10	31.90
2-pole 20 amp	1E@0.50	Ea	41.90	13.40	55.30
3-way 20 amp	1E@0.50	Ea	21.70	13.40	35.10
4-way 20 amp	1E@0.60	Ea	42.40	16.10	58.50

Use these figures to estimate the cost of installing assemblies under the conditions described on pages 5 and 6. Costs listed are for each assembly installed. The crew is one electrician working at a labor cost of $26.88 per manhour. These costs include layout, material handling and normal waste. Add for the sales tax, delivery, supervision, mobilization, demobilization, cleanup, overhead and profit.

2-1/8" Deep Handy Box Switch Assemblies

Material	Craft@Hrs	Unit	Material Cost	Labor Cost	Installed Cost
Switches in 2-1/8" deep handy boxes with handy box covers					
1-pole 15 amp brown	1E@0.48	Ea	3.87	12.90	16.77
2-pole 15 amp brown	1E@0.53	Ea	7.51	14.20	21.71
3-way 15 amp brown	1E@0.53	Ea	5.27	14.20	19.47
4-way 15 amp brown	1E@0.63	Ea	10.10	16.90	27.00
1-pole 15 amp brown 277V	1E@0.48	Ea	3.87	12.90	16.77
2-pole 15 amp brown 277V	1E@0.53	Ea	7.51	14.20	21.71
3-way 15 amp brown 277V	1E@0.53	Ea	5.27	14.20	19.47
4-way 15 amp brown 277V	1E@0.63	Ea	10.10	16.90	27.00
1-pole 15 amp ivory	1E@0.48	Ea	3.87	12.90	16.77
2-pole 15 amp ivory	1E@0.53	Ea	7.51	14.20	21.71
3-way 15 amp ivory	1E@0.53	Ea	5.27	14.20	19.47
4-way 15 amp ivory	1E@0.63	Ea	10.10	16.90	27.00
1-pole 15 amp ivory 277V	1E@0.48	Ea	3.87	12.90	16.77
2-pole 15 amp ivory 277V	1E@0.53	Ea	7.51	14.20	21.71
3-way 15 amp ivory 277V	1E@0.53	Ea	5.21	14.20	19.41
4-way 15 amp ivory 277V	1E@0.63	Ea	10.10	16.90	27.00
1-pole 20 amp brown	1E@0.48	Ea	4.85	12.90	17.75
2-pole 20 amp brown	1E@0.53	Ea	10.40	14.20	24.60
3-way 20 amp brown	1E@0.53	Ea	6.09	14.20	20.29
4-way 20 amp brown	1E@0.63	Ea	17.80	16.90	34.70
1-pole 20 amp brown 277V	1E@0.48	Ea	4.85	12.90	17.75
2-pole 20 amp brown 277V	1E@0.53	Ea	10.40	14.20	24.60
3-way 20 amp brown 277V	1E@0.53	Ea	6.09	14.20	20.29
4-way 20 amp brown 277V	1E@0.63	Ea	17.80	16.90	34.70
1-pole 20 amp ivory	1E@0.48	Ea	4.85	12.90	17.75
2-pole 20 amp ivory	1E@0.53	Ea	10.40	14.20	24.60
3-way 20 amp ivory	1E@0.53	Ea	6.09	14.20	20.29
4-way 20 amp ivory	1E@0.63	Ea	17.80	16.90	34.70
1-pole 20 amp ivory 277V	1E@0.48	Ea	4.85	12.90	17.75
2-pole 20 amp ivory 277V	1E@0.53	Ea	10.40	14.20	24.60
3-way 20 amp ivory 277V	1E@0.53	Ea	6.09	14.20	20.29
4-way 20 amp ivory 277V	1E@0.63	Ea	17.80	16.90	34.70

S

Material	Craft@Hrs	Unit	Material Cost	Labor Cost	Installed Cost
Key switches in 2-1/8" deep handy boxes with handy box covers					
1-pole 15 amp	1E@0.48	Ea	10.30	12.90	23.20
2-pole 15 amp	1E@0.53	Ea	20.50	14.20	34.70
3-way 15 amp	1E@0.53	Ea	14.10	14.20	28.30
4-way 15 amp	1E@0.63	Ea	40.00	16.90	56.90
1-pole 20 amp	1E@0.48	Ea	20.10	12.90	33.00
2-pole 20 amp	1E@0.53	Ea	42.20	14.20	56.40
3-way 20 amp	1E@0.53	Ea	22.00	14.20	36.20
4-way 20 amp	1E@0.63	Ea	42.70	16.90	59.60

S$_K$

Use these figures to estimate the cost of installing assemblies under the conditions described on pages 5 and 6. Costs listed are for each assembly installed. The crew is one electrician working at a labor cost of $26.88 per manhour. These costs include layout, material handling and normal waste. Add for the sales tax, delivery, supervision, mobilization, demobilization, cleanup, overhead and profit.

2-13/16" Deep Handy Box Switch Assemblies

Material	Craft@Hrs	Unit	Material Cost	Labor Cost	Installed Cost
Switches in 2-13/16" deep handy boxes with handy box covers					
1-pole 15 amp brown	1E@0.53	Ea	4.57	14.20	18.77
2-pole 15 amp brown	1E@0.58	Ea	8.21	15.60	23.81
3-way 15 amp brown	1E@0.58	Ea	5.98	15.60	21.58
4-way 15 amp brown	1E@0.68	Ea	10.80	18.30	29.10
1-pole 15 amp brown 277V	1E@0.53	Ea	4.57	14.20	18.77
2-pole 15 amp brown 277V	1E@0.58	Ea	8.21	15.60	23.81
3-way 15 amp brown 277V	1E@0.58	Ea	5.98	15.60	21.58
4-way 15 amp brown 277V	1E@0.68	Ea	10.80	18.30	29.10
1-pole 15 amp ivory	1E@0.53	Ea	4.57	14.20	18.77
2-pole 15 amp ivory	1E@0.58	Ea	8.21	15.60	23.81
3-way 15 amp ivory	1E@0.58	Ea	5.98	15.60	21.58
4-way 15 amp ivory	1E@0.68	Ea	10.80	18.30	29.10
1-pole 15 amp ivory 277V	1E@0.53	Ea	4.57	14.20	18.77
2-pole 15 amp ivory 277V	1E@0.58	Ea	8.21	15.60	23.81
3-way 15 amp ivory 277V	1E@0.58	Ea	5.98	15.60	21.58
4-way 15 amp ivory 277V	1E@0.68	Ea	10.80	18.30	29.10
1-pole 20 amp brown	1E@0.53	Ea	5.55	14.20	19.75
2-pole 20 amp brown	1E@0.58	Ea	11.10	15.60	26.70
3-way 20 amp brown	1E@0.58	Ea	6.79	15.60	22.39
4-way 20 amp brown	1E@0.68	Ea	18.50	18.30	36.80
1-pole 20 amp brown 277V	1E@0.53	Ea	5.55	14.20	19.75
2-pole 20 amp brown 277V	1E@0.58	Ea	11.10	15.60	26.70
3-way 20 amp brown 277V	1E@0.58	Ea	6.79	15.60	22.39
4-way 20 amp brown 277V	1E@0.68	Ea	18.50	18.30	36.80
1-pole 20 amp ivory	1E@0.53	Ea	5.55	14.20	19.75
2-pole 20 amp ivory	1E@0.58	Ea	11.10	15.60	26.70
3-way 20 amp ivory	1E@0.58	Ea	6.79	15.60	22.39
4-way 20 amp ivory	1E@0.68	Ea	18.50	18.30	36.80
1-pole 20 amp ivory 277V	1E@0.53	Ea	5.55	14.20	19.75
2-pole 20 amp ivory 277V	1E@0.58	Ea	11.10	15.60	26.70
3-way 20 amp ivory 277V	1E@0.58	Ea	6.79	15.60	22.39
4-way 20 amp ivory 277V	1E@0.68	Ea	18.50	18.30	36.80
Key switches in 2-13/16" deep handy boxes with handy box covers					
1-pole 15 amp	1E@0.53	Ea	11.00	14.20	25.20
2-pole 15 amp	1E@0.58	Ea	21.20	15.60	36.80
3-way 15 amp	1E@0.58	Ea	14.80	15.60	30.40
4-way 15 amp	1E@0.68	Ea	40.70	18.30	59.00
1-pole 20 amp	1E@0.53	Ea	20.80	14.20	35.00
2-pole 20 amp	1E@0.58	Ea	42.90	15.60	58.50
3-way 20 amp	1E@0.58	Ea	22.70	15.60	38.30
4-way 20 amp	1E@0.68	Ea	43.40	18.30	61.70

Use these figures to estimate the cost of installing assemblies under the conditions described on pages 5 and 6. Costs listed are for each assembly installed. The crew is one electrician working at a labor cost of $26.88 per manhour. These costs include layout, material handling and normal waste. Add for the sales tax, delivery, supervision, mobilization, demobilization, cleanup, overhead and profit.

2" Deep 15 Amp Sectional Box Switch Assemblies

Material	Craft@Hrs	Unit	Material Cost	Labor Cost	Installed Cost	
Switches, 15 amp 1-pole in 2" deep sectional box and plastic plate						
1 gang brown	1E@0.45	Ea	3.81	12.10	15.91	
2 gang brown	1E@0.90	Ea	7.64	24.20	31.84	
3 gang brown	1E@1.35	Ea	9.52	36.30	45.82	S
4 gang brown	1E@1.80	Ea	12.80	48.40	61.20	
5 gang brown	1E@2.25	Ea	16.80	60.50	77.30	
6 gang brown	1E@2.70	Ea	20.10	72.60	92.70	
Switches, 15 amp 2-pole in 2" deep sectional box and plastic plate						
1 gang brown	1E@0.50	Ea	6.81	13.40	20.21	
2 gang brown	1E@1.00	Ea	13.60	26.90	40.50	
3 gang brown	1E@1.50	Ea	20.40	40.30	60.70	S_2
4 gang brown	1E@2.00	Ea	27.40	53.80	81.20	
5 gang brown	1E@2.50	Ea	35.00	67.20	102.20	
6 gang brown	1E@3.00	Ea	41.90	80.60	122.50	
Switches, 15 amp 3-way in 2" deep sectional box and plastic plate						
1 gang brown	1E@0.50	Ea	4.58	13.40	17.98	
2 gang brown	1E@1.00	Ea	9.18	26.90	36.08	
3 gang brown	1E@1.50	Ea	13.80	40.30	54.10	S_3
4 gang brown	1E@2.00	Ea	18.50	53.80	72.30	
5 gang brown	1E@2.50	Ea	23.80	67.20	91.00	
6 gang brown	1E@3.00	Ea	28.50	80.60	109.10	
Switches, 15 amp 4-way in 2" deep sectional box and plastic plate						
1 gang brown	1E@0.60	Ea	9.34	16.10	25.44	
2 gang brown	1E@1.20	Ea	18.80	32.30	51.10	
3 gang brown	1E@1.80	Ea	28.20	48.40	76.60	S_4
4 gang brown	1E@2.40	Ea	37.70	64.50	102.20	
5 gang brown	1E@3.00	Ea	47.90	80.60	128.50	
6 gang brown	1E@3.60	Ea	57.40	96.80	154.20	
Switches, 15 amp key in 2" deep sectional box and plastic plate						
1 gang 1-pole	1E@0.45	Ea	9.59	12.10	21.69	
1 gang 2-pole	1E@0.50	Ea	19.80	13.40	33.20	S_K
1 gang 3-way	1E@0.50	Ea	13.40	13.40	26.80	
1 gang 4-way	1E@0.60	Ea	39.30	16.10	55.40	

Use these figures to estimate the cost of installing assemblies under the conditions described on pages 5 and 6. Costs listed are for each assembly installed. The crew is one electrician working at a labor cost of $26.88 per manhour. These costs include layout, material handling and normal waste. Add for the sales tax, delivery, supervision, mobilization, demobilization, cleanup, overhead and profit.

2-1/2" Deep 15 Amp Sectional Box Switch Assemblies

Material	Craft@Hrs	Unit	Material Cost	Labor Cost	Installed Cost

Switches, 15 amp 1-pole in 2-1/2" deep sectional box and plastic plate

Material	Craft@Hrs	Unit	Material Cost	Labor Cost	Installed Cost
1 gang brown	1E@0.47	Ea	3.45	12.60	16.05
2 gang brown	1E@0.94	Ea	7.28	25.30	32.58
3 gang brown	1E@1.41	Ea	9.16	37.90	47.06
4 gang brown	1E@1.88	Ea	12.50	50.50	63.00
5 gang brown	1E@2.35	Ea	16.40	63.20	79.60
6 gang brown	1E@2.82	Ea	19.70	75.80	95.50

S

Switches, 15 amp 2-pole in 2-1/2" deep sectional box and plastic plate

Material	Craft@Hrs	Unit	Material Cost	Labor Cost	Installed Cost
1 gang brown	1E@0.52	Ea	6.45	14.00	20.45
2 gang brown	1E@1.04	Ea	13.30	28.00	41.30
3 gang brown	1E@1.56	Ea	20.10	41.90	62.00
4 gang brown	1E@2.08	Ea	27.00	55.90	82.90
5 gang brown	1E@2.60	Ea	34.60	69.90	104.50
6 gang brown	1E@3.12	Ea	41.60	83.90	125.50

S_2

Switches, 15 amp 3-way in 2-1/2" deep sectional box and plastic plate

Material	Craft@Hrs	Unit	Material Cost	Labor Cost	Installed Cost
1 gang brown	1E@0.52	Ea	4.22	14.00	18.22
2 gang brown	1E@1.04	Ea	8.82	28.00	36.82
3 gang brown	1E@1.56	Ea	13.40	41.90	55.30
4 gang brown	1E@2.08	Ea	18.10	55.90	74.00
5 gang brown	1E@2.60	Ea	23.50	69.90	93.40
6 gang brown	1E@3.12	Ea	28.20	83.90	112.10

S_3

Switches, 15 amp 4-way in 2-1/2" deep sectional box and plastic plate

Material	Craft@Hrs	Unit	Material Cost	Labor Cost	Installed Cost
1 gang brown	1E@0.62	Ea	9.05	16.70	25.75
2 gang brown	1E@1.24	Ea	18.40	33.30	51.70
3 gang brown	1E@1.86	Ea	27.80	50.00	77.80
4 gang brown	1E@2.48	Ea	37.40	66.70	104.10
5 gang brown	1E@3.10	Ea	47.50	83.30	130.80
6 gang brown	1E@3.72	Ea	57.00	100.00	157.00

S_4

Switches, 15 amp key in 2-1/2" deep sectional box and plastic plate

Material	Craft@Hrs	Unit	Material Cost	Labor Cost	Installed Cost
1 gang 1-pole	1E@0.47	Ea	9.23	12.60	21.83
1 gang 2-pole	1E@0.52	Ea	19.40	14.00	33.40
1 gang 3-way	1E@0.52	Ea	13.00	14.00	27.00
1 gang 4-way	1E@0.62	Ea	38.90	16.70	55.60

S_K

Use these figures to estimate the cost of installing assemblies under the conditions described on pages 5 and 6. Costs listed are for each assembly installed. The crew is one electrician working at a labor cost of $26.88 per manhour. These costs include layout, material handling and normal waste. Add for the sales tax, delivery, supervision, mobilization, demobilization, cleanup, overhead and profit.

2-3/4" Deep 15 Amp Sectional Box Switch Assemblies

Material	Craft@Hrs	Unit	Material Cost	Labor Cost	Installed Cost

Switches, 15 amp 1-pole in 2-3/4" deep sectional box and plastic plate

Material	Craft@Hrs	Unit	Material Cost	Labor Cost	Installed Cost	
1 gang brown	1E@0.50	Ea	3.60	13.40	17.00	
2 gang brown	1E@1.00	Ea	7.43	26.90	34.33	
3 gang brown	1E@1.50	Ea	9.31	40.30	49.61	S
4 gang brown	1E@2.00	Ea	12.60	53.80	66.40	
5 gang brown	1E@2.50	Ea	16.60	67.20	83.80	
6 gang brown	1E@3.00	Ea	19.90	80.60	100.50	

Switches, 15 amp 2-pole in 2-3/4" deep sectional box and plastic plate

Material	Craft@Hrs	Unit	Material Cost	Labor Cost	Installed Cost	
1 gang brown	1E@0.55	Ea	6.60	14.80	21.40	
2 gang brown	1E@1.10	Ea	13.40	29.60	43.00	
3 gang brown	1E@1.65	Ea	20.20	44.40	64.60	S_2
4 gang brown	1E@2.20	Ea	27.20	59.10	86.30	
5 gang brown	1E@2.75	Ea	34.80	73.90	108.70	
6 gang brown	1E@3.30	Ea	41.70	88.70	130.40	

Switches, 15 amp 3-way in 2-3/4" deep sectional box and plastic plate

Material	Craft@Hrs	Unit	Material Cost	Labor Cost	Installed Cost	
1 gang brown	1E@0.55	Ea	4.37	14.80	19.17	
2 gang brown	1E@1.10	Ea	8.97	29.60	38.57	
3 gang brown	1E@1.65	Ea	13.50	44.40	57.90	S_3
4 gang brown	1E@2.20	Ea	18.30	59.10	77.40	
5 gang brown	1E@2.75	Ea	23.60	73.90	97.50	
6 gang brown	1E@3.30	Ea	28.30	88.70	117.00	

Switches, 15 amp 4-way in 2-3/4" deep sectional box and plastic plate

Material	Craft@Hrs	Unit	Material Cost	Labor Cost	Installed Cost	
1 gang brown	1E@0.65	Ea	9.18	17.50	26.68	
2 gang brown	1E@1.30	Ea	18.60	34.90	53.50	
3 gang brown	1E@1.95	Ea	28.00	52.40	80.40	S_4
4 gang brown	1E@2.60	Ea	37.50	69.90	107.40	
5 gang brown	1E@3.25	Ea	47.70	87.40	135.10	
6 gang brown	1E@3.90	Ea	57.20	105.00	162.20	

Switches, 15 amp key in 2-3/4" deep sectional box and plastic plate

Material	Craft@Hrs	Unit	Material Cost	Labor Cost	Installed Cost	
1 gang 1-pole	1E@0.50	Ea	9.38	13.40	22.78	
1 gang 2-pole	1E@0.55	Ea	19.60	14.80	34.40	S_K
1 gang 3-way	1E@0.55	Ea	13.20	14.80	28.00	
1 gang 4-way	1E@0.65	Ea	39.10	17.50	56.60	

Use these figures to estimate the cost of installing assemblies under the conditions described on pages 5 and 6. Costs listed are for each assembly installed. The crew is one electrician working at a labor cost of $26.88 per manhour. These costs include layout, material handling and normal waste. Add for the sales tax, delivery, supervision, mobilization, demobilization, cleanup, overhead and profit.

3-1/2" Deep 15 Amp Sectional Box Switch Assemblies

Material	Craft@Hrs	Unit	Material Cost	Labor Cost	Installed Cost
Switches, 15 amp 1-pole in 3-1/2" deep sectional box and plastic plate					
1 gang brown	1E@0.52	Ea	3.91	14.00	17.91
2 gang brown	1E@1.04	Ea	7.74	28.00	35.74
3 gang brown	1E@1.56	Ea	9.62	41.90	51.52
4 gang brown	1E@2.08	Ea	12.90	55.90	68.80
5 gang brown	1E@2.60	Ea	16.90	69.90	86.80
6 gang brown	1E@3.12	Ea	20.20	83.90	104.10
Switches, 15 amp 2-pole in 3-1/2" deep sectional box and plastic plate					
1 gang brown	1E@0.57	Ea	6.91	15.30	22.21
2 gang brown	1E@1.14	Ea	13.70	30.60	44.30
3 gang brown	1E@1.71	Ea	20.50	46.00	66.50
4 gang brown	1E@2.28	Ea	27.50	61.30	88.80
5 gang brown	1E@2.85	Ea	35.10	76.60	111.70
6 gang brown	1E@3.42	Ea	42.00	91.90	133.90
Switches, 15 amp 3-way in 3-1/2" deep sectional box and plastic plate					
1 gang brown	1E@0.57	Ea	4.68	15.30	19.98
2 gang brown	1E@1.14	Ea	9.28	30.60	39.88
3 gang brown	1E@1.71	Ea	13.90	46.00	59.90
4 gang brown	1E@2.28	Ea	18.60	61.30	79.90
5 gang brown	1E@2.85	Ea	23.90	76.60	100.50
6 gang brown	1E@3.42	Ea	28.60	91.90	120.50
Switches, 15 amp 4-way in 3-1/2" deep sectional box and plastic plate					
1 gang brown	1E@0.67	Ea	9.49	18.00	27.49
2 gang brown	1E@1.34	Ea	18.90	36.00	54.90
3 gang brown	1E@2.01	Ea	28.30	54.00	82.30
4 gang brown	1E@2.68	Ea	37.80	72.00	109.80
5 gang brown	1E@3.35	Ea	48.00	90.00	138.00
6 gang brown	1E@4.02	Ea	57.50	108.00	165.50
Switches, 15 amp key in 3-1/2" deep sectional box and plastic plate					
1 gang 1-pole	1E@0.52	Ea	9.69	14.00	23.69
1 gang 2-pole	1E@0.57	Ea	19.90	15.30	35.20
1 gang 3-way	1E@0.57	Ea	13.50	15.30	28.80
1 gang 4-way	1E@0.67	Ea	39.40	18.00	57.40

Use these figures to estimate the cost of installing assemblies under the conditions described on pages 5 and 6. Costs listed are for each assembly installed. The crew is one electrician working at a labor cost of $26.88 per manhour. These costs include layout, material handling and normal waste. Add for the sales tax, delivery, supervision, mobilization, demobilization, cleanup, overhead and profit.

2" Deep 15 Amp Sectional Box Switch Assemblies

Material	Craft@Hrs	Unit	Material Cost	Labor Cost	Installed Cost
Switches, 15 amp 1-pole in 2" deep sectional box and plastic plate					
1 gang ivory	1E@0.45	Ea	3.81	12.10	15.91
2 gang ivory	1E@0.90	Ea	7.64	24.20	31.84
3 gang ivory	1E@1.35	Ea	9.52	36.30	45.82
4 gang ivory	1E@1.80	Ea	12.80	48.40	61.20
5 gang ivory	1E@2.25	Ea	16.80	60.50	77.30
6 gang ivory	1E@2.70	Ea	20.10	72.60	92.70

S

Material	Craft@Hrs	Unit	Material Cost	Labor Cost	Installed Cost
Switches, 15 amp 2-pole in 2" deep sectional box and plastic plate					
1 gang ivory	1E@0.50	Ea	6.81	13.40	20.21
2 gang ivory	1E@1.00	Ea	13.60	26.90	40.50
3 gang ivory	1E@1.50	Ea	20.40	40.30	60.70
4 gang ivory	1E@2.00	Ea	27.40	53.80	81.20
5 gang ivory	1E@2.50	Ea	35.00	67.20	102.20
6 gang ivory	1E@3.00	Ea	41.90	80.60	122.50

S_2

Material	Craft@Hrs	Unit	Material Cost	Labor Cost	Installed Cost
Switches, 15 amp 3-way in 2" deep sectional box and plastic plate					
1 gang ivory	1E@0.50	Ea	4.58	13.40	17.98
2 gang ivory	1E@1.00	Ea	9.18	26.90	36.08
3 gang ivory	1E@1.50	Ea	13.80	40.30	54.10
4 gang ivory	1E@2.00	Ea	18.50	53.80	72.30
5 gang ivory	1E@2.50	Ea	23.80	67.20	91.00
6 gang ivory	1E@3.00	Ea	28.50	80.60	109.10

S_3

Material	Craft@Hrs	Unit	Material Cost	Labor Cost	Installed Cost
Switches, 15 amp 4-way in 2" deep sectional box and plastic plate					
1 gang ivory	1E@0.60	Ea	9.39	16.10	25.49
2 gang ivory	1E@1.20	Ea	18.90	32.30	51.20
3 gang ivory	1E@1.80	Ea	28.20	48.40	76.60
4 gang ivory	1E@2.40	Ea	39.70	64.50	104.20
5 gang ivory	1E@3.00	Ea	47.90	80.60	128.50
6 gang ivory	1E@3.60	Ea	57.40	96.80	154.20

S_4

Use these figures to estimate the cost of installing assemblies under the conditions described on pages 5 and 6. Costs listed are for each assembly installed. The crew is one electrician working at a labor cost of $26.88 per manhour. These costs include layout, material handling and normal waste. Add for the sales tax, delivery, supervision, mobilization, demobilization, cleanup, overhead and profit.

2-1/2" Deep 15 Amp Sectional Box Switch Assemblies

Material	Craft@Hrs	Unit	Material Cost	Labor Cost	Installed Cost
Switches, 15 amp 1-pole in 2-1/2" deep sectional box and plastic plate					
1 gang ivory	1E@0.47	Ea	3.45	12.60	16.05
2 gang ivory	1E@0.94	Ea	7.28	25.30	32.58
3 gang ivory	1E@1.41	Ea	9.16	37.90	47.06
4 gang ivory	1E@1.88	Ea	12.50	50.50	63.00
5 gang ivory	1E@2.35	Ea	16.40	63.20	79.60
6 gang ivory	1E@2.82	Ea	19.70	75.80	95.50
Switches, 15 amp 2-pole in 2-1/2" deep sectional box and plastic plate					
1 gang ivory	1E@0.52	Ea	6.45	14.00	20.45
2 gang ivory	1E@1.04	Ea	13.30	28.00	41.30
3 gang ivory	1E@1.56	Ea	20.10	41.90	62.00
4 gang ivory	1E@2.08	Ea	27.00	55.90	82.90
5 gang ivory	1E@2.60	Ea	34.60	69.90	104.50
6 gang ivory	1E@3.12	Ea	41.60	83.90	125.50
Switches, 15 amp 3-way in 2-1/2" deep sectional box and plastic plate					
1 gang ivory	1E@0.52	Ea	4.22	14.00	18.22
2 gang ivory	1E@1.04	Ea	8.82	28.00	36.82
3 gang ivory	1E@1.56	Ea	13.40	41.90	55.30
4 gang ivory	1E@2.08	Ea	18.10	55.90	74.00
5 gang ivory	1E@2.60	Ea	23.50	69.90	93.40
6 gang ivory	1E@3.12	Ea	28.20	83.90	112.10
Switches, 15 amp 4-way in 2-1/2" deep sectional box and plastic plate					
1 gang ivory	1E@0.62	Ea	9.03	16.70	25.73
2 gang ivory	1E@1.24	Ea	18.40	33.30	51.70
3 gang ivory	1E@1.86	Ea	27.80	50.00	77.80
4 gang ivory	1E@2.48	Ea	37.40	66.70	104.10
5 gang ivory	1E@3.10	Ea	47.50	83.30	130.80
6 gang ivory	1E@3.72	Ea	57.00	100.00	157.00

Use these figures to estimate the cost of installing assemblies under the conditions described on pages 5 and 6. Costs listed are for each assembly installed. The crew is one electrician working at a labor cost of $26.88 per manhour. These costs include layout, material handling and normal waste. Add for the sales tax, delivery, supervision, mobilization, demobilization, cleanup, overhead and profit.

2-3/4" Deep 15 Amp Sectional Box Switch Assemblies

Material	Craft@Hrs	Unit	Material Cost	Labor Cost	Installed Cost	

Switches, 15 amp 1-pole in 2-3/4" deep sectional box and plastic plate

Material	Craft@Hrs	Unit	Material Cost	Labor Cost	Installed Cost	
1 gang ivory	1E@0.50	Ea	3.60	13.40	17.00	
2 gang ivory	1E@1.00	Ea	7.43	26.90	34.33	
3 gang ivory	1E@1.50	Ea	9.31	40.30	49.61	S
4 gang ivory	1E@2.00	Ea	12.60	53.80	66.40	
5 gang ivory	1E@2.50	Ea	16.60	67.20	83.80	
6 gang ivory	1E@3.00	Ea	19.90	80.60	100.50	

Switches, 15 amp 2-pole in 2-3/4" deep sectional box and plastic plate

Material	Craft@Hrs	Unit	Material Cost	Labor Cost	Installed Cost	
1 gang ivory	1E@0.55	Ea	6.60	14.80	21.40	
2 gang ivory	1E@1.10	Ea	13.40	29.60	43.00	
3 gang ivory	1E@1.65	Ea	20.20	44.40	64.60	S_2
4 gang ivory	1E@2.20	Ea	27.20	59.10	86.30	
5 gang ivory	1E@2.75	Ea	34.80	73.90	108.70	
6 gang ivory	1E@3.30	Ea	41.70	88.70	130.40	

Switches, 15 amp 3-way in 2-3/4" deep sectional box and plastic plate

Material	Craft@Hrs	Unit	Material Cost	Labor Cost	Installed Cost	
1 gang ivory	1E@0.55	Ea	4.37	14.80	19.17	
2 gang ivory	1E@1.10	Ea	8.97	29.60	38.57	
3 gang ivory	1E@1.65	Ea	13.50	44.40	57.90	S_3
4 gang ivory	1E@2.20	Ea	18.30	59.10	77.40	
5 gang ivory	1E@2.75	Ea	23.60	73.90	97.50	
6 gang ivory	1E@3.30	Ea	28.30	88.70	117.00	

Switches, 15 amp 4-way in 2-3/4" deep sectional box and plastic plate

Material	Craft@Hrs	Unit	Material Cost	Labor Cost	Installed Cost	
1 gang ivory	1E@0.65	Ea	9.18	17.50	26.68	
2 gang ivory	1E@1.30	Ea	18.60	34.90	53.50	S_4
3 gang ivory	1E@1.95	Ea	28.00	52.40	80.40	
4 gang ivory	1E@2.60	Ea	37.50	69.90	107.40	
5 gang ivory	1E@3.25	Ea	47.70	87.40	135.10	
6 gang ivory	1E@3.90	Ea	57.70	105.00	162.70	

Use these figures to estimate the cost of installing assemblies under the conditions described on pages 5 and 6. Costs listed are for each assembly installed. The crew is one electrician working at a labor cost of $26.88 per manhour. These costs include layout, material handling and normal waste. Add for the sales tax, delivery, supervision, mobilization, demobilization, cleanup, overhead and profit.

3-1/2" Deep 15 Amp Sectional Box Switch Assemblies

Material	Craft@Hrs	Unit	Material Cost	Labor Cost	Installed Cost

Switches, 15 amp 1-pole in 3-1/2" deep sectional box and plastic plate

Material	Craft@Hrs	Unit	Material Cost	Labor Cost	Installed Cost
1 gang ivory	1E@0.52	Ea	3.91	14.00	17.91
2 gang ivory	1E@1.04	Ea	7.74	28.00	35.74
3 gang ivory	1E@1.56	Ea	9.62	41.90	51.52
4 gang ivory	1E@2.08	Ea	12.90	55.90	68.80
5 gang ivory	1E@2.60	Ea	16.90	69.90	86.80
6 gang ivory	1E@3.12	Ea	20.20	83.90	104.10

S

Switches, 15 amp 2-pole in 3-1/2" deep sectional box and plastic plate

Material	Craft@Hrs	Unit	Material Cost	Labor Cost	Installed Cost
1 gang ivory	1E@0.57	Ea	6.91	15.30	22.21
2 gang ivory	1E@1.14	Ea	13.70	30.60	44.30
3 gang ivory	1E@1.71	Ea	20.50	46.00	66.50
4 gang ivory	1E@2.28	Ea	27.50	61.30	88.80
5 gang ivory	1E@2.85	Ea	35.10	76.60	111.70
6 gang ivory	1E@3.42	Ea	42.00	91.90	133.90

S_2

Switches, 15 amp 3-way in 3-1/2" deep sectional box and plastic plate

Material	Craft@Hrs	Unit	Material Cost	Labor Cost	Installed Cost
1 gang ivory	1E@0.57	Ea	4.68	15.30	19.98
2 gang ivory	1E@1.14	Ea	9.28	30.60	39.88
3 gang ivory	1E@1.71	Ea	13.90	46.00	59.90
4 gang ivory	1E@2.28	Ea	18.60	61.30	79.90
5 gang ivory	1E@2.85	Ea	23.90	76.60	100.50
6 gang ivory	1E@3.42	Ea	28.60	91.90	120.50

S_3

Switches, 15 amp 4-way in 3-1/2" deep sectional box and plastic plate

Material	Craft@Hrs	Unit	Material Cost	Labor Cost	Installed Cost
1 gang ivory	1E@0.67	Ea	9.49	18.00	27.49
2 gang ivory	1E@1.34	Ea	18.90	36.00	54.90
3 gang ivory	1E@2.01	Ea	28.30	54.00	82.30
4 gang ivory	1E@2.68	Ea	37.80	72.00	109.80
5 gang ivory	1E@3.35	Ea	48.00	90.00	138.00
6 gang ivory	1E@4.02	Ea	57.50	108.00	165.50

S_4

Use these figures to estimate the cost of installing assemblies under the conditions described on pages 5 and 6. Costs listed are for each assembly installed. The crew is one electrician working at a labor cost of $26.88 per manhour. These costs include layout, material handling and normal waste. Add for the sales tax, delivery, supervision, mobilization, demobilization, cleanup, overhead and profit.

2" Deep 20 Amp Sectional Box Switch Assemblies

Material	Craft@Hrs	Unit	Material Cost	Labor Cost	Installed Cost
Switches, 20 amp 1-pole in 2" deep sectional box and plastic plate					
1 gang brown	1E@0.45	Ea	4.79	12.10	16.89
2 gang brown	1E@0.90	Ea	9.60	24.20	33.80
3 gang brown	1E@1.35	Ea	12.50	36.30	48.80
4 gang brown	1E@1.80	Ea	16.80	48.40	65.20
5 gang brown	1E@2.25	Ea	21.70	60.50	82.20
6 gang brown	1E@2.70	Ea	26.00	72.60	98.60

S

Material	Craft@Hrs	Unit	Material Cost	Labor Cost	Installed Cost
Switches, 20 amp 2-pole in 2" deep sectional box and plastic plate					
1 gang brown	1E@0.50	Ea	9.70	13.40	23.10
2 gang brown	1E@1.00	Ea	19.40	26.90	46.30
3 gang brown	1E@1.50	Ea	29.10	40.30	69.40
4 gang brown	1E@2.00	Ea	39.00	53.80	92.80
5 gang brown	1E@2.50	Ea	49.40	67.20	116.60
6 gang brown	1E@3.00	Ea	59.30	80.60	139.90

S_2

Material	Craft@Hrs	Unit	Material Cost	Labor Cost	Installed Cost
Switches, 20 amp 3-way in 2" deep sectional box and plastic plate					
1 gang brown	1E@0.50	Ea	5.39	13.40	18.79
2 gang brown	1E@1.00	Ea	10.80	26.90	37.70
3 gang brown	1E@1.50	Ea	16.20	40.30	56.50
4 gang brown	1E@2.00	Ea	21.70	53.80	75.50
5 gang brown	1E@2.50	Ea	27.90	67.20	95.10
6 gang brown	1E@3.00	Ea	33.40	80.60	114.00

S_3

Material	Craft@Hrs	Unit	Material Cost	Labor Cost	Installed Cost
Switches, 20 amp 4-way in 2" deep sectional box and plastic plate					
1 gang brown	1E@0.60	Ea	17.10	16.10	33.20
2 gang brown	1E@1.20	Ea	34.20	32.30	66.50
3 gang brown	1E@1.80	Ea	51.20	48.40	99.60
4 gang brown	1E@2.40	Ea	68.40	64.50	132.90
5 gang brown	1E@3.00	Ea	86.30	80.60	166.90
6 gang brown	1E@3.60	Ea	103.00	96.80	199.80

S_4

Material	Craft@Hrs	Unit	Material Cost	Labor Cost	Installed Cost
Switches, 20 amp key in 2" deep sectional box and plastic plate					
1-pole 20 amp	1E@0.43	Ea	19.40	11.60	31.00
2-pole 20 amp	1E@0.48	Ea	21.80	12.90	34.70
3-way 20 amp	1E@0.48	Ea	21.30	12.90	34.20
4-way 20 amp	1E@0.58	Ea	64.20	15.60	79.80

S_K

Use these figures to estimate the cost of installing assemblies under the conditions described on pages 5 and 6. Costs listed are for each assembly installed. The crew is one electrician working at a labor cost of $26.88 per manhour. These costs include layout, material handling and normal waste. Add for the sales tax, delivery, supervision, mobilization, demobilization, cleanup, overhead and profit.

2-1/2" Deep 20 Amp Sectional Box Switch Assemblies

Material	Craft@Hrs	Unit	Material Cost	Labor Cost	Installed Cost
Switches, 20 amp 1-pole in 2-1/2" deep sectional box and plastic plate					
1 gang brown	1E@0.47	Ea	4.43	12.60	17.03
2 gang brown	1E@0.94	Ea	5.32	25.30	30.62
3 gang brown	1E@1.41	Ea	12.10	37.90	50.00
4 gang brown	1E@1.88	Ea	17.10	50.50	67.60
5 gang brown	1E@2.35	Ea	21.30	63.20	84.50
6 gang brown	1E@2.82	Ea	25.60	75.80	101.40
Switches, 20 amp 2-pole in 2-1/2" deep sectional box and plastic plate					
1 gang brown	1E@0.52	Ea	9.34	14.00	23.34
2 gang brown	1E@1.04	Ea	19.10	28.00	47.10
3 gang brown	1E@1.56	Ea	28.80	41.90	70.70
4 gang brown	1E@2.08	Ea	38.60	55.90	94.50
5 gang brown	1E@2.60	Ea	49.10	69.90	119.00
6 gang brown	1E@3.12	Ea	58.90	83.90	142.80
Switches, 20 amp 3-way in 2-1/2" deep sectional box and plastic plate					
1 gang brown	1E@0.52	Ea	5.03	14.00	19.03
2 gang brown	1E@1.04	Ea	10.40	28.00	38.40
3 gang brown	1E@1.56	Ea	15.80	41.90	57.70
4 gang brown	1E@2.08	Ea	21.40	55.90	77.30
5 gang brown	1E@2.60	Ea	27.50	69.90	97.40
6 gang brown	1E@3.12	Ea	33.00	83.90	116.90
Switches, 20 amp 4-way in 2-1/2" deep sectional box and plastic plate					
1 gang brown	1E@0.62	Ea	16.70	16.70	33.40
2 gang brown	1E@1.24	Ea	33.80	33.30	67.10
3 gang brown	1E@1.86	Ea	50.90	50.00	100.90
4 gang brown	1E@2.48	Ea	68.10	66.70	134.80
5 gang brown	1E@3.10	Ea	85.90	83.30	169.20
6 gang brown	1E@3.72	Ea	103.00	100.00	203.00
Switches, 20 amp key in 2-1/2" deep sectional box and plastic plate					
1 gang 1-pole	1E@0.47	Ea	19.00	12.60	31.60
1 gang 2-pole	1E@0.52	Ea	21.40	14.00	35.40
1 gang 3-way	1E@0.52	Ea	20.90	14.00	34.90
1 gang 4-way	1E@0.62	Ea	63.80	16.70	80.50

Use these figures to estimate the cost of installing assemblies under the conditions described on pages 5 and 6. Costs listed are for each assembly installed. The crew is one electrician working at a labor cost of $26.88 per manhour. These costs include layout, material handling and normal waste. Add for the sales tax, delivery, supervision, mobilization, demobilization, cleanup, overhead and profit.

2-3/4" Deep 20 Amp Sectional Box Switch Assemblies

Material	Craft@Hrs	Unit	Material Cost	Labor Cost	Installed Cost

Switches, 20 amp 1-pole in 2-3/4" deep sectional box and plastic plate

Material	Craft@Hrs	Unit	Material Cost	Labor Cost	Installed Cost
1 gang brown	1E@0.50	Ea	4.58	13.40	17.98
2 gang brown	1E@1.00	Ea	5.47	26.90	32.37
3 gang brown	1E@1.50	Ea	12.30	40.30	52.60
4 gang brown	1E@2.00	Ea	16.60	53.80	70.40
5 gang brown	1E@2.50	Ea	21.50	67.20	88.70
6 gang brown	1E@3.00	Ea	25.80	80.60	106.40

S

Switches, 20 amp 2-pole in 2-3/4" deep sectional box and plastic plate

Material	Craft@Hrs	Unit	Material Cost	Labor Cost	Installed Cost
1 gang brown	1E@0.55	Ea	9.49	14.80	24.29
2 gang brown	1E@1.10	Ea	19.20	29.60	48.80
3 gang brown	1E@1.65	Ea	28.90	44.40	73.30
4 gang brown	1E@2.20	Ea	38.80	59.10	97.90
5 gang brown	1E@2.75	Ea	49.20	73.90	123.10
6 gang brown	1E@3.30	Ea	59.10	88.70	147.80

S_2

Switches, 20 amp 3-way in 2-3/4" deep sectional box and plastic plate

Material	Craft@Hrs	Unit	Material Cost	Labor Cost	Installed Cost
1 gang brown	1E@0.55	Ea	5.18	14.80	19.98
2 gang brown	1E@1.10	Ea	10.60	29.60	40.20
3 gang brown	1E@1.65	Ea	16.00	44.40	60.40
4 gang brown	1E@2.20	Ea	21.50	59.10	80.60
5 gang brown	1E@2.75	Ea	27.70	73.90	101.60
6 gang brown	1E@3.30	Ea	33.20	88.70	121.90

S_3

Switches, 20 amp 4-way in 2-3/4" deep sectional box and plastic plate

Material	Craft@Hrs	Unit	Material Cost	Labor Cost	Installed Cost
1 gang brown	1E@0.65	Ea	16.90	17.50	34.40
2 gang brown	1E@1.30	Ea	34.00	34.90	68.90
3 gang brown	1E@1.95	Ea	51.00	52.40	103.40
4 gang brown	1E@2.60	Ea	68.20	69.90	138.10
5 gang brown	1E@3.25	Ea	86.10	87.40	173.50
6 gang brown	1E@3.90	Ea	103.00	105.00	208.00

S_4

Switches, 20 amp key in 2-3/4" deep sectional box and plastic plate

Material	Craft@Hrs	Unit	Material Cost	Labor Cost	Installed Cost
1 gang 1-pole	1E@0.50	Ea	19.20	13.40	32.60
1 gang 2-pole	1E@0.55	Ea	21.60	14.80	36.40
1 gang 3-way	1E@0.55	Ea	21.10	14.80	35.90
1 gang 4-way	1E@0.65	Ea	64.00	17.50	81.50

S_K

Use these figures to estimate the cost of installing assemblies under the conditions described on pages 5 and 6. Costs listed are for each assembly installed. The crew is one electrician working at a labor cost of $26.88 per manhour. These costs include layout, material handling and normal waste. Add for the sales tax, delivery, supervision, mobilization, demobilization, cleanup, overhead and profit.

3-1/2" Deep 20 Amp Sectional Box Switch Assemblies

Material	Craft@Hrs	Unit	Material Cost	Labor Cost	Installed Cost
Switches, 20 amp 1-pole in 3-1/2" deep sectional box and plastic plate					
1 gang brown	1E@0.52	Ea	4.89	14.00	18.89
2 gang brown	1E@1.04	Ea	9.70	28.00	37.70
3 gang brown	1E@1.56	Ea	12.60	41.90	54.50
4 gang brown	1E@2.08	Ea	16.90	55.90	72.80
5 gang brown	1E@2.60	Ea	21.80	69.90	91.70
6 gang brown	1E@3.12	Ea	26.10	83.90	110.00
Switches, 20 amp 2-pole in 3-1/2" deep sectional box and plastic plate					
1 gang brown	1E@0.57	Ea	9.80	15.30	25.10
2 gang brown	1E@1.14	Ea	19.50	30.60	50.10
3 gang brown	1E@1.71	Ea	29.20	46.00	75.20
4 gang brown	1E@2.28	Ea	39.10	61.30	100.40
5 gang brown	1E@2.85	Ea	49.50	76.60	126.10
6 gang brown	1E@3.42	Ea	59.40	91.90	151.30
Switches, 20 amp 3-way in 3-1/2" deep sectional box and plastic plate					
1 gang brown	1E@0.57	Ea	5.49	15.30	20.79
2 gang brown	1E@1.14	Ea	10.90	30.60	41.50
3 gang brown	1E@1.71	Ea	16.30	46.00	62.30
4 gang brown	1E@2.28	Ea	21.80	61.30	83.10
5 gang brown	1E@2.85	Ea	28.00	76.60	104.60
6 gang brown	1E@3.42	Ea	33.50	91.90	125.40
Switches, 20 amp 4-way in 3-1/2" deep sectional box and plastic plate					
1 gang brown	1E@0.67	Ea	17.20	18.00	35.20
2 gang brown	1E@1.34	Ea	34.30	36.00	70.30
3 gang brown	1E@2.01	Ea	51.30	54.00	105.30
4 gang brown	1E@2.68	Ea	68.50	72.00	140.50
5 gang brown	1E@3.35	Ea	86.40	90.00	176.40
6 gang brown	1E@4.02	Ea	104.00	108.00	212.00
Switches, 20 amp key in 3-1/2" deep sectional box and plastic plate					
1 gang 1-pole	1E@0.52	Ea	19.50	14.00	33.50
1 gang 2-pole	1E@0.57	Ea	21.90	15.30	37.20
1 gang 3-way	1E@0.57	Ea	21.40	15.30	36.70
1 gang 4-way	1E@0.67	Ea	64.30	18.00	82.30

The row-label symbols in the left margin are: S, S_2, S_3, S_4, S_K.

Use these figures to estimate the cost of installing assemblies under the conditions described on pages 5 and 6. Costs listed are for each assembly installed. The crew is one electrician working at a labor cost of $26.88 per manhour. These costs include layout, material handling and normal waste. Add for the sales tax, delivery, supervision, mobilization, demobilization, cleanup, overhead and profit.

2" Deep 20 Amp Sectional Box Switch Assemblies

Material	Craft@Hrs	Unit	Material Cost	Labor Cost	Installed Cost
Switches, 20 amp 1-pole in 2" deep sectional box and plastic plate					
1 gang ivory	1E@0.45	Ea	4.74	12.10	16.84
2 gang ivory	1E@0.90	Ea	9.60	24.20	33.80
3 gang ivory	1E@1.35	Ea	12.50	36.30	48.80
4 gang ivory	1E@1.80	Ea	16.80	48.40	65.20
5 gang ivory	1E@2.25	Ea	21.70	60.50	82.20
6 gang ivory	1E@2.70	Ea	26.00	72.60	98.60

S

Material	Craft@Hrs	Unit	Material Cost	Labor Cost	Installed Cost
Switches, 20 amp 2-pole in 2" deep sectional box and plastic plate					
1 gang ivory	1E@0.50	Ea	9.70	13.40	23.10
2 gang ivory	1E@1.00	Ea	19.40	26.90	46.30
3 gang ivory	1E@1.50	Ea	29.10	40.30	69.40
4 gang ivory	1E@2.00	Ea	39.00	53.80	92.80
5 gang ivory	1E@2.50	Ea	49.40	67.20	116.60
6 gang ivory	1E@3.00	Ea	59.30	80.60	139.90

S_2

Material	Craft@Hrs	Unit	Material Cost	Labor Cost	Installed Cost
Switches, 20 amp 3-way in 2" deep sectional box and plastic plate					
1 gang ivory	1E@0.50	Ea	5.39	13.40	18.79
2 gang ivory	1E@1.00	Ea	10.80	26.90	37.70
3 gang ivory	1E@1.50	Ea	16.20	40.30	56.50
4 gang ivory	1E@2.00	Ea	21.70	53.80	75.50
5 gang ivory	1E@2.50	Ea	21.90	67.20	89.10
6 gang ivory	1E@3.00	Ea	33.40	80.60	114.00

S_3

Material	Craft@Hrs	Unit	Material Cost	Labor Cost	Installed Cost
Switches, 20 amp 4-way in 2" deep sectional box and plastic plate					
1 gang ivory	1E@0.60	Ea	17.10	16.10	33.20
2 gang ivory	1E@1.20	Ea	34.20	32.30	66.50
3 gang ivory	1E@1.80	Ea	51.20	48.40	99.60
4 gang ivory	1E@2.40	Ea	68.40	64.50	132.90
5 gang ivory	1E@3.00	Ea	86.30	80.60	166.90
6 gang ivory	1E@3.60	Ea	103.00	96.80	199.80

S_4

Use these figures to estimate the cost of installing assemblies under the conditions described on pages 5 and 6. Costs listed are for each assembly installed. The crew is one electrician working at a labor cost of $26.88 per manhour. These costs include layout, material handling and normal waste. Add for the sales tax, delivery, supervision, mobilization, demobilization, cleanup, overhead and profit.

2-1/2" Deep 20 Amp Sectional Box Switch Assemblies

Material	Craft@Hrs	Unit	Material Cost	Labor Cost	Installed Cost

Switches, 20 amp 1-pole in 2-1/2" deep sectional box and plastic plate

Material	Craft@Hrs	Unit	Material Cost	Labor Cost	Installed Cost
1 gang ivory	1E@0.47	Ea	4.43	12.60	17.03
2 gang ivory	1E@0.94	Ea	5.32	25.30	30.62
3 gang ivory	1E@1.41	Ea	12.10	37.90	50.00
4 gang ivory	1E@1.88	Ea	17.10	50.50	67.60
5 gang ivory	1E@2.35	Ea	21.30	63.20	84.50
6 gang ivory	1E@2.82	Ea	25.60	75.80	101.40

S

Switches, 20 amp 2-pole in 2-1/2" deep sectional box and plastic plate

Material	Craft@Hrs	Unit	Material Cost	Labor Cost	Installed Cost
1 gang ivory	1E@0.52	Ea	9.34	14.00	23.34
2 gang ivory	1E@1.04	Ea	19.10	28.00	47.10
3 gang ivory	1E@1.56	Ea	28.80	41.90	70.70
4 gang ivory	1E@2.08	Ea	38.60	55.90	94.50
5 gang ivory	1E@2.60	Ea	49.10	69.90	119.00
6 gang ivory	1E@3.12	Ea	58.90	83.90	142.80

S_2

Switches, 20 amp 3-way in 2-1/2" deep sectional box and plastic plate

Material	Craft@Hrs	Unit	Material Cost	Labor Cost	Installed Cost
1 gang ivory	1E@0.52	Ea	5.03	14.00	19.03
2 gang ivory	1E@1.04	Ea	10.40	28.00	38.40
3 gang ivory	1E@1.56	Ea	15.80	41.90	57.70
4 gang ivory	1E@2.08	Ea	21.40	55.90	77.30
5 gang ivory	1E@2.60	Ea	27.50	69.90	97.40
6 gang ivory	1E@3.12	Ea	33.00	83.90	116.90

S_3

Switches, 20 amp 4-way in 2-1/2" deep sectional box and plastic plate

Material	Craft@Hrs	Unit	Material Cost	Labor Cost	Installed Cost
1 gang ivory	1E@0.62	Ea	16.70	16.70	33.40
2 gang ivory	1E@1.24	Ea	33.80	33.30	67.10
3 gang ivory	1E@1.86	Ea	50.90	50.00	100.90
4 gang ivory	1E@2.48	Ea	68.10	66.70	134.80
5 gang ivory	1E@3.10	Ea	85.90	83.30	169.20
6 gang ivory	1E@3.72	Ea	103.00	100.00	203.00

S_4

Use these figures to estimate the cost of installing assemblies under the conditions described on pages 5 and 6. Costs listed are for each assembly installed. The crew is one electrician working at a labor cost of $26.88 per manhour. These costs include layout, material handling and normal waste. Add for the sales tax, delivery, supervision, mobilization, demobilization, cleanup, overhead and profit.

2-3/4" Deep 20 Amp Sectional Box Switch Assemblies

Material	Craft@Hrs	Unit	Material Cost	Labor Cost	Installed Cost

Switches, 20 amp 1-pole in 2-3/4" deep sectional box and plastic plate

Material	Craft@Hrs	Unit	Material Cost	Labor Cost	Installed Cost
1 gang ivory	1E@0.50	Ea	4.58	13.40	17.98
2 gang ivory	1E@1.00	Ea	5.47	26.90	32.37
3 gang ivory	1E@1.50	Ea	12.30	40.30	52.60
4 gang ivory	1E@2.00	Ea	16.60	53.80	70.40
5 gang ivory	1E@2.50	Ea	21.50	67.20	88.70
6 gang ivory	1E@3.00	Ea	25.80	80.60	106.40

S

Switches, 20 amp 2-pole in 2-3/4" deep sectional box and plastic plate

Material	Craft@Hrs	Unit	Material Cost	Labor Cost	Installed Cost
1 gang ivory	1E@0.55	Ea	9.49	14.80	24.29
2 gang ivory	1E@1.10	Ea	19.20	29.60	48.80
3 gang ivory	1E@1.65	Ea	28.90	44.40	73.30
4 gang ivory	1E@2.20	Ea	38.80	59.10	97.90
5 gang ivory	1E@2.75	Ea	49.20	73.90	123.10
6 gang ivory	1E@3.30	Ea	59.10	88.70	147.80

S_2

Switches, 20 amp 3-way in 2-3/4" deep sectional box and plastic plate

Material	Craft@Hrs	Unit	Material Cost	Labor Cost	Installed Cost
1 gang ivory	1E@0.55	Ea	5.18	14.80	19.98
2 gang ivory	1E@1.10	Ea	10.60	29.60	40.20
3 gang ivory	1E@1.65	Ea	16.00	44.40	60.40
4 gang ivory	1E@2.20	Ea	21.50	59.10	80.60
5 gang ivory	1E@2.75	Ea	27.70	73.90	101.60
6 gang ivory	1E@3.30	Ea	33.20	88.70	121.90

S_3

Switches, 20 amp 4-way in 2-3/4" deep sectional box and plastic plate

Material	Craft@Hrs	Unit	Material Cost	Labor Cost	Installed Cost
1 gang ivory	1E@0.65	Ea	16.90	17.50	34.40
2 gang ivory	1E@1.30	Ea	34.00	34.90	68.90
3 gang ivory	1E@1.95	Ea	51.00	52.40	103.40
4 gang ivory	1E@2.60	Ea	68.20	69.90	138.10
5 gang ivory	1E@3.25	Ea	86.10	87.40	173.50
6 gang ivory	1E@3.90	Ea	103.00	105.00	208.00

S_4

Use these figures to estimate the cost of installing assemblies under the conditions described on pages 5 and 6. Costs listed are for each assembly installed. The crew is one electrician working at a labor cost of $26.88 per manhour. These costs include layout, material handling and normal waste. Add for the sales tax, delivery, supervision, mobilization, demobilization, cleanup, overhead and profit.

3-1/2" Deep 20 Amp Sectional Box Switch Assemblies

Material	Craft@Hrs	Unit	Material Cost	Labor Cost	Installed Cost
Switches, 20 amp 1-pole in 3-1/2" deep sectional box and plastic plate					
1 gang ivory	1E@0.52	Ea	4.89	14.00	18.89
2 gang ivory	1E@1.04	Ea	9.70	28.00	37.70
3 gang ivory	1E@1.56	Ea	12.60	41.90	54.50
4 gang ivory	1E@2.08	Ea	16.90	55.90	72.80
5 gang ivory	1E@2.60	Ea	21.80	69.90	91.70
6 gang ivory	1E@3.12	Ea	26.10	83.90	110.00
Switches, 20 amp 2-pole in 3-1/2" deep sectional box and plastic plate					
1 gang ivory	1E@0.57	Ea	9.80	15.30	25.10
2 gang ivory	1E@1.14	Ea	19.50	30.60	50.10
3 gang ivory	1E@1.71	Ea	29.20	46.00	75.20
4 gang ivory	1E@2.28	Ea	39.10	61.30	100.40
5 gang ivory	1E@2.85	Ea	49.50	76.60	126.10
6 gang ivory	1E@3.42	Ea	59.40	91.90	151.30
Switches, 20 amp 3-way in 3-1/2" deep sectional box and plastic plate					
1 gang ivory	1E@0.57	Ea	5.49	15.30	20.79
2 gang ivory	1E@1.14	Ea	10.90	30.60	41.50
3 gang ivory	1E@1.71	Ea	16.30	46.00	62.30
4 gang ivory	1E@2.28	Ea	21.80	61.30	83.10
5 gang ivory	1E@2.85	Ea	28.00	76.60	104.60
6 gang ivory	1E@3.42	Ea	33.50	91.90	125.40
Switches, 20 amp 4-way in 3-1/2" deep sectional box and plastic plate					
1 gang ivory	1E@0.67	Ea	17.20	18.00	35.20
2 gang ivory	1E@1.34	Ea	34.30	36.00	70.30
3 gang ivory	1E@2.01	Ea	51.30	54.00	105.30
4 gang ivory	1E@2.68	Ea	68.50	72.00	140.50
5 gang ivory	1E@3.35	Ea	86.40	90.00	176.40
6 gang ivory	1E@4.02	Ea	104.00	108.00	212.00

The row-label symbols printed in the left margin are: S (1-pole), S₂ (2-pole), S₃ (3-way), S₄ (4-way).

Use these figures to estimate the cost of installing assemblies under the conditions described on pages 5 and 6. Costs listed are for each assembly installed. The crew is one electrician working at a labor cost of $26.88 per manhour. These costs include layout, material handling and normal waste. Add for the sales tax, delivery, supervision, mobilization, demobilization, cleanup, overhead and profit.

4/S x 1-1/4"-Deep, 15 Amp, 1 Gang Switch Assemblies

Material	Craft@Hrs	Unit	Material Cost	Labor Cost	Installed Cost

1 gang brown switches, 15 amp, 1-pole, in 4/S x 1-1/4" deep box, ring and plastic plate

Material	Craft@Hrs	Unit	Material Cost	Labor Cost	Installed Cost
4-S-ring flat	1E@0.60	Ea	3.97	16.10	20.07
4-S-ring 1/4"	1E@0.60	Ea	3.89	16.10	19.99
4-S-ring 1/2"	1E@0.60	Ea	3.70	16.10	19.80
4-S-ring 5/8"	1E@0.60	Ea	3.80	16.10	19.90
4-S-ring 3/4"	1E@0.60	Ea	3.97	16.10	20.07
4-S-ring 1"	1E@0.60	Ea	4.29	16.10	20.39
4-S-ring 1-1/4"	1E@0.60	Ea	4.54	16.10	20.64
4-S-tile ring 1/2"	1E@0.60	Ea	5.16	16.10	21.26
4-S-tile ring 3/4"	1E@0.60	Ea	5.25	16.10	21.35
4-S-tile ring 1"	1E@0.60	Ea	5.67	16.10	21.77
4-S-tile ring 1-1/4"	1E@0.61	Ea	5.81	16.40	22.21
4-S-tile ring 1-1/2"	1E@0.61	Ea	5.25	16.40	21.65
4-S-tile ring 2"	1E@0.61	Ea	5.37	16.40	21.77

S

1 gang brown switches, 15 amp, 2-pole, in 4/S x 1-1/4" deep box, ring and plastic plate

Material	Craft@Hrs	Unit	Material Cost	Labor Cost	Installed Cost
4-S-ring flat	1E@0.65	Ea	8.74	17.50	26.24
4-S-ring 1/4"	1E@0.65	Ea	8.66	17.50	26.16
4-S-ring 1/2"	1E@0.65	Ea	8.47	17.50	25.97
4-S-ring 5/8"	1E@0.65	Ea	8.57	17.50	26.07
4-S-ring 3/4"	1E@0.65	Ea	8.74	17.50	26.24
4-S-ring 1"	1E@0.65	Ea	9.06	17.50	26.56
4-S-ring 1-1/4"	1E@0.65	Ea	9.31	17.50	26.81
4-S-tile ring 1/2"	1E@0.65	Ea	10.60	17.50	28.10
4-S-tile ring 3/4"	1E@0.65	Ea	10.00	17.50	27.50
4-S-tile ring 1"	1E@0.65	Ea	10.40	17.50	27.90
4-S-tile ring 1-1/4"	1E@0.66	Ea	9.18	17.70	26.88
4-S-tile ring 1-1/2"	1E@0.66	Ea	9.54	17.70	27.24
4-S-tile ring 2"	1E@0.66	Ea	10.10	17.70	27.80

S_2

1 gang brown switches, 15 amp, 3-way, in 4/S x 1-1/4" deep box, ring and plastic plate

Material	Craft@Hrs	Unit	Material Cost	Labor Cost	Installed Cost
4-S-ring flat	1E@0.65	Ea	4.42	17.50	21.92
4-S-ring 1/4"	1E@0.65	Ea	4.34	17.50	21.84
4-S-ring 1/2"	1E@0.65	Ea	4.15	17.50	21.65
4-S-ring 5/8"	1E@0.65	Ea	4.25	17.50	21.75
4-S-ring 3/4"	1E@0.65	Ea	4.42	17.50	21.92
4-S-ring 1"	1E@0.65	Ea	4.74	17.50	22.24
4-S-ring 1-1/4"	1E@0.65	Ea	4.99	17.50	22.49
4-S-tile ring 1/2"	1E@0.65	Ea	5.61	17.50	23.11
4-S-tile ring 3/4"	1E@0.65	Ea	5.70	17.50	23.20
4-S-tile ring 1"	1E@0.65	Ea	6.12	17.50	23.62
4-S-tile ring 1-1/4"	1E@0.66	Ea	6.26	17.70	23.96
4-S-tile ring 1-1/2"	1E@0.66	Ea	5.70	17.70	23.40
4-S-tile ring 2"	1E@0.66	Ea	5.82	17.70	23.52

S_3

Use these figures to estimate the cost of installing assemblies under the conditions described on pages 5 and 6. Costs listed are for each assembly installed. The crew is one electrician working at a labor cost of $26.88 per manhour. These costs include layout, material handling and normal waste. Add for the sales tax, delivery, supervision, mobilization, demobilization, cleanup, overhead and profit.

4/S x 1-1/4"-Deep, 15 Amp, 1 Gang Switch Assemblies

Material	Craft@Hrs	Unit	Material Cost	Labor Cost	Installed Cost
1 gang brown switches, 15 amp, 4-way, in 4/S x 1-1/4" deep box, ring and plastic plate					
4-S-ring flat	1E@0.70	Ea	11.30	18.80	30.10
4-S-ring 1/4"	1E@0.70	Ea	11.20	18.80	30.00
4-S-ring 1/2"	1E@0.70	Ea	11.10	18.80	29.90
4-S-ring 5/8"	1E@0.70	Ea	11.20	18.80	30.00
4-S-ring 3/4"	1E@0.70	Ea	11.30	18.80	30.10
4-S-ring 1"	1E@0.70	Ea	11.60	18.80	30.40
4-S-ring 1-1/4"	1E@0.70	Ea	11.90	18.80	30.70
4-S-tile ring 1/2"	1E@0.70	Ea	12.50	18.80	31.30
4-S-tile ring 3/4"	1E@0.70	Ea	12.60	18.80	31.40
4-S-tile ring 1"	1E@0.70	Ea	13.00	18.80	31.80
4-S-tile ring 1-1/4"	1E@0.71	Ea	13.20	19.10	32.30
4-S-tile ring 1-1/2"	1E@0.71	Ea	12.60	19.10	31.70
4-S-tile ring 2"	1E@0.71	Ea	12.70	19.10	31.80
1 gang key switches, 15 amp, 1-pole, in 4/S x 1-1/4" deep box, ring and plastic plate					
4-S-ring flat	1E@0.60	Ea	11.50	16.10	27.60
4-S-ring 1/4"	1E@0.60	Ea	11.40	16.10	27.50
4-S-ring 1/2"	1E@0.60	Ea	11.30	16.10	27.40
4-S-ring 5/8"	1E@0.60	Ea	11.40	16.10	27.50
4-S-ring 3/4"	1E@0.60	Ea	11.50	16.10	27.60
4-S-ring 1"	1E@0.60	Ea	11.80	16.10	27.90
4-S-ring 1-1/4"	1E@0.60	Ea	12.10	16.10	28.20
4-S-tile ring 1/2"	1E@0.60	Ea	12.70	16.10	28.80
4-S-tile ring 3/4"	1E@0.60	Ea	12.80	16.10	28.90
4-S-tile ring 1"	1E@0.60	Ea	13.20	16.10	29.30
4-S-tile ring 1-1/4"	1E@0.61	Ea	13.40	16.40	29.80
4-S-tile ring 1-1/2"	1E@0.61	Ea	12.80	16.40	29.20
4-S-tile ring 2"	1E@0.61	Ea	12.90	16.40	29.30
1 gang key switches, 15 amp, 2-pole, in 4/S x 1-1/4" deep box, ring and plastic plate					
4-S-ring flat	1E@0.65	Ea	21.70	17.50	39.20
4-S-ring 1/4"	1E@0.65	Ea	21.60	17.50	39.10
4-S-ring 1/2"	1E@0.65	Ea	21.40	17.50	38.90
4-S-ring 5/8"	1E@0.65	Ea	21.50	17.50	39.00
4-S-ring 3/4"	1E@0.65	Ea	21.70	17.50	39.20
4-S-ring 1"	1E@0.65	Ea	22.00	17.50	39.50
4-S-ring 1-1/4"	1E@0.65	Ea	22.30	17.50	39.80
4-S-tile ring 1/2"	1E@0.65	Ea	22.90	17.50	40.40
4-S-tile ring 3/4"	1E@0.65	Ea	23.00	17.50	40.50
4-S-tile ring 1"	1E@0.65	Ea	23.40	17.50	40.90
4-S-tile ring 1-1/4"	1E@0.66	Ea	23.60	17.70	41.30
4-S-tile ring 1-1/2"	1E@0.66	Ea	23.00	17.70	40.70
4-S-tile ring 2"	1E@0.66	Ea	23.10	17.70	40.80

The row-label groups are marked in the left margin as S_4, S_K, and S_{K2} respectively.

Use these figures to estimate the cost of installing assemblies under the conditions described on pages 5 and 6. Costs listed are for each assembly installed. The crew is one electrician working at a labor cost of $26.88 per manhour. These costs include layout, material handling and normal waste. Add for the sales tax, delivery, supervision, mobilization, demobilization, cleanup, overhead and profit.

4/S x 1-1/4"-Deep, 15 Amp, 1 and 2 Gang Switch Assemblies

Material	Craft@Hrs	Unit	Material Cost	Labor Cost	Installed Cost	

1 gang key switches, 15 amp, 3-way, in 4/S x 1-1/4" deep box, ring and plastic plate

Material	Craft@Hrs	Unit	Material Cost	Labor Cost	Installed Cost	
4-S-ring flat	1E@0.65	Ea	15.30	17.50	32.80	
4-S-ring 1/4"	1E@0.65	Ea	15.20	17.50	32.70	
4-S-ring 1/2"	1E@0.65	Ea	15.00	17.50	32.50	
4-S-ring 5/8"	1E@0.65	Ea	15.10	17.50	32.60	
4-S-ring 3/4"	1E@0.65	Ea	15.30	17.50	32.80	
4-S-ring 1"	1E@0.65	Ea	15.60	17.50	33.10	S_{K3}
4-S-ring 1-1/4"	1E@0.65	Ea	15.90	17.50	33.40	
4-S-tile ring 1/2"	1E@0.65	Ea	16.50	17.50	34.00	
4-S-tile ring 3/4"	1E@0.65	Ea	16.60	17.50	34.10	
4-S-tile ring 1"	1E@0.65	Ea	17.00	17.50	34.50	
4-S-tile ring 1-1/4"	1E@0.66	Ea	17.20	17.70	34.90	
4-S-tile ring 1-1/2"	1E@0.66	Ea	16.60	17.70	34.30	
4-S-tile ring 2"	1E@0.66	Ea	16.70	17.70	34.40	

1 gang key switches, 15 amp, 4-way, in 4/S x 1-1/4" deep box, ring and plastic plate

Material	Craft@Hrs	Unit	Material Cost	Labor Cost	Installed Cost	
4-S-ring flat	1E@0.70	Ea	41.20	18.80	60.00	
4-S-ring 1/4	1E@0.70	Ea	41.10	18.80	59.90	
4-S-ring 1/2"	1E@0.70	Ea	40.90	18.80	59.70	
4-S-ring 5/8"	1E@0.70	Ea	41.00	18.80	59.80	
4-S-ring 3/4"	1E@0.70	Ea	41.20	18.80	60.00	
4-S-ring 1"	1E@0.70	Ea	41.50	18.80	60.30	S_{K4}
4-S-ring 1-1/4"	1E@0.70	Ea	41.80	18.80	60.60	
4-S-tile ring 1/2"	1E@0.70	Ea	42.40	18.80	61.20	
4-S-tile ring 3/4"	1E@0.70	Ea	42.50	18.80	61.30	
4-S-tile ring 1"	1E@0.70	Ea	42.90	18.80	61.70	
4-S-tile ring 1-1/4"	1E@0.71	Ea	43.10	19.10	62.20	
4-S-tile ring 1-1/2"	1E@0.71	Ea	42.50	19.10	61.60	
4-S-tile ring 2"	1E@0.71	Ea	42.60	19.10	61.70	

2 gang brown switches, 15 amp, 1-pole, 4/S x 1-1/4" deep box, ring and plastic plate

Material	Craft@Hrs	Unit	Material Cost	Labor Cost	Installed Cost	
4-S-ring flat	1E@0.91	Ea	5.36	24.50	29.86	
4-S-ring 1/4	1E@0.91	Ea	5.72	24.50	30.22	
4-S-ring 1/2"	1E@0.91	Ea	5.11	24.50	29.61	
4-S-ring 5/8"	1E@0.91	Ea	5.24	24.50	29.74	
4-S-ring 3/4"	1E@0.91	Ea	5.44	24.50	29.94	
4-S-ring 1"	1E@0.91	Ea	6.01	24.50	30.51	SS
4-S-ring 1-1/4"	1E@0.91	Ea	6.12	24.50	30.62	
4-S-tile ring 1/2"	1E@0.91	Ea	6.87	24.50	31.37	
4-S-tile ring 3/4"	1E@0.91	Ea	7.01	24.50	31.51	
4-S-tile ring 1"	1E@0.91	Ea	7.08	24.50	31.58	
4-S-tile ring 1-1/4"	1E@0.93	Ea	7.03	25.00	32.03	
4-S-tile ring 1-1/2"	1E@0.93	Ea	6.73	25.00	31.73	
4-S-tile ring 2"	1E@0.93	Ea	6.90	25.00	31.90	

Use these figures to estimate the cost of installing assemblies under the conditions described on pages 5 and 6. Costs listed are for each assembly installed. The crew is one electrician working at a labor cost of $26.88 per manhour. These costs include layout, material handling and normal waste. Add for the sales tax, delivery, supervision, mobilization, demobilization, cleanup, overhead and profit.

4/S x 1-1/4"-Deep, 15 Amp, 2 Gang Switch Assemblies

Material	Craft@Hrs	Unit	Material Cost	Labor Cost	Installed Cost

2 gang brown switches, 15 amp, 2-pole, in 4/S x 1-1/4" deep box, ring and plastic plate

S_2S_2

Material	Craft@Hrs	Unit	Material Cost	Labor Cost	Installed Cost
4-S-ring flat	1E@1.01	Ea	14.90	27.10	42.00
4-S-ring 1/4"	1E@1.01	Ea	15.30	27.10	42.40
4-S-ring 1/2"	1E@1.01	Ea	14.70	27.10	41.80
4-S-ring 5/8"	1E@1.01	Ea	14.80	27.10	41.90
4-S-ring 3/4"	1E@1.01	Ea	15.00	27.10	42.10
4-S-ring 1"	1E@1.01	Ea	15.60	27.10	42.70
4-S-ring 1-1/4"	1E@1.01	Ea	15.70	27.10	42.80
4-S-tile ring 1/2"	1E@1.01	Ea	16.40	27.10	43.50
4-S-tile ring 3/4"	1E@1.01	Ea	16.60	27.10	43.70
4-S-tile ring 1"	1E@1.01	Ea	16.60	27.10	43.70
4-S-tile ring 1-1/4"	1E@1.03	Ea	16.60	27.70	44.30
4-S-tile ring 1-1/2"	1E@1.03	Ea	16.30	27.70	44.00
4-S-tile ring 2"	1E@1.03	Ea	16.40	27.70	44.10

2 gang brown switches, 15 amp, 3-way, in 4/S x 1-1/4" deep box, ring and plastic plate

S_3S_3

Material	Craft@Hrs	Unit	Material Cost	Labor Cost	Installed Cost
4-S-ring flat	1E@1.01	Ea	6.26	27.10	33.36
4-S-ring 1/4"	1E@1.01	Ea	6.62	27.10	33.72
4-S-ring 1/2"	1E@1.01	Ea	6.01	27.10	33.11
4-S-ring 5/8"	1E@1.01	Ea	6.14	27.10	33.24
4-S-ring 3/4"	1E@1.01	Ea	6.34	27.10	33.44
4-S-ring 1"	1E@1.01	Ea	6.91	27.10	34.01
4-S-ring 1-1/4"	1E@1.01	Ea	7.02	27.10	34.12
4-S-tile ring 1/2"	1E@1.01	Ea	7.77	27.10	34.87
4-S-tile ring 3/4"	1E@1.01	Ea	7.91	27.10	35.01
4-S-tile ring 1"	1E@1.01	Ea	7.98	27.10	35.08
4-S-tile ring 1-1/4"	1E@1.03	Ea	7.93	27.70	35.63
4-S-tile ring 1-1/2"	1E@1.03	Ea	7.63	27.70	35.33
4-S-tile ring 2"	1E@1.03	Ea	7.80	27.70	35.50

2 gang brown switches, 15 amp, 4-way, in 4/S x 1-1/4" deep box, ring and plastic plate

S_4S_4

Material	Craft@Hrs	Unit	Material Cost	Labor Cost	Installed Cost
4-S-ring flat	1E@1.06	Ea	20.10	28.50	48.60
4-S-ring 1/4"	1E@1.06	Ea	20.40	28.50	48.90
4-S-ring 1/2"	1E@1.06	Ea	19.80	28.50	48.30
4-S-ring 5/8"	1E@1.06	Ea	19.90	28.50	48.40
4-S-ring 3/4"	1E@1.06	Ea	20.10	28.50	48.60
4-S-ring 1"	1E@1.06	Ea	20.70	28.50	49.20
4-S-ring 1-1/4"	1E@1.06	Ea	20.80	28.50	49.30
4-S-tile ring 1/2"	1E@1.06	Ea	21.60	28.50	50.10
4-S-tile ring 3/4"	1E@1.06	Ea	21.70	28.50	50.20
4-S-tile ring 1"	1E@1.06	Ea	21.80	28.50	50.30
4-S-tile ring 1-1/4"	1E@1.08	Ea	21.70	29.00	50.70
4-S-tile ring 1-1/2"	1E@1.08	Ea	21.40	29.00	50.40
4-S-tile ring 2"	1E@1.08	Ea	21.60	29.00	50.60

Use these figures to estimate the cost of installing assemblies under the conditions described on pages 5 and 6. Cost listed are for each assembly installed. The crew is one electrician working at a labor cost of $26.88 per manhour. These costs include layout, material handling and normal waste. Add for the sales tax, delivery, supervision, mobilization, demobilization, cleanup, overhead and profit.

4/S x 1-1/4"-Deep, 15 Amp, 1 Gang Switch Assemblies

Material	Craft@Hrs	Unit	Material Cost	Labor Cost	Installed Cost

1 gang ivory switches, 15 amp, 1-pole, in 4/S x 1-1/4" deep box, ring and plastic plate

Material	Craft@Hrs	Unit	Material Cost	Labor Cost	Installed Cost
4-S-ring flat	1E@0.60	Ea	3.97	16.10	20.07
4-S-ring 1/4"	1E@0.60	Ea	3.89	16.10	19.99
4-S-ring 1/2"	1E@0.60	Ea	3.70	16.10	19.80
4-S-ring 5/8"	1E@0.60	Ea	3.80	16.10	19.90
4-S-ring 3/4"	1E@0.60	Ea	3.97	16.10	20.07
4-S-ring 1"	1E@0.60	Ea	4.29	16.10	20.39
4-S-ring 1-1/4"	1E@0.60	Ea	4.54	16.10	20.64
4-S-tile ring 1/2"	1E@0.60	Ea	5.16	16.10	21.26
4-S-tile ring 3/4"	1E@0.60	Ea	5.25	16.10	21.35
4-S-tile ring 1"	1E@0.60	Ea	5.67	16.10	21.77
4-S-tile ring 1-1/4"	1E@0.61	Ea	5.81	16.40	22.21
4-S-tile ring 1-1/2"	1E@0.61	Ea	5.25	16.40	21.65
4-S-tile ring 2"	1E@0.61	Ea	5.37	16.40	21.77

S

1 gang ivory switches, 15 amp, 2-pole, in 4/S x 1-1/4" deep box, ring and plastic plate

Material	Craft@Hrs	Unit	Material Cost	Labor Cost	Installed Cost
4-S-ring flat	1E@0.65	Ea	8.74	17.50	26.24
4-S-ring 1/4"	1E@0.65	Ea	8.66	17.50	26.16
4-S-ring 1/2"	1E@0.65	Ea	8.47	17.50	25.97
4-S-ring 5/8"	1E@0.65	Ea	8.57	17.50	26.07
4-S-ring 3/4"	1E@0.65	Ea	8.74	17.50	26.24
4-S-ring 1"	1E@0.65	Ea	9.06	17.50	26.56
4-S-ring 1-1/4"	1E@0.65	Ea	9.31	17.50	26.81
4-S-tile ring 1/2"	1E@0.65	Ea	10.60	17.50	28.10
4-S-tile ring 3/4"	1E@0.65	Ea	10.00	17.50	27.50
4-S-tile ring 1"	1E@0.65	Ea	10.40	17.50	27.90
4-S-tile ring 1-1/4"	1E@0.66	Ea	9.18	17.70	26.88
4-S-tile ring 1-1/2"	1E@0.66	Ea	9.54	17.70	27.24
4-S-tile ring 2"	1E@0.66	Ea	10.10	17.70	27.80

S_2

1 gang ivory switches, 15 amp, 3-way, in 4/S x 1-1/4" deep box, ring and plastic plate

Material	Craft@Hrs	Unit	Material Cost	Labor Cost	Installed Cost
4-S-ring flat	1E@0.65	Ea	4.42	17.50	21.92
4-S-ring 1/4"	1E@0.65	Ea	4.34	17.50	21.84
4-S-ring 1/2"	1E@0.65	Ea	4.15	17.50	21.65
4-S-ring 5/8"	1E@0.65	Ea	4.25	17.50	21.75
4-S-ring 3/4"	1E@0.65	Ea	4.42	17.50	21.92
4-S-ring 1"	1E@0.65	Ea	4.74	17.50	22.24
4-S-ring 1-1/4"	1E@0.65	Ea	4.99	17.50	22.49
4-S-tile ring 1/2"	1E@0.65	Ea	5.61	17.50	23.11
4-S-tile ring 3/4"	1E@0.65	Ea	5.70	17.50	23.20
4-S-tile ring 1"	1E@0.65	Ea	6.12	17.50	23.62
4-S-tile ring 1-1/4"	1E@0.66	Ea	6.26	17.70	23.96
4-S-tile ring 1-1/2"	1E@0.66	Ea	5.70	17.70	23.40
4-S-tile ring 2"	1E@0.66	Ea	5.82	17.70	23.52

S_3

Use these figures to estimate the cost of installing assemblies under the conditions described on pages 5 and 6. Costs listed are for each assembly installed. The crew is one electrician working at a labor cost of $26.88 per manhour. These costs include layout, material handling and normal waste. Add for the sales tax, delivery, supervision, mobilization, demobilization, cleanup, overhead and profit.

4/S x 1-1/4"-Deep, 15 Amp, 1 and 2 Gang Switch Assemblies

Material	Craft@Hrs	Unit	Material Cost	Labor Cost	Installed Cost
1 gang ivory switches, 15 amp, 4-way, in 4/S x 1-1/4" deep box, ring and plastic plate					
4-S-ring flat	1E@0.70	Ea	11.30	18.80	30.10
4-S-ring 1/4"	1E@0.70	Ea	11.20	18.80	30.00
4-S-ring 1/2"	1E@0.70	Ea	11.10	18.80	29.90
4-S-ring 5/8"	1E@0.70	Ea	11.20	18.80	30.00
4-S-ring 3/4"	1E@0.70	Ea	11.30	18.80	30.10
4-S-ring 1"	1E@0.70	Ea	11.60	18.80	30.40
4-S-ring 1-1/4"	1E@0.70	Ea	11.90	18.80	30.70
4-S-tile ring 1/2"	1E@0.70	Ea	12.50	18.80	31.30
4-S-tile ring 3/4"	1E@0.70	Ea	12.60	18.80	31.40
4-S-tile ring 1"	1E@0.70	Ea	13.00	18.80	31.80
4-S-tile ring 1-1/4"	1E@0.71	Ea	13.20	19.10	32.30
4-S-tile ring 1-1/2"	1E@0.71	Ea	12.60	19.10	31.70
4-S-tile ring 2"	1E@0.71	Ea	12.70	19.10	31.80
2 gang ivory switches, 15 amp, 1-pole, in 4/S x 1-1/4" deep box, ring and plastic plate					
4-S-ring flat	1E@0.91	Ea	5.36	24.50	29.86
4-S-ring 1/4"	1E@0.91	Ea	5.72	24.50	30.22
4-S-ring 1/2"	1E@0.91	Ea	5.11	24.50	29.61
4-S-ring 5/8"	1E@0.91	Ea	5.24	24.50	29.74
4-S-ring 3/4"	1E@0.91	Ea	5.44	24.50	29.94
4-S-ring 1"	1E@0.91	Ea	6.01	24.50	30.51
4-S-ring 1-1/4"	1E@0.91	Ea	6.12	24.50	30.62
4-S-tile ring 1/2"	1E@0.91	Ea	6.87	24.50	31.37
4-S-tile ring 3/4"	1E@0.91	Ea	7.01	24.50	31.51
4-S-tile ring 1"	1E@0.91	Ea	7.08	24.50	31.58
4-S-tile ring 1-1/4"	1E@0.93	Ea	7.03	25.00	32.03
4-S-tile ring 1-1/2"	1E@0.93	Ea	6.73	25.00	31.73
4-S-tile ring 2"	1E@0.93	Ea	6.90	25.00	31.90

Use these figures to estimate the cost of installing assemblies under the conditions described on pages 5 and 6. Costs listed are for each assembly installed. The crew is one electrician working at a labor cost of $26.88 per manhour. These costs include layout, material handling and normal waste. Add for the sales tax, delivery, supervision, mobilization, demobilization, cleanup, overhead and profit.

4/S x 1-1/4"-Deep, 15 Amp, 2 Gang Switch Assemblies

Material	Craft@Hrs	Unit	Material Cost	Labor Cost	Installed Cost
2 gang ivory switches, 15 amp, 2-pole, in 4/S x 1-1/4" deep box, ring and plastic plate					
4-S-ring flat	1E@1.01	Ea	14.90	27.10	42.00
4-S-ring 1/4"	1E@1.01	Ea	15.30	27.10	42.40
4-S-ring 1/2"	1E@1.01	Ea	14.70	27.10	41.80
4-S-ring 5/8"	1E@1.01	Ea	14.80	27.10	41.90
4-S-ring 3/4"	1E@1.01	Ea	15.00	27.10	42.10
4-S-ring 1"	1E@1.01	Ea	15.60	27.10	42.70
4-S-ring 1-1/4"	1E@1.01	Ea	15.70	27.10	42.80
4-S-tile ring 1/2"	1E@1.01	Ea	16.40	27.10	43.50
4-S-tile ring 3/4"	1E@1.01	Ea	16.60	27.10	43.70
4-S-tile ring 1"	1E@1.01	Ea	16.60	27.10	43.70
4-S-tile ring 1-1/4"	1E@1.03	Ea	16.60	27.70	44.30
4-S-tile ring 1-1/2"	1E@1.03	Ea	16.30	27.70	44.00
4-S-tile ring 2"	1E@1.03	Ea	16.40	27.70	44.10

S_2S_2

Material	Craft@Hrs	Unit	Material Cost	Labor Cost	Installed Cost
2 gang ivory switches, 15 amp, 3-way, in 4/S x 1-1/4" deep box, ring and plastic plate					
4-S-ring flat	1E@1.01	Ea	6.26	27.10	33.36
4-S-ring 1/4"	1E@1.01	Ea	6.62	27.10	33.72
4-S-ring 1/2"	1E@1.01	Ea	6.01	27.10	33.11
4-S-ring 5/8"	1E@1.01	Ea	6.14	27.10	33.24
4-S-ring 3/4"	1E@1.01	Ea	6.34	27.10	33.44
4-S-ring 1"	1E@1.01	Ea	6.91	27.10	34.01
4-S-ring 1-1/4"	1E@1.01	Ea	7.02	27.10	34.12
4-S-tile ring 1/2"	1E@1.01	Ea	7.77	27.10	34.87
4-S-tile ring 3/4"	1E@1.01	Ea	7.91	27.10	35.01
4-S-tile ring 1"	1E@1.01	Ea	7.98	27.10	35.08
4-S-tile ring 1-1/4"	1E@1.03	Ea	7.93	27.70	35.63
4-S-tile ring 1-1/2"	1E@1.03	Ea	7.63	27.70	35.33
4-S-tile ring 2"	1E@1.03	Ea	7.80	27.70	35.50

S_3S_3

Material	Craft@Hrs	Unit	Material Cost	Labor Cost	Installed Cost
2 gang ivory switches, 15 amp, 4-way, in 4/S x 1-1/4" deep box, ring and plastic plate					
4-S-ring flat	1E@1.06	Ea	20.10	28.50	48.60
4-S-ring 1/4"	1E@1.06	Ea	20.40	28.50	48.90
4-S-ring 1/2"	1E@1.06	Ea	19.80	28.50	48.30
4-S-ring 5/8"	1E@1.06	Ea	19.90	28.50	48.40
4-S-ring 3/4"	1E@1.06	Ea	20.10	28.50	48.60
4-S-ring 1"	1E@1.06	Ea	20.70	28.50	49.20
4-S-ring 1-1/4"	1E@1.06	Ea	20.80	28.50	49.30
4-S-tile ring 1/2"	1E@1.06	Ea	21.60	28.50	50.10
4-S-tile ring 3/4"	1E@1.06	Ea	21.70	28.50	50.20
4-S-tile ring 1"	1E@1.06	Ea	21.80	28.50	50.30
4-S-tile ring 1-1/4"	1E@1.08	Ea	21.70	29.00	50.70
4-S-tile ring 1-1/2"	1E@1.08	Ea	21.40	29.00	50.40
4-S-tile ring 2"	1E@1.08	Ea	21.60	29.00	50.60

S_4S_4

Use these figures to estimate the cost of installing assemblies under the conditions described on pages 5 and 6. Costs listed are for each assembly installed. The crew is one electrician working at a labor cost of $26.88 per manhour. These costs include layout, material handling and normal waste. Add for the sales tax, delivery, supervision, mobilization, demobilization, cleanup, overhead and profit.

4/S x 1-1/2"-Deep, 15 Amp, 1 Gang Switch Assemblies

Material	Craft@Hrs	Unit	Material Cost	Labor Cost	Installed Cost
1 gang brown switches, 15 amp, 1-pole, in 4/S x 1-1/2" deep box, ring and plastic plate					
4-S-ring flat	1E@0.60	Ea	3.38	16.10	19.48
4-S-ring 1/4"	1E@0.60	Ea	3.30	16.10	19.40
4-S-ring 1/2"	1E@0.60	Ea	3.11	16.10	19.21
4-S-ring 5/8"	1E@0.60	Ea	3.21	16.10	19.31
4-S-ring 3/4"	1E@0.60	Ea	3.38	16.10	19.48
4-S-ring 1"	1E@0.60	Ea	3.70	16.10	19.80
4-S-ring 1-1/4"	1E@0.60	Ea	3.95	16.10	20.05
4-S-tile ring 1/2"	1E@0.60	Ea	4.57	16.10	20.67
4-S-tile ring 3/4"	1E@0.60	Ea	4.66	16.10	20.76
4-S-tile ring 1"	1E@0.60	Ea	5.08	16.10	21.18
4-S-tile ring 1-1/4"	1E@0.61	Ea	5.22	16.40	21.62
4-S-tile ring 1-1/2"	1E@0.61	Ea	4.66	16.40	21.06
4-S-tile ring 2"	1E@0.61	Ea	4.78	16.40	21.18
1 gang brown switches, 15 amp, 2-pole, in 4/S x 1-1/2" deep box, ring and plastic plate					
4-S-ring flat	1E@0.65	Ea	8.15	17.50	25.65
4-S-ring 1/4"	1E@0.65	Ea	8.07	17.50	25.57
4-S-ring 1/2"	1E@0.65	Ea	7.88	17.50	25.38
4-S-ring 5/8"	1E@0.65	Ea	7.98	17.50	25.48
4-S-ring 3/4"	1E@0.65	Ea	8.15	17.50	25.65
4-S-ring 1"	1E@0.65	Ea	8.47	17.50	25.97
4-S-ring 1-1/4"	1E@0.65	Ea	8.72	17.50	26.22
4-S-tile ring 1/2"	1E@0.65	Ea	9.99	17.50	27.49
4-S-tile ring 3/4"	1E@0.65	Ea	9.43	17.50	26.93
4-S-tile ring 1"	1E@0.65	Ea	9.82	17.50	27.32
4-S-tile ring 1-1/4"	1E@0.66	Ea	8.59	17.70	26.29
4-S-tile ring 1-1/2"	1E@0.66	Ea	8.95	17.70	26.65
4-S-tile ring 2"	1E@0.66	Ea	9.55	17.70	27.25
1 gang brown switches, 15 amp, 3-way, in 4/S x 1-1/2" deep box, ring and plastic plate					
4-S-ring flat	1E@0.65	Ea	3.83	17.50	21.33
4-S-ring 1/4"	1E@0.65	Ea	3.75	17.50	21.25
4-S-ring 1/2"	1E@0.65	Ea	3.56	17.50	21.06
4-S-ring 5/8"	1E@0.65	Ea	3.66	17.50	21.16
4-S-ring 3/4"	1E@0.65	Ea	3.83	17.50	21.33
4-S-ring 1"	1E@0.65	Ea	4.15	17.50	21.65
4-S-ring 1-1/4"	1E@0.65	Ea	4.40	17.50	21.90
4-S-tile ring 1/2"	1E@0.65	Ea	5.02	17.50	22.52
4-S-tile ring 3/4"	1E@0.65	Ea	5.11	17.50	22.61
4-S-tile ring 1"	1E@0.65	Ea	5.53	17.50	23.03
4-S-tile ring 1-1/4"	1E@0.66	Ea	5.67	17.70	23.37
4-S-tile ring 1-1/2"	1E@0.66	Ea	5.11	17.70	22.81
4-S-tile ring 2"	1E@0.66	Ea	5.23	17.70	22.93

The symbols S, S_2, and S_3 label the three groups above, respectively.

Use these figures to estimate the cost of installing assemblies under the conditions described on pages 5 and 6. Costs listed are for each assembly installed. The crew is one electrician working at a labor cost of $26.88 per manhour. These costs include layout, material handling and normal waste. Add for the sales tax, delivery, supervision, mobilization, demobilization, cleanup, overhead and profit.

4/S x 1-1/2"-Deep, 15 Amp, 1 Gang Switch Assemblies

Material	Craft@Hrs	Unit	Material Cost	Labor Cost	Installed Cost
1 gang brown switches, 15 amp, 4-way, in 4/S x 1-1/2" deep box, ring and plastic plate					
4-S-ring flat	1E@0.70	Ea	10.70	18.80	29.50
4-S-ring 1/4"	1E@0.70	Ea	10.70	18.80	29.50
4-S-ring 1/2"	1E@0.70	Ea	10.50	18.80	29.30
4-S-ring 5/8"	1E@0.70	Ea	10.60	18.80	29.40
4-S-ring 3/4"	1E@0.70	Ea	10.70	18.80	29.50
4-S-ring 1"	1E@0.70	Ea	11.10	18.80	29.90
4-S-ring 1-1/4"	1E@0.70	Ea	11.30	18.80	30.10
4-S-tile ring 1/2"	1E@0.70	Ea	11.90	18.80	30.70
4-S-tile ring 3/4"	1E@0.70	Ea	12.00	18.80	30.80
4-S-tile ring 1"	1E@0.70	Ea	12.40	18.80	31.20
4-S-tile ring 1-1/4"	1E@0.71	Ea	12.60	19.10	31.70
4-S-tile ring 1-1/2"	1E@0.71	Ea	12.00	19.10	31.10
4-S-tile ring 2"	1E@0.71	Ea	12.10	19.10	31.20

S_4

Material	Craft@Hrs	Unit	Material Cost	Labor Cost	Installed Cost
1 gang key switches, 15 amp, 1-pole, in 4/S x 1-1/2" deep box, ring and plastic plate					
4-S-ring flat	1E@0.60	Ea	10.90	16.10	27.00
4-S-ring 1/4"	1E@0.60	Ea	10.90	16.10	27.00
4-S-ring 1/2"	1E@0.60	Ea	10.70	16.10	26.80
4-S-ring 5/8"	1E@0.60	Ea	10.80	16.10	26.90
4-S-ring 3/4"	1E@0.60	Ea	10.90	16.10	27.00
4-S-ring 1"	1E@0.60	Ea	11.30	16.10	27.40
4-S-ring 1-1/4"	1E@0.60	Ea	11.50	16.10	27.60
4-S-tile ring 1/2"	1E@0.60	Ea	12.10	16.10	28.20
4-S-tile ring 3/4"	1E@0.60	Ea	12.20	16.10	28.30
4-S-tile ring 1"	1E@0.60	Ea	12.60	16.10	28.70
4-S-tile ring 1-1/4"	1E@0.61	Ea	12.80	16.40	29.20
4-S-tile ring 1-1/2"	1E@0.61	Ea	12.20	16.40	28.60
4-S-tile ring 2"	1E@0.61	Ea	12.30	16.40	28.70

S_K

Material	Craft@Hrs	Unit	Material Cost	Labor Cost	Installed Cost
1 gang key switches, 15 amp, 2-pole, in 4/S x 1-1/2" deep box, ring and plastic plate					
4-S-ring flat	1E@0.65	Ea	21.10	17.50	38.60
4-S-ring 1/4"	1E@0.65	Ea	21.00	17.50	38.50
4-S-ring 1/2"	1E@0.65	Ea	20.90	17.50	38.40
4-S-ring 5/8"	1E@0.65	Ea	21.00	17.50	38.50
4-S-ring 3/4"	1E@0.65	Ea	21.10	17.50	38.60
4-S-ring 1"	1E@0.65	Ea	21.40	17.50	38.90
4-S-ring 1-1/4"	1E@0.65	Ea	21.70	17.50	39.20
4-S-tile ring 1/2"	1E@0.65	Ea	22.30	17.50	39.80
4-S-tile ring 3/4"	1E@0.65	Ea	22.40	17.50	39.90
4-S-tile ring 1"	1E@0.65	Ea	22.80	17.50	40.30
4-S-tile ring 1-1/4"	1E@0.66	Ea	23.00	17.70	40.70
4-S-tile ring 1-1/2"	1E@0.66	Ea	22.40	17.70	40.10
4-S-tile ring 2"	1E@0.66	Ea	22.50	17.70	40.20

S_{K2}

Use these figures to estimate the cost of installing assemblies under the conditions described on pages 5 and 6. Costs listed are for each assembly installed. The crew is one electrician working at a labor cost of $26.88 per manhour. These costs include layout, material handling and normal waste. Add for the sales tax, delivery, supervision, mobilization, demobilization, cleanup, overhead and profit.

4/S x 1-1/2"-Deep, 15 Amp, 1 and 2 Gang Switch Assemblies

Material	Craft@Hrs	Unit	Material Cost	Labor Cost	Installed Cost
1 gang key switches, 15 amp, 3-way, in 4/S x 1-1/2" deep box, ring and plastic plate					
4-S-ring flat	1E@0.65	Ea	14.70	17.50	32.20
4-S-ring 1/4"	1E@0.65	Ea	14.60	17.50	32.10
4-S-ring 1/2"	1E@0.65	Ea	14.50	17.50	32.00
4-S-ring 5/8"	1E@0.65	Ea	14.60	17.50	32.10
4-S-ring 3/4"	1E@0.65	Ea	14.70	17.50	32.20
4-S-ring 1"	1E@0.65	Ea	15.00	17.50	32.50
4-S-ring 1-1/4"	1E@0.65	Ea	15.30	17.50	32.80
4-S-tile ring 1/2"	1E@0.65	Ea	15.90	17.50	33.40
4-S-tile ring 3/4"	1E@0.65	Ea	16.00	17.50	33.50
4-S-tile ring 1"	1E@0.65	Ea	16.40	17.50	33.90
4-S-tile ring 1-1/4"	1E@0.66	Ea	15.60	17.70	33.30
4-S-tile ring 1-1/2"	1E@0.66	Ea	16.00	17.70	33.70
4-S-tile ring 2"	1E@0.66	Ea	16.10	17.70	33.80
1 gang key switches, 15 amp, 4-way, in 4/S x 1-1/2" deep box, ring and plastic plate					
4-S-ring flat	1E@0.70	Ea	40.60	18.80	59.40
4-S-ring 1/4"	1E@0.70	Ea	40.50	18.80	59.30
4-S-ring 1/2"	1E@0.70	Ea	40.40	18.80	59.20
4-S-ring 5/8"	1E@0.70	Ea	40.50	18.80	59.30
4-S-ring 3/4"	1E@0.70	Ea	40.60	18.80	59.40
4-S-ring 1"	1E@0.70	Ea	40.90	18.80	59.70
4-S-ring 1-1/4"	1E@0.70	Ea	41.20	18.80	60.00
4-S-tile ring 1/2"	1E@0.70	Ea	41.80	18.80	60.60
4-S-tile ring 3/4"	1E@0.70	Ea	41.90	18.80	60.70
4-S-tile ring 1"	1E@0.70	Ea	42.30	18.80	61.10
4-S-tile ring 1-1/4"	1E@0.71	Ea	42.50	19.10	61.60
4-S-tile ring 1-1/2"	1E@0.71	Ea	41.90	19.10	61.00
4-S-tile ring 2"	1E@0.71	Ea	42.00	19.10	61.10
2 gang brown switches, 15 amp, 1-pole, in 4/S x 1-1/2" deep box, ring and plastic plate					
4-S-ring flat	1E@0.91	Ea	4.77	24.50	29.27
4-S-ring 1/4"	1E@0.91	Ea	5.13	24.50	29.63
4-S-ring 1/2"	1E@0.91	Ea	4.52	24.50	29.02
4-S-ring 5/8"	1E@0.91	Ea	4.65	24.50	29.15
4-S-ring 3/4"	1E@0.91	Ea	4.85	24.50	29.35
4-S-ring 1"	1E@0.91	Ea	5.42	24.50	29.92
4-S-ring 1-1/4"	1E@0.91	Ea	5.53	24.50	30.03
4-S-tile ring 1/2"	1E@0.91	Ea	6.28	24.50	30.78
4-S-tile ring 3/4"	1E@0.91	Ea	6.42	24.50	30.92
4-S-tile ring 1"	1E@0.91	Ea	6.49	24.50	30.99
4-S-tile ring 1-1/4"	1E@0.93	Ea	6.44	25.00	31.44
4-S-tile ring 1-1/2"	1E@0.93	Ea	6.14	25.00	31.14
4-S-tile ring 2"	1E@0.93	Ea	6.31	25.00	31.31

S_{K3} S_{K4} SS

Use these figures to estimate the cost of installing assemblies under the conditions described on pages 5 and 6. Costs listed are for each assembly installed. The crew is one electrician working at a labor cost of $26.88 per manhour. These costs include layout, material handling and normal waste. Add for the sales tax, delivery, supervision, mobilization, demobilization, cleanup, overhead and profit.

4/S x 1-1/2"-Deep, 15 Amp, 2 Gang Switch Assemblies

Material	Craft@Hrs	Unit	Material Cost	Labor Cost	Installed Cost

2 gang brown switches, 15 amp, 2-pole, in 4/S x 1-1/2" deep box, ring and plastic plate

Material	Craft@Hrs	Unit	Material Cost	Labor Cost	Installed Cost
4-S-ring flat	1E@1.01	Ea	14.30	27.10	41.40
4-S-ring 1/4"	1E@1.01	Ea	14.70	27.10	41.80
4-S-ring 1/2"	1E@1.01	Ea	14.10	27.10	41.20
4-S-ring 5/8"	1E@1.01	Ea	14.20	27.10	41.30
4-S-ring 3/4"	1E@1.01	Ea	14.40	27.10	41.50
4-S-ring 1"	1E@1.01	Ea	15.00	27.10	42.10
4-S-ring 1-1/4"	1E@1.01	Ea	15.10	27.10	42.20
4-S-tile ring 1/2"	1E@1.01	Ea	15.80	27.10	42.90
4-S-tile ring 3/4"	1E@1.01	Ea	16.00	27.10	43.10
4-S-tile ring 1"	1E@1.01	Ea	16.00	27.10	43.10
4-S-tile ring 1-1/4"	1E@1.03	Ea	16.00	27.70	43.70
4-S-tile ring 1-1/2"	1E@1.03	Ea	15.70	27.70	43.40
4-S-tile ring 2"	1E@1.03	Ea	15.90	27.70	43.60

S_2S_2

2 gang brown switches, 15 amp, 3-way, in 4/S x 1-1/2" deep box, ring and plastic plate

Material	Craft@Hrs	Unit	Material Cost	Labor Cost	Installed Cost
4-S-ring flat	1E@1.01	Ea	5.67	27.10	32.77
4-S-ring 1/4"	1E@1.01	Ea	6.03	27.10	33.13
4-S-ring 1/2"	1E@1.01	Ea	5.42	27.10	32.52
4-S-ring 5/8"	1E@1.01	Ea	5.55	27.10	32.65
4-S-ring 3/4"	1E@1.01	Ea	5.75	27.10	32.85
4-S-ring 1"	1E@1.01	Ea	6.32	27.10	33.42
4-S-ring 1-1/4"	1E@1.01	Ea	6.43	27.10	33.53
4-S-tile ring 1/2"	1E@1.01	Ea	7.18	27.10	34.28
4-S-tile ring 3/4"	1E@1.01	Ea	7.32	27.10	34.42
4-S-tile ring 1"	1E@1.01	Ea	7.39	27.10	34.49
4-S-tile ring 1-1/4"	1E@1.03	Ea	7.34	27.70	35.04
4-S-tile ring 1-1/2"	1E@1.03	Ea	7.04	27.70	34.74
4-S-tile ring 2"	1E@1.03	Ea	7.21	27.70	34.91

S_3S_3

2 gang brown switches, 15 amp, 4-way, in 4/S x 1-1/2" deep box, ring and plastic plate

Material	Craft@Hrs	Unit	Material Cost	Labor Cost	Installed Cost
4-S-ring flat	1E@1.06	Ea	19.50	28.50	48.00
4-S-ring 1/4"	1E@1.06	Ea	19.80	28.50	48.30
4-S-ring 1/2"	1E@1.06	Ea	19.20	28.50	47.70
4-S-ring 5/8"	1E@1.06	Ea	19.40	28.50	47.90
4-S-ring 3/4"	1E@1.06	Ea	19.60	28.50	48.10
4-S-ring 1"	1E@1.06	Ea	20.10	28.50	48.60
4-S-ring 1-1/4"	1E@1.06	Ea	20.20	28.50	48.70
4-S-tile ring 1/2"	1E@1.06	Ea	21.00	28.50	49.50
4-S-tile ring 3/4"	1E@1.06	Ea	21.10	28.50	49.60
4-S-tile ring 1"	1E@1.06	Ea	21.20	28.50	49.70
4-S-tile ring 1-1/4"	1E@1.08	Ea	21.10	29.00	50.10
4-S-tile ring 1-1/2"	1E@1.08	Ea	20.80	29.00	49.80
4-S-tile ring 2"	1E@1.08	Ea	21.00	29.00	50.00

S_4S_4

Use these figures to estimate the cost of installing assemblies under the conditions described on pages 5 and 6. Costs listed are for each assembly installed. The crew is one electrician working at a labor cost of $26.88 per manhour. These costs include layout, material handling and normal waste. Add for the sales tax, delivery, supervision, mobilization, demobilization, cleanup, overhead and profit.

4/S x 1-1/2" Deep, 15 Amp, 1 Gang Switch Assemblies

Material	Craft@Hrs	Unit	Material Cost	Labor Cost	Installed Cost
1 gang ivory switches, 15 amp, 1-pole, in 4/S x 1-1/2" deep box, ring and plastic plate					
4-S-ring flat	1E@0.60	Ea	3.38	16.10	19.48
4-S-ring 1/4"	1E@0.60	Ea	3.30	16.10	19.40
4-S-ring 1/2"	1E@0.60	Ea	3.11	16.10	19.21
4-S-ring 5/8"	1E@0.60	Ea	3.21	16.10	19.31
4-S-ring 3/4"	1E@0.60	Ea	3.38	16.10	19.48
4-S-ring 1"	1E@0.60	Ea	3.70	16.10	19.80
4-S-ring 1-1/4"	1E@0.60	Ea	3.95	16.10	20.05
4-S-tile ring 1/2"	1E@0.60	Ea	4.57	16.10	20.67
4-S-tile ring 3/4"	1E@0.60	Ea	4.66	16.10	20.76
4-S-tile ring 1"	1E@0.60	Ea	5.08	16.10	21.18
4-S-tile ring 1-1/4"	1E@0.61	Ea	5.22	16.40	21.62
4-S-tile ring 1-1/2"	1E@0.61	Ea	4.66	16.40	21.06
4-S-tile ring 2"	1E@0.61	Ea	4.78	16.40	21.18
1 gang ivory switches, 15 amp, 2-pole, in 4/S x 1-1/2" deep box, ring and plastic plate					
4-S-ring flat	1E@0.65	Ea	8.15	17.50	25.65
4-S-ring 1/4"	1E@0.65	Ea	8.07	17.50	25.57
4-S-ring 1/2"	1E@0.65	Ea	7.88	17.50	25.38
4-S-ring 5/8"	1E@0.65	Ea	7.98	17.50	25.48
4-S-ring 3/4"	1E@0.65	Ea	8.15	17.50	25.65
4-S-ring 1"	1E@0.65	Ea	8.47	17.50	25.97
4-S-ring 1-1/4"	1E@0.65	Ea	8.72	17.50	26.22
4-S-tile ring 1/2"	1E@0.65	Ea	9.99	17.50	27.49
4-S-tile ring 3/4"	1E@0.65	Ea	9.43	17.50	26.93
4-S-tile ring 1"	1E@0.65	Ea	9.82	17.50	27.32
4-S-tile ring 1-1/4"	1E@0.66	Ea	8.59	17.70	26.29
4-S-tile ring 1-1/2"	1E@0.66	Ea	8.95	17.70	26.65
4-S-tile ring 2"	1E@0.66	Ea	9.55	17.70	27.25
1 gang ivory switches, 15 amp, 3-way, in 4/S x 1-1/2" deep box, ring and plastic plate					
4-S-ring flat	1E@0.65	Ea	3.83	17.50	21.33
4-S-ring 1/4"	1E@0.65	Ea	3.75	17.50	21.25
4-S-ring 1/2"	1E@0.65	Ea	3.56	17.50	21.06
4-S-ring 5/8"	1E@0.65	Ea	3.66	17.50	21.16
4-S-ring 3/4"	1E@0.65	Ea	3.83	17.50	21.33
4-S-ring 1"	1E@0.65	Ea	4.15	17.50	21.65
4-S-ring 1-1/4"	1E@0.65	Ea	4.40	17.50	21.90
4-S-tile ring 1/2"	1E@0.65	Ea	5.02	17.50	22.52
4-S-tile ring 3/4"	1E@0.65	Ea	5.11	17.50	22.61
4-S-tile ring 1"	1E@0.65	Ea	5.53	17.50	23.03
4-S-tile ring 1-1/4"	1E@0.66	Ea	5.67	17.70	23.37
4-S-tile ring 1-1/2"	1E@0.66	Ea	5.11	17.70	22.81
4-S-tile ring 2"	1E@0.66	Ea	5.23	17.70	22.93

The left margin marks the three sections: **S** (1-pole), **S₂** (2-pole), **S₃** (3-way).

Use these figures to estimate the cost of installing assemblies under the conditions described on pages 5 and 6. Costs listed are for each assembly installed. The crew is one electrician working at a labor cost of $26.88 per manhour. These costs include layout, material handling and normal waste. Add for the sales tax, delivery, supervision, mobilization, demobilization, cleanup, overhead and profit.

4/S x 1-1/2" Deep, 15 Amp, 1 Gang Switch Assemblies

Material	Craft@Hrs	Unit	Material Cost	Labor Cost	Installed Cost
1 gang ivory switches, 15 amp, 4-way, in 4/S x 1-1/2" deep box, ring and plastic plate					
4-S-ring flat	1E@0.70	Ea	10.70	18.80	29.50
4-S-ring 1/4"	1E@0.70	Ea	10.70	18.80	29.50
4-S-ring 1/2"	1E@0.70	Ea	10.50	18.80	29.30
4-S-ring 5/8"	1E@0.70	Ea	10.60	18.80	29.40
4-S-ring 3/4"	1E@0.70	Ea	10.70	18.80	29.50
4-S-ring 1"	1E@0.70	Ea	11.10	18.80	29.90
4-S-ring 1-1/4"	1E@0.70	Ea	11.30	18.80	30.10
4-S-tile ring 1/2"	1E@0.70	Ea	11.90	18.80	30.70
4-S-tile ring 3/4"	1E@0.70	Ea	12.00	18.80	30.80
4-S-tile ring 1"	1E@0.70	Ea	12.40	18.80	31.20
4-S-tile ring 1-1/4"	1E@0.71	Ea	12.60	19.10	31.70
4-S-tile ring 1-1/2"	1E@0.71	Ea	12.00	19.10	31.10
4-S-tile ring 2"	1E@0.71	Ea	12.10	19.10	31.20

S_4

Material	Craft@Hrs	Unit	Material Cost	Labor Cost	Installed Cost
2 gang ivory switches, 15 amp, 1-pole, in 4/S x 1-1/2" deep box, ring and plastic plate					
4-S-ring flat	1E@0.91	Ea	5.39	24.50	29.89
4-S-ring 1/4"	1E@0.91	Ea	5.75	24.50	30.25
4-S-ring 1/2"	1E@0.91	Ea	5.14	24.50	29.64
4-S-ring 5/8"	1E@0.91	Ea	5.27	24.50	29.77
4-S-ring 3/4"	1E@0.91	Ea	5.47	24.50	29.97
4-S-ring 1"	1E@0.91	Ea	6.04	24.50	30.54
4-S-ring 1-1/4"	1E@0.91	Ea	6.15	24.50	30.65
4-S-tile ring 1/2"	1E@0.91	Ea	6.90	24.50	31.40
4-S-tile ring 3/4"	1E@0.91	Ea	7.04	24.50	31.54
4-S-tile ring 1"	1E@0.91	Ea	7.11	24.50	31.61
4-S-tile ring 1-1/4"	1E@0.93	Ea	7.06	25.00	32.06
4-S-tile ring 1-1/2"	1E@0.93	Ea	6.76	25.00	31.76
4-S-tile ring 2"	1E@0.93	Ea	6.93	25.00	31.93

SS

Use these figures to estimate the cost of installing assemblies under the conditions described on pages 5 and 6. Costs listed are for each assembly installed. The crew is one electrician working at a labor cost of $26.88 per manhour. These costs include layout, material handling and normal waste. Add for the sales tax, delivery, supervision, mobilization, demobilization, cleanup, overhead and profit.

4/S x 1-1/2" Deep, 15 Amp, 2 Gang Switch Assemblies

Material	Craft@Hrs	Unit	Material Cost	Labor Cost	Installed Cost
2 gang ivory switches, 15 amp, 2-pole, in 4/S x 1-1/2" deep box, ring and plastic plate					
4-S-ring flat	1E@1.01	Ea	14.30	27.10	41.40
4-S-ring 1/4"	1E@1.01	Ea	14.70	27.10	41.80
4-S-ring 1/2"	1E@1.01	Ea	14.10	27.10	41.20
4-S-ring 5/8"	1E@1.01	Ea	14.20	27.10	41.30
4-S-ring 3/4"	1E@1.01	Ea	14.40	27.10	41.50
4-S-ring 1"	1E@1.01	Ea	15.00	27.10	42.10
4-S-ring 1-1/4"	1E@1.01	Ea	15.10	27.10	42.20
4-S-tile ring 1/2"	1E@1.01	Ea	15.80	27.10	42.90
4-S-tile ring 3/4"	1E@1.01	Ea	16.00	27.10	43.10
4-S-tile ring 1"	1E@1.01	Ea	16.00	27.10	43.10
4-S-tile ring 1-1/4"	1E@1.03	Ea	16.00	27.70	43.70
4-S-tile ring 1-1/2"	1E@1.03	Ea	15.70	27.70	43.40
4-S-tile ring 2"	1E@1.03	Ea	15.90	27.70	43.60
2 gang ivory switches, 15 amp, 3-way, in 4/S x 1-1/2" deep box, ring and plastic plate					
4-S-ring flat	1E@1.01	Ea	5.67	27.10	32.77
4-S-ring 1/4"	1E@1.01	Ea	6.03	27.10	33.13
4-S-ring 1/2"	1E@1.01	Ea	5.42	27.10	32.52
4-S-ring 5/8"	1E@1.01	Ea	5.55	27.10	32.65
4-S-ring 3/4"	1E@1.01	Ea	5.75	27.10	32.85
4-S-ring 1"	1E@1.01	Ea	6.32	27.10	33.42
4-S-ring 1-1/4"	1E@1.01	Ea	6.43	27.10	33.53
4-S-tile ring 1/2"	1E@1.01	Ea	7.18	27.10	34.28
4-S-tile ring 3/4"	1E@1.01	Ea	7.32	27.10	34.42
4-S-tile ring 1"	1E@1.01	Ea	7.39	27.10	34.49
4-S-tile ring 1-1/4"	1E@1.03	Ea	7.34	27.70	35.04
4-S-tile ring 1-1/2"	1E@1.03	Ea	7.04	27.70	34.74
4-S-tile ring 2"	1E@1.03	Ea	7.21	27.70	34.91
2 gang ivory switches, 15 amp, 4-way, in 4/S x 1-1/2" deep box, ring and plastic plate					
4-S-ring flat	1E@1.06	Ea	19.40	28.50	47.90
4-S-ring 1/4"	1E@1.06	Ea	19.80	28.50	48.30
4-S-ring 1/2"	1E@1.06	Ea	19.20	28.50	47.70
4-S-ring 5/8"	1E@1.06	Ea	19.40	28.50	47.90
4-S-ring 3/4"	1E@1.06	Ea	19.60	28.50	48.10
4-S-ring 1"	1E@1.06	Ea	20.10	28.50	48.60
4-S-ring 1-1/4"	1E@1.06	Ea	20.20	28.50	48.70
4-S-tile ring 1/2"	1E@1.06	Ea	21.00	28.50	49.50
4-S-tile ring 3/4"	1E@1.06	Ea	21.10	28.50	49.60
4-S-tile ring 1"	1E@1.06	Ea	21.20	28.50	49.70
4-S-tile ring 1-1/4"	1E@1.08	Ea	21.10	29.00	50.10
4-S-tile ring 1-1/2"	1E@1.08	Ea	20.80	29.00	49.80
4-S-tile ring 2"	1E@1.08	Ea	21.00	29.00	50.00

The row labels S_2S_2, S_3S_3 and S_4S_4 appear at the left of the three sections respectively.

Use these figures to estimate the cost of installing assemblies under the conditions described on pages 5 and 6. Costs listed are for each assembly installed. The crew is one electrician working at a labor cost of $26.88 per manhour. These costs include layout, material handling and normal waste. Add for the sales tax, delivery, supervision, mobilization, demobilization, cleanup, overhead and profit.

4/S x 2-1/8" Deep, 15 Amp, 1 Gang Switch Assemblies

Material	Craft@Hrs	Unit	Material Cost	Labor Cost	Installed Cost
1 gang brown switches, 15 amp, 1-pole, in 4/S x 2-1/8" deep box, ring and plastic plate					
4-S-ring flat	1E@0.62	Ea	4.00	16.70	20.70
4-S-ring 1/4"	1E@0.62	Ea	3.92	16.70	20.62
4-S-ring 1/2"	1E@0.62	Ea	3.73	16.70	20.43
4-S-ring 5/8"	1E@0.62	Ea	3.83	16.70	20.53
4-S-ring 3/4"	1E@0.62	Ea	4.00	16.70	20.70
4-S-ring 1"	1E@0.62	Ea	4.32	16.70	21.02
4-S-ring 1-1/4"	1E@0.62	Ea	4.57	16.70	21.27
4-S-tile ring 1/2"	1E@0.62	Ea	5.19	16.70	21.89
4-S-tile ring 3/4"	1E@0.62	Ea	5.28	16.70	21.98
4-S-tile ring 1"	1E@0.62	Ea	5.70	16.70	22.40
4-S-tile ring 1-1/4"	1E@0.63	Ea	5.84	16.90	22.74
4-S-tile ring 1-1/2"	1E@0.63	Ea	5.28	16.90	22.18
4-S-tile ring 2"	1E@0.63	Ea	5.40	16.90	22.30

S

Material	Craft@Hrs	Unit	Material Cost	Labor Cost	Installed Cost
1 gang brown switches, 15 amp, 2-pole, in 4/S x 2-1/8" deep box, ring and plastic plate					
4-S-ring flat	1E@0.67	Ea	8.77	18.00	26.77
4-S-ring 1/4"	1E@0.67	Ea	8.69	18.00	26.69
4-S-ring 1/2"	1E@0.67	Ea	8.50	18.00	26.50
4-S-ring 5/8"	1E@0.67	Ea	8.60	18.00	26.60
4-S-ring 3/4"	1E@0.67	Ea	8.77	18.00	26.77
4-S-ring 1"	1E@0.67	Ea	9.09	18.00	27.09
4-S-ring 1-1/4"	1E@0.67	Ea	9.34	18.00	27.34
4-S-tile ring 1/2"	1E@0.67	Ea	10.60	18.00	28.60
4-S-tile ring 3/4"	1E@0.67	Ea	10.10	18.00	28.10
4-S-tile ring 1"	1E@0.67	Ea	10.40	18.00	28.40
4-S-tile ring 1-1/4"	1E@0.68	Ea	9.21	18.30	27.51
4-S-tile ring 1-1/2"	1E@0.68	Ea	9.57	18.30	27.87
4-S-tile ring 2"	1E@0.68	Ea	10.20	18.30	28.50

S_2

Material	Craft@Hrs	Unit	Material Cost	Labor Cost	Installed Cost
1 gang brown switches, 15 amp, 2-pole, in 4/S x 2-1/8" deep box, ring and plastic plate					
4-S-ring flat	1E@0.67	Ea	4.45	18.00	22.45
4-S-ring 1/4	1E@0.67	Ea	4.37	18.00	22.37
4-S-ring 1/2"	1E@0.67	Ea	4.18	18.00	22.18
4-S-ring 5/8"	1E@0.67	Ea	4.28	18.00	22.28
4-S-ring 3/4"	1E@0.67	Ea	4.45	18.00	22.45
4-S-ring 1"	1E@0.67	Ea	4.77	18.00	22.77
4-S-ring 1-1/4"	1E@0.67	Ea	5.02	18.00	23.02
4-S-tile ring 1/2"	1E@0.67	Ea	5.64	18.00	23.64
4-S-tile ring 3/4"	1E@0.67	Ea	5.73	18.00	23.73
4-S-tile ring 1"	1E@0.67	Ea	6.15	18.00	24.15
4-S-tile ring 1-1/4"	1E@0.68	Ea	6.29	18.30	24.59
4-S-tile ring 1-1/2"	1E@0.68	Ea	5.73	18.30	24.03
4-S-tile ring 2"	1E@0.68	Ea	5.85	18.30	24.15

S_3

Use these figures to estimate the cost of installing assemblies under the conditions described on pages 5 and 6. Costs listed are for each assembly installed. The crew is one electrician working at a labor cost of $26.88 per manhour. These costs include layout, material handling and normal waste. Add for the sales tax, delivery, supervision, mobilization, demobilization, cleanup, overhead and profit.

4/S x 2-1/8" Deep, 15 Amp, 1 Gang Switch Assemblies

Material	Craft@Hrs	Unit	Material Cost	Labor Cost	Installed Cost

1 gang brown switches, 15 amp, 4-way, in 4/S x 2-1/8" deep box, ring and plastic plate

Material	Craft@Hrs	Unit	Material Cost	Labor Cost	Installed Cost
4-S-ring flat	1E@0.72	Ea	11.40	19.40	30.80
4-S-ring 1/4"	1E@0.72	Ea	11.30	19.40	30.70
4-S-ring 1/2"	1E@0.72	Ea	11.10	19.40	30.50
4-S-ring 5/8"	1E@0.72	Ea	11.20	19.40	30.60
4-S-ring 3/4"	1E@0.72	Ea	11.40	19.40	30.80
4-S-ring 1"	1E@0.72	Ea	11.70	19.40	31.10
4-S-ring 1-1/4"	1E@0.72	Ea	11.90	19.40	31.30
4-S-tile ring 1/2"	1E@0.72	Ea	12.50	19.40	31.90
4-S-tile ring 3/4"	1E@0.72	Ea	12.60	19.40	32.00
4-S-tile ring 1"	1E@0.72	Ea	13.10	19.40	32.50
4-S-tile ring 1-1/4"	1E@0.73	Ea	13.20	19.60	32.80
4-S-tile ring 1-1/2"	1E@0.73	Ea	12.60	19.60	32.20
4-S-tile ring 2"	1E@0.73	Ea	12.80	19.60	32.40

S_4

1 gang key switches, 15 amp, 1-pole, in 4/S x 2-1/8" deep box, ring and plastic plate

Material	Craft@Hrs	Unit	Material Cost	Labor Cost	Installed Cost
4-S-ring flat	1E@0.62	Ea	11.60	16.70	28.30
4-S-ring 1/4"	1E@0.62	Ea	11.50	16.70	28.20
4-S-ring 1/2"	1E@0.62	Ea	11.30	16.70	28.00
4-S-ring 5/8"	1E@0.62	Ea	11.40	16.70	28.10
4-S-ring 3/4"	1E@0.62	Ea	11.60	16.70	28.30
4-S-ring 1"	1E@0.62	Ea	11.90	16.70	28.60
4-S-ring 1-1/4"	1E@0.62	Ea	12.10	16.70	28.80
4-S-tile ring 1/2"	1E@0.62	Ea	12.70	16.70	29.40
4-S-tile ring 3/4"	1E@0.62	Ea	12.80	16.70	29.50
4-S-tile ring 1"	1E@0.62	Ea	13.30	16.70	30.00
4-S-tile ring 1-1/4"	1E@0.63	Ea	13.40	16.90	30.30
4-S-tile ring 1-1/2"	1E@0.63	Ea	12.80	16.90	29.70
4-S-tile ring 2"	1E@0.63	Ea	13.00	16.90	29.90

S_K

1 gang key switches, 15 amp, 2-pole, in 4/S x 2-1/8" deep box, ring and plastic plate

Material	Craft@Hrs	Unit	Material Cost	Labor Cost	Installed Cost
4-S-ring flat	1E@0.67	Ea	21.70	18.00	39.70
4-S-ring 1/4"	1E@0.67	Ea	21.70	18.00	39.70
4-S-ring 1/2"	1E@0.67	Ea	21.50	18.00	39.50
4-S-ring 5/8"	1E@0.67	Ea	21.60	18.00	39.60
4-S-ring 3/4"	1E@0.67	Ea	21.70	18.00	39.70
4-S-ring 1"	1E@0.67	Ea	22.10	18.00	40.10
4-S-ring 1-1/4"	1E@0.67	Ea	22.30	18.00	40.30
4-S-tile ring 1/2"	1E@0.67	Ea	22.90	18.00	40.90
4-S-tile ring 3/4"	1E@0.67	Ea	23.00	18.00	41.00
4-S-tile ring 1"	1E@0.67	Ea	23.40	18.00	41.40
4-S-tile ring 1-1/4"	1E@0.68	Ea	23.60	18.30	41.90
4-S-tile ring 1-1/2"	1E@0.68	Ea	23.00	18.30	41.30
4-S-tile ring 2"	1E@0.68	Ea	23.10	18.30	41.40

S_{K2}

Use these figures to estimate the cost of installing assemblies under the conditions described on pages 5 and 6. Costs listed are for each assembly installed. The crew is one electrician working at a labor cost of $26.88 per manhour. These costs include layout, material handling and normal waste. Add for the sales tax, delivery, supervision, mobilization, demobilization, cleanup, overhead and profit.

4/S x 2-1/8" Deep, 15 Amp, 1 and 2 Gang Switch Assemblies

Material	Craft@Hrs	Unit	Material Cost	Labor Cost	Installed Cost
1 gang key switches, 15 amp, 3-way, in 4/S x 2-1/8" deep box, ring and plastic plate					
4-S-ring flat	1E@0.67	Ea	15.30	18.00	33.30
4-S-ring 1/4"	1E@0.67	Ea	15.30	18.00	33.30
4-S-ring 1/2"	1E@0.67	Ea	15.10	18.00	33.10
4-S-ring 5/8"	1E@0.67	Ea	15.20	18.00	33.20
4-S-ring 3/4"	1E@0.67	Ea	15.30	18.00	33.30
4-S-ring 1"	1E@0.67	Ea	15.70	18.00	33.70
4-S-ring 1-1/4"	1E@0.67	Ea	15.90	18.00	33.90
4-S-tile ring 1/2"	1E@0.67	Ea	16.50	18.00	34.50
4-S-tile ring 3/4"	1E@0.67	Ea	16.60	18.00	34.60
4-S-tile ring 1"	1E@0.67	Ea	17.00	18.00	35.00
4-S-tile ring 1-1/4"	1E@0.68	Ea	17.20	18.30	35.50
4-S-tile ring 1-1/2"	1E@0.68	Ea	16.60	18.30	34.90
4-S-tile ring 2"	1E@0.68	Ea	16.70	18.30	35.00

S_{K3}

Material	Craft@Hrs	Unit	Material Cost	Labor Cost	Installed Cost
1 gang key switches, 15 amp, 4-way, in 4/S x 2-1/8" deep box, ring and plastic plate					
4-S-ring flat	1E@0.72	Ea	41.20	19.40	60.60
4-S-ring 1/4"	1E@0.72	Ea	41.20	19.40	60.60
4-S-ring 1/2"	1E@0.72	Ea	41.00	19.40	60.40
4-S-ring 5/8"	1E@0.72	Ea	41.10	19.40	60.50
4-S-ring 3/4"	1E@0.72	Ea	41.20	19.40	60.60
4-S-ring 1"	1E@0.72	Ea	41.60	19.40	61.00
4-S-ring 1-1/4"	1E@0.72	Ea	41.80	19.40	61.20
4-S-tile ring 1/2"	1E@0.72	Ea	42.40	19.40	61.80
4-S-tile ring 3/4"	1E@0.72	Ea	42.50	19.40	61.90
4-S-tile ring 1"	1E@0.72	Ea	42.90	19.40	62.30
4-S-tile ring 1-1/4"	1E@0.73	Ea	43.10	19.60	62.70
4-S-tile ring 1-1/2"	1E@0.73	Ea	42.50	19.60	62.10
4-S-tile ring 2"	1E@0.73	Ea	42.60	19.60	62.20

S_{K4}

Material	Craft@Hrs	Unit	Material Cost	Labor Cost	Installed Cost
2 gang brown switches, 15 amp, 1-pole, in 4/S x 2-1/8" deep box, ring and plastic plate					
4-S-ring flat	1E@0.93	Ea	5.39	25.00	30.39
4-S-ring 1/4"	1E@0.93	Ea	5.75	25.00	30.75
4-S-ring 1/2"	1E@0.93	Ea	5.14	25.00	30.14
4-S-ring 5/8"	1E@0.93	Ea	5.27	25.00	30.27
4-S-ring 3/4"	1E@0.93	Ea	5.47	25.00	30.47
4-S-ring 1"	1E@0.93	Ea	6.04	25.00	31.04
4-S-ring 1-1/4"	1E@0.93	Ea	6.15	25.00	31.15
4-S-tile ring 1/2"	1E@0.93	Ea	6.90	25.00	31.90
4-S-tile ring 3/4"	1E@0.93	Ea	7.04	25.00	32.04
4-S-tile ring 1"	1E@0.93	Ea	7.11	25.00	32.11
4-S-tile ring 1-1/4"	1E@0.95	Ea	7.07	25.50	32.57
4-S-tile ring 1-1/2"	1E@0.95	Ea	6.76	25.50	32.26
4-S-tile ring 2"	1E@0.95	Ea	6.93	25.50	32.43

SS

Use these figures to estimate the cost of installing assemblies under the conditions described on pages 5 and 6. Costs listed are for each assembly installed. The crew is one electrician working at a labor cost of $26.88 per manhour. These costs include layout, material handling and normal waste. Add for the sales tax, delivery, supervision, mobilization, demobilization, cleanup, overhead and profit.

4/S x 2-1/8" Deep, 15 Amp, 2 Gang Switch Assemblies

Material	Craft@Hrs	Unit	Material Cost	Labor Cost	Installed Cost
2 gang brown switches, 15 amp, 2-pole, in 4/S x 2-1/8" deep box, ring and plastic plate					
4-S-ring flat	1E@1.03	Ea	14.90	27.70	42.60
4-S-ring 1/4"	1E@1.03	Ea	15.30	27.70	43.00
4-S-ring 1/2"	1E@1.03	Ea	14.70	27.70	42.40
4-S-ring 5/8"	1E@1.03	Ea	14.80	27.70	42.50
4-S-ring 3/4"	1E@1.03	Ea	15.00	27.70	42.70
4-S-ring 1"	1E@1.03	Ea	15.60	27.70	43.30
4-S-ring 1-1/4"	1E@1.03	Ea	15.70	27.70	43.40
4-S-tile ring 1/2"	1E@1.03	Ea	16.40	27.70	44.10
4-S-tile ring 3/4"	1E@1.03	Ea	16.60	27.70	44.30
4-S-tile ring 1"	1E@1.03	Ea	16.70	27.70	44.40
4-S-tile ring 1-1/4"	1E@1.05	Ea	16.60	28.20	44.80
4-S-tile ring 1-1/2"	1E@1.05	Ea	16.30	28.20	44.50
4-S-tile ring 2"	1E@1.05	Ea	16.50	28.20	44.70
2 gang brown switches, 15 amp, 3-way, in 4/S x 2-1/8" deep box, ring and plastic plate					
4-S-ring flat	1E@1.03	Ea	6.29	27.70	33.99
4-S-ring 1/4"	1E@1.03	Ea	6.65	27.70	34.35
4-S-ring 1/2"	1E@1.03	Ea	6.04	27.70	33.74
4-S-ring 5/8"	1E@1.03	Ea	6.17	27.70	33.87
4-S-ring 3/4"	1E@1.03	Ea	6.37	27.70	34.07
4-S-ring 1"	1E@1.03	Ea	6.94	27.70	34.64
4-S-ring 1-1/4"	1E@1.03	Ea	7.05	27.70	34.75
4-S-tile ring 1/2"	1E@1.03	Ea	7.80	27.70	35.50
4-S-tile ring 3/4"	1E@1.03	Ea	7.94	27.70	35.64
4-S-tile ring 1"	1E@1.03	Ea	8.01	27.70	35.71
4-S-tile ring 1-1/4"	1E@1.05	Ea	7.96	28.20	36.16
4-S-tile ring 1-1/2"	1E@1.05	Ea	7.66	28.20	35.86
4-S-tile ring 2"	1E@1.05	Ea	7.83	28.20	36.03
2 gang brown switches, 15 amp, 4-way, in 4/S x 2-1/8" deep box, ring and plastic plate					
4-S-ring flat	1E@1.08	Ea	20.10	29.00	49.10
4-S-ring 1/4"	1E@1.08	Ea	20.50	29.00	49.50
4-S-ring 1/2"	1E@1.08	Ea	19.80	29.00	48.80
4-S-ring 5/8"	1E@1.08	Ea	20.00	29.00	49.00
4-S-ring 3/4"	1E@1.08	Ea	20.20	29.00	49.20
4-S-ring 1"	1E@1.08	Ea	20.70	29.00	49.70
4-S-ring 1-1/4"	1E@1.08	Ea	20.90	29.00	49.90
4-S-tile ring 1/2"	1E@1.08	Ea	21.60	29.00	50.60
4-S-tile ring 3/4"	1E@1.08	Ea	21.70	29.00	50.70
4-S-tile ring 1"	1E@1.08	Ea	21.80	29.00	50.80
4-S-tile ring 1-1/4"	1E@1.10	Ea	21.80	29.60	51.40
4-S-tile ring 1-1/2"	1E@1.10	Ea	21.50	29.60	51.10
4-S-tile ring 2"	1E@1.10	Ea	21.60	29.60	51.20

Use these figures to estimate the cost of installing assemblies under the conditions described on pages 5 land 6. Costs listed are for each assembly installed. The crew is one electrician working at a labor cost of $26.88 per manhour. These costs include layout, material handling and normal waste. Add for the sales tax, delivery, supervision, mobilization, demobilization, cleanup, overhead and profit.

4/S x 2-1/8" Deep, 15 Amp, 1 Gang Switch Assemblies

Material	Craft@Hrs	Unit	Material Cost	Labor Cost	Installed Cost

1 gang ivory switches, 15 amp, 1-pole, in 4/S x 2-1/8" deep box, ring and plastic plate

Material	Craft@Hrs	Unit	Material Cost	Labor Cost	Installed Cost
4-S-ring flat	1E@0.62	Ea	4.00	16.70	20.70
4-S-ring 1/4"	1E@0.62	Ea	3.92	16.70	20.62
4-S-ring 1/2"	1E@0.62	Ea	3.73	16.70	20.43
4-S-ring 5/8"	1E@0.62	Ea	3.83	16.70	20.53
4-S-ring 3/4"	1E@0.62	Ea	4.00	16.70	20.70
4-S-ring 1"	1E@0.62	Ea	4.32	16.70	21.02
4-S-ring 1-1/4"	1E@0.62	Ea	4.57	16.70	21.27
4-S-tile ring 1/2"	1E@0.62	Ea	5.19	16.70	21.89
4-S-tile ring 3/4"	1E@0.62	Ea	5.28	16.70	21.98
4-S-tile ring 1"	1E@0.62	Ea	5.70	16.70	22.40
4-S-tile ring 1-1/4"	1E@0.63	Ea	5.84	16.90	22.74
4-S-tile ring 1-1/2"	1E@0.63	Ea	5.28	16.90	22.18
4-S-tile ring 2"	1E@0.63	Ea	5.40	16.90	22.30

S

1 gang ivory switches, 15 amp, 2-pole, in 4/S x 2-1/8" deep box, ring and plastic plate

Material	Craft@Hrs	Unit	Material Cost	Labor Cost	Installed Cost
4-S-ring flat	1E@0.67	Ea	8.77	18.00	26.77
4-S-ring 1/4"	1E@0.67	Ea	8.69	18.00	26.69
4-S-ring 1/2"	1E@0.67	Ea	8.50	18.00	26.50
4-S-ring 5/8"	1E@0.67	Ea	8.60	18.00	26.60
4-S-ring 3/4"	1E@0.67	Ea	8.77	18.00	26.77
4-S-ring 1"	1E@0.67	Ea	9.09	18.00	27.09
4-S-ring 1-1/4"	1E@0.67	Ea	9.34	18.00	27.34
4-S-tile ring 1/2"	1E@0.67	Ea	10.60	18.00	28.60
4-S-tile ring 3/4"	1E@0.67	Ea	10.10	18.00	28.10
4-S-tile ring 1"	1E@0.67	Ea	10.40	18.00	28.40
4-S-tile ring 1-1/4"	1E@0.68	Ea	9.21	18.30	27.51
4-S-tile ring 1-1/2"	1E@0.68	Ea	9.57	18.30	27.87
4-S-tile ring 2"	1E@0.68	Ea	10.20	18.30	28.50

S_2

1 gang ivory switches, 15 amp, 3-way, in 4/S x 2-1/8" deep box, ring and plastic plate

Material	Craft@Hrs	Unit	Material Cost	Labor Cost	Installed Cost
4-S-ring flat	1E@0.67	Ea	4.45	18.00	22.45
4-S-ring 1/4"	1E@0.67	Ea	4.37	18.00	22.37
4-S-ring 1/2"	1E@0.67	Ea	4.18	18.00	22.18
4-S-ring 5/8"	1E@0.67	Ea	4.28	18.00	22.28
4-S-ring 3/4"	1E@0.67	Ea	4.45	18.00	22.45
4-S-ring 1"	1E@0.67	Ea	4.77	18.00	22.77
4-S-ring 1-1/4"	1E@0.67	Ea	5.02	18.00	23.02
4-S-tile ring 1/2"	1E@0.67	Ea	5.64	18.00	23.64
4-S-tile ring 3/4"	1E@0.67	Ea	5.73	18.00	23.73
4-S-tile ring 1"	1E@0.67	Ea	6.15	18.00	24.15
4-S-tile ring 1-1/4"	1E@0.68	Ea	6.29	18.30	24.59
4-S-tile ring 1-1/2"	1E@0.68	Ea	5.73	18.30	24.03
4-S-tile ring 2"	1E@0.68	Ea	5.85	18.30	24.15

S_3

Use these figures to estimate the cost of installing assemblies under the conditions described on pages 5 and 6. Costs listed are for each assembly installed. The crew is one electrician working at a labor cost of $26.88 per manhour. These costs include layout, material handling and normal waste. Add for the sales tax, delivery, supervision, mobilization, demobilization, cleanup, overhead and profit.

4/S x 2-1/8" Deep, 15 Amp, 1 and 2 Gang Switch Assemblies

Material	Craft@Hrs	Unit	Material Cost	Labor Cost	Installed Cost
1 gang ivory switches, 15 amp, 4-way, in 4/S x 2-1/8" deep box, ring and plastic plate					
4-S-ring flat	1E@0.72	Ea	11.40	19.40	30.80
4-S-ring 1/4"	1E@0.72	Ea	11.30	19.40	30.70
4-S-ring 1/2"	1E@0.72	Ea	11.10	19.40	30.50
4-S-ring 5/8"	1E@0.72	Ea	11.20	19.40	30.60
4-S-ring 3/4"	1E@0.72	Ea	11.40	19.40	30.80
4-S-ring 1"	1E@0.72	Ea	11.70	19.40	31.10
4-S-ring 1-1/4"	1E@0.72	Ea	11.90	19.40	31.30
4-S-tile ring 1/2"	1E@0.72	Ea	12.50	19.40	31.90
4-S-tile ring 3/4"	1E@0.72	Ea	12.60	19.40	32.00
4-S-tile ring 1"	1E@0.72	Ea	13.10	19.40	32.50
4-S-tile ring 1-1/4"	1E@0.73	Ea	13.20	19.60	32.80
4-S-tile ring 1-1/2"	1E@0.73	Ea	12.60	19.60	32.20
4-S-tile ring 2"	1E@0.73	Ea	12.80	19.60	32.40
2 gang ivory switches, 15 amp, 1-pole, in 4/S x 2-1/8" deep box, ring and plastic plate					
4-S-ring flat	1E@0.93	Ea	5.34	25.00	30.34
4-S-ring 1/4"	1E@0.93	Ea	5.75	25.00	30.75
4-S-ring 1/2"	1E@0.93	Ea	5.14	25.00	30.14
4-S-ring 5/8"	1E@0.93	Ea	5.27	25.00	30.27
4-S-ring 3/4"	1E@0.93	Ea	5.47	25.00	30.47
4-S-ring 1"	1E@0.93	Ea	6.04	25.00	31.04
4-S-ring 1-1/4"	1E@0.93	Ea	6.15	25.00	31.15
4-S-tile ring 1/2"	1E@0.93	Ea	6.90	25.00	31.90
4-S-tile ring 3/4"	1E@0.93	Ea	7.04	25.00	32.04
4-S-tile ring 1"	1E@0.93	Ea	7.11	25.00	32.11
4-S-tile ring 1-1/4"	1E@0.95	Ea	7.06	25.50	32.56
4-S-tile ring 1-1/2"	1E@0.95	Ea	6.76	25.50	32.26
4-S-tile ring 2"	1E@0.95	Ea	6.93	25.50	32.43

S_4 (1 gang section) · SS (2 gang section)

Use these figures to estimate the cost of installing assemblies under the conditions described on pages 5 and 6. Costs listed are for each assembly installed. The crew is one electrician working at a labor cost of $26.88 per manhour. These costs include layout, material handling and normal waste. Add for the sales tax, delivery, supervision, mobilization, demobilization, cleanup, overhead and profit.

4/S x 2-1/8" Deep, 15 Amp, 2 Gang Switch Assemblies

Material	Craft@Hrs	Unit	Material Cost	Labor Cost	Installed Cost
2 gang ivory switches, 15 amp, 2-pole, in 4/S x 2-1/8" deep box, ring and plastic plate					
4-S-ring flat	1E@1.03	Ea	14.90	27.70	42.60
4-S-ring 1/4"	1E@1.03	Ea	15.30	27.70	43.00
4-S-ring 1/2"	1E@1.03	Ea	14.70	27.70	42.40
4-S-ring 5/8"	1E@1.03	Ea	14.90	27.70	42.60
4-S-ring 3/4"	1E@1.03	Ea	15.00	27.70	42.70
4-S-ring 1"	1E@1.03	Ea	15.60	27.70	43.30
4-S-ring 1-1/4"	1E@1.03	Ea	15.70	27.70	43.40
4-S-tile ring 1/2"	1E@1.03	Ea	16.40	27.70	44.10
4-S-tile ring 3/4"	1E@1.03	Ea	16.60	27.70	44.30
4-S-tile ring 1"	1E@1.03	Ea	16.70	27.70	44.40
4-S-tile ring 1-1/4"	1E@1.05	Ea	16.60	28.20	44.80
4-S-tile ring 1-1/2"	1E@1.05	Ea	16.30	28.20	44.50
4-S-tile ring 2"	1E@1.05	Ea	16.50	28.20	44.70

$S_2 S_2$

Material	Craft@Hrs	Unit	Material Cost	Labor Cost	Installed Cost
2 gang ivory switches, 15 amp, 3-way, in 4/S x 2-1/8" deep box, ring and plastic plate					
4-S-ring flat	1E@1.03	Ea	6.29	27.70	33.99
4-S-ring 1/4"	1E@1.03	Ea	6.65	27.70	34.35
4-S-ring 1/2"	1E@1.03	Ea	6.04	27.70	33.74
4-S-ring 5/8"	1E@1.03	Ea	6.17	27.70	33.87
4-S-ring 3/4"	1E@1.03	Ea	6.37	27.70	34.07
4-S-ring 1"	1E@1.03	Ea	6.94	27.70	34.64
4-S-ring 1-1/4"	1E@1.03	Ea	7.05	27.70	34.75
4-S-tile ring 1/2"	1E@1.03	Ea	7.80	27.70	35.50
4-S-tile ring 3/4"	1E@1.03	Ea	7.94	27.70	35.64
4-S-tile ring 1"	1E@1.03	Ea	8.01	27.70	35.71
4-S-tile ring 1-1/4"	1E@1.05	Ea	7.96	28.20	36.16
4-S-tile ring 1-1/2"	1E@1.05	Ea	7.66	28.20	35.86
4-S-tile ring 2"	1E@1.05	Ea	7.83	28.20	36.03

$S_3 S_3$

Material	Craft@Hrs	Unit	Material Cost	Labor Cost	Installed Cost
2 gang ivory switches, 15 amp, 4-way, in 4/S x 2-1/8" deep box, ring and plastic plate					
4-S-ring flat	1E@1.08	Ea	20.10	29.00	49.10
4-S-ring 1/4"	1E@1.08	Ea	20.50	29.00	49.50
4-S-ring 1/2"	1E@1.08	Ea	19.80	29.00	48.80
4-S-ring 5/8"	1E@1.08	Ea	20.00	29.00	49.00
4-S-ring 3/4"	1E@1.08	Ea	20.20	29.00	49.20
4-S-ring 1"	1E@1.08	Ea	20.70	29.00	49.70
4-S-ring 1-1/4"	1E@1.08	Ea	20.90	29.00	49.90
4-S-tile ring 1/2"	1E@1.08	Ea	21.60	29.00	50.60
4-S-tile ring 3/4"	1E@1.08	Ea	21.70	29.00	50.70
4-S-tile ring 1"	1E@1.08	Ea	21.80	29.00	50.80
4-S-tile ring 1-1/4"	1E@1.10	Ea	21.80	29.60	51.40
4-S-tile ring 1-1/2"	1E@1.10	Ea	21.50	29.60	51.10
4-S-tile ring 2"	1E@1.10	Ea	21.60	29.60	51.20

$S_4 S_4$

Use these figures to estimate the cost of installing assemblies under the conditions described on pages 5 and 6. Costs listed are for each assembly installed. The crew is one electrician working at a labor cost of $26.88 per manhour. These costs include layout, material handling and normal waste. Add for the sales tax, delivery, supervision, mobilization, demobilization, cleanup, overhead and profit.

Handy Box and Receptacle Assemblies

Material	Craft@Hrs	Unit	Material Cost	Labor Cost	Installed Cost
Receptacles in 1-1/2" deep handy boxes with handy box cover					
Single 15 amp brown	1E@0.43	Ea	4.33	11.60	15.93
Single 15 amp ivory	1E@0.43	Ea	3.90	11.60	15.50
Single 15 amp orange	1E@0.43	Ea	18.03	11.60	29.63
Duplex 15 amp brown	1E@0.43	Ea	2.02	11.60	13.62
Duplex 15 amp ivory	1E@0.43	Ea	2.02	11.60	13.62
Duplex 15 amp orange	1E@0.43	Ea	13.22	11.60	24.82
Single 20 amp brown	1E@0.43	Ea	5.69	11.60	17.29
Single 20 amp ivory	1E@0.43	Ea	5.73	11.60	17.33
Single 20 amp orange	1E@0.43	Ea	15.34	11.60	26.94
Duplex 20 amp brown	1E@0.43	Ea	3.41	11.60	15.01
Duplex 20 amp ivory	1E@0.43	Ea	3.41	11.60	15.01
Duplex 20 amp orange	1E@0.43	Ea	16.10	11.60	27.70
Receptacles in 1-7/8" deep handy boxes with handy box cover					
Single 15 amp brown	1E@0.45	Ea	4.35	12.10	16.45
Single 15 amp ivory	1E@0.45	Ea	3.92	12.10	16.02
Single 15 amp orange	1E@0.45	Ea	18.05	12.10	30.15
Duplex 15 amp brown	1E@0.45	Ea	2.04	12.10	14.14
Duplex 15 amp ivory	1E@0.45	Ea	2.04	12.10	14.14
Duplex 15 amp orange	1E@0.45	Ea	13.24	12.10	25.34
Single 20 amp brown	1E@0.45	Ea	5.71	12.10	17.81
Single 20 amp ivory	1E@0.45	Ea	5.75	12.10	17.85
Single 20 amp orange	1E@0.45	Ea	15.36	12.10	27.46
Duplex 20 amp brown	1E@0.45	Ea	3.43	12.10	15.53
Duplex 20 amp ivory	1E@0.45	Ea	3.43	12.10	15.53
Duplex 20 amp orange	1E@0.45	Ea	16.12	12.10	28.22
Receptacles in 2-1/8" deep handy boxes with handy box cover					
Single 15 amp brown	1F@0.48	Ea	4.68	12.90	17.58
Single 15 amp ivory	1E@0.48	Ea	4.25	12.90	17.15
Single 15 amp orange	1E@0.48	Ea	18.38	12.90	31.28
Duplex 15 amp brown	1E@0.48	Ea	2.37	12.90	15.27
Duplex 15 amp ivory	1E@0.48	Ea	2.37	12.90	15.27
Duplex 15 amp orange	1E@0.48	Ea	13.57	12.90	26.47

Use these figures to estimate the cost of installing assemblies under the conditions described on pages 5 and 6. Costs listed are for each assembly installed. The crew is one electrician working at a labor cost of $26.88 per manhour. These costs include layout, material handling and normal waste. Add for the sales tax, delivery, supervision, mobilization, demobilization, cleanup, overhead and profit.

Material	Craft@Hrs	Unit	Material Cost	Labor Cost	Installed Cost
Receptacles in 2-1/8" deep handy boxes with handy box cover					
Single 20 amp brown	1E@0.48	Ea	6.04	12.90	18.94
Single 20 amp ivory	1E@0.48	Ea	6.08	12.90	18.98
Single 20 amp orange	1E@0.48	Ea	15.69	12.90	28.59
Duplex 20 amp brown	1E@0.48	Ea	3.76	12.90	16.66
Duplex 20 amp ivory	1E@0.48	Ea	3.76	12.90	16.66
Duplex 20 amp orange	1E@0.48	Ea	16.45	12.90	29.35
Receptacles in 1-7/8" deep handy boxes with handy box cover and flat bracket					
Single 15 amp brown	1E@0.45	Ea	5.09	12.10	17.19
Single 15 amp ivory	1E@0.45	Ea	4.66	12.10	16.76
Single 15 amp orange	1E@0.45	Ea	18.76	12.10	30.86
Duplex 15 amp brown	1E@0.45	Ea	2.78	12.10	14.88
Duplex 15 amp ivory	1E@0.45	Ea	2.78	12.10	14.88
Duplex 15 amp orange	1E@0.45	Ea	13.98	12.10	26.08
Single 20 amp brown	1E@0.45	Ea	6.45	12.10	18.55
Single 20 amp ivory	1E@0.45	Ea	6.45	12.10	18.55
Single 20 amp orange	1E@0.45	Ea	16.10	12.10	28.20
Duplex 20 amp brown	1E@0.45	Ea	4.17	12.10	16.27
Duplex 20 amp ivory	1E@0.45	Ea	4.17	12.10	16.27
Duplex 20 amp orange	1E@0.45	Ea	16.86	12.10	28.96
Receptacles in 2-1/8" deep handy boxes with handy box cover and flat bracket					
Single 15 amp brown	1E@0.48	Ea	5.25	12.90	18.15
Single 15 amp ivory	1E@0.48	Ea	4.82	12.90	17.72
Single 15 amp orange	1E@0.48	Ea	18.95	12.90	31.85
Duplex 15 amp brown	1E@0.48	Ea	2.94	12.90	15.84
Duplex 15 amp ivory	1E@0.48	Ea	2.94	12.90	15.84
Duplex 15 amp orange	1E@0.48	Ea	14.14	12.90	27.04
Single 20 amp brown	1E@0.48	Ea	6.61	12.90	19.51
Single 20 amp ivory	1E@0.48	Ea	6.65	12.90	19.55
Single 20 amp orange	1E@0.48	Ea	16.26	12.90	29.16
Duplex 20 amp brown	1E@0.48	Ea	4.33	12.90	17.23
Duplex 20 amp ivory	1E@0.48	Ea	4.33	12.90	17.23
Duplex 20 amp orange	1E@0.48	Ea	17.02	12.90	29.92

Use these figures to estimate the cost of installing assemblies under the conditions described on pages 5 and 6. Costs listed are for each assembly installed. The crew is one electrician working at a labor cost of $26.88 per manhour. These costs include layout, material handling and normal waste. Add for the sales tax, delivery, supervision, mobilization, demobilization, cleanup, overhead and profit.

Sectional Box and Receptacle Assemblies

Material	Craft@Hrs	Unit	Material Cost	Labor Cost	Installed Cost
Receptacles in 2" deep sectional box and plastic plate					
Single 15 amp brown	1E@0.40	Ea	4.63	10.80	15.43
Single 15 amp ivory	1E@0.40	Ea	4.20	10.80	15.00
Single 15 amp orange	1E@0.40	Ea	18.33	10.80	29.13
Duplex 15 amp brown	1E@0.40	Ea	2.55	10.80	13.35
Duplex 15 amp ivory	1E@0.40	Ea	2.55	10.80	13.35
Duplex 15 amp orange	1E@0.40	Ea	13.75	10.80	24.55
Single 20 amp brown	1E@0.40	Ea	4.37	10.80	15.17
Single 20 amp ivory	1E@0.40	Ea	4.47	10.80	15.27
Single 20 amp orange	1E@0.40	Ea	15.64	10.80	26.44
Duplex 20 amp brown	1E@0.40	Ea	3.94	10.80	14.74
Duplex 20 amp ivory	1E@0.40	Ea	3.94	10.80	14.74
Duplex 20 amp orange	1E@0.40	Ea	16.63	10.80	27.43
Receptacles in 2" deep 2 gang sectional box and plastic plate					
Single 15 amp brown	1E@0.80	Ea	9.35	21.50	30.85
Single 15 amp ivory	1E@0.80	Ea	8.49	21.50	29.99
Single 15 amp orange	1E@0.80	Ea	36.75	21.50	58.25
Duplex 15 amp brown	1E@0.80	Ea	5.17	21.50	26.67
Duplex 15 amp ivory	1E@0.80	Ea	5.17	21.50	26.67
Duplex 15 amp orange	1E@0.80	Ea	31.61	21.50	53.11
Single 20 amp brown	1E@0.80	Ea	8.83	21.50	30.33
Single 20 amp ivory	1E@0.80	Ea	9.03	21.50	30.53
Single 20 amp orange	1E@0.80	Ea	31.37	21.50	52.87
Duplex 20 amp brown	1E@0.80	Ea	7.95	21.50	29.45
Duplex 20 amp ivory	1E@0.80	Ea	7.95	21.50	29.45
Duplex 20 amp orange	1E@0.80	Ea	33.33	21.50	54.83

Use these figures to estimate the cost of installing assemblies under the conditions described on pages 5 and 6. Costs listed are for each assembly installed. The crew is one electrician working at a labor cost of $26.88 per manhour. These costs include layout, material handling and normal waste. Add for the sales tax, delivery, supervision, mobilization, demobilization, cleanup, overhead and profit.

Sectional Box and Receptacle Assemblies

Material	Craft@Hrs	Unit	Material Cost	Labor Cost	Installed Cost
Receptacles in 2-1/2" deep sectional box and plastic plate					
Single 15 amp brown	1E@0.42	Ea	4.27	11.30	15.57
Single 15 amp ivory	1E@0.42	Ea	3.84	11.30	15.14
Single 15 amp orange	1E@0.42	Ea	17.97	11.30	29.27
Duplex 15 amp brown	1E@0.42	Ea	2.19	11.30	13.49
Duplex 15 amp ivory	1E@0.42	Ea	2.19	11.30	13.49
Duplex 15 amp orange	1E@0.42	Ea	13.39	11.30	24.69
Single 20 amp brown	1E@0.42	Ea	4.01	11.30	15.31
Single 20 amp ivory	1E@0.42	Ea	4.11	11.30	15.41
Single 20 amp orange	1E@0.42	Ea	15.28	11.30	26.58
Duplex 20 amp brown	1E@0.42	Ea	3.58	11.30	14.88
Duplex 20 amp ivory	1E@0.42	Ea	3.58	11.30	14.88
Duplex 20 amp orange	1E@0.42	Ea	16.27	11.30	27.57
Receptacles in 2-1/2" deep 2 gang sectional box and plastic plate					
Single 15 amp brown	1E@0.84	Ea	8.63	22.60	31.23
Single 15 amp ivory	1E@0.84	Ea	7.77	22.60	30.37
Single 15 amp orange	1E@0.84	Ea	36.03	22.60	58.63
Duplex 15 amp brown	1E@0.84	Ea	4.45	22.60	27.05
Duplex 15 amp ivory	1E@0.84	Ea	4.45	22.60	27.05
Duplex 15 amp orange	1E@0.84	Ea	30.89	22.60	53.49
Single 20 amp brown	1E@0.84	Ea	8.11	22.60	30.71
Single 20 amp ivory	1E@0.84	Ea	8.31	22.60	30.91
Single 20 amp orange	1E@0.84	Ea	30.65	22.60	53.25
Duplex 20 amp brown	1E@0.84	Ea	7.23	22.60	29.83
Duplex 20 amp ivory	1E@0.84	Ea	7.23	22.60	29.83
Duplex 20 amp orange	1E@0.84	Ea	32.61	22.60	55.21

Use these figures to estimate the cost of installing assemblies under the conditions described on pages 5 and 6. Costs listed are for each assembly installed. The crew is one electrician working at a labor cost of $26.88 per manhour. These costs include layout, material handling and normal waste. Add for the sales tax, delivery, supervision, mobilization, demobilization, cleanup, overhead and profit.

Sectional Box and Receptacle Assemblies

Material	Craft@Hrs	Unit	Material Cost	Labor Cost	Installed Cost
Receptacles in 2-3/4" deep sectional box and plastic plate					
Single 15 amp brown	1E@0.45	Ea	4.42	12.10	16.52
Single 15 amp ivory	1E@0.45	Ea	3.99	12.10	16.09
Single 15 amp orange	1E@0.45	Ea	18.12	12.10	30.22
Duplex 15 amp brown	1E@0.45	Ea	2.34	12.10	14.44
Duplex 15 amp ivory	1E@0.45	Ea	2.34	12.10	14.44
Duplex 15 amp orange	1E@0.45	Ea	13.54	12.10	25.64
Single 20 amp brown	1E@0.45	Ea	4.16	12.10	16.26
Single 20 amp ivory	1E@0.45	Ea	4.26	12.10	16.36
Single 20 amp orange	1E@0.45	Ea	15.43	12.10	27.53
Duplex 20 amp brown	1E@0.45	Ea	3.73	12.10	15.83
Duplex 20 amp ivory	1E@0.45	Ea	3.73	12.10	15.83
Duplex 20 amp orange	1E@0.45	Ea	16.42	12.10	28.52
Receptacles in 2-3/4" deep 2 gang sectional box and plastic plate					
Single 15 amp brown	1E@0.90	Ea	8.93	24.20	33.13
Single 15 amp ivory	1E@0.90	Ea	8.07	24.20	32.27
Single 15 amp orange	1E@0.90	Ea	36.33	24.20	60.53
Duplex 15 amp brown	1E@0.90	Ea	4.75	24.20	28.95
Duplex 15 amp ivory	1E@0.90	Ea	4.75	24.20	28.95
Duplex 15 amp orange	1E@0.90	Ea	31.19	24.20	55.39
Single 20 amp brown	1E@0.90	Ea	8.41	24.20	32.61
Single 20 amp ivory	1E@0.90	Ea	8.61	24.20	32.81
Single 20 amp orange	1E@0.90	Ea	30.95	24.20	55.15
Duplex 20 amp brown	1E@0.90	Ea	7.53	24.20	31.73
Duplex 20 amp ivory	1E@0.90	Ea	7.53	24.20	31.73
Duplex 20 amp orange	1E@0.90	Ea	32.91	24.20	57.11

Use these figures to estimate the cost of installing assemblies under the conditions described on pages 5 and 6. Costs listed are for each assembly installed. The crew is one electrician working at a labor cost of $26.88 per manhour. These costs include layout, material handling and normal waste. Add for the sales tax, delivery, supervision, mobilization, demobilization, cleanup, overhead and profit.

Sectional Box and Receptacle Assemblies

Material	Craft@Hrs	Unit	Material Cost	Labor Cost	Installed Cost
Receptacles in 3-1/2" deep sectional box and plastic plate					
Single 15 amp brown	1E@0.47	Ea	4.73	12.60	17.33
Single 15 amp ivory	1E@0.47	Ea	4.30	12.60	16.90
Single 15 amp orange	1E@0.47	Ea	18.43	12.60	31.03
Duplex 15 amp brown	1E@0.47	Ea	2.65	12.60	15.25
Duplex 15 amp ivory	1E@0.47	Ea	2.65	12.60	15.25
Duplex 15 amp orange	1E@0.47	Ea	13.85	12.60	26.45
Single 20 amp brown	1E@0.47	Ea	4.47	12.60	17.07
Single 20 amp ivory	1E@0.47	Ea	4.57	12.60	17.17
Single 20 amp orange	1E@0.47	Ea	15.74	12.60	28.34
Duplex 20 amp brown	1E@0.47	Ea	4.04	12.60	16.64
Duplex 20 amp ivory	1E@0.47	Ea	4.04	12.60	16.64
Duplex 20 amp orange	1E@0.47	Ea	16.73	12.60	29.33
Receptacles in 3-1/2" deep 2 gang sectional box and plastic plate					
Single 15 amp brown	1E@0.94	Ea	9.55	25.30	34.85
Single 15 amp ivory	1E@0.94	Ea	8.69	25.30	33.99
Single 15 amp orange	1E@0.94	Ea	36.95	25.30	62.25
Duplex 15 amp brown	1E@0.94	Ea	5.37	25.30	30.67
Duplex 15 amp ivory	1E@0.94	Ea	5.37	25.30	30.67
Duplex 15 amp orange	1E@0.94	Ea	31.51	25.30	56.81
Single 20 amp brown	1E@0.94	Ea	9.03	25.30	34.33
Single 20 amp ivory	1E@0.94	Ea	9.23	25.30	34.53
Single 20 amp orange	1E@0.94	Ea	31.57	25.30	56.87
Duplex 20 amp brown	1E@0.94	Ea	7.46	25.30	32.76
Duplex 20 amp ivory	1E@0.94	Ea	7.46	25.30	32.76
Duplex 20 amp orange	1E@0.94	Ea	36.16	25.30	61.46

Use these figures to estimate the cost of installing assemblies under the conditions described on pages 5 and 6. Costs listed are for each assembly installed. The crew is one electrician working at a labor cost of $26.88 per manhour. These costs include layout, material handling and normal waste. Add for the sales tax, delivery, supervision, mobilization, demobilization, cleanup, overhead and profit.

4/S x 1-1/2" Deep, 15 Amp, 1 Gang
Single Receptacle Assemblies

1 gang brown single receptacle, 15 amp, in 4/S x 1-1/2" deep box, ring and plastic plate

Material	Craft@Hrs	Unit	Material Cost	Labor Cost	Installed Cost
4-S-ring flat	1E@0.55	Ea	5.33	14.80	20.13
4-S-ring 1/4"	1E@0.55	Ea	5.25	14.80	20.05
4-S-ring 1/2"	1E@0.55	Ea	5.06	14.80	19.86
4-S-ring 5/8"	1E@0.55	Ea	5.16	14.80	19.96
4-S-ring 3/4"	1E@0.55	Ea	5.33	14.80	20.13
4-S-ring 1"	1E@0.55	Ea	5.65	14.80	20.45
4-S-ring 1-1/4"	1E@0.55	Ea	5.90	14.80	20.70
4-S-tile ring 1/2"	1E@0.55	Ea	6.52	14.80	21.32
4-S-tile ring 3/4"	1E@0.55	Ea	6.61	14.80	21.41
4-S-tile ring 1"	1E@0.55	Ea	7.03	14.80	21.83
4-S-tile ring 1-1/4"	1E@0.56	Ea	7.17	15.10	22.27
4-S-tile ring 1-1/2"	1E@0.56	Ea	6.61	15.10	21.71
4-S-tile ring 2"	1E@0.56	Ea	6.73	15.10	21.83

1 gang ivory single receptacle, 15 amp, in 4/S x 1-1/2" deep box, ring and plastic plate

Material	Craft@Hrs	Unit	Material Cost	Labor Cost	Installed Cost
4-S-ring flat	1E@0.55	Ea	4.90	14.80	19.70
4-S-ring 1/4"	1E@0.55	Ea	4.82	14.80	19.62
4-S-ring 1/2"	1E@0.55	Ea	4.63	14.80	19.43
4-S-ring 5/8"	1E@0.55	Ea	4.73	14.80	19.53
4-S-ring 3/4"	1E@0.55	Ea	4.90	14.80	19.70
4-S-ring 1"	1E@0.55	Ea	5.22	14.80	20.02
4-S-ring 1-1/4"	1E@0.55	Ea	5.47	14.80	20.27
4-S-tile ring 1/2"	1E@0.55	Ea	6.09	14.80	20.89
4-S-tile ring 3/4"	1E@0.55	Ea	6.18	14.80	20.98
4-S-tile ring 1"	1E@0.55	Ea	6.60	14.80	21.40
4-S-tile ring 1-1/4"	1E@0.56	Ea	6.74	15.10	21.84
4-S-tile ring 1-1/2"	1E@0.56	Ea	6.18	15.10	21.28
4-S-tile ring 2"	1E@0.56	Ea	6.30	15.10	21.40

1 gang orange single receptacle, 15 amp, in 4/S x 1-1/2" deep box, ring and plastic plate

Material	Craft@Hrs	Unit	Material Cost	Labor Cost	Installed Cost
4-S-ring flat	1E@0.55	Ea	19.03	14.80	33.83
4-S-ring 1/4"	1E@0.55	Ea	18.95	14.80	33.75
4-S-ring 1/2"	1E@0.55	Ea	18.76	14.80	33.56
4-S-ring 5/8"	1E@0.55	Ea	18.86	14.80	33.66
4-S-ring 3/4"	1E@0.55	Ea	19.03	14.80	33.83
4-S-ring 1"	1E@0.55	Ea	19.35	14.80	34.15
4-S-ring 1-1/4"	1E@0.55	Ea	19.60	14.80	34.40
4-S-tile ring 1/2"	1E@0.55	Ea	20.22	14.80	35.02
4-S-tile ring 3/4"	1E@0.55	Ea	20.31	14.80	35.11
4-S-tile ring 1"	1E@0.55	Ea	20.73	14.80	35.53
4-S-tile ring 1-1/4"	1E@0.56	Ea	20.87	15.10	35.97
4-S-tile ring 1-1/2"	1E@0.56	Ea	20.31	15.10	35.41
4-S-tile ring 2"	1E@0.56	Ea	20.43	15.10	35.53

Use these figures to estimate the cost of installing assemblies under the conditions described on pages 5 and 6. Costs listed are for each assembly installed. The crew is one electrician working at a labor cost of $26.88 per manhour. These costs include layout, material handling and normal waste. Add for the sales tax, delivery, supervision, mobilization, demobilization, cleanup, overhead and profit.

Material	Craft@Hrs	Unit	Material Cost	Labor Cost	Installed Cost
1 gang brown duplex receptacle, 15 amp, in 4/S x 1-1/2" deep box, ring and plastic plate					
4-S-ring flat	1E@0.55	Ea	3.25	14.80	18.05
4-S-ring 1/4"	1E@0.55	Ea	3.17	14.80	17.97
4-S-ring 1/2"	1E@0.55	Ea	2.98	14.80	17.78
4-S-ring 5/8"	1E@0.55	Ea	3.08	14.80	17.88
4-S-ring 3/4"	1E@0.55	Ea	3.25	14.80	18.05
4-S-ring 1"	1E@0.55	Ea	3.57	14.80	18.37
4-S-ring 1-1/4"	1E@0.55	Ea	3.82	14.80	18.62
4-S-tile ring 1/2"	1E@0.55	Ea	4.44	14.80	19.24
4-S-tile ring 3/4"	1E@0.55	Ea	4.53	14.80	19.33
4-S-tile ring 1"	1E@0.55	Ea	4.95	14.80	19.75
4-S-tile ring 1-1/4"	1E@0.56	Ea	5.09	15.10	20.19
4-S-tile ring 1-1/2"	1E@0.56	Ea	4.53	15.10	19.63
4-S-tile ring 2"	1E@0.56	Ea	4.65	15.10	19.75
1 gang ivory duplex receptacle, 15 amp, in 4/S x 1-1/2" deep box, ring and plastic plate					
4-S-ring flat	1E@0.55	Ea	3.25	14.80	18.05
4-S-ring 1/4"	1E@0.55	Ea	3.17	14.80	17.97
4-S-ring 1/2"	1E@0.55	Ea	2.98	14.80	17.78
4-S-ring 5/8"	1E@0.55	Ea	3.08	14.80	17.88
4-S-ring 3/4"	1E@0.55	Ea	3.25	14.80	18.05
4-S-ring 1"	1E@0.55	Ea	3.57	14.80	18.37
4-S-ring 1-1/4"	1E@0.55	Ea	3.82	14.80	18.62
4-S-tile ring 1/2"	1E@0.55	Ea	4.44	14.80	19.24
4-S-tile ring 3/4"	1E@0.55	Ea	4.53	14.80	19.33
4-S-tile ring 1"	1E@0.55	Ea	4.95	14.80	19.75
4-S-tile ring 1-1/4"	1E@0.56	Ea	5.09	15.10	20.19
4-S-tile ring 1-1/2"	1E@0.56	Ea	4.53	15.10	19.63
4-S-tile ring 2"	1E@0.56	Ea	4.65	15.10	19.75
1 gang orange duplex receptacle, 15 amp, in 4/S x 1-1/2" deep box, ring and plastic plate					
4-S-ring flat	1E@0.55	Ea	14.45	14.80	29.25
4-S-ring 1/4"	1E@0.55	Ea	14.37	14.80	29.17
4-S-ring 1/2"	1E@0.55	Ea	14.18	14.80	28.98
4-S-ring 5/8"	1E@0.55	Ea	14.28	14.80	29.08
4-S-ring 3/4"	1E@0.55	Ea	14.45	14.80	29.25
4-S-ring 1"	1E@0.55	Ea	14.77	14.80	29.57
4-S-ring 1-1/4"	1E@0.55	Ea	15.02	14.80	29.82
4-S-tile ring 1/2"	1E@0.55	Ea	15.64	14.80	30.44
4-S-tile ring 3/4"	1E@0.55	Ea	15.73	14.80	30.53
4-S-tile ring 1"	1E@0.55	Ea	16.15	14.80	30.95
4-S-tile ring 1-1/4"	1E@0.56	Ea	16.29	15.10	31.39
4-S-tile ring 1-1/2"	1E@0.56	Ea	15.73	15.10	30.83
4-S-tile ring 2"	1E@0.56	Ea	15.85	15.10	30.95

Use these figures to estimate the cost of installing assemblies under the conditions described on pages 5 and 6. Costs listed are for each assembly installed. The crew is one electrician working at a labor cost of $26.88 per manhour. These costs include layout, material handling and normal waste. Add for the sales tax, delivery, supervision, mobilization, demobilization, cleanup, overhead and profit.

4/S x 1-1/2" Deep, 20 Amp, 1 Gang
Single Receptacle Assemblies

Material	Craft@Hrs	Unit	Material Cost	Labor Cost	Installed Cost

1 gang brown single receptacle, 20 amp, in 4/S x 1-1/2" deep box, ring and plastic plate

Material	Craft@Hrs	Unit	Material Cost	Labor Cost	Installed Cost
4-S-ring flat	1E@0.55	Ea	5.07	14.80	19.87
4-S-ring 1/4"	1E@0.55	Ea	4.99	14.80	19.79
4-S-ring 1/2"	1E@0.55	Ea	4.80	14.80	19.60
4-S-ring 5/8"	1E@0.55	Ea	4.90	14.80	19.70
4-S-ring 3/4"	1E@0.55	Ea	5.07	14.80	19.87
4-S-ring 1"	1E@0.55	Ea	5.39	14.80	20.19
4-S-ring 1-1/4"	1E@0.55	Ea	5.64	14.80	20.44
4-S-tile ring 1/2"	1E@0.55	Ea	6.26	14.80	21.06
4-S-tile ring 3/4"	1E@0.55	Ea	6.35	14.80	21.15
4-S-ring 1"	1E@0.55	Ea	6.77	14.80	21.57
4-S-ring 1-1/4"	1E@0.56	Ea	6.91	15.10	22.01
4-S-tile ring 1-1/2"	1E@0.56	Ea	6.35	15.10	21.45
4-S-tile ring 2"	1E@0.56	Ea	6.47	15.10	21.57

1 gang ivory single receptacle, 20 amp, in 4/S x 1-1/2" deep box, ring and plastic plate

Material	Craft@Hrs	Unit	Material Cost	Labor Cost	Installed Cost
4-S-ring flat	1E@0.55	Ea	4.64	14.80	19.44
4-S-ring 1/4"	1E@0.55	Ea	4.56	14.80	19.36
4-S-ring 1/2"	1E@0.55	Ea	4.37	14.80	19.17
4-S-ring 5/8"	1E@0.55	Ea	4.47	14.80	19.27
4-S-ring 3/4"	1E@0.55	Ea	4.64	14.80	19.44
4-S-ring 1"	1E@0.55	Ea	4.96	14.80	19.76
4-S-ring 1-1/4"	1E@0.55	Ea	5.21	14.80	20.01
4-S-tile ring 1/2"	1E@0.55	Ea	5.83	14.80	20.63
4-S-tile ring 3/4"	1E@0.55	Ea	5.92	14.80	20.72
4-S-tile ring 1"	1E@0.55	Ea	6.34	14.80	21.14
4-S-tile ring 1-1/4"	1E@0.56	Ea	6.48	15.10	21.58
4-S-tile ring 1-1/2"	1E@0.56	Ea	5.92	15.10	21.02
4-S-tile ring 2"	1E@0.56	Ea	6.04	15.10	21.14

1 gang orange single receptacle, 20 amp, in 4/S x 1-1/2" deep box, ring and plastic plate

Material	Craft@Hrs	Unit	Material Cost	Labor Cost	Installed Cost
4-S-ring flat	1E@0.55	Ea	16.34	14.80	31.14
4-S-ring 1/4"	1E@0.55	Ea	16.26	14.80	31.06
4-S-ring 1/2"	1E@0.55	Ea	16.07	14.80	30.87
4-S-ring 5/8"	1E@0.55	Ea	16.17	14.80	30.97
4-S-ring 3/4"	1E@0.55	Ea	16.34	14.80	31.14
4-S-ring 1"	1E@0.55	Ea	16.66	14.80	31.46
4-S-ring 1-1/4"	1E@0.55	Ea	16.91	14.80	31.71
4-S-tile ring 1/2"	1E@0.55	Ea	17.53	14.80	32.33
4-S-tile ring 3/4"	1E@0.55	Ea	17.62	14.80	32.42
4-S-tile ring 1"	1E@0.55	Ea	18.04	14.80	32.84
4-S-tile ring 1-1/4"	1E@0.56	Ea	18.18	15.10	33.28
4-S-tile ring 1-1/2"	1E@0.56	Ea	17.62	15.10	32.72
4-S-tile ring 2"	1E@0.56	Ea	17.74	15.10	32.84

Use these figures to estimate the cost of installing assemblies under the conditions described on pages 5 and 6. Costs listed are for each assembly installed. The crew is one electrician working at a labor cost of $26.88 per manhour. These costs include layout, material handling and normal waste. Add for the sales tax, delivery, supervision, mobilization, demobilization, cleanup, overhead and profit.

Material	Craft@Hrs	Unit	Material Cost	Labor Cost	Installed Cost

1 gang brown duplex receptacle, 20 amp, in 4/S x 1-1/2" deep box, ring and plastic plate

Material	Craft@Hrs	Unit	Material Cost	Labor Cost	Installed Cost
4-S-ring flat	1E@0.55	Ea	4.64	14.80	19.44
4-S-ring 1/4"	1E@0.55	Ea	4.56	14.80	19.36
4-S-ring 1/2"	1E@0.55	Ea	4.37	14.80	19.17
4-S-ring 5/8"	1E@0.55	Ea	4.47	14.80	19.27
4-S-ring 3/4"	1E@0.55	Ea	4.64	14.80	19.44
4-S-ring 1"	1E@0.55	Ea	4.96	14.80	19.76
4-S-ring 1-1/4"	1E@0.55	Ea	5.21	14.80	20.01
4-S-tile ring 1/2"	1E@0.55	Ea	5.83	14.80	20.63
4-S-tile ring 3/4"	1E@0.55	Ea	5.92	14.80	20.72
4-S-tile ring 1"	1E@0.55	Ea	6.34	14.80	21.14
4-S-tile ring 1-1/4"	1E@0.56	Ea	6.48	15.10	21.58
4-S-tile ring 1-1/2"	1E@0.56	Ea	5.92	15.10	21.02
4-S-tile ring 2"	1E@0.56	Ea	6.04	15.10	21.14

1 gang ivory duplex receptacle, 20 amp, in 4/S x 1-1/2" deep box, ring and plastic plate

Material	Craft@Hrs	Unit	Material Cost	Labor Cost	Installed Cost
4-S-ring flat	1E@0.55	Ea	4.64	14.80	19.44
4-S-ring 1/4"	1E@0.55	Ea	4.56	14.80	19.36
4-S-ring 1/2"	1E@0.55	Ea	4.37	14.80	19.17
4-S-ring 5/8"	1E@0.55	Ea	4.47	14.80	19.27
4-S-ring 3/4"	1E@0.55	Ea	4.64	14.80	19.44
4-S-ring 1"	1E@0.55	Ea	4.96	14.80	19.76
4-S-ring 1-1/4"	1E@0.55	Ea	5.21	14.80	20.01
4-S-tile ring 1/2"	1E@0.55	Ea	5.83	14.80	20.63
4-S-tile ring 3/4"	1E@0.55	Ea	5.92	14.80	20.72
4-S-tile ring 1"	1E@0.55	Ea	6.34	14.80	21.14
4-S-tile ring 1-1/4"	1E@0.56	Ea	6.48	15.10	21.58
4-S-tile ring 1-1/2"	1E@0.56	Ea	5.92	15.10	21.02
4-S-tile ring 2"	1E@0.56	Ea	6.04	15.10	21.14

1 gang orange duplex receptacle, 20 amp, in 4/S x 1-1/2" deep box, ring and plastic plate

Material	Craft@Hrs	Unit	Material Cost	Labor Cost	Installed Cost
4-S-ring flat	1E@0.55	Ea	17.33	14.80	32.13
4-S-ring 1/4"	1E@0.55	Ea	17.25	14.80	32.05
4-S-ring 1/2"	1E@0.55	Ea	17.06	14.80	31.86
4-S-ring 5/8"	1E@0.55	Ea	17.16	14.80	31.96
4-S-ring 3/4"	1E@0.55	Ea	17.33	14.80	32.13
4-S-ring 1"	1E@0.55	Ea	17.65	14.80	32.45
4-S-ring 1-1/4"	1E@0.55	Ea	17.90	14.80	32.70
4-S-tile ring 1/2"	1E@0.55	Ea	18.52	14.80	33.32
4-S-tile ring 3/4"	1E@0.55	Ea	18.61	14.80	33.41
4-S-tile ring 1"	1E@0.55	Ea	19.03	14.80	33.83
4-S-tile ring 1-1/4"	1E@0.56	Ea	19.17	15.10	34.27
4-S-tile ring 1-1/2"	1E@0.56	Ea	18.61	15.10	33.71
4-S-tile ring 2"	1E@0.56	Ea	18.73	15.10	33.83

Use these figures to estimate the cost of installing assemblies under the conditions described on pages 5 and 6. Costs listed are for each assembly installed. The crew is one electrician working at a labor cost of $26.88 per manhour. These costs include layout, material handling and normal waste. Add for the sales tax, delivery, supervision, mobilization, demobilization, cleanup, overhead and profit.

Troffler Fluorescent Assemblies

Material	Craft@Hrs	Unit	Material Cost	Labor Cost	Installed Cost

Framed troffer lay-in T-bar fluorescent fixtures with lamps, 5' aluminum flex whip, 3-#12 THHN solid copper wire and wire connectors

24" wide steel frame troffer lay-in T-bar fixtures, 5' flex whip, 3-#12 THHN solid copper wire

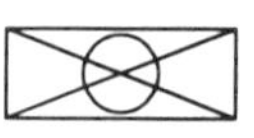

Material	Craft@Hrs	Unit	Material Cost	Labor Cost	Installed Cost
48" 2 lamp	1E@0.75	Ea	71.78	20.20	91.98
48" 3 lamp	1E@0.80	Ea	80.37	21.50	101.87
48" 4 lamp	1E@0.85	Ea	80.06	22.80	102.86
48" 2 lamp energy saver	1E@0.75	Ea	82.88	20.20	103.08
48" 3 lamp energy saver	1E@0.80	Ea	100.57	21.50	122.07
48" 4 lamp energy saver	1E@0.85	Ea	99.16	22.80	121.96

Regressed 24" wide steel frame troffer T-bar fixtures, 5' flex whip, 3-#12 THHN solid copper wire

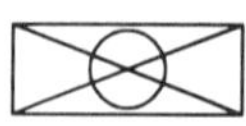

Material	Craft@Hrs	Unit	Material Cost	Labor Cost	Installed Cost
48" 2 lamp	1E@0.75	Ea	73.58	20.20	93.78
48" 3 lamp	1E@0.80	Ea	82.17	21.50	103.67
48" 4 lamp	1E@0.85	Ea	81.76	22.80	104.56
48" 2 lamp energy saver	1E@0.75	Ea	84.68	20.20	104.88
48" 3 lamp energy saver	1E@0.80	Ea	102.77	21.50	124.27
48" 4 lamp energy saver	1E@0.85	Ea	100.96	22.80	123.76

24" wide aluminum frame troffer T-bar fixtures, 5' flex whip, 3-#12 THHN solid copper wire

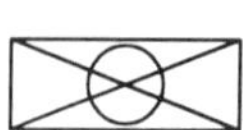

Material	Craft@Hrs	Unit	Material Cost	Labor Cost	Installed Cost
48" 2 lamp	1E@0.75	Ea	73.68	20.20	93.88
48" 3 lamp	1E@0.80	Ea	83.17	21.50	104.67
48" 4 lamp	1E@0.85	Ea	88.86	22.80	111.66
48" 2 lamp energy saver	1E@0.75	Ea	87.68	20.20	107.88
48" 3 lamp energy saver	1E@0.80	Ea	104.77	21.50	126.27
48" 4 lamp energy saver	1E@0.85	Ea	103.36	22.80	126.16

Regressed 24" wide aluminum frame troffer T-bar fixtures, 5' flex whip, 3-#12 THHN solid copper wire

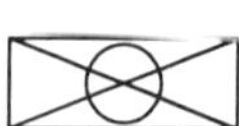

Material	Craft@Hrs	Unit	Material Cost	Labor Cost	Installed Cost
48" 2 lamp	1E@0.75	Ea	78.48	20.20	98.68
48" 3 lamp	1E@0.80	Ea	88.27	21.50	109.77
48" 4 lamp	1E@0.85	Ea	88.86	22.80	111.66
48" 2 lamp energy saver	1E@0.75	Ea	89.58	20.20	109.78
48" 3 lamp energy saver	1E@0.80	Ea	108.77	21.50	130.27
48" 4 lamp energy saver	1E@0.85	Ea	107.36	22.80	130.16

Use these figures to estimate the cost of installing assemblies under the conditions described on pages 5 and 6. Costs listed are for each assembly installed. The crew is one electrician working at a labor cost of $26.88 per manhour. These costs include layout, material handling and normal waste. Add for the sales tax, delivery, supervision, mobilization, demobilization, cleanup, overhead and profit.

Simple Conversion Table

For insulated conductors rated 0-2000 volts 60 degrees C (140 degree F), not more than three conductors in a raceway or cable or earth (direct burial) on ambient temperature of 30 degree C (86 degree F)

| 60 degree C | | 90 degree C | |
Copper	Aluminum	Copper	Aluminum
# 10	# 8	# 10	# 8
# 8	# 6	# 8	# 6
# 6	# 4	# 6	# 4
# 4	# 2	# 4	# 2
# 2	# 1/0	# 2	# 1/0
# 1	# 2/0	# 1	# 2/0
# 1/0	# 3/0	# 1/0	# 3/0
# 2/0	# 4/0	# 2/0	# 4/0
# 3/0	# 250 kcmil	# 3/0	# 250 kcmil
# 4/0	# 350 kcmil	# 4/0	# 350 kcmil
# 250 kcmil	# 400 kcmil	# 250 kcmil	# 400 kcmil
# 300 kcmil	# 500 kcmil	# 300 kcmil	# 500 kcmil
# 350 kcmil	# 500 kcmil	# 350 kcmil	# 500 kcmil
# 400 kcmil	# 600 kcmil	# 400 kcmil	# 600 kcmil
# 500 kcmil	# 750 kcmil	# 500 kcmil	# 750 kcmil
# 600 kcmil	# 900 kcmil	# 600 kcmil	# 900 kcmil
# 750 kcmil	# 1250 kcmil	# 750 kcmil	# 1250 kcmil

See NEC Tables 310-16, 310-17, 310-18 for conductor ampacities for general wiring.

Electrical Abbreviations

A	Ammeter	OSCG	Oscillograph
AH	Ampere-hour meter	PH	Phase meter
CMA	Contact-making (or breaking) ammeter	PI	Position indicator
CMC	Contact-making (or breaking) clock	PF	Power-factor meter
CMV	Contact-making (or breaking) voltmeter	RD	Recording demand meter
CRO	Oscilloscope or cathode-ray oscillograph	REC	Recording
D	Demand meter	RF	Reactive-factor meter
DB	Decibel meter	S	Synchroscope
DTR	Demand-totalizing relay	TLM	Telemeter
F	Frequency meter	T	Temperature meter
G	Galvanometer	TT	Total time
GD	Ground detector	VH	Varhour meter
I	Indicating	V	Voltmeter
M	Integrating	VA	Voltammeter
UA	Microammeter	VAR	Varmeter
MA	Milliammeter	VI	Volume indicator
N	Noise meter	VU	Standard volume indicator
OHM	Ohmmeter	W	Wattmeter
OP	Oil pressure	WH	Watthour meter

National Estimator Quick Start

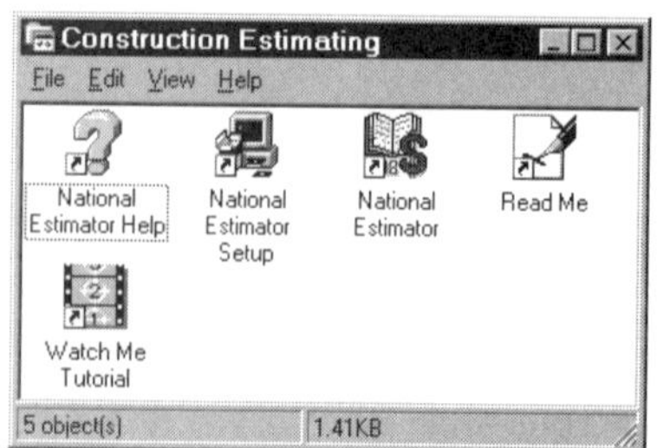

Construction Estimating program group

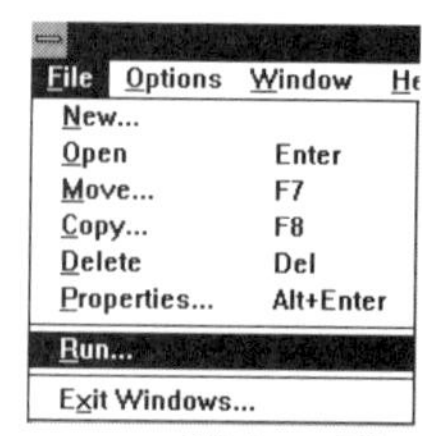

*Click on File.
Click on Run in Windows 3.1*

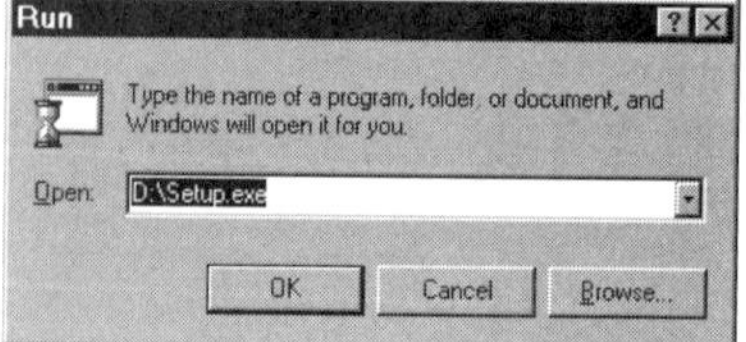

*Type D:\SETUP
in Windows 95 or NT*

National Estimator is a construction cost estimating program with all the cost estimates in this book. You can page through these costs one screen at a time or use the electronic index to search by keyword for exactly the information you need. When you find what you're looking for, split the screen in two so your estimate is on the bottom half of the screen and the database is on the top half. Then copy and paste into your estimate. Type the estimated quantity. The program extends prices and totals columns automatically. Add your overhead and profit as a percentage and then print the estimate. It's as easy as that.

If you're a Windows user, you know most of the National Estimator program already. But even computer novices can be writing labor and material cost estimates in an hour or two. The first step is installing the National Estimator disk.

The installation program creates a Construction Estimating program group and puts several icons in that group. If you have trouble installing National Estimator, call Craftsman tech support at 760-438-7828. If your computer can't use the CD, exchange it for National Estimator on 3½" disk. Send the CD to Craftsman, Box 6500, Carlsbad, CA 92018. There's no charge for this exchange.

Installing National Estimator

Put the National Estimator disk in your CD drive (such as D:). Start Windows.

In Windows 3.1 or 3.11
go to the Program Manager.

1. Click on File
2. Click on Run . . .
3. Type D:\SETUP
4. Press [Enter ↵]

In Windows 95 or NT

1. Click on ⊞ Start
2. Click on Run . . .
3. Type D:\SETUP
4. Press [Enter ↵]

We recommend accepting the installation defaults:

1. National directory
2. "Typical" setup

When installation is complete, click on *Finish*.

Uninstalling National Estimator

In Windows 3.1 or 3.11: Select the Construction Estimating program group. Then click on the Uninstall icon.

In Windows 95 or NT: Click on *Start*, *Settings*, *Control Panel*. Double click on *Add/Remove Programs*. Then click on *National Estimator* and *Add/Remove*.

Learn the Easy Way!

Estimating with Jay and Julie

Watch Me icon

The **Watch Me** tutorial is an interactive video guide to National Estimator. Let Jay and Julie help you get started using the National Estimator program. Sit back and relax and let the video run — or click on buttons and menus to jump from topic to topic inside the video. Exit any time you want. Then go back to Watch Me later to brush up on some topic. Watch Me takes up very little space on your hard drive and is part of the "Typical" installation. Watch Me requires 8Mb of RAM, a Windows-compatible sound system and at least a 33 MHz 486 computer. Click on the Watch Me icon to begin Watch Me.

Using National Estimator

National Estimator icon

National Estimator begins when you click on the National Estimator icon (in Windows 3.1 or 3.11) or click on *Start*, *Programs*, the *Construction Estimating* group and then *National Estimator* (in Windows 95 or NT).

On the title bar at the top of the screen you see the program name, National Estimator, and [Estimate1], showing that Estimate1 is on the screen. Let's take a closer look at the other information at the top of your screen.

Your estimating form ready to begin the first estimate.

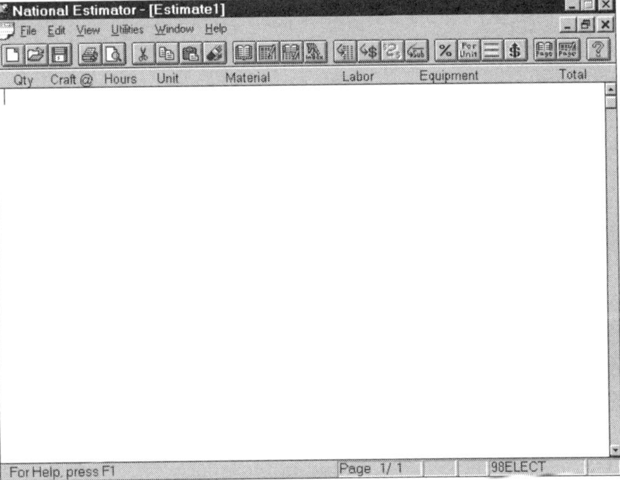

The Menu Bar

Below the title bar you see the menu bar: Every option in National Estimator is available on the menu bar. Click with your left mouse button on any item on the menu to open a list of available commands.

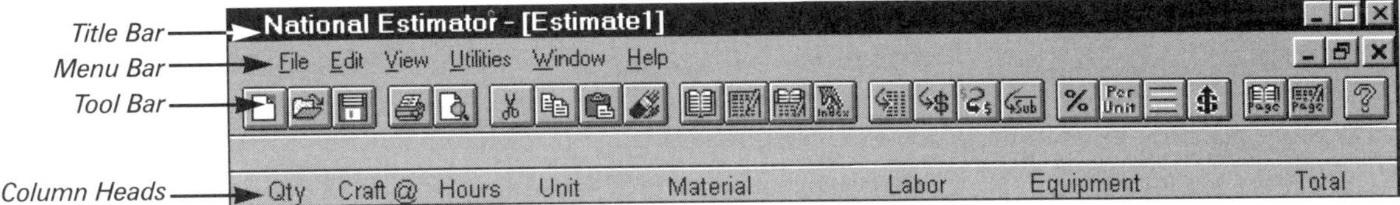

Buttons on the Tool Bar

Below the menu bar you see 24 buttons that make up the tool bar. The options you use most in National Estimator are only a mouse click away on the tool bar.

Column Headings

Below the tool bar you'll see column headings for your estimate form:

Qty for quantity

Craft@Hours for craft (the crew doing the work) and manhours (to complete the task)

Unit for unit of measure, such as per linear foot or per square foot

Material for material cost

Labor for labor cost

Equipment for equipment cost

Total for the total of all cost columns

The Status Bar

The bottom line on your screen is the status bar. Here you'll find helpful information about the choices available. Notice "Page 1/1" near the right end of the status line. That's a clue that you're looking at page 1 of a one-page estimate. When you see PICT at the right of the status bar, a picture is available in the costbook. Click on PICT to show the picture. Click on PICT again to hide the picture.

Check the status bar occasionally for helpful tips and explanations of what you see on screen.

Beginning an Estimate

Let's start by putting a heading on this estimate.

The Blinking Cursor (insert point)

Mouse Pointer

Mouse Pointer

1. Press [Enter ↵] once to space down one line.

2. Press [Tab ↹] four times (or hold the space bar down) to move the Blinking Cursor (the insert point) near the middle of the line.

3. Type "Estimate One" and press [Enter ↵]. That's the title of this estimate, "Estimate One."

4. Press [Enter ↵] again to move the cursor down a line. That opens up a little space below the title.

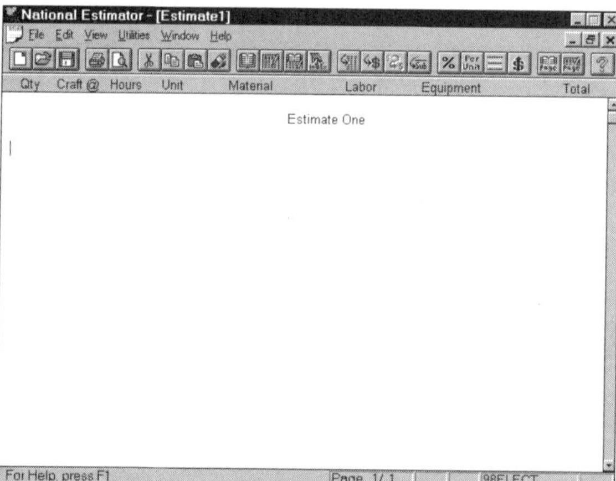

Begin by putting a title on your estimate, such as "Estimate One."

The Costbook

Let's leave your estimating form for a moment to take a look at estimates stored in the costbook. To switch to the costbook, either:

■ Click on the (Open Costbook) button, -or-

■ Click on View on the menu bar. Then click on Costbook Window, -or-

- Tap the [Alt] key, tap the letter V (for View) and tap the letter C (for Costbook Window), -or-

- Press [Esc]. Press the [↓] arrow key to highlight Costbook Window. Then press [Enter ↵].

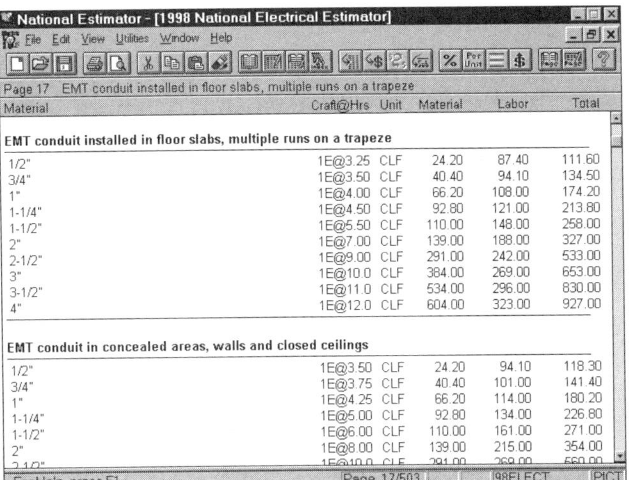

The costbook window has the entire National Electrical Estimator.

The entire 1998 National Electrical Estimator is available in the Costbook Window. Notice the words *Page 17 EMT conduit* at the left side of the screen just below the tool bar. That's your clue that the conduit section of page 17 is on the screen. To turn to the next page, either:

- Press [PgDn] (with Num Lock off), -or-

- Click on the lower half of the scroll bar at the right edge of the screen.

To move down one line at a time, either:

- Press the [↓] arrow key (with Num Lock off), -or-

- Click on the arrow on the down scroll bar at the lower right corner of the screen.

Press [PgDn] about 1,400 times and you'll page through the entire National Electrical Estimator. Obviously, there's a better way. To turn quickly to any page, either:

- Click on the [icon] (Turn to Costbook Page) button near the right end of the tool bar, -or-

- Click on View on the menu bar. Then click on Turn to Costbook Page, -or-

- Tap the [Alt] key, tap the letter V (for View) and tap the letter T (for Turn to Costbook Page), -or-

- Press [Esc]. Press the [↓] arrow key to highlight Turn to Costbook Page. Then press [Enter ↵].

Type the page number you want to see.

Type the number of the page you want to see and press Enter. National Estimator will turn to the top of the page you requested.

An Even Better Way

Find the small square in the slide bar at the right side of the Costbook Window. Click and hold on that square while rolling the mouse up or down. Keep dragging the square until you see the page you want in the Page: box. Release the mouse button to turn to the top of that page.

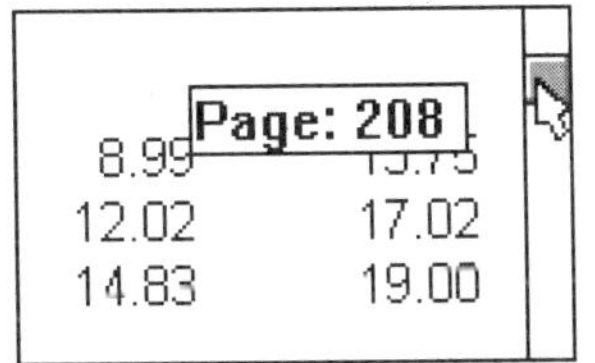

Drag the square to see any page.

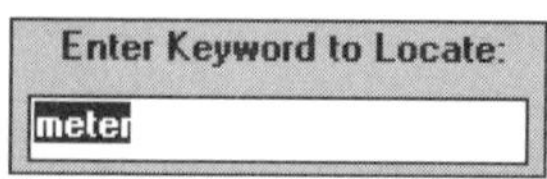

10,000 AIC DC breakers: 286
10,000 amp interrupt capacity: 278
120 volt fixtures: 173, 177
120 volt smoke detectors: 343
120/277 volt switches: 219
14,000 AIC breakers: 284
2" range expansion PVC couplings: 39
208 volt fixtures: 173, 177
22,000 AIC DC breakers: 286
240 volt fixtures: 173, 177
240 volt safety switches: 247-249

*Use the electronic index
to find cost estimates
for any item.*

Enter Keyword to Locate:

meter

Enter keyword to locate.

Meter centers: 290-293
Meter sockets: 244, 289
Mobilization: 7
Modulating valves, hookup: 352
Mogul aluminum conduit bodies: 65
Moisture-resistant thermoplastic: 88
Momentary contact switches: 208, 219
Momentary control switch: 354

The index jumps to Meter centers.

*If costs appear on several pages,
click on the page you prefer.*

Costs for meter sockets on page 289.

A Still Better Way: Keyword Search

To find any cost estimate in seconds, search by keyword in the index. To go to the index, either:

❚ Click on the (Index) button near the center of the tool bar, -or-

❚ Press Esc. Press Enter↵, -or-

❚ Tap the Alt key, tap the letter V (for View) and press Enter↵, -or-

❚ Click on View on the menu bar. Then press Enter↵.

Notice that the cursor is blinking in the Enter Keyword to Locate box at the right of the screen. Obviously, the index is ready to begin a search.

Your First Estimate

Suppose we're estimating the cost of surface mounted electric meter sockets. Let's put the index to work with a search for meter sockets. In the box under Enter Keyword to Locate, type *meter*. The index jumps to the heading *Meter*.

The second item is *Meter sockets: 244, 289*. Either:

❚ Click once on that line and press Enter↵, -or-

❚ Double click on that line, -or-

❚ Press Tab⇆ and the ↓ arrow key to move the highlight to *Meter sockets: 244, 289*. Then press Enter↵.

To select the page you want to see (page 289 in this case), either:

❚ Click on the number 289, -or-

❚ Press Tab⇆ to highlight 289. Then press Enter↵.

National Estimator turns to the top of page 289. See the example below. Notice that estimates at the top of this page are for 120/240 volt meter sockets.

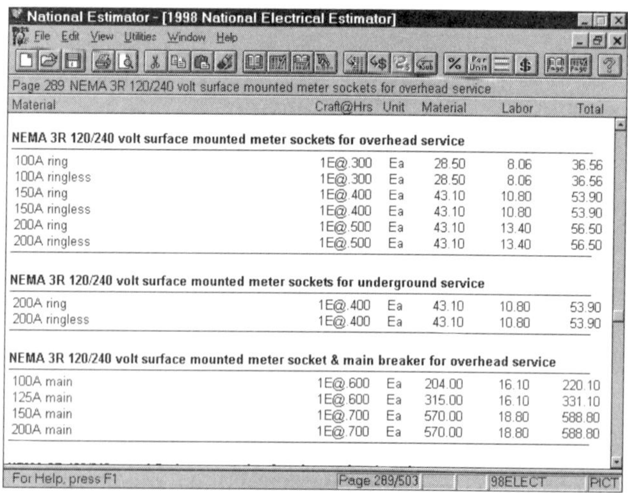

NEMA 3R 120/240 volt surface mounted meter sockets for overhead service					
100A ring	1E@.300	Ea	28.50	8.06	36.56
100A ringless	1E@.300	Ea	28.50	8.06	36.56
150A ring	1E@.400	Ea	43.10	10.80	53.90
150A ringless	1E@.400	Ea	43.10	10.80	53.90
200A ring	1E@.500	Ea	43.10	13.40	56.50
200A ringless	1E@.500	Ea	43.10	13.40	56.50
NEMA 3R 120/240 volt surface mounted meter sockets for underground service					
200A ring	1E@.400	Ea	43.10	10.80	53.90
200A ringless	1E@.400	Ea	43.10	10.80	53.90
NEMA 3R 120/240 volt surface mounted meter socket & main breaker for overhead service					
100A main	1E@.600	Ea	204.00	16.10	220.10
125A main	1E@.600	Ea	315.00	16.10	331.10
150A main	1E@.700	Ea	570.00	18.80	588.80
200A main	1E@.700	Ea	570.00	18.80	588.80

Splitting the Screen

Most of the time you'll want to see what's in both the costbook and your estimate. To split the screen into two halves, either:

- Click on the ▣ (Split Window) button near the center of the tool bar, -or-
- Press [Esc] and the [↓] arrow key to move the selection bar to Split Window, -or-
- Tap the [Alt] key, tap the letter V (for View), tap the letter S (for Split Window), -or-
- Click on View on the menu bar. Then click on Split Window and your screen should look like the example below.

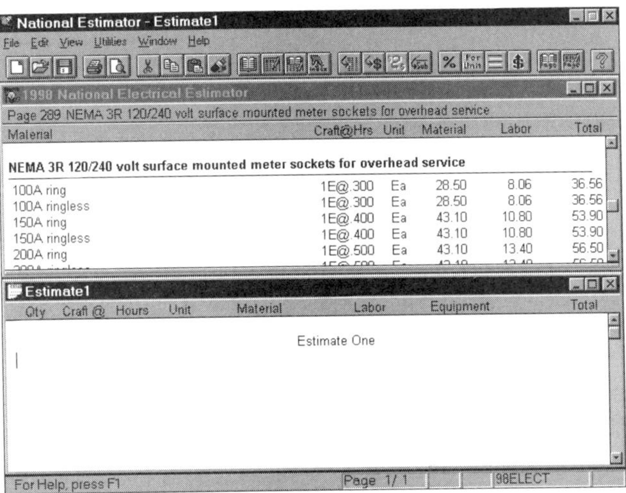

The split window: Costbook above and estimate below.

Notice the eight lines of the costbook are at the top of the screen and your estimate is at the bottom. You should recognize "Estimate One." It's your title for the estimate. Column headings are at the top of the costbook and across the middle of the screen (for your estimate).

To Switch from Window to Window

- Click in the window of your choice, -or-
- Hold the [Ctrl] key down and press [Tab↹].

Notice that a window title bar turns dark when that window is selected. The selected window is where keystrokes appear as you type.

Copying Costs to Your Estimate

Next, we'll estimate the cost of three meter sockets. Click the ▣ (Split Window) button on the tool bar to be sure you're in the split window. Click anywhere in the costbook (the top half of your screen). Then press the [↓] arrow key until the cursor is on the line:

100A ring	1E@.300	Ea	28.50	8.06	36.56

To copy this line to your estimate:

1. Click on the line.
2. Click the ▣ (Copy) button.
3. Click on the ▣ (Paste) button to open the Enter Cost Information dialog box.

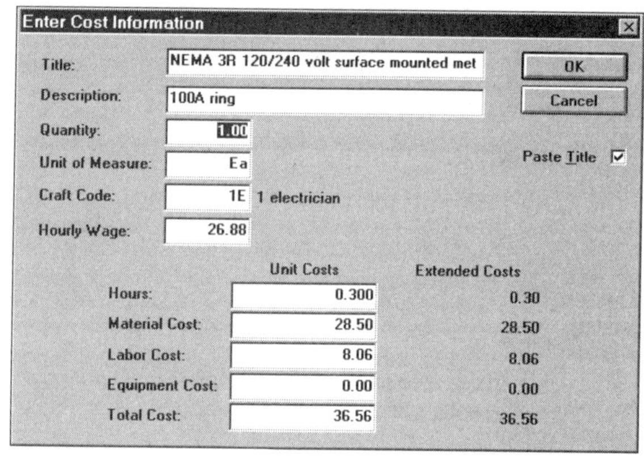 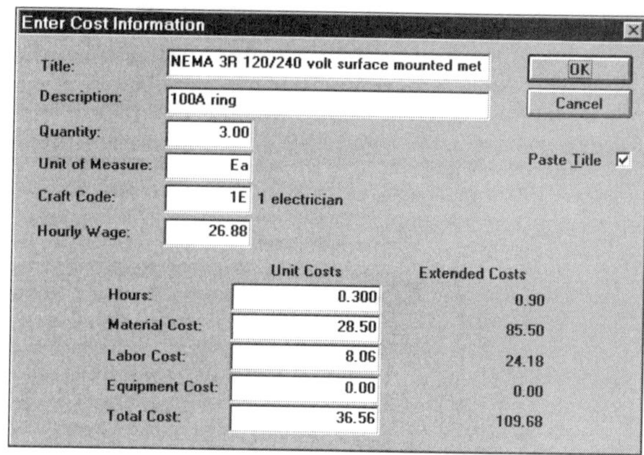

Use the Enter Cost Information dialog box to copy or change costs.

Costs for 3 meter sockets (extended costs) are on the right.

Notice that the blinking cursor is in the Quantity box:

1. Type a quantity of 3 because the job has 3 meter sockets.

2. Press `Tab` and check the estimate for accuracy.

3. Notice that the column headed Unit Costs shows costs per unit, per "Ea" (each) in this case.

4. The column headed Extended Costs shows costs for 3 sockets.

5. The lines opposite Title and Description show what's getting installed. You can change the words in either of these boxes. Just click on what you want to change and start typing or deleting.

6. You can also change any numbers in the Unit Cost column. Just click and start typing.

7. When the words and costs are exactly right, press `Enter` or click on OK to copy these figures to the end of your estimate.

The new line at the bottom of your estimate shows:

Extended costs for three meter sockets as they appear on your estimate form.

NEMA 3R 120/240 volt surface mounted meter sockets for overhead service
100A ring

| 3.00 | 1E@.9000 | Ea | 85.50 | 24.18 | 0.00 | 109.68 |

3.00 is the quantity of meter sockets

1E is the recommended crew, an electrician

@.9000 shows the manhours required for the work

Ea is the unit of measure, each in this case

85.50 is the material cost (the meter sockets)

24.18 is labor cost for the job

0.00 shows there is no equipment cost

109.68 is the total of material, labor and equipment columns

Copy Anything to Anywhere in Your Estimate

Anything in the costbook can be copied to your estimate. Just click on the line (or select the words) you want to copy and press the `F8` key. It's copied to the last line of your estimating form. If your selection includes costs, you'll have a chance to enter the quantity. To copy to the *middle* of your estimate:

1. Select what you want to copy.
2. Click on the ▣ (Copy) button.
3. Click in the estimate where you want to paste.
4. Click on the ▣ (Paste) button.

Changing Wage Rates

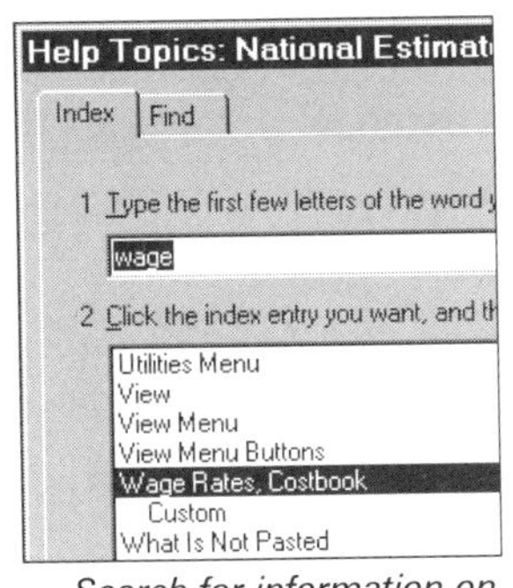

Search for information on setting wage rates.

The labor cost in the example above is based on an electrician working at a cost of $26.88 per hour. (See page 5 in the National Electrical Estimator for crew rates used in the costbook.) Suppose $26.88 per hour isn't right for your estimate. What then? No problem! It's easy to use your own wage rate for any crew or even make up your own crew codes. To get more information on setting wage rates, press F1. At National Estimator Help Contents, click on the Search button. Type *wage* then double click on *"custom"* under Wage Rates. To return to your estimate, click on File on the National Estimator Help menu bar. Then click on Exit.

Changing Cost Estimates

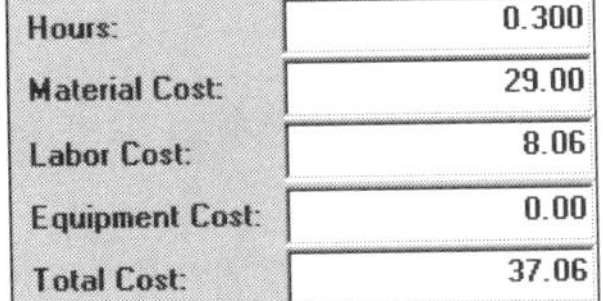

Change the material cost to 29.00.

With Num Lock off, use the ↑ or ↓ arrow key to move the cursor to the line you want to change (or click on that line). In this case, move to the line that begins with a quantity of 3. To open the Enter Cost Information Dialog box, either:

▌ Press Enter↵, -or-

▌ Click on the 🔧 (Change Cost) button on the tool bar.

To make a change, either

▌ Click on what you want to change, -or-

▌ Press Tab⇆ until the cursor advances to what you want to change.

Then type the correct figure. In this case, change the material cost to 29.00.

Press Tab⇆ and check the Extended Costs column. If it looks OK, press Enter↵ and the change is made on your estimating form.

Changing Text (Descriptions)

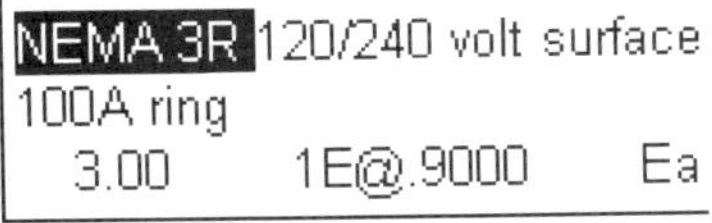

To select, click and hold the mouse button while dragging the mouse.

Click on the 🖽 (Estimate Window) button on the tool bar to be sure you're in the estimate. With Num Lock off, use the ↑ or ↓ arrow key or click the mouse button to put the cursor where you want to make a change. In this case, we're going to make a change on the line that begins "NEMA 3R . . ."

To make a change, click where the change is needed. Then either:

▌ Press the Del or ←Bksp key to erase what needs deleting, -or-

▌ Select what needs deleting and click on the ✂ (Cut) button on the tool bar.

▌ Type what needs to be added.

In this case, click just before "NEMA 3R." Then hold the left mouse button down and drag the mouse to the right until you've put a dark background behind "NEMA 3R." The dark background shows that this word is selected and ready for editing.

Press the Del key, or click on the ✂ (Cut) button on the tool bar, and the selection is cut from the estimate. If that's not what you wanted, click on the 🖌 (Undo) button and "NEMA 3R" is back again.

Adding Text (Descriptions)

Some of your estimates will require descriptions (text) and costs that can't be found in the National Electrical Estimator. What then? With National Estimator it's easy to add descriptions and costs of your choice anywhere in the estimate. For practice, let's add an estimate for four meter panels to Estimate One.

Click on the ▓ (Estimate Window) button to be sure the estimate window is maximized. We can add lines anywhere on the estimate. But in this case, let's make the addition at the end. Press the ⬇ arrow key to move the cursor down until it's just above the horizontal line that separates estimate detail lines from estimate totals. To open a blank line, either:

- Press [Enter ↵], -or-

- Click on the ▓ (Insert Text) button on the tool bar, -or-

- Click on Edit on the menu bar. Then click on Insert a Text Line.

Type "Meters with panel" and press [Enter ↵].

Adding a Cost Estimate Line

Now let's add a cost for "Meters with panel" to your estimate. Begin by opening the Enter Cost Information dialog box. Either:

- Click on the ▓ (Insert Cost) button on the tool bar, -or-

- Click on Edit on the menu bar. Then click on Insert a Cost Line.

1. The cursor is in the Quantity box. Type the number of units (4 in this case) and press [Tab ⇆].

2. The cursor moves to the next box, Unit of Measure.

3. In the Unit of Measure box, type *Each* and press [Tab ⇆].

4. Press [Tab ⇆] twice to leave the Craft Code blank and Hourly Wage at zero.

5. Since these meters will be installed by the supplier, there's no material, labor or equipment cost. So press [Tab ⇆] four times to skip over the Hours, Material Cost, Labor Cost and Equipment Cost boxes.

6. In the Total Cost box, type 20.00. That's the cost per meter quoted by your supplier.

7. Press [Tab ⇆] once more to advance to OK.

8. Press [Enter ↵] and the cost of four meters is written to your estimate.

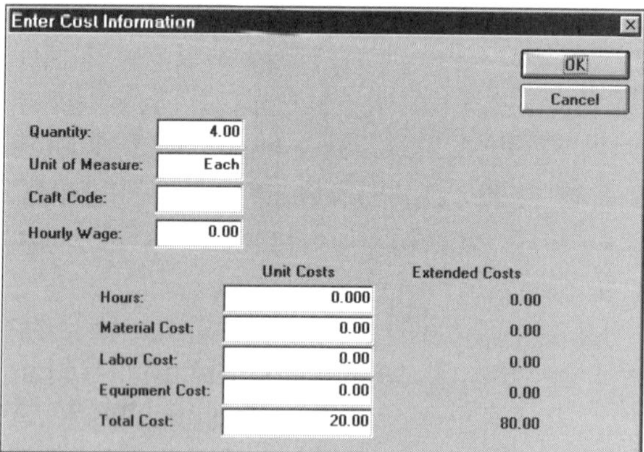

Unit and extended costs for four meter panels.

Note: The sum of material, labor and equipment costs appears automatically in the Total Cost box. If there's no cost entered in the Material Cost, Labor Cost or Equipment Cost boxes (such as for a subcontracted item), you can enter any figure in the Total Cost box.

120/240 volt surface mount
100A ring
   3.00   1E@.9000
Meters with panel

Adding "Meters with panel"

Adding Lines to the Costbook

Add lines or make changes in the costbook the same way you add lines or make changes in an estimate. The additions and changes you make become part of the user costbook. For more information on user costbooks, press F1. Click on Search. Type "user" and press Enter ←.

Adding Tax

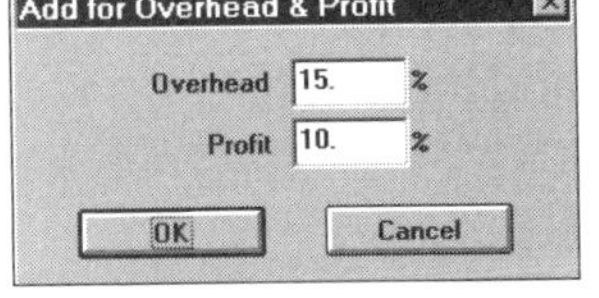

Type the tax rate that applies.

To include sales tax in your estimate:

1. Click on Edit.

2. Click on Tax Rates.

3. Type the tax rate in the appropriate box.

4. Press Tab⇥ to advance to the next box.

5. Press Enter ← or click on OK when done.

In this case, the tax rate is 7.25% on materials only. Tax will appear as the last line of the estimate.

Adding Overhead and Profit

Set markup percentages in the Add for Overhead & Profit dialog box. To open the box, either:

- Click on the [$] (Markup) button on the tool bar, -or-
- Click on Edit on the menu bar. Then click on Markup.

Type the percentages you want to add for overhead. For this estimate:

1. Type 15 on the Overhead line.

2. Press Tab⇥ to advance to Profit.

3. Type 10 on the Profit line.

4. Press Enter ←.

Adding overhead & profit.

Markup percentages can be changed at any time. Just reopen the Add for Overhead & Profit dialog box and type the correct figure.

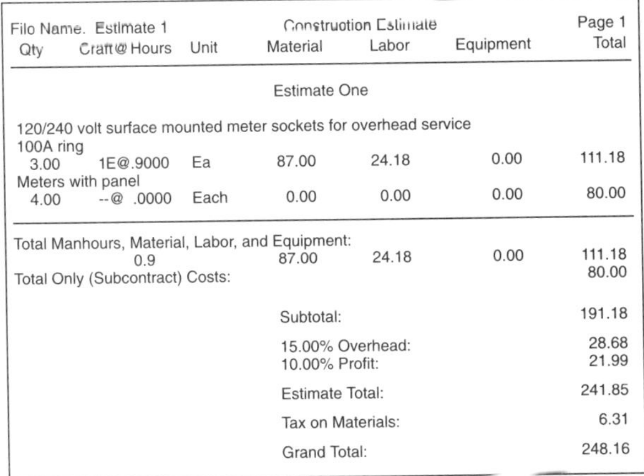

| File Name: Estimate 1 | | | Construction Estimate | | | Page 1 |
| Qty | Craft @ Hours | Unit | Material | Labor | Equipment | Total |

Estimate One

120/240 volt surface mounted meter sockets for overhead service
100A ring

| 3.00 | 1E@.9000 | Ea | 87.00 | 24.18 | 0.00 | 111.18 |

Meters with panel

| 4.00 | --@ .0000 | Each | 0.00 | 0.00 | 0.00 | 80.00 |

Total Manhours, Material, Labor, and Equipment:

| 0.9 | | | 87.00 | 24.18 | 0.00 | 111.18 |

Total Only (Subcontract) Costs: 80.00

	Subtotal:	191.18
	15.00% Overhead:	28.68
	10.00% Profit:	21.99
	Estimate Total:	241.85
	Tax on Materials:	6.31
	Grand Total:	248.16

A preview of Estimate One.

Preview Your Estimate

You can display an estimate on screen just the way it will look when printed on paper. To preview your estimate, either:

- Click on the [🔍] (Print Preview) button on the tool bar, -or-
- Click on File on the menu bar. Then click on Print Preview.

In print preview:

- Click on Next Page or Prev Page to turn pages.
- Click on Two Page to see two estimate pages side by side.
- Click on Zoom In to get a closer look.
- Click on Close when you've seen enough.

Use buttons on Print Preview to see your estimate as it will look when printed.

501

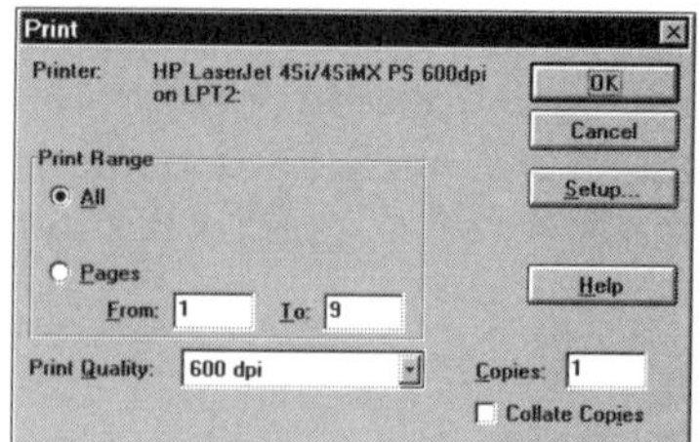

Options available depend on the printer you're using.

Printing Your Estimate

When you're ready to print the estimate, either:

▮ Click on the (Print) button on the tool bar, -or-

▮ Click on File on the menu bar. Then click on Print, -or-

▮ Hold the [Ctrl] key down and type the letter P.

Press [Enter ◄┘] or click on OK to begin printing.

Save Your Estimate to Disk

To store your estimate on the hard disk where it can be re-opened and changed at any time, either:

▮ Click on the (Save) button on the tool bar, -or-

▮ Click on File on the menu bar. Then click on Save, -or-

▮ Hold the [Ctrl] key down and type the letter S.

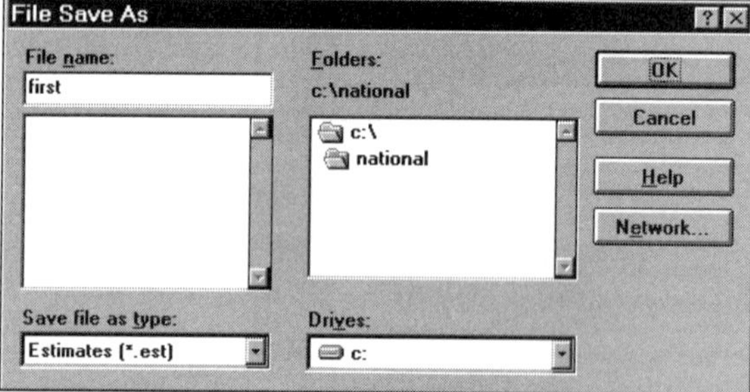

Type the estimate name in the File Name box to assign a file name.

The cursor is in the File Name box. Type the name you want to give this estimate, such as *First*. The name can be up to eight letters and numbers, but don't use symbols or spaces. Press [Enter ◄┘] or click on OK and the estimate is written to disk.

Opening Other Costbooks

Many construction cost estimating databases are available for the National Estimator program. For example, a National Estimator costbook comes with each of these estimating manuals:

National Construction Estimator

National Electrical Estimator

National Plumbing & HVAC Estimator

National Painting Cost Estimator

National Renovation & Insurance Repair Estimator

National Repair & Remodeling Estimator

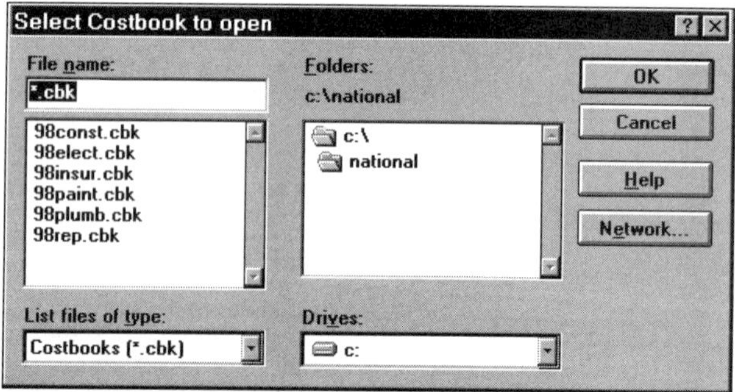

Open the costbook of your choice.

If you own several of these manuals and have installed the database that comes with each, you can open several costbooks at the same time. To open another costbook:

▋ Click on File.

▋ Click on Open Costbook.

▋ Be sure the drive and directory are correct, usually *c:\national*.

▋ Double click on the costbook of your choice.

To see a list of the costbooks open, click on Window. The name of the current estimate or costbook will be checked. Click on any other costbook name to display that costbook. Click on Window, then click on Tile to display all open costbooks and estimates.

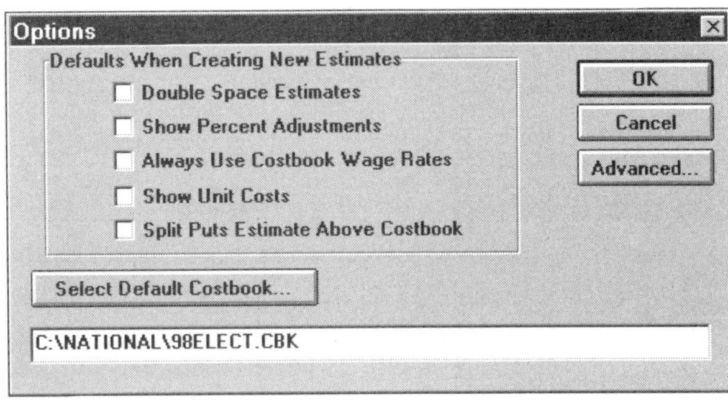

Click to switch costbooks.

Select Your Default Costbook

Your default costbook opens automatically every time you begin using National Estimator. Save time by making the default costbook the one you use most.

To change your default costbook, click on Utilities on the menu bar. Then click on Options. Next, click on Select Default Costbook. Click on the costbook of your choice. Click on OK. Then click on OK again.

Select the default costbook.

Use National Estimator Help

That completes the basics of National Estimator. You've learned enough to complete most estimates. When you need more information about the fine points, use National Estimator Help. Click on the [?] (Help) button to see Help Contents. Then click on the menu selection of your choice. To print 28 pages of instructions for National Estimator, go to Help Contents. Click on Print All Topics (at the bottom of Help Contents). Click on File on the Help menu bar. Then click on Print Topic. Click on OK.

Click on File and then Print Topic.

Index

Practical References for Builders

Rough Framing Carpentry

If you'd like to make good money working outdoors as a framer, this is the book for you. Here you'll find shortcuts to laying out studs; speed cutting blocks, trimmers and plates by eye; quickly building and blocking rake walls; installing ceiling backing, ceiling joists, and truss joists; cutting and assembling hip trusses and California fills; arches and drop ceilings — all with production line procedures that save you time and help you make more money. Over 100 on-the-job photos of how to do it right and what can go wrong. **304 pages, 8¹/2 x 11, $26.50**

Commercial Electrical Wiring

Make the transition from residential to commercial electrical work. Here are wiring methods, spec reading tips, load calculations and everything you need for making the transition to commercial work: commercial construction documents, load calculations, electric services, transformers, overcurrent protection, wiring methods, raceway, boxes and fittings, wiring devices, conductors, electric motors, relays and motor controllers, special occupancies, and safety requirements. This book is written to help any electrician break into the lucrative field of commercial electrical work. **320 pages, 8¹/2 x 11, $27.50**

Residential Electrical Design Revised

If you've ever had to draw up an electrical plan for an addition, or add corrections to an existing plan, you know how complicated it can get. And how many electrical plans - no matter how well designed - fit the reality of what the homeowner wants? Here you'll find everything you need to know about blueprints, what the *NEC* requires, how to size electric service, calculate and size loads and conductors, install ground-fault circuit interrupters, ground service entrances, and recommeded wiring methods, It covers branch circuite layout, how to analyze existing lighting layouts and install outdoor lighting, methods for remote-control switching, residential HVAC systems and controls, and more. **256 pages, 8¹/2 x 11, $22.50**

Electrical Blueprint Reading Revised

Shows how to read and interpret electrical drawings, wiring diagrams, and specifications for constructing electrical systems. Shows how a typical lighting and power layout would appear on a plan, and explains what to do to execute the plan. Describes how to use a panelboard or heating schedule, and includes typical electrical specifications. **208 pages, 8¹/2 x 11, $18.00**

CD Estimator

If your computer has *Windows*™ and a CD-ROM drive, CD Estimator puts at your fingertips 85,000 construction costs for new construction, remodeling, renovation & insurance repair, electrical, plumbing, HVAC and painting. You'll also have the National Estimator program — a stand-alone estimating program for *Windows*™ that *Remodeling* magazine called a "computer wiz." Quarterly cost updates are available at no charge on the Internet. To help you create professional-looking estimates, the disk includes over 40 construction estimating and bidding forms in a format that's perfect for nearly any word processing or spreadsheet program for *Windows*™. And to top it off, a 70-minute interactive video teaches you how to use this CD-ROM to estimate construction costs. **CD Estimator is $68.50**

Construction Forms & Contracts

125 forms you can copy and use — or load into your computer (from the FREE disk enclosed). Then you can customize the forms to fit your company, fill them out, and print. Loads into *Word for Windows, Lotus 1-2-3, WordPerfect,* or *Excel* programs. You'll find forms covering accounting, estimating, fieldwork, contracts, and general office. Each form comes with complete Instructions on when to use it and how to fill it out. These forms were designed, tested and used by contractors, and will help keep your business organized, profitable and out of legal, accounting and collection troubles. Includes a 3¹/2" disk for your PC. For Macintosh disks, add $15. **432 pages, 8¹/2 x 11, $39.75**

Illustrated Guide to the 1996 *National Electrical Code*

This fully-illustrated guide offers a quick and easy visual reference for installing eiectrical systems. Whether you're installing a new system or repairing an old one, you'll appreciate the simple explanations written by a code expert, and the detailed, intricately-drawn and labeled diagrams. A real time-saver when it comes to deciphering the current *NEC*. **384 pages, 8¹/2 x 11, $34.75**

Audio: Electrician's Exam Preparation Guide

These tapes are made to order for the busy electrician looking for a better-paying career as a licensed apprentice, journeyman, or master electrician. This two-audiotape set asks you over 150 often-used exam questions, waits for your answer, then gives you the correct answer and an explanation. This is the easiest way to study for the exam. **Two 50-minute audiotapes, $26.50**

Electrician's Exam Preparation Guide

Need help in passing the apprentice, journeyman, or master electrician's exam? This is a book of questions and answers based on actual electrician's exams over the last few years. Almost a thousand multiple-choice questions — exactly the type you'll find on the exam — cover every area of electrical installation: electrical drawings, services and systems, transformers, capacitors, distribution equipment, branch circuits, feeders, calculations, measuring and testing, and more. It gives you the correct answer, an explanation, and where to find it in the latest *NEC*. Also tells how to apply for the test, how best to study, and what to expect on examination day. **352 pages, 8¹/2 x 11, $28.00**

1996 National Electrical Code Interpretive Diagrams

Based on the 1996 *NEC*, this new edition visually and graphically explains and interprets the key provisions of the *National Electrical Code* as seen through the eyes of an electrical inspector. All drawings have been carefully prepared and reviewed by inspectors and the Codes and Standards Committee. Here you'll find approved installations of just about every type of electrical part. **270 pages, 5 x 8, $34.95** 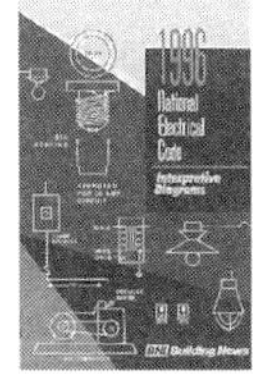

Basic Engineering for Builders

If you've ever been stumped by an engineering problem on the job, yet wanted to avoid the expense of hiring a qualified engineer, you should have this book. Here you'll find engineering principles explained in non-technical language and practical methods for applying them on the job. With the help of this book you'll be able to understand engineering functions in the plans and how to meet the requirements, how to get permits issued without the help of an engineer, and anticipate requirements for concrete, steel, wood and masonry. See why you sometimes have to hire an engineer and what you can undertake yourself: surveying, concrete, lumber loads and stresses, steel, masonry, plumbing, and HVAC systems. This book is designed to help the builder save money by understanding engineering principles that you can incorporate into the jobs you bid. **400 pages, 8¹/2 x 11, $34.00**

Vest Pocket Guide to HVAC Electricity

This handy guide will be a constant source of useful information for anyone working with electrical systems for heating, ventilating, refrigeration, and air conditioning. Includes essential tables and diagrams for calculating and installing electrical systems for powering and controlling motors, fans, heating elements, compressors, transformers and every electrical part of an HVAC system. **304 pages, 3¹/2 x 5¹/2, $18.00**

How to Succeed With Your Own Construction Business

Everything you need to start your own construction business: setting up the paperwork, finding the work, advertising, using contracts, dealing with lenders, estimating, scheduling, finding and keeping good employees, keeping the books, and coping with success. If you're considering starting your own construction business, all the knowledge, tips, and blank forms you need are here. **336 pages, 8¹/2 x 11, $24.25**

Craftsman Book Company
6058 Corte del Cedro
P.O. Box 6500
Carlsbad, CA 92018

24 hour order line
1-800-829-8123 Fax (760) 438-0398
Order online: http://www.craftsman-book.com
In A Hurry? We accept phone orders charged to your
Visa, MasterCard, Discover or American Express

Name______________________________

Company___________________________

Address____________________________

City/State/Zip_______________________

Total enclosed _______________ (In California add 7.25% tax)

☐ Send check or money order and we pay postage, or use your
☐ Visa ☐ MasterCard ☐ Discover or ☐ Amex

Card#______________________________

Expiration Date _____________ Initials__________

10-Day Money Back Guarantee

☐ 34.00 Basic Engineering for Builders
☐ 38.00 Basic Lumber Engineering for Builders
☐ 27.25 Builder's Guide to Room Additions
☐ 35.00 Building Contractor's Exam Preparation Guide
☐ 35.50 Carpentry Estimating with FREE estimating program on disk.
☐ 68.50 CD Estimator
☐ 69.00 CD Estimator — Heavy
☐ 25.00 Concrete Construction & Estimating
☐ 39.50 Construction Estimating Reference Data with FREE stand-alone *Windows* estimating program on a 3½" disk.
☐ 39.75 Construction Forms & Contracts with a 3½" disk. Add $15.00 for ☐ Macintosh disks.
☐ 39.00 Contractor's Guide to Building Code Revised
☐ 22.25 Contractor's Survival Manual
☐ 27.50 Drafting House Plans
☐ 17.00 Estimating Home Building Costs
☐ 21.50 Estimating Tables for Home Building
☐ 22.75 Excavation & Grading Handbook Revised
☐ 22.50 Finish Carpenter's Manual
☐ 23.00 National Building Cost Manual

☐ 47.75 National Electrical Estimator with FREE stand-alone *Windows* estimating program on CD ROM.
☐ 48.00 National Painting Cost Estimator with FREE stand-alone *Windows* estimating program on CD ROM.
☐ 48.25 National Plumbing & HVAC Estimator with FREE stand-alone *Windows* estimating program on CD ROM.
☐ 49.50 National Renovation & Insurance Repair Estimator with FREE stand-alone *Windows* estimating program on CD ROM.
☐ 48.50 National Repair & Remodeling Estimator with FREE stand-alone *Windows* estimating program on CD ROM.
☐ 28.25 Plumbing & HVAC Manhour Estimates
☐ 29.00 Residential Electrical Estimating
☐ 38.80 Residential Steel Framing Guide
☐ 35.00 Roofing Construction & Estimating
☐ 26.50 Rough Framing Carpentry
☐ 47.50 National Construction Estimator with FREE stand-alone *Windows* estimating program on CD ROM
☐ FREE **Full Color Catalog**

Receive Fresh Cost Data Every Year – Automatically
Join Craftsman's Standing Order Club and automatically receive special membership discounts on annual cost books!

   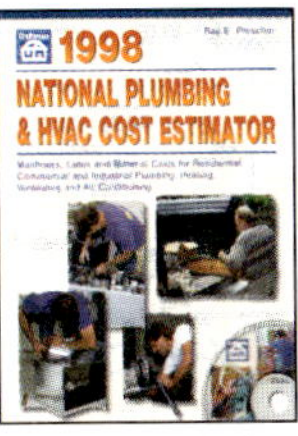 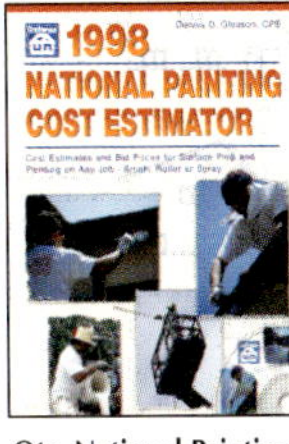  

Qty. National Building Cost Manual
—— Standing Order price $19.55

Qty. National Construction Estimator
—— Standing Order price $40.37

Qty. National Electrical Estimator
—— Standing Order price $40.58

Qty. National Plumbing & HVAC Estimator
—— Standing Order price $41.01

Qty. National Painting Cost Estimator
—— Standing Order price $40.80

Qty. National Renovation & Insurance Repair Est.
—— Standing Order price $42.07

Qty. National Repair & Remodeling Est.
—— Standing Order price $41.22

How many times have you missed one of Craftsman's pre-publication discounts, or procrastinated about buying the updated book and ended up bidding your jobs using obsolete cost data that you "updated" on your own? As a Standing Order Member you never have to worry about ordering every year. Instead, you will receive a confirmation of your standing order each September. If the order is correct, do nothing, and your order will be shipped and billed as listed on the confirmation card. If you wish to change your address, or your order, you can do so at that time. Your standing order for the books you selected will be shipped as soon as the publications are printed. You'll automatically receive any pre-publication discount price.

Bill to:
Name ___________________

Company ___________________

Address ___________________

City/State/Zip ___________________

Ship to:
Name ___________________

Company ___________________

Address ___________________

City/State/Zip ___________________

Purchase order# ___________________

Phone # (___) ___________________

Signature ___________________

Mail This Card Today for a Free Full Color Catalog

Over 100 construction references at your fingertips with information that can save you time and money. Here you'll find information on carpentry, contracting, estimating, remodeling, electrical work and plumbing.

All items come with an unconditional 10-day money-back guarantee. If they don't save you money, mail them back for a full refund.

Name ___________________

Company ___________________

Address ___________________

City/State/Zip ___________________

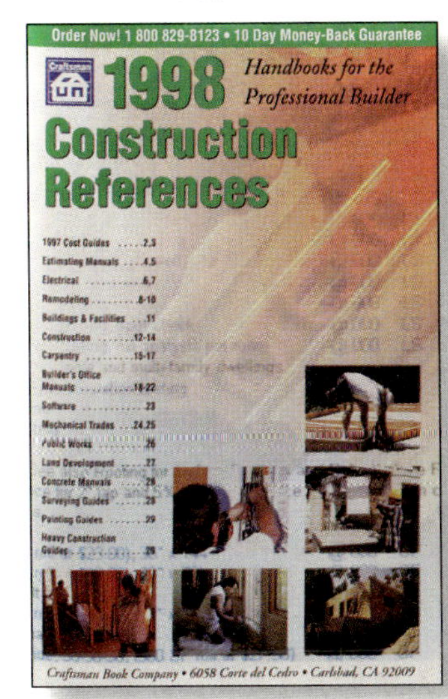

Craftsman Book Company
6058 Corte del Cedro
P.O. Box 6500 • Carlsbad, CA 92018

BUSINESS REPLY MAIL
FIRST CLASS MAIL PERMIT NO. 271 CARLSBAD, CA

POSTAGE WILL BE PAID BY ADDRESSEE

Craftsman Book Company

6058 Corte del Cedro
P.O. Box 6500
Carlsbad, CA 92018-9974

BUSINESS REPLY MAIL
FIRST CLASS MAIL PERMIT NO. 271 CARLSBAD, CA

POSTAGE WILL BE PAID BY ADDRESSEE

Craftsman Book Company

6058 Corte del Cedro
P.O. Box 6500
Carlsbad, CA 92018-9974

NO POSTAGE
NECESSARY
IF MAILED
IN THE
UNITED STATES

BUSINESS REPLY MAIL
FIRST CLASS MAIL PERMIT NO. 271 CARLSBAD, CA

POSTAGE WILL BE PAID BY ADDRESSEE

Craftsman Book Company

6058 Corte del Cedro
P.O. Box 6500
Carlsbad, CA 92018-9974

NO POSTAGE
NECESSARY
IF MAILED
IN THE
UNITED STATES